# MEDICINAL CHEMISTRY
# OF BIOACTIVE
# NATURAL PRODUCTS

# MEDICINAL CHEMISTRY OF BIOACTIVE NATURAL PRODUCTS

Edited by

**XIAO-TIAN LIANG**
**WEI-SHUO FANG**
Chinese Academy of Medical Sciences Beijing, China

WILEY-INTERSCIENCE

A JOHN WILEY & SONS, INC., PUBLICATION

*Library of Congress Cataloging-in-Publication Data:*

Medicinal chemistry of bioactive natural products / [edited by] Xiao-Tian Liang, Wei-Shuo Fang.
    p.   cm.
    Includes index.
    ISBN-13   978-0-471-66007-1 (cloth)
    ISBN-10   0-471-66007-8 (cloth)
    1. Materia medica, Vegetable.   2. Pharmaceutical chemistry.   3. Natural products.   4. Bioactive
compounds.   I. Liang, Xiaotian.   II. Fang, Wei-Shuo
RS431.M37M43 2006
615′ .321 - - dc22                                                    2006000882

10 9 8 7 6 5 4 3 2 1

# CONTENTS

# PREFACE

Although the use of bioactive natural products as herbal drug preparations dates back hundreds, even thousands, of years ago, their application as isolated and characterized compounds to modern drug discovery and development only started in the 19th century, the dawn of the chemotherapy era. It has been well documented that natural products played critical roles in modern drug development, especially for antibacterial and antitumor agents.[1] More importantly, natural products presented scientists with unique chemical structures, which are beyond human imagination most of the time, and inspired scientists to pursue new chemical entities with completely different structures from known drugs.

Medicinal chemistry has evolved from the chemistry of bioactive compounds in early days to works at the interface of chemistry and biology nowadays. Medicinal chemistry of bioactive natural products spans a wide range of fields, including isolation and characterization of bioactive compounds from natural sources, structure modification for optimization of their activity and other physical properties, and total and semi-synthesis for a thorough scrutiny of structure activity relationship (SAR). In addition, synthesis of natural products also provides a powerful means in solving supply problems in clinical trails and marketing of the drug, for obtaining natural products in bulk amounts is often very difficult.

Since the 1980s, the rapid progress in molecular biology, computational chemistry, combinatorial chemistry (combichem), and high throughput screening (HTS) technologies has begun to reshape the pharmaceutical industry and changed their views on natural products. People once thought that natural products discovery was of less value because it is time-consuming, and thus, uneconomic. However, natural products survived as a result of disappointing outcomes of combichem and HTS after a decade of explosive investments. People began to once more appreciate the value of natural products and revived natural products research by integrating rapid isolation and identification with hyphenated technologies, parallel synthesis, computations and may other new techniques into medicinal chemistry of natural products. People also again stressed the unique properties of natural products including their disobedience to Lipinsky's "Rule of Five" which has been widely recognized as the most useful "drug-like" compounds selection criteria.[2,3]

In this single volume entitled *Medicinal Chemistry of Bioactive Natural Products*, discovery, structure elucidation, and elegant synthetic strategies are described, with an emphasis on structure activity relationships for bioactive natural products.

The topics in this book were carefully selected—all classes of bioactive natural products are either clinically useful pharmaceuticals or leading compounds under extensive exploration. Our primary intention in writing such a book is to attract graduate students and spur their interests in bioactive natural products research by providing them with illustrative examples. The real practice of medicinal chemistry of these natural products may teach them how to optimize a target molecule of interest and what kind of techniques they could apply to address the most important issues in medicinal chemistry. In addition, we hope that experts in natural products may find this book's information useful to them, in terms of updated research results on several classes of renowned natural products.

It has been reported that most natural products were used as antibiotics and anticancer agents.[1] However, their uses in the treatment of other epidemics such as AIDS, cardiovascular and neurodegenerative diseases have also been extensively explored. In Chapters 1 and 3, two important anticancer drugs, epothilones and taxanes based on microtuble function inhibition are depicted in a wide range of topics, including syntheses of their analogs, SAR and pharmacophore studies, their pharmacology and mechanism researches and discovery of other antimicrotubule compounds. In Chapter 6, cembranoids from soft coral represents a class of marine organism derived natural product with potent cytotoxicities and other activities. Many synthetic efforts have been made to prepare versatile structural analogs of cembranoids in order to facilitate their SAR research. Chapter 10 also discussed another class of highly cytotoxic natural products, acetogenins, although mainly regarding the synthetic efforts of these molecules. Chapter 2 gave a comprehensive description about glycopeptide antibiotics. As star molecule, vancomycin, belongs to this important family of antibiotics with versatile activities. Chapter 4 and 5 dealt with two unique natural products first discovered by Chinese researchers from traditional Chinese medicines, namely Huperzine A and artemisinin. The former is a cholinesterase inhibitor which has been used in anti-Alzheimer's disease clinics, and the latter is an antimalarial drug and maybe the best known natural product of Chinese origin. Chapter 7 not only discussed the chemistry of Ginkgolides and their actions as PAF receptor inhibitors, but also reported the recent progress in their new target—glycine receptor. Anti-HIV agents calanolides as well as other plant-derived natural products are topics of Chapters 8 and 9. Many different kinds of natural products, coumarins, biphenyl lignans and triterpenes, have been found to be active in anti-HIV models and thus are undergoing further investigations.

It should be pointed out that natural products research will benefit from the integration of many new technologies into this field. One of these technologies is combichem which is still not common in today's practice in medicinal chemistry of natural products, probably due to the complex nature and unpredictable reactivity of natural products. However, many strategies evolved in combichem, such as focused library,[4] libraries based on privileged structures,[5] and fragment based method,[6] are compatible with natural products based libraries. Natural products or its substructures have been utilized as scaffolds by many pioneer researchers.[7] Although very few examples were cited in this book, we believe that we will see rapid progress in the application of combichem and other techniques to natural products research in the future.

Finally, we would like to think all contributors to this book. They have been working on natural products for many years and many of them are active in this field. It is their participation that makes our efforts to organize such a book possible. We also should express our grateful appreciation to the editorial help of Jonathan Rose and Rosalyn Farkas of John Wiley & Sons, who helped us to "polish" the language in contexts and design the impressive book cover.

Xiao-Tian Liang
Wei-Shuo Fang
*Beijing, China*

## REFERENCES

1. Newman D. J., Cragg G. M., Snader K. M. *J. Nat. Prod.* **2003,** 66: 1002-1037.
2. Koehn F. E., Carter G. T. *Nature Rev. Drug Disc.* **2005,** 4: 206-220.
3. Lipinski C. A., Lombardo F., Dominy B. W., Feeney P. J. *Adv. Drug Deliv. Rev.* **1997,** 23: 3-25.
4. Breinbauer R., Vetter I. R., Waldmann H. *Angew. Chem. Int. Ed.* **2002,** 41: 2878-2890.
5. Horton D. A., Bourne G. T., Smythe M. L. *Chem. Rev.* **2003,** 103: 893-930.
6. Erlanson D. A., McDowell R. S., O'Brien T. *J. Med. Chem.* **2004,** 47: 3463-3482.
7. Nicolaou K. C., Pfefferkorn J. A., Roecker A. J., Cao G.-Q., Barluenga S. *J. Am. Chem. Soc.* **2000,** 122: 9939-9953.

# CONTRIBUTORS

*Karl-Heinz Altmann  Institute of Pharmaceutical Sciences, ETH Hönggerberg, HCI H 405, CH-8093 Zürich, Switzerland (e-mail: karl-heinz.altmann@pharma.ethz.ch)

Qi-Cheng Fang  Institute of Materia Medica, Chinese Academy of Medical Sciences, 1 Xian Nong Tan St., Beijing 100050, China

*Wei-Shuo Fang  Institute of Materia Medica, Chinese Academy of Medical Sciences, 1 Xian Nong Tan St., Beijing 100050, China (e-mail: wfang@imm.ac.cn)

Tai-Shan Hu  State Key Laboratory of Bioorganic and Natural Products Chemistry, Shanghai Institute of Organic Chemistry, Chinese Academy of Sciences, 354 Fenglin Road, Shanghai 200032, China

Hao Huang  State Key Laboratory of Bioorganic and Natural Products Chemistry, Shanghai Institute of Organic Chemistry, Chinese Academy of Sciences, 354 Fenglin Road, Shanghai 200032, China

*Kuo-Hsiung Lee  Natural Products Laboratory, School of Pharmacy, University of North Carolina, Chapel Hill, NC 27599, USA (e-mail: khlee@unc.edu)

*Ying Li  Shanghai Institute of Materia Medica, Shanghai Institutes for Biological Sciences, Chinese Academy of Sciences, 555 Zuchongzhi Road, Shanghai 201203, China (e-mail: yli@mail.shcnc.ac.cn)

*Yulin Li  State Key Laboratory of Applied Organic Chemistry, Institute of Organic Chemistry, Lanzhou University, Lanzhou 730000, China (e-mail: liyl@lzu.edu.cn)

*Corresponding author.

**Yi-Ming Li**   State Key Laboratory of Drug Research, Shanghai Institute of Materia Medica, Shanghai Institutes for Biological Scinces, Chinese Academy of Sciences 555 ZuChongZhi Road, Shanghai 201023, China

**Xiao-Tian Liang**   Institute of Materia Medica, Chinese Academy of Medical Sciences, 1 Xian Nong Tan St., Beijing 100050, China (e-mail: xtliang@public. bta.net.cn)

**Gang Liu**   Department of Medicinal Chemistry, Institute of Materia Medica, Chinese Academy of Medical Sciences and Peking Union Medical College, 1 Xian Nong Tan Road, Beijing 100050, China

**Tao Ma**   Department of Medicinal Chemistry, Institute of Materia Medica, Chinese Academy of Medical Sciences and Peking Union Medical College, 1 Xian Nong Tan Road, Beijing 100050, China

**Lizeng Peng**   State Key Laboratory of Applied Organic Chemistry, Institute of Organic Chemistry, Lanzhou University, Lanzhou 730000, China

*__Kristian Strømgaard__   Department of Medicinal Chemistry, The Danish University of Pharmaceutical Sciences, Universitetsparken, 2 DK-2100 Copenhagen, Denmark (e-mail: krst@dfuni.dk)

*__Roaderich Süssmuth__   Rudolf-Wiechert-Professor für Biologische Chemie, Technische Universitäat Berlin, Institut Für Chemie/FG Organische Chemie, Strasse des 17. Juni 124, D-10623 Berlin, Germany (e-mail: suessmuth@ chem.tu-berlin.de)

**Chang-Heng Tan**   State Key Laboratory of Drug Research, Shanghai Institute of Materia Medica, Shanghai Institutes for Biological Sciences, Chinese Academy of Sceinces, 555 ZuChongzhi Road, Shanghai 201023, China

*__Lin Wang__   Department of Medicinal Chemistry, Institute of Materia Medica, Chinese Academy of Medical Sciences & Peking Union medical college, 1 Xian Nong Tan Road Beijing 100050, China (e-mail: wanglin1933@163.com)

**Yu-Lin Wu**   State Key Laboratory of Bioorganic and Natural Products Chemistry, Shanghai Institute of Organic Chemistry, Chinese Academy of Sciences, 354 Fenglin Road, Shanghai 200032, China (e-mail: ylwu@mail.sioc.ac.cn)

*__Zhu-Jun Yao__   State Key Laboratory of Bioorganic and Natrual Products Chemistry, Shanghai Institute of Organic Chemistry, Chinese Academy of Sciences, 354 Fenglin Road, Shanghai 200032, China (e-mail: yaoz@mail.sioc.ac.cn)

**Donglei Yu**   Natural Products Laboratory, School of Pharmacy, University of North Carolina, Chapel Hill, NC 27599

**Tao Zhang**   State Key Laboratory of Applied Organic Chemistry, Institute of Organic Chemistry, Lanzhou University, Lanzhou 730000, China

*Corresponding author.

*Da-Yuan Zhu   State Key Laboratory of Drug Research, Shanghai Institute of Materia Medica, Shanghai Institutes for Biological Sciences, Chinese Academy of Sciences, 555 ZuChongZhi Road, Shanghai 201023, China (e-mail: dyzhu@mail.shcnc.ac.cn)

*Corresponding author.

# 1

# THE CHEMISTRY AND BIOLOGY OF EPOTHILONES—LEAD STRUCTURES FOR THE DISCOVERY OF IMPROVED MICROTUBULE INHIBITORS

KARL-HEINZ ALTMANN

*Department of Chemistry and Applied BioSciences, Institute of Pharmaceutical Sciences, Swiss Federal Institute of Technology (ETH), Zürich, Switzerland*

## 1.1  INTRODUCTION

Cancer represents one of the most severe health problems worldwide, and the development of new anticancer drugs and more effective treatment strategies are fields of utmost importance in drug discovery and clinical therapy. Much of the research in these areas is currently focused on cancer-specific mechanisms and the corresponding molecular targets (e.g., kinases related to cell cycle progression or signal transduction),[1] but the search for improved cytotoxic agents (acting on ubiquitous targets such as DNA or tubulin) still constitutes an important part of modern anticancer drug discovery. As the major types of solid human tumors (breast, lung, prostate, and colon), which represent most cancer cases today, are multicausal in nature, there is a growing recognition that the treatment of solid tumors with "mechanism-based" agents alone is unlikely to be successful. Instead, improved treatment strategies are likely to involve combinations of signal transduction inhibitors with new and better cytotoxic drugs.

Microtubule inhibitors are an important class of anticancer agents,[2] with clinical applications in the treatment of a variety of cancer types, either as single agents or as part of different combination regimens.[3] Microtubule-interacting agents can be

*Medicinal Chemistry of Bioactive Natural Products*   Edited by Xiao-Tian Liang and Wei-Shuo Fang
Copyright © 2006 John Wiley & Sons, Inc.

1

grouped into two distinct functional classes, namely compounds that inhibit the assembly of tubulin heterodimers into microtubule polymers ("tubulin polymerization inhibitors") and those that stabilize microtubules under normally destabilizing conditions ("microtubule stabilizers").[4] The latter will also promote the assembly of tubulin heterodimers into microtubule polymers. While the use of tubulin polymerization inhibitors such as vincristine and vinblastine in cancer therapy dates back around 40 years (vincristine and vinblastine received U.S. Food and Drug Administration (FDA) approval in 1963 and 1965, respectively), the introduction of microtubule stabilizers into clinical practice constitutes a relatively recent development, which took place only in 1993. The first agent of this type to obtain FDA approval was paclitaxel (Taxol) in 1992, which was followed by its closely related analog docetaxel (Taxotere) in 1996 and the emergence of microtubule-stabilizing anticancer drugs clearly marks a significant advance in cancer chemotherapy.[5]

While several small synthetic molecules are known, which act as efficient tubulin polymerization inhibitors,[6] it is intriguing to note that all potent microtubule-stabilizing agents identified to date are natural products or natural product-derived (for a recent review, see Altmann[7]). Historically, more than a decade passed after the elucidation of paclitaxel's mode of action in 1979[8] before alternative microtubule-stabilizing agents were discovered, bearing no structural resemblance to paclitaxel or other taxanes. Most prominent among these new microtubule stabilizers is a group of bacteria-derived macrolides, which were discovered in 1993 by Reichenbach and Höfle and have been termed "epothilones" by their discoverers[9,10]. Although not immediately recognized, these compounds were subsequently demonstrated by a group at Merck Research Laboratories to possess a paclitaxel-like mechanism of action.[11]

Paclitaxel                R = H: Epothilone A
                          R = CH₃: Epothilone B

The major products originally isolated from the myxobacterium *Sorangium cellulosum Sc 90* are epothilone A and epothilone B (Epo A and B), but numerous related members of this natural products family have subsequently been isolated as minor components from fermentations of myxobacteria.[12] The relative and absolute stereochemistry of Epo B was determined by Höfle et al. in 1996 based on a combination of x-ray crystallography and chemical degradation studies,[13] and the availability of this information shortly after the discovery of their mechanism of action has provided an important impetus for the extensive synthetic chemistry efforts on Epo A and B and their analogs over the subsequent years. Although they are not part of this chapter, it is worth noting that a growing number of additional natural products have been recognized over the last few years to be microtubule stabilizers,[7,14–18] thus providing a whole new set of diverse lead structures for anticancer drug discovery.

While exerting their antiproliferative activity through interference with the same molecular target, a major distinction between paclitaxel and epothilones is the ability of the latter to inhibit the growth of multidrug-resistant cancer cell lines.[11,19–21] In addtition, epothilones have also been shown to be active in vitro against cancer cells, whose paclitaxel resistance originates from specific tubulin mutations.[19,22] At the same time, epothilones possess more favorable biopharmaceutical properties than paclitaxel, such as improveded water-solubility,[13] which enables the use of clinical formulation vehicles less problematic than Cremophor EL. (For a discussion of the clinical side-effects of Taxol believed to originate in this particular formulation vehicle, see Ref. 5). Epo B and several of its analogs have been demonstrated to possess potent in vivo antitumor activity, and at least five compounds based on the epothilone structural scaffold are currently undergoing clinical evaluation in humans. These compounds include Epo B (EPO906; developed by Novartis), Epo D (deoxyEpo B, KOS-862; Kosan/Sloan-Kettering/Roche), BMS-247550 (the lactam analog of Epo B; BMS), BMS-310705 (C21-amino-Epo B; BMS), and ABJ879 (C20-desmethyl-C20-methylsulfanyl-Epo B; Novartis).

The combination of an attractive biological profile and comparatively limited structural complexity (at least for a natural product) has made epothilones attractive targets for total chemical synthesis. Thus, numerous syntheses of Epo A and B have been published in the literature (for reviews of work up to 2001, see Refs. 23–26; for more recent work, see Refs. 27–36) since the first disclosure of their absolute stereochemistry in 1996.[13] At the same time, the methodology developed in the course of those studies has been exploited for the synthesis of a host of synthetic analogs (reviewed in Refs. 23, 24, 37–40); although structural information on complexes between epothilones and their target protein β-tubulin (or microtubules) at atomic resolution is still lacking (vide infra), this has allowed the empirical elucidation of the most important structural parameters required for biological activity. The chemistry developed for the preparation of some of these analogs should even allow the production of amounts of material sufficient for clinical trials,[24,41] thus highlighting the difference in structural complexity (which is reflected in synthetic accessibility) between epothilone-type structures and paclitaxel, for which an industrial scale synthesis is clearly out of reach.

The chemistry, biology, and structure activity relationship (SAR) of epothilones have been extensively discussed in recent review articles.[20,23,24,37–40] It is thus not the intention of this chapter to provide a detailed review of these different facets of epothilone-related research. Rather, this chapter will focus on some selected aspects of the chemistry, biology, and clinical evaluation of natural epothilones and their synthetic analogs, with particular emphasis on SAR work performed in our laboratories. Details of the organic chemistry of epothilones and their analogs will not be discussed. Likewise, the impressive advances in the elucidation of epothilone biosynthesis and the development of heterologous expression systems is largely outside of the scope of this chapter,[42] except for a few selected modified analogs produced in heterologous expression systems. (For a recent review on the biosynthesis of epothilones see Ref. 42). These systems will be covered in Section 1.3.

## 1.2 BIOLOGICAL EFFECTS OF Epo B

### 1.2.1 In Vitro Activity

The basic biology and pharmacology of Epo B (as the most potent and most widely studied natural epothilone) have been summarized in several previous review articles.[15,23,37,39,43,44] As indicated in Section 1.1, the biological effects of the compound are based on its ability to bind to microtubules and alter the intrinsic stability and dynamic properties of these supramolecular structures. In cell-free in vitro systems, this is demonstrated by the prevention of $Ca^{2+}$- or cold-induced depolymerization of preformed microtubule polymers[19] as well as by the promotion of tubulin polymerization (to form microtubule-like polymers) in the absence of either microtubule-associated proteins (MAPs) and/or guanosine triphosphate (GTP), at temperatures significantly below 37 °C, and in the presence of $Ca^{2+}$.[11,19] The latter phenomenon, that is, the induction of tubulin polymerization, is frequently used as a biochemical readout for the assessment of the interaction of microtubule-stabilizing agents with tubulin in a quantitative fashion. Epo B is a more efficient tubulin-polymerizing agent than paclitaxel, which in turn polymerizes tubulin with about the same potency as Epo A. (e.g., $EC_{50}$ values for the polymerization of microtubule protein by Epo A, Epo B, and paclitaxel have been determined as 1.12, 0.67, and 1.88 μM, respectively).[45] However, it should be noted that the exact magnitude of tubulin-polymerizing effects in vitro (absolute and even relative polymerization rates, extent of tubulin polymer formation) strongly depends on the assay conditions employed (e.g., biological source and purity of tubulin, concentration of microtubule-stabilizing buffer components, and reaction temperature).[45] Epothilones can displace [$^3$H]-paclitaxel from microtubules with efficiencies similar or superior to those of unlabeled paclitaxel or docetaxel.[11,19] Inhibition of paclitaxel binding occurs in a competitive fashion [with apparent $K_i$ values of 1.4 μM (Epo A) and 0.7 μM (Epo B)], which thus suggests that the microtubule binding sites of paclitaxel and Epo A and B are largely overlapping or even identical (vide infra). More recently, the binding constants of Epo A and B to stabilized microtubules in vitro have been determined as $2.93 \times 10^7 M^{-1}$ (Epo A) and $6.08 \times 10^8 M^{-1}$ (37 °C) with a fluorescence-based displacement assay.[46]

In line with its effects on tubulin polymerization in vitro (i.e., in an excellular context), the prevention of cold-induced depolymerization of microtubules by epothilones has also been demonstrated in cells.[37] Microtubule stabilization in intact cells (as well as cancer cell growth inhibition, vide infra), however, is observed at strikingly lower concentrations than those required for the induction of tubulin polymerization in vitro. This apparent discrepancy has been resolved by careful uptake experiments in HeLa cells,[20,45] which have shown that Epo A and B, like paclitaxel,[47] accumulate several-hundred-fold inside cells over external medium concentrations. Similar findings have been reported for a close analog of Epo B in MCF-7 cells.[48]

Early experiments investigating the effects of epothilones on microtubule bundling in intact cells had demonstrated that treatment of cultured cells (RAT1, HeLa, Hs578T, Hs578Bst, PtK$_2$) with high ($10^{-6}$–$10^{-4}$ M) concentrations of epothilones resulted in the formation of characteristic, extensive microtubule bundles (lateral association of microtubules) throughout the cytoplasm of interphase cells.[11,19] These bundles develop independently of the centrosome, which indicates that epothilones override the microtubule-nucleating activity of the centrosome in interphase cells. In contrast, at lower epothilone concentrations ($10^{-7}$–$10^{-8}$ M), interphase microtubule arrays were reported to remain largely unaffected, with the primary effect occurring on cells entering mitosis.[11] However, employing live fluorescence microscopy of HeLa cells ectopically expressing mouse β6-tubulin fused to enhanced green fluorescent protein (EGFP), more recent experiments conducted in our laboratories have demonstrated that even low nM concentrations of Epo B lead to the bundling of interphase microtubules after 24 h of drug exposure.[20]

In general, the growth inhibitory effect of epothilones (and other microtubule-interacting agents) is assumed to be a consequence of the suppression of microtubule dynamics rather than an overall increase in microtubule polymer mass caused by massive induction of tubulin polymerization.[49] Indeed, using time-lapse microscopy in MCF-7 cells stably transfected with GFP-α-tubulin, Kamath and Jordan[50] have recently demonstrated that the inhibition of the dynamics of interphase microtubules by Epo B occurs in a concentration-dependent manner and correlates well with the extent of mitotic arrest (G2/M block; vide infra). These data provide a direct demonstration of suppression of cellular microtubule dynamics by Epo B, although it is still unclear how effects on *interphase* microtubules relate to the dynamics of *spindle* microtubules (which could not be measured) and thus to mitotic arrest. Likewise, it remains to be established how mitotic entry and the associated rearrangement of the entire microtubule network can occur from a largely static state of this network. It seems likely, however, that mitotic signals lead to profound changes in microtubule dynamics, which may cause the dynamics of spindle microtubules to be suppressed less efficiently than is the case for interphase microtubules.

Treatment of human cancer cells with low nM concentrations of Epo B leads to profound growth inhibition and cell death (Table 1-1). In line with the effects on tubulin polymerization in vitro, Epo B is a more potent antiproliferative agent than Epo A, which in turn is about equipotent with paclitaxel. As observed for paclitaxel, Epo B treatment produces aberrant mitotic spindles, results in cell cycle arrest in mitosis, and eventually leads to apoptotic cell death.[11,19] It is often assumed that apoptosis is a direct consequence of G2/M arrest, which in turn would be a prerequisite for growth inhibition and cell death. However, as has been elegantly demonstrated by Chen et al. in a series of recent experiments, the situation is clearly more complex,[51,52] such that low concentrations of Epo B (and paclitaxel

**TABLE 1-1. IC$_{50}$ Values [nM] for Net Growth Inhibition of Human Cancer Cell Lines by Epo A and B in Comparison With Paclitaxel[a,b]**

| | Cell Line | | | | | | |
|---|---|---|---|---|---|---|---|
| | HCT-116 (colon) | PC-3M (prostate) | A549 (lung) | MCF-7 (breast) | MCF-7/ ADR[c] | KB-31 (epidermoid) | KB-8511[d] |
| Epo A | 2.51 | 4.27 | 2.67 | 1.49 | 27.5 | 2.1 | 1.9 |
| Epo B | 0.32 | 0.52 | 0.23 | 0.18 | 2.92 | 0.19 | 0.19 |
| Paclitaxel | 2.79 | 4.77 | 3.19 | 1.80 | 9105 | 2.31 | 533 |

[a]Cells were exposed to drugs for 3–5 days, allowing for at least two population doublings. Cell numbers were estimated by quantification of protein content of fixed cells by methylene blue staining. Data from Ref. 38.
[b]Multidrug-resistant cell lines are underlined.
[c]Multiple resistance mechanisms/MDR.
[d]P-gp overexpression/ MDR.

or discodermolide) produce a large aneuploid cell population in A549 lung carcinoma cells in the absence of a mitotic block. These cells, which are arrested in the G1 phase of the cell cycle, originate from aberrant mitosis after formation of multipolar spindles and eventually will undergo apoptosis. On the other hand, higher drug concentrations lead to a protracted mitotic block from which the cells exit without division, thus forming tetraploid G1 cells.[52] In contrast to this differential behavior, tubulin polymerization inhibitors such as colchicine, nocodazole, and vinblastine do not give rise to aneuploid cells, but they always lead to mitotic arrest followed by apoptosis.[51] This result suggests that the suppression of microtubule dynamics, which is common to both types of tubulin-interacting agents, cannot fully account for the complex array of biological effects displayed by Epo B or paclitaxel. The above results from the Horwitz laboratory clearly demonstrate that entry of cells into mitosis is a fundamental prerequisite for cell killing by microtubule-stabilizing agents. At the same time, however, cell death does not necessarily require prior mitotic arrest, but at low concentrations of Epo B, it can simply be a consequence of mitotic slippage (aberrant mitosis) and subsequent cell cycle arrest in G1.

The notion of apoptosis as the predominant mechanism of cell killing in response to Epo B treatment has recently been questioned based on the observation that both caspase inhibition and blockage of the death receptor pathway fail to reduce the cytotoxic effects of Epo B in non-small cell lung cancer (NSCLC) cells (as measured by the appearance of cells with a hypodiploid DNA content).[53] At the same time, it was found that a specific inhibitor of cathepsin B completely inhibits cell killing by Epo B, which thus implicates this cysteine protease as a central player in the execution of cell death in response to Epo B treatment of NSCLC cells. Although caspase activity does not seem to be an essential prerequisite for Epo B-induced cell kill, activation of these proteases still occurs after prolonged exposure of NSCLC cells to Epo B, as indicated by PARP cleavage, chromatin condensation, and phosphatidylserine externalization.[53] Thus, the overall effects described by Bröker et al.[53,54] are similar to those reported by Chen et al.,[51,52] including insensitivity of Epo B-induced cell kill to caspase inhibition, which had been previously observed in HeLa cells by McDaid and Horwitz.[55]

As is generally the case for anticancer drugs, the cellular response to microtubule-stabilizing agents can be modulated by adaptive changes of the cell that lead to acquired drug resistance. Alternatively, cells may be inherently protected from the antiproliferative effects of cytotoxic agents by a variety of mechanisms. In contrast to paclitaxel (as well as other standard cytotoxic anticancer agents), Epo A/B are not susceptible to phosphoglyco-protein-170 (P-gp)-mediated drug efflux and thus retain full antiproliferative activity against the corresponding multidrug-resistant cell lines in vitro (Table 1-1).[11,19,20,45] This characteristic may provide a distinctive advantage of epothilones over current taxane-based therapy, but the clinical significance of P-gp-mediated drug resistance is a matter of significant debate.[56] On the other hand, recent discoveries from various laboratories demonstrate that cancer cells can become resistant to epothilones through alternative mechanisms, such as tubulin mutations. For example, Wartmann and Altmann have isolated an epothilone-resistant subline of the KB-31 epidermoid carcinoma cell line (termed KB-31/C5), which carries a single point-mutation (Thr274 Pro) in the HM40 tubulin gene (the major β-tubulin isoform expressed in these cells).[37] Similar findings have been independently reported by Giannakakou et al., who have produced an Epo A-resistant cell line, 1A9/A8, in which Thr274 is mutated to Ile rather than to Pro.[57] Thr274 maps to the taxane-binding site on β-tubulin,[58] which based on competition studies, is likely to be targeted also by epothilones (vide supra). Consistent with the notion of a shared binding site between epothilones and paclitaxel, both KB-31/C5 as well as 1A9/A8 cells are cross-resistant to paclitaxel, albeit to varying degrees. More recently, He et al. have generated three different Epo-resistant cell lines, A549B40, HeLa.EpoA9, and HeLa.EpoB1.8, each of which is characterized by a specific β-tubulin mutation, namely Gln292Glu in A549B40 cells, Pro173Ala in HeLa.EpoA9 cells, and Tyr422Cys in HeLa.EpoB1.8 cells.[59] The highest degree of resistance is associated with A549.EpoB40 cells, which are 95-fold resistant to Epo B and exhibit marked cross-resistance with other microtubule-stabilizing agents, with the notable exception of discodermolide. The β-tubulin mutations identified by the Horwitz laboratory map to sites on β-tubulin that have been suggested to be involved in paclitaxel binding and lateral protofilament interaction (Gln292), GTP hydrolysis (Pro173), and binding of MAPs (Tyr422), respectively.

The above Gln292Glu mutation in combination with a second mutation at position 231 of β-tubulin (Thr → Ala) was also identified by Verrills et al.[60] in a highly resistant subline of the human T-cell acute leukemia cell line CCRF-CEM (termed dEpoB300), which had been selected with Epo D (deoxyEpo B). dEpo300 cells are 307-fold resistant to the selecting agent Epo D and exhibit 77-fold and 467-fold cross-resistance with Epo B and paclitaxel, respectively. Epo D failed to induce any measurable tubulin polymerization in cell lysates prepared from dEpoB300 cells at 8 μM compound concentration, which thus illustrates that impaired growth inhibition is indeed paralleled by diminished effects on tubulin polymerization.

In summary, all tubulin mutations identified to date in epothilone-resistant cells are found in regions of the tubulin structure, which are predicted to be important for tubulin polymerization and/or microtubule stability (including those that may additionally affect drug binding). Thus, these mutations may not only affect drug–target interactions, but they may also (or alternatively) impair intrinsic tubulin functions in a way that could result in hypostable microtubules.[37,59] Consistent with this hypothesis (previously formulated by Cabral and Barlow based on observations made with paclitaxel-resistant CHO cells[61]), these cell lines are hypersensitive to tubulin-depolymerizing drugs, such as vincristine or colchicine.

Epothilones retain significant activity against paclitaxel-selected cell lines that harbor a distinct set of tubulin mutations,[22] and again this could perhaps translate into clinical utility in the treatment of Taxol-resistant tumors. However, any such predictions must be treated with great caution, as the clinical significance of individual resistance mechanisms identified in vitro has not been established.

### 1.2.2 In Vivo Antitumor Activity

The in vivo effects of Epo B have been investigated in some detail by a group at the Sloan-Kettering Cancer Center as well as by our group at Novartis. Initial experiments by the Sloan-Kettering group in xenograft models of human leukemia [CCRF-CEM and CCRF-CEM/VBL (MDR)] in CB-SCID mice (drug-sensitive as well as multidrug-resistant tumors) suggested promising antitumor activity but also a narrow therapeutic window.[62] In subsequent experiments, the compound was found to exhibit considerable toxicity, while having only limited effects on tumor growth in human MX-1 breast or SKOV-3 ovarian tumors in mice. These data led to the conclusion that Epo B might simply be too toxic to become a clinically useful anticancer agent.[63]

In contrast to these findings, studies in our laboratory have demonstrated potent antitumor activity of Epo B in several drug-sensitive human tumor models (nude mice) upon intravenous administration, despite the compound's limited plasma stability in rodents.[20] Activity was observed for models encompassing all four major types of solid human tumors (lung, breast, colon, and prostate) and was manifest either as profound growth inhibition (stable disease) or significant tumor regression. In addition, Epo B was found to be a potent inhibitor of tumor growth in P-gp-overexpressing multidrug-resistant human tumor models. Regressions were observed in two such models (KB-8511 (epidermoid carcinoma)[20] and HCT-15 (colon carcinoma)[64]), where tumors were either poorly responsive or completely nonresponsive to treatment with Taxol. In general, therapeutic effects could be achieved at tolerated dose levels, but significant body weight loss was observed in several experiments (which was generally reversible after cessation of treatment), which indicates a relatively narrow therapeutic window. As demonstrated in a recent study by Pietras et al., the antitumor effect of Epo B in a model of human anaplastic thyroid carcinoma can be potentiated by coadministration with the tyrosine kinase inhibitor STI571 (Gleevec) without any obvious decrease in tolerability.[65] The enhanced antitumor activity of the combination is presumed to be a consequence of a selective enhancement in drug uptake by the tumor because of inhibition of platelet-derived growth factor receptor (PDGF-R) by STI571.

To conclude the discussion on the in vivo antitumor activity of Epo B, it should be noted that the disparate results of in vivo experiments by the Sloan-Kettering and the Novartis groups are not necessarily incompatible, but they may simply reflect differences in the experimental setups, such as tumor models, formulation, and/or dosing regimens. The results of the preclinical evaluation of Epo B at Novartis have led to the initiation of clinical trials with the compound in 1999 (vide infra).

## 1.3 EPOTHILONE ANALOGS AND SAR STUDIES

The chemistry of epothilones has been extensively explored, and a wealth of SAR information has been generated for this family of structures over the last several years. Most synthetic analogs have originated from the groups of Nicoloau (cf., e.g., Ref. 23) and Danishefsky (cf., e.g., Ref. 24) and to a lesser extent from the groups at Novartis[20,45] and Schering AG.[66] Semisynthetic work has been reported by the groups at the "Gesellschaft für Biotechnologische Forschung" in Braunschweig, Germany (GBF; cf., e.g., Refs. 67 and 68) and Bristol-Myers Squibb (BMS; cf., e.g., Ref. 69). The SAR data that have emerged from this research have been summarized in several recent review articles.[20,23,24,37-40] In Sections 1.3.1–1.3.6, the most significant features of the epothilone SAR will be discussed, with particular emphasis on the work conducted in our laboratory.

### 1.3.1 Lactam-Based Analogs

The replacement of the lactone oxygen by nitrogen, that is, converting the macro-*lactone* into a macro*lactam* ring,[70,71] has emerged as one of the most important scaffold modifications in epothilones reported so far. This strategy was spearheaded by the group at BMS, and it has led to the identification of the lactam analog of Epo B (**1** = BMS-247550) as a highly promising antitumor agent, which is currently undergoing extensive clinical evaluation by BMS (vide infra).

1

Lactam-based analogs of epothilones were conceived by the BMS group as metabolically more stable alternatives to the lactone-based natural products (which exhibit limited metabolic stability in rodent plasma). It is worth noting, however, that despite its short plasma half-life in rodent species, Epo B shows potent antitumor activity in a variety of nude mouse human tumor models,[20] and the same is true for Epo D (vide infra). In addition, Epo D has subsequently been demonstrated to be significantly more stable in human than in rodent plasma,[72] which is in line with our studies on Epo B and clearly is a result of the well-known difference in plasma esterase activity between humans and rats or mice. BMS-247550 (**1**) is a potent inducer of tubulin polymerization, but its antiproliferative activity is circa one order of magnitude lower than that of Epo B[37,70] (e.g., $IC_{50}$ values against the human colon carcinoma cell line HCT-116 are 3.6 nM and 0.42 nM, respectively, for **1** and Epo B[70]). Methylation of the lactam nitrogen has been shown to result in a substantial loss in potency.[73]

In contrast to Epo B, **1** exhibits a significant activity differential between the drug-sensitive KB-31 cell line and its P-gp overexpressing multidrug-resistant KB-8511 variant ($IC_{50}$'s of 2.85 nM and 128 nM against KB-31 and KB-8511 cells, respectively[37]), thus indicating that the compound is a substrate for the P-gp efflux pump. Similar differences between P-gp-overexpressing and drug-sensitive human cancer cell lines have been observed for lactam-based analogs of Epo C and D.[73,74] Like other types of microtubule inhibitors, BMS-247550 (**1**) was found by Yamaguchi et al. to induce G2/M arrest and apoptosis in human cancer cell lines.[75] Interestingly, BMS-247550-induced cell death, at least in MDA-MB 468 breast carcinoma cells, seems to be caspase-dependent,[75] contrary to what has been reported for Epo B in other cell lines.[53–55] It is also worth noting that Yamaguchi et al. strongly emphasize that [BMS-247550 (**1**)-induced] "apoptosis of A2780-1A9 cells follows mitotic arrest, which is not associated with a marked increase in the levels of survivin" (cf., however, recent work by Chen et al.[51,52]).

BMS247550 (**1**) exhibits antitumor activity similar to that of Taxol in Taxol-sensitive tumor models (i.e., A2780 human ovarian carcinoma, HCT116 and LS174T human colon carcinomas) when each drug is given at its optimal dose.[71] Despite its limited effects against highly multidrug-resistant cell lines in vitro, **1** was also shown to be superior to Taxol in Taxol-resistant tumor models (i.e., Pat-7 and A2780Tax human ovarian carcinomas, Pat-21 human breast carcinoma, and Pat-26 human pancreatic carcinoma, and M5076 murine sarcoma). Furthermore, the compound showed remarkable antitumor activity against Pat-7 ovarian and HCT-116 colon carcinoma xenografts after oral administration.[71]

### 1.3.2  Modifications in the C9–C12 Region

Initial reports on structural modifications in the C9–C11 trimethylene fragment adjacent to the epoxide moiety were rare, and the corresponding analogs were generally found to exhibit diminished biological activity. Early work by the Nicolaou and Danishefsky groups had shown that ring contraction or expansion via the removal of existing or the incorporation of additional $CH_2$-groups in the C9–C11 region causes a substantial loss in biological potency.[62,76] An alternative approach pursued in our laboratory for modifications situated in the Northern hemisphere of epothilones was based on an epothilone pharmacophore model, which was derived from a comparative analysis of the x-ray crystal structure of Epo B[13] with those of paclitaxel and discodermolide (P. Furet and N. Van Campenhout, unpublished results) (see Refs. 57 and 77–80). According to this model, the bioactive conformation of

Epo B is closely related to its x-ray crystal structure; in particular, the three bonds between C8/C9, C9/C10, and C10/C11 all adopt an *anti*-periplanar conformation. The model suggested that the incorporation of *meta*-substituted phenyl rings in the C9 to C12 segment, such as in compounds **2** or **3**, should lead to a stabilization of the purported bioactive conformation in this potentially flexible region of natural epothilones. While the synthesis and biological evaluation of **3** has been reported by the Schering AG group,[66] we have recently described the synthesis of analog **8**.[81] Unfortunately, **2** and **3** were found to be substantially less active than Epo B or D (**2**; Ref. 81) or to exhibit "reduced" activity (**11**; Ref. 66). More specific data are not available from Ref. 66.

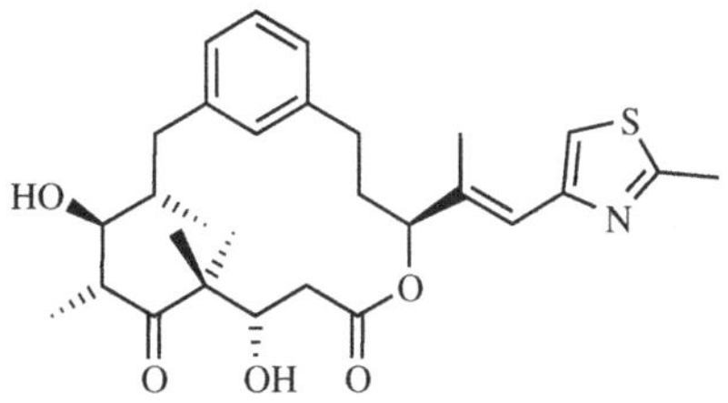

2

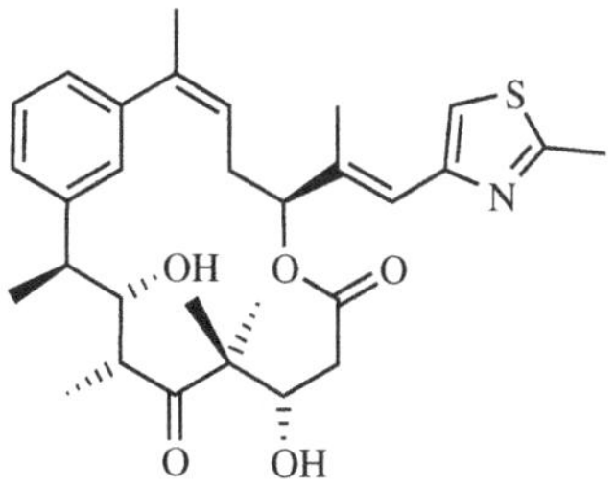

3

In contrast to these disappointing early findings, several highly potent epothilone analogs with structural variations in the C9–C11 trimethylene region have been described more recently, some of which have also been found to exhibit favorable in vivo pharmacological properties.

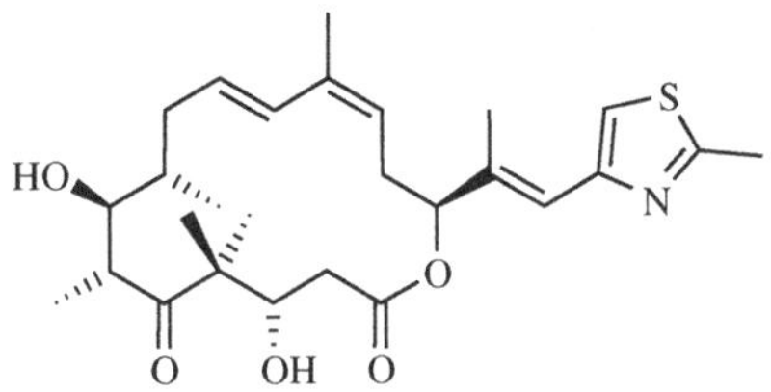

4

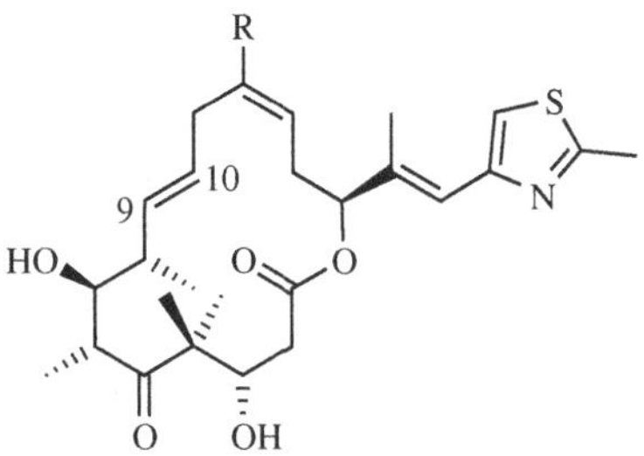

R = CH$_3$: **5**
R = CF$_3$: **6**

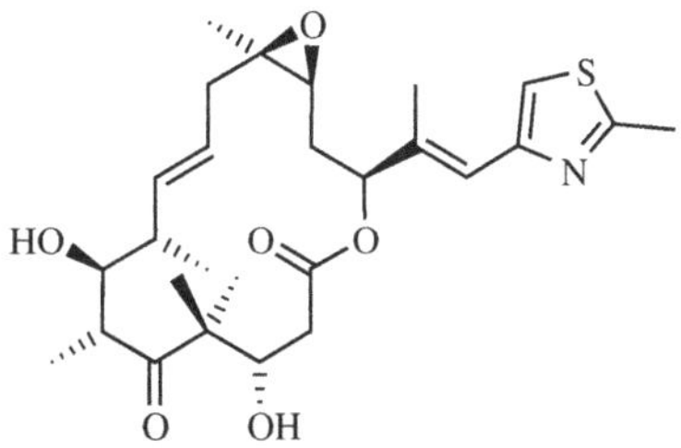

7

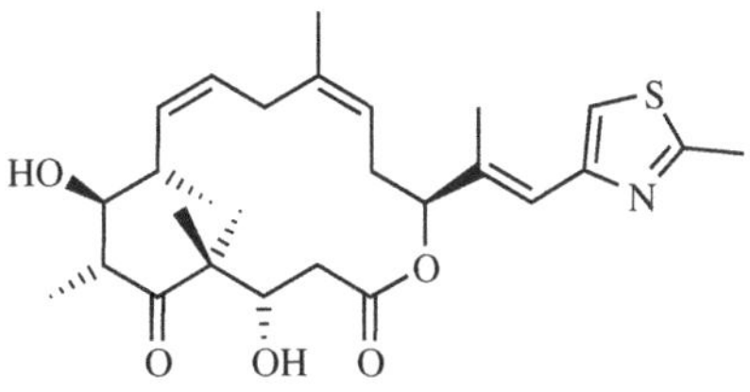

8

These analogs were obtained through several different approaches, including heterologous expression of the modified epothilone polyketide synthases in *Myccococus xanthus*,[82] total chemical synthesis (spearheaded by the Danishefsky group),[83-88] or biotransformation of Epo B.[89,90] For example, the in vitro antiproliferative activity of epothilone 490 **4** is only three- to four-fold lower than that of the parent compound Epo D against the MCF7 breast, SF268 glioma, NCI-H460 lung cancer, and HL60 promyeolocytic leukemia cell lines, and the compound is equipotent with Epo B against the human T-cell leukemia cell lines CCRM-VEM and CCRM-VEM/VBL.[82] These findings corroborate and extend earlier results reported by Hardt et al. for C10/C11 dehydro-Epo C, which had been isolated as a minor fermentation product from cultures of the myxobacterium *S. cellulosum*.[12] Subsequent to these findings, Rivkin et al. have shown that the presence of a *trans*-double bond between C9 and C10, such as in **5**, similarly results in a marked increase in antiproliferative activity over Epo D (the $IC_{50}$ value of **4** against the human leukemia cell line CCRF-CEM is 0.9 nM vs. 3.6 nM for Epo D).[86] Likewise, the C12/C13 epoxide corresponding to **5**, that is, **7**, is three- to four-fold more potent than Epo B.[87] In contrast, *cis*-analog **8** has been reported by White et al. to be circa 30-fold less active than Epo D against the human epidermoid cancer cell line KB-31.[91] (Note that the compound assumed to be *trans*-analog **5** in Ref. 91 later was found to be in fact *cis*-isomer **8**[92]). These data support findings from recent spectroscopic studies,[93] which suggest that the bioactive conformation of epothilones is characterized by *anti*-periplanar conformations about the C9/C10 and C10/C11 bonds. For analogs **5** and **7**, it has also been suggested that the presence of a C9/C10 *trans*-double bond favors the bioactive conformation of the macrocycle in the C5–C8 polyketide region.[87] In light of these recent findings, it seems likely that the lack of biological activity in analog **3** is related to the increase in steric bulk associated with the presence of the phenylene moiety.

*trans*-9,10-Didehydro epothilone analogs **5** and **7** were found to possess markedly improved in vivo antitumor activity over their respective parent structures Epo D and Epo B in a mouse model of human breast cancer MX-1. For **5**, this effect was specifically ascribed to a combination of enhanced antiproliferative activity and improved plasma stability in mice,[86] but unfortunately, the compound is also associated with significantly enhanced toxicity.[88] In contrast, the corresponding C26-trifluoro derivative **6** exhibits exquisite antitumor activity in mouse models of human breast (MX-1) and lung (A549) carcinoma as well as in a model of Taxol-resistant lung carcinoma (A549/Taxol) in the absence of unacceptable overt toxicity.[88] The in vitro antiproliferative of **6** is comparable with that of Epo D, and the enhanced in vivo activity of the compound, as for the nonfluorinated analog **5**, may be a consequence of improved pharmacokinetic properties. The discovery of this compound could mark a major milestone in epothilone-based anticancer drug discovery, and it represents the preliminary culmination of the extensive efforts of the Danishefsky group in this area.

Apart from the discovery of the potent in vivo activity of **6**, recent work of the Danishefsky laboratory has also shown that the presence of a *trans*-double bond between C10 and C11 allows the insertion of an additional methylene group between C11 and C12 (thus creating a 17-membered ring) without substantial loss in antiproliferative activity. Thus, in contrast to previously studied ring-expanded analogs (vide supra; Ref. 76), compound **9** is only four-fold less active against the human leukemia cell line CCRF-CEM than the parent compound Epo D.[84]

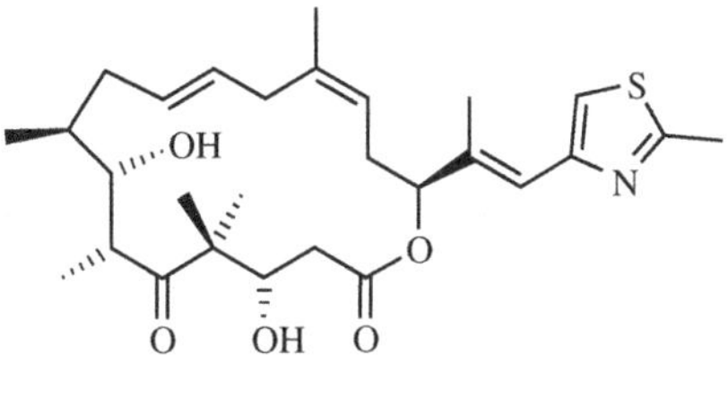

**9**

As for other modifications in the Northern part of the epothilone macrocycle, the replacement of C10 by oxygen has recently been shown to be detrimental for biological activity,[94] whereas the incorporation of a furan moiety incorporating C8, C9, and C10 seems to be better tolerated.[95]

### 1.3.3 Modifications of the Epoxide Moiety

A large part of the early SAR work on epothilones has focused on modifications of the epoxide moiety at positions 12/13 of the macrolactone ring. These studies have demonstrateded that the presence of the epoxide ring is not an indispensible prerequisite for efficient microtubule stabilization and potent antiproliferative activity. Thus, Epo C (**10**) and D (**11**) (Figure 1-1) are virtually equipotent inducers of tubulin polymerization as Epo A and B, respectively. They are also potent inhibitors of human cancer cell growth in vitro,[20,45,62,96–99] although antiproliferative activity is somewhat reduced in comparison with the corresponding parent epoxides. For

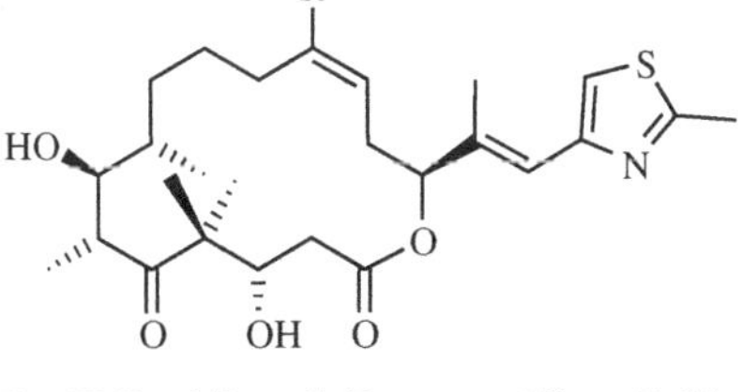

R = H: Epothilone C (Deoxyepothilone A) **10**
(IC$_{50}$ KB-31: 25.9 nM)

R = CH$_3$: Epothilone D (Deoxyepothilone B) **11**
(IC$_{50}$ KB-31: 2.70 nM)

**Figure 1-1.** Molecular structures of deoxyepothilones. Numbers in parentheses are IC$_{50}$-value for growth inhibition of the human epidermoid carcinoma cell line KB-31. Data are from Altmann et al.[45]

example, Epo D inhibits the growth of the human epidermoid cancer cell line KB-31 and the leukemia cell line CCRF-CEM with $IC_{50}$-values of 2.7 nM[45] and 9.5 nM,[63] respectively, versus $IC_{50}$'s of 0.19 nM and 0.35 nM for Epo B. The reduced antiproliferative activity of Epo D compared with Epo B may be related to differences in cellular uptake between the two compounds.[48] Like Epo B, Epo D is equally active against drug-sensitive and multidrug-resistant human cancer cell lines, which indicates that it, too, is a poor substrate for the P-gp efflux pump.

Epo D has been extensively characterized in vivo by the group at the Sloan-Kettering Cancer Center. Employing a specifically optimized intravenous dosing regimen (30 mg/kg, 6-h infusion, q2d × 5), the toxicity and efficacy of Epo D was shown to be comparable with that of paclitaxel when tested against MX-1 breast carcinoma and HT-29 colon tumors.[100] However, the compound was found far superior to paclitaxel when tested in two multidrug-resistant models, MCF-7/Adr and CCRF-CEM/paclitaxel. Therapeutic effects ranged from tumor stasis to complete tumor regressions at the end of the treatment period, despite the fact that Epo D exhibits a very short half-life in rodent plasma, similar to what has been reported for the parent compound Epo B[70] (vide supra). In contrast, the compound is significantly more stable in human plasma (in vitro),[72] thus indicating that plasma stability is unlikely to be limiting for therapeutic applications of lactone-based epothilone analogs in humans.

The replacement of the oxirane ring in epothilones by a cyclopropane[101–103] or variously N-substituted aziridine[104] moieties is generally well tolerated and can even lead to enhanced cellular potency (Figure 1-2). Moreover, replacement of the epoxide oxygen by a methylene group was recently shown to produce enhanced binding to stabilized microtubules in a study by Buey et al.[46] Together with the earlier data on Epo C and D, these findings indicate that the oxirane ring system in epothilones simply serves to stabilize the proper bioactive conformation of the macrocycle rather than to act as a reactive electrophile or a hydrogen bond acceptor. The C12/C13 cyclobutyl analog of Epo A is less potent than **12** (Figure 1-2), but the magnitude of the activity loss seems to be cell line-dependent.[103]

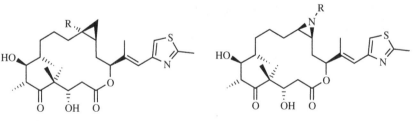

R = H: **12** ($IC_{50}$ HCT-116: 1.4 nM)
R = CH₃: **13** ($IC_{50}$ HCT-116: 0.7 nM)

R = H: **14** ($IC_{50}$ HCT-116: 2.7 nM)
R = CH₃: **15** ($IC_{50}$ HCT-116: 0.13 nM)

**Figure 1-2.** Molecular structures of cyclopropane- and aziridine-based analogs of epothilones. Numbers in parentheses are $IC_{50}$-values for growth inhibition of the human colon carcinoma cell line HCT-116. Data are from Johnson et al.[101] (**12** and **13**) and Regueiro-Ren et al.[104] (**14** and **15**).

Our work in the area of C12/C13-modified epothilone analogs was initially guided by the potent biological activity associated with the deoxyepothilone structural framework; this approach will be discussed in Section 1.3.6. In addition, we have investigated a series of semisynthetic derivatives of Epo A, which were obtained through nucleophilic ring opening of the epoxide moiety.[45] Acid-catalyzed hydrolytic epoxide opening in Epo A leads to an inseparable mixture of *trans*-diols, which were elaborated into the corresponding acetonides **16a** and **16b**.[27–36,67] The corresponding *cis*-analogs **17a** and **17b** were obtained via Epo C and standard *cis*-dihydroxylationn of the double bond.[45] Very similar chemistry has been independently reported by Sefkow et al.[67]

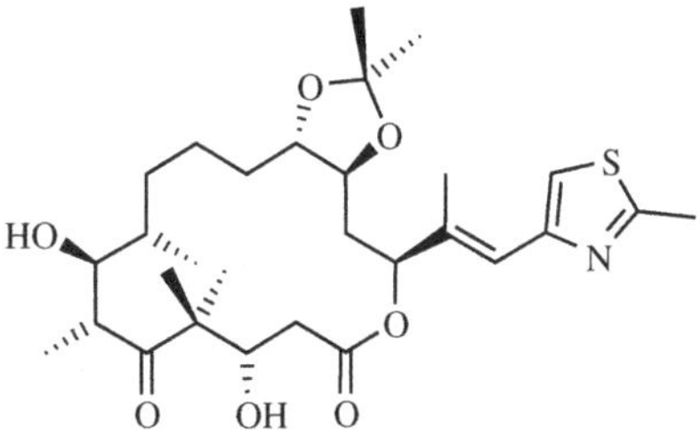

16a

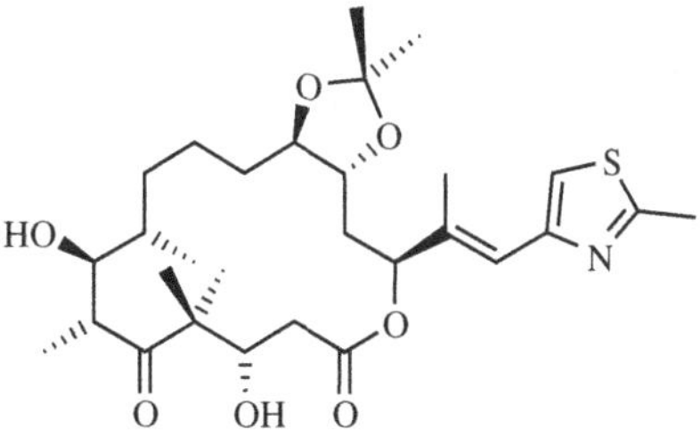

16b

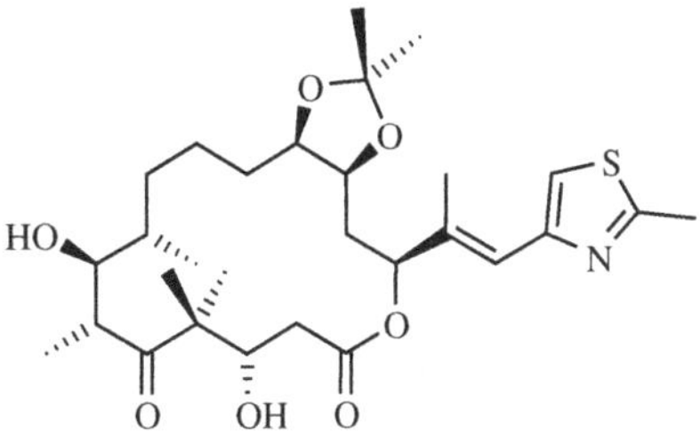

17a

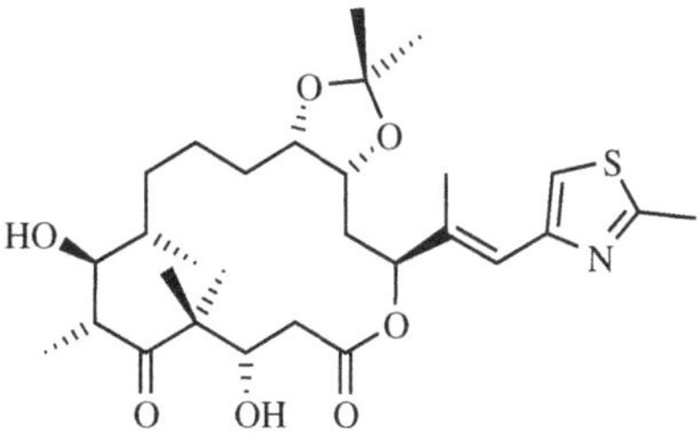

17b

None of the various diols or a related amino alcohol (structure not shown) showed any appreciable biological activity, with IC$_{50}$'s for cancer cell growth inhibition being above 1 μM in all cases. In contrast, azido alcohol **18** (obtained through epoxide ring opening with NaN$_3$) is significantly more potent (e.g., IC$_{50}$'s of **18** against the human epidermoid cancer cell lines KB-31 and KB-8511 are 61 nM and 64 nM, respectively).[45]

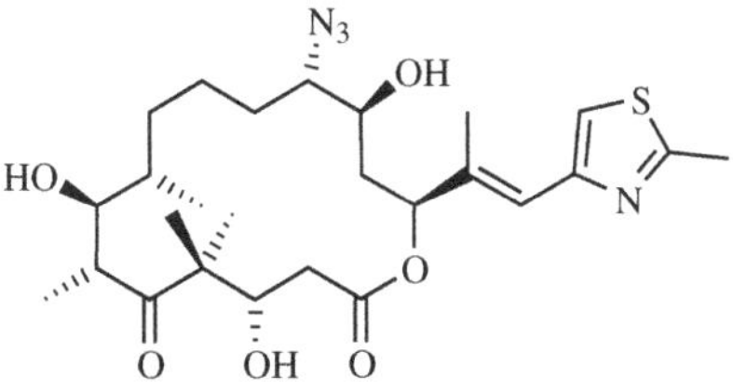

18

This compound indicates that the loss in activity for these diols cannot be simply ascribed to increased conformational flexibility. However, the interpretation of changes in cellular activity is not straightforward, as these may be caused by a combination of changes in target affinity, cellular penetration, and metabolic stability. Analog **18** showed no measurable (>10%) induction of tubulin polymerization at 2 μM compound concentration (vs. 69% for Epo A), but we have not determined the $EC_{50}$-value for induction of tubulin polymerization. Interestingly, acetonides **16a** and **17a** are highly active antiproliferative agents, which are only 10–15-fold less potent than Epo A [$IC_{50}$-values against the KB-31 (KB-8511) line are 23 nM (10 nM) and 30 nM (17 nM) for **16a** and **17a**, respectively], whereas the respective diastereoisomers **16b** and **17b** are 30–100-fold less potent. These data suggest that for a tetrahedral geometry at C12 and C13, the size of the ring fused to the C12-C13 single bond can be significantly increased without substantial loss in biological potency (in contrast to analogs with a planar geometry of the C12-C13 bond; vide infra). Moreover, the data for **16b** also illustrate that, given the proper absolute stereochemistry at C12 and C13, activity is retained even upon moving from a *cis*- to a *trans*-fused system (vide infra).

Another intriguing feature of the epothilone SAR revealed during early SAR studies was that even C12/C13 *trans*-analogs of epothilones exhibit potent tubulin-polymerizing as well as antiproliferative activity.[62,96,98,99] The literature data available at the outset of our work in this area indicated that *trans*-deoxyepothilone A was only slightly less active than deoxyepothilone A (Epo C), whereas in the B series, the activity difference seemed to be more pronounced. At the same time, *trans*-epothilone A was reported by Nicolaou et al. to be virtually equipotent with Epo A on an ovarian (1A9) and a breast cancer (MCF-7) cell line.[98] However, the absolute stereochemistry of the active epoxide isomer was not reported, and the *trans* isomers were obtained as minor components during the synthesis of the natural *cis* isomers rather than being the result of a directed synthetic effort. In view of the interesting biological features of *trans*-(deoxy)epothilones, we embarked on a project directed at the stereoselective synthesis of *trans*-epothilones A, the determination of the absolute stereochemistry of the bioactive isomer, and a more exhaustive biological characterization of this compound.[105] The results of these studies clearly demonstrate that compound **19**, which retains the natural stereochemistry at C13, is a strong inducer of tubulin polymerization in vitro and exhibits potent antiproliferative activity, whereas its (12R,13R)-isomer **19a** is at least 500-fold less active than **19**.[105]

19

19a

(12$S$,13$S$) *trans*-epothilone A (**19**) in fact shows slightly higher growth inhibitory activity than Epo A, and we have observed this rank order of activity across a wide range of human cancer cell lines, which makes this compound an interesting candidate for in vivo profiling (e.g., average $IC_{50}$ values across a panel of seven human cancer cell lines of 2.72, 0.30, and 1.32 nM have been reported for Epo A, Epo B, and **19**, respectively[105]). Whether the *cis/trans* equivalence observed for Epo A and *trans*-epothilone A (**19**) also occurs in the Epo B series is unclear at this point, as the literature data on this question are somewhat contradictory (an 88-fold activity difference between Epo B and *trans*-Epo B against the human ovarian cancer cell line 1A9 is reported in,[98] whereas a difference of only 8-fold is reported in Ref. 99 for the same cell line; cf. also Ref. 106). On the other hand, our finding that the *trans*-Epo A scaffold of **19** is associated with potent biological activity has recently been confirmed and expanded by Nicolaou et al. for a series of highly potent C12/C13 cyclopropane-based analogs of **19**.[103,106]

### 1.3.4   C-26-Modified Analogs

In addition to the changes in the epoxide structure, a variety of modifications of the 26-methyl group in Epo B or D have been reported. These studies have shown that the replacement of one hydrogen atom of this methyl group by relatively small and apolar substituents such as F, Cl, $CH_3$, or $C_2H_5$ (Figure 1-3, X = $CH_2F$, $CH_2Cl$, $C_2H_5$, n-$C_3H_7$), is well tolerated, thus producing analogs that are only slightly less potent in vitro than Epo B or Epo D.[62,99,107]

In general, activity decreases with increasing size of the C26-substituents,[62,99] but exceptions from this general theme have been reported recently. Thus,

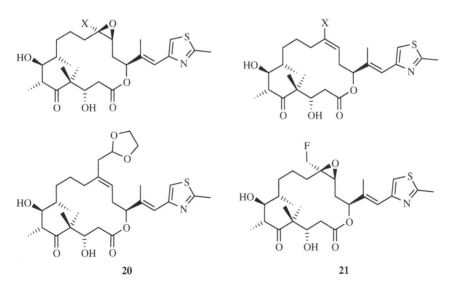

**20**                                    **21**

**Figure 1-3.** Molecular structures of C-26-modified analogs of Epo B and D. For meaning of X, see the text.

C26-(1,3-dioxolanyl)-12,13-Epo D **20** (Figure 1-3) exhibits enhanced in vitro anti-proliferative activity over Epo D,[108] but it was found to be significantly less effica-cious than Epo D in vivo. In contrast, C26-fluoro-Epo B **21** (Figure 1-3), which exhibits comparable in vitro antiproliferative activity with Epo B,[107] was demon-strated to possess significantly better antitumor activity than paclitaxel in a human prostate xenograft model when both compounds were administered at equitoxic doses.[109] No comparison with Epo B was included in this work, but data from our laboratory indicate that the in vivo profile of C26-fluoro-Epo B is similar to that of Epo B.[110]

### 1.3.5 Side-Chain Modifications

Not too surprisingly the second part of the epothilone structure that has been tar-geted for SAR studies most extensively, apart from the epoxide moiety, is the unsa-turated heterocycle-bearing side chain. Structural changes in this area, in particular involving the pendant heterocycle, hold the potential to modulate the physico-chemical, and perhaps the pharmacokinetic, properties of the natural products. The corresponding SAR studies include modifications of the thiazole moiety at the 2- and 4-positions,[99,106,111–113] the replacement of the thiazole ring by other heterocyclic structures[62,99,114] or a simple phenyl group,[62,106,115] and the synthesis of C16-desmethyl-Epo B.[115,116] For example, these studies have shown that the allylic methyl group attached to C16 can be removed with only a minor change in biological activity (see Ref. 45). Likewise, the substitution of oxygen for sulphur in the heterocycle (to produce oxazole-derived epothilone analogs) does not affect biological potency.[62,99] Replacement of the 2-methyl group on the thiazole ring by relatively small substituents such as $CH_2OH$, $CH_2F$, $SCH_3$, or $CH_2CH_3$ is well tol-erated, whereas more bulky substituents result in a substantial loss in potency.[37,105]

22                                          23

Out of this latter family of analogs, in vivo data have been reported by the Sloan-Kettering group for deoxyEpo F [C21-hydroxy-Epo D (**22**)]. The compound was found to exhibit comparable in vivo efficacy as Epo D,[74,117] but by virtue of its enhanced water-solubility, it may be a more attractive drug development candidate. Employing a 6-h continuous intravenous infusion regimen for both compounds, **22** was found to have significantly superior antitumor effects over BMS-247550 (**1**) in a CCRF-CEM as well as a MX-1 tumor model.[72,74] It should be emphasized, how-ever, that although the 6-h continuous infusion schedule may be optimal for Epo D and **22**, this is not necessarily the case for BMS-247550 (**1**), which thus renders the interpretation of these comparisons problematic.

C21-Amino-Epo B [BMS-310705 (**23**)] is undergoing clinical evaluation in humans, but only limited biological data are currently available in the literature for this compound. Thus, an $IC_{50}$ value of 0.8 nM for growth inhibition of the human epidermoid cancer cell line KB.31 has been reported in a patent application[118] (vs 1.2 nM for Epo B under comparable experimental conditions[111]). More recently, Uyar et al.[119] have demonstrated that 50 nM BMS-310705 induces substantial apoptosis in early passage ovarian cancer cells (OC-2), which were derived from a clinical tumor sample and were refractory to paclitaxel and platinum treatment. A concentration of 50 nM of BMS-310705 (**23**) is clinically achievable at a dose of 10 mg/m$^2$, which is below the phase I maximum tolerated dose (MTD) for the compound.[39,119] BMS-310705 (**23**) exhibits improved water-solubility over BMS-247550 (**1**), which enables the use of clinical formulations not containing Cremophor-EL.[39]

Major contributions to the area of heterocycle modifications in epothilones have come from the collaborative work of the Nicolaou group at The Scripps Research Institute (TSRI) in La Jolla, CA, and our group at Novartis. One of the most significant findings of this research is that pyridine-based analog **24** (Figure 1-4) and methyl-substituted variants thereof are basically equipotent with Epo B,[114] which clearly demonstrates that the presence of a five-membered heterocycle attached to C17 is not a prerequisite for highly potent biological activity.

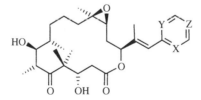

X, Y, Z = N, CH, CH: **24** ($IC_{50}$ KB-31: 0.30 nM)
X, Y, Z = CH, N, CH: **25** ($IC_{50}$ KB-31: 4.32 nM)
X, Y, Z = CH, CH, N: **26** ($IC_{50}$ KB-31: 11.8 nM)

X, Y, Z = N, N, CH: **27** ($IC_{50}$ KB-31: 8.78 nM)
X, Y, Z = N, CH, N: **28** ($IC_{50}$ KB-31: 14.9 nM)
X, Y, Z = CH, CH, CH: **29** ($IC_{50}$ KB-31: 2.88 nM)

**Figure 1-4.** Molecular structures of pyridinyl-, pyrimidinyl-, and phenyl-based Epo B analogs. Numbers in parentheses are $IC_{50}$-values for growth inhibition of the human epidermoid carcinoma cell line KB-31. Data are from Nicolaou et al.[114] and Altmann et al.[120]

At the same time, it was shown that the occurrence of Epo B-like activity in pyridine-based Epo B analogs (i.e., sub-nM $IC_{50}$'s for growth inhibition) strongly depends on the proper position of the ring N-atom, which needs to be located *ortho* to the attachment point of the linker between the heterocycle and the macrocyclic skeleton. Different positioning of the ring N-atom such as in **25** and **26** (Figure 1-4) leads to a significant decrease in cellular potency.[114] Moreover, the incorporation of a second nitrogen atom either at the 3- or the 4- position of the six-membered ring (**27** or **28**; Figure 1-4) results in a profound decrease in antiproliferative activity, even with one N-atom in the obligatory position.[120] In fact, the corresponding analogs are even less potent than phenyl-derived analog **29** (Figure 1-4).[120] The underlying reasons for these differences have not been determined at this point, and their understanding will require structural information on complexes between β-tubulin and various types of epothilone analogs.

In addition to analogs incorporating an olefinic double bond as a linker between the macrolactone ring and different types of heterocycles, we have also studied a new family of side-chain modified structures, which are characterized by rigidification of the entire side-chain manifold (exemplified for quinoline-based analogs **30** and **31**).[66,121]

30                                      31

The design of these analogs was guided by preliminary nuclear magnetic resonance (NMR) data on the bioactive (tubulin-bound) conformation of epothilones, which indicated that the C16/C17 double bond and the aromatic C18-N bond were present in a transoid arrangement (corresponding to a ~180° C16-C17-C18-N22 torsion angle). (These data have subsequently been consolidated and have recently appeared in the literature[93]).

In general, analogs of this type are more potent inhibitors of human cancer cell proliferation than the respective parent compounds Epo D and Epo B.[121] For example, compounds **30** and **31** inhibit the growth of the human epidermoid carcinoma cell line KB-31 with $IC_{50}$ values of 0.11 nM and 0.59 nM, respectively, versus 0.19 nM and 2.77 nM for Epo B and Epo D. As observed for other analogs of this type (incorporating benzothiazole-, benzoxazole-, or benzimidazole-type side chains), the activity difference is more pronounced in the deoxy case, with **30** being an almost five-fold more potent antiproliferative agent than Epo D.[121] Only a few other analogs with enhanced in vitro activity over natural epothilones have been described in the literature so far, recent examples being compounds **5** and **7**.[86,87] Interestingly, however, the observed increase in antiproliferative activity does not seem to be a consequence of more effective interactions with tubulin (data not shown), but it may be related to parameters such as cell penetration or intracellular accumulation. Given that Epo D is currently undergoing phase II clinical trials, the improved antiproliferative activity of analog **30** and related structures could make these compounds interesting candidates for continued development.

Most recently, the collaborative work between the Nicolaou group at TSRI and our group at Novartis has resulted in the discovery of 20-desmethyl-20-methyl-sulfanyl-Epo B (**32** = ABJ879) as a highly promising antitumor agent,[106,122] which has recently entered phase I clinical trials sponsored by Novartis.

**32**                                         **33**

ABJ879 (**32**) induces tubulin polymerization in vitro with slightly higher potency than Epo B or paclitaxel. At the same time, the compound is a markedly more potent antiproliferative agent, with an average $IC_{50}$ for growth inhibition across a panel of drug-sensitive human cancer cell lines of 0.09 nM versus 0.24 nM for Epo B and 4.7 nM for paclitaxel.[122] ABJ879 (**32**) retains full activity against cancer cells overexpressing the drug efflux pump P-gp or harboring tubulin mutations.

The binding of **32** to stabilized microtubules has been carefully evaluated in a recent study by Buey et al.[46] At 35 °C, a change in binding free energy of $-2.8$ kJ/mole is observed upon replacement of the C20-methyl group in Epo B by a methylsulfanyl group, which corresponds to an increase in the binding constant from $7.5 \times 10^8$ for Epo B to $2.5 \times 10^9$ for ABJ879 (**32**). Interestingly, binding enthalpy is less favorable for ABJ879 than for Epo B and the increase in binding free energy for ABJ879 (**32**) is in fact entropy driven. In the same study, an increase in binding free energy was also observed for C12/C13-cyclopropane-based epothi-lone analogs, and the energetic effects of the replacement of the C20-methyl group by a methylsulfanyl group and the substitution of a methylene group for the epoxide oxygen are in fact additive. Thus, analog **33** binds to stabilized microtubules with 27.4-fold enhanced affinity over Epo B ($\Delta\Delta G^{35°C} = -8.2$ kJ/mole).[46] Further-more, this compound in some cases has been found to be a more potent antiproli-ferative agent in vitro than either Epo B or ABJ879 (**32**) (e.g., $IC_{50}$-values for growth inhibition of the human ovarian carcinoma cell line 1A9 are 0.3 (0.6) nM, 0.17 nM, and 0.10 n for Epo B, **32**, and **33**, respectively[106,123]). ABJ879 (**32**) has demonstrated potent antitumor activity[122] in experimental human tumor models in mice, where it produced transient regressions and inhibition of tumor growth of slow-growing NCI H-596 lung adenocarcinomas and HT-29 colon tumors. Inhi-bition of tumor growth was observed in fast-growing, difficult-to-treat NCI H-460 large cell lung tumors. Finally, single-dose administration of ABJ879 (**32**) resulted in long-lasting regressions and cures in a Taxol-resistant KB-8511 epidermoid carcinoma model.

### 1.3.6  Aza-Epothilones

Of the numerous epothilone analogs recently reported in the literature, only a few are characterized by the replacement of carbon atoms in the macrocyclic skeleton by heteroatoms, and those investigated to date were found to be poorly active (for examples, see Refs. 69, 94, and 120). Overall, however, the potential of such modifications remains largely unexplored, which is despite the fact that the replacement of carbon by heteroatoms in complex structures could lead to improved synthetic accessibility and offer the potential to generate large sets of diverse analogs in a straightforward manner (e.g., through amide bond formation or reductive amination in the case of nitrogen). In light of this fact and guided by the potent biological activity associated with the deoxyepothilone structural framework (vide supra), some of our initial work in the area of epothilone modifications was directed at the replacement of the C12/C13 olefinic double bond in Epo D by *N*-alkyl amides and 1,2-disubstituted heterocycles, such as imidazole. Structural units of this type were hypothesized to act as *cis* C=C double bond mimetics and thus to result in a similar conformation of the macrocycle as for Epo D (assuming a preference of the C–N partial double bond for a *cis* conformation). As a consequence, such analogs were expected to exhibit similar antiproliferative activities as the parent deoxyepothilones. At the same time, these polar double bond substitutes were assumed to lead to improved aqueous solubility of the corresponding analogs over the very lipophilic Epo D.

R = CH$_3$: (**34**)
R = C$_2$H$_5$: (**35**)
R = H: (**36**)

(**37**)

Unfortunately, none of the analogs **34**, **35**, **37**, or (*N*-unsubstituted secondary amide) **36**, which would mimic a *trans*-olefin geometry) showed any appreciable tubulin-polymerizing or antiproliferative activity, despite the fact that preliminary NMR studies with compound (**34**) in DMSO/water indicate that the preferred conformation about the 12/13 *N*-methyl amide bond is indeed *cis*, that is, the methyl group and the carbonyl oxygen are located on the same side of the partial C–N double bond (*cis/trans*-ratio ~4/1; **34** may thus be considered a direct structural mimetic of Epo D). The underlying reasons for the lack of biological activity of analogs **34–37** have not been elucidated, but subsequent data obtained in various laboratories,[124,125] including ours, for other (non-amide-based) structures suggested that increasing the steric bulk at C13 was generally associated with reduced potency. In light of these findings, we decided to continue exploration of the potential utility of nitrogen incorporation at position 12 of the macrocycle, as a functional handle for additional substitution, without concomitant modification of C13. At the most straightforward level, this approach involved simple acylation of the 12-nitrogen atom, which thus lead to amide- and carbamate-based analogs of type **38**, whose carbonyl oxygen could potentially assume the role of the epoxide oxygen in natural epothilones.

**38**

Analogs **38** were tested for their ability to promote in vitro tubulin polymerization, and their antiproliferative activity was assessed against the human epidermoid cancer cell lines KB-31 and KB-8511, which serve as representative examples of drug-sensitive and P-gp-overexpressing, multidrug-resistant human cancer cell lines, respectively (see, e.g., Refs. 20 and 37).

As illustrated by the data summarized in Table 1-2, compounds of type **38**, although less active inhibitors of cancer cell growth than Epo A or B, can indeed

**TABLE 1-2. Induction of Tubulin Polymerization and Growth Inhibition of Human Carcinoma Cell Lines by 12-Aza-Epothilones 38**

| Compound | R | % Tubulin Polymerization[a] | IC$_{50}$ KB-31 [nM][b] | IC$_{50}$ KB-8511 [nM][b] |
|---|---|---|---|---|
| **38a** | O-*tert*-C$_4$H$_9$ | 27 | 31 | 105 |
| **38b** | OCH$_2$Ph | <10 | 297 | 703 |
| **38c** | OC$_2$H$_5$ | 17 | 85 | 465 |
| **38d** | O-*iso*-C$_3$H$_7$ | <10 | 297 | 737 |
| **38e** | CH$_3$ | <10 | 116 | N. D.[c] |
| **38f** | C$_2$H$_5$ | <10 | 71 | 1352 |
| **38g** | *tert*-C$_4$H$_9$ | <10 | 206 | >1000 |
| **38h** | C$_6$H$_5$ | <10 | >1000 | >1000 |

[a]Induction of polymerization of porcine brain microtubule protein by 5 µM of test compound relative to the effect of 25 µM of Epo B, which gave maximal polymerization (85% of protein input).
[b]IC$_{50}$ values for growth inhibition of human epidermoid carcinoma cell lines KB-31 and KB-8511. Values generally represent the average of two independent experiments.
[c]Not determined.

be potent antiproliferative agents. Interestingly, however, some of these analogs are significantly less active against the multidrug-resistant KB-8511 line than the drug-sensitive KB-31 parental line, thus indicating that compounds **38** are better P-gp substrates than natural epothilones. The structural basis for this phenomenon is not understood at this point, but the finding is in line with a more general tendency for polar epothilone analogs (e.g., compounds incorporating (other) amide bonds or additional hydroxyl groups) to exhibit increased resistance factors in the KB-31/ KB-8511 cell line pair (i.e., increased ratios of $IC_{50}$ (KB-8511)/$IC_{50}$ (KB-31); M. Wartmann and K.-H. Altmann, unpublished observations). The most interesting of the compounds included in Table 1-2 is analog **38a**, which is only circa 15-fold less active against the drug-sensitive KB-31 line than Epo A (and thus roughly equipotent with Epo C) and is characterized by an only modest resistance factor of ~3. This analog also shows measurable induction of tubulin polymerization in vitro, but it remains to be determined whether the antiproliferative activity of **38a** is mainly related to interference with microtubule functionality or whether other/additional mechanisms may also be operative.

A third type of aza-epothilones, which we have investigated as part of our program on backbone-modified hetero-analogs of epothilones, is characterized by the replacement of C4 by nitrogen and the presence of a C5/N4 amide group rather than a C5-ketone (**39**; Ref. 126). These analogs were inspired by the fact that one of the characteristic features of the tubulin-bound structure of Epo A[93] is the presence of a *syn*-periplanar conformation about the C4–C5 bond. The same geometry would be enforced in analogs of type **39**, provided that the amide bond between N4 and C5 would be present in a *cis* conformation. At the same time, preliminary modeling studies indicated that the presence of a *cis* amide bond in this position should allow replacement of the C1–C4 segment by various types of β-amino acids without causing significant distortions in the bioactive conformation of the C5–O16 segment. Apart from these structural considerations, structures of type **39** also appeared attractive for chemical reasons, as they would lend themselves to an efficient combinatorial chemistry approach employing a single advanced intermediate (i.e., a C5-carboxylic acid encompassing the C21–C5 fragment **40**; epothilone numbering).

**39**                                         **40**

So far, only a few examples of analogs of type **39** have been investigated, all of which were found to lack any meaningful tubulin polymerizing or antiproliferative activity.[126] However, many of these structures (incorporating different types of α-, β-, and γ-amino acids) will need to be investigated before allowing a final conclusion on the (pharmaceutical) validity of this modification approach. Based on the chemistry developed in our laboratory for these compounds,[126] the synthesis of such additional analogs should be a straightforward undertaking.

## 1.4  PHARMACOPHORE MODELING AND CONFORMATIONAL STUDIES

Our current understanding of the three-dimensional conformation of tubulin is largely based on the structure of a tubulin/docetaxel complex within a two-dimensional tubulin polymer sheet, which has been solved by electron crystallography at 3.7-Å resolution.[58] The availability of this information has significantly improved our gross understanding of paclitaxel binding to β-tubulin, but the structure-based design of epothilone analogs or mimics so far has been hampered by the lack of high (atomic level)-resolution structural data either for tubulin or tubulin/microtubule-epothilone complexes. However, recent studies on the tubulin-bound conformation of Epo A either by NMR spectroscopy on a soluble β-tubulin/Epo A complex[93] or by a combination of electron crystallography, NMR spectroscopic conformational analysis, and molecular modeling of a complex between Epo A and a $Zn^{2+}$-stabilized two-dimensional α,β-tubulin sheet (solved at 2.89 Å resolution)[127] have provided completely new insights into the bioactive conformation of the epothilone-class of microtubule inhibitors (vide infra). Before the availability of these data, various attempts have been described to develop a predictive pharmacophore model for epothilones, which would be of substantial value for the design of new analogs. Different approaches have been followed to address this problem, which were generally based on the assumption of a common tubulin binding site between epothilones and paclitaxel.[57,77–80] For example, the common paclitaxel/epothilone pharmacophore model presented by Giannakakou et al.[57] is based on an energy-refined model of the 3.7-Å density map of docetaxel bound to β-tubulin.[58] According to this model, the position of the epoxide oxygen in epothilones within the microtubule binding pocket corresponds with that of the oxetane oxygen in paclitaxel, whereas the epothilone side chain is located in the same region as either the C3′-phenyl group or, alternatively, the C2-benzoyloxy moiety of paclitaxel. The model also suggests that the methyl group attached to C12 in Epo B is involved in hydrophobic interactions with the side chains of Leu-β273, Leu-β215, Leu-β228, and Phe-β270, and that this may account for the higher activity of Epo B versus Epo A. Different conclusions with regard to the relative positioning of paclitaxel and epothilones within the microtubule binding site have been reached by Wang et al.[78] In their model, the position of the thiazole moiety in epothilones within the microtubule binding pocket matches the position of the phenyl group of the C-3′-benzamido substituent in paclitaxel. Furthermore, the epoxide oxygen is concluded not to be involved in interactions with the protein, which is in line with the experimental data discussed here for cyclopropane-based epothilone analogs. A model similar to that of Wang et al. has recently been proposed by Manetti et al.[79] Although these computational models by their very nature are of limited accuracy, some of them can reproduce at least part of the published epothilone SAR with reasonable accuracy.[78,79] They may thus provide a useful basis for the design of new analogs, and experimental data generated for such compounds will then help to additionally refine the model. The notion of a common pharmacophore between paclitaxel and epothilones, however, has been seriously questioned by recent data on the structure of a complex between a $Zn^{2+}$-stabilized two- dimensional tubulin sheet and Epo A.[127] Although paclitaxel and Epo A are located within the same gross binding pocket on the protein, the structure indicates that the compounds exploit this pocket in different ways, that is, through distinctively different sets of hydrogen bonding and hydrophobic interactions. Interestingly, the tubulin-bound conformation of Epo A derived from the electron crystallographic data at 2.89 Å is different from any of the computational models mentioned here, but also from the recent NMR structure of tubulin-bound Epo A as had been determined by Carlomagno et al. by means of magnetization transfer NMR techniques.[93] Although similar to the X-ray crystal structure of Epo A in the C5–C15 part of the macrocycle, a distinct difference between the NMR-derived and X-ray crystal structures exists in the C1–C4 region. Furthermore, the conformation of the side chain attached to C15 in the tubulin-bound conformation is characterized by a *transoid* arrangement of the olefinic double bond between C16 and C17 and the C–N bond of the thiazole ring (i.e., by a C16–C17–C18–N dihedral angle of 180°). In contrast, a dihedral angle C16–C17–C18–N of $-7.6°$ is observed in the X-ray crystal structure of Epo A, and both conformational states are in rapid equilibrium for the unbound compound free in solution.[93] Although agreement exists between Carlomagno et al.'s NMR-derived structure of tubulin-bound Epo A and the more recent structure derived from the $\alpha,\beta$-tubulin sheet/Epo A complex, the structures clearly differ with regard to the conformation in the C1–C9 region. Whether these differences may reflect differences in the basic experimental conditions (soluble tubulin vs. insoluble tubulin-polymer sheets) or whether they may perhaps point to conformational heterogeneity in the tubulin-bound state of epothilones remains to be determined. In any case, this new information on the bioactive conformation of epothilones should provide valuable guidance for the design of structurally new (and hopefully diverse) analogs for biological testing. Apart from the potential for the discovery of new therapeutic agents, such rationally designed analogs should also represent useful tools to probe the topology of the epothilone (paclitaxel) binding site on microtubules in more detail. At the same time, however, it should be kept in mind that even the availability of precise structural information on a $\beta$-tubulin/epothilone complexes might not have an immediate impact on our ability to create analogs with an *improved therapeutic window*, which is the fundamental issue associated with all epothilone-based drug discovery work.

## 1.5   EPOTHILONE ANALOGS IN CLINICAL DEVELOPMENT

As indicated above, the first compound of the epothilone class of microtubule inhibitors to enter clinical trials was Epo B (*EPO906*, Novartis; vide supra). This involved initial testing in two phase I studies, employing either a weekly or a once-every-three-weeks dosing regimen.[64] MTDs of 2.5 mg/m$^2$ and 6.0 mg/m$^2$, respectively, were observed in the two studies, with diarrhoea being the dose-limiting toxicity in both cases. All other toxicities observed at MTD were mild to moderate, and no significant myelosuppression was found. The drug seemed to be particularly active in patients with colorectal cancer, with additional responses in patients with breast, ovarian, and NSCLC cancers, and carcinoid tumors, whose disease had progressed on several other therapies. Phase II trials with Epo B in colorectal cancer are currently ongoing (second-line treatment), and preliminary data seem to indicate that the drug is active.[64]

Subsequent of the initiation of clinical studies with Epo B, a variety of modified analogs have also proceeded to clinical evaluation in humans. The most advanced of these compounds is *BMS-247550* (**1**), and the current state of the clinical pharmacology for this compound has been summarized recently in an excellent review by Lin et al.[128] (see also Ref. 39). (As for all other agents of this class, most of the original information related to clinical trial results with BMS-247550 (**1**) is only available in the form of meeting abstracts and posters). Briefly, clinical trials with BMS-247550 (**1**) were initiated in early 2000, and several objective responses to single-agent treatment with the compound were observed in these studies in breast, ovarian, cervical, prostate, colon, lung, and renal cancers as well as in squamous cell cancers of the head and neck, lymphoma, and angiosarcoma. Dose-limiting toxicities included fatigue, prolonged neutropenia, and peripheral neuropathy.[129] The compound was also shown to be orally bioavailable in humans,[39] which thus confirms previous results from animal studies (vide supra).[71] Very importantly, BMS-247550 (**1**) has been demonstrated to induce microtubule bundling in peripheral blood monocytes (PBMCs) of treated persons, and a good correlation was observed between the magnitude of this effect and plasma areas under the curve.[130] These findings validate the in vitro pharmacodynamic findings with BMS-247550 (**1**) in the clinical setting.

Phase II trials with BMS-247550 (**1**) have produced objective responses for a variety of tumor types, including tumors that had been refractory to treatment with platinum-based drugs or taxanes.[39,128] Based on these highly promising results, the compound has been advanced to phase III studies, which are currently ongoing in parallel with several phase II trials (including combination studies).[39]

Phase I clinical studies have recently been completed with Epo D, and phase II trials with the compound are now ongoing together with an additional phase I study [sponsored by Kosan Biosciences (as KOS-862)].[131,132] As for BMS-247550, KOS-862 was demonstrated to induce microtubule bundling in patient PBMCs, and several tumor responses have been noted in the phase I studies.[131]

The most recent additions to the portfolio of epothilone-type clinical development compounds are the semisynthetic derivatives BMS-310705 (**23**),[39,133] which differs from Epo B by the presence of a primary amino group at C21, and C20-desmethyl-C20-methylsulfanyl-Epo B (**32**, ABJ879).[122] As indicated above, BMS-310705 (**23**) exhibits improved water-solubility over BMS-247550 (**1**), thus allowing the use of clinical formulations not containing Cremophor-EL.[39] Accordingly, no hypersensitivity reactions were observed in phase I studies with BMS-310705 (**23**) in contrast to BMS-247550 (**1**) (in the absence of premedication).

## 1.6 CONCLUSIONS

The discovery by Bollag et al.[11] that Epo A and B represent a new class of micro-tubule depolymerization inhibitors, which are not subject to P-gp-mediated efflux mechanisms, immediately established these compounds as highly interesting lead structures for the development of a new and improved generation of Taxol-like anticancer drugs. This notion was enforced by the fact that Epo A and B are more water-soluble than paclitaxel, thus raising the prospect of improved formulability and the absence of formulation-associated side-effects as they are characteristic for Taxol. At the same time, and in contrast to many other natural product-based drug discovery efforts, total chemical synthesis has proven to be a perfectly feasible approach to access a wide variety of structural analogs of epothilones. This chapter has tried to summarize the basic biology and pharmacology of Epo B as the most potent and most widely studied natural epothilone. It has also outlined the most important structural modifications, which have been investigated for epothilones and the most significant SAR features that have emerged from these studies. These efforts have led to a sizable number of analogs with antiproliferative activities comparable with those of Epo A/B; in addition, many other compounds with epothilone-type structures have been discovered to be highly potent antiproliferative agents, even if they are less active than the corresponding parent epothilones.

To enable the rational design of improved epothilone analogs, different pharmacophore models have been proposed over the past several years, which accommodate and rationalize at least some aspects of the epothilone SAR. These computational efforts have recently been complemented (and perhaps superseded) by the elucidation of the tubulin-bound structure of Epo A by means of NMR spectroscopy and by electron-crystallography. However, notwithstanding these significant advances, it is clear that our current understanding of the interrelationship between structure and biological activity for epothilone-type structures at the molecular level is far from complete, and it remains to be seen how the recent gain in structural information will impact the design of improved epothilone analogs. So far, at least five compounds of the epothilone class have been advanced to clinical development and more are likely to follow, which includes Epo B (EPO906, developed by Novartis), Epo B lactam [BMS-247550, BMS], C21-amino-Epo B (BMS-310705, BMS), Epo D (KOS-862, Kosan/Sloan-Kettering/Roche), and most recently, C20-desmethyl-C20-methylsulfanyl-Epo B (ABJ879, Novartis). None of these compounds has yet successfully completed clinical development, but the available published data for the most advanced compounds EPO906 and BMS-247550 suggest that the highly promising preclinical profile of these agents may indeed translate into therapeutic utility at the clinical level. Interestingly, the clinical profiles of EPO906 and BMS-247550 seem to be distinctly different, which highlights the importance and usefulness of clinical testing even of structurally closely related analogs.

## ACKNOWLEDGMENTS

The author would like to thank all former collaborators at Novartis for their help and inspiration over the last several years. In particular, I want to acknowledge the support by Dr. A. Flörsheimer, Dr. T. O'Reilly, and Dr. M. Wartmann, who had worked with me on the epothilone program at Novartis since its very beginning. My sincere thanks also goes to Prof. K. C. Nicolaou and his group at the The Scripps Research Institute for a very productive collaboration.

## REFERENCES

1. McLaughlin, F.; Finn, P.; La Thangue, N. B. *Drug Discovery Today*, **2003**, 8: 793.

2. Lu, M. C. In Foye, W. O. editor. *Cancer Chemotherapeutic Agents*. Washington DC: American Chemical Society, **1995**, 345.

3. Mekhail, T. M.; Markman, M. *Expert Opin. Pharmacother.*, **2002**, 3: 755. Obasaju, C.; Hudes, G. R. *Hematol./oncol. Clin. N. Am.*, **2001**, 15: 525.

4. Hamel, E. *Med. Res. Rev.*, **1996**, 16: 207.

5. Rowinsky, E. K. *Ann. Rev. Med.*, **1997**, 48: 353.

6. Prinz, H. *Exp. Rev. Anticancer Ther.*, **2002**, 2: 695.

7. Altmann, K.-H. *Curr. Opin. Chem. Biol.*, **2001**, 5: 424.

8. Schiff, P. B.; Fant, J.; Horwitz, S. B. *Nature*, **1979**, 277: 665.

9. Höfle, G.; Bedorf, N.; Gerth, K.; Reichenbach, H. German Patent Disclosure, DE 4,138,042, 1993; Chem. Abstr., **1993**, 120: 52841.

10. Gerth, K.; Bedorf, N.; Höfle, G.; Irschik, H.; Reichenbach, H. *J. Antibiotics*, **1996**, 49: 560.

11. Bollag, D. M.; McQueney, P. A.; Zhu, J.; Hensens, O.; Koupal, L.; Liesch, J.; Goetz, M.; Lazarides, E.; Woods, C. M. *Cancer Res.*, **1995**, 55: 2325.

12. Hardt, I. H.; Steinmetz, H.; Gerth, K.; Sasse, F.; Reichenbach, H.; Höfle, G. *J. Nat. Prod.*, **2001**, 64: 847.

13. Höfle, G.; Bedorf, N.; Steinmetz, H.; Schomburg, D.; Gerth, K.; Reichenbach, H. *Angew. Chem. Int. Ed. Engl.*, **1996**, 35: 1567.

14. ter Haar, E.; Kowalski, R. J.; Hamel, E.; Lin, C. M.; Longley, R. E.; Gunasekara, S. P.; Rosenkranz, H. S.; Day, B. W. *Biochemistry*, **1996**, 35: 243.

15. Long, B. H.; Carboni, J. M.; Wasserman, A. J.; Cornell, L. A.; Cassaza, A. M.; Jensen, P. R.; Lindel, T.; Fenical, W.; Fairchild, C. R. *Cancer Res.*, **1998**, 58: 1111.

16. Mooberry, S. L.; Tien, G.; Hernandez, A. H.; Plubrurkan, A.; Davidson, B. S. *Cancer Res.*, **1999**, 59: 653.

17. Hood, K. A.; West, L. M.; Rouwe, B.; Northcote, P. T.; Berridge, M. V.; Wakefield, St. J.; Miller, J. H. *Cancer Res.*, **2002**, 62: 3356.

18. Isbrucker, A.; Cummins, J.; Pomponi, S.; Longley, R. E.; Wright, A. E. *Biochem. Pharmacol.*, **2003**, 66: 75.

19. Kowalski, R. J.; Giannakakou, P.; Hamel, E. *J. Biol. Chem.*, **1999**, 272: 2534.

20. Altmann, K.-H.; Wartmann, M.; O'Reilly, T. *Biochim. Biophys. Acta*, **2000**, 1470: M79.

21. Wolff, A.; Technau, A.; Brandner, G. *Int. J. Oncol.*, **1997**, 11: 123.

22. Giannakakou, P.; Sackett, D. L.; Kang, Y. K.; Zhan, Z.; Buters, J. T.; Fojo, T.; Poruchynsky, M. S. *J. Biol. Chem.*, **1997**, 272: 17118.

23. Nicolaou, K. C.; Roschangar, F.; Vourloumis, D. *Angew. Chem. Int. Ed. Engl.*, **1998**, 37: 2014.

24. Harris, C. R.; Danishefsky, S. J. *J. Org. Chem.*, **1999**, 64: 8434.

25. Mulzer, J.; Martin, H. J.; Berger, M. *J. Heterocycl. Chem.*, **1999**, 36: 1421.

26. Nicolaou, K. C.; Ritzén, A.; Namoto, K. *JCS Chem. Commun.*, **2001**, 1523.

27. Storer, R. I.; Takemoto, T.; Jackson, P. S.; Ley, S. V. *Angew. Chem. Int. Ed. Engl.*, **2003**, 42: 2521.

28. Liu, Z.-Y.; Chen, Z.-C.; Yu, C.-Z.; Wang, R.-F.; Zhang, R. Z.; Huang, C.-S.; Yan, Z.; Cao, D.-R.; Sun, J.-B.; Li, G. *Chemistry–Eur. J.* **2002**, 8: 3747.

29. Liu, J.; Wong, C.-H. *Angew. Chem. Intl. Ed. Engl.*, **2002**, 41: 1404.

30. Sun, J.; Sinha, S. C. *Angew. Chem. Intl. Ed. Engl.*, **2002**, 41: 1381.

31. Ermolenko, M. S.; Potier, P. *Tetrahedron Lett.*, **2002**, 43: 2895.

32. Martin, N.; Thomas, E. J. *Tetrahedron Lett.*, **2001**, 42: 8373.

33. Martin, H. J.; Pojarliev, P.; Kahlig, H.; Mulzer, J. *Chem. – Eur. J.*, **2001**, 7: 2261.

34. Valluri, M.; Hindupur, R. M.; Bijoy, P.; Labadie, G.; Jung, J.-C.; Avery, M. A. *Org. Lett.*, **2001**, 3: 3607.

35. Hindupur, R. M.; Panicker, B.; Valluri, M.; Avery, M. A. *Tetrahedron Lett.*, **2001**, 42: 7341.

36. Zhu, B.; Panek, J. S. *Eur. J. Org. Chem.*, **2001**, 9: 1701.

37. Wartmann, M.; Altmann, K.-H. *Curr. Med. Chem: Anti-Cancer Agents*, **2002**, 2: 123.

38. Altmann, K.-H. *Mini-Rev. Med. Chem.*, **2003**, 3: 149.

39. Borzilleri, R. M.; Vite, G. D. *Drugs of the Future*, **2003**, 27: 1149.

40. Altmann, K.-H. *Org. Biomol. Chem.*, **2004**, 2: 2137.

41. Chappell, M. D.; Stachel, S. J.; Lee, C. B.; Danishefsky, S. J. *Org. Lett.*, **2000**, 2: 1633.

42. Walsh, C. T.; O'Connor, S.; Schneider, T. L. *J. Ind. Microbiol. Biotechnol.*, **2003**, 30: 448.

43. He, L.; Orr, G. A.; Horwitz, S. B. *Drug Discovery Today*, **2001**, 6: 1153.

44. Altaha, R.; Fojo, T.; Reed, E.; Abraham, J. *Curr. Pharm. Des.*, **2002**, 8: 1707.

45. Altmann, K.-H.; Bold, G.; Caravatti, G.; End, N.; Flörsheimer, A.; Guagnano, V.; O'Reilly, T.; Wartmann, M. *Chimia*, **2000**, 54: 612.

46. Buey, R. M.; Diaz, J. F.; Andreu, J. M.; O'Brate, A.; Giannakakou, P.; Nicolaou, K. C.; Sasmal, P. K.; Ritzén, A.; Namoto, K. *Chem. Biol.*, **2004**, 11: 225.

47. Jordan, M. A.; Wendell, K.; Gardiner, S.; Derry, W. B.; Copp, H.; Wilson, L. *Cancer Res.*, **1996**, 56: 816.

48. Lichtner, R. B.; Rotgeri, A.; Bunte, T.; Buchmann, B.; Hoffmann, J.; Schwede, W.; Skuballa, W.; Klar, U. *Proc. Natl. Acad. Sci. USA*, **2001**, 98: 11743.

49. Jordan, M. A. *Curr. Med. Chem: Anti-Cancer Agents*, **2002**, 2: 1.

50. Kamath, K.; Jordan, M. A. *Cancer Res.*, **2003**, 63: 6026.

51. Chen, J.-G. C.; Horwitz, S. B. *Cancer Res.*, **2002**, 62: 1935.

52. Chen, J.-G. C.; Yang, C.-P. H. Y.; Cammer, M.; Horwitz, S. B. *Cancer Res.*, **2003**, 63: 7891.

53. Bröker, L. E.; Huisman, C.; Ferreira, C. G.; Rodriguez, J. A.; Kruyt, F. A. E.; Giaccone, G. *Cancer Res.*, **2002**, 62: 4081.

54. Bröker, L. E.; Huisman, C.; Span, S. W.; Rodriguez, J. A.; Kruyt, F. A. E.; Giaccone, G. *Cancer Res.*, **2004**, 64: 27.

55. McDaid, H. M.; Horwitz, S. B. *Mol. Pharmacol.*, **2001**, 60: 290.

56. Garraway, L. A.; Chabner, B. *Eur. J. Cancer*, **2002**, 38: 2337.

57. Giannakakou, P.; Gussio, R.; Nogales, E.; Downing, K. H.; Zaharevitz, D.; Bollbuck, B.; Poy, G.; Sackett, D.; Nicolaou, K. C.; Fojo, T. *Proc. Natl. Acad. Sci. USA*, **2000**, 97: 2904.

58. Nogales, E.; Wolf, S. G.; Downing, K. H. *Nature*, **1998**, 391: 199.

59. He, L.; Yang, C.-P.; Horwitz, S. B. *Molecular Cancer Ther.*, **2001**, 1: 3.

60. Verrills, N. M.; Flemming, C. L.; Liu, M.; Ivery, M. T.; Cobon, G. S.; Norris, M. D.; Haber, M.; Kavallaris, M. *Chem. Biol.*, **2003**, 10: 597.

61. Cabral, F.; Barlow, S. B. *Pharmacol. Ther.*, **1991**, 52: 159.

62. Su, D. S.; Balog, A.; Meng, D.; Bertinato, P.; Danishefsky, S. J.; Zheng, Y. H.; Chou, T. C.; He, L.; Horwitz, S. B. *Angew. Chem. Int. Ed. Engl.*, **1997**, 36: 2093.

63. Chou, T. C.; Zhang, X. G.; Balog, A.; Su, D. S.; Meng, D.; Savin, K.; Bertino, J. R.; Danishefsky, S. J. *Proc. Natl. Acad. Sci. USA*, **1998**, 95: 9642.

64. Rothermel, J.; Wartmann, M.; Chen, T.; Hohneker, T. *Sem. Oncol.*, **2003**, 30(Suppl 6): 51.

65. Pietras, K.; Stumm, M.; Hubert, M.; Buchdunger, E.; Rubin, K.; Heldin, C.-K.; McSheehy, P.; Wartmann, M.; Oestman, A. *Clin. Cancer Res.*, **2003**, 9: 3779.

66. Klar, U.; Skuballa, W.; Buchmann, B.; Schwede, W.; Bunte, T.; Hoffmann, J.; Lichtner, R. In Ojima, I.; Vite, G. D.; Altmann, K.-H. editors. *Anticancer Agents – Frontiers in Cancer Chemotherapy*. Washington, DC: American Chemical Society; ACS Symposium Series 796, **2001**, 131.

67. Sefkow, M.; Kiffe, M.; Schummer, D.; Höfle, G.; *Bioorg. Med. Chem. Lett.*, **1998**, 8: 3025.

68. Sefkow, M.; Kiffe, M.; Höfle, G. *Bioorg. Med. Chem. Lett.*, **1998**, 8: 3031.

69. Vite, G. D.; Borzilleri, R. M.; Kim, S. H.; Regueiro-Rin, A.; Humphreys, W. G.; Lee, F. Y. F. In Ojima, I.; Vite, G. D.; Altmann, K.-H. editors. *Anticancer Agents – Frontiers in Cancer Chemotherapy*. Washington, DC: American Chemical Society; ACS Symposium Series 796, **2001**, 148.

70. Borzilleri, R. M.; Zheng, X.; Schmidt, R. J.; Johnson, J. A.; Kim, S. H.; DiMarco, J. D.; Fairchild, C. R.; Gougoutas, J. Z.; Lee, F. Y. F.; Long, B. H.; Vite, G. D. *J. Am. Chem. Soc.*, **2000**, 122: 8890.

71. Lee, F. Y.; Borzilleri, R. M.; Fairchild, C. R.; Kim, S. H.; Long, B. H.; Reventos-Suarez, C.; Vite, G. D.; Rose, W. C.; Kramer, R. A. *Clin. Cancer Res.*, **2001**, 7: 1429.

72. Chou, T. C.; O'Connor, O. A.; Tong, W. P.; Guan, Y.; Zhang, Z. G.; Stachel, S. J.; Lee, C.; Danishefsky, S. J. *Proc. Natl. Acad. Sci. USA*, **2001**, 98: 8113.

73. Schinzer, D.; Altmann, K.-H.; Stuhlmann, F.; Bauer, A.; Wartmann, M. *ChemBioChem.*, **2000**, 1: 67.

74. Stachel, S. J.; Lee, C. B.; Spassova, M.; Chappell, M. D.; Bornmann, W. G.; Danishefsky, S. J.; Chou, T.-C.; Guan, Y. *J. Org. Chem.*, **2001**, 66: 4369.

75. Yamaguchi, H.; Paranawithana, S. R.; Lee, M. W.; Huang, Z.; Bhalla, K. N.; Wang, H. G. *Cancer Res.*, **2002**, 62: 466.

76. Nicolaou, K. C.; Sarabia, F.; Ninkovic, S.; Finlay, M. R.; Boddy, C. N. C. *Angew. Chem. Int. Ed. Engl.*, **1998**, 37: 81.

77. Ojima, I.; Chakravarty, S.; Inoue, T.; Lin, S.; He, L.; Horwitz, S. B.; Kuduk, S. D.; Danishefsky, S. J. *Proc. Natl. Acad. Sci. USA*, **1999**, 96: 4256.

78. Wang, M.; Xia, X.; Kim, Y.; Hwang, D.; Jansen, J. M.; Botta, M.; Liotta, D. C.; Snyder, J. P. *Org. Lett.*, **1999**, 1: 43.

79. Manetti, F.; Forli, S.; Maccari, L.; Corelli, F.; Botta, M. *Il Farmaco*, **2003**, 58: 357.

80. Manetti, F.; Maccari, L.; Corelli, F.; Botta, M. *Curr. Topics Med. Chem.*, **2004**, 4: 203.

81. End, N.; Wartmann, M.; Altmann, K.-H. *Chem. & Biodiversity*, **2004**, 1: 1771.

82. Arslanian, R. L.; Tang, L.; Blough, S.; Ma, W.; Qiu, R.-G.; Katz, L.; Carney, J. R. *J. Nat. Prod.*, **2002**, 65: 1061.

83. Biswas, K.; Lin, H.; Njardson, J. T.; Chappell, M. D.; Chou, T.-C.; Guan, Y.; Tong, W. P.; He, L.; Horwitz, S. B.; Danishefsky, S. J. *J. Am. Chem. Soc.*, **2002**, 124: 9825.

84. Rivkin, A.; Njardson, J. T.; Biswas, K.; Chou, T.-C.; Danishefsky, S. J. *J. Org. Chem.*, **2002**, 67: 7737.

85. Rivkin, A.; Biswas, K.; Chou, T.-C.; Danishefsky, S. J. *Org. Lett.*, **2002**, 4: 4081.

86. Rivkin, A.; Yoshimura, F.; Gabarda, A. E.; Chou, T.-C.; Dong, H.; Tong, W. P.; Danishefsky, S. J. *J. Am. Chem. Soc.*, **2003**, 125: 2899.

87. Yoshimura, F.; Rivkin, A.; Gabarda, A. E.; Chou, T.-C.; Dong, H.; Sukenick, H.; Morel, F. F.; Taylor, R. E.; Danishefsky, S. J. *Angew. Chem. Int. Ed. Engl.*, **2003**, 42: 2518.

88. Chou, T.-C.; Dong, H.; Rivkin, A.; Yoshimura, F.; Gabarda, A. E.; Cho, Y. S.; Tong, W. P.; Danishefsky, S. J. *Angew. Chem. Int. Ed. Engl.*, **2003**, 42: 4762.

89. Tang, L.; Li, R.-G.; Yong, L.; Katz, L. *J. Antibiot.*, **2003**, 56: 16.

90. Starks, C. M.; Zhou, Y.; Liu, F.; Licari, P. *J. Nat. Prod.*, **2003**, 66: 1313.

91. White, J. D.; Carter, R. G.; Sundermann, K. F.; Wartmann, M. *J. Am. Chem. Soc.*, **2001**, 123: 5407.

92. White, J. D.; Carter, R. G.; Sundermann, K. F.; Wartmann, M. *J. Am. Chem. Soc.*, **2003**, 125: 3190.

93. Carlomagno, T.; Blommers, M. J. J.; Meiler, J.; Jahnke, W.; Schupp, T.; Petersen, F.; Schinzer, D.; Altmann, K.-H.; Griesinger, C. *Angew. Chem. Int. Ed. Engl.*, **2003**, 42: 2511.

94. Quintard, D.; Bertrand, P.; Vielle, S.; Raimbaud, E.; Renard, P.; Pfeiffer, B.; Gesson, J.-P. *Synlett*, **2003**, 13: 2033.

95. Schinzer, D.; Böhm, O. M.; Altmann, K.-H.; Wartmann, M. *Synlett* **2004**, 14: 1375.

96. Meng, D.; Su, D. S.; Balog, A.; Bertinato, P.; Sorensen, E. J.; Danishefsky, S. J.; Zheng, Y. H.; Chou, T. C.; He, L.; Horwitz, S. B. *J. Am. Chem. Soc.*, **1997**, 119: 2733.

97. Su, D. S.; Meng, D.; Bertinato, P.; Balog, A.; Sorensen, E. J.; Danishefsky, S. J.; Zheng, Y. H.; Chou, T. C.; He, L.; Horwitz, S. B. *Angew. Chem. Int. Ed. Engl.*, **1997**, 36: 757.

98. Nicolaou, K. C.; Winssinger, N.; Pastor, J.; Ninkovic, S.; Sarabia, F.; He, Y.; Vourloumis, D.; Yang, Z.; Li, T.; Giannakakou, P.; Hamel, E. *Nature*, **1997**, 387: 268.

99. Nicolaou, K. C.; Vourloumis, D.; Li, T.; Pastor, J.; Winssinger, N.; He, Y.; Ninkovic, S.; Sarabia, F.; Vallberg, H.; Roschangar, F.; King, N. P.; Finlay, M. R.; Giannakakou, P.; Verdier-Pinard, P.; Hamel, E. *Angew. Chem. Int. Ed. Engl.*, **1997**, 36: 2097.

100. Chou, T. C.; Zhang, X. G.; Harris, C. R.; Kuduk, S. D.; Balog, A.; Savin, K. A.; Bertino, J. R.; Danishefsky, S. J. *Proc. Natl. Acad. Sci. USA*, **1998**, 95: 15798.

101. Johnson, J.; Kim, S. H.; Bifano, M.; DiMarco, J.; Fairchild, C.; Gougoutas, J.; Lee, F.; Long, B.; Tokarski, J.; Vite, G. D. *Org. Lett.*, **2000**, 2: 1537.

102. Nicolaou, K. C.; Namoto, K.; Li, J.; Ritzén, A.; Ulven, T.; Shoji, M.; Zaharevitz, D.; Gussio, R.; Sackett, D. L.; Ward, R. D.; Hensler, A.; Fojo, T.; Giannakakou, P. *ChemBioChem*, **2001**, 2: 69.

103. Nicolaou, K. C.; Namoto, K.; Ritzén, A.; Ulven, T.; Shoji, M.; Li, J.; D'Amico, G.; Liotta, D.; French, C. T.; Wartmann, M.; Altmann, K.-H.; Giannakakou, P. *J. Am. Chem. Soc.*, **2001**, 123: 9313.

104. Regueiro-Ren, A.; Borzilleri, R. M.; Zheng, X.; Kim, S. H.; Johnson, J. A.; Fairchild, C. R.; Lee, F. Y.; Long, B. H.; Vite, G. D. *Org. Lett.*, **2001**, 3: 2693.

105. Altmann, K.-H.; Bold, G.; Caravatti, G.; Denni, D.; Flörsheimer, A.; Schmidt, A.; Rihs, G.; Wartmann, M. *Helv. Chim. Acta*, **2002**, 85: 4086.

106. Nicolaou, K. C.; Ritzén, A.; Namoto, K.; Ruben, M. B.; Diaz, F.; Andreu, J. M.; Wartmann, M.; Altmann, K.-H.; O'Brate, A.; Giannakakou, P. *Tetrahedron*, **2002**, 58: 6413.

107. Nicolaou, K. C.; Ninkovic, S.; Finlay, M. R.; Sarabia, F.; Li, T. *JCS Chem. Commun.*, **1997**, 2343.

108. Chappell, M. D.; Harris, C. R.; Kuduk, S. D.; Balog, A.; Wu, Z.; Zhang, F.; Lee, C. B.; Stachel, S. J.; Danishefsky, S. J.; Chou, T.-C.; Guan, Y. *J. Org. Chem.*, **2002**, 67: 7730.

109. Newman, R. A.; Yang, J.; Finlay, M. R. V.; Cabral, F.; Vourloumis, D.; Stevens, L. C.; Troncoso, L. P.; Wu, X.; Logothetis, C. J.; Nicolaou, K. C.; Navone, N. M. *Cancer Chemother. Pharmacol.*, **2001**, 48: 319.

110. Altmann, K.-H.; Nicolaou, K. C.; Wartmann, M.; O'Reilly, T. *Proc. American Association for Cancer Research*, **2001**, 42: Abstract #1979.

111. Höfle, G.; Glaser, N.; Kiffe, M.; Hecht, H.-J.; Sasse, F.; Reichenbach, H. *Angew. Chem. Int. Ed. Engl.*, **1999**, 38: 1971.

112. Sefkow, M.; Höfle, G. *Heterocycles*, **1998**, 48: 2485.

113. Nicolaou, K. C.; King, N. P.; Finlay, M. R. V.; He, Y.; Roschangar, F.; Vourloumis, D.; Vallberg, H.; Sarabia, F.; Ninkovich, S.; Hepworth, D. *Bioorg. Med. Chem.*, **1999**, 7: 665.

114. Nicolaou, K. C.; Scarpelli, R.; Bollbuck, B.; Werschkun, B.; Pereira, M. M.; Wartmann, M.; Altmann, K.-H.; Zaharevitz, D.; Gussio, R.; Giannakakou, P. *Chem. Biol.*, **2000**, 7: 593.

115. Nicolaou, K. C.; Hepworth, D.; King, N. P.; Finlay, M. R.; Scarpelli, R.; Pereira, M. M.; Bollbuck, B.; Bigot, A.; Werschkun, B. *Chem. Eur. J.*, **2000**, 6: 2783.

116. Nicolaou, K. C.; Hepworth, D.; Finlay, M. R.; King, N. P.; Werschkun, B.; Bigot, A. *JCS Chem. Commun.*, **1999**, 519.

117. Lee, C. B.; Chou, T.-C.; Zhang, X.-G.; Wang, Z.-G.; Kuduk, S. D.; Chappell, M. D.; Stachel, S. J.; Danishefsky, S. J. *J. Org. Chem.*, **2000**, 65: 6525.

118. Höfle, G.; Glaser, N.; Leibold, T. *Ger. Offen.*, **2000**; DE 19907588 A1 20000824.

119. Uyar, D.; Takigawa, N.; Mekhail, T.; Grabowski, D.; Markman, M.; Lee, F.; Canetta, R.; Peck, R.; Bukowski, R.; Ganapathi, R. *Gynecolog. Oncol.*, **2003**, 91: 173.

120. Altmann, K.-H.; Blommers, M. J. J.; Caravatti, G.; Flörsheimer, A.; Nicolaou, K. C.; O'Reilly, T.; Schmidt, A.; Schinzer, D.; Wartmann, M., In Ojima, I.; Vite, G. D.; Altmann, K.-H. editors. *Anticancer Agents—Frontiers in Cancer Chemotherapy.* Washington, DC: American Chemical Society; ACS Symposium Series 796, **2001**, 112.

121. Altmann, K.-H.; Bold, G.; Caravatti, G.; Flörsheimer, A.; Guagnano, V.; Wartmann, M. *Bioorg. Med. Chem. Lett.*, **2000**, 10: 2765.

122. Wartmann, M.; Loretan, J.; Reuter, R.; Hattenberger, M.; Muller, M.; Vaxelaire, J.; Maira, S.-M.; Flörsheimer, A.; O'Reilly, T.; Nicolaou, K. C.; Altmann, K.-H. *Proc. American Association for Cancer Research*, **2004**, 45: Abstract #5440.

123. Nicolaou, K. C.; Sasmal, P. K.; Rassias, G.; Reddy, M. V.; Altmann, K.-H.; Wartmann, M. *Angew. Chem. Int. Ed. Engl.*, **2003**, 42: 3515.

124. Glunz, P. W.; He, L.; Horwitz, S. B.; Chakravarty, S.; Ojima, I.; Chou, T.-C.; Danishefsky, S. J. *Tetrahedron Lett.*, **1999**, 40: 6895.

125. Sinha, S. C.; Sun, J.; Wartmann, M.; Lerner, R. A. *ChemBioChem.*, **2001**, 2: 656.

126. Cachoux, F.; Schaal, F.; Teichert, A.; Wagner, T.; Altmann, K.-H. *Synlett.*, **2004**, 2709.

127. Nettles, J. H.; Li, H.; Cornett, B.; Krahn, J. M.; Snyder, J. P.; Downing, K. H. *Science*, **2004**, 305: 866.

128. Lin, N.; Brakora, K.; Seiden, M. *Curr. Opin. Investigational Drugs*, **2003**, 4: 746.

129. Mani, S.; McDaid, H.; Hamilton, A.; Hochster, H.; Cohen, M. B.; Khabelle, D.; Griffin, T.; Lebwohl, D. E.; Liebes, L.; Muggia, F.; Horwitz, S. B. *Clin. Canc. Res.*, **2004**, 10: 1289.

130. McDaid, H. M.; Mani, S.; Shen, H.-J.; Muggia, F.; Sonnichsen, D.; Horwitz, S. B. *Clin. Canc. Res.*, **2002**, 8: 2035.

131. Spriggs, D.; Dupont, J.; Pezzulli, S.; Larkin, J.; Cropp, J.; Johnson, R.; Hannah, A. L. AACR-NCI-EORTC International Conference: Molecular Targets and Cancer Therapeutics, Nov. 17–21, 2003, Boston, MA. In *Clin. Cancer Res.*, **2003**, 9(16): Abstract #A248.

132. Holen, K.; Hannah, A.; Zhou, Y.; Cropp, G.; Johnson, R.; Volkman, J.; Binger, K.; Alberti, D.; Wilding, G. AACR-NCI-EORTC International Conference: Molecular Targets and Cancer Therapeutics, Nov. 17–21, 2003, Boston, MA. In *Clin. Cancer Res.*, **2003**, 9(16): Abstract #A261.

133. Vite, G.; Höfle, G.; Bifano, M.; Fairchild, C.; Glaser, N.; Johnston, K.; Kamath, A.; Kim, S.-H.; Leavitt, K.; Lee, F. F.-Y.; Leibold, T.; Long, B.; Peterson, R.; Raghavan, K.; Regueiro-Ren, A. Abstracts of Papers, 223rd ACS National Meeting, Apr. 7–11, 2002, Orlando, FL, MEDI-018.

# 2

# THE CHEMISTRY AND BIOLOGY OF VANCOMYCIN AND OTHER GLYCOPEPTIDE ANTIBIOTIC DERIVATIVES

RODERICH D. SÜSSMUTH

*Institut für Chemie, Berlin, Germany*

## 2.1 INTRODUCTION

Vancomycin (Scheme 2-1) is the most prominent representative of the family of glycopeptide antibiotics. It was isolated from a culture of the gram-positive bacterial strain *Amycolatopsis orientalis* in the mid 1950s in a screening program of Eli Lilly and Company (Indianapolis, IN).[1] Since then, many structurally related glycopeptides have been isolated from bacterial glycopeptide producing strains. Soon after its discovery, vancomycin served clinically as an antibiotic. However, the impurities of byproducts caused toxic side effects that were overcome with enhanced purification protocols. Since then, vancomycin has been used over 30 years, without the observation of significant bacterial resistances. The first clinical-resistant isolates were reported in the late 1980s,[2–4] and since then, this number has continuously increased. Nowadays, vancomycin is still widely used by clinicians as an antibiotic of last resort, especially against methicillin-resistant *Staphylococcus aureus* (MRSA) strains. Because of the emergence of vancomycin-resistant strains over the past 10 years, researchers urgently seek for alternative antibiotics to counter the expected threat of public health. From various strategies to combat vancomycin resistance, one is the screening and evaluation of compound classes other than glycopeptides. However, accepting glycopeptides as validated biological lead

*Medicinal Chemistry of Bioactive Natural Products* Edited by Xiao-Tian Liang and Wei-Shuo Fang
Copyright © 2006 John Wiley & Sons, Inc.

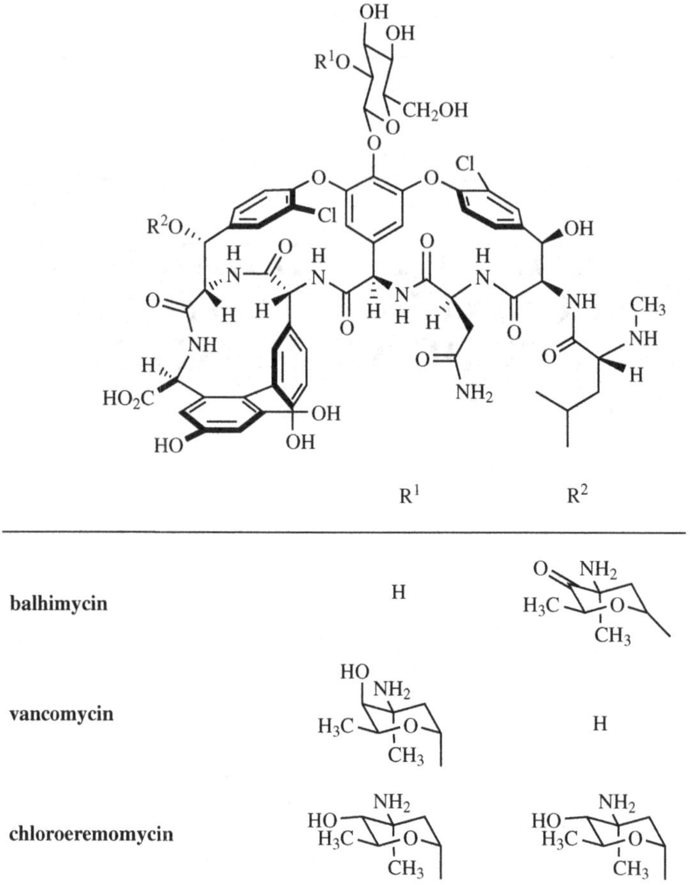

**Scheme 2-1.** Structures of important vancomycin-type glycopeptide antibiotics.

structures, the modification by semisynthetic or biotechnological approaches still offers interesting and promising alternatives. Since its discovery, the molecular architecture and the properties of vancomycin have fascinated many researchers from various disciplines. Because of their significance, all scientific aspects of glycopeptides are continuously highlighted in review articles. A selection of recent reviews or articles can be found on the discovery and mode of action,[5] the total synthesis,[6,7] the biosynthesis,[8,9] as well as the novel antibiotics and strategies to overcome glycopeptide resistance.[10,11] Besides these topics, glycopeptides play a key role in the field of enantioseparation[12] and as a model system for ligand–receptor interactions.

    Remarkably, it took almost 25 years to unravel the structure of glycopeptide antibiotics. The reasons are based on the highly complex structure of glycopeptides and the limited power of analytical methods, especially of nuclear magnetic resonance (NMR) methods, by that time. With the progress in the development of NMR methods, partial structures were determined.[13] Paralleled crystallization experiments yielded x-ray structures of antibiotically inactive CDP-1, a degradation product of vancomycin.[14] The final and commonly accepted structures were published by Williamson and Williams[15] and Harris and Harris.[16] In 1995, Sheldrick et al. presented the first x-ray structure of a naturally occuring glycopeptide antibiotic, balhimycin (Scheme 2-1).[17] On the basis of these contributions, the structures of many other glycopeptides have been elucidated.

## 2.2 CLASSIFICATION OF GLYCOPEPTIDE ANTIBIOTICS

The high structural diversity within the glycopeptide family led to a classification into five subtypes (Scheme 2-2). Types I–IV show antibacterial activity, whereas type V shows antiviral activity, e.g., against the human immunodeficiency virus (HIV).[18,19] The basic structural motif of types I–IV are three side-chain cyclizations of the aromatic amino acids of the heptapeptide backbone. These ring systems are called AB-ring (biaryl) and C-$O$-D- and D-$O$-E-rings (diarylethers), respectively. They are formed by cross-linking of the non-proteinogenic amino acids β-hydroxytyrosine/tyrosine (Hty/Tyr), 4-hydroxyphenylglycine (Hpg), and 3,5-dihydroxyphenylglycine (Dpg).

From the aglycon portion of glycopeptide antibiotics, the amino acids in positions 1 and 3 are the criterion for the classification into types I–III. The vancomycin-type (type I) glycopeptides have aliphatic amino acids in postions 1 and 3. In contrast, the actinoidin-type (type II) glycopeptides have aromatic amino acids in these positions, which are linked in the ristocetin A-type (type III) by one arylether bond. This additional ring system is commonly assigned as the F-$O$-G-ring. The rings of the teicoplanin-type (type IV) correspond to those of the ristocetin A-type. However, the classification of teicoplanin as another subtype is based on the acylation of an aminosugar with a fatty acid. Other structural features, e.g., the glycosylation pattern, the halogenation pattern, and the number of $N$-terminal methyl groups vary widely within each subtype.

The type-V glycopeptide antibiotics with complestatin or chloropeptin as representatives show no antibacterial activity. Instead, the inhibition of binding of viral glycoprotein gp120 to cellular CD4-receptors was found.[19] Characteristic features for this subclass are a DE-biaryl ring, which is formed by 4-hydroxyphenylglycine (AA4) and tryptophan (AA2).

To highlight the structural features of glycopeptides in more detail, in Scheme 2-3, representatively the structure of vancomycin is shown. The glycopeptide consists of a heptapeptide backbone with the sequence $(R)^1$MeLeu-$(2R,3R)^2$Cht-$(S)^3$ Asn-$(R)^4$Hpg-$(R)^5$Hpg-$(2S,3R)^6$Cht-$(S)^7$Dpg. Vancomycin has a total number of 18 stereocenters, with 9 stereocenters located in the aglycon and the remaining stereocenters located in the carbohydrate residues. The AB-ring formed by $^5$Hpg and $^7$Dpg is an atropoisomer with axial chirality. The C-$O$-D- and D-$O$-E-rings formed by $^2$Cht, $^4$Hpg, and $^6$Cht have planar chirality. The β-hydroxy groups of $^2$Cht and $^6$Cht are in *anti*-position to the chorine substituents attached to the aromatic rings.

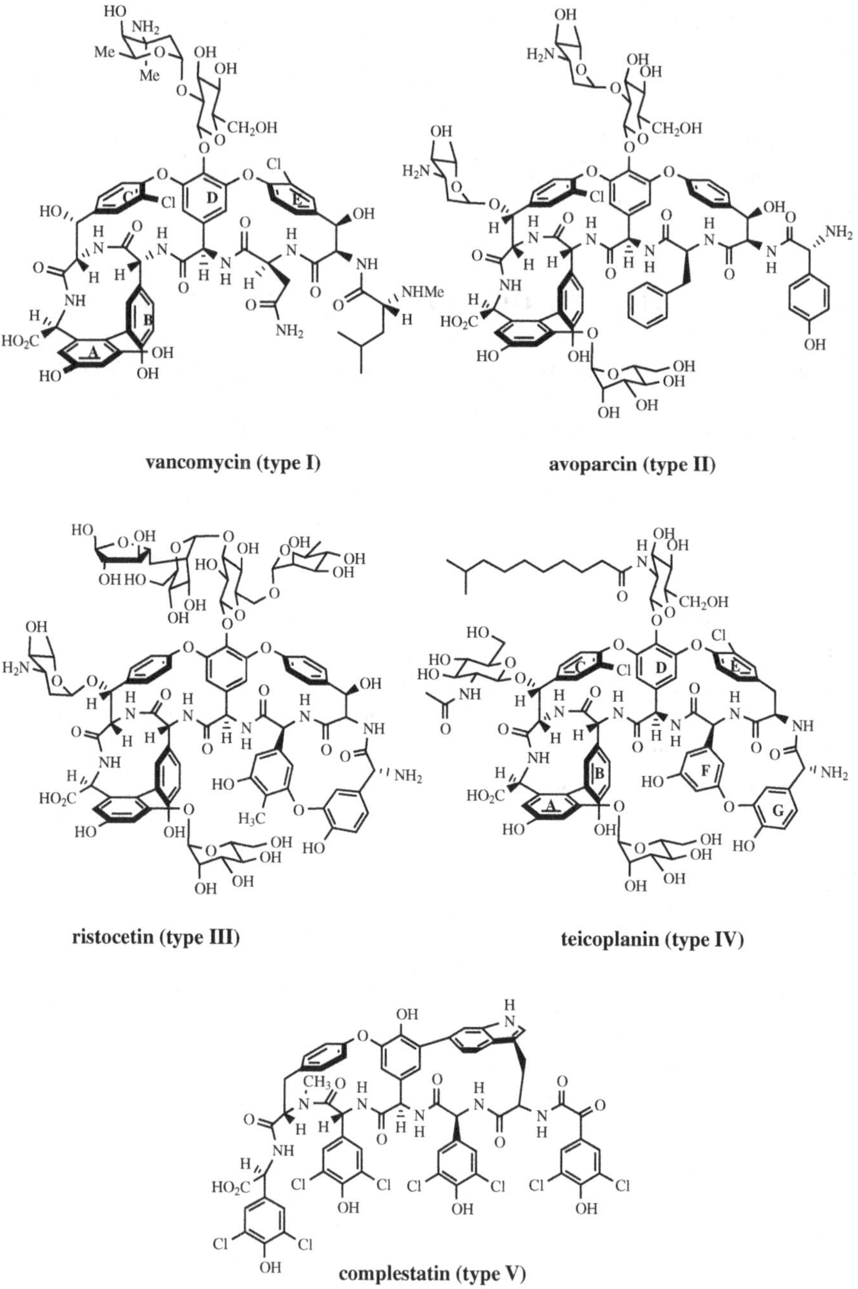

vancomycin (type I)

avoparcin (type II)

ristocetin (type III)

teicoplanin (type IV)

complestatin (type V)

**Scheme 2-2.**  Structures of the five naturally occurring glycopeptide antibiotic subtypes.

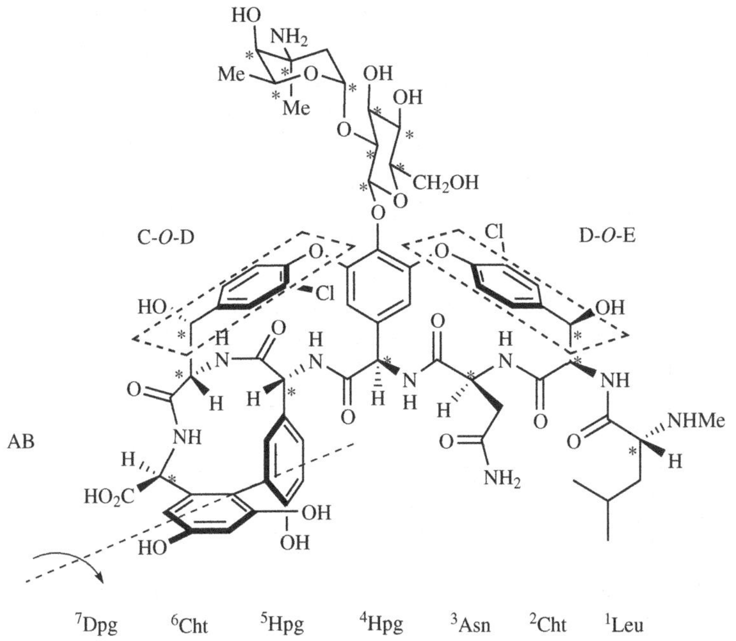

C-*O*-D     Cl     D-*O*-E

AB

$^7$Dpg    $^6$Cht    $^5$Hpg    $^4$Hpg    $^3$Asn    $^2$Cht    $^1$Leu

**Scheme 2-3.** Stereochemical features of the vancomycin structure. The AB-ring (chiral axis) and the C-*O*-D-*O*-E-rings (chiral planes) fixate the heptapeptide aglycon in a rigid comformation.

The side-chain cyclizations are the basis for conformative rigidity of the molecules and thus for the antibiotic activity of glycopeptide antibiotics.

## 2.3 MODE OF ACTION

Various antibacterial targets exist, in which antibiotics interfere with the essential pathways of the bacterial metabolism. These targets are the interaction with the cytoplasmic membrane, the inhibition of cell wall biosynthesis, or the inhibition of replicational, transcriptional, and translational processes. Like the penicillins and the cephalosporins, glycopeptide antibiotics also inhibit the cell wall biosynthesis. According to the features of the bacterial cell wall, bacteria are divided into gram-positive and gram-negative organisms. Gram-negative bacteria (e.g., *Escherichia coli*) have a thin layer of peptidoglycan that is covered by an outer membrane. Gram-positive bacteria (e.g., *S. aureus*) lack this outer membrane but have a thicker peptidoglycan layer compared with gram-negative organisms. Because of their size and polarity, glycopeptide antibiotics cannot cross the outer membrane of gram-negative bacteria, and thus, their antibiotic effects are restricted to gram-positive bacteria. As a consequence, the most important bacterial strains, which are combated with glycopeptides, are gram-positive enterococci, staphylococci, and streptococci.

The peptidoglycan layer confers mechanical stability to the cell wall of the bacteria. An important intermediate of the peptidoglycan biosynthesis is the GlcNAc- MurNAc-L-Ala-D-$\gamma$-Gln-L-Lys-D-Ala-D-Ala peptide (muramyl-pentapeptide), which is in its lipid-carrrier bound form transglycosylated to a linear polysaccharide. The linear polysaccharide is then cross-linked to peptidoglycan by transpeptidation reactions. Perkins observed[20] that vancomycin binds to the Lys-D-Ala-D-Ala peptide motif of bacterial cell wall intermediates. This observation was later investigated on a molecular level by NMR[21,22] and by x-ray crystallographic studies.[23,24]

As mentioned, the primary antibiotic effect of glycopeptide antibiotics is based on the binding to the D-Ala-D-Ala dipeptide motive of the bacterial cell wall biosynthesis. In contrast to penicillin, which covalently binds to an enzymatic target, glycopeptide antibiotics represent substrate binders that shield the substrate from transpeptidation but also from transglycosylation reactions. On the molecular level, five hydrogen bonds between the peptidic backbones of the D-Ala-D-Ala ligand and the glycopetide receptor (Scheme 2-4) contribute to a tight binding with binding constants in the range of $10^5$ M$^{-1}$ to $10^6$ M$^{-1}$.[25] The microheterogenicity found for glycopeptides, that is structural variations in the degree of glycosylation, N-terminal methylation, chlorination, and differences in the length of fatty acid side chains, result in varying antibiotic activities of these derivatives.

Of some importance is the ability of most glycopeptides to form dimers (e.g., eremomycin)[26] or to insert into bacterial membranes (e.g., teicoplanin).[27] Dimer formation is strongly dependent on the nature of the carbohydrates attached to the aglycon and on the attachment site of these residues. Chloroeremomycin (Scheme 2-1), which contains the amino sugar 4-*epi*-vancosamine bound to AA6, forms dimers with six hydrogen bonds (Scheme 2-4), whereas vancomycin shows a weak dimerization tendency by the formation of only four hydrogen bonds. The dimerization behavior originally observed with NMR has also been confirmed by x-ray crystallography.[17,23] Furthermore, cooperativity effects of ligand bound glycopeptides in dimerization have been found as well as a stronger binding of ligands through glycopeptide dimers.[28] In constrast, for type IV-glycopeptide antibiotics, membrane anchoring is assumed, which forms an "intramolecular" complex with its target peptide on the cell surface.[27–29] An excellent review, which highlights details of the mode of action of glycopeptide antibiotics on a molecular level, has been published by Williams and Bardsley.[5] In summary, the binding of D-Ala-D-Ala peptides can be considered as the primary and main effect for antibiotic activity of naturally occuring glycopeptide antibiotics. Dimerization and membrane anchoring mechanisms are secondary effects, which only modulate the antibiotic activity. This consideration is true, if no other inhibiting effects have to be taken into account.

## 2.4   GLYCOPEPTIDE RESISTANCE

Over the past decade the emergence of resistant enterococci and *S. aureus* strains has been observed in clinics. Already, nowadays, the increase of glycopeptide-resistant

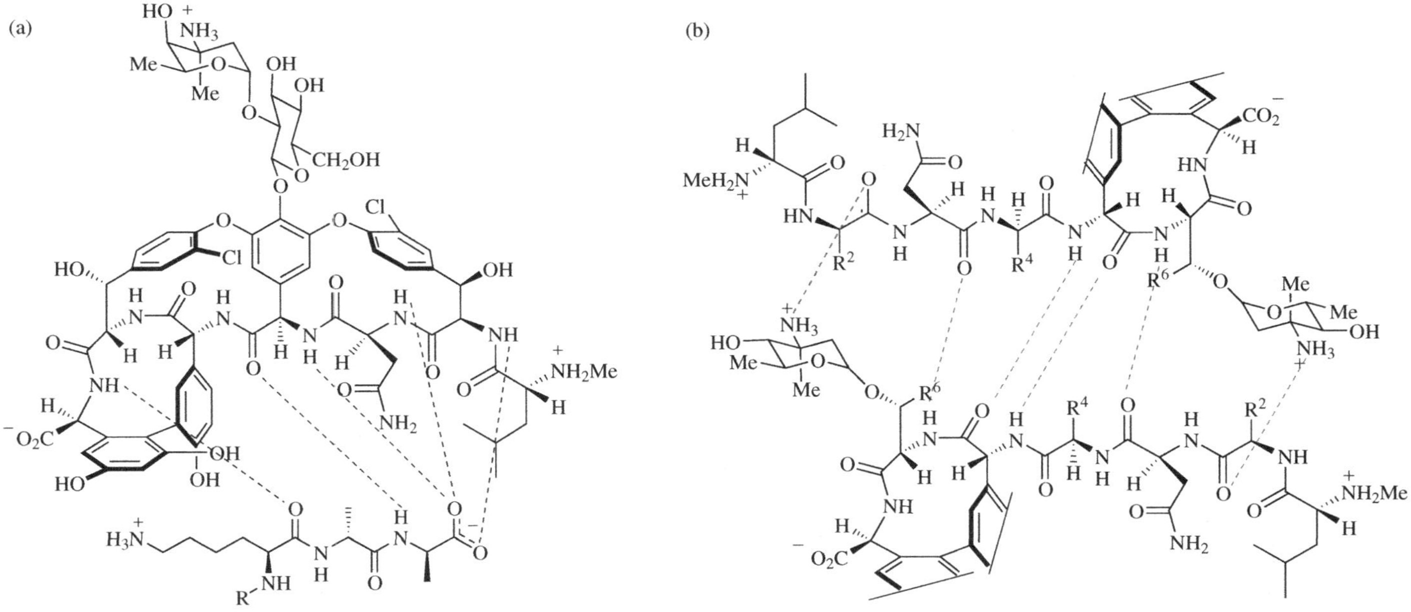

**Scheme 2-4.** (a) Binding of vancomycin to the Lys-D-Ala-D-Ala peptide motive with five hydrogen bonds. (b) Dimerization of vancomycin-type glycopeptide antibiotics (e.g., chloroeremomycin) over six hydrogen bonds. Because vancomycin lacks a vancosamine sugar at AA6, it interacts only over four hydrogen bonds.

bacterial isolates causes severe treatment problems especially with elderly people and immunosuppressed patients. The molecular mechanism of enterococcal resistance is based on the change of the D-Ala-D-Ala peptide of the bacterial cell wall biosynthesis to D-Ala-D-Lac (vanA/B resistance) or D-Ala-D-Ser (vanC resistance). The well-investigated vanA/B phenotypes[30–34] are based on five genes (vanS, vanR, vanH, vanA, and vanX). VanS is a vancomycin-dependant sensor kinase, which induces the cleavage of D-Ala-D-Ala-dipeptides (vanX) and at the same time promotes the formation of D-Ala-D-Lac (vanA). The increased amounts of lactate necessary for cell wall biosynthesis are formed from pyruvate by the action of ketoreductase vanH. The glycopeptide resistance on the molecular level is based on the loss of one hydrogen bond combined with the electronic repulsion of two oxygen atoms between the lactate oxygen and the glycopeptide carbonyl (Scheme 2-5). As a result of the alteration of cell wall biosynthesis, the affinity of glycopeptides to D-Ala-D-Lac is ~1000-fold decreased[35] and thus is no longer useful as an antibiotic.[36] For the D-Ala-D-Ser-modification (vanC resistance),[37,38] it is assumed that glycopeptide binding is less tight because of steric reasons.[39] As a consequence, both cell wall biosynthesis alterations lead to a markedly reduced susceptibiliy toward glycopeptide antibiotics, whereas some strains remain sensitive to teicoplanin.

Thus far, the reasons for glycopeptide-resistance of *S. aureus* strains are not entirely clear. It is assumed that an increased biosynthesis of cell wall precursors combined with a thickened cell wall are the likely reasons.[40] However, also the transferance of enterococcal resistance to *S. aureus* has been shown in the

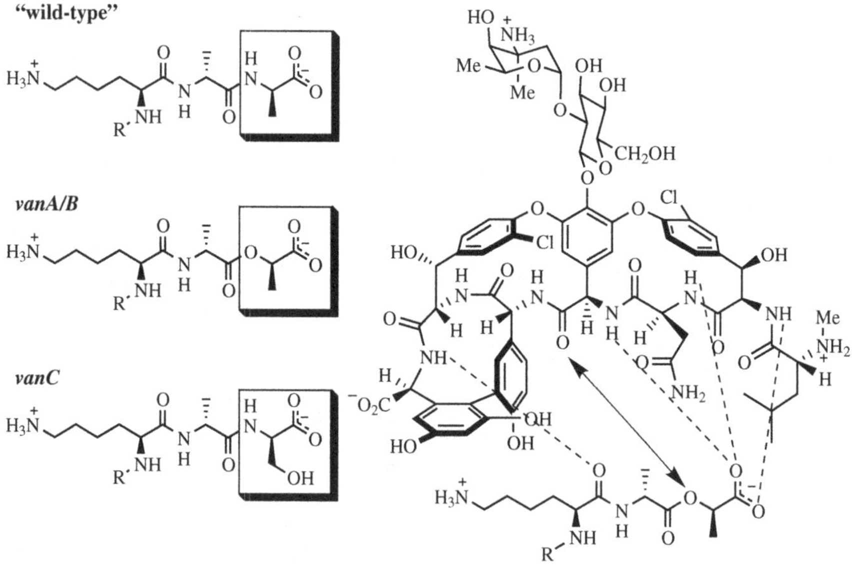

**Scheme 2-5.** Molecular basis of enterococcal vanA/B- and vanC-glycopeptide resistance. Interaction of vancomycin with D-Ala-D-Lac. The lactate ester bond leads to a lowered ligand binding affinity by loss of one hydrogen bond and electronic repulsion (solid arrow).

laboratory,[41] and even more recently, enterococcal vanA-resistance has been found in clinical isolates of *S. aureus* strains.[42]

## 2.5  BIOSYNTHESIS

The investigation of the glycopeptide biosynthesis has been a main focus of many research groups over the past 5 years, and currently, a relatively detailed picture already exists of the biosynthetic assembly. As a consequence, this topic has also been the subject of several review articles, which give an overview of the current status of research.[8,9]

The sequencing of the chloroeremomycin biosynthesis gene cluster in 1998 allowed the putative gene functions for the assembly of glycopeptide antibiotics to be deduced for the first time.[43] Until now the biosynthesis gene clusters of several other glycopeptide producers have been sequenced: balhimycin (*Amycolatopsis balhimycina*),[44] teicoplanin (*Actinoplanes teichomyceticus*),[45–47] A40926 (*Nonomuraea* species),[48] A47934 (*Streptomyces toyocaensis*),[49] and complestatin (*Streptomyces lavendulae*).[50] The peptide synthetase genes found in the gene clusters of all sequenced glycopeptide producers indicate a peptide assembly by nonribosomal peptide synthetases (NRPS).[51,52] The biosynthetic assembly of glycopeptide antibiotics can be subdivided into three parts. The first stage comprises the assembly of building blocks, which are the nonproteinogenic amino acids and the vancosamine sugars. The second stage is the peptide assembly on the NRPS and finally the tailoring by P450-dependant oxygenases, glycosylation, and *N*-methylation. In the case of teicoplanin, the amino sugar is acylated with fatty acids. The biosynthesis of vancomycin-type glycopeptides has been investigated by two different approaches. One approach is the overexpression of biosynthetic proteins followed by protein characterization and the conversion of substrates. The other approach is the inactivation of biosynthetic genes followed by the chemical characterization of accumulated biosynthesis intermediates. Both approaches have been used in a complementary way to shed light on the different stages of glycopeptide assembly. The protein overexpression approach was successful for the early and late stages, namely the building block assembly, the glycosylation, and *N*-methylation reactions.[53] The biosyntheses of the five aromatic core amino acids 4-hydroxyphenylglycine (Hpg),[54,55] 3,5-dihydroxyphenylglycine (Dpg),[54,56–59] and β-hydroxytyrosine (Hty)[60–62] are well investigated. However, the central structural features exclusively characteristic for glycopeptide antibiotics are the three side-chain cyclizations with the AB, C-*O*-D, and D-*O*-E rings. The characterization of a linear heptapeptide, which was obtained from gene inactivation of P450-monooxygenases of the balhimycin (Scheme 2-1) producer *Amycolatopsis balhimycina*, shed light on the nature of putative peptide intermediates and the likely course of aglycon formation.[44,63] This heptapeptide intermediate aready showed β-hydroxylation, chlorination, as well as D-Leu, D-Hpg, and L-Dpg as the structural features being present before side-chain cyclization. Subsequent contributions determined a sequence of ring closure reactions (Scheme 2-6) in the order: (1) C-*O*-D, (2) D-*O*-E, and (3)

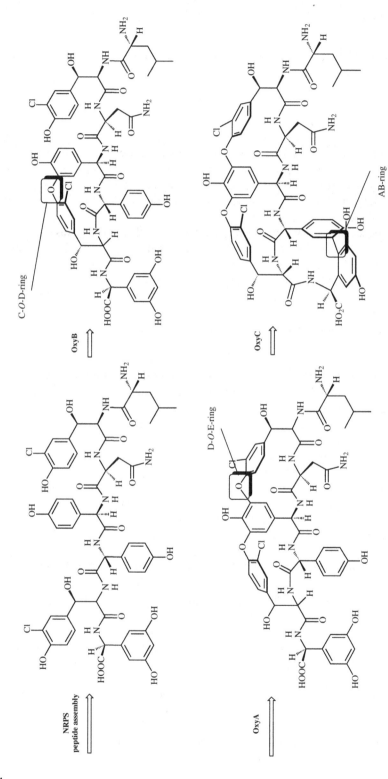

**Scheme 2-6.** The order of the side-chain cyclization reactions performed by three P450-dependent monooxygenases (OxyA/B/C).

AB.[64,65] Similarly, with this approach, it could be shown that *N*-methylation and glycosylation are biosynthetic steps that follow the oxidative ring closure reactions. Currently, the knowledge of biosynthetic glycopeptide assembly is used for the generation of novel glycopeptide antibiotics with modifications in the aglycon and of the carbohydrate residues.

## 2.6  TOTAL SYNTHESIS

The total synthesis of the vancomycin aglycon was accomplished by Evans et al. (Harvard University, Cambridge, MA)[66,67] and Nicolaou et al. (Scripps Research Institute, La Jolla, CA)[68–74] nearly at the same time. A short time later, Boger et al. (Scripps Research Institute) contributed their total synthesis of the aglycon.[75–78] With the attachment of the carbohydrate residues, the total synthesis was formally completed by Nicolaou et al.[79] The various strategies and the number of synthetic steps, which were tested for the synthetic assembly show, that only a few groups worldwide can conduct research efforts on such a synthetic problem. The achievement of the total synthesis is certainly one milestone in the art of peptide synthesis, if not in organic synthesis. The main challenge of the synthetic strategy was the atropisomerism of the diphenylether and of the biaryl ring systems. From eight possible atropoisomers, there is only found one in nature. Additional problems included the synthetic access to the amino acid building blocks and the attachment of the carbohydrate residues to the aglycon, both of which require a well-planned protecting group strategy. Accordingly, preceding the achievement of the total synthesis, there was a decade of synthetic studies at model systems also by other research groups, which are documented in several review articles.[6,7,80–82] In Schemes 2-7 and 2-8, some characteristics of the total synthesis strategies are representatively highlighted. For detailed descriptions and discussions of the total synthesis, it is recommended to review the literature cited here.

The syntheses of nonproteinogenic amino acids have been performed by three different enantioselective reactions. These were the use of the Evans auxiliaries (Evans et al.), the Sharpless hydroxylation/aminohydroxylation (Nicolaou et al.), and the Schöllkopf bislactimether synthesis (Boger et al.). The second characteristic of the total synthesis was the underlying chemistry of the side-chain cyclizations and the sequence of the ring closure reactions. Evans et al. group used a tripeptide amide with $VOF_3$-mediated AB-biaryl formation (Scheme 2-7). The *C*-terminal amide protection at [7]Dpg had to be introduced to prevent racemization of this amino acid. The next steps were condensation of a 3,4,5-trihydroxyphenylglycine and ring closure of the C-*O*-D-ring via nucleophilic aromatic substitution ($S_NAr$). The remaining aromatic nitro group in the C-*O*-D ring was then converted into a H-substituent. The AB/C-*O*-D-tetrapeptide was *N*-terminally coupled with a tripeptide to the heptapeptide, which was then again submitted to a nucleophilic aromatic substitution to yield the AB/C-*O*-D/D-*O*-E ring system. The atropostereoselectivity of the D-*O*-E ring was 5:1 of the desired protected aglycon derivative.

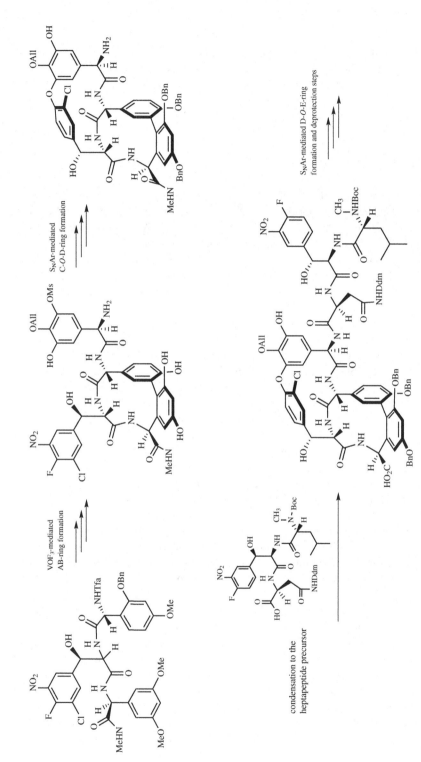

**Scheme 2-7.** Synthetic strategy of ring closure reactions according to Evans et al.[83]

The key compound of the synthetic route of Nicolaou et al. was the triazene-phenylglycine derivative (Scheme 2-8) used for Cu-mediated C-$O$-D and D-$O$-E ring formation. The first ring closure reaction was the formation of the C-$O$-D ring, which was followed by macrolactamization of a preformed AB-biaryl amino acid to yield the AB/C-$O$-D ring system. The mixture of atropoisomers obtained from this step afforded a chromatographic separation. To protect $^7$Dpg sensitive to racemization, the amino acid was masked as an amino alcohol, which was oxidized to the carboxy function at a late stage of the aglycon synthesis.

Both total syntheses of Evans et al. and Nicolaou et al. first perform synthesis of the AB/C-$O$-D-ring systems before condensation with the D-$O$-E ring system. The Evans et al. route was extended by several synthetic steps by the conversion of the 2-OH-group of $^5$Hpg and the conversion of nitro substituents after $S_N$Ar into hydrogen substituents. Problems developed with the transformation of the C-terminal amide into the carboxy function during deprotection steps. In contrast, Nicolaou et al. had to face the relatively moderate atroposelectivities and had to solve the somewhat obscure conversion of the triazene into a phenolic group. Both syntheses have been compared and discussed in the literature.[6,7,83]

## 2.7 GLYCOPEPTIDES AS CHIRAL SELECTORS IN CHROMATOGRAPHY AND CAPILLARY ELECTROPHORESIS

The significance of glycopeptide antibiotics as chiral selectors for enantiomeric separations is often neglected. Among the glycopeptides, vancomycin and ristocetin are the most important chiral selectors used for thin-layer chromatography (TLC),[84] high-performance liquid chromatography (HPLC), capillary electrochromatography (CEC),[85,86] and capillary electrophoresis (CE).[12,87–89] Armstrong et al. used for the first time vancomycin as a chiral stationary phase for HPLC.[90] The advantages of glycopeptides compared with amino acids, proteins, and cyclodextrines are the high stability, the commercial availability, and the broad application range to various separation problems. The glycopeptides used for HPLC are covalently bound to the packing material. They have a high stability during column packing, and unlike proteins, no denaturation occurs. They can be used for normal-phase and for reversed-phase separations. For separation with CE, glycopeptides are added as a chiral selector to the running buffer in a concentration of 1–4 mmol. In a representative example, vancomycin was used for separations of more than 100 racemates of structurally highly diverse compounds.[91] The use of micelles of vancomycin and sodium dodecylsulfate resulted in enhanced separations of racemates and extended the applicability toward neutral analytes.[92,93]

The broad applicability of glycopeptides is assumed to be based on the interactions with the analyte of various functional groups being part of the glycopeptide molecule. These comprise hydrophobic, dipole–dipole-, π–π-, and ionic interactions as well as the formation of hydrogen bonds and steric factors.[94–96] Hydrophilic and ionic groups contribute to a good solubility in buffer systems, whereas the number of stereocenters apparently plays a role for the quality of the separations

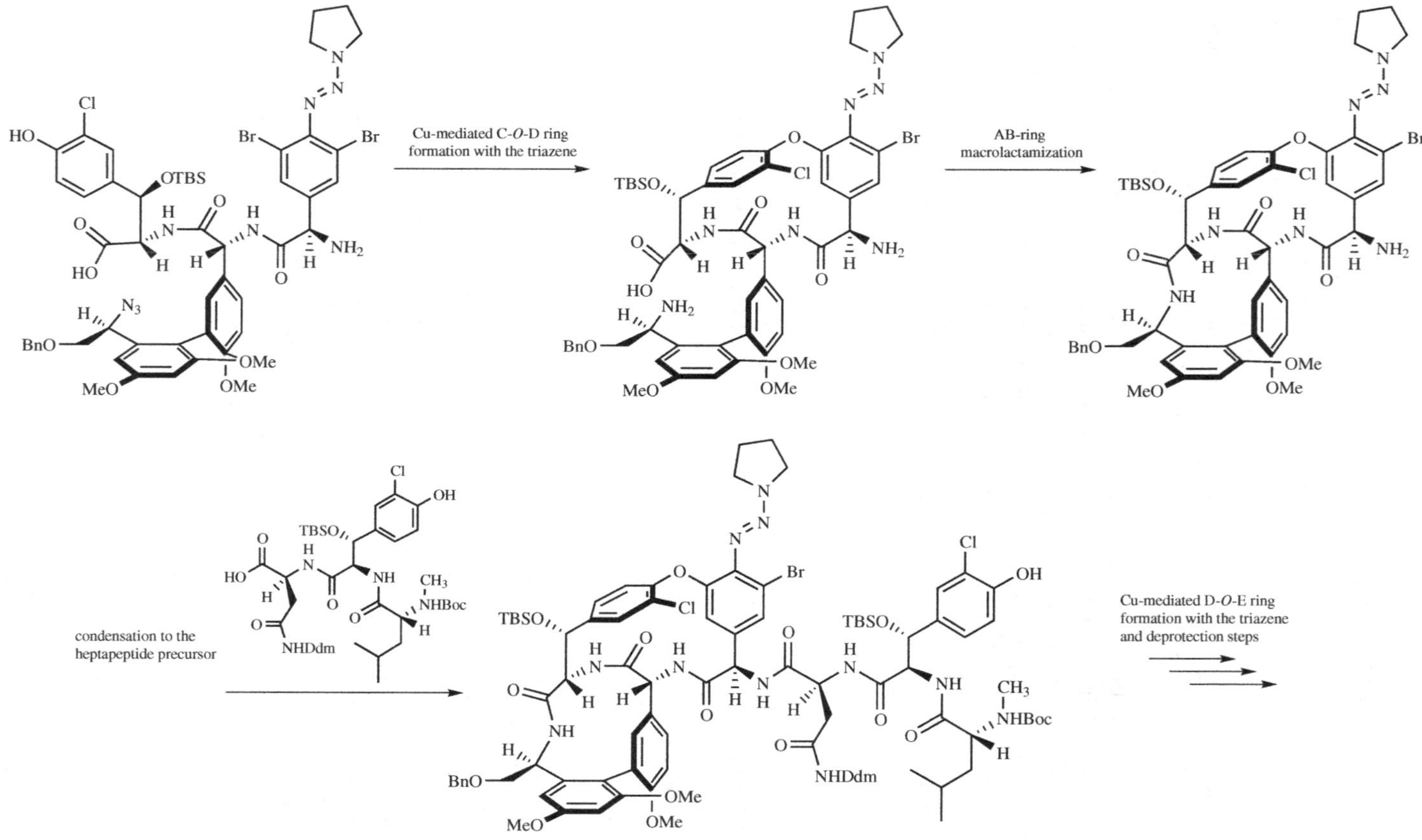

**Scheme 2-8.** Synthetic strategy of ring closure reactions according to Nicolaou et al.[83]

condensation to the heptapeptide precursor

Cu-mediated C-O-D ring formation with the triazene

AB-ring macrolactamization

Cu-mediated D-O-E ring formation with the triazene and deprotection steps

**TABLE 2-1. Physicochemical Properties of Glycopeptide Antibiotics Used for Enantioseparation**

|                       | Vancomycin   | Ristocetin A | Teicoplanin  |
|-----------------------|--------------|--------------|--------------|
| Molecular mass        | 1449 amu     | 2066 amu     | 1877 amu     |
| No. of stereocenters  | 18           | 38           | 23           |
| No. of ring systems   | 3            | 4            | 4            |
| Carbohydrate residues | 2            | 6            | 3            |
| Aromatic rings        | 5            | 7            | 7            |
| OH-groups (phenols)   | 9 (3)        | 21 (4)       | 15 (4)       |
| Amide bonds           | 7            | 6            | 7            |
| Amines (sec. amines)  | 2 (1)        | 2 (0)        | 1 (0)        |
| pI                    | 7.2          | 7.5          | 4.2, 6.5     |
| Stability             | 1–2 weeks    | 3–4 weeks    | 2–3 weeks    |
| Separation method     | HPLC, CE, DC | HPLC, CE     | HPLC, CE     |

*Source*: Modified from Ref. 97.

(Table 2-1). It has to be expected that glycopeptides will establish next to cyclodextrines for a routine use in enantioseparations.

## 2.8 STRUCTURAL MODIFICATIONS OF GLYCOPEPTIDE ANTIBIOTICS AND STRUCTURE ACTIVITY RELATIONSHIP (SAR) STUDIES

The SARs of glycopeptide antibiotics have mostly been studied with vancomycin and teicoplanin. Some other examples also comprise the glycopeptides eremomycin, balhimycin, ristocetin, and avoparcin. The reasons for the predominance of vancomycin and teicoplanin are that, particularly vancomycin (Vancomycin, Eli Lilly and Company) and teicoplanin (Targocid, Lepetit, Italy), are industrial large-scale fermentation products, and most of these studies were performed within research programs or with the participation of researchers from Eli Lilly and Lepetit.

The structure of glycopeptide antibiotics can roughly be divided into a D-Ala-D-Ala-binding site and the periphery of the molecule, which is not directly involved in binding of the cell wall biosynthesis component (Scheme 2-4). This recognition site that is essential for D-Ala-D-Ala-binding is constituted by the five aromatic amino acids, which are cross-linked to AB, C-*O*-D, and D-*O*-E rings in the aromatic side chains and thus held in a fixed conformation. In contrast, the sugar moieties as well as the *C*-terminus are not directly involved in D-Ala-D-Ala-binding.

Because of the complexity of the molecular architecture of glycopeptides, total synthesis as a tool for SAR studies is devoid of applicability. Better chances offer semisynthetic approaches, albeit a relatively limited number of modifications can be introduced. An attempt to subdivide these modifications is the alteration or modification of amino acid residues, or alternatively the attachment of molecules to the *C*- terminus, the *N*-terminus, or variations at the carbohydrate moieties (Scheme 2-9). The latter modifications mostly comprise deglycosylation reactions and alterations of

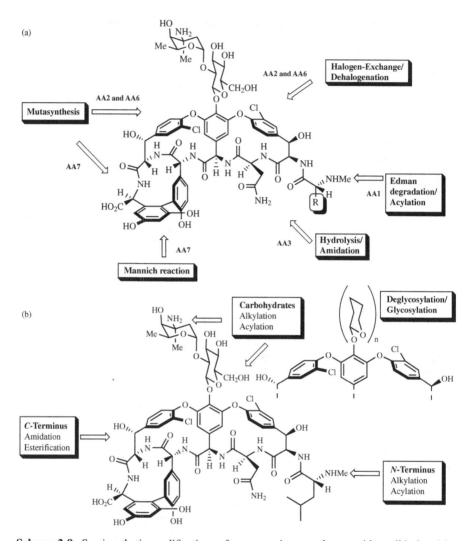

**Scheme 2-9.** Semisynthetic modifications of vancomycin-type glycopeptide antibiotics. (a) Alterations and modifications of amino acids. (b) Attachment of molecules to the amino groups, to the carboxy groups, and to phenolic carbohydrate functionalities. Similar modifications have been performed for antibiotics of the teicoplanin-type.

easily modifiable functional groups, for example, amino and carboxy groups. However, over the past few years, chemists have introduced more sophisticated modifications that are closer to the D-Ala-D-Ala binding pocket. As a consequence of the understanding of glycopeptide biosynthesis over the recent years, biotechnological approaches have been performed. One example is mutasynthesis,[98,99] which is the feeding of modified amino acid building blocks to glycopeptide producer mutants, which results in a restored production of altered glycopeptide antibiotics. In the future, combinatorial biosynthesis approaches, such as the exchange of biosynthesis genes of glycopeptide producers, are expected to be developed. In the following sections, semisynthetic SAR studies and rational semisynthetic concepts for the development of novel glycopeptides will be presented.

### 2.8.1 Modifications of Glycopeptide Antibiotics

#### 2.8.1.1 Deglycosylated Glycopeptide Derivatives

The complete or sequential deglycosylation of glycopeptides is one basic reaction to dissect contributions of glycosyl residues to the antibiotic activity. Vancomycin and teicoplanin can be converted to desvancosaminyl vancomycin in trifluoroacetic acid (TFA) at $-15\,^\circ$C and completely deglycosylated in TFA at $50\,^\circ$C according to previously published procedures.[100–102] Using TFA-based protocols, considerable loss of glycopeptide has been observed, and therefore, less harsh procedures have been developed with an HF-mediated cleavage of carbohydrate residues.[103] Previous publications on the dimerization of glycopeptide antibiotics (see Section 2.3) have shown that glycosylation represents more than merely the enhancement of water solubility and pharmacological properties. An important finding in this context is that the absence of amino sugars leads to a reduced dimerization and in most cases to a reduced antibiotic activity. Selectively deglycosylated teicoplanines have also been investigated for their antibiotic activity.[104] As in vitro assays show, only slight variations exist in the activity against certain bacterial strains for these derivatives. Most significantly, the loss of the N-acylglucosamine moiety has a less favorable influence on the pharmacokinetic performance. In general, the partially or fully deglycosylated derivatives commonly serve as starting material for other carbohydrate modifications or substitution reactions by chemical or enzymatic approaches.

#### 2.8.1.2 Glycosylation of Glycopeptide Derivatives

For the substitution of the carbohydrate portion of glycopeptides or for the coupling of novel glycosyl residues, the corresponding aglycones serve as starting materials for semisynthetic modifications. To introduce glycosylations by chemical methods, in one example, all phenolic groups and the C-terminus were allylated, followed by the protection of the N-terminus with Alloc.[105,106] Remaining hydroxy groups were acetylated, and the phenolic side chain of ⁴Hpg, which was previously protected, was then glycosylated with sulfoxide chemistry to yield vancomycines.

A similar example, which demonstrates the feasability of aglycon glycosylation on solid-support is the C-terminal immobilization of a fully protected vancomycin to a proallyl selenium-resin.[107] After deglycosylation and semisynthetic reglycosylation, the glycopeptide was cleaved by $H_2O_2$-mediated oxidation from the resin under simultanous conversion to the allyl ester. With this approach, several variously monoglycosylated and/or AA1-altered glycopeptides have been synthesized and evaluated for their antibiotic activity.[108] Wong et al. reported on the glycosylation of a protected vancomycin with non-natural disaccharides and trisaccharides.[109] In this contribution, acetylated saccharides were presynthesized and coupled with trichloroacetimidate chemistry to the vancomycin aglycon at the phenole of [4]Hpg. The general disadvantages of these chemical aproaches are the lenghty protection/deprotection of glycopeptides and the relatively poor yields.

Besides chemical glycosylations, the evaluation of biotechnological methods has also been performed. An early example is the heterologous expression of glycosyl-transferases in *E. coli* followed by the in vitro conversion of aglycon substrates with NDP-sugars to hybrid glycopeptides.[110] The authors of this contribution also used a genetically modified glycopeptide producer strain with a glycosyltransferase gene, which produced a glucosylated glycopeptide. Novel hybrid teicoplanins have been synthesized by enzymatic conversion of aglycones with heterologously expressed and purified glycosyltransferases.[111] Currently, the substrate specificity and the availability of novel glycosyltransferases together with the availability of a broad substrate range of NDP-sugars poses some restrictions for a broad applicability of this approach. However, with the future developments in this field, such as the availability of novel glycosyltransferase genes and carbohydrate biosynthesis genes from genome mining, considerable progress has to be expected. The next stage would then represent the cloning of glycosyltransferase genes and their carbohy-drate biosynthesis genes into glycopeptide producing strains to generate novel glycopeptides by combinatorial biosynthesis.

### 2.8.1.3 Variations and Modifications of Amino Acids of the Aglycon Heptapeptide

From the aglycon portion of the naturally occurring glycopeptides (types I–IV), five amino acids ([2]Hty/[2]Tyr, [4]Hpg, [5]Hpg, [6]Hty, and [7]Dpg) are highly conserved and cross-linked in the aromatic side chains by one biaryl and two diarylethers. The conformational rigidity that is established by these cross-links is mandatory for tight D-Ala-D-Ala-binding. The hydrolytic cleavage of amide bonds of the peptide backbone,[102] their acidic rearrangement to CDP-1,[112] or the lack of one biayl/diarylether cross-link[63–65] immediately leads to a significant decrease in D-Ala-D-Ala binding mostly cocommitant with the complete loss of antibiotic activity.

Unfortunately, the D-Ala-D-Ala binding region of the aglycon core is hardly accessible to chemical modification reactions. As a consequence, the options for the introduction of synthetic variations in the peptide core region are limited. We herein currently present established modifications of amino acids of the heptapep-tide aglycon by chemical and biotechnological approaches.

*Modification of AA1*   The NMR- and x-ray data of glycopeptide-D-Ala-D-Ala-complexes strongly underline a crucial role of a positive charge located at the *N*-terminus of AA1, for D-Ala-D-Ala binding.[113,114] In this context, a variation in *N*-methylation does not significantly alter the antibiotic activity.[115] The exchange of side chains of AA1 can be achieved by Edman-degradation to the antibiotically inactive hexapeptide aglycon,[100,116,117] which is then aminoacylated with other amino acids. Early aminoacylation experiments of vancomycin-derived or eremomycin-derived hexapeptides with other amino acids, however, did not yield compounds of increased antibiotic activity compared with vancomycin.[118–120] Similarly, subsequent aminoacylation studies of the Nicolaou et al. with, for example, L-Asn, L-Phe, and L-Arg, did not result in more active vancomycin derivatives.[108]

*Modification of AA2 and AA6*　The naturally occurring glycopeptides mostly contain β-hydroytyrosines in positions 2 and 6 of the aglycon, and a varying degree of chlorination is observed. In some cases, such as for teicoplanin but also for vancomycin derivatives, AA2 can also be a tyrosine. The replacement of the β-hydroxy group of AA2 by hydrogen, which has been found for a vancomycin analog, leads to a two-fold decrease in antibiotic acitivity.[121] The contribution of the chlorine atoms to antibiotic activity has been investigated already at an early stage of glycopeptide research. Chlorine atoms can be removed by catalytic dehalogenation with $Pd/H_2$, but these dechloro-glycopeptides have also been found in culture filtrates of bacterial type I glycopeptide producers. The example of orienticin A ($X^{AA2}$=H; $X^{AA6}$=Cl), a vancomycin-type glycopeptide, as a reference compound compared with chloroorienticin A ($X^{AA2}$=$X^{AA6}$=Cl) and A82846B ($X^{AA2}$=Cl; $X^{AA6}$=H), shows a five- to ten-fold reduced antibacterial activity for orienticin A in MIC tests.[122,123] Chloroorienticin A and A82846B have comparable antibiotic activity, which indicates a significant role for chlorine at AA2 for antibiotic activity compared with the less important chlorination of AA6.[17,124] A mechanistic explanation suggests a role of chlorine substituents supporting dimerization and D-Ala-D-Ala-binding.[5,28,125] In this context, a co-crystal of vancomycin and *N*-Acetyl-D-Ala shows that the chlorine substituent of AA6 is directed toward the putative pocket formed by the other glycopeptide dimerization partner, which thus contributes to an enhanced dimerization.[17,24]

This aspect of substitutions at AA2 and AA6 other than chlorine has been further investigated by Bister et al. with the example of the vancomycin-type glycopeptide balhimycin (Scheme 2-1). The replacement of chloride salts in the culture media of the balhimycin producer *Amycolatopsis balhimycina* with bromide salts rendered the corresponding bromobalhimycin (Scheme 2-10). The use of 1:1 molar ratios in fermentation media rendered a statistic distribution of chlorine and bromine at AA2 and AA6, respectively. Comparative MIC tests showed similar antibiotic activities for the bromobalhimycines compared with balhimycin.[126] Furthermore, it was shown that this approach could be transferred to other glycopeptide producers (types II and III) to yield the corresponding bromo-glycopeptides. Because of the toxicity of fluorine and iodine salts to glycopeptide-producing bacterial strains, this approach could not be tested for the generation of fluorinated or iodinated glycopeptides.

Performing mutasynthesis with the balhimycin producer *Amycolatopsis balhimycina*, several fluorinated glycopeptides finally could be obtained.[98] The mutasynthesis principle was established by Rinehart et al. with the example of neomycin.[127] The experimental approach is based on the generation of directed or undirected mutants of a secondary metabolite-producing bacterial strain, which

54

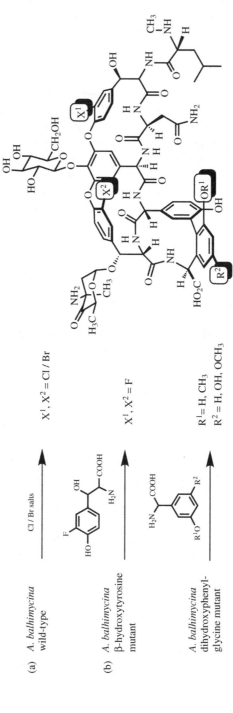

(a) *A. balhimycina* wild-type $\xrightarrow{\text{Cl / Br salts}}$ $X^1, X^2 = Cl / Br$

(b) *A. balhimycina* β-hydroxytyrosine mutant $\longrightarrow$ $X^1, X^2 = F$

*A. balhimycina* dihydroxyphenyl-glycine mutant $\longrightarrow$ $R^1 = H, CH_3$
$R^2 = H, OH, OCH_3$

**Scheme 2-10.** Biotechnological generation of novel type I glycopeptide aglycones modified at AA2/AA6 and AA7 with (a) media supplementation[126] and (b) mutasynthesis of modified amino acids.[98,99]

carries a gene inactivation in the biosynthesis of a building block of this secondary metabolite. By supplementing the media with analogs of this building block, the gene inactivation can be bypassed in certain cases to yield a modified secondary metabolite. Prerequesites are mostly the use of structurally similar modifications. This approach was transferred to a balhimycin mutant, which was inactivated in the biosynthesis of β-hydroxytyrosine. The supplementation of 3-fluoro-β-hydroxytyrosine (Scheme 2-10), 3,5-difluoro-β-hydroxytyrosine, and 2-fluoro-β-hydroxytyrosine rendered the corresponding fluorobalhimycines, which were detected by HPLC-ESI-MS. The qualitative agar diffusion test showed in all cases antibacterial activity against the indicator strain *B. subtilis*. However, because of the low efficiency of this method, the amounts obtained did not suffice for comparative MIC tests.

*Modification of AA3*   Glycopeptides of the vancomycin-type mostly bear Asn in position 3 of the aglycon but also the aspartic acid analogs as well as the glutamine analogs are known.[128] However, these derivatives are antibiotically less active that their Asn-analog. Reasons may include the negative charge of the carboxylate side chain or a less favorable binding to D-Ala-D-Ala by the Gln-derivative. The above-mentioned glycopeptide CDP-1, which is an isoaspartic acid analog of vancomycin, is derived from an acidic rearrangement and has no antibiotic activity because of an extended D-*O*-E-ring combined with an altered conformation. With semisynthetic approaches, Asn can be selectively hydrolyzed to Asp, which was shown at the example of Ba(OH)$_2$-mediated hydrolysis of eremomycin to yield [3]Asp-eremomycin.[129] The coupling of several amines gave the corresponding *bis*-amides, which are modified at [3]Asp as well as at the *C*-terminus of eremomycin.

Based on extensive degradation studies, Malabarba et al. synthesized AA1- and AA3-substituted teicoplanin aglycones with excellent activity against staphylococci.[130] Some of these derivatives had some moderate activity also against vanA-resistant enterococci. The novel glycopeptides were the D-[1]Lys-L-[3]Phe, D-[1]Lys-L-[3]Lys, and D-[1]MeLeu-L-[3]Lys derivatives, and thus, they resembled the vancomycin aglycon more than the teicoplanin aglycon. A significant drawback of this approach is that a considerable amount of synthetic steps had to be performed to degrade the teicoplanin to a tetrapeptide precursor, which then helped to rebuild the glycopeptide.[131,132] As a consequence of the laborious synthesis of those derivatives, the generation of such derivatives will only be feasible if biosynthetic processes of glycopeptide antibiotics can be designed for a biotechnological large-scale production.

*Modification of AA4 and AA5*   The phenolic group of [4]Hpg serves as the carbohydrate attachment site, and in semisynthetic approaches, this group has been derivatized by protecting groups. Besides these modifications, other glycopeptide derivatives do not exist. This is similarly the case for [5]Hpg, where some naturally occuring derivatives are known, which are chlorinated in the *o*-position of the phenole. The restricted conformational flexibility and steric hindrance impede semisynthetic modification of these amino acids.

*Modification of AA7*    Besides the *C*-terminal derivatization or coupling of other molecules, which is subject to the Section 2.8.1.4, there are two phenoles and two electrophilic attachment sites in the aromatic side chain of $^7$Dpg. In the context of the derivatization of the glycopeptide aglycon with protecting groups, the phenoles have been alkylated or acylated. However, this approach does not provide any selectivity for the hydroxy groups of AA7 in favor of the phenoles of AA5 and AA4. With mutasynthesis, some vancomycin-type glycopeptide antibiotics have been generated, which were selectively modified at AA7 (Scheme 2-10).[99] These were 3,5-dimethoxyphenylglycines as well as 3-methoxy- and 3-hydroxyphenylglycines. Because of the low amounts obtained with this approach, no MIC tests were performed for these compounds.

Alternatives for the modification of $^7$Dpg are electrophilic substitution reactions at the aromatic ring, which has been demonstrated by a contribution of Pavlov et al.[133] A set of 15 novel glycopeptides has been synthesized with a Mannich-reaction (Scheme 2-9) with formaldehyde and various amines. Amines with long alkyl chains were the preferred substrates because these modifications previously showed promising antibacterial acivity in other glycopeptide derivatives. The antibiotic activities of these derivatives were good in the case of streptococci but less active against *S. aureus* strains, and they only showed weak activity against vanA-resistant enterococci.

### 2.8.1.4   Coupling Reactions at the C-Terminus, the N-Terminus, and the Carbohydrate Residues

In this chapter, modifications are discussed that concern easily accessible functional groups of glycopeptides and that are based on the coupling of bigger molecular entities rather than exchange reactions of carbohydrates, amino acids, or the substitution of functional groups. These modifications comprise mainly the amidation of the *C*-terminus, as well as the alkylation and acylation reactions of the *N*-terminus and the amino groups of the carbohydrate residues (Scheme 2-9). In fact, this approach rendered the majority of all semisynthetic glycopeptide derivatives and has been most successful in the generation of glycopeptides active against vancomycin-resistant enterococci and staphylococcci.

*Derivatization of the C-Terminus*    The derivatization of glycopeptides has been performed with various amines to yield the corresponding amides. An early example is the synthesis of vancomycin propanamide, histamide, and 3-aminopropanamide with DCC/HOBt.[134] Solution and solid-phase synthesis protocols were developed to yield similar derivatives as mentioned above, but also *C*-terminally elongated nonapeptide and decapeptide derivatives have been obtained.[135] The glycopeptide eremomycin was *C*-terminally converted to the methylester, the hydrazide, some urea derivatives, and the amide along with the methylamide and the benzylamide. The latter two derivatives were more active than eremomycin, however, they did not show antibiotic activity against vancomycin-resistant enterococci (VRE).[136] In a more representative study by Miroshnikova et al.,[137] a series of more than 25 eremomycin carboxamides were synthesized and tested for their antibiotic activity. As a result, small substituents (C0–C4) had comparable activity with the parent antibiotic but no activity against vancomycin-resistant strains. Tryptamine carboxamides and linear lipophilic substituents had activity against both vancomycin-sensitive and vancomycin-resistant strains. Among the teicoplanin analogs, C-terminal amide modifications have been tested with several dimethylamines. These compounds partially showed excellent activities against VRE and staphylococci.[138]

*Derivatization of the N-Terminus and of the Amino Groups of Vancosamine Sugars* Amino groups are good nucleophiles, and thus, they provide an excellent target for modification with a range of carbonyl and acyl compounds, also in the presence of other functional groups, such as phenoles and hydroxy groups. Most glycopeptides bear two amino groups, with the N-terminus and an amino sugar, which both compete in the reaction with derivatizing agents. Alkylation reactions and especially acylation reactions, which preferably occur at the N-terminus mostly led to a significantly reduced antibiotic activity. This result can be explained by the importance of a positive charge at the N-terminus for D-Ala-D-Ala-binding[113,114] and because of steric reasons, which might interfere with D-Ala-D-Ala-binding. As a consequence, the modification of the amino groups of vancosamine sugars was the preferred target of semisynthetic modifications. The performance of alkylation reactions with vancomycin yielded a set of more than 80 modified vancomycines.[139] The comparison of alkyl with alkanoyl derivatives showed that a better activity of alkyl derivatives, also against vancomycin-resistant strains, is obtained. To raise the yields in favor of amino sugar modifications, researchers developed synthesis protocols for their selective derivatization. Early modifications were the N-acylation of vancosamines, which also resulted in N-terminally acylated compounds.[140] From these semisynthetic glycopeptide derivatives, modifications with aryl residues showed better antibacterial activity than did alkyl residues. These early contributions were the groundwork for the continued development of other, more potent glycopeptide antibiotics. The alkylation reactions were mostly performed by reductive alkylation, for example, of eremomycin-related glycopeptides with benzaldehydes using boron hydrides as the reducing agents. These regioselectively carbohydrate-alkylated derivatives with lipophilic substituents showed good antibiotic activity against VRE.[123] The study also confirmed the previously observed trend of the importance of the N-alkylation sites resulting in either enhanced or lowered antibiotic activity. N'-monoalkylated compounds had higher antibiotic activity than did N-terminally alkylated compounds or multiple alkylated compounds against vancomycin-sensitive and vancomycin-resistant strains. In other extensive structure activity studies with chloroeremomycin (LY264826) derivatives, the significance of N'-alkylated benzyl substituents at the disaccharide portion of glycopeptides for antibiotic activity had been recognized.[141,142] Their activities were excellent against both vancomycin-susceptible and vancomycin-resistant antibiotics. The most important representative of this group is LY-333328, which is a N'-p-(p-chlorobiphenyl)benzyl-chloroeremomycin and which displays an excellent activity against vancomycin-susceptible and vancomycin-resistant strains (see Section 2.8.2.2).[143] Other studies were performed with acylation reactions of eremomycin, which preferably occur with an unoptimized synthesis protocol regioselectively at the N-terminus.[144] Because of the low antibiotic activity compared with N'-carbohydrate-modified glycopeptides, the approach selectively protected the N-terminus with Boc or Fmoc and then derivatized the N'- and N''-positions of the vancosamine sugars in an subsequent step.[145] Starting with these N-terminally protected eremomycins, N'-mono- and N',N''-dialkylated eremomycins have been synthesized by reductive alkylation with biphenyl and decyl residues. The double alkylation led to a decrease in antibiotic acitivity in contrast to alkylated desacyl-teicoplanins with two lipophilic substituents.[146] Other systematic studies have been performed to evaluate decylcarboxamides and chlorobiphenylbenzylcarboxamides of vancomycines, teicoplanines, and eremomycines, and of their aglycones and their *des*-leucylderivatives for their antibiotic activities.[147]

## 2.8.2   Rational Concepts for the Design of Novel Glycopeptides

The originally nonbiased modifications of glycopeptide molecules used in SAR studies together with the knowledge on the mode of action, stimulated researchers to develop more rational and sophisticated approaches and concepts to raise antibiotic activity of glycopeptide derivatives also against glycopeptide-resistant bacterial strains. A selection of these approaches is discussed in the subsequent sections.

### 2.8.2.1   Dimer, Trimer, and Multimer Approaches
The observation of glycopeptide self-assembly to dimers by Williams et al.[27,28] gave impulses for the chemical synthesis of covalently linked glycopeptide dimers (Figure 2-1). This approach was first introduced by Sundram et al.,[148] with *bis*-(vancomycin)carboxamides linked via 1,6-diaminohexane, cystamine, homocystamine, and triethylenetetramine. These derivatives showed moderate activity against VRE. C-terminally linked dimers with different tethers have been also synthesized by Jain et al.[149] The antibacterial activity of these derivatives against vanA-resistant enterococci has been tested and surprisingly, the desleucyl dimers still showed considerable antibiotic activity.

Besides C-terminal tethers, Nicolaou et al. synthesized dimers that were tethered via the amino group of the vancosamine sugar (Figure 2-1).[150] The underlying concept to this work was the dynamic target-accelerated combinatorial synthesis of vancomycin dimers. In the presence of the D-Ala-D-Ala ligand, the vancomycin monomers were allowed to assemble, whereas a previously introduced functionality covalently dimerized the glycopeptide monomers. Ligation was performed by olefin metathesis or by disulfide bond formation. A rate acceleration of the ligation reactions to the homo-dimer was observed in the presence of $Ac_2$-L-Lys-D-Ala-D-Ala compared with the dimerization reaction without ligand. From an eight-component combinatorial synthesis with 36 expected products, three derivatives were

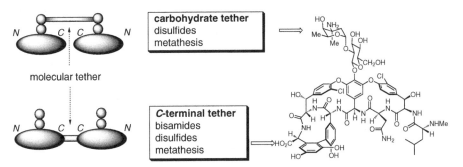

**Figure 2-1.** Scheme of the glycopeptide (big ellipse) dimer concept with tethers attached to the carbohydrate residues (small circles) or to the *C*-terminus.

preferentially formed. These derivatives showed significantly higher antibacterial activity than did other dimers and vancomycin alone. This approach was followed by a broader study with different tethers using the disulfide and the olefine metathesis coupling as well as a several amino acids other than leucine at the *N*-terminus.[151]

It is important to realize that vancomycin is not only an antibiotic of considerable significance but also the smallest known ligand–receptor system. Other well-known receptor systems, such as siderophores and crown ethers, specifically bind inorganic ions. However, as the only representative of molecules in the mass range of ~1000 Da, only glycopeptide antibiotics show clearly defined receptor properties for small molecules. As a consequence, with regard to ligand–receptor interactions, vancomycin is the most cited example for the chemical design of novel model receptors applicable also in aqueous solutions. The specificity of vancomycin for peptide ligands was investigated by Whitesides et al. with affinity capillary electrophoresis (ACE).[152,153] The Lys-D-Ala-D-Ala peptide was confirmed as the best binder from several peptides investigated. Whitesides et al. also synthesized a divalent vancomycin receptor model and determined a 1000-fold increase of D-Ala-D-Ala binding with ACE.[154] A *tris*(vancomycincarboxamide)[155] was designed as a trivalent receptor that binds a trivalent D-Ala-D-Ala ligand with a binding constant of $\approx 4 \times 10^{-17}$ M (Figure 2-2). This system has a 25-fold increased binding constant compared with biotin-avidin, which accounts for the strongest known binders in biological systems.[156] A detailed characterization of the *tris*(vancomycincarboxamide) has been performed in a subsequent publication.[157]

### 2.8.2.2 *Glycopeptides and Glycopeptide-Based Approaches with Antibiotic Activity Against Vancomycin-Resistant Bacteria*

As mentioned, semisynthetic approaches resulted in glycopeptides active against VRE. From these derivatives, the most promising results in SAR studies showed amino sugar-modified *N*-alkylated glycopeptide derivatives.[123,139,142,143] The most prominent member LY-333328 (oritavancin) (Scheme 2-11) showed a high antibiotic activity against vanA/B-resistant enterococci, MRSA, and streptococci.[143,158] LY-333328 is a chloroeremomycin derivative, which is modified at

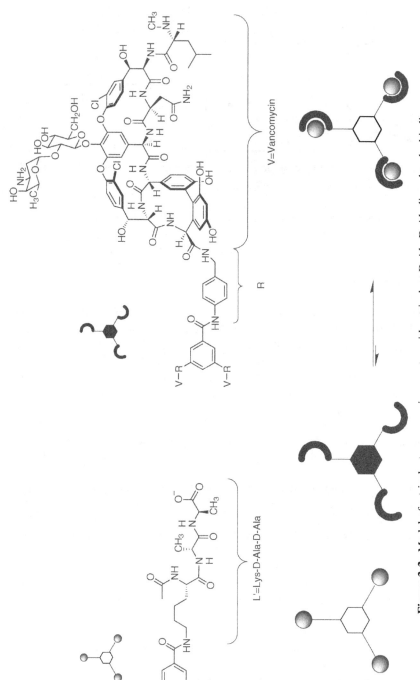

**Figure 2-2.** Model of a trivalent vancomycin-receptor with a trivalent D-Ala-D-Ala-ligand and a binding constant of $\approx 4 \times 10^{-17}$ M.[133] (a) Structure formulas of the ligands and the receptors. (b) Model of the interaction.

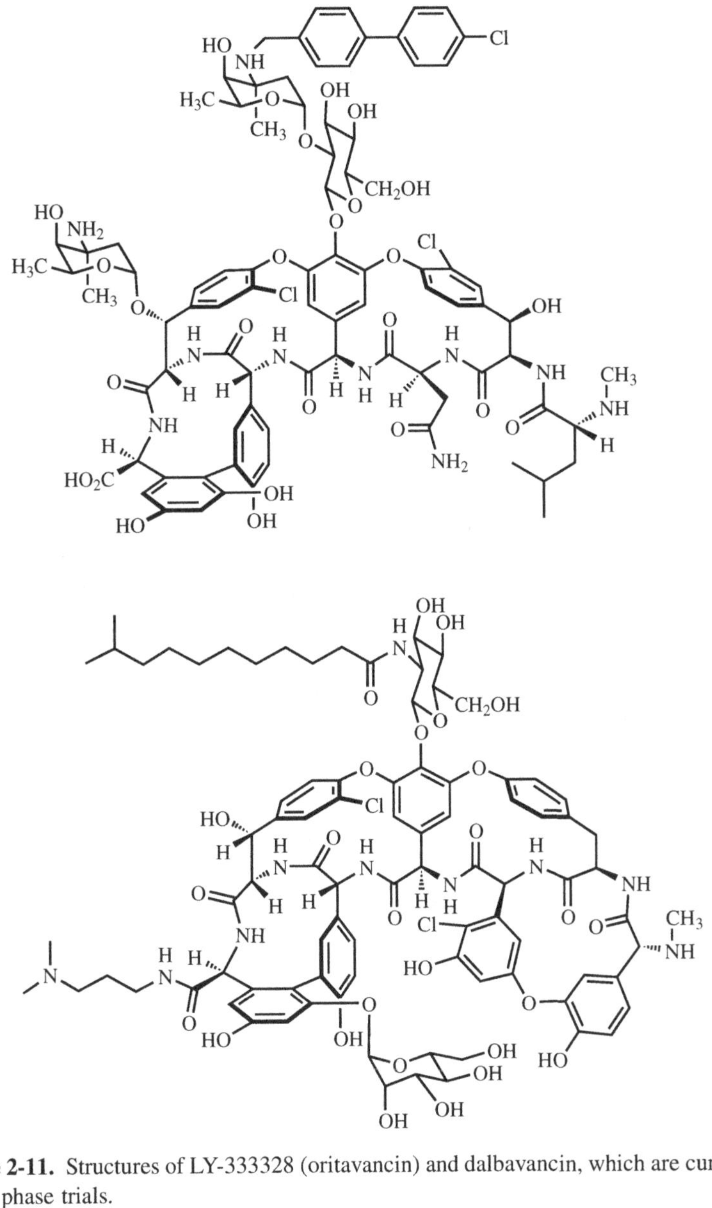

**Scheme 2-11.** Structures of LY-333328 (oritavancin) and dalbavancin, which are currently in clinical phase trials.

the amino group of 4-*epi*-vancosamine with a *N'*-*p*-(*p*-chlorobiphenyl)benzyl-residue. The antibacterial activity spectrum of LY-333328 has been investigated in various studies, which are summarized in a review article.[159] In analogy to the suggested membrane insertion mechanism of teicoplanin,[28] also for of the *N*-alkyl glycopeptide derivatives, a similar mechanism is assumed. This mechanism had been shown with model lipid monolayers[160,161] and membrane vesicles[162,163] and was supported by NMR studies.[164]

However, the recent contributions of Ge et al.[165] strongly support an additional mechanism that is not based on D-Ala-D-Ala binding or membrane insertion. According to previously established synthesis protocols,[100,116] desleucyl-vancomycin and its chlorobiphenylbenzyl derivative (Scheme 2-12) were synthesized. Surprisingly, from both compounds, which do not bind D-Ala-D-Ala, the chlorobiphenylbenzyl derivative showed good antibacterial activity against vancomycin-sensitive and vancomycin-resistant strains. The synthesis of disaccaride compounds with a chlorobiphenylbenzyl residue (Scheme 2-12) lacking the vancomycin aglycon also showed a significantly stronger antibiotic activity than did vancomycin. From these experiments, it was suggested that the chlorobiphenylbenzyl moiety inhibits transglycosylating enzymes. In a subsequent publication by the same group,[166] chlorobiphenylbenzyl derivatives attached to C6 of glucose were synthesized. These derivatives were also more active than vancomycin against vancomycin-sensitive bacterial strains. Moreover, they displayed good antibiotic activity against vancomycin-resistant strains. However, the corresponding desleucyl derivatives and a damaged teicoplanin incapable of D-Ala-D-Ala binding showed a complete loss of activity against vancomycin-resistant strains. As a consequence, for these C6-alkyl derivatives, a membrane insertion mechanism according to Williams rather than an inhibition of transglycosylating enzymes was suggested. The most recent results also support the hypothesis that chlorobiphenylbenzyl derivatives inhibit transglycosylases.[167] Although this hypothesis was only shown with penicillin binding protein 1b (PBP1b) and lipid II analogs of gram-negative E. coli, it seems justified to extrapolate these results to gram-positive bacteria. The above-mentioned hypothesis of Ge et al. is in the meantime also supported with experiments by other groups.[168] A very interesting approach is the design of hybrid glycopeptides combining the aglycon portion of vancomycin with the carbohydrate motives of the transglycosylase inhibitor moenomycin.[169,170] These hybrid derivatives showed an increased antibiotic activity against vancomycin-sensitive and vanA-resistant E. faecium strains compared with vancomycin. In the same contribution, the authors show that the introduction of one polyethyleneglycol unit linking the carbohydrate with the aglycon is not essential for antibiotic activity.

Many publications deal with the synthesis of vancomycin mimetics or partial structures to design novel peptide binding receptors. In an example by Xu et al., a combinatorial library was synthesized on solid support.[171] In this library, the D-O-E ring as well as AA1 and AA3 were conserved, whereas structural variety was introduced at AA5–AA7 with a combination of more than 30 amino acids. The screening was performed against fluorophore-labeled ligands with D-Ala-D-Ala or D-Ala-D-Lac peptides. The best binding molecules had binding affinities to D-Ala-D-Ala of about one magnitude less than vancomycin and a five-fold increased binding of D-Ala-D-Lac compared with vancomycin.

In another conceptually different approach, the ester linkage of D-Ala-D-Lac of VRE was considered as a hydrolytically cleavable functional group.[172] The screening of a nonbiased peptide library rendered an ε-aminopentanoylated prolinol

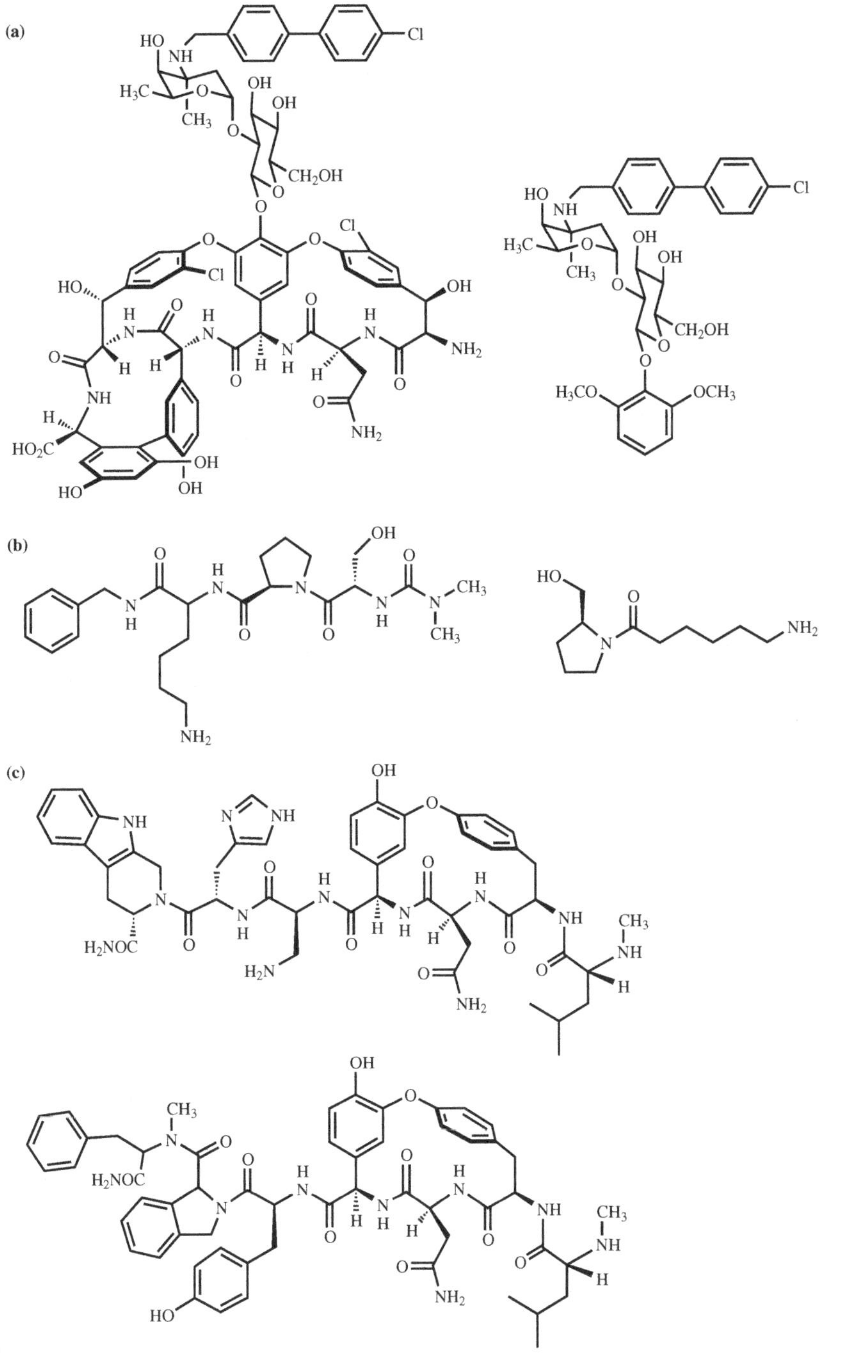

**Scheme 2-12.** Approaches for the design of glycopeptide-related molecules with also activity against vancomycin-resistant bacteria. (a) Chlorobiphenylbenzyl derivatives with activity against VRE;[165,166] (b) D-Ala-D-Lac small molecule cleavers;[172] (c) Synthetic receptors from a combinatorial library with binding properties to D-Ala-D-Ala and D-Ala-D-Lac.[171]

derivative (Scheme 2-12), which showed antibiotic activity in combination with vancomycin against vanA-resistant enterococci. In contrast, the separate use of both componds in antibacterial assays rendered no effects.

### 2.8.3 Conclusions

As can be observed from this discussion, many semisynthetic glycopeptide derivatives have been synthesized over the past few decades. The modifications, which were contributed particularly from academic research groups, seem sometimes less systematic and mostly contain only a few glycopeptide derivatives. The antibacterial testings of only a few compounds often reflect a tendency for enhanced or diminished activity, albeit they do not give a full picture. Therefore, a more systematic evaluation of SAR studies mostly guided or conducted by companies, with at least 20 to even 80 compounds per set, are of much bigger value to judge the antibiotic potential of site-specific glycopeptide modifications. One important basis for these systematic SAR studies has been the elaboration of chemoselective and regioselective synthesis protocols to obtain the desired compounds (e.g., N-alkylation) and high yields. Although on the first view the chemistry developed for the semisynthesis of glycopeptide derivatives does not seem to reach beyond well-established standard chemical reactions, glycopeptides considerably show degradation during modification reactions and yields can be dramatically decreased. Sensitive structural features are the carbohydrates residues and the asparagine residue of vancomycin-type glycopeptides, which tend to undergo hydrolysis and rearrangement reactions, respectively. Furthermore, it has to be assumed that the phenolic residues also contribute to the degradation of glycopeptides. A disadvantage of SAR studies performed by companies is that they are less "visible" to the scientific community. A good overview of patents on recent developments in the field of semisynthetic glycopeptides is given by Preobrazhenskaja and Olsufyeva.[173]

The attachment of lipophilic residues has brought a major breakthrough in obtaining derivatives also active against vancomycin-resistant strains. Remarkably, by modification of the oritavancin-type, an additional independent mode of action has been introduced by this semisynthetic modification. As a conclusion, the major trends of successful glycopeptide derivatizations are reflected by either carbohydrate N-alkylated or C-terminally amidated glycopeptides, represented by LY-333328 (oritavancin) and dalbavancin (Figure 2-11), which both are currently investigated in clinical studies. Dalbavancin shows excellent antibiotic activity against MRSA strains and was active against vanB-resistant enterococci.[174] However, only oritavancin remained active against vanA- and vanB-resistant enterococci and is thus preferably developed for VRE infections.[174]

From the current view of the author, few chances remain for any more enhancement of antibiotic activity based on the derivatization with lipophilic residues. This view is also reflected by other approaches, which tested covalently tethered dimers and trimers and vancomycin-based libraries. An apparent disadvantage of the dimer approach for clinical use, however, is the relatively high molecular masses of such derivatives combined with unfavorable pharmacokinetic properties. The modulation of D-Ala-D-Ala peptide binding site remains an approach with a high potential for novel glycopeptides with enhanced antibacterial properties. However, manipulations of the aglycon, such as the exchange of AA1 and AA3, as it has already been demonstrated, are extremely difficult and laborious. This characteristic presents a dilemma, because the potential still is widely unexplored. Continued semisynthetic experiments in this direction require large amounts of relatively expensive glycopeptides as starting materials. In this context, one elegant alternative approach would be the cyclization of linear heptapeptides with overexpressed P450-dependent oxygenase enzymes from vancomycin and teicoplanin biosynthesis gene clusters. The synthesis of heptapeptides can be easily performed with solid-phase peptide synthesis, and diversity can be introduced in a combinatorial approach with heptapeptide libraries. Additional interesting alternatives to the semisynthesis of glycopeptides with the potential for large-scale production can be observed in biotechnological approaches, which are expected to gain importance in the coming years. However, a lot of research still has to be performed to make such approaches attractive for industrial processes.

Finally, it has to be noted that glycopeptides only represent one option to combat infections by gram-positive bacteria. Current research is focused on other cell wall biosynthesis inhibitors (e.g., β-lactams, cephalosporins) or even on the development of antibacterial agents (e.g., tetracyclines, ketolides, and quinolone antibiotics) against other targets.[11,175] An important drug candidate in this context is linezolid (Zyvox), which is an entirely synthetic oxazolidinone antibiotic with in vitro and in vivo efficiency against MRSA and VRE.[175]

## ACKNOWLEDGMENT

The work of R. D. Süssmuth is supported by an Emmy-Nother-Fellowship (SU 239/2) of the Deutsche Forschungsgemeinschaft (DFG) and a grant from the European Community (COMBIG-TOP, LSHG-CT-2003–503491).

## REFERENCES

1. McCormick, M. H.; Stark, W. M.; Pittenger, G. E.; Pittenger, R. C.; McGuire, J. M. *Antibiot. Annu.*, **1955–1956**, 606–611.

2. Courvalin, P. *Antimicrob. Agents Chemother.*, **1990**, 34: 2291–2296.

3. Leclerq, R.; Derlot, E.; Duval, J.; Courvalin, P., *New Engl. J. Med.*, **1988**, 319: 157–161.

4. Uttley, A. H. C.; Collins, C. H.; Naidon, J.; George, R. C. *Lancet*, **1988**, 1: 57–58.

5. Williams, D. H.; Bardsley, B. *Angew. Chem.*, **1999**, 111: 1264–1286; *Angew. Chem. Int. Ed. Engl.*, **1999**, 38: 1172–1193.

6. Nicolaou, K. C.; Boddy, C. N. C.; Bräse, S.; Winssinger, N. *Angew. Chem. Int. Ed.*, **1999**, 38: 2096–2152.

7. Nicolaou, K. C.; Snyder, S. A. In *Classics in Total Synthesis II*. Weinheim, Germany: Wiley-VCH, **2003**, 239.

8. Hubbard, B. K.; Walsh, C. T. *Angew. Chem. Int. Ed.*, **2003**, 42: 730–765.

9. Süssmuth, R. D.; Wohlleben, W. *Appl. Microbiol. Biotechnol.*, **2004**, 63: 344–350.

10. Green, D. W. *Expert Opin. Ther. Targets*, **2002**, 6: 1–19.

11. Boneca, I. G.; Chiosis, G. *Expert Opin. Ther. Targets*, **2003**, 7: 311–328.

12. Ward, T. J.; Farris, A. B., III. *J. Chromatogr. A*, **2001**, 906: 73–89.

13. Williams, D. H.; Kalman, J. R. *J. Am. Chem. Soc.*, **1977**, 99: 2768–2774.

14. Sheldrick, G. M.; Jones, P. G.; Kennard, O.; Williams, D. H.; Smith, G. A. *Nature*, **1978**, 271: 223–225.

15. Williamson, M. P.; Williams, D. H. *J. Am. Chem. Soc.*, **1981**, 103: 6580–6585.

16. Harris, C. M.; Harris, T. M. *J. Am. Chem. Soc.*, **1982**, 104: 4293–4295.

17. Sheldrick, G. M.; Paulus, E.; Vertesy, L.; Hahn, F. *Acta Crystallogr. B Struct. Sci.*, **1995**, B51: 89–98.

18. Singh, S. B.; Jayasuriya, H.; Salituro, G. M.; Zink, D. L.; Shafiee, A.; Heimbuch, B.; Silverman, K. C.; Lingham, R. B.; Genilloud, O.; Teran, A.; Vilella, D.; Felock, P.; Hazuda, D. *J. Nat. Prod.*, **2001**, 64: 874–882.

19. Matsuzaki, K.; Ikeda, H.; Ogino, T.; Matsumoto, A.; Woodruff, H. B.; Tanaka, H.; Omura, S. *J. Antibiot.*, **1994**, 47: 1173–1174.

20. Perkins, H. R. *Biochem. J.*, **1969**, 111: 195–205.

21. Kalman, J. R.; Williams, D. H. *J. Am. Chem. Soc.*, **1980**, 102: 906–912.

22. Williams, D. H.; Williamson, M. P.; Butcher, D. W.; Hammond, S. J. *J. Am. Chem. Soc.*, **1983**, 105: 1332–1339.

23. Schäfer, M.; Schneider, T. R.; Sheldrick, G. M. *Structure*, **1996**, 4: 1509–1515.

24. Loll, P. J.; Bevivino, A. E.; Korty, B. D.; Axelsen, P. H. *J. Am. Chem. Soc.*, **1997**, 119: 1516–1522.

25. Nieto, M.; Perkins, H. R. *Biochem. J.*, **1971**, 123: 780–803.

26. Waltho, J. P.; Williams, D. H. *J. Am. Chem. Soc.*, **1989**, 111: 2475–2480.

27. Beauregard, D. A.; Williams, D. H.; Gwynn, M. N.; Knowles, D. J. C. *Antimicrob. Agents Chemother.*, **1995**, 39: 781–785.

28. Mackay, J. P.; Gerhard, U.; Beauregard, D. A.; Westwell, M. S.; Searle, M. S.; Williams, D. H. *J. Am. Chem. Soc.*, **1994**, 116: 4581–4590.

29. Cooper, M. A.; Wiliams, D. H. *Chem. Biol.*, **1999**, 6: 891–899.

30. Arthur, M.; Molinas, C.; Bugg, T. D. H.; Wright, G. D.; Walsh, C. T.; Courvalin, P. *Antimicrob. Agents Chemother.*, **1992**, 36: 867–869.

31. Arthur, M.; Molinas, C.; Courvalin. P. *J. Bacteriol.*, **1992**, 174: 2582–2591.

32. Arthur, M.; Reynolds, P.; Courvalin, P. *Trends Microbiol.*, **1996**, 4: 401–407.

33. Walsh, C. T. *Science*, **1993**, 261: 308–309.

34. Walsh, C. T.; Fisher, S. L.; Park, I.-S.; Prahalad, M.; Wu, Z. *Chem. Biol.*, **1996**, 3: 21–28.

35. Allen, N. E.; LeTourneau, D. L.; Hobbs, J. N. Jr., *J. Antibiot.*, **1997**, 50: 677–684.

36. Nicas, T. I.; Cole, C. T.; Preston, D. A.; Schabel, A. A.; Nagarajan, R. *Antimicrob. Agents Chemother.*, **1989**, 33: 1477–1481.

37. Reynolds, P. E.; Snaith, H. A.; Maguire, A. J.; Dutka-Malen, S.; Courvalin, P. *Biochem. J.*, **1994**, 301: 5–8.

38. Billot-Klein, D.; Gutmann, L.; Sable, S.; Guittet, E.; van Heijenoort, J. *J. Bacteriol.*, **1994**, 176: 2398–2405.

39. Billot-Klein, D.; Blanot, D.; Gutmann, L.; van Heijenoort, J. *Biochem. J. Lett.*, **1994**, 304: 1021–1022.

40. Hiramatsu, K.; Hanaki, H.; Ino, T.; Yabuta, K.; Oguri, T.; Tenover, F. C. *J. Antimicrob. Chemother.*, **1997**, 97: 1233–1250.

41. Noble, W. C.; Virani, Z.; Cree, R. G. A. *FEMS Microbiol. Lett.*, **1992**, 93: 195–198.

42. Chang, S.; Sievert, D. M.; Hageman, J. C.; Boulton, M. L.; Tenover, F. C.; Downes, F. P.; Shah, S.; Rudrik, J. T.; Pupp, G. R.; Brown, W. J.; Cardo, D.; Fridkin, S. K. *New Engl. J. Med.*, **2003**, 348: 1342–3147.

43. Van Wageningen, A. M. A.; Kirkpatrick, P. N.; Williams, D. H.; Harris, B. R.; Kershaw, J. K.; Lennard, N. J.; Jones, M.; Jones, S. J. M.; Solenberg, P. J. *Chem. Biol.*, **1998**, 5: 155–162.

44. Pelzer, S.; Süssmuth, R.; Heckmann, D.; Recktenwald, J.; Huber, P.; Jung, G.; Wohlleben, W. *Antimicrob. Agents Chemother.*, **1999**, 43: 1565–1573.

45. Sosio, M.; Bianchi, A.; Bossi, E.; Donadio, S. *Antonie van Leeuwenhoek*, **2000**, 78: 379–384.

46. Li, T.-L.; Huang, F.; Haydock, S. F.; Mironenko, T.; Leadley, P. F.; Spencer, J. B. *Chem. Biol.*, **2004**, 11: 107–119.

47. Sosio, M.; Kloosterman, H.; Bianchi, A.; deVreugd, P.; Dijkhuizen, L.; Donadio, S. *Microbiology*, **2004**, 150: 95–102.

48. Sosio, M.; Stinchi, S.; Beltrametti, F.; Lazzarini, A.; Donadio, S. *Chem. Biol.*, **2003**, 10: 1–20.

49. Pootoolal, J.; Thomas, M. G.; Marshall, C. G.; Neu, J. M.; Hubbard, B. K.; Walsh, C. T.; Wright, G. D. *Proc. Natl. Acad. Sci. USA*, **2002**, 99: 8962–8967.

50. Chiu, H.-T.; Hubbard, B. K.; Shah, A. N.; Eide, J.; Fredenburg, R. A.; Walsh, C. T.; Koshla, C. *Proc. Natl. Acad. Sci. USA*, **2001**, 98: 8548–8553.

51. Marahiel, M. A.; Stachelhaus, T.; Mootz, H. D. *Chem. Rev.*, **1997**, 97: 2651–2673.

52. Mootz, H. D.; Schwarzer, D.; Marahiel, M. A. *Chem. Bio. Chem.*, **2002**, 3: 490–504.

53. O'Brian, D. P.; Kirkpatrick, P. N.; O'Brian, S. W.; Staroske, T.; Richardson, T. I.; Evans, D. A.; Hopkinson, A.; Spencer, J. B.; Williams, D. H. *Chem. Commun.*, **2000**, 103–104.

54. Choroba, O. W., Williams, D. H.; Spencer, J. B. *J. Amer. Chem. Soc.*, **2000**, 122: 5389–5390.

55. Hubbard, B. K.; Thomas, M. G.; Walsh, C. T. *Chem. Biol.*, **2000**, 7: 931–942.

56. Pfeifer, V.; Nicholson, G. J.; Ries, J.; Recktenwald, J.; Schefer, A. B.; Shawky, R. M.; Schröder, J.; Wohlleben, W.; Pelzer, S. *J. Biol. Chem.*, **2001**, 276: 38370–38377.

57. Sandercock, A. M.; Charles, E. H.; Scaife, W.; Kirkpatrick, P. N.; O'Brien, S. W.; Papageorgiou, E. A.; Spencer, J. B.; Williams, D. H. *Chem. Commun.*, **2001**, 14: 1252–1253.

58. Chen, H.; Tseng, C. C.; Hubbard, B. K.; Walsh, C. T. *Proc. Natl. Acad. Sci.*, **2001**, 98: 14901–14906.

59. Li, T. L.; Choroba, O. W.; Hong, H.; Williams, D. H.; Spencer, J. B. *J. Chem. Soc. Chem. Commun.*, **2001**, 2156–2157.

60. Chen, H.; Walsh, C. T. *Chem. Biol.*, **2001**, 8: 301–312.

61. Puk, O.; Huber, P.; Bischoff, D.; Recktenwald, J.; Jung, G.; Süssmuth, R. D.; van Pee, K.-H.; Wohlleben, W.; Pelzer, S. *Chem. Biol.*, **2002**, 9: 225–235.

62. Puk, O.; Bischoff, D.; Kittel, C.; Pelzer, S.; Weist, S.; Stegmann, E.; Süßmuth, R.; Wohlleben, W. *J. Bacteriol.*, **2004**, 86: 6093–6100.

63. Süssmuth, R.; Süssmuth, R. D.; Pelzer, S.; Nicholson, G.; Walk, T.; Wohlleben, W.; Jung, G. *Angew. Chem.*, **1999**, 111: 2096–2099; *Angew. Chem. Int. Ed.*, **1999**, 38: 1976–1979.

64. Bischoff, D.; Pelzer, S.; Höltzel, A.; Nicholson, G.; Stockert, S.; Wohlleben, W.; Jung, G.; Süssmuth, R. D.; *Angew. Chem.*, **2001**, 113: 1736–1739; *Angew. Chem. Int. Ed.*, **2001**, 40: 1693–1696.

65. Bischoff, D.; Pelzer, S.; Bister, B.; Nicholson, G. J.; Stockert, S.; Schirle, M.; Wohlleben, W.; Jung, G.; Süssmuth, R. D. *Angew. Chem.*, **2001**, 113: 4824–4827; *Angew. Chem. Int. Ed.*, **2001**, 40: 4688–4691.

66. Evans, D. A.; Wood, M. R.; Trotter, B. W.; Richardson, T. I.; Barrow, J. C.; Katz, J. L. *Angew. Chem.*, **1998**, 110: 2864–2868; *Angew. Chem. Int. Ed.*, **1998**, 37: 2700–2704.

67. Evans, D. A.; Dinsmore, C. J.; Watson, P. S.; Wood, M. R.; Richardson, T. I.; Trotter, B. W.; Katz, J. L. *Angew. Chem.*, **1998**, 110: 2868–2872; *Angew. Chem. Int. Ed.*, **1998**, 37: 2704–2708.

68. Nicolaou, K. C.; Natarajan, S.; Li, H.; Jain, N. F.; Hughes, R.; Solomon, M. E.; Ramanjulu, J. M.; Boddy, C. N. C.; Takayanagi, M. *Angew. Chem.*, **1998**, 110: 2872–2878; *Angew. Chem. Int. Ed.*, **1998**, 37: 2708–2714.

69. Nicolaou, K. C.; Jain, N. F.; Natarajan, S.; Hughes, R.; Solomon, M. E.; Li, H.; Ramanjulu, J. M.; Takayanagi, M.; Koumbis, A. E.; Bando, T. *Angew. Chem.*, **1998**, 110: 2879–2881; *Angew. Chem. Int. Ed.*, **1998**, 37: 2714–2716.

70. Nicolaou, K. C.; Takayanagi, M.; Jain, N. F.; Natarajan, S.; Koumbis, A. E.; Bando, T.; Ramanjulu, J. M. *Angew. Chem.*, **1998**, 110: 2881–2883; *Angew. Chem. Int. Ed.*, **1998**, 37: 2717–2719.

71. Nicolaou, K. C.; Li, H.; Boddy, C. N. C.; Ramanjulu, J. M.; T.-Y. Yue, Natarajan, S.; Chu, X.-J.; Bräse, S.; Rübsam, F. *Chem. Eur. J.*, **1999**, 5: 2584–2601.

72. Nicolaou, K. C.; Boddy, C. N. C.; Li, H.; Koumbis, A. E.; Hughes, R.; Natarajan, S.; Jain, N. F.; Ramanjulu, J. M.; Bräse, S.; Solomon, M. E. *Chem. Eur. J.*, **1999**, 5: 2602–2621.

73. Nicolaou, K. C.; Koumbis, A. E.; Takayanagi, M.; Natarajan, S.; Jain, N. F.; Bando, T.; Li, H.; Hughes, R. *Chem. Eur. J.*, **1999**, 5: 2622–2647.

74. Nicolaou, K. C.; Mitchell, H. J.; Jain, N. F.; Bando, T.; Hughes, R.; Winssinger, N.; Natarajan, S.; Koumbis, A. E. *Chem. Eur. J.*, **1999**, 5: 2648–2667.

75. Boger, D. L.; Miyazaki, S.; Kim, S. H.; Wu, J. H.; Castle, S. L.; Loiseleur, O.; Jin, Q. *J. Am. Chem. Soc.*, **1999**, 121: 10004–10011.

76. Boger, D. L. *Medicinal Res. Rev.*, **2001**, 21: 356–381.

77. Boger, D. L.; Kim, S. H.; Miyazaki, S.; Strittmatter, H.; Weng, J. H.; Mori, Y.; Rogel, O.; Castle, S. L.; McAtee, J. J. *J. Am. Chem. Soc.*, **2000**, 122: 7416–7417.

78. Boger, D. L.; Miyazaki, S.; Kim, S. H.; Wu, J. H.; Loiseleur, O.; Castle, S. L. *J. Am. Chem. Soc.*, **1999**, 121: 3226–3227.

79. Nicolaou, K. C.; Mitchell, H. J.; Jain, N. F.; Winssinger, N.; Hughes, R.; Bando, T. *Angew. Chem.*, **1999**, 111: 253–255; *Angew. Chem. Int. Ed.*, **1999**, 38: 240–244.

80. Rama Rao, A. V.; Gurjar, M. K.; Reddy, K. L.; Rao, A. S. *Chem. Rev.*, **1995**, 95: 2135–2167.

81. Evans, D. E.; DeVries, K. M. In Nagarajan, R., editor. *Glycopeptide Antibiotics.*, New York: Marcel Dekker; **1994**, 63–104.

82. Sztaricskai, F.; Pelyvas-Ferenczik, I. In Nagarajan, R., editor. *Glycopeptide Antibiotics*, New York: Marcel Dekker; **1994**, 105–195.

83. Zhang, A. J.; Burgess, K. *Angew. Chem.*, **1999**, 111: 666–669; *Angew. Chem. Int. Ed.*, **1999**, 38: 634–636.

84. Armstrong, D. W.; Zhou, Y. *J. Liq. Chromatogr.*, **1994**, 17: 1695–1707.

85. Dermaux, A.; Lynen, F.; Sandra, P. *J. High Resolut. Chromatogr.*, **1998**, 21: 575–576.

86. Karlsson, C.; Karlsson, L.; Armstrong, D. W.; Owens, P. K. *Anal. Chem.*, **2000**, 72: 4394–4401.

87. Gasper, M. P.; Berthod, A.; Nair, U. B.; Armstrong, D. W. *Anal. Chem.*, **1996**, 68: 2501–2514.

88. Aboul-Enein, H. Y.; Ali, I. *Chromatographia*, **2000**, 52: 679–691.

89. Desiderio, C.; Fanali, S. *J. Chromatogr. A.*, **1998**, 807: 37–56.

90. Armstrong, D. W.; Tang, Y. B.; Chen, S. S.; Zhou, Y.; Bagwill, C.; Chen, J.-R. *Anal. Chem.*, **1994**, 66: 1473–1484.

91. Armstrong, D. W.; Rundlett, K. L.; Chen, J. *Chirality*, **1994**, 6: 496–509.

92. Rundlett, K. L.; Armstrong, D. W. *Anal. Chem.*, **1995**, 67: 2088–2095.

93. Armstrong, D. W.; Rundlett, K. L. *J. Liq. Chromatogr.*, **1995**, 18: 3659–3674.

94. Armstrong, D. W.; Nair, U. B. *Electrophoresis*, **1997**, 18: 2331–2342.

95. Ward, T. J.; Oswald, T. M. *J. Chromatogr. A*, **1997**, 792: 309–325.

96. Ward, T. J.; Dann, C. III; Blaylock, A. *J. Chromatogr. A.*, **1995**, 715: 337–344

97. Ward, T. J.; Farris, A. B. III. *J. Chromatogr. A.*, **2001**, 906, 73–89.

98. Weist, S.; Bister, B.; Puk, O.; Bischoff, D.; Nicholson, G.; Stockert, S.; Wohlleben, W.; Jung, G.; Süssmuth, R. D. *Angew. Chem.*, **2002**, 114: 3531–3534; *Angew. Chem. Int. Ed.*, **2002**, 40: 3383–3385.

99. Weist, S.; Kittel, C.; Bischoff, D.; Bister, B.; Pfeifer, V.; Nicholson, G. J.; Wohlleben, W.; Süssmuth, R. D. *J. Am. Chem. Soc.*, **2004**, 126: 5942–5943.

100. Nagarajan, R.; Schabel, A. A. *J. Chem. Soc. Chem. Comm.*, **1988,** 23: 1306–1307.

101. Nagarajan, R.; Berry, D. M.; Schabel, A. A. *J. Antibiot.*, **1989**, 42: 1438–1440.

102. Cavalleri, B.; Ferrari, P.; Malabarba, A.; Magni, A.; Pallanza, R.; Gallo, G. *J. Antibiot.*, **1987**, 40: 49–59.

103. Wanner, J.; Tang, D.; McComas, C. C.; Crowley, B. M.; Jiang, W.; Moss, J.; Boger, D. L. *Bioorg. Med. Chem. Lett.*, **2003**, 13: 1169–1173.

104. Malabarba, A.; Nicas, T. I.; Thompson, R. C. *Med. Chem. Rev.*, **1997**, 17: 69–1997.

105. Ge, M.; Thompson, C.; Kahne, D. *J. Am. Chem. Soc.*, **1998**, 120: 11014–11015.

106. Thompson, C.; Ge, M.; Kahne, D. *J. Am. Chem. Soc.*, **1999**, 121: 1237–1244.

107. Nicolaou, K. C.; Winssinger, N.; Hughes, R.; Smethurst, C.; Cho, S. Y. *Angew. Chem.*, **2000**, 112: 1126–1130; *Angew. Chem. Int. Ed.*, **2000**, 39: 1084–1088.

108. Nicolaou, K. C.; Cho, S. Y.; Hughes, R.; Winssinger, N.; Smethurst, C.; Labischinski, H.; Endermann, R. *Chem. Eur. J.*, **2001**, 7: 3799–3823.

109. Ritter, T. K.; Mong, K.-K. T.; Liu, H.; Nakatani, T.; Wong, C.-H. *Angew. Chem.*, **2003**, 115: 4805–4808; *Angew. Chem. Int. Ed.*, **2003**, 42: 4657–4660.

110. Solenberg, P. J.; Matsushima, P.; Stack, D. R.; Wilkie, S. C.; Thompson, R. C.; Baltz, R. H. *Chem. Biol.*, **1997**, 5: 195–202.

111. Losey, H. C.; Peczuh, M. W.; Chen, Z.; Eggert, U. S.; Dong, S. D.; Pelczer, I.; Kahne, D.; Walsh, C. T. *Biochemistry*, **2001**, 40: 4745–4755.

112. Harris, C. M.; Kopecka, H.; Harris, T. M. *J. Am. Chem. Soc.*, **1983**, 105: 6915–6922.

113. Kannan, R.; Harris, C. M.; Harris, T. M.; Waltho, J. P.; Skelton, N. J.; Williams, D. H. *J. Am. Chem. Soc.*, **1988**, 110: 2946–2953.

114. Convert, O.; Bongini, A.; Feeney, J. *J. Chem. Soc. Perkin. Trans. I*, **1980**, 1262–1270.

115. Nagarajan, R. *J. Antibiot.*, **1993**, 46: 1181–1195.

116. Booth, P. M.; Stone, D. J. M.; Williams, D. H. *J. Chem. Soc. Chem. Comm.*, **1987**, 22: 1694–1695.

117. Booth, P. M.; Williams, D. H. *J. Chem. Soc. Perkin. Trans. 1*, **1989**, 12: 2335–2339.

118. Nagarajan, R. In Nagarajan, R., editor. *Glycopeptide Antibiotics*, vol. 5. New York: Marcel Dekker; **1994**, 199.

119. Christofaro, N. F.; Beauregard, D. A.; Yan, H.; Osborn, N. J.; Williams, D. H. *J. Antibiot.*, **1995**, 48: 805–810.

120. Miroshnikova, O. V.; Berdnikova, T. F.; Olsufyeva, E. N.; Pavlov, A. Y.; Reznikova, M. I.; Preobrazhenkaya, M. N.; Ciabatti, R.; Malabarba, A.; Colombo, L. *J. Antibiot.*, **1996**, 49: 1157–1161.

121. Nagarajan, R. In Nagarajan, R., editor. *Glycopeptide Antibiotics*, vol. 5. New York: Marcel Dekker; **1994**, 201.

122. Nagarajan, R. In Nagarajan, R., editor. *Glycopeptide Antibiotics*, vol. 5. New York: Marcel Dekker; **1994**, 206.

123. Cooper, R. D. G.; Snyder, N. J.; Zweifel, M. J.; Staszak, M. A.; Wilkie, S. C.; Nicas, T. I.; Mullen, D. L.; Butler, T. F.; Rodriguez, M. J.; Huff, B. E.; Thompson, R. C. *J. Antibiot.*, **1996**, 49: 575–581.

124. Harris, C. M.; Kannan, R.; Kopecka, H.; Harris, T. M. *J. Am. Chem. Soc.*, **1985**, 107: 6652–6658.

125. Gerhard, U.; Mackay, J. P.; Maplestone, R. A.; Williams, D. H. *J. Am. Chem. Soc.*, **1993**, 115: 232–237.

126. Bister, B.; Bischoff, D.; Nicholson, G.; Stockert, S.; Wink, J.; Brunati, C.; Donadio, S.; Pelzer, S.; Wohlleben, W.; Süssmuth, R. D. *ChemBioChem*, **2003**, 4: 658–662.

127. Rinehart K. L. *Pure Appl. Chem.*, **1977**, 49: 1361–1384.

128. Nagarajan, R. In: Nagarajan, R., editor. *Glycopeptide Antibiotics*, vol. 5. New York: Marcel Dekker; **1994**, 203.

129. Olsufyeva, E. N.; Berdnikova, T. F.; Miroshnikova, O. V.; Reznikova, M. I.; Preobrazhenskaya, M. N. *J. Antibiot.*, **1999**, 52: 319–324.

130. Malabarba, A.; Ciabatti, R.; Gerli, E.; Ripamonti, F.; Ferrari, P.; Colombo, L.; Olsufyeva, E. N.; Pavlov, A. Y.; Reznikova, M. I.; Lazhko, E. I.; Preobrazhenkaya, M. N. *J. Antibiot.*, **1997**, 50: 70–81.

131. Malabarba, A.; Ciabatti, R.; Kettenring, J.; Ferrari, P.; Vekey, K.; Bellasio, E.; Denaro, M. *J. Org. Chem.*, **1996**, 61: 2137–2150.

132. Malabarba, A.; Ciabatti, R.; Maggini, M.; Ferrari, P.; Colombo, L.; Denaro, M. *J. Org. Chem.*, **1996**, 61: 2151–2157.

133. Pavlov, A. Y.; Lazhko, E. I.; Preobrazhenskaya, M. N. *J. Antibiot.*, **1997**, 50: 509–513.

134. Shi, Z.; Griffin, J. H. *J. Am. Chem. Soc.*, **1993**, 115: 6482–6486.

135. Sundram, U. N.; Griffin, J. H. *J. Org. Chem.*, **1995**, 60, 1102–1103.

136. Pavlov, A. Y.; Berdnikova, T. F.; Olsufyeva, E. N.; Miroshnikova, O. V.; Filiposyanz, S. T.; Preobrazhenskaya, M. N.; Scottani, C.; Colombo, L.; Goldstein, B. P. *J. Antibiot.*, **1996**, 49: 194–198.

137. Miroshnikova, O. V.; Printsevskaya, S. S.; Olsufyeva, E. N.; Pavlov, A. Y.; Nilius, A.; Hensey-Rudloff, D.; Preobrazhenkaya, M. N. *J. Antibiot.*, **2000**, 53: 286–293.

138. Malabarba, A.; Ciabatti, R.; Scotti, R.; Goldstein, B. P.; Ferrari, P.; Kurz, M.; Andreini, B. P.; Denaro, M. *J. Antibiot.*, **1995**, 48: 869–882.

139. Nagarajan, R.; Schabel, A. A.; Occolowitz, J. L.; Counter, F. T.; Ott, J. L.; Felty-Duckworth, A. M. *J. Antibiot.*, **1989**, 42: 63–72.

140. Nagarajan, R.; Schabel, A. A.; Occolowitz, J. L.; Counter, F. T.; Ott, J. L. *J. Antibiot.*, **1988**, 41: 1430–1438.

141. Nicas, T. I.; Mullen, D. L.; Flokowitsch, J. E.; Preston, D. A.; Snyder, N. J.; Stratford, R. E.; Cooper, R. D. G. *Antimicrob. Agents Chemother.*, **1995**, 39: 2585–2587.

142. Rodriguez, M. J.; Snyder, N. J.; Zweifel, M. J.; Wilkie, S. C.; Stack, D. R.; Cooper, R. D. G.; Nicas, T. I.; Mullen, D. L.; Butler, T. F.; Thompson, R. C. *J. Antibiot.*, **1998**, 51: 560–569.

143. Nicas, T. I.; Mullen, D. L.; Flokowitsch, J. E.; Preston, D. A.; Snyder, N. J.; Wilkie, S. C.; Rodriguez, M. J.; Thompson, R. C.; Cooper, R. D. G. *Antimicrob. Agents Chemother.*, **1996**, 40: 2194–2199.

144. Pavlov, A. Y.; Berdnikova, T. F.; Olsufyeva, E. N.; Lazhko, E. I.; Malkova, I. V.; Preobrazhenskaya, M. N.; Testa, R. T.; Petersen, P. J. *J. Antibiot.*, **1993**, 46: 1731–1739.

145. Pavlov, A. Y.; Miroshnikova, O. V.; Printsevskaya, S. S.; Olsufyeva, E. N.; Preobrazhenskaya, M. N. *J. Antibiot.*, **2001**, 54: 455–459.

146. Snyder, N. J.; Cooper, R. D. G.; Briggs, B. S.; Zmijewski, M.; Mullen, D. L.; Kaiser, R. E.; Nicas, T. I. *J. Antibiot.*, **1998**, 51: 945–951.

147. Printsevskaya, S. S.; Pavlov, A. Y.; Olsufyeva, E. N.; Mirchink, E. P.; Preobrazhenskaya, M. N. *J. Med. Chem.*, **2003**, 46: 1204–1209.

148. Sundram, U. N.; Griffin, J. H.; Nicas, T. I. *J. Am. Chem. Soc.*, **1996**, 118: 13107–13108.

149. Jain, R. K.; Trias, J.; Ellman, J. A. *J. Am. Chem. Soc.*, **2003**, 125: 8740–8741.

150. Nicolaou, K. C.; Hughes, R.; Cho, S. Y.; Winssinger, N.; Smethurst, C.; Labischinski, H.; Endermann, R. *Angew. Chem.*, **2000**, 112: 3981–3986; *Angew. Chem. Int. Ed.*, **2000**, 39: 3823–3828.

151. Nicolaou, K. C.; Hughes, R.; Cho, S. Y.; Winssinger, N.; Labischinski, H.; Endermann, R. *Chem. Eur. J.*, **2001**, 7: 3824–3843.

152. Chu, Y. H.; Whitesides, G. M. *J. Org. Chem.*, **1992**, 57: 3524–3525.

153. Chu, Y. H.; Avila, L. Z.; Biebuyck, H. A.; Whitesides, G. M. *J. Org. Chem.*, **1993**, 58: 648–652.

154. Rao, J.; Whitesides, G. M. *J. Am. Chem. Soc.*, **1997**, 119: 10286–10290.

155. Rao, J.; Lahiri, J.; Isaacs, L.; Weis, R. M.; Whitesides, G. M. *Science.*, **1998**, 280: 708–711.
156. Green, N. M. *Biochem. J.*, **1963**, 89: 585–591.
157. Rao, J.; Lahiri, J.; Weis, R. M.; Whitesides, G. M. *J. Am. Chem. Soc.*, **2000**, 122: 2698–2710.
158. Jones, R. N.; Barrett, M. S.; Erwin, M. E. *Antimicrob. Agents Chemother.*, **1997**, 41: 488–493.
159. Allen, N. E.; Nicas, T. I. *FEMS Microbiol. Rev.*, **2003**, 26: 511–532.
160. Cooper, M. A.; Williams, D. H.; Cho, Y. R. *J.Chem. Soc. Chem. Commun.*, **1997**, 1625–1626.
161. Cooper, M. A.; Williams, D. H. *Chem. Biol.*, **1999**, 6: 891–899.
162. Allen, N. E.; LeTourneau, D. L.; Hobbs, J. N. Jr., *Antimicrob. Agents Chemother.*, **1997**, 41: 66–71.
163. Allen, N. E.; LeTourneau, D. L.; Hobbs, J. N. Jr., *J. Antibiot.*, **1997**, 50: 677–684.
164. Sharman, G. J.; Try, A. C.; Dancer, R. J.; Cho, Y. R.; Staroske, T.; Bardsley, B.; Maguire, A. J.; Cooper, M. A.; O'Brien, D. P.; Williams, D. H. *J. Am. Chem. Soc.*, **1997**, 119: 12041–12047.
165. Ge, M.; Chen, Z.; Onishi, H. R.; Kohler, J.; Silver, L. L.; Kerns, R.; Fukuzawa, S.; Thompson, C.; Kahne, D. *Science*, **1999**, 284: 507–511.
166. Kerns, R.; Dong, S. D.; Fukuzawa, S.; Carbeck, J.; Kohler, J.; Silver, L.; Kahne, D. *J. Am. Chem. Soc.*, **2000**, 122: 12608–12609.
167. Chen, L.; Walker, D.; Sun, B.; Hu, Y.; Walker, S.; Kahne, D. *Proc. Natl. Acad. Sci. USA*, **2003**, 100: 5658–5663.
168. Printsevskaya, S. S.; Pavlov, A. Y.; Olsufyeva, E. N.; Mirchink, E. P.; Isakova, E. B.; Reznikova, M. I.; Goldman, R. C.; Branstrom, A. A.; Baizman, E. R.; Longley, C. B.; Sztaricskai, F.; Batta, G.; Preobrazhenskaya, M. N. *J. Med. Chem.*, **2002**, 45: 1340–1347.
169. Sun, B.; Chen, Z.; Eggert, U. S.; Shaw, S. J.; LaTour, J. V.; Kahne, D. *J. Am. Chem. Soc.*, **2001**, 123: 12722–12723.
170. Chen, Z.; Eggert, U. S.; Dong, S. D.; Shaw, S. J.; Sun, B.; LaTour, J. V.; Kahne, D. *Tetrahedron*, **2002**, 58: 6585–6594.
171. Xu, R.; Greiveldinger, G.; Marenus, L. E.; Cooper, A.; Ellman, J. A. *J. Am. Chem. Soc.*, **1999**, 121: 4898–4899.
172. Chiosis, G.; Boneca, I. G. *Science*, **2001**, 293: 1484–1487.
173. Preobrazhenskaja, M. N.; Olsufyeva, E. N. *Expert Opin. Ther. Patents*, **2004**, 14: 141–173.
174. Candiani, G.; Abbondini, M.; Borgonovi, M.; Romano, G.; Parenti, F. *J. Antimicrob. Chemother.*, **1999**, 44: 179–192.
175. Abbanat, D.; Macielag, M.; Bush, K. *Expert Opin. Invest. Drugs*, **2003**, 12: 379–399.

# 3

# STRUCTURE MODIFICATIONS AND THEIR INFLUENCES ON ANTITUMOR AND OTHER RELATED ACTIVITIES OF TAXOL AND ITS ANALOGS

WEI-SHUO FANG, QI-CHENG FANG, AND XIAO-TIAN LIANG

*Institute of Materia Medica, Chinese Academy of Medical Sciences, Beijing, P. R. China*

## 3.1 DISCOVERY AND RESEARCH AND DEVELOPMENT OF TAXOL

Taxol **1a** is an antitumor plant diterpene isolated from the yew tree *Taxus* spp. in the late 1960s. It was named taxol at the time of its first isolation by Wani et al.[1] In the 1990s, when Bristol-Myers Squibb (BMS) launched it on the market, Taxol was registered as an anticancer product, and Taxol was assigned a generic name, paclitaxel. Both taxol and paclitaxel are referred to in this chapter to describe the same compound.

The discovery of Taxol is a fruit of a National Cancer Institute (NCI)-sponsored project on identification of antitumor agents from natural resources. Bioassay-guided fractionation led to the isolation of this unique compound from *Taxus brevifolia* (pacific yew). Wani et al. also identified another famous antitumor natural product camptothecin. Unlike camptothecin, which was abandoned in the phase of clinical trial because of its severe toxicity, Taxol was almost discarded at the preliminary phase because it only exhibited moderate in vitro activity toward P388, a murine leukemia cell line that was used in the standard evaluation system by NCI researchers at that time. However, it was rescued by a subsequent finding of its strong and selective antitumor activities toward several solid tumors, and more

*Medicinal Chemistry of Bioactive Natural Products* Edited by Xiao-Tian Liang and Wei-Shuo Fang
Copyright © 2006 John Wiley & Sons, Inc.

importantly, its unique mechanism for interference with the microtubule polymerization–depolymerization equilibrium, that is, to promote the polymerization of tubulins into the microtubule and to stabilize it to prevent its depolymerization, led to the cell cycle arrest at $G_2/M$ phase. The subsequent clinical trials starting in the early 1980s proved its extraordinary efficacy against some solid tumors.

However, during the clinical tests and commercialization of Taxol, a supply crisis developed because of its scarce origin—the bark of Pacific yew. To extract enough Taxol for one patient, stem bark from three 50 year old trees had to be stripped. Then BMS supported NCI financially for collaborative efforts to overcome the supply crisis. In return, BMS obtained exclusive rights to commercialize Taxol in the U.S. market for 10 years.

Taxol is now a well-known anticancer drug for the treatment of several kinds of late stage, reconcurrent tumors. Despite its success in chemotherapy, there are demands to improve its efficacy and lower its toxicity. Since the mid-1980s, the structure activity relationship (SAR) of Taxol and its analogs was thoroughly explored to find more active analogs. Also, mechanistic studies of paclitaxel were conducted not only to reveal the molecular basis of its action, but also to provide clues to the rational design of new analogs of paclitaxel and even the molecules with different structures. The most important and updated results will be described in the following sections.

Other efforts to maximize the use of Taxol by clinicians include synthesis of conjugates or prodrugs with better bioavailability and specificity, preparation of new formulations with improved physical properties, and combined use with other drugs. Those efforts have been the topics of some reviews, and they will also be discussed in this chapter.

## 3.2 PACLITAXEL ANALOGS ACTIVE AGAINST NORMAL TUMOR CELLS

Reaching a peak in the 1990s, SAR studies of paclitaxel analogs are still active today. Many reviews written by leading researchers in this area have appeared in scientific journals and books.[2–4] Instead of a systematic retrospect, we will mainly concentrate on recent progress in this review, together with some of the most important SAR results known to date. It should be pointed out that these results were mainly obtained by traditional medicinal chemistry methods, with little knowledge on the drug–receptor interaction.

Our discussion will follow a left to right route in the molecular framework, that is, C-13 side chain, A, B, C, and D rings. Other results that are difficult to be categorized are shown at the end of this section.

### 3.2.1 C-13 Side Chain

The incorporation of the C-13 side chain into Taxol and its analogs, roughly speaking, includes two stages—syntheses of side chains with various substituents, and attachment of them to the properly protected baccatin core structures and subsequent deprotection to furnish the target compounds. The side chains were usually used in enantiopure forms, which are prepared by asymmetric synthesis or resolution. The β-lactam side chains were commericially available for the semisynthesis of paclitaxel, and those in acid forms usually for the synthesis of another semisynthetic taxane, Taxotere **1b** (generic name docetaxel), both from 10-deacetyl baccatin III **2** (10-DAB), a natural taxane that is abundant in the regenerated resources—twigs and needles of yew trees. The illustrative schemes for the semisynthesis of taxanes from 10-DAB are shown below. The latter finding[5] on the usefulness of the side chain in oxazoline form to the synthesis of paclitaxel can be categorized into the "acid approach," in contrast to the "β-lactam approach" (Scheme 3-1).

For the semisynthesis of paclitaxel and its analogs, selective modification of 10-DAB is an important issue to be addressed, that is, the 7-OH and 10-OH should

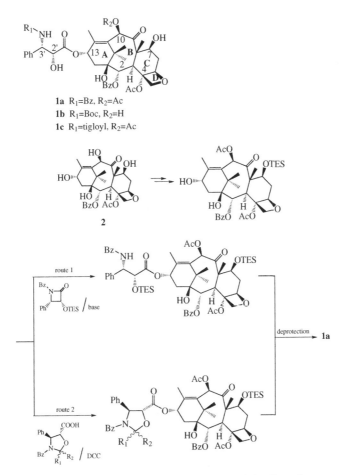

**1a** R$_1$=Bz, R$_2$=Ac
**1b** R$_1$=Boc, R$_2$=H
**1c** R$_1$=tigloyl, R$_2$=Ac

**Scheme 3-1.** Illustrative synthesis of Taxol (paclitaxel).

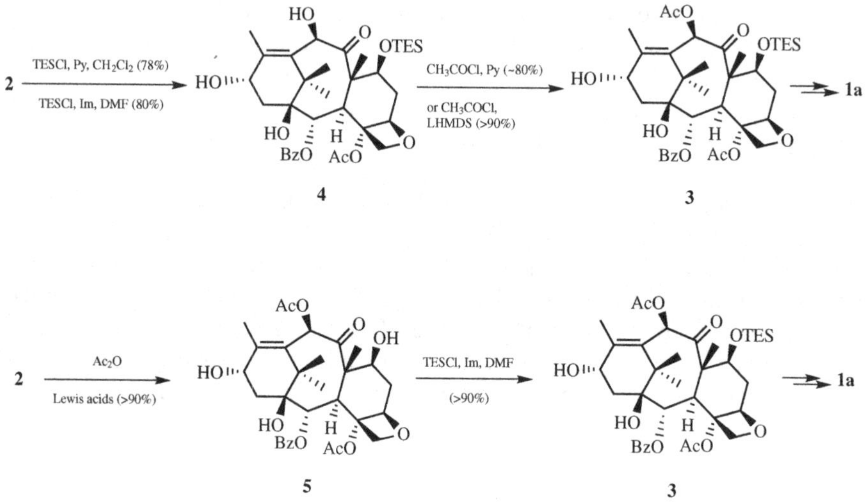

**Scheme 3-2.** Selective protection of hydroxyls in 10-DAB (**3**) for the semisynthesis of Taxol.

be protected appropriately before the incorporation of the C-13 side chain. The acylation order in the presence of acyl chloride and base was 7-OH > 10-OH > 13-OH.[6,7] As early as 1988, a prominent semisynthesis of Taxol was attributed from Greene's group, in which 7-OH in 10-DAB was protected in selective manner by TESCl and pyridine, and the acetyl group was incorporated subsequently in the C-10 position to furnish **3**.[8] Although the semisynthetic routes for paclitaxel and related taxanes have been improved, the protection of 10-DAB was almost not changed for a long time and only slightly modified by a BMS group.[9] Two groups then found independently the Lewis acid catalysed 10-OH acylation of 10-DAB with anhydride in a highly regioselective manner to prepare baccatin III **5**,[10,11] which is different from the acylation order under the basic condition (Scheme 3-2). Another attempt to reverse the reactivity of 7-, 10-, and 13-OH is enzymatic acylation,[12] in which preferential acylation of 10-OH in 10-DAB (10-OH > 7-OH) and 13-OH in 7-TES-10-DAB **4** (13-OH > 10-OH) was observed.

The importance of the C-13 substituted phenylisoserine side chain to the bioactivity of paclitaxel has been acknowledged for a long time. It has been shown that the substitutions and/or stereochemistry of C-2′, 3′, and 3′-N contribute to its activity in different ways. An early report has demonstrated the baccatin core bearing the C-13 side chain in 2′R,3′S form as naturally occurring taxane—paclitaxel are active, whereas three other stereogenic forms are all but inactive.[13–15]

The extension of the C-13 side chain to its homologated one furnished poorly active paclitaxel and docetaxel analogs in tubulin assembly assay.[16] The authors assumed that the poor activity of the analogs may have originated from their different conformations from that of paclitaxel in water.

### 3.2.1.1 C-2' Position

Numerous results have demonstrated that the free 2'-OH is crucial to the activity. Its replacement with other bioisosteric atoms/groups, for example, $NH_2$, F, $OCH_3$, or deoxygenation at this position led to complete loss of activity (two to three orders of magnitude less cytotoxic).[13,15] It was then deduced that 2'-OH may get involved in hydrogen bonding to the receptor, which was confirmed later by x-ray crystallography of docetaxel and its receptor tubulin complex.

Steric hindrance, that is, the introduction of (S)-Me to C-2', while 2'(R)-OH is retained, makes a positive contribution to the antitumor activity as well as to the tubulin binding ability. This result may have come from the reduced rotation of the side chain, which thus enhances the ratios of bioactive conformers in all conformers.[17–19] The preparation of 2'-Me analog was usually undertaken by β-lactam approach, in which 3-keto-β-lactam was attacked by a nucleophile to yield stereoselectively 3-methyl-3-OH (equivalent to the 2' position in paclitaxel) β-lactam ready for attachment to baccatin core structures. Battaglia et al. prepared a series of 2'(S)-Me of paclitaxel analogs from 10-DAB and 14β-OH-10-DAB with different C-3' and 3'-N substitutents, and all compounds 6a–e are comparable with or more active than paclitaxel toward A2780 human lung carcinoma in vitro.[20]

6a R=$C_6H_5$
6b R=$Me_3CO$

6c R=2-furyl
6d R=$C_6H_5$
6e R=$CF_3$

Génisson et al. also prepared two diastereomers of 2'-hydroxymethyl analogs through an asymmetric Baylis–Hillman reaction-like sequence to prepare a 3'(S)-N-substituted 2'-methylene C-13 side chain (Scheme 3-3). After incorporation of the side chain to C-13 of 7,10-di-Troc-10-DAB, the product was then subjected to osmylation to yield 2'(R)- and 2'(S)-hydroxymethyl docetaxel 7a–b analogs stereoselectively. Both taxoids displayed less tubulin polymerization abilities than did paclitaxel, but the major product 2'R-isomer 7a is more active than the minor one 2'S-isomer 7b.[21]

### 3.2.1.2 C-3' Position

Replacement of 3'-Ph with other aromatic and aliphatic groups have been investigated, among which three to four carbon alkyl or alkenyl substitutions, especially 3'-isobutenyl and 3'-isobutyl groups, could improve the activity to a great extent. In 1996, Ojima et al. reported that such Taxol analogs, butitaxels, exhibited stronger activity than paclitaxel. In combination with changes at C-10 acyl substitutions,[22]

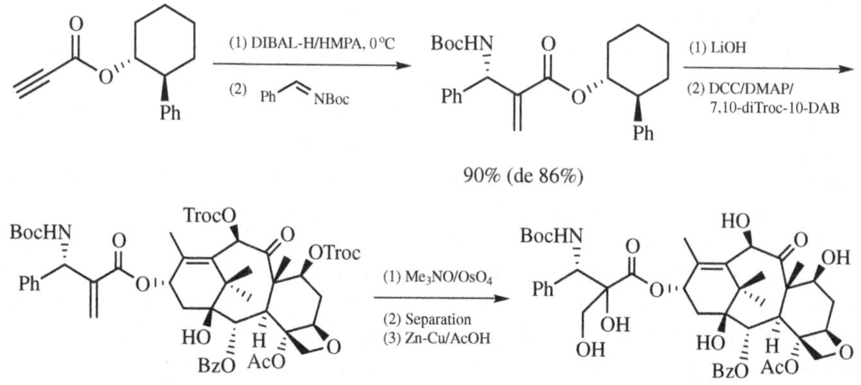

**Scheme 3-3.** Preparation of 2′-hydroxymethyl doctaxel.

as well as C-2 *meta*-substituted benzoyl groups[23] that had been recognized to enhance the in vitro activity of paclitaxel analogs, some promising taxoids were prepared. Additional modification on the C-3′ alkenyl group (isobutenyl), that is, preparation of taxoids bearing 3′-cyclopropane and 3′-epoxide moieties, is also encouraging.[24] The cyclopropantion of isobutenyl-substitued lactams occurred stereospecifically, and epoxidation occurred in a highly stereoselective manner to furnish the lactams ready for C-13 side-chain incorporation. Two 3′-cyclopropane **8a-b** and one 3′(*R*)-epoxide taxanes **9a**, all with $IC_{50}$ less than 1 nM, are among the most potent analogs in this series. Although 10 to 30 times less active than **9a**, the cytotoxicity of 3′(*S*)-epoxide **9b** is still comparable with that of paclitaxel.

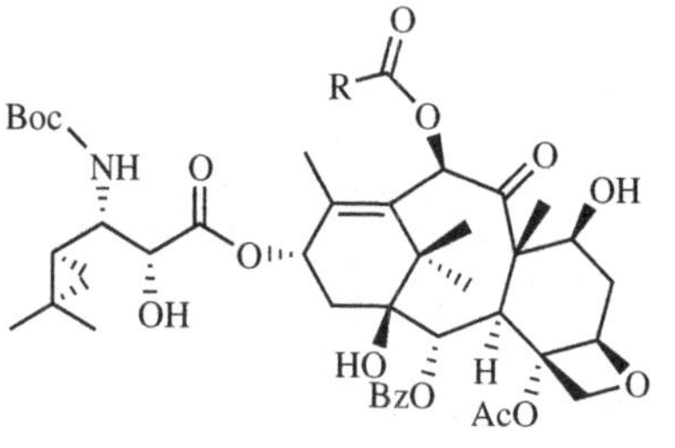

**8a** R=Et
**8b** R=cyclopropane

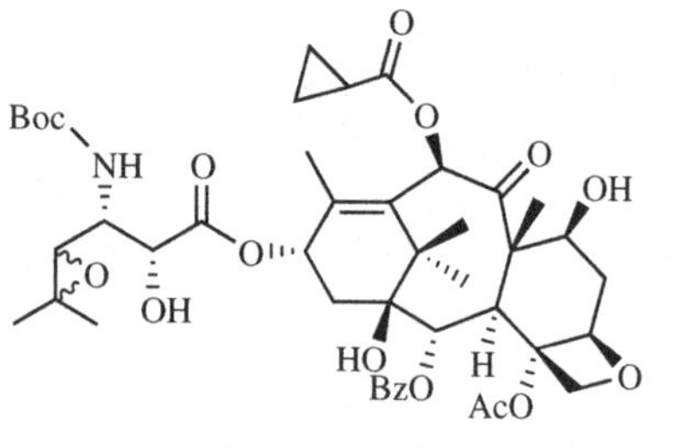

**9a** (R)-epoxide
**9b** (S)-epoxide

Recently, Ojima et al.[25-27] described the synthesis of some taxanes with C-13 fluorine-substituted isoserine side chains. In pharmaceutical practices, the fluorine atom is usually introduced as an isosteric atom of the hydrogen atom, but it showed higher or sometimes unique activity against its hydrogen-containing counterparts. Flourine also blocks the metabolism of the parent molecule, which led to the improvement of its therapeutic potential. Introduction of fluorine atom to the *para* position of 3′-phenyl decreased activity in most cases, except in **10**, which are comparably cytotoxic with docetaxel in all tested cell lines. For 3′-CF$_3$ docetaxel analogs **11a–h**, in combination with or without the change of 10-acetate to other ester, carbonate, and carbamate groups, an enhancement of activity against sensitive tumor cells of several times and two to three orders of magnitude against MDR tumor cells were observed.[25,26] For 3′-difluoromethyl docetaxels, most derivatives with changes at 10-OH to 10-esters and 10-*N,N*-dimethylcarbamate were comparable with or modestly more active than docetaxel, similar to their 3′-CF$_3$ counterparts, whereas their 14β-OH counterparts were less active.[27] Briefly, for 3′-CF$_3$ and 3′-CF$_2$H docetaxel series, taxoids with greater potencies can be expected, whereas fluorine substitution in 3′-Ph may not be beneficial to the antitumor activities of paclitaxel analogs. Interested readers may refer to a recent review by Ojima.[28]

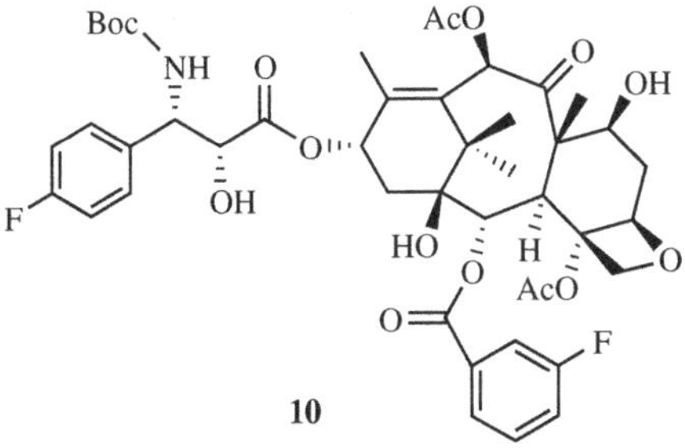

**10**

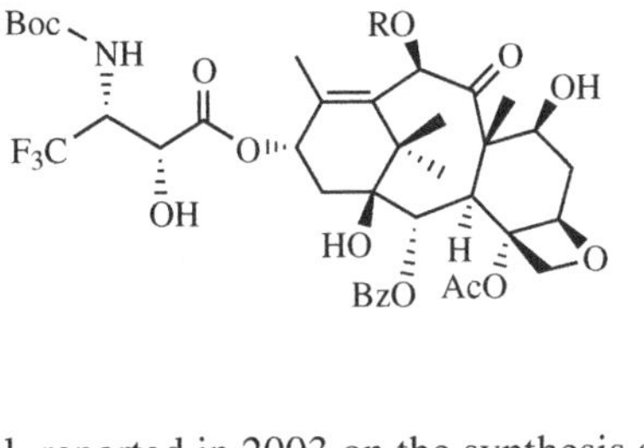

**11a**: R = Ac
**11b**: R = (CH$_3$)$_2$N-CO
**11c**: R = morpholine-CO
**11d**: R = cyclopropane-CO
**11e**: R = CH$_3$CH$_2$-CO
**11f**: R = CH$_3$(CH$_2$)$_3$-CO
**11g**: R = (CH$_3$)$_3$CCH$_2$-CO
**11h**: R = MeO-CO

Liu et al. reported in 2003 on the synthesis and cytotoxicities of 3′-cyclopropane analogs with both 7α- and 7β-OH and 2′(R)- and 2′(S)-OH functionalities. Both 2′-(R) isomers were 400 times less active than paclitaxel in A2780 cancer cell assays, and 2′-(S) isomers displayed even weaker cytotoxicities.[29] Unsatisfactory results may have developed from small volumes of the 3′-cycloprapane, which is similar to 3′-methyl in 9(R)-dihydro paclitaxels as reported earlier.[30] As mentioned, larger groups such as isobutyl or isobutenyl may interact with the receptor better.

The introduction of steric hindrance through a 3′(R)-Me group, however, did not prove to be as favorable for C-2′. The 3′(R)-methyldocetaxel exhibited no activity even at a concentration of 20 μM in microtubule assays, whereas docetaxel displayed 100% activity at 5 μM.[31]

### 3.2.1.3  *C3'-N Substitutions*

Although some C3'-*N*-debenzoyl *N*-acyl paclitaxels were evaluated, the impact of C3'-*N* substituents on antitumor activity seems to be somewhat complicated— neither electronic effects nor steric effects alone can be applied to explain the contributions of the acyl groups.

A Korean group extensively studied such analogs with aliphatic acyl groups and found that those with conjugated double and triple bonds displayed higher activities against both sensitive and resistant tumor cells. When α-substitutions or β-substitutions were attached to the double bonds, reduced or enhanced activities were observed, respectively. More importantly, they demonstrated the essential role of the size of the acyl group. Most analogs bearing three to six carbons displayed high potency.[32] Comparative molecular field analysis (COMFA) was then applied to create a rational explanation and prediction protocol of SAR of C3'-*N*-acyl analogs by this Korean group.[33] Three-dimensional contour maps were provided to predict the distribution of seemingly scattered antitumor activities in a precise manner, and the steric effect is regarded as the dominating factor that made up about 80% of contributions, whereas the electrostatic effect made up about 20% of contributions.

Ali et al. reported[34] the synthesis of a series of C3'-*t*-butyl paclitaxel analogs with C3'-*N* amides and carbamates, among which *N*-debenzoyl-*N*-(2-thienoyl) analog **12** was the most potent. Although equipotent to docetaxel, and about 25 times more water soluble than paclitaxel, this taxane was not superior to analog **13**[35] reported earlier.

A BMS research group found that 3'-*t*-butylaminocarbonyloxy paclitaxel analogs **14a** and **14b** (regioisomers of docetaxel and its 10-Ac derivative) were several times less active than paclitaxel in vitro, but **14b** was equipotent to paclitaxel in vivo.[36] They also prepared 3'-*N*-thiocarbamate and C3'-*N*-thiourea bearing analogs. Although C3'-*N*-thiocarbamate was found to be more potent than paclitaxel and docetaxel in both tubulin polymerization and cytotoxicity assays, thioureas are usually less active.[37]

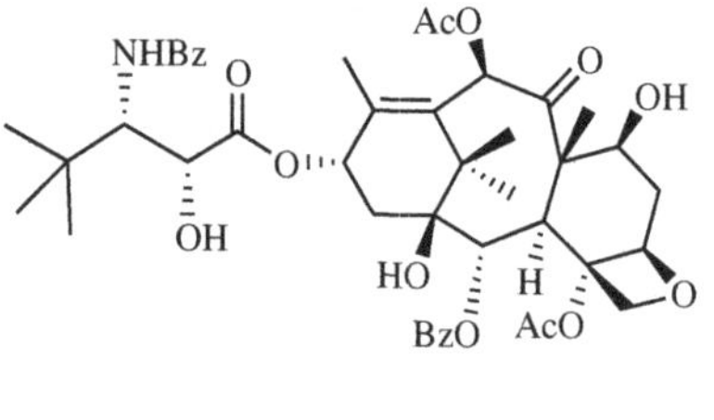

12

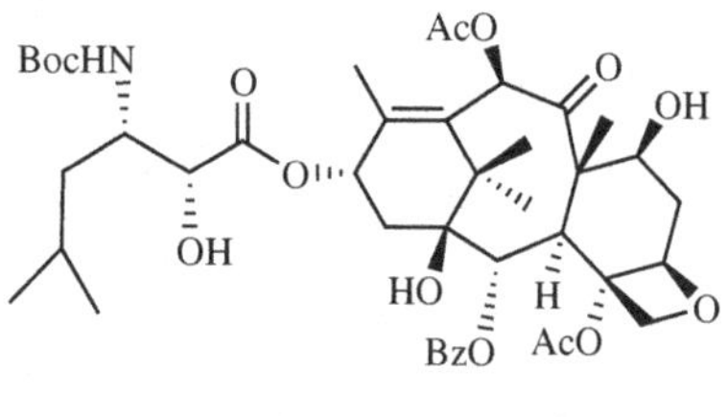

13

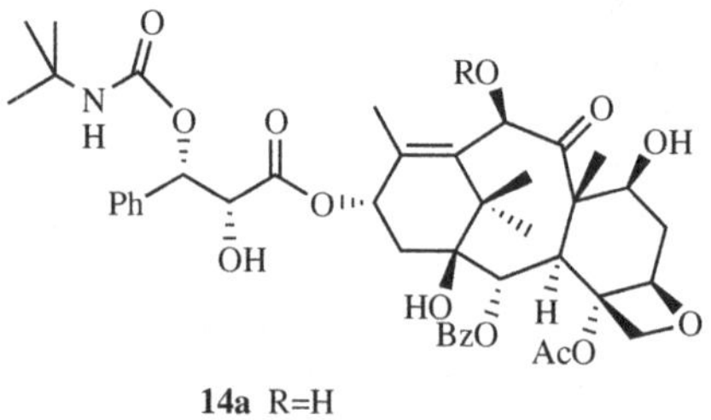

**14a**  R=H
**14b**  R=Ac

In 1997, Xiao et al. prepared the first $C3'$-$N$ modified taxane library with 400 compounds by a radio-frequency encoded, solid-phase synthesis method.[38] However, they did not report the biological assay data. Although combichem is a powerful tool in medicinal chemistry, only a few reports[39-41] have appeared in this field to date, maybe because of the complexity in taxane structure and chemistry.

Cephalomannine (1c) is a congener of paclitaxel in several *Taxus* sp. plants, which showed comparable cytotoxicity with paclitaxel. Several bromine and chlorine adducts to the double bond of $C3'$-$N$-tigloyl in cephalomannine were prepared. The dichlorocephalomannine derivatives were one order of magnitude less active, whereas the dibromocephalomannines were better than paclitaxel against several colon, ovarian, and breast cancer cell lines.[42] Epoxidation products of the double bond on $3'$-$N$-tigloyl exhibited comparable activity.[43]

### 3.2.1.4  C2′–C3′ Linkage

The C-2$'$ and C-3$'$ substituents can be rotated along the C2$'$–C3$'$ axis in palictaxel, and various conformers, including biologically active species, were observed in nonpolar and polar solvents. When C2$'$–C3$'$ is tethered appropriately, one can expect a more active analog. However, during the semisynthesis of paclitaxel and its analogs, the intermediates with oxazoline-protected C-13 side chains are usually less active. After the oxazoline rings were opened after deprotection, paclitaxel or more active analogs were obtained.

Barboni et al.[44] prepared a conformationally strained paclitaxel analog 15a, in which C-2$'$ and *ortho* position of 3$'$-phenyl is tethered with a methylene group. It exhibited comparable cytotoxicity with that of paclitaxel. After synthesizing several tethered analogs including docetaxel analog 15b, they observed ethylene linkage between C2$'$ and C3$'$ led to a drastic decrease in activity. Furthermore, analogs with reversed C2$'$ and C3$'$ configurations were totally inactive.[45] Although this result was disappointing, it may provide an insight into the conformational requirement for taxane-tubulin bindings.

15a  $R_1$-Bz, $R_2$=Ac
15b  $R_1$=Boc, $R_2$=H

## 3.2.2  A Ring and Its Substitutions

### 3.2.2.1  C-13 Linkage

When C-13$\alpha$ ester linkage was substituted by amide or epimerized to $\beta$ form, reduced or loss of activity was observed. Chen et al. initially failed to transform the 13-keto group in 13-*keto*-7-TES-baccatin III (**16a**) to C-13α *N*-substitutions by directly reductive amination conditions. After many tests, they found that when all hydroxyls in baccatin III were protected by silyl groups and the 4-acetyl group was removed, 13α-OH in the starting material **17a** can be transformed into desired 13α-azido baccatin **17b** by double $S_N2$ conversion with $CBr_4$/$PPh_3$ in 50% yield and $NaN_3$/DMF in 63% yield. The reason for the necessity to remove 4-acetyl is that the reactivity of 13α-OH is reduced because of its intramolecular hydrogen bonding to the 4-acetyl group. Reduction of the 13α-azido group was also found to be difficult, unless heating it with $PhSeH$/$Et_3N$ at 60 °C to furnish 13α-amino baccatin **17c** in 80% yield. Besides the C-13 amide-linked paclitaxel analog **18a** being inactive in both tubulin polymerization and cytotoxicity assays, the C-4 methyl carbonate and 3'-furyl analog **18b** were found to be inactive (these changes in paclitaxel have been proven to enhance activity).[46]

4-OH in 13-*keto*-4-deacetyl-7-TES baccatin III (**16b**) was used for transannular hydride transfer during reduction of the C-13 keto group in this compound, because these two groups are in proximity in space. In contrast to the hydride's attack of the C-13 keto group in baccatin III-like taxanes, which usually occured from the β-face because of the rigid convex conformation of baccatins, $Me_4NBH(OAc)_3$ reduced C-13 keto in **16b** to 13β-OH in **19**, which demonstrates that the hydride was transferred from the α-face of the C-13 keto group. It was disappointing to find that the 13-epi paclitaxel and docetaxel analogs could not promote tubulin polymerization.[47]

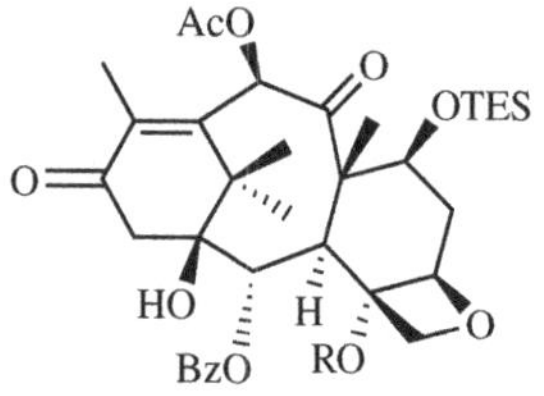

**16a** R = Ac
**16b** R = H

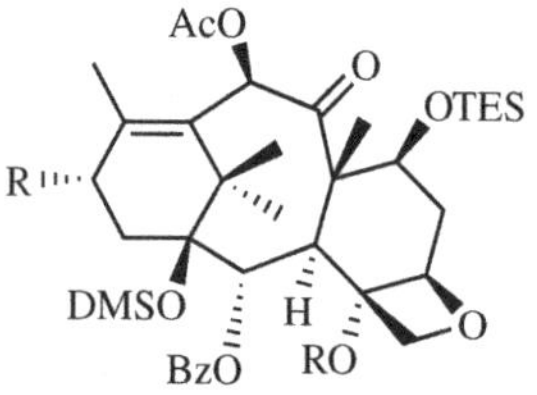

**17a** R = OH
**17b** R = $N_3$
**17c** R = $NH_2$

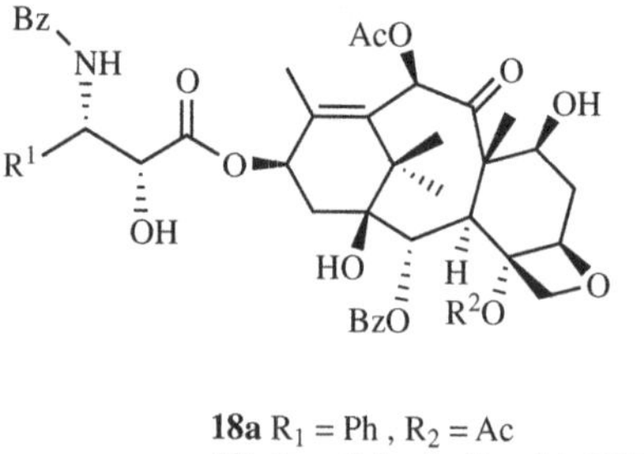

**18a** $R_1$ = Ph , $R_2$ = Ac
**18b** $R_1$ = 2-furyl , $R_2$ = MeOCO

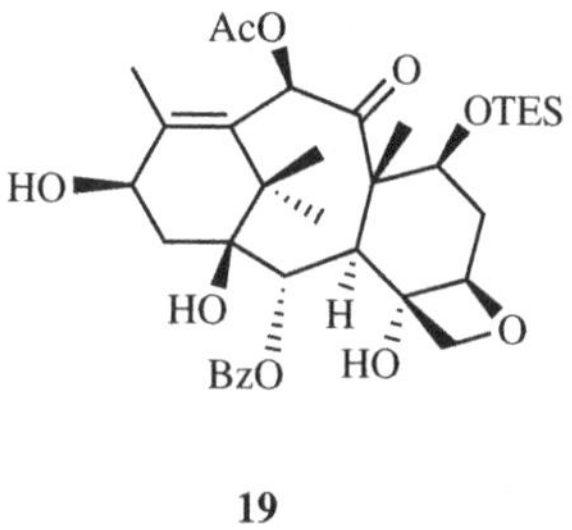

**19**

During the reduction of 11,12-olefin in **16a** with Zn in acetic acid, baccatin **20** in stable enol form was unexpectedly obtained (Scheme 3-4). However, this baccatin was tautomerized to more stable keto form **21a** with silica gel. It was also

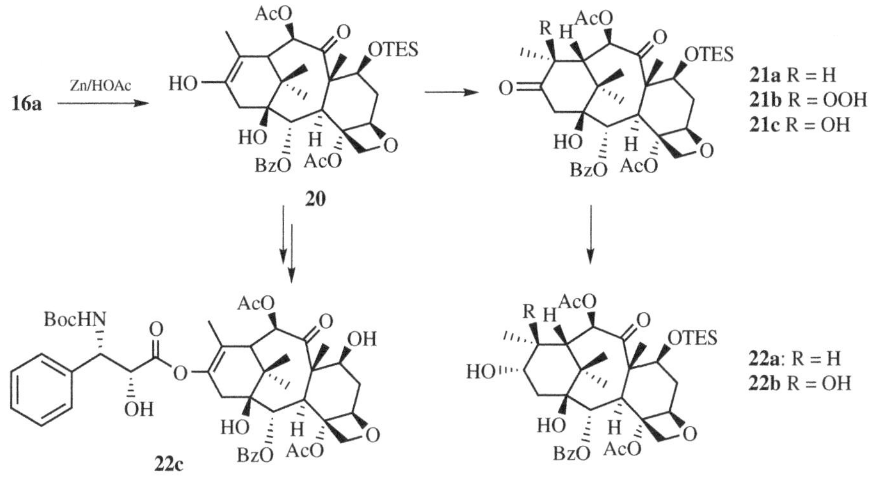

**Scheme 3-4.** Preparation of baccatin in enol form and its transformation.

peroxidized to **21b** when shaken up with dichloromethane (oxidized by oxygen), and hydroxylated product **21c** was obtained after treatment of mCPBA under nitrogen atmosphere. Analog **22c** was prepared by side-chain attachment to baccatin **20**, which showed equipotency to that of docetaxel. When reduced with sodium borohydride, **21a** and **21c** was transformed into 13α-OH baccatins **22a–b**. The docetaxel analog derived from 12-hydroxylated baccatin **22b** exhibited high potency, although about one order of magnitude weaker than docetaxel, whereas those from **22a** were much less active by two orders of magnitudes.[48]

### 3.2.2.2   A Ring

Except for the C-13 substitutions, the most extensively studied in A ring modifications are contracted ring (A-nor), opened ring (A-seco), and the 11,12-dihydro analogs. Although some derivatives showed comparable activities, most of them were less active than paclitaxel.

The A-nor taxoid **23a** was first prepared serendipitously by Samaranayake et al. when they tried to prepare 1-mesylate of 2',7-*di*-TES paclitaxel, and its desilyated product **23b** displayed no cytotoxicity while retaining tubulin stablization activity.[49] Chordia et al. then synthesized many compounds in this series, including different 2-aroyl esters and A-nor taxoids through hydrogenation, epoxidation, chlorohydrination, and oxidation cleavage of the isopropenyl radical attached to C-1. Most of them were far more less active than paclitaxel in both cytotoxicity and tubulin polymerization tests but more active than **23b**.[50] Preliminary in vivo tests for this series of compounds were also disappointing.

A C-11,12 double-bond rearranged product **24** was observed when 2',7-*di*-Troc-10-deacetyl paclitaxel was treated with Yarovenko reagent (Et$_2$NCF$_2$CHFCl), which can be then transformed into 2'-Troc-10-deoxy paclitaxel (**25**) by catalytic hydrogenation, along with the formation of a small amount of 12-fluorinated **26**. All taxoids, after deprotection, showed reduced cytotoxicity seven to eight times.[51]

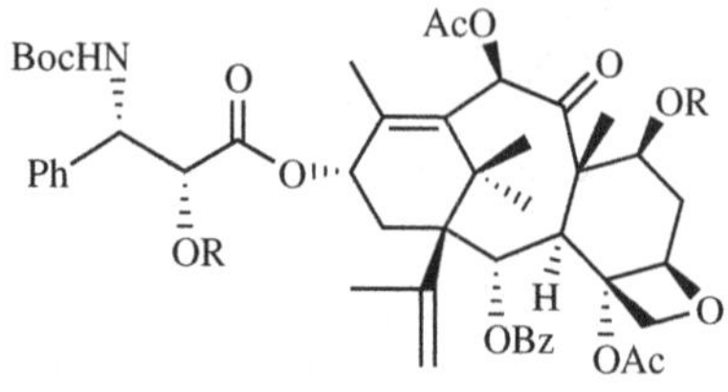

23a R=TES
23b R=H

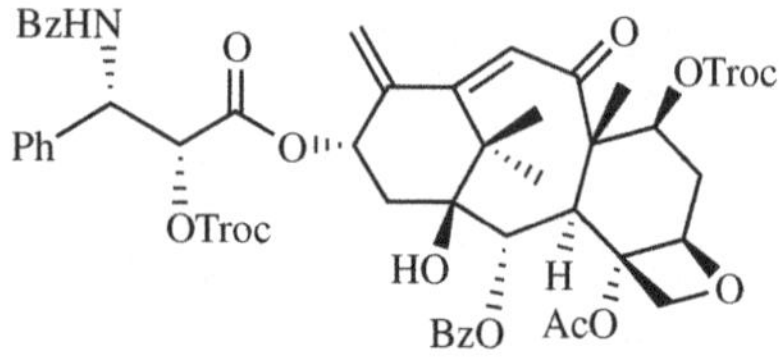

24

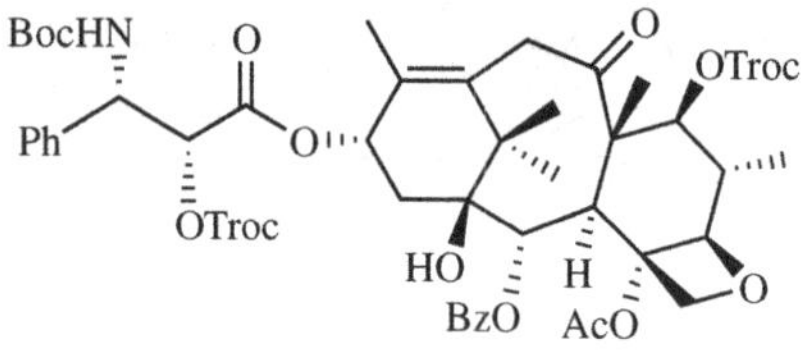

25 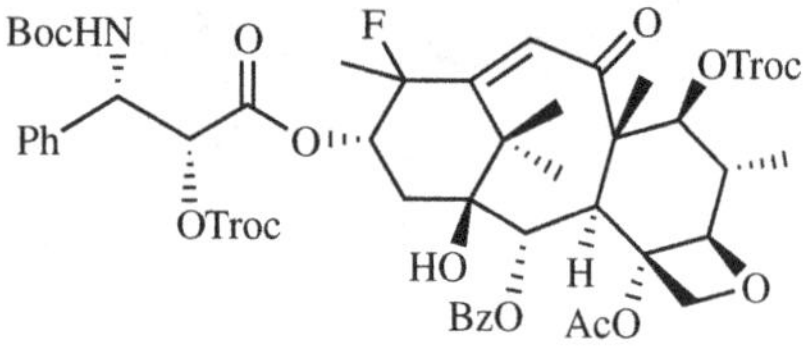                                                          26

It has been found that the C-11,12 double bond in paclitaxel was resistant to many reaction conditions, including catalytic hydrogenation and ozonolysis. Harriman et al. reasoned that C-10 acetate may hamper the epoxidation from the β-face and the cup-shaped core structure prevents the epoxidation from the α-face. By removal of 10-acetate in paclitaxel, the 11,12-β-epoxide was formed quantitatively. The 11,12-epoxidized taxoid **27** was only one third as cytotoxic as paclitaxel against B16 melanoma cells.[52]

The 11,12 double bond in baccatin was reducible by zinc in acidic conditions. Since treatment of 11,12-dihydro baccatin **28b** with base only yielded 13-acetyl-4-

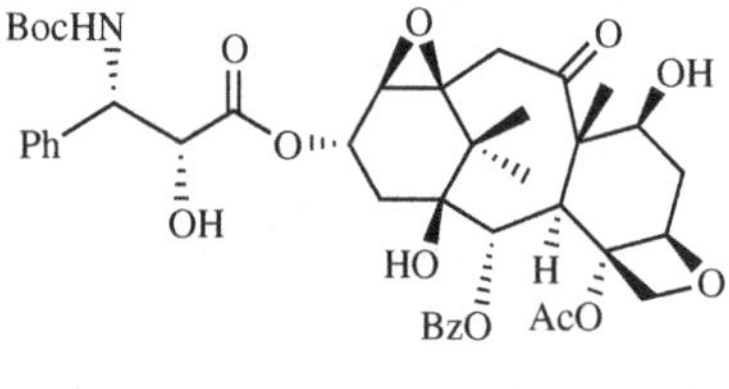

27

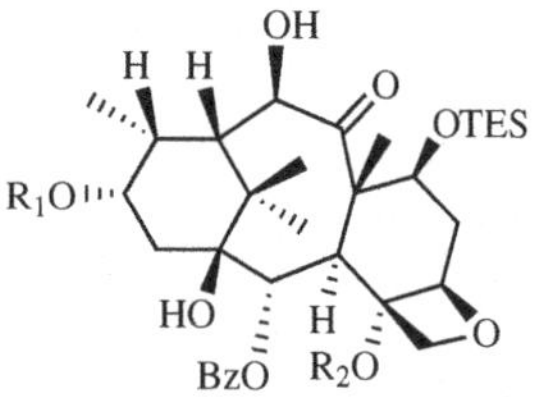

28a R₁=H, R₂=H
28b R₁=H, R₂=Ac
28c R₁=Ac, R₂=H

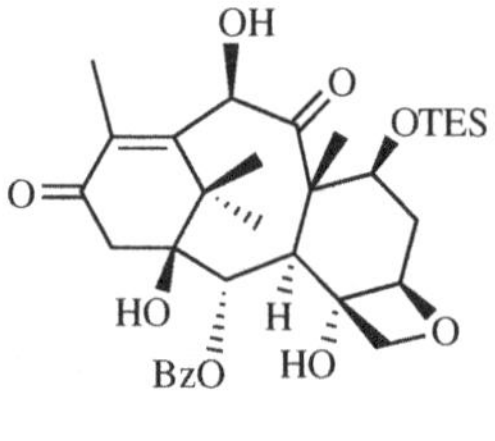

29

deacetyl baccatin **28c**, presumably through transannular acyl migration in proximity, Marder et al. attempted[53] to prepare 11,12-dihydropaclitaxel by attachment of the C-13 side chain to 4-deacetyl-11,12-dihydrobaccatin **28a**, which was prepared by reduction of the 11,12 double bond with zinc and C-13 keto group with sodium borohydride in baccatin **29**, and then 4-acetylation. However, they found that the docetaxel analogs from **28a** were resistant to 4-acetylation, and the 4-deacetyl-11,12-dihydrodocetaxel was almost inactive. Because of the poor activity of the documented 4-deacetyl taxoids, it is not sure if the inactivity has developed from the reduction of 11,12-olefin. Considering another report[48] (see Section 3.2.2.1), the 11,12-dihydropaclitaxel was indeed inactive even if 4-OH is acetylated.

### 3.2.2.3  14β and 1β Substitutions

From 14β-OH-10-DAB, a taxoid isolated from the needles of *T. wallichiana* Zucc., A-nor-seco baccatin **31** was derived from the oxidative cleavage of 14β-OH-10-DAB **30**, and additional reduction of C-13 aldehyde with sodium cyanoborohydride and incorporation of the side chain furnished A-seco paclitaxel and docetaxel analogs.[54,55] The C-13 ester A-nor-seco analogs **32** were 20–40 times weaker in activity,[54] and those taxoids **33** (R$_3$=H) with C-13 amide side chain almost inactive (more than three orders of magnitudes weaker activity). The authors thus reasoned that it may have originated from the mobility of C-13 amides. When the amide was methylated, the taxoids' **33** (R$_3$=Me) activity were enhanced to be comparable with that of the esters. Based on computer simulations, they found that this series of taxanes possessed convex conformations similar to that of paclitaxel and docetaxel. Those with higher activity overlapped with hydrophobic clustering conformation of paclitaxel better than those poorly active ones. Recently, Appendino et al. prepared more 14-nor-A-seco analogs, all of which were almost inactive against MCF-7 cells.[56]

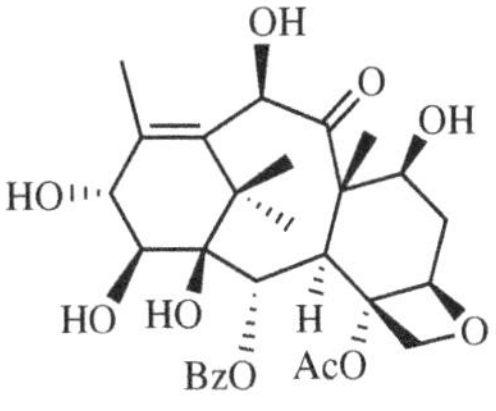

30

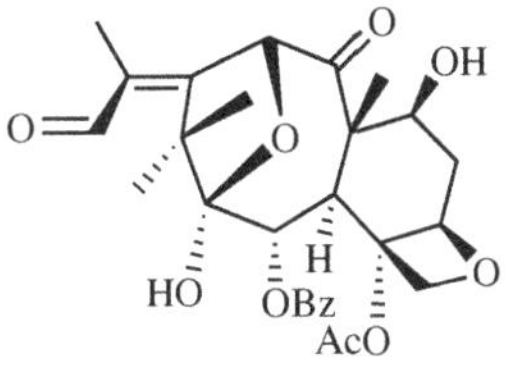

31

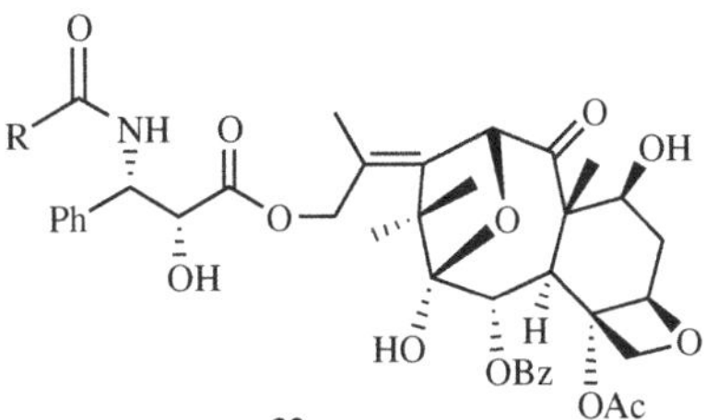

32

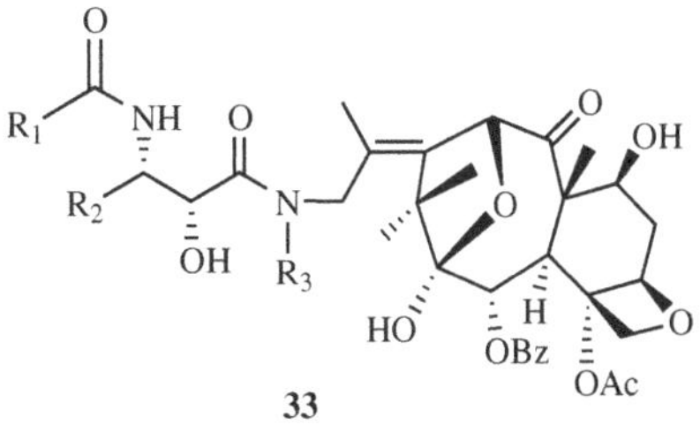

33

The SAR results for many other 14β-OH paclitaxel and docetaxel analogs were scattered in different parts in Section 3.2, for example, the results in Refs. 20, 54, and 55. Most of the works for the 14β-OH series were conducted by Ojima et al. and their Italian collaborators. An antitumor agent effective against MDR tumors, IND5109, has been prepared from **30** and is currently under clinical trials.[58]

A series of 14β-hydroxy and 14β-acyloxy taxoids **34** with 10β, 5α, 2α-acetoxy, or hydroxy substitutions without C-13 oxygenations and oxetane D ring have been isolated from the cell culture of *T. yunnanensis*. Attachment of the *N*-benzoylphenylisoserine side chain to C-14, and incorporation of the 4(20), 5-oxetane ring and change of 2-acetate did not improve their activity.[59–61] The fact that all derivatives are far less active than paclitaxel prompted the authors to suggest that C-14 isoserine isomers did not bind to tubulin properly.

The role of 1β-OH in SAR of paclitaxel was not clear until Kingston et al. finished the synthesis of the 1-deoxypaclitaxel analog and examined its activity. They have made such efforts previously with Barton's deoxygenation protocol. However, when 7,10-*di*-TES paclitaxel was treated with the standard Barton protocol, the 1-benzoyl-2-debenzoyloxy derivative **35b** was prepared.[62,63] Taxoid **35b** was prepared by deoxygenation of the 1-BzO-2-xanthate intermediate **35c**, which was obtained through transesterification of 1-OH and 2-benzoate in 7,10-*di*-TES paclitaxel. Deprotection of **35b** led to the formation of inactive taxoid **35a**. The 1-deoxy-9(*R*)-dihydro analog **36a** was prepared from a naturally occurring taxane 1-deoxybaccatin VI.[64] After comparison of **36a** with 9(*R*)-dihydropaclitaxel analog **36b** and other 1-deoxy analogs with their paclitaxel counterparts, they concluded that 1-deoxygenation caused only slightly reduced activity in both cytotoxicity and tubulin assembly assays.

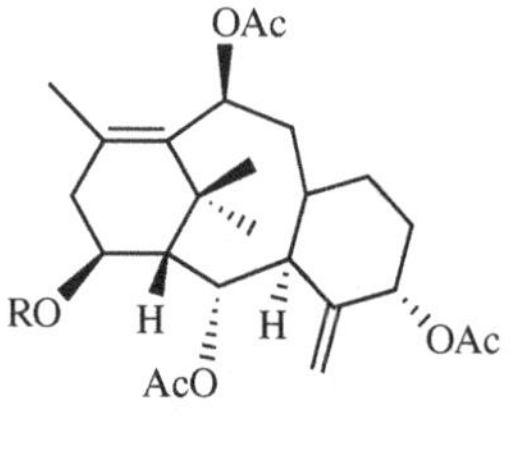

**34**

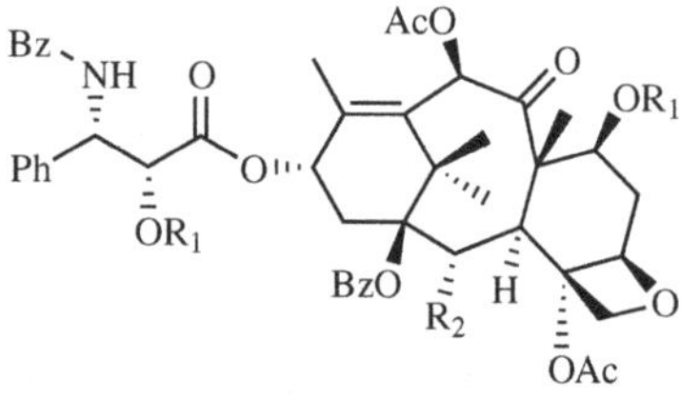

**35a** R$_1$=H, R$_2$=H
**35b** R$_1$=TES, R$_2$=H
**35c** R$_1$=TES, R$_2$=OCSSMe

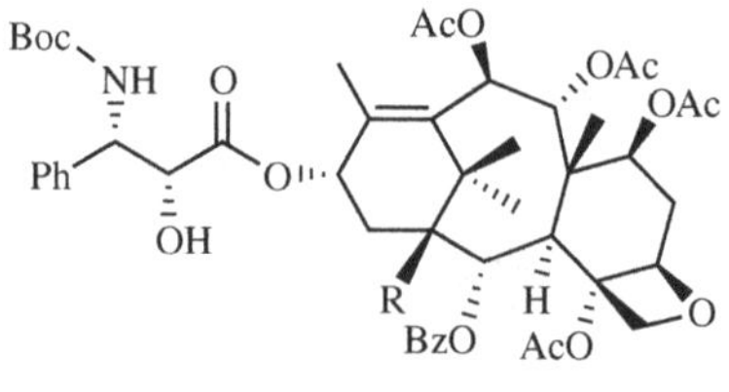

**36a** R=H
**36b** R=OH

### 3.2.3   B Ring and Its Substitutions

#### 3.2.3.1   C-10 Substitutions

Generally, C-10 deacetylation or deoxygenation made an insignificantly negative and positive contribution to the cytotoxicity against sensitive tumor cells, respectively, by comparison with paclitaxel.

For different acyl groups at the C-10 position, it was thought previously that change from acetyl in paclitaxel to other acyl groups usually did not affect the activity significantly and even improve the activity sometimes. Recently, Liu et al. constructed a library with C-10-modified paclitaxel analogs, in which aliphatic, heteroatom-containing aliphatic, alicyclic, aromatic, heteroaromatic acyl groups were introduced.[40] These taxoids were less active in both tubulin assembly and B16 melanoma cytotoxicity assays. Quite different from Liu et al. results, Kingston et al. found C-10 propionate, isobutyrate, and butyrate of paclitaxel were more active against A2780 human ovarian cancer.[64] They investigated the effects of simultaneous modification of analogs at C3'-$N$/C-2 and C-10/C-2 positions.[65] Interestingly, either reduction or the synergistic effect was observed in this study. For example, simultaneous introduction of C-10 substitutions with positive effects, that is, propionate or cortonate, to the C-2 $m$-substituted benzoyl taxoids did not improve or even reduce activity in comparison with C-2 $m$-substituted benzoyl taxoids (**37b–c** vs. **37a**). In addition, taxoid **38** with the C-10 propanoyl and C-3'-$N$ furoyl group increased activity unexpectedly, which infers a synergistic effect.

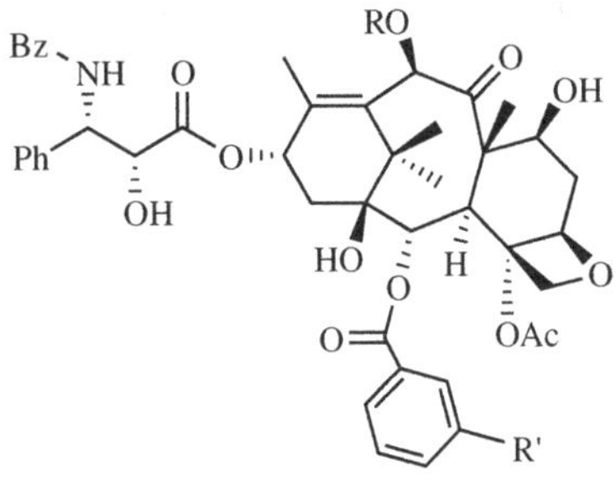

**37a** R=Ac
**37b** R=$i$-PrCO
**37c** R=Crotonyl

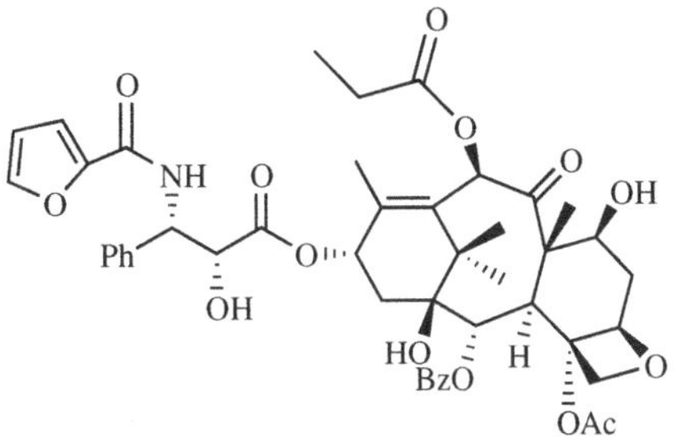

**38**

Datta et al. synthesized 10-$epi$ and 9$\alpha$-OH paclitaxel analogs by iterative oxidation–reduction transformations, furnishing **39–41**. 10-$Epi$ paclitaxel **39a** and 10-deacetyl paclitaxel **39b** are slightly more active than paclitaxel in both cytotoxicity and tubulin binding assays, whereas their 9($R$) counterparts **40** are comparable or slightly less active. 10-Keto analog **41** is also comparable with paclitaxel in both assays.[66]

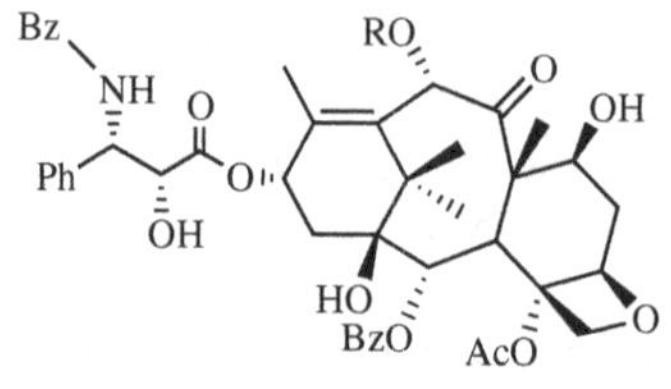

**39a** R=Ac
**39b** R=H

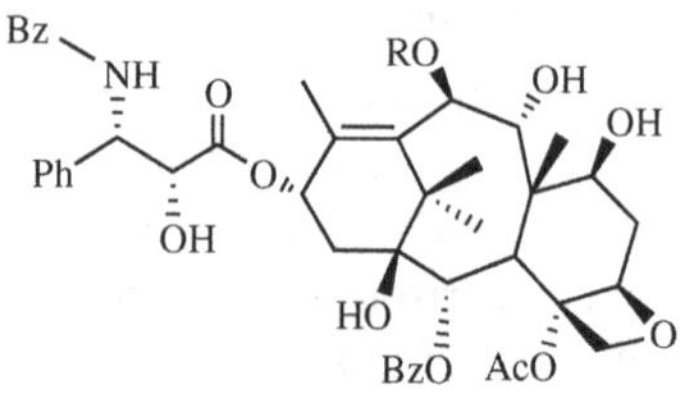

**40a** R=Ac
**40b** R=H

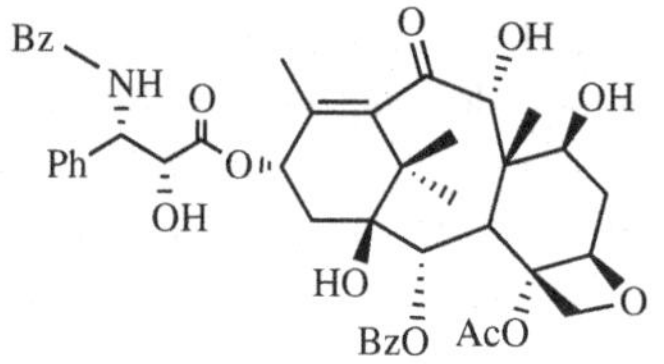

**41**

Walker et al.[67] prepared 10α-spiro epoxide (**42a**) and its 7-methoxymethyl (MOM) ether (**42b**), which exhibited comparable cytotoxicity and tubulin assembly activity with paclitaxel.

10-Deoxygenation was realized by either Barton's method[68,69] or samarium iodide-mediated deoxygenation.[70,71] The latter one is direct and chemoselective, and function group protection is not needed. Treatment of paclitaxel with $SmI_2$ led to the formation of 10-deacetoxy product **43a** in 5 minutes, which was reduced to 10-deacetoxy-9β-OH paclitaxel **43b** with prolonged treatment of $SmI_2$. For 10-deacetyl paclitaxel, 9β-OH derivatives **43b** and **43c** were obtained in a ratio of 50:40. All taxoids were biologically evaluated, and two of them, 10-deacetoxy and 9β-OH-10-deacetylpaclitaxel, were comparable with paclitaxel, whereas the other is less active in tubulin assembly and cytotoxicity assays.[68,72,73]

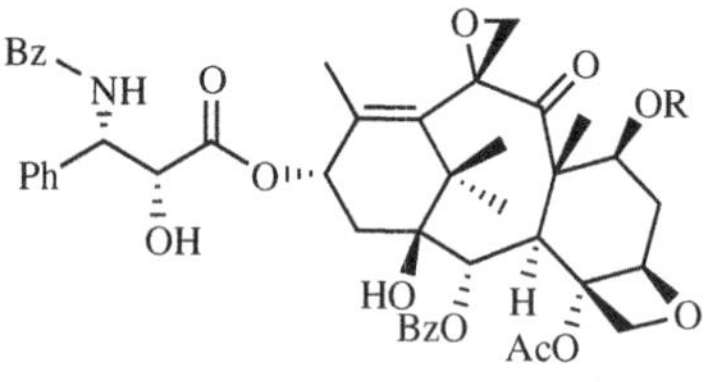

**42a** R=H
**42b** R=CH$_2$OCH$_3$

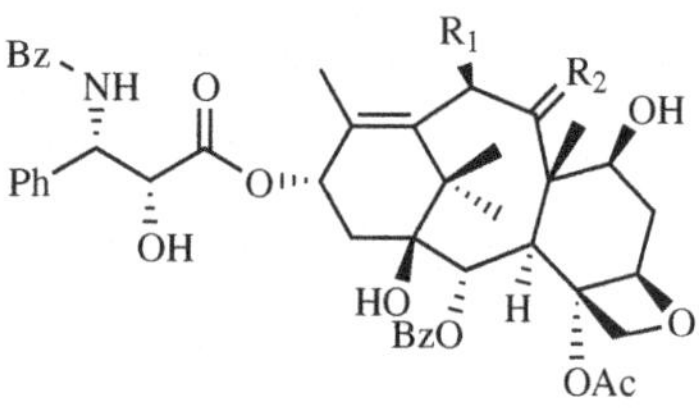

**43a** R$_1$=H, R$_2$=O
**43b** R$_1$=H, R$_2$=β–OH
**43c** R$_1$=OH, R$_2$=β–OH

### 3.2.3.2  C-9 Substitutions

When the C-9 keto group in paclitaxel was replaced by hydroxyl, either α or β-OH analogs displayed slightly higher potencies. The 9α-OH, that is, 9(R)-OH, analog of

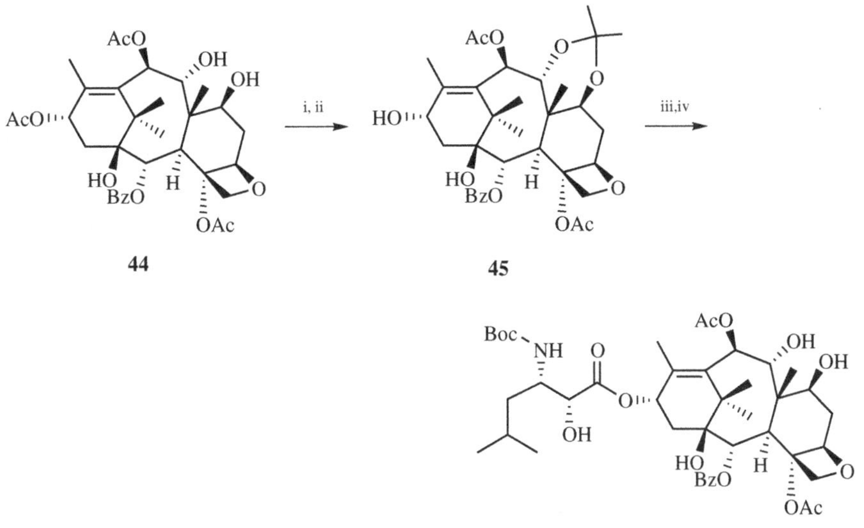

Reagents: (i) Me$_2$C(OMe)$_2$, CSA; (ii) MeLi, THF; (iii) LHMDS, β-lactam; (iv) 1% HCl-EtOH.

**Scheme 3-5.** Synthesis of 9(R)-dihydrotaxol ABT-271.

paclitaxel has been isolated from *Taxus* sp. plants and has exhibited slightly more activity than paclitaxel. At the same time, the discovery of 13-acetyl-9(R)-OH-baccatin III **44** from *T. canadensis* in high content enabled the SAR research of 9(R)-OH taxoids. As shown in Scheme 3-5, the starting material **44** was first selectively deacetylated with BuLi, and then more reactive 9α- and 7β-OH protected by isopropylidene to prepare **45**, which was ready to the attachment of different C-13 side chains, and the 3'-dephenyl-3'-isobutyl-10-acetyl docetaxel analog **46** was found to be the most active in this series, with one order of magnitude higher potency.[30,74,75] After many efforts, researchers from Abott Laboratories demonstrated that a 9(R)-dihydropaclitaxel analog ABT-271 (**46**) was more potent than paclitaxel in several anticancer bioassays. In 2001, they reported on an efficient synthesis of **46** on the 600 g scale, as a part of an effort to file a new drug application for its clinical trial.

9β-OH paclitaxel and docetaxel analogs can be prepared by SmI$_2$ reduction of 10-OH taxoids. Their in vitro activity was comparable with that of paclitaxel and docetaxel.[73,76] A group from Daiichi Co. reported recently the synthesis of highly active and water-soluble 9β-OH-dihydropaclitaxels. Based on their findings on a more active analog, 3'-(2-pyridyl)-7-deoxy- 9β-dihydrodocetaxel analog **47**,[77,78] and a nonhydrolyzed hydrophilic C-10 analog **48**,[79,80] they designed to combine these two substructures into the new taxoid DJ-927 (**49**). Taxoid **50a** and its 2'-(S)-Me analog **50b** exhibited significant antitumor effects in vitro and in vivo, and they were also three orders of magnitude more water soluble. In addition, they were superior to docetaxel in terms of oral bioavailability.[81]

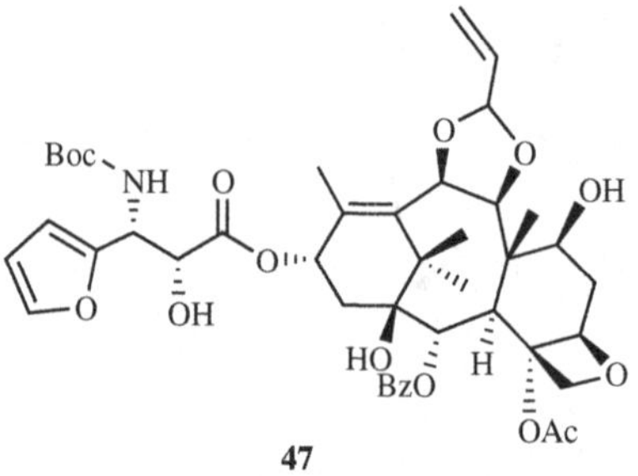

47

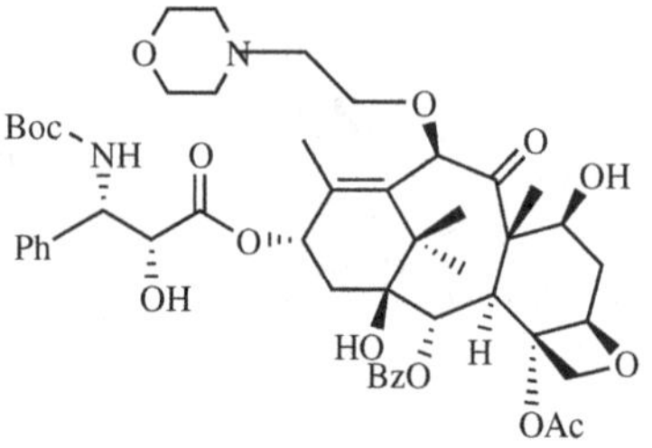

48

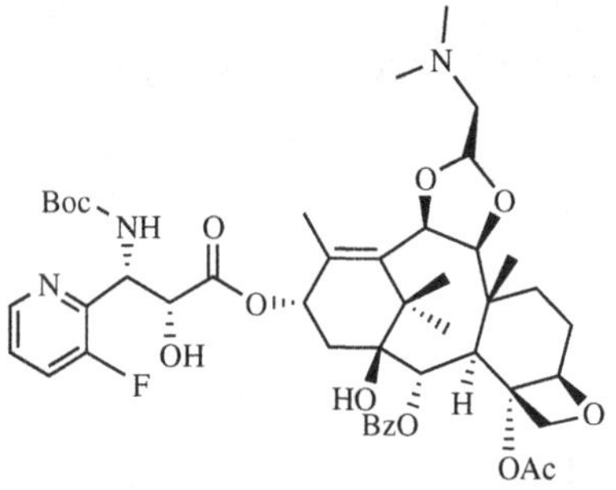

49

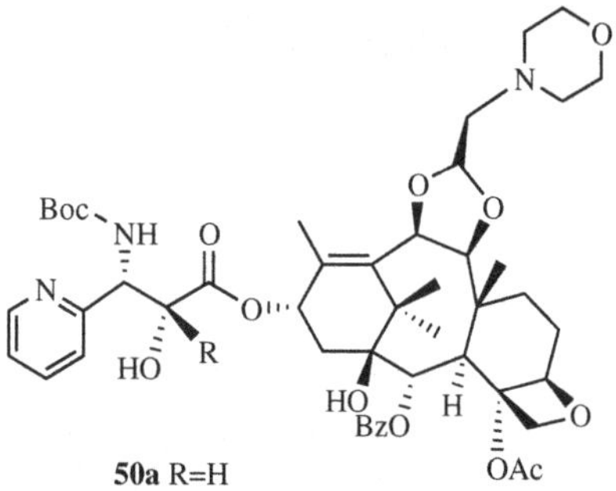

50a R=H
50b R=Me

Cheng et al.[82] conjugated cyclic adenosine monophosphate (cAMP) with 9-OH in the 7-deoxy-9-(R)-dihydro paclitaxel analog, with the hope that cAMP can be converted to ATP in vivo to promote tubulin polymerization actions of paclitaxel, and they found conjugate **51a** was more cytotoxic for two to three times, and the cAMP 2′-conjugate **51b** exhibited reduced cytotoxicity. They also coupled different purine and pyrimidine ribosides through a succinyl linker to the 9α-OH in 7-deoxy-9(R)-dihydropaclitaxel.[83] Those derivatives were generally less active toward all five human normal tumor cells in the assays, whereas enhancement of cytotoxicity was also observed for two of these analogs toward Bel-7402 human liver and Eca-109 human esophagus cancer cells.

A series of 9-deoxy analogues was prepared from 13-acetyl-9(R)-OH-baccatin III (**44**) by an Abbott Lab group.[84] They found that the 9-deoxypaclitaxel **52a**

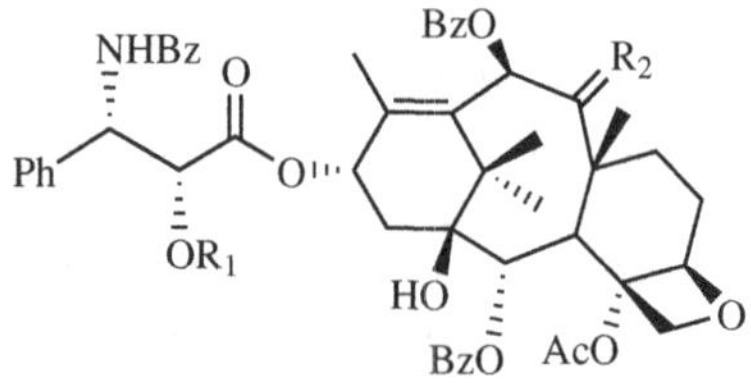

| **51a** | R₁=H, R₂= 9α–O–cAMP |
| **51b** | R₁=cAMP, R₂=O |

| **52a** | R₁=OAc, R₂=OH |
| **52b** | R₁=OAc, R₂=H |
| **52c** | R₁=H, R₂=H |

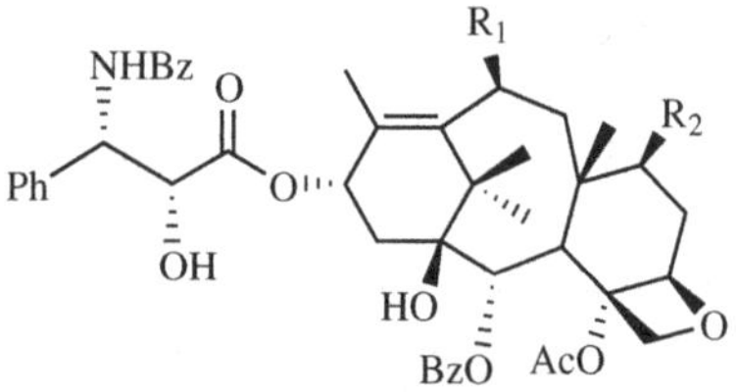

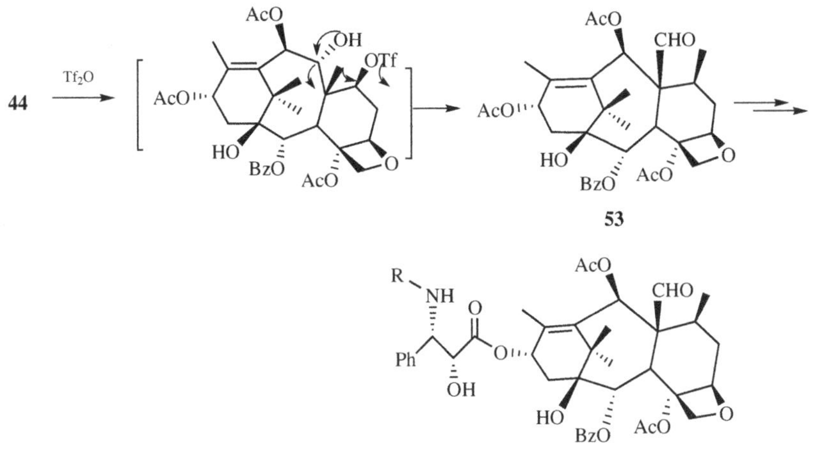


**Scheme 3-6.** Synthesis of B-ring contracted taxoids.

did not affect the in vitro activity at all. Continued deoxygenation at the C-9 and C-10 positions yield the less-active 7,9-dideoxypaclitaxel **52b** and 7,9,10-trideoxy-paclitaxel **52c** within one order of magnitude. The results showed that deoxygenation at these positions may be not of great importance to the cytotoxicity of taxoids.

When **44** was treated with trifluoromethanesulfonic anhydride, a B-ring contracted product was obtained unexpectedly through a plausible Wagner–Meerwein rearrangement (Scheme 3-6). The initially formed 7-triflate was easily disassociated to C-7 cation. Subsequent migration of carbon bonds led to the formation of a B-ring rearranged baccatin **53**.[85] The B-ring contracted paclitaxel analog **54a** was about one order of magnitude less active than paclitaxel, whereas the docetaxel analog **54b** was comparable with that of paclitaxel.

When treated with hydrazine, 10-oxo baccatin **55** was transformated into 7,9-pyrazoline derivatives **56**. Cytotoxicities of those C-13 phenylisoserine derivatives **57** were comparable with that of corresponding taxoids, paclitaxel, docetaxel, and butitaxel.[86] This example is the first successful one in heteroatom substitution at the C-9 position.

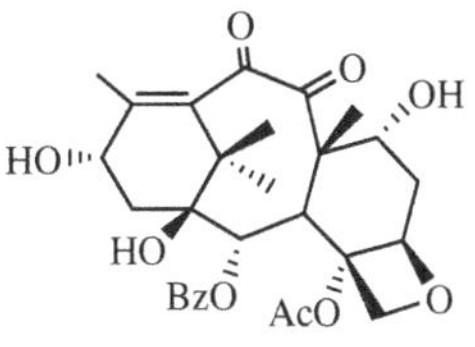


**55**

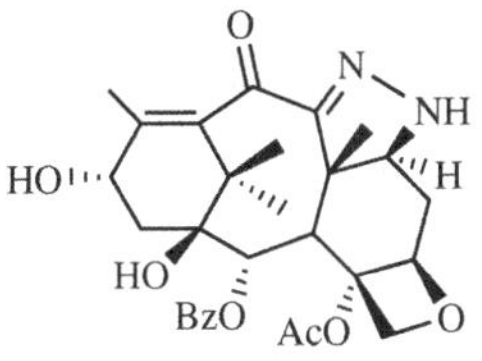


**56**

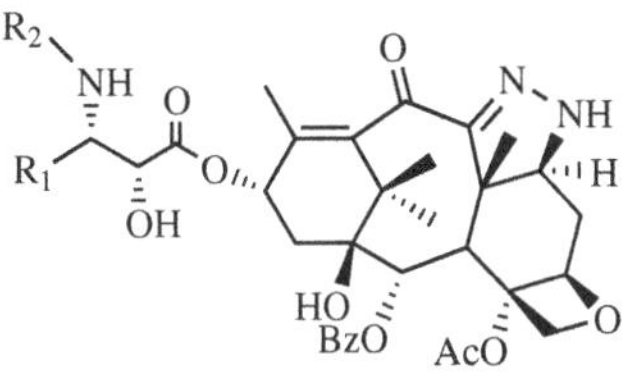


**57a** $R_1$=Ph, $R_2$=Bz
**57b** $R_1$=Ph, $R_2$=Boc
**57c** $R_1$=*i*-Bu, $R_2$=Boc

From properly protected 19-hydroxy-10-DAB **58**, a natural taxane, its docetaxel analog **59** was obtained by the "acid approach," which was proven to be active in promoting tubulin polymerization ability as well as cytotoxicity.[87]

**58**                                                                    **59**

### 3.2.3.3   C-2 Substitutions

As early as 1994, Chaudhary et al. enabled the facile SAR exploration at C-2, which is possible by establishment of the selective C-2 debenzoylation method of 2′-TBS-paclitaxel by phase-transfer catalysed basic hydrolysis,[88] and, later on, basic hydrolysis with Triton B.[91] Datta et al. also found potassium *tert*-butoxide as a selective deacylating reagent in 7,13-*di*-TES-baccatin III.[89] The C-2 debenzoylation can also be realized by the electrochemistry method.[90] Based on their retrosynthetic studies for total synthesis of taxol, Nicolaou et al. also made the incorporation of various acyl groups at C-2 realized in a different way, that is, through treatment of 1,2-carbonate with organolithium reagents.[91]

For the benzoate series, Kingston's group found *meta*-substitution is usually superior to *ortho*- or *para*-substituted phenyls, among which *m*-azido substituted analog **60a** is the best, with two to three orders of magnitude enhancement of cytotoxicity.[88] Recently, his group conducted systematic exploration[92] on the acyl substitutions at C-2, and it found reasonable agreement in the correlation of cytotoxicity and tubulin polymerization activity for those active analogs. In general, *meta*-substituted compounds are more cytotoxic than paclitaxel. The *para*- and *ortho*-substituents usually showed negative impact on activity, except for some specific compounds, for example, 2-(*o*-azido)benzoyl analog. Disubstituted benzoyl analogs were generally less active than their monosubstituted counterparts. For other heteroaromatic analogs tested, only thiophene analogs showed improved activity, in accordance with Nicolaou's finding. The 2,4-diacyl paclitaxel analogs were also prepared. Taxoids **60b–d** gave the best data in tubulin assembly and cytotoxicity assays.[93]

For C-2 heteroaromatic esters, promising results are seldom reported, except the analog with certain groups (e.g., 2-thienoyl) **61**, which retained comparable or superior activities.[91,94]

Boge et al. found that hydrogenation of phenyl of C-2 benzoate in paclitaxel with a ruthenium catalyst led to the formation of a cyclohexanoate analog with reduced activity,[95] and Kingston et al. claimed that nonaromatic analogs, for example, acetyl and valeryl taxoids, were significantly less active.[92] Ojima et al. investigated the impact of combined C-2 and C-3′ modifications.[96] When phenyl groups

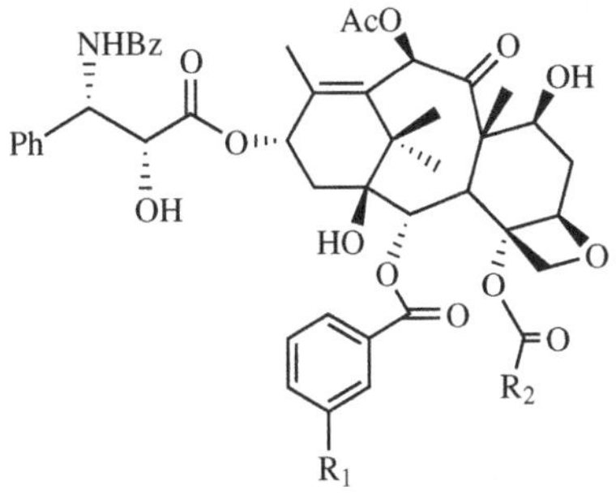

**60a** R$_1$=N$_3$, R$_2$=Ac
**60b** R$_1$=N$_3$, R$_2$=cycloprapane
**60c** R$_1$=Cl, R$_2$=cycloprapane

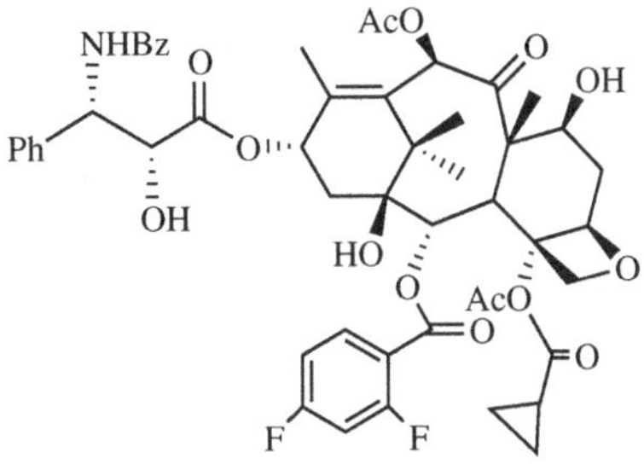

**60d**

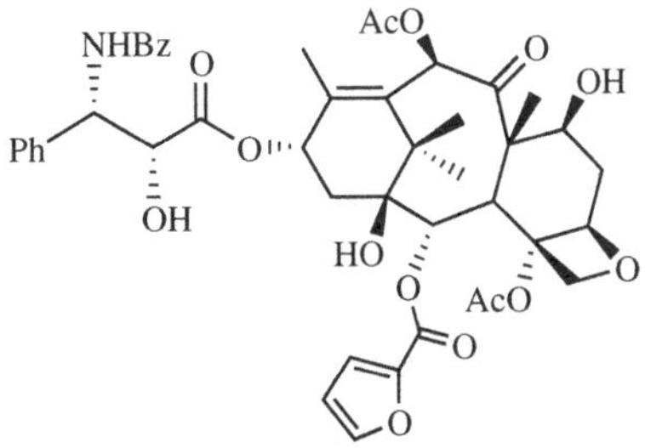

**61**

at these two positions were replaced with nonaromatic groups, for example, cyclo-hexane, isobutene, and *trans*-prop-1-ene, more rigid groups seem to be better than freely rotating ones. However, even the best one in this series, the 2-debenzoyl-2-isobutenoyl-3'-N-debenzoyl-N-isobutenoyl analog **62a**,[96] was not superior to SB-T-1212 (**62b**).

Chen et al. synthesized C-2-acetoxy-C-4-benzoate, so-called *iso*-paclitaxel **63**. This compound was totally inactive either in cytotoxicity or in tubulin polymerization assays.[97] It is in agreement with previous observations that only small C-4 substituents were tolerated.

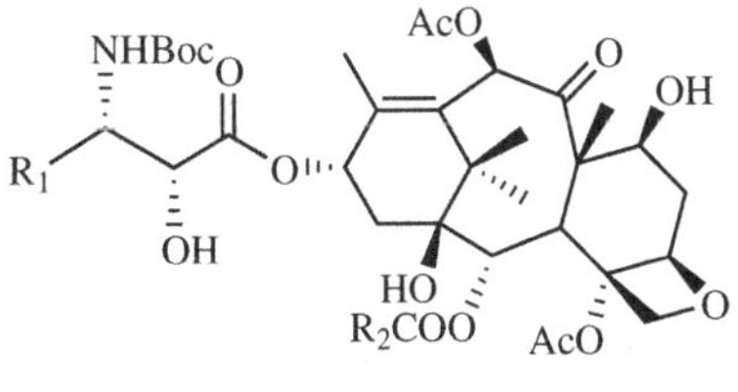

**62a** R$_1$=*i*-butene, R$_2$=*i*-Bu
**62b** R$_1$=*i*-butene, R$_2$=Ph

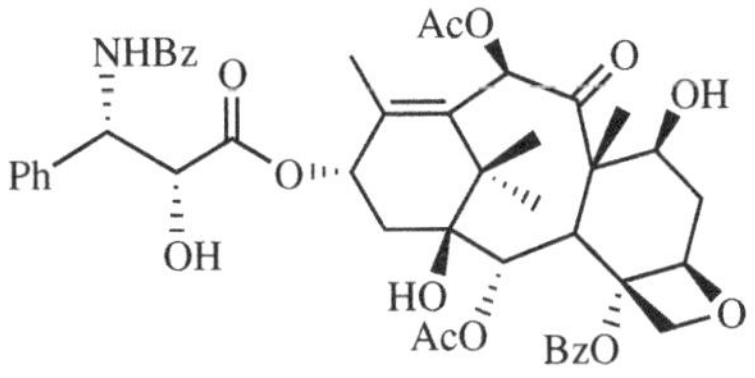

**63**

Synthesis of a 2α-*N* substituted analog 2-debenzoyloxy-2α-benzamido docetaxel analog **65a** was realized as the first example in the preparation of 2α-heteroatom linkages in our laboratory recently. The key step for the synthesis is transformation of 2α-baccatin **64a** to 2α-azido baccatin **64b** through a double S$_N$2 conversion. The taxoid **65a** is comparably active with paclitaxel against human lung cancer A-549 and less active in the other two tumor cell lines.[98] A series of C-2 *meta*- and *para*-substituted benzamido analogs were also prepared, among which 2-*m*-methoxy and 2-*m*-chloro benzamide taxoids **65b-c** are the most active, but not more than paclitaxel and docetaxel.[99] Besides, a 2α-phenylthio analog of docetaxel exhibited almost no activity.[100]

2-Deoxygenation was realized in the early 1990s, and it was found to be detrimental to the antitumor activity (see also Section 3.2.2.3).[101] A 2-epimerized derivative was prepared by an oxidation–reduction sequence for the transformation of 2α-OH into 2β-OH through the 2-keto group.[102] The 2-keto group was stereoselectively reduced to 2β-OH because of steric hindrance of C-16 and C-19 from the β-face to prevent hydride transfer from this face. Unfortunately, 2-*epi*-paclitaxel was inactive, which demonstrates the importance of stereochemistry of C-2 substitutions in its activity.

### 3.2.4   C Ring and Its Substitutions

#### 3.2.4.1   C-7 Substitutions

7-OH, because of its location at β-position to the 9-keto group, is easily epimerized from β orientation in taxanes to α in 7-*epi* taxanes through a retro-aldol reaction. This reversible reaction was accelerated under basic conditions. Although 7α-OH is thermodynamically more stable and thus predominant in a mixture consisting of 7α- and 7β-OH taxanes, 7α taxanes can be converted into a mixture of 7α- and 7β isomers kinetically.[103] Being frequently observed in organic solvents and biological fluids, 7-epimerization did not make much contribution to its activity, either positively or negatively.

Acylation of free hydroxyl at C-7 in paclitaxel usually led to the reduction and even loss of cytotoxicity of the derivatives, when steric hindrance of the acyl groups increases. Bhat et al. have used a parallel solution phase synthetic method to construct a 26-membered library of C-7 esters,[39] and concluded that modification at C-7 were detrimental to cytotoxicity against the MCF-7 cell line. Only a few exceptions, including 10-deacetyl-10-propionyl-7-chloroacetyl paclitaxel (**66a**),[37] were reported. However, the 7-esters are not stable in biological fluids and the parent compound paclitaxel is released slowly after its injection, and **66a**'s superior activity could be attributed to 10-deacetyl-10-propionyl paclitaxel **66b**, a taxoid that is more potent than paclitaxel. Hence, the 7-esters or other 7-OH derivatives with hydrolysable linkages have been frequently used as prodrugs.

Such an observation was also documented in many efforts to prepare prodrugs and fluorescent or photoreactive probes at the C-7 position. A 7-{[$^3$H$_2$]-3-(4-benzoyl)-phenylpropanolyl} paclitaxel analog (**67a**) and its 10-deacetyl derivative (**67b**) as photoreactive probes to explore paclitaxel binding sites on tubulin and P-glycoprotein.[104] Taxoid **67a** was less cytotoxic than paclitaxel against either normal or MDR tumor cells, and **67b** is much less active. 7-(β-Alanyl)-biotin derivative (**68**) of paclitaxel is comparably active against human leukemia U937 cell (4.0 nM for **68** vs. 4.5 nM for paclitaxel).[105] Another derviative 7-(*p*-azidobenzoyl)taxol (**69**) prepared as a photoaffinity label by Georg et al.[106] and a series of 7-esters during the preparation of water-soluble analogs[107] also demenstrated that different 7-esters may exert different bioactivity according to its steric hindrance and orientation of the C-7 ester side chain. The fluorescent derivatives (**70**) of paclitaxel, also active against tumor cells, although bearing a large group at C-7.[108,109]

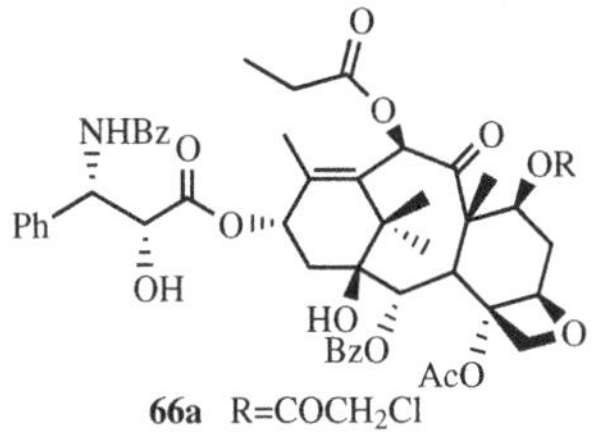

66a  R=COCH$_2$Cl
66b  R=H

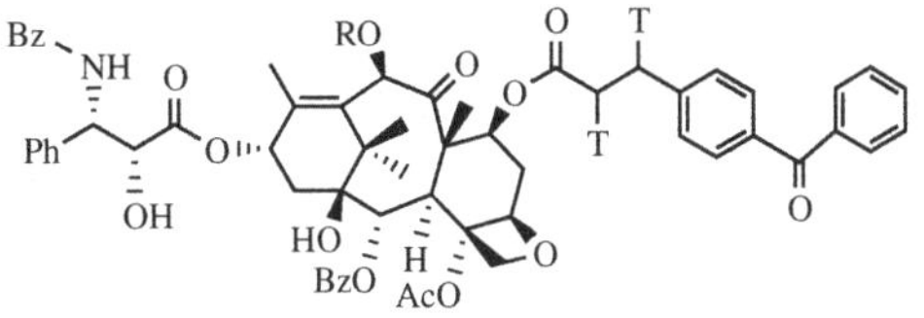

67a  R=Ac
67b  R=H

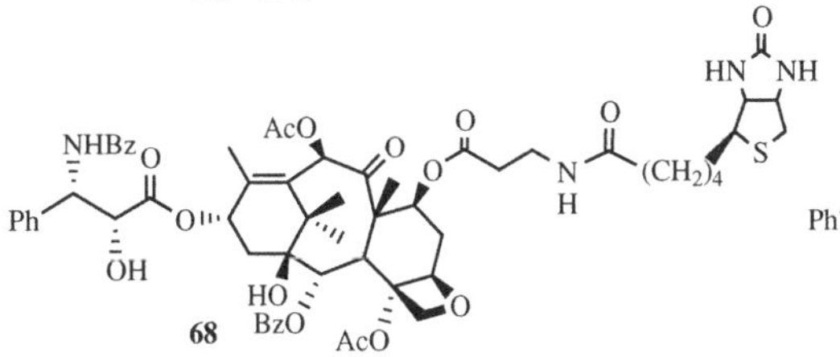

68

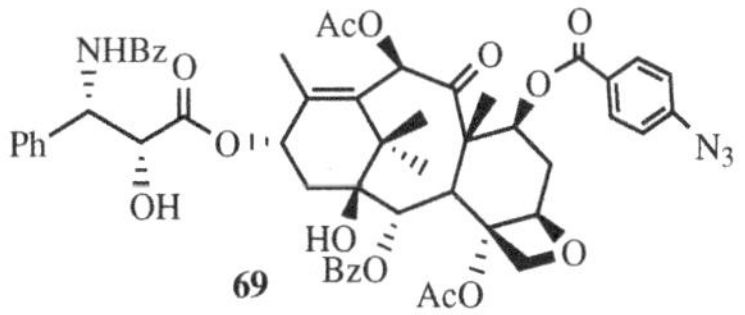

69

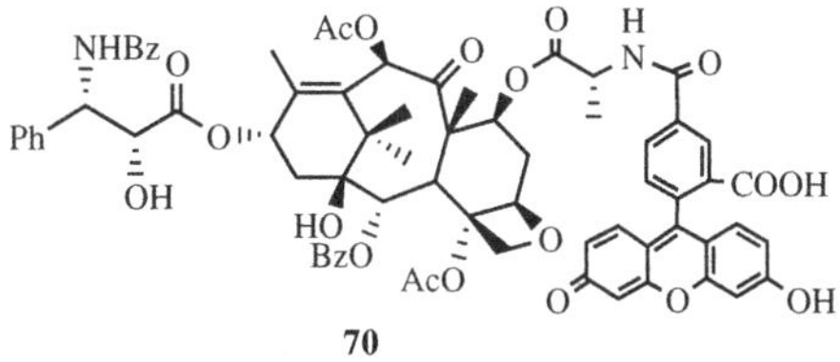

70

Previous results have shown that 10-deacetyl-7-acyl paclitaxels or docetaxels were superior to most 7-acyl paclitaxel.[110] A series of tertiary amine-containing

10-deacetyl-7-acyl analogs was prepared from **66a** by nucleophilic substitution to enhance their water solubility.[41] In cytotoxicity assays, all compounds prepared in this series were comparable with paclitaxel, among which **71** was most active. In addition, hydrochloride salt of **71** was nine times more soluble in water.

Many paclitaxel 7-ethers were found to be comparable or more active. The 7-methylthiomethyl (MTM) ether of paclitaxel (**73**) was prepared and found to be comparably active with paclitaxel.[111] In recent research efforts disclosed by BMS scientists, a series of MOM ethers and MTM ethers was prepared. BMS-184476 (**73**) was chosen to be the starting point in systematic evaluation of C-7 ether analogs of paclitaxel with general formula **72**, in which phenyl, 2-furyl, and $i$-butenyl were selected as $R_1$; phenyl and $t$-Boc as $R_2$; and MTM, MOM, $CH_2O(CH_2)_2OH$, and Me as $R_3$ moieties. BMS-184476, although scored behind many competitors in vitro, exhibited superior activity in several in vivo tumor-bearing animal models, including most paclitaxel-resistant tumor HCC79 model in BMS.[112]

71

72 general formula
73 $R_1$=Bz, $R_2$=ph, $R_3$=$CH_2$SMe

Several 7-$O$-glycosylated taxanes have been found in *Taxus*, and one of them, 7-xylosyl-10-deacetylpaclitaxel **74**, is abundant (about 0.2–0.3% dry weight) in biomass from *T. yunnanensis* and hence regarded as starting material for the semi-synthesis of paclitaxel in industry. Because the glycoside is more hydrophilic than most taxanes in hydrophobic nature, it is reasonable to think that if the 7-$O$-glyco-side is cytotoxic, it may be a superior antitumor agent than paclitaxel for water solubility. A semisynthetic 7-$O$-glucopyranosyl docetaxel analog **75** that was obtained from **74** by four steps of chemical conversions displayed comparable cyto-toxicity and tubulin binding ability with docetaxel and was twice as soluble as paclitaxel.[113]

7-Deoxy-7β-sulfur analogs were prepared by epimerization with DBU from 7α-thiol, the latter one from 7β-triflate upon treatment of LiSMe or KSAc.[114] In contrast to 7α-SH and 7β-SH analogs, which are less toxic, 7β-MeS (**76a**) and 7β-MeOCH$_2$S (**76b**) analogs as well as **73** are superior to paclitaxel.

Treatment of 2′-$CO_2$Bn-paclitaxel with DAST reagent yielded 7α-F **77** and 7,19-cyclopropane **78** products,[115,116] and 6,7-olefin derivative **79** derivatives, which can also be obtained through 7-triflate intermediate when treated with

DBU and silica gel in 1,2-dichloroethane, respectively.[117] Paclitaxel and docetaxel analogs with 7α-F are not superior to the parent compounds in both in vitro and in vivo, and those with 7,19-cyclopropane and 6,7-olefin were comparably active.[118]

In the early 1990s, the Bartons protocol was widely applied to the preparation of deoxygenated derivatives at many sites on taxanes. Enhancement of cytotoxicity was observed for many 7-deoxy paclitaxel and docetaxel analogs, along with a reduction of cytotoxicity for 7,10-dideoxy analogs.[119,120] An efficient synthetic route of 7-deoxypaclitaxel from taxine, a mixture containing several structurally related taxanes with 4(20)-exocyclic methylenes,[121] as well as synthesis of 7-deoxytaxane from $\Delta^{6,7}$ taxane were also reported.[122]

In general, all above-mentioned modifications at C-7 did not change their activities to a great extent, which indicates that C-7 radicals may not interact with tubulin binding site significantly. This hypothesis was confirmed by x-ray crystallography and molecular simulation.

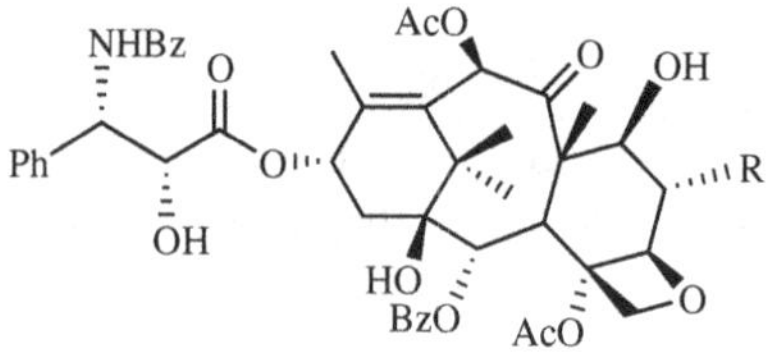

**80a**  R=Ac
**80b**  R=H
**80c**  R=Cl
**80d**  R=Br

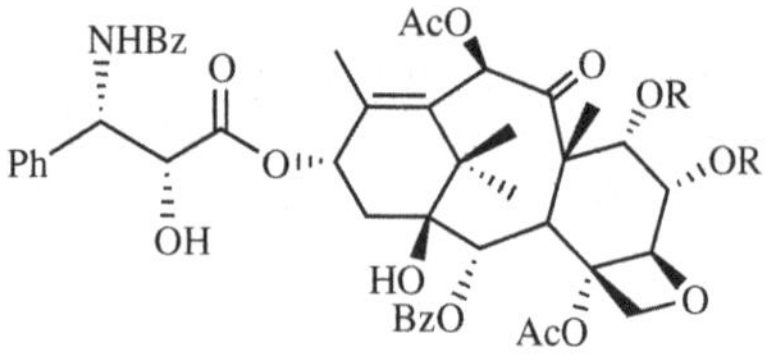

**81a**  R=H
**81b**  R=SO
**81c**  R=SO$_2$

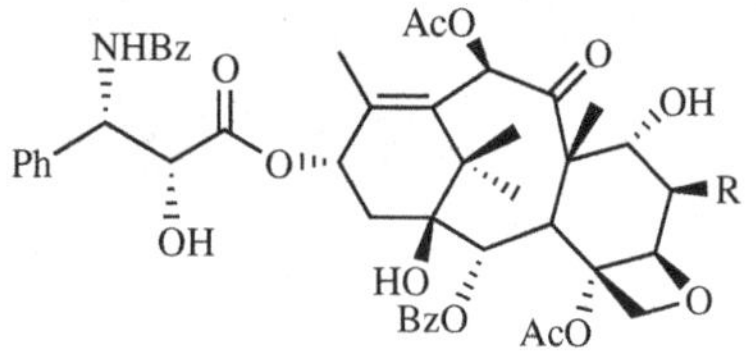

**82a**  R=N$_3$
**82b**  R=NH$_2$

### 3.2.4.2  C-6 Substitutions

6α-OH paclitaxel **80a**, the major metabolite of paclitaxel in human that has been known to be less cytotoxic than paclitaxel (both 6-hydroxylation and 3′-*p*-phenyl-hydroxylation were detected in mice, whereas 6α-hydroxylation predominated in humans), can be prepared via epimerization of 6α-OH-7-*epi*-paclitaxel **81a**.[123,124] Taxoid **81a**, which was prepared via dihydroxylation of Δ$^{6,7}$ taxane, was slightly less active than paclitaxel. 6α-F, Cl, and Br paclitaxels **80b–d** were designed and prepared as the metabolic site blocked analogs, but they could not alter their in vitro and in vivo efficacies significantly.[125] 6α-Hydroxy-7-*epi*-paclitaxel 6,7-*O,O′*-cyclosulfite **81b** and 6,7-*O,O′*-cyclosulfate **81c** were obtained from **81a**, and 6β-azido- (**82a**) and 6β-amino-7-*epi*-paclitaxel (**82b**) were also prepared from the same intermediate. Taxoid **82a** was two to three times more cytotoxic than paclitaxel, but **82b** and **81b** were less active, and **81c** was essentially inactive.[126]

Synthesis of 7-deoxy-6α-hydroxypaclitaxel was realized through the regiospecific reduction of 6,7-α-thiocarbonate **83** as the key step. 2′-TES-7-deoxy-6α-hydro-

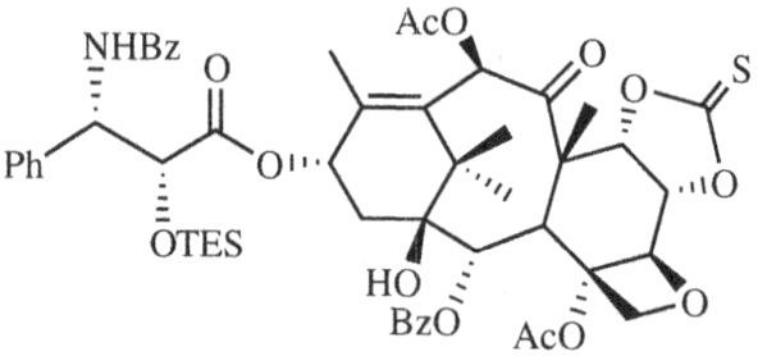

**83**

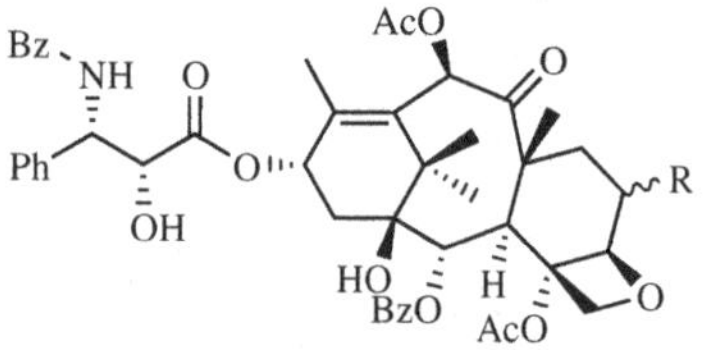

**84a** R=α–OH
**84b** R=β–OH

xypaclitaxel was transformed into its C-6β epimer by oxidation–reduction manipulation. Both isomers (**84a-b**) are equipotent to paclitaxel in tubulin assembly assay and less cytotoxic by about one order of magnitude.[127]

### 3.2.4.3 C-4 Substitutions

Selective C-4 deacetylation reactions were reported by several groups independently. In 1994, Chen et al. found that 7,13-di-TES-1-DMS-baccatin III **85a** can be selectively deacetylated with Red-Al in 66% yield. The C-4-OH taxoid **85b** is an ideal starting material for 4-acylation under LHMDS/RCOCl acylation conditions. C-4 cyclopropanoyl analog **86a** showed better in vitro activity than paclitaxel, whereas the benzoyl analog **86b** did not.[128] C-4 OMe derivative was also obtained with similar manipulation.[129] Georg et al. discovered that t-BuO⁻K⁺ can exert selective deacylation at C-4 in 7-TES-baccatin III (**3**) in 58% yield, possibly by assistance of 13α-OH in close proximity. When 13α-OH was protected by TES, the 2-debenoyl product was obtained instead.[89,130] 4-Isobutyric paclitaxel **86c** exhibited strong activity, although three to five times less active than paclitaxel.[125]

In a systematic study of C-4 ester, carbonate, and carbamate analogs, some aliphatic esters and carbonates as highly cytotoxic taxoids, several times better than paclitaxel, were reported.[131] The cyclopropanoyl and methylcarbonate analogs displayed the strongest in vitro and in vivo activity in this series of taxoids, and the latter one **86d** underwent a phase I clinical trial. For 4-aziridine analogs (general formulas **87**), change of 3′-Ph and 3′N-Bz to 2-furyl and Boc, respectively, did not improve the activity; **86a** was still the most potent one.[132] Kingston et al. found[93] that C-2/C-4 modification was in agreement with previous results; only methoxylcarbonyl and cyclopropylcarbonyl analogs were active, whereas other larger groups at C-4 were detrimental to activity. It is noteworthy that most of these active analogs share common structure characterstics, that is, relatively small substitution at C-4.

C-4 OMe paclitaxel analog **88** exhibited 10 times the reduced activity, whereas a change of 3′-Ph and 3′N-Bz to 2-furyl and Boc, respectively, did improve the activity by about ten times.[124] This result demonstrated the importance of the C-4 carbonyl group.

|  |  |
|---|---|
| **85a** | R=Ac |
| **85b** | R=H |

|  |  |
|---|---|
| **86a** | R=cyclopropane |
| **86b** | R=Ph |
| **86c** | R=s-Pr |
| **86d** | R=OMe |

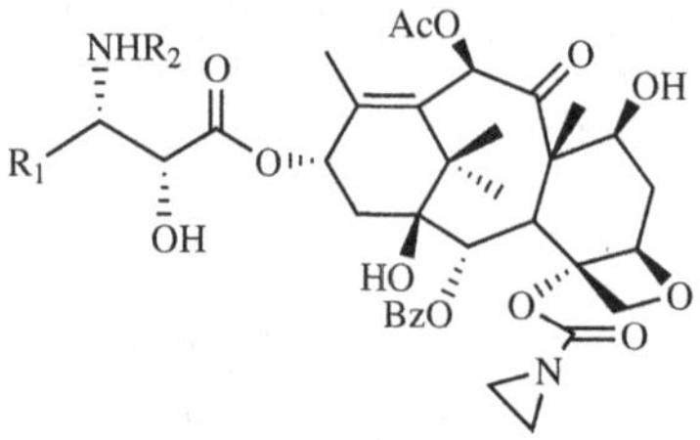

87 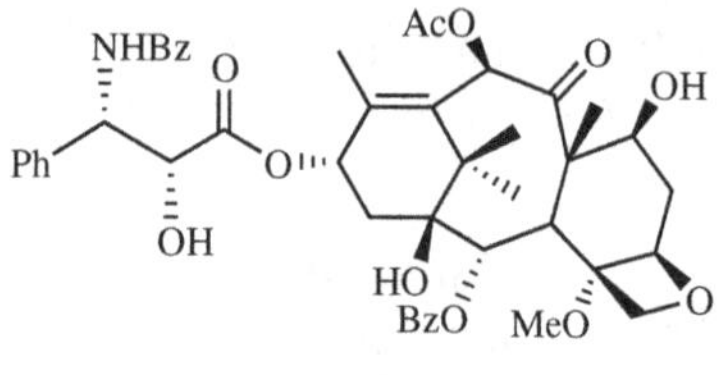                                                88

A paclitaxel analog with a C4–C6 bridge, built on the connection of a carboxyl group of 4-glutarate and the hydroxyl group of 6α-hydroxyacetate, was found almost inactive.[133] This observation was in accordance with the hypothesis that the Southern Hemisphere of paclitaxel binds to the tubulin receptor.

### 3.2.4.4  C Ring Contraction and Expansion

Yuan et al. have reported on the synthesis of a C-ring contracted analog (**90**) from 2'-TBS-6α-OH-7-*epi*-paclitaxel (**89**), upon the treatment of lead tetraacetate. After 2'-desilylation, the analog **91a** showed 10 times reduced activity, and its 7-Ac derivative **91b** was even less active.[134]

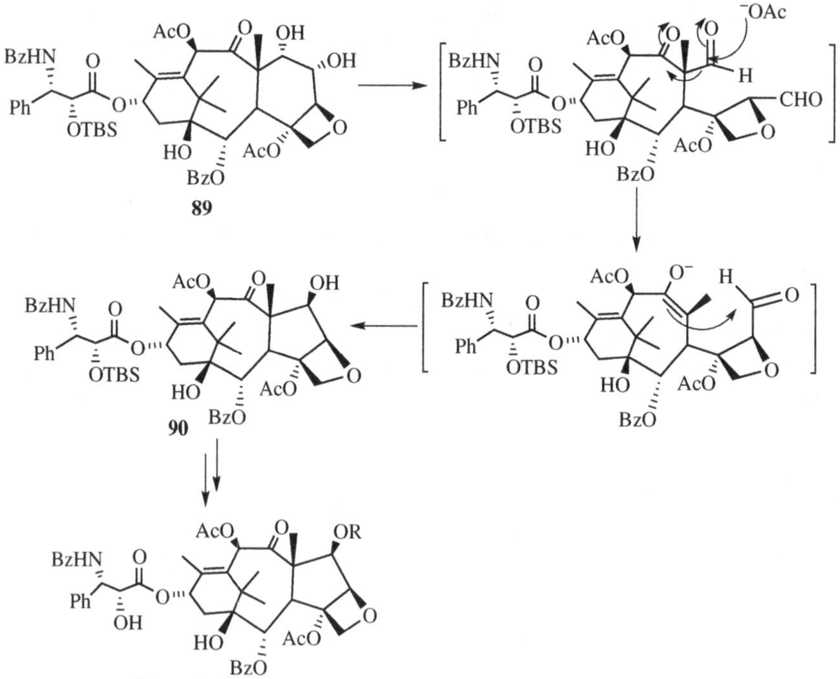

**Scheme 3-7.** Formation of C-ring contracted analogues.

Bourchard et al. attempted to reduce the 7,19-cyclopropane docetaxel analog **92a** electrochemically. Besides the major 10-deoxy product **92b**, C-ring expanded 19-nor analog **93** was obtained as a minor product. Taxoid **93** did not exhibit any activity in both cytotoxicity and tubulin binding assays, although its conformation is analogous to that of docetaxel.[135]

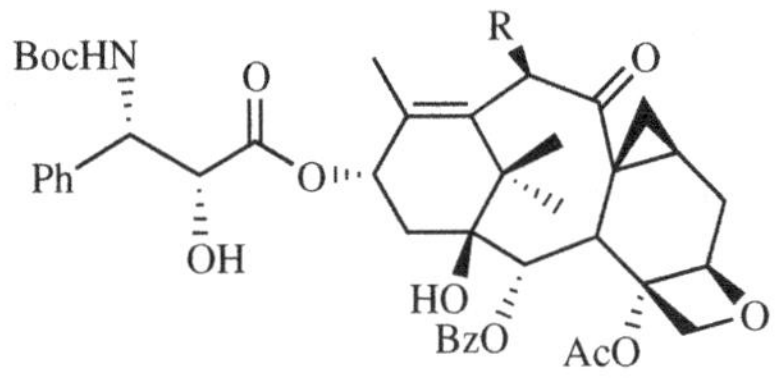

92a  R=OH
92b  R=H

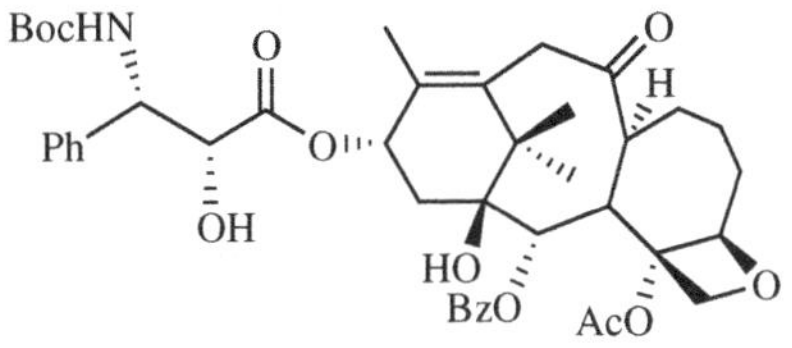

93

### 3.2.5  D Ring

Although numerous researchers have claimed the crucial role of oxetane D ring in cytotoxicity of taxoids, its role was still not well understood structurally. Constraint and hydrogen bond acceptor are assumed to be two major roles of D ring in paclitaxel's binding to its receptor tubulin. Moreover, recent successful syntheses of aza-, thia-, and selena-substituted, as well as deoxy oxetane ring (cyclopropane) analogs, were helpful in clarifying its role.

Low activity or inactivity of azetidine D-ring **94a** and **94b**,[136] very poor activity of 4-carbonate-thia-paclitaxel **95** in tubulin polymerization and inactivity in cytotoxicity tests, together with inactivity of 4-deacetyl-selena-paclitaxel **96** in both

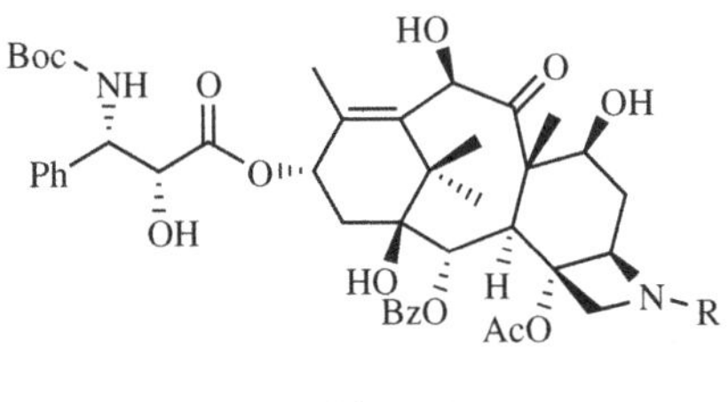

94a  R=Bn
94b  R=H

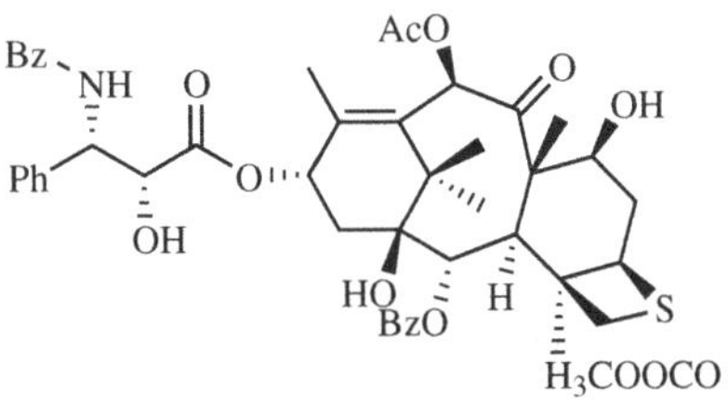

95

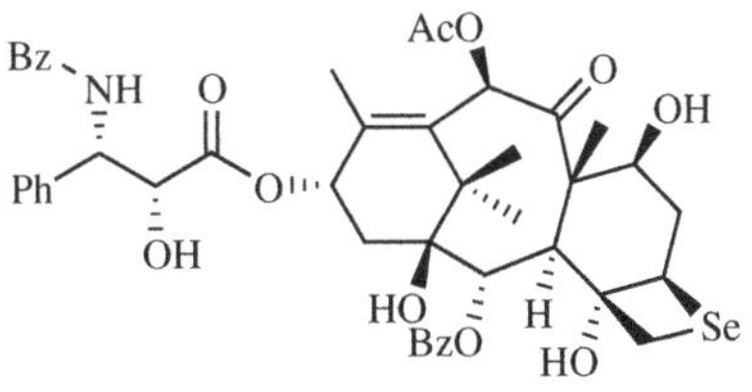

96

assays[137] suggested that (1) the oxygen atom may be involved in the interaction with an amino acid residue of tubulin and this interaction may not be replaced by NH; (2) the region surrouding the oxetane ring is sensitive to steric effects (the inactivity of 4-deacetyl-selena analog **96** may also be from the absence of 4-acyl substitution that is crucial to the activity). A more recent study conducted by a Japanese group on 4-deacetoxy-1,7-dideoxy D-azetidine analogs also verified that, after the replacement of the oxygen atom in the D ring with the nitrogen atom, the analogs largely retained activities.[138]

In 2001, two D-thia analogs, 5(20)-thia-docetaxel **97a** and 7-deoxy-5(20)-thia-docetaxel **97b**, were synthesized.[139] Compound **97a** is two orders of magnitude less cytotoxic than paclitaxel, but it retained tubulin assembly ability stronger than in a previous report.[137] Meanwhile, **97b** is inactive in tubulin assay and 3 times more cytotoxic than **97a** against KB cells, but it is still much less active than both 7-deoxy-10-acetyl-docetaxel and docetaxel.

Cyclopropane analog[140] **98** showed microtubule disassembly inhibitory activity comparable with paclitaxel, but lower than docetaxel. The author thus demonstrated that the oxetane ring is not essential for the interaction of paclitaxel analogs with microtubules when the C-ring conformation is locked by cyclopropane, but the oxygen atom in the D ring of paclitaxel may participate in the stabilization of a drug-tubulin complex.

**97a** R$_1$=OH, R$_2$=OH
**97b** R$_1$=OAc, R$_2$=H

**98**

It has been found that a 4-deacetoxy D-seco analog was about ten times less active than paclitaxel, but it was somewhat more active than 4-deacetoxyopaclitaxel.[141] Barboni et al. synthesized some D-seco paclitaxel analogs without the 5α-oxygenated group, usually presented in those D-seco analogs obtained from the oxetane ring cleavage. Jones' oxidation of 2′-TBS-paclitaxel led to a 5,6-unsaturated-7-*keto* D-seco analog, which was then converted into 7α-OH and 7β-OH analogs. Although it is disappointing to find that these compounds did not exhibit any activities in biological assays, it provided the authors an opportunity to review and revise the "hydrophobic collapse" pharmacophore model.[142]

Based on the experimental results of azetidine, thia- and cyclopropane analogs of D-ring in paclitaxel, as well as a computational prediction of a taxol minireceptor model, some taxoids devoid of the D-oxetane ring, including D-seco analogs, were proposed to bind to tubulin in similar free energies to that of paclitaxel, which infers comparable activities. The design and syntheses of several D-seco, oxirane, and cyclopropane analogs were realized and biologically tested by a Dutch group.[143] The lower activities of those analogs than paclitaxel by three orders of magnitude may also be partly attributed to the absence of the 4-acetoxy group, which was demonstrated earlier[141] in addition to the absence or inappropriate arrangement of the oxygen atom of the oxetane D ring in space.

### 3.2.6 Macrocyclic Analogs

Ojima et al. proposed a common pharmacophore for several anticancer natural products targeting microtubules, including paclitaxel, epothilones, eleutherobin, and discodermolide.[144] It was suggested that macrocyclic taxoids such as **99a** may represent hybrid constructs of paclitaxel and epothilone. A group of these macrocyclic taxoids with unsaturated and saturated linkages between C-2 and C-3' were prepared through ring-closure metathesis (RCM) catalysed by the Grubbs catalyst and subsequent hydrogenation.[145] Among 30 macrocyclic taxoids, only three of them, including **99a**, retained strong cytotoxicity, although they were less active than paclitaxel by two orders of magnitude. Taxoid **99a** was also one of three active taxoids in tubulin polymerization assay with 36% relative activity to paclitaxel.

Boge et al. also prepared a group of macrocyclic taxoids, in which C-2 benzoate and C-3' phenyl were tethered by alkenyl, alkyl, and ester linkers. The alkenyl linked taxoids were synthesized through Heck reaction, and additional hydrogenation formed alkyl linkage. All compounds subjected to tubulin polymerization tests were inactive. Unfortunately, the authors did not report the biological evaluation and molecular modeling results of the precursors for those C2–C3' tethered analogs, which may provide insight to their questions on the "hydrophobic collapse" hypothesis.[146]

Ojima et al. also completed the synthesis of a series of C3'$N$–C2 linked macrocyclic analogs of 3'-isobutyl analogs by RCM strategy.[147] Interestingly, most analogs in this series were more potent cytotoxic agents against LCC6 human breast carcinoma and its drug-resistant counterpart LCC6-MDR than were the C3'–C2 linked macrocyclic taxoids through a *meta*-substitutent on C-2 benzoate, although they were still less active than paclitaxel by one to two orders of magnitude. Taxoids **99b** and **99c** were among the most potent inhibitors for both wild-type and drug-resistant cancer cells. A CNRS group also demonstrated the impact of ring size for C3'$N$–C2 tethered taxoids through C2 aliphatic ester groups by RCM and sulfide at the end of C3'-$N$ amides and C-2 aliphatic esters,[148] and they compared the cytotoxicities of the macrocyclic taxoids with their open-chain counterparts. They concluded that the sulfide linkage is deleterious to the activity[148] and that the 22-membered ring taxoid was active, whereas 18-, 20-, and 21-membered ring taxoids were inactive in tubulin binding and cytotoxicity assays.[149] Also, taxoids with both C-3'$N$ and C-2 alkyl acyl groups (open-chain taxoids), although they exhibited several times weaker binding affinity and cytotoxicity than docetaxel, they were still one to two orders of magnitude more active than their macrocyclic derivatives.

Metaferia et al. reported on the synthesis and cytotoxicity of C3'-C4 linked macrocyclic taxoids in 2001. The synthetic strategy was also based on RCM. Similar to Ojima et al.'s finding, these compounds (**100**) are less active than paclitaxel in both cytotoxicity and tubulin assays.[150] However, after careful inspection of the bioactive conformations of taxoids with computer simulation and nuclear magnetic resonance (NMR) experiments, they designed several C-4 and C3'-Ph *ortho*-position linked taxoids (**101**) that were highly active, several times better than paclitaxel.[151]

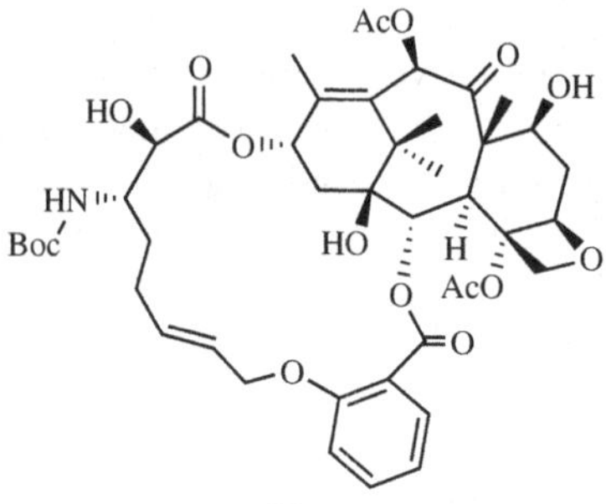

**99a**

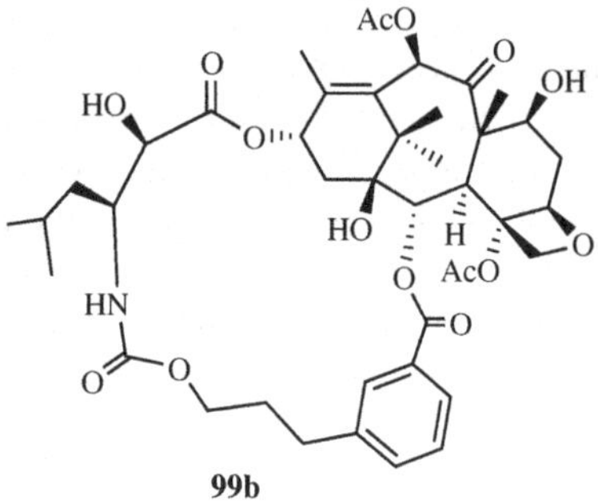

**99b**

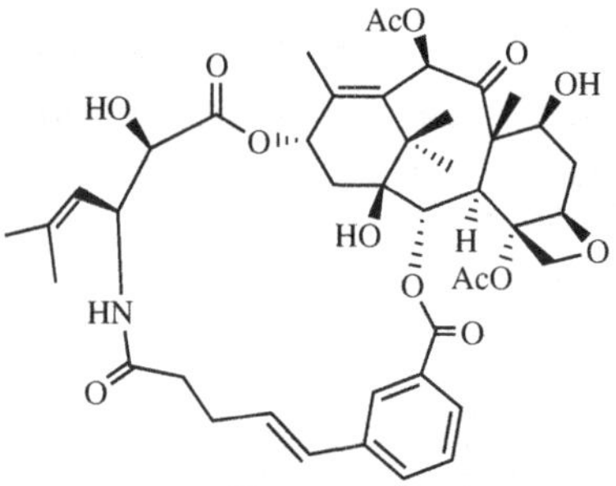

**99c**

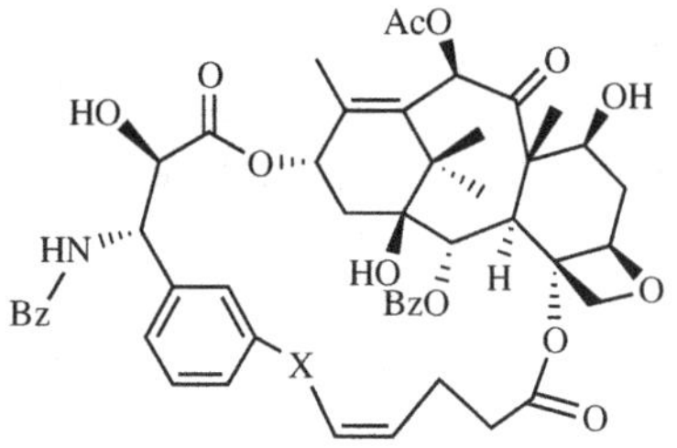

**100a** X=CH$_2$
**100b** X=OCH$_2$

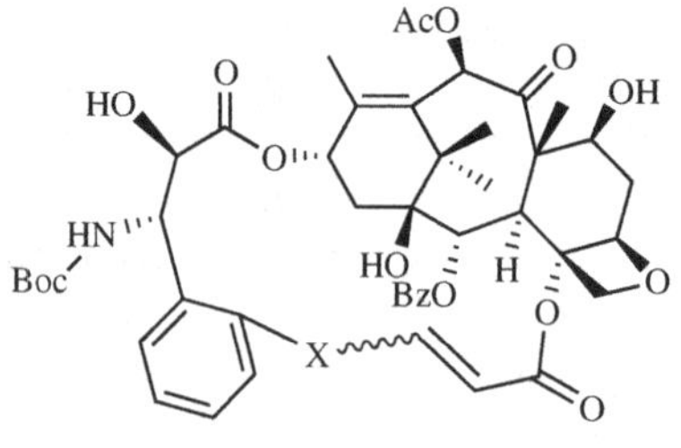

**101a** X=OCH$_2$, trans
**101b** X=CH$_2$, cis

### 3.2.7  Miscellaneous

The conjugate of two molecules with different functions or mechanisms of action often displayed dual functions or enhanced activities. Based on this rationale, Shi et al. prepared five taxoid-epipodophyllotoxin dimers. All dimers showed cytotoxicity but were less active than paclitaxel and cephalomannine in most cases and better than etoposide. Some dimers showed a little enhancement in cytotoxicity against drug-resistant tumor cells, compared with both precursors. In topoisomerase assays, two paclitaxel conjugates are active topo II inhibitor in vitro and intracellular poisons.[152] Shi et al. also prepared conjugates of paclitaxel and camptothecin through amino acid and imine linkage, and they found that the conjugates' activities were distinct from either the two drugs alone or a simple 1:1 mixture of the two drugs. Whereas most conjugates were usually less active, **102a–c** were more active than paclitaxel against HCT-8.[153]

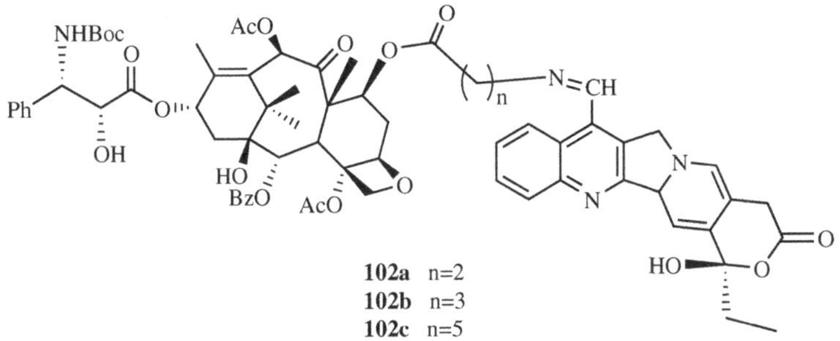

**102a** n=2
**102b** n=3
**102c** n=5

Sometimes natural taxanes or synthetic taxoids exhibited unexpected cytotoxicities, which may provide new clues for SAR. Because *m*-azido baccatin III **103** was found to be almost as comparably active as paclitaxel, 7-deoxy-9,10-*O*-acetylbaccatin **104** was prepared from taxinine **105**, a plain and abundant inactive taxane from *T. cuspidata*. Both **104a** and **104b** were found to be inactive (IC$_{50}$ at $10^{-5}$M level).[154]

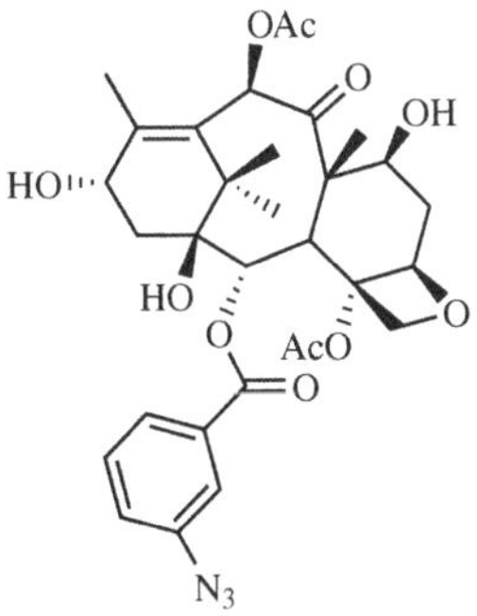

**103**

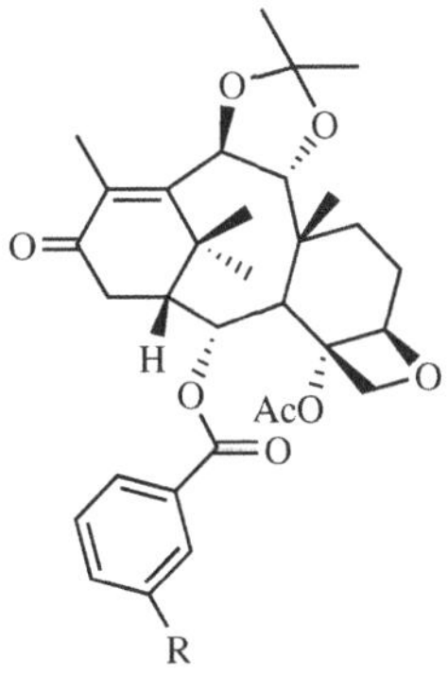

**104a** R=N$_3$
**104b** R=Cl

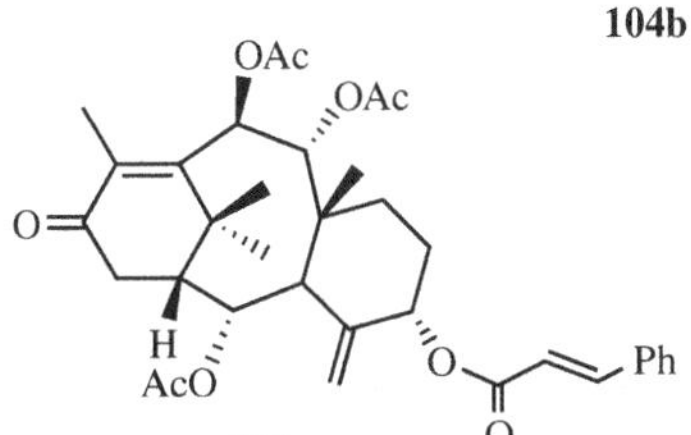

**105**

Wu and Zamir tried to explain similar tubulin binding activity of Taxuspine D (**106**) with that of paclitaxel by molecular simulation. Based on their rationale that the C-5 cinnamoyl in Taxuspine D mimics the C-13 side chain in paclitaxel,[155] they reasoned that some Taxuspine D derivatives with paclitaxel's side chain at C-5 are worthy of investigation. However, neither of the derivatives **107a–b** with C-5 or C-20 phenylisoserine side chains displayed tubulin binding ability even at 10 μM.[156]

Taxoid **108** was prepared from a 2(3→20)*abeo*taxane compound deaminoacyltaxine A after incoporation of 2-benzoate and paclitaxel side chain at C-13. It was more active than the parent compound, but it was still much less potent than paclitaxel.[157]

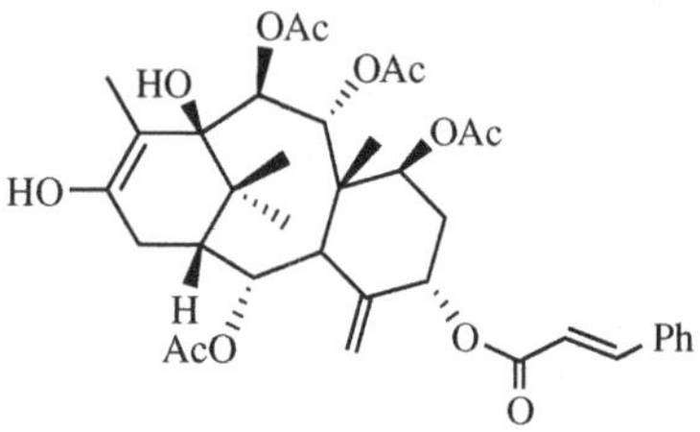

**106**

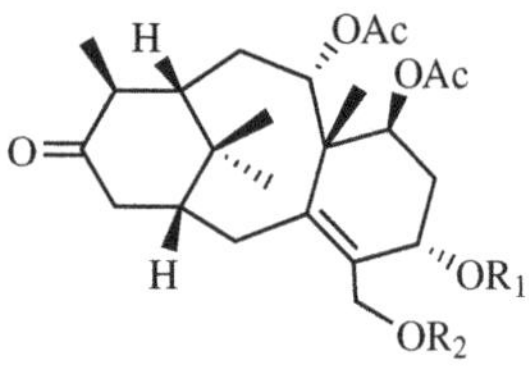

**107a**  R₁=side chain, R₂=Ac
**107b**  R₁=H, R₂=side chain

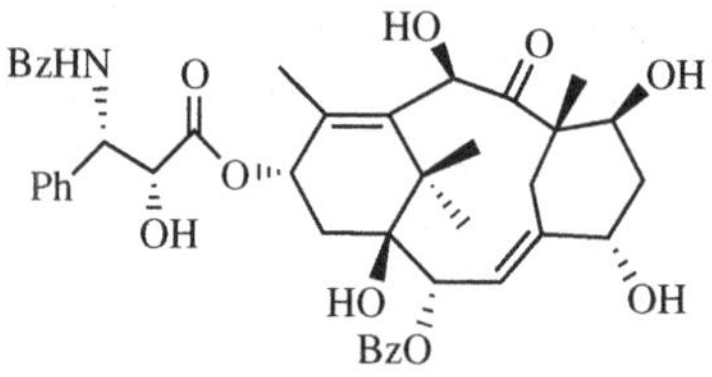

**108**

The 9-keto group in 10-DAB was converted into thiosemicarbazone and then complexed with copper ion. The complex was almost as active as 10-DAB below 12.5 μM. Beyond that concentration, the cytotoxicity of the complex increased dramatically.[158] Although the complex is still much less active than paclitaxel, it may provide an alternative way to enhance the cytotoxicity of inactive taxoids. However, the complex is probably involved in a different cytotoxic mechnism from 10-DAB.

## 3.3   EXPLORATION ON MECHANISM OF PACLITAXEL RELATED TO TUBULIN BINDING AND QUEST FOR ITS PHARMACOPHORE

### 3.3.1   Biochemical Mechanism of Paclitaxel Related to Tubulin Binding

It has been believed that the anticancer mechanism of paclitaxel promotes tubulin polymerization and stabilizes the polymer since the pioneering work done by

Horwitz's group two decades ago. Recent studies stressed the importance of the dynamics for tubulin assembly. It is even proposed that paclitaxel exerts its effect by affecting the dynamic of microtubules rather than its mass.[159]

Horwitz's group reported two photoaffinity labeling experiment results before the crystal structure of β-tubulin dimer was solved by Nogales et al.[160] Employing [³H]-labeled 3′-(p-azidobenzamido)paclitaxel and 2-(m-azidobenzoyl)paclitaxel, they found that the amino acid residues 1–31 and 217–233 were photolabeled, but that the precise position of photoincorporation was not determined. Recently, Rao et al. used another photoreactive probe, [³H]-7-(p-benzoyl)dihydrocinnamate, to explore the binding site of paclitaxel on β-tubulin.[161] Residues 277–293 were attributed to the photolabeling domain, and Arg282 was found to directly cross-link to the probe molecule. Their photoaffinity results are compatible with the electronic crystallographic structure of β-tubulin. A ligand-tubulin model proposed by Li et al.[162] was consistent with photolabeling results but inconsistent to that proposed by Nogales et al.[160] on the basis of electronic spectroscopy.

Chatterjee et al. compared the dynamic properties of paclitaxel and one of its "inactive" analogs, baccatin III,[163] and concluded that they behave similarly in their interactions with tubulin. These results supported the hypothesis that baccatin, the core structure of paclitaxel, is responsible for most of its interaction with tubulin at the binding site. However, one should be aware that the interaction cannot be translated into the cytotoxicity of taxoids directly. Andreu and Barasoain also noted the importance of the baccatin III core in the binding process.[164] They estimated that the C-2 and C-4 substitutions on the core structure account for about 75% of free energy change during the taxol binding process. Without the assistance of the C-13 side chain, the binding of the core structure is sufficient to initiate those pharmacological events induced by paclitaxel. During an attempt to create a common pharmacophore for paclitaxel and epothilones, He et al. proposed that C-2 benzoate is placed in the pocket formed by His-227 and Asp-224 and that the C-13 side chain and C-2 benzoate act as "anchors" for the binding of the taxane ring to tubulin.[165]

Bane's group also explored the binding of paclitaxel to its receptor quantitatively by employing a fluorescent paclitaxel analog.[166] They proposed that there are two types of binding sites, each as a single site on microtubules assembled from a different nucleotide-tubulin complex. Before GTP hydrolysis, paclitaxel-tubulin binding has a high affinity with a dissociation constant at the nM level. The affinity decreased sharply to a μM level after GTP hydrolysis to GDP. Diaz et al. probed the binding site of paclitaxel on microtubules with two of its fluorescent derivatives Flutax-1 and -2.[167] They found that paclitaxel binds rapidly, a fact that is difficult to explain with the current model. So they suggested a rotated or structure-modified microtubule model, in which the binding site is located between protofilaments and easily accessed from the surface of the microtubule, in contrast to the Nogales et al. model,[168] in which the location facing the microtubule lumen was proposed. However, Lillo et al. proposed[169] another model for taxoid binding to the β-unit of tubulin after completing a picosecond laser study with two C-7 fluorescein conjugated paclitaxels. In that model, the paclitaxel binding site is located at the inner wall of the microtubule, and C-7 is close to a positively charged peptide segment of M-loop in the β-unit of the microtubule.

In 2001, several articles revealed a microtubule structure with improved resolution.[170,171] In the refined tubulin structure, the binding pocket of paclitaxel was modified slightly.[170] These structures will provide a good starting point in the construction of the pharmacophores of paclitaxel and other antitubulin drugs.

Although many studies concentrated on β-tubulin, the role of α-tubulin in the binding process was still scarcely known. In a recent report, the authors found that the assembly of different α-tubulin isoforms differs greatly in the presence of paclitaxel, and thus they proposed at least partial involvement of α-tubulin in the binding process.[172]

### 3.3.2  Identification of Bioactive Conformations and Quest for a Pharmacophore for Paclitaxel

What conformation paclitaxel adopts when it binds to its receptor tubulin is an important question to be answered. There are many efforts on the construction of common pharmacophore for several other antitubulin natural products sharing the same binding site with paclitaxel.

In a hypothetical common pharmacophore for a nonaromatic analog of paclitaxel and other antimicrotubule agents, it was proposed that the baccatin core structure is not essential to activity but it acts as a scaffold for the substituents.[173] A more recent report noted the importance of the baccatin structure and the usefulness of C-13 and C-2 side chains in enhancing the binding of taxoid to its receptor.[163]

Because of the rigid structure of the tetracyclic core of taxane with less change during binding, people focused on the side chain of paclitaxel, especially C-13 isoserine and substitutions at other sites. In 1993, the first hypothesis on "active" conformation, "hydrophobic collapse," was established on the basis of NOESY data of paclitaxel in DMSO-$d_6$/$D_2O$ solution. It was proposed that 3′-Ph, phenyl rings of 3′-NH and 2-benzoate as well as 4-OAc were close to each other in the hydrophilic environment. Major differences in paclitaxel conformations in nonpolar and polar solvents were found, and those conformations were assigned as "nonpolar" and "polar" as two representative groups, respectively. Despite the difference, both "nonpolar" and "polar" conformations showed to some extent the "hydrophobic collapse" property. In 2001, a noncollapsed "T-shaped" conformation on the basis of molecular simulation was proposed as the binding conformation of paclitaxel to tubulin.[174]

The "nonpolar" conformation, also termed the "extended" conformation, was established on the basis of NMR data of paclitaxel and docetaxel in nonpolar solvent such as chloroform, as well as crystallographic data of docetaxel. Its presence was experimentally confirmed by fluorescence and solid-state NMR spectroscopies (REDOR).[163] Wang et al. selected 20 amino acid residues and paclitaxel conformation in CDCl$_3$ as a starting point to construct a "mini-receptor" model[57] for both paclitaxel and epothilone, a family of macrocyclic antimitotic agents that was assumed to bind to the same site on tubulin as paclitaxel. This model has been applied to predict binding of some D-seco analogs. Unfortunately, two D-seco analogs with a saturated C ring, which is predicted to be similar to paclitaxel in the free energy of drug binding, were inactive in bioassay.[142] This result prompted the authors to improve their model. They changed the amino acid H-bonding to the oxetane ring from Arg to Thr, and they incoparted more amino acid residues close to the oxetane ring so that the inactivity of those D-seco analogs can be explained.[142]

The "polar" conformation, also called the "hydrophobic collapse" conformation since it was named in 1993, was proposed on the basis of NMR data of taxoids in polar solvent and the x-ray structure of paclitaxel. Recently, Ojima et al. proposed a common pharmcophore for paclitaxel and several other antimotic natural products on the basis of NMR data recorded in DMSO-$d_6$/$D_2O$ of a macrocyclic pacliaxel analog, nonataxel, and molecular modeling results.[144] Later on, supportive evidences were collected in photoaffinity labeling experiments,[161] and fluorescence spectroscopy/REDOR NMR by using C3'-N-(p-aminobenzoyl)pacli- taxel as a fluorescence probe and $^{13}C$, $^{15}N$-radiolabeled 2-debenzoyl-2-(m-F-ben- zoyl)- paclitaxel.[174] In the latter report,[174] the distance between C-3'N carbonyl carbon and fluorine at C-2 benzoate was determined to be 9.8 Å and that between C-3' methane and C-2 fluorine to be 10.3 Å, in agreement with "hydrophobic collapse" conformation. Ojima et al. have suggested that two major "collapsed" conformations of paclitaxel, one with a H2'–C2'–C3'–H3' dihedral angle of 180° (the characteristics for the "polar" conformation mentioned earlier) and another with the angle of 124° (which is believed to be the third "active" conformation of paclitaxel at that time), are in equilibrium in the aqueous environment using "fluorine-probe approach."[175] They explained the bioactivity of a series of A-seco analogs[55] and fluorine-substituted analogs[25] in compliance with the "polar" conformation hypothesis. However, "hydrophobic collapse" conformation was questioned as to whether it was an "active" conformation because some macrocyc- lic tethered paclitaxel analogs mimicking "hydrophobic collapse" were found inactive.[146]

Snyder et al. proposed that the NMR reflect probably the dynamic averages of large sets of conformers rather than one or two major conformers. Ojima et al. have recognized the dynamic equilibrium behavior of paclitaxel conformers, but they have not found the "T-shaped" conformer subsets. After reanalyzing paclitaxel ROESY data published earlier with NMR analysis of molecular flexibility in solu- tion (NAMFIS) techniques, Snyder et al. identified eight energy optimized confor- mers, among which four represented 33% of the whole conformer mixture belonging to neither "nonpolar" nor "polar" conformations but what they called the "open" conformer subfamily.[176] Other studies led to the discovery of the unique "T-shaped" ("open") conformation of paclitaxel, which does not resemble either of the above-mentioned conformations.[174] In this model constructed on the electronic crystallographic data of β-tubulin and subsequent molecular simulation, C-2 benzoate and C-3' of paclitaxel cannot collapse because of the prevention of His-229 of the receptor protein. NAMFIS revealed that all three major groups of conformers (nonpolar, polar, and T-shaped) existed in the solution for a group

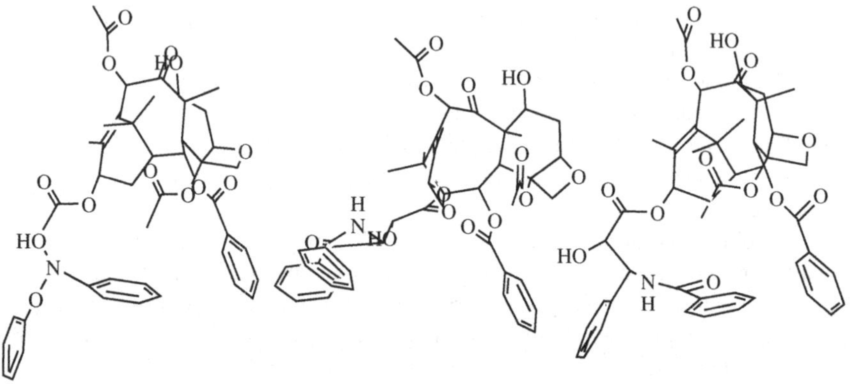

| T-conformation | "polar" conformation | "nonpolar" conformation |

**Figure 3-1.** Active conformations proposed for paclitaxel.

of C2′–C3′–Ph tethered analogs of paclitaxel (Figure 3-1).[45] In fact, extended conformations predominate in the mixture of conformers, and three T-shaped conformers constitute 59% of the extended conformers. In another picosecond fluorescence spectroscopy experiment, the data supported the binding of paclitaxel in "T-shaped" conformation to tubulin.[169] In a conformation study conducted in 2003[149] for C3′-N–C2 linked macrocyclic analogs, a conformer situated between "nonpolar" and "T-shaped" forms was identified as the bioactive conformation.

It should be noted that in 2004, Ganesh et al.[151] and in 2000, Snyder et al.[176] claimed in their articles that only the T-conformation was the conformation to be adopted by paclitaxel when binding to tubulin, whereas collapsed "polar" and "nonpolar" conformations do not work. Their conclusion was supported by computer simulation and NMR experiments of semisynthetic taxoids, in which the C-4 alkyl terminal and the C3′-Ph *ortho*-position were linked.[151]

Some reports focused on conformation of the C-13 phenylisoserine side chain of paclitaxel.[177] From the point of view of setting up an "active" conformation model for paclitaxel, a drug–receptor complex rather than a part of the drug, for example, the C-13 side chain, will be more informative and meaningful.

The oxetane D-ring in paclitaxel also attracts a lot of attention. Previous SAR studies have suggested it plays a critical role in binding, either through taxane skeleton rigidification or a weak hydrogen bonding acceptor. But its essence in binding is not acknowledged in some reports.[140,178] Wang et al. tried to reveal the role of the D-ring in paclitaxel in 2000, showing that the binding energies of some D-seco analogs of paclitaxel are comparable with that of paclitaxel. They predicted that some analogs without an intact oxetane D ring can still bind to tubulin very well.[178] Barboni et al.[142] and Boge et al.[179] found that conformational changes are relayed from ring C to A in the D-seco analogs. Another consequence of the SAR study of D-seco analogs[142] is that the second generation of the paclitaxel–epothilone minireceptor[57,178] was revised because of its inconsistency with experimental results.

## 3.4 NATURAL AND SEMISYNTHETIC TAXOIDS OVERCOMING MULTIDRUG RESISTANCE (MDR)

MDR is one major reason for the failure of chemotherapy.[180] Overexpression of P-glycoprotein (P-gP), which results in massive transport of anticancer drugs and other hydrophobic substrates out of cells, is the best known contributing factor to MDR. Also, other factors attributing to the failure of taxane anticancer drugs, such as inherently insensitive isotypes of tubulin and amino acid mutations in tubulin, will also be discussed in this section.

### 3.4.1 Structure-Modified Taxoids With Better Activity Toward MDR Tumors

Although paclitaxel was reported to be effective against ovarian and breast cancers resistant to other first-line anticancer drugs in clinics, the patients often relapsed and did not respond to these antitumor agents, including paclitaxel, anymore. Unfortunately, docetaxel was also inactive toward paclitaxel-resistant tumors.

In recent years, during research and development of new antitumor taxoids, scientists have begun to shift their attention from begun drug-sensitive tumors to drug-resistant tumors, especially paclitaxel-resistant tumors. It was found that subtle changes in structure led to new taxoids with much more potent activities against MDR tumors. It is worth noting that a series of taxoids with C-2, C-3', and C-10 modifications prepared by Ojima's group can serve this purpose. It has been reasoned that these taxoids are not good P-gP substrates and thus exhibited potent activities against drug-resistant tumors.

In 1996, Ojima et al. reported that the introduction of carbonate and carbamate to C-10 and the simultaneous replacement of 3'-phenyl with an alkenyl or alkyl group provide the taxoids that exhibit one to two orders of magnitude higher potency against drug-resistant cancer cells.[22] Among them, SB-T-1213 (**109a**) was selected for additional research efforts because of highest potencies in this series. Other paclitaxel analogs, IDN5109 (**113**) and ABT-271 (**46**), have shown their efficacies against MDR tumors and have been subject to clinical evaluation as well.

Subsequently, new taxoids bearing 3'-cyclopropane and 3'-epoxide moieties were synthesized. The R/S ratio (ratio of $IC_{50}$ in a drug-resistant cell to that in a sensitive cell) was 2.48 for the 3'-cyclopropane/10-PrCO compound (**8a**).[23] Later on, the same group discovered more potent analogs with modifications on the C-2 as well as the C-3' and C-10 positions; three of them (**109a–c**) showed the best R/S ratios at 0.89–1.3 in LCC6 (breast) and 0.92–1.2 in MCF-7 (breast) cell lines, whereas for paclitaxel and docetaxel, 112 and 130 in LCC6 and 300 and 235 in MCF-7.[23] C-3'-difluoromethyl docetaxel analogs were found to be one to two orders of magnitude more potent in MDR LCC6 cell lines.[27] Those C-3'-$CF_2H$ taxoids prepared from 14β-hydroxyl-10-DAB also exhibited comparable activity in both normal and MDR cell lines, but they were generally less active than their counterparts from 10-DAB. Several paclitaxel analogs bearing C-3'-(p-F)-substitution in combination with 2-m-F, 2-difluoro, and 2-m-$CF_3$-p-F benzoates were found to be less active in most cases, whereas C-3'-CF$_3$ docetaxel analogs were more potent in either normal or MDR tumor cell lines.[25] C-3'-thiocarbamate paclitaxel analogs exhibited superior activitiy to that of the parent compound in HCT-116 drug-resistant tumors.[34] Conjugated double and triple bonds in C3'-*N* analogs make positive contributions to the activity against MDR tumor cells, as depicted in Section 3.2.1.3.[32]

109a R$_1$=CH$_2$CH(CH$_3$)$_2$, R$_2$=OMe
109b R$_1$=CH=C(CH$_3$)$_2$, R$_2$=OMe
109c R$_1$=CH=C(CH$_3$)$_2$, R$_2$=N$_3$

110

Battaglia et al. prepared a series of 2'-(*S*)-Me of paclitaxel analogs, and most compounds are more active toward drug-resistant A2780 human lung carcinoma and drug-resistant MCF-7 human breast carcinoma.[20] For the best compound **6c** among them, it exerted better antitumor effects on drug-resistant tumors than paclitaxel, docetaxel, and IDN5109 (**113**).

A Korean group, Roh et al., found that some C3'-*N*-debenzoyl *N*-acyl analogs, without any change of substitutions at other sites, can exhibit high potencies toward MDR tumors. Even if less active against sensitive tumor cells, all *N*-acyl analogs bearing conjugated bonds tested in bioassays were as much as seven times more active than paclitaxel against MDR tumor cells.[32]

In a library with C-10-modified paclitaxel analogs with various acyl groups, some of them, including short-chain aliphatic, most alicyclic, and some nitrogen-containing aromatic and heteroaromatic analogs, were more effective toward drug-resistant MCF-7R breast cancer cell lines.[40] Two C-10 spiro epoxides were prepared and biologically evaluated.[67] Taxoids **42b** is more active against the MDR tumor cell line HCT-VM46 than is paclitaxel, and **42a** is almost one order of magnitude less active.

In 2002, it was found that a 2-difluorobenzoyl paclitaxel analog (**60d**) exhibited comparable activity with paclitaxel, are best in C-2 mono- and di-substituted benzoyl analogs and better than **9a** and **9b** in paclitaxel-resistant HCT-116/VM46 cancer cell lines.[77,78] It was also reported that some 2-debenzoyloxy-2α-benzamido docetaxel analogs were comparably cytotoxic with paclitaxel toward some drug-resistant tumor cell lines.[99]

A group from Daiichi Co. reported many 9β-dihydro-paclitaxel and docetaxel analogs, including **47**,[77] **49**,[78] and **50a–b**,[81] which exhibited significant antitumor effects in vitro and in vivo against MDR tumors. Some of them were also highly water soluble as well as orally active.

Distefano et al. reported the activity of 7,9-pyrazoline (general formula **111**) and C-seco (general formula **112**) analogs. The pyrazoline analogs of docetaxel were better than paclitaxel but less active than docetaxel against adariamycin-resistant MCF-7 cells.[181] C-seco analogs were less potent than pyrazoline analogs. Because these pyrazoline analogs could arrest a cell cycle at $G_2/M$ phase and DNA fragmentation, their activities are probably related to apoptosis.

An interesting finding on taxoid-related MDR is that the inactivity of orally administered paclitaxel originated from overexpressing P-gP in the gastrointestinal tract. Combined use of paclitaxel and a P-gP inhibitor will improve bioavailability to a great extent. Taxoid **113** was found to be a poor substrate of P-gP, which thus showed good oral bioavailability (48% p.o./i.v.) and significant efficacy in clinical trial.[58] 10-Deoxy-10-C-morpholinoethyl deocetaxel analogs are also orally active taxoids, and 10-(C-morpholinoethyl)-7-MeO docetaxel (**114**) is the best in this series.[80]

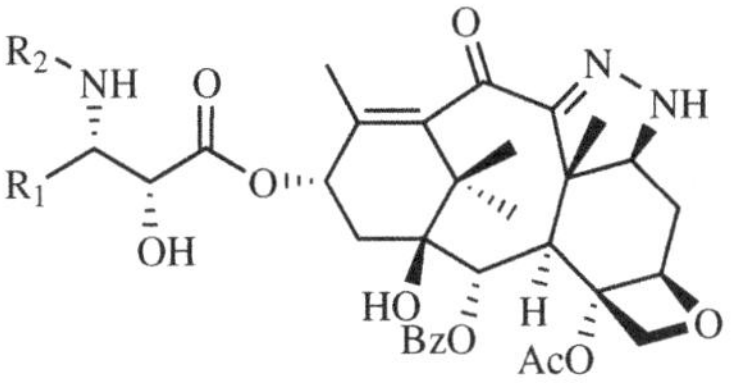

111

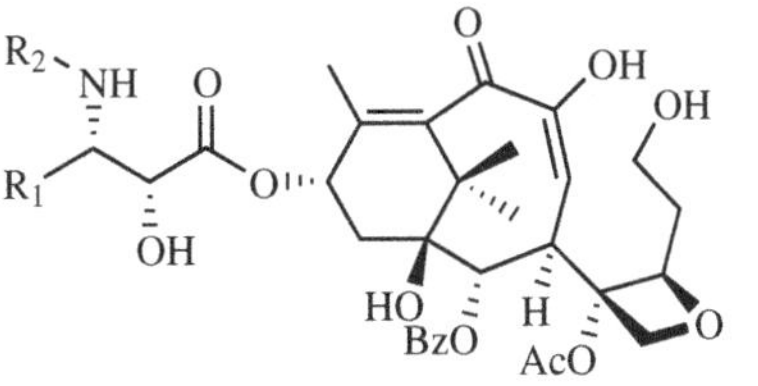

112

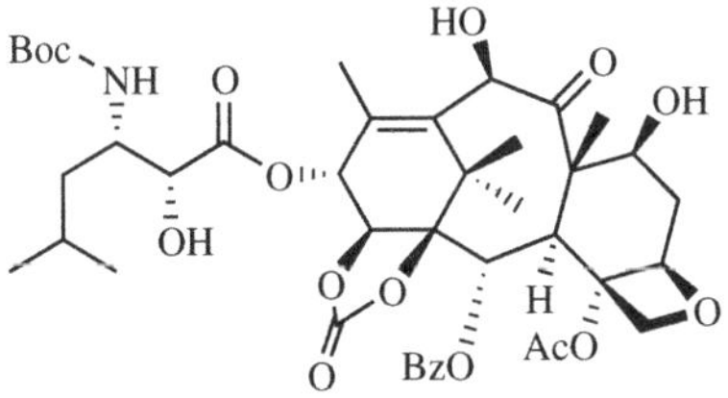

113

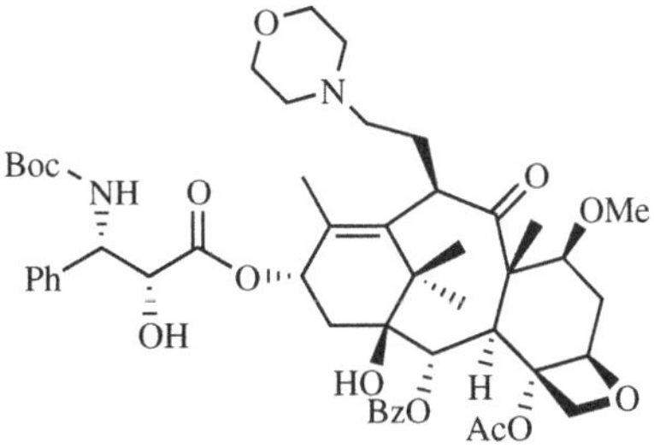

114

DJ-927 (**49**), a 9-dihydro-7-deoxy docetaxel analog under its phase I clinical trial, was effective against various tumors, especially P-gP overexpressing MDR tumors in vivo. Additional investigation showed that it is not a P-gP substrate and its cytotoxicity is not influenced by P-gP modulators. Although the authors proposed that the effectiveness of DJ-927 may be partly from higher intracellular accumulation, its mechanism against MDR tumors should be addressed in the future.[182]

Some macrocyclic taxoids such as **99b** and **99c** were among the most potent inhibitors for drug-resistant cancer cells.[147]

Dimers of paclitaxel or docetaxel with 2-deacetoxytaxinine J were designed and prepared for their dual role in cytotoxicity as well as MDR reversal activity, because 2-deacetoxytaxinine J exhibited strong MDR reversal activity. However, biological evaluation results are disappointing.[183]

Interestingly, some conjugates of paclitaxel and camptothecin through amino acid and imine linkage exhibited better R/S ratios against drug-resistant tumor cells induced by the two drugs, KB-CPT and 1A9-PTX10.[153]

### 3.4.2 Nonpaclitaxel-Type Taxoids With MDR Reversal Activities

Some naturally occurring or semisynthetic non-Taxol taxoids can restore MDR tumor cells sensitivity toward paclitaxel and other anticancer drugs. These taxoids are usually weakly cytotoxic; thus, they are ideal candidates for combined use with cytotoxic agents.

Ojima et al. prepared 23 taxoids with hydrophobic side chains at different positions of 10-DAB.[184] Taxoids with mono-hydrophobic ester substitution could be grouped into two categories. One group including taxoids with C-7 and C-10 modifications showed strong reversal activity (>95%) in most cases at the concentration of 1~3 μM, and another group with C-13 and C-2 modifications exhibited less or no activity. The effects of introducing two or three hydrophobic groups seemed to be complicated. Baccatin III-7-(*trans*-1-naphthanyl-acrylic acid) ester (**115**) is the best among those semisynthetic taxoids, which does not increase paclitaxel accumulation in sensitive tumor cell MCF-7, but it drastically increases the paclitaxel accumulation in drug-resistant cell MCF-7-R with overexpression of P-gP.

Kobayashi et al. first reported the effects of nonpaclitaxel-type taxoids from *Taxus cuspidata* on vincristine (VCR) accumulation in the adriamycin-resistant human leukemia K562/ADM cell.[185] Seven taxoids belonging to different subtypes are as potent as Verapamil for MDR reversal, and some of them can competitively bind to P-gP. Among them, taxinine (**105**) and taxuspine C (**117**) were chosen for additional studies.[186–189] For taxinine, hydrophobic groups attached to C-2, C-5, and C-13 enhance MDR reversal activity, and to C-9 and C-10 reduce the activity.[187] A hydrogenated product (**116**) of taxinine retaining 4(20)-exomethylene is the most potent MDR-reversal taxoid reported to date.[181] Deacylations at C-2, 5, 9, and 10 in taxuspine C result in a drastic reduction of activity.[189] Two taxoids 2-deacetyl taxinine and 1-hydorxy Taxuspine C were isolated from *T. cuspidata*, both of which showed better activity than Verapamil in vitro.[190] A common observation for taxinine and taxuspine C derivatives is that a phenyl containing hydrophobic group attached to C-5 apparently increases the accumulation of paclitaxel in MDR. However, comparison of these results with those obtained from baccatin III derivatives is difficult because of reversed C-5 configuration.

In 2002, Kobaysahi et al. reviewed their work on taxoids, including MDR reversal and other biological activities of these taxoids.[191] Interested readers can refer to it for a comprehensive description.

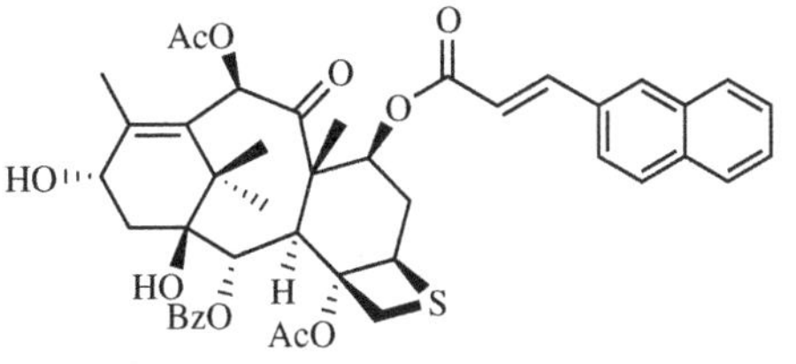

**115**

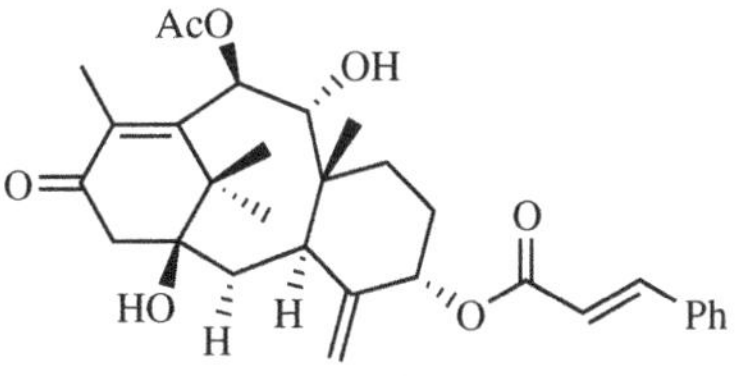

**116**

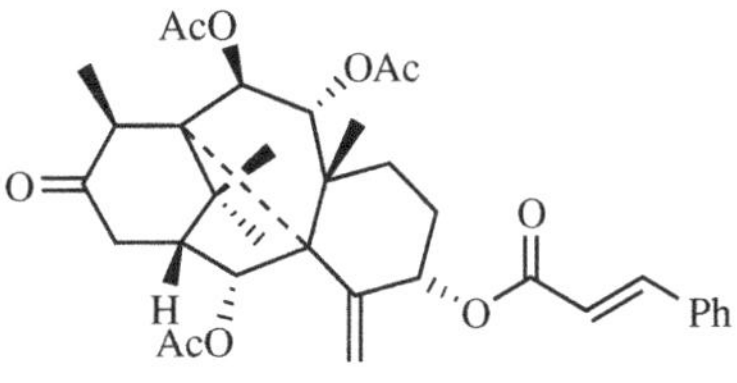

**117**

In addition, some tricyclic C-aromatic taxoid intermediates also exhibited MDR reversal activity, one of which is comparable with Verapamil. Incorporation of the Taxol side chain resulted in the reduction of the activity.[192,193]

Some nontaxoid MDR reversal compounds may share common structure features with the above-mentioned MDR reversal taxoids. For example, Chibale et al. attached hydrophobic moities to the antimalarial drugs chloroquine and primaquine, and they found that chloroquine derivatives are superior to primaquine derivatives against MDR in vitro and in vivo when coinjected with paclitaxel. Unexpectedly, they observed that those chloroquines fit very well to two baccatin III-based MDR reversal compounds in silico.[194]

### 3.4.3 Factors Contributing to the Resistance to Paclitaxel

Structural changes in P-gPs can affect their response to their substrates (anticancer agents) and modulators (MDR reversal agents). Groul et al. isolated six mutants of mdr1b in mice, equivalent to MDR1 (a subtype of P-gP) in humans, with the treatment of **115** in combination with colchine as the selection pressure. Five of the six mutants can reduce the MDR reversal efficacy of **115**, and the resistance to paclitaxel, which demonstrates the possibility of **115** as the competitive inhibitor of P-gP to paclitaxel. In the second round of selection, a double mutant enables a complete loss of resistance to paclitaxel and a five-fold reduction of efficacy of **115**.[195] Most of these mutants located within the tenth and twelfth transmemberane spanning (TMS) segments of the second half of the protein, which is consistent with the previous results obtained with photolabeling paclitaxel analogs. Wu et al.[196] used tritium-labeled benzoyldihydrocinnamoyl analogs (BzDC) of paclitaxel at the C-7 and C-3' positions to photoincorporate into the segments of MDR1b P-gP. They demonstrated 7-BzDC incorporated into the twelfth and 3'-BzDC into the seventh to eighth TMS region. Because the tertiary structure of MDR1 or mdr1b is still unknown, the above results may help to identify the binding domain of hydrophobic molecules on the P-gP, which thus leads to the rationale of the MDR mechanism and the structure-based design of new P-gP inhibitors.

Besides P-gP overexpression, alteration of tubulin isotypes, amino acid mutations in tubulin, and microtubule dynamic changes are also involved in resistance to paclitaxel and other anti-microtubule agents. Here we concentrate on some recent results. Comprehensive descriptions can be found in reviews by Burkhart et al.[197] and Sangrajrang et al.[198]

Drug-induced alteration of tubulin isotype expression in resistant cells is considered to be a general mechanism of antimitotic drugs resistance.[198] Research on the distribution of different tubulin subtypes showed that in MDR tumor cells, β-II and β-IVa isotype tubulins were absent, whereas the level of β-III tubulin increased.[199] Direct evidence for the involvement of such alteration is that antisense oligonucleotides to class III β-tubulin isotype did enhance sentivity about 30% to paclitaxel in resistant lung cancer cells.[200] These results indicated that β-III tubulin isotype is a biomarker of resistance. Other cellular factors binding to tubulin, e.g., microtubule associating protein 4 (MAP4),[201] rather than tubulin itself, are also responsible for the resistance.

Amino acid mutations also make contributions. Gonzalez-Garay et al. identified a cluster of mutations in class I β-tubulin isotype.[202] All six mutants had substitutions at leucines, such as Leu215, Leu217, and Leu228, which may lead to destablization of a microtubule. This finding also claimed the importance of the leucine cluster in microtubule assembly. Giannakakou et al. found some β-tubulin mutations in paclitaxel and epothilone-resistant tumor cells. These mutations are Phe270 to Val, Ala364 to Thr, Thr274 to Ile, and Arg282 to Gln. The first two mutations cause paclitaxel resistance, whereas the latter two cause resistance to both paclitaxel and epothilone.[203] In a recent report, two mutants were isolated from tumor cells resistant to both paclitaxel and desoxyepothilone B. The mutant of Ala231 to Thr locates at helix-7 within the binding pocket of paclitaxel derived from crystallographic data, whereas Gln292 to Glu in helix-9 near the M loop, outside of the taxol binding site.[201] Recently, a clinical report demenostrated the correlation between tubulin mutant and paclitaxel resistance in non-small-cell lung cancer patients. The group of patients without mutations had longer median survival time and higher 1-, 3-, and 5-year survival rates.[204] Also, α-tubulin mutations are widely distributed in paclitaxel-resistant cells, and the evidence for their roles to confer resistance is still indirect. More studies are needed to figure out whether α- and β-tubulin are involved in resistance in a synergic way.

Microtubule dynamics alterations also showed impacts on resistance. Goncalves et al. found a 57% increase of microtubule instability in paclitaxel-resistant cell A549-T12 as compared with parental sensitive cell A549, and a 167% overall increase in the more resistant cell A549-T24.[205] It is interesting to note that the resistant cells, in the absence of paclitaxel, suffered mitotic block as well, which suggests that both increased or suppressed microtubule dynamics can impair cell function and proliferation.

## 3.5  DESIGN, SYNTHESIS, AND PHARMACOLOGICAL ACTIVITY OF PRODRUGS OF PACLITAXEL

Currently the paclitaxel formulation contains a surfactant, Cremophor EL, to improve the poor water solubility of the drug. Some adverse effects including hypersensitivity have been attributed to Cremophor. Other severe adverse effects, such as neutropenia and dose-dependent neurotoxicity, also occur at a high dosage of paclitaxel administration. Improved water solubility may lower the dosage of paclitaxel because of effective transportation of the drug to the active sites, which thus reduces high dosage-related toxicity. Several alternative formulations, such as emulsions[206] and liposomes[207,208] have been developed to improve efficacy and minimize the toxicity of paclitaxel. Another approach, design and preparation of water-soluble prodrugs will be discussed here. Also, the "smart" prodrugs aimed at specific sites or kinds of tumors have emerged recently. Some prodrugs with both improved water solubility and enhanced effectiveness and specificity were also documented.

### 3.5.1  Prodrugs Prepared to Improve Water Solubility

Most water-soluble prodrugs were realized by the derivatization of 2'-OH and 7-OH positions, the two most liable positions in paclitaxel. Some prodrugs are as active as the parental drug both in vitro and in vivo.

α-amino acids have been applied to the preparation of water-soluble taxanes as early as the late 1980s. A series of amino acids was conjugated to 2'-OH through a glutaryl linker, and the asparagine- and glutamine-glutarylpaclitaxels improved the water solubility as much as three orders of magnitude.[209] These two derivatives as well as serine- and glycine-derivatives showed strong cytotoxicity against several sensitive cancer cell lines, and no activity against paclitaxel-resistant cells. Poly(L-glutamic acid)-paclitaxel (PG-TXL),[210,211] a derivative about five orders of magnitude more soluble than paclitaxel, was active against several kinds of tumors in vivo, including those not responsive to both paclitaxel and a combination of paclitaxel and polyglutamic acid.

Hydroxy acid esters were also applied to the synthesis of prodrugs. Damen et al.[212] prepared 2'- and 7-malic esters of paclitaxel and found that 2'-ester behaved as the prodrug, whereas 7-ester and 2', 7-diesters did not. The 2'-malyl paclitaxel and its sodium salt are more water soluble than paclitaxel by 20 and 60 times, and both are stable at pH 7.4 PBS buffer for 48 hours at 37 °C. The prodrug, because of its two-fold higher maximum tolerance dose (MTD) than its parent drug, exhibited more significant antitumor activity at a higher dosage. Niethammer et al. reported a 7-glycerolyl carbonate of paclitaxel as a prodrug with improved antitumor activity and hydrophilicity.[213] The prodrug, with 50-fold higher solubility in water, possessed 2.5-fold higher MTD, reduced toxicity to stem cells by 100 times, and exhibited almost equal activity in vitro as compared with paclitaxel. In vivo results are also promising—tumor growth regression in all prodrug treating groups (40 mg/kg) were greater than that in paclitaxel groups (16 mg/kg). A series of polyol-carbonate prodrug of paclitaxel at 2′ and 7 postitions were synthesized, and 7-(2″, 3″-dihy-droxypropylcarbonato) paclitaxel, named protaxel, was the best in solubilty and stablity assays in human serum.[214] Protaxel is actually the same compound as the one in Ref. 213. The 7-phosphate/2′-aminoester and 2′-phosphoamidate deriva-tives of paclitaxel were also prepared to enhance the water solubility.[215]

Takahashi et al. conjugated sialic acid to 7-OH of paclitaxel through an oligo ethylene glycol linker to prepare the potential neuraminidase cleavable, water-soluble prodrug.[216] Like extremely water-soluble PEG-paclitaxel,[217,218] the conjugate is also highly soluble in water (about 28 mg/mL), which is improved by more than four orders of magnitude. To improve the pharmacokinetic properties of PEG-taxol, α-, β-, and ω-amino acids were used as linkages between PEG and the parent drug. The most cytotoxic one among them is the proline linked conjugate. Unfortunately, its antitumor efficacies are not better than palictaxel in vivo, although it possesses a higher MTD.[219] Other saccharide and PEG-linked prodrugs included conjugates of glucuronide,[220] of PEG-glycinate,[221] of PEG-HSA.[222] The conjugates with 5KD of PEG and HSA showed comparable cytotoxicity in vitro and reduced blood clear-ance and disposition in the liver and spleen.[222] These changes in pharmacokinetics may have a positive contribution to its improved in vivo antitumor activity. Suga-hara et al. prepared an amino acid linked carboxymethyldextran prodrug for pacli-taxel to enhance its solubility as well as improve its pharmacokinetics. It is observed that those conjugates releasing the highest amounts of paclitaxel were most active toward Colon 26, a paclitaxel-resistant tumor xenograft in mice.[223] The acid-sensitive linked PEG conjugates of paclitaxel were prepared, and their in vitro antitumor activity was evaluated. All three are less cytotoxic than paclitaxel, and release was less than 10% of the parent drug after 48 hours at pH 7.4.[224]

Wrasidlo et al.[225] demonstrated the impact of nature and position of different substitutents on the activity with the 2′- and 7-pyridinium and 2′-sulfonate paclitaxel derivatives. They found that the 2′-(N-methyl-pyridinium acetate) paclitaxel (118) behaved more likely as a prodrug, in comparison with 7-pyridinium and 2′-sulfo-nate derivatives, because it showed almost no cytotoxicity and tubulin binding abil-ity in the absence of plasma. However, it exhibited higher in vivo activity and reduced system toxicitiy than those of paclitaxel, whereas 7-pyridinium and 2′-sul-fonate derivatives showed little activities in nude mice although they retained strong cytotoxicities.

**118**

Scientists from BMS applied the disulfide linkage to the preparation of several prodrugs of paclitaxel, containing glucose, GSH, or captopril. Under reductive activation conditions, the captopril conjugate was most stable in vitro and exhibited the most enhanced activity (50 times). More importantly, a 60% tumor regression rate against L2987 lung carcinoma mice model was observed at a 125-mg/kg dose for the compound, whereas no activity occurred for paclitaxel at its MTD of 30 mg/kg.[226]

Contrary to the usual preparation of hydrophilic conjugates, Ali et al. prepared a series of hydrophobic α-bromoacyl prodrugs of paclitaxel with 6, 8, 10, 12, 14, and 16 carbon chains[227] on the basis of their discovery that the association of paclitaxel prodrugs with lipid bilayers was influenced by the chain length of the bromoacyl paclitaxel analogs.[228] In the absence of α-bromo substitution, the prodrugs were 50- to 250-fold less active, which implies the assistance of bromine in the hydrolysis of 2′-carbonate. Interestingly, it was found that the longer the chain, the stronger the growth inhibitory activity, probably developing "the slow hydrolysis of the prodrug followed by sustained delivery of paclitaxel to the tumor," according to the authors.

Non-prodrug water-soluble paclitaxel analogs were also reported.[79] Because their C-10 positions were covalently attached to secondary amines instead of acetate in paclitaxel, they cannot convert into paclitaxel under physiological conditions.

### 3.5.2  Prodrugs Designed for Enhancing Specificity

Antibody-directed enzyme prodrug therapy (ADEPT) is one promising strategy in prodrug design. In this strategy, the conjugate consisted of a monoclonal antibody of a tumor cell surface receptor for recognition and the prodrug of an antitumor agent subjected to specifically enzymatic cleavage to release the drug at the tumor site. Rodrigues et al.[229] reported for the first time the application of the ADEPT method to paclitaxel, using the conjugate of β-lactamase and MAb, and the cephalosporin-paclitaxel prodrug. As an improvement to this brilliant strategy, BMS scientists changed the structures of linkers and antibodies to ensure a faster release, more stable to unspecific hydrolysis and specificity of prodrug activation. The 3,3-dimethyl-4-aminobutyric acid linked conjugate demonstrated the fastest release of paclitaxel, and its specificity was also observed as expected when activated by the melanotrasferrin mAb-fused protein L-49-sFv-β-lactamase in human 3677 melanoma cells.[230] To avoid immunogenicity caused by the non-human enzyme, glucuronidase was chosen as the enzyme in ADEPT.[231] Unfortunately, although the prodrug was hydrolyzed by glucoronidase to exhibit similar cytotoxicity to paclitaxel, activation of the prodrug with enzyme-MAb was not realized after 24 hours probably because of an insufficient amount of enzyme bound to cells. Schmidt et al. prepared another ADEPT candidate in 2001.[232] They chose 2′-carbamate instead of 2′-esters, and the *para*-nitro group on the benzene ring in the linker can facilitate the attack of the phenol ion to 2′-carbamate to liberate the parental drug. However, this prodrug may also suffer from similar problems because 100 μg/mL of enzyme was needed for fast release of the parental drug.

The lower toxicity, improved efficacy and water solubility, as well as tumor specificity make the MAb-paclitaxel conjugate a promising candidate for the treatment of tumors. The MAb for the p75 tyrosine kinase low-affinity receptor has been developed previously, and one of them, anti-p75 MAb MC192, was chosen to target p75-overexpressing tumor cells. The in vivo efficacy of the conjugate is higher than free paclitaxel and coinjection of paclitaxel and MC192. Also, the "all-purpose" prodrug, which is supposed to target any kind of tumors wanted, was prepared by the conjugation of paclitaxel with an anti-immunoglobulin secondary antibody.[233]

Attachment of small peptides capable of recognizing tumor cell surface receptors to anticancer drug is also useful to enhance specificity. Safavy et al. reported on the conjugation of paclitaxel-2'-succinate, PEG linker, and a bombesin (BBN) fragment BBN [7-13], a hepta-peptide recognizing binding site on the BBN/gastri-releasing peptide (GRP) receptor. The binding ability of BBN in this highly water-soluble conjugate is comparable with the free peptide, and the cytotoxicity of the conjugate is stronger than free paclitaxel after 24 and 96 hours of administration at dosages of 15 nM and 30 nM against the human non-small-cell lung cancer NCI-H1299 cell line with the BBN/GRP receptor.[234]

Erbitux (C225), an anti-epidermal growth factor receptor (anti-EGFR) mAb, was attached to paclitaxel through suitable linkers. The conjugate using 2'-succinate linkage was more active than C225, paclitaxel, and a mixture of the two in vitro, but it exhibited similar spectra and activities to C225.[235] Ojima et al. incorporated methyldisufanyl (MDS) alkanoyl groups at C-10, 7, and 2' positions and conjugated three kinds of mAbs of human EGFR via disulfide bonds. It is promising to find that those nontoxic conjugates are highly effective and specific in vivo against EGFR overexpressing tumors.[236]

Hyaluronic acid (HA) is a linear polysaccharide and one of several glycosaminoglycan components of the extracellular matrix (ECM). Some HA receptors are overexpressed in human breast epithelial cells and other cancer cells. Luo and Prestwich prepared a series of conjugates with different adipic dihydrarzide (ADH) loading from paclitaxel-2'-succinate and ADH-modified HA. Those conjugates showed selective activity toward human ovarian SK-OV-3, colon HCT-116, and breast HBL-100 cell lines, whereas there was no activity against the untransformed murine fibroblast NIH 3T3 cell line. The conjugates with either highest or lowest loading of paclitaxel did not show the best activity.[237]

Folate-PEG-modified prodrugs of paclitaxel were recently prepared. Although the selected prodrug taxol-7-PEG-folate increased the survival in mice, it was not better than paclitaxel.[238,239]

A group from BMS prepared several cathepsin B cleavable dipeptide (Phe-Lys) conjugates through the p-aminobenzylcarbonyl linker as prodrugs for paclitaxel and mitomycin C and doxorubicin.[240] Unfortunately, the prodrug did not release the parent drug in human plasma.

De Groot et al. reported on the synthesis of tumor-associated protease cleavable prodrugs of paclitaxel.[241] The carbamate and carbonate linkers between 2'-OH of paclitaxel and tripeptides D-Ala-Phe-Lys and D-Val-Leu-Lys were used instead of ester in these prodrugs to avoid nonspecific hydrolysis of 2'-ester by widely distributed proteases or esterases in vivo. These prodrugs are nontoxic, maybe the least toxic paclitaxel prodrugs hitherto reported. They did release the parental drug upon treatment of human plasmin, although at relatively high concentrations (100 μg/mL of plasmin and 200 μM of prodrug).

Considering the reductive condition in the anaerobic environment in the center of tumor tissue, Damen et al. designed and prepared 2'-carbonate and 3'-*N*-carbamate prodrugs of paclitaxel that release the parental drug targeting hypoxic tumor tissue. Two of 11 prodrugs are selected for additional investigation.[242]

In 2004, Liu et al. made an effort to incorporate estradiol into C-2', 7, and 10 positions in Taxol to target the drug to estrogen receptor (ER) positive breast cancer. For C-2' and C-7 conjugates, no satisfactory results were obtained for either activity or selectivity. But a 7-*epi*-10-conjugated taxoid (**119**) did exhibit some selectivity between ER-positive and negative cancer cells, and ER-β (MDA-MB-231 cell) and ER-α expressing (MCF-7) cancer cells.[243]

**119**

## 3.6  OTHER BIOLOGICAL ACTIONS OF PACLITAXEL

Apoptosis is another important mechanism for cell poisoning, in addition to tubulin assembly promotion and microtubule stabilization for paclitaxel, especially at a high concentration (sub μM to μM level). It is also observed that in higher concentrations (5 to 50 μM), paclitaxel induced tumor necrosis through microtubules rather than apoptosis.[244] Note that paclitaxel concentration in the clinic is lower than μM.

The involvement of many biochemical pathways and cytokines in paclitaxel-induced apoptosis has been extensively investigated. Among them, bcl-2, the well-known apoptosis inhibitor, was found to be phosphorylated in the presence of paclitaxel, with its inhibitory effect on apoptosis downregulated. One phosphorylation position is Ser-70, and the cells with Ser70 to Ala mutant showed lower reponse to paclitaxel-induced apoptosis. Deletion of the bcl-2 loop area comprising

60 amino acids suppressed its anti-apoptotic effect completely.[245] Other investigations indicated that phosphorylation is a marker of mitotic arrest rather than a determinant of paclitaxel-induced apoptosis.[246] Interested readers can refer to several recent reviews, especially on bcl-2 related apoptosis.[247–250]

At a higher concentration, paclitaxel exhibits immunostimulating effects, e.g., lipopolysaccharide (LPS)-like and tumor necrosis factor (TNF)-like activity.[251,252] The LPS-mimetic and subsequent immune factors releasing action induced by paclitaxel were found in mouse and scarcely in human. Nakano et al. found that an A-nor-B-seco taxoid with a C-13 amide chain, SB-T-2022, a nonactive analog of paclitaxel, enhanced LPS and paclitaxel-induced nitric oxide (NO) production, which did not affect TNF production.[253] Kirikae et al. explored the SAR of a series of 3′-N-benzoyl and aroyl analogs of paclitaxel in murine macrophage (Mφ) activation and NO and TNF production. They claimed that p-substitution of benzoyl affects potencies of taxoids in Mφ activation, and p-Cl-benzoyl is even stronger than paclitaxel. Also, these analogs only showed marginal activity in LPS-induced TNF production in humans.[254] During a systematical evaluation of C-2, 7, and 10 substitutions of paclitaxel, they also found that several compounds possessing different substitutions on all three positions showed stronger Mφ activation activity, and A-nor taxoid exhibited no activity.[255] Also, none of these compounds induced TNF production in humans.

An antiproliferative property of paclitaxel has also found another clinical usage in reduction of restenosis. The marketing of paclitaxel-coated stents have been approved in Europe and will be approved in the United States in 2004. An effort has been made to prepare such a conjugate by combing paclitaxel and a nitric oxide donor.[256] C-7 nitroso paclitaxel derivative **120** as its NO donor conjugate exhibited both strong antitumor (20 nM) and antiplatelet (10 μM) activities in vitro, and anti-stenotic activity in the rabbit model, which indicates the beneficial effect of such a conjugation for the treatment of stenotic vessel disease.

120

## 3.7 NEW ANTIMICROTUBULE MOLECULES MIMICKING ACTION OF PACLITAXEL

Several natural products, such as epothilones, discodermolide, and eleutherobin, were found to have a similar mechanism of action as paclitaxel. A recent review outlined many tubulin stabilization natural products and their analogs as anticancer agents.[257] Some small synthetic molecules were designed to simplify the complex structure of paclitaxel while retaining its activity. A series of compounds with the C-13 isoserine side chain of paclitaxel attached to a borneol derivative, an intermediate of total synthesis of paclitaxel, showed good microtubule stabilizing effects. The compound **121** was chosen for cytotoxicity test, and it is far less cytotoxic than paclitaxel.[258] Another interesting example is GS-164 (**122**) with a simple structure. This compound exhibits many aspects similar to those of paclitaxel in microtubule polymerization, but with three orders of magnitude reduced activity in cytotoxic assays.[259] Recently, Haggart et al. discovered a small molecule, namely synstab A (**123**), using reversed genetics or chemical genetics methodology.[260] At first, they used an antibody-based high-throughput screening method to fish out those compounds, which can cause mitotic arrest from a library of 16,320 compounds, and then they evaluated the tubulin assembly activity of them. Those hits were divided into a colchicine-like group, which destabilizes microtubules, and a paclitaxel-like group, which stabilizes them. Synstab A may bind to the same site on a microtubule as paclitaxel or change the conformation of the microtubule to prevent paclitaxel from binding. This approach may be useful in the discovery of antimitotic leading compounds.

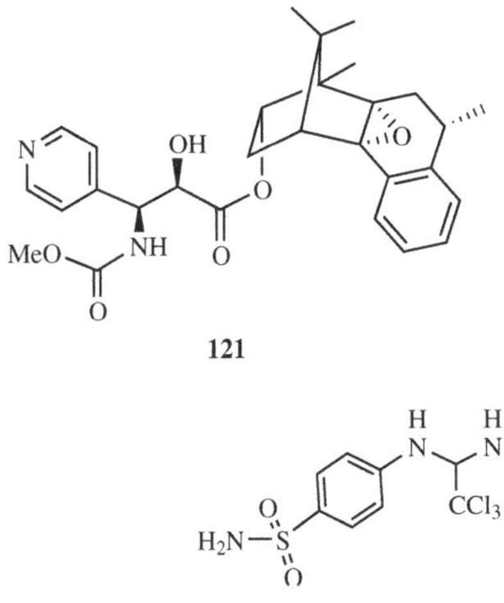

121

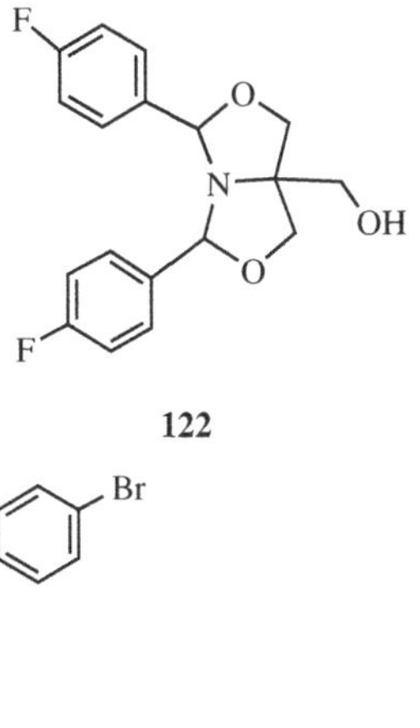

122

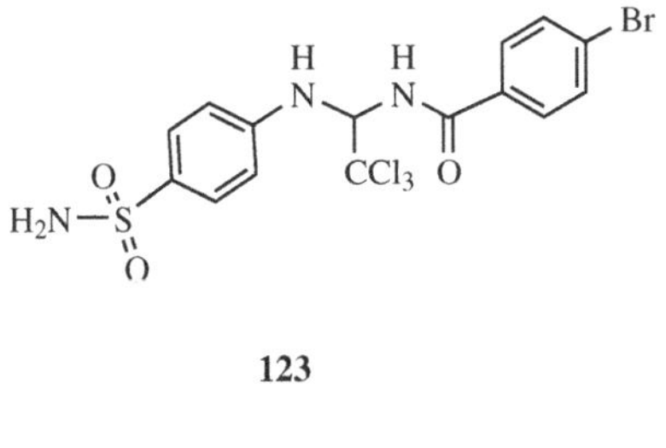

123

## 3.8 CONCLUSION

Most research has focused on the development of paclitaxel analogs or prodrugs with enhanced specificity; MDR reversal and orally effective taxoids have also been developed recently. Meanwhile, scientists have gained insights into the mechanism of action of taxoids at molecular level, that is, binding sites on tubulin and dynamics of tubulin polymerization. It is worth pointing out that the SAR results derived from traditional medicinal chemistry[2-4] have shown the essential role of the C-13 side chain, but some pharmacophore studies have suggested that the C-13 isoserine chain only contributes a small part compared with the baccatin core structure, to binding and triggering subsequent physiological response.[160,164,165] Replacement with a much more simple C-13 side chain is expected to furnish a new generation of antitumor taxanes. On the basis of common pharmacophore established for paclitaxel and several other tubulin-targeting molecules, people tried to apply SAR results of one drug, such as paclitaxel, to another molecule sharing the same pharmacophore. But such efforts were usually unsuccessful (it has been shown recently that these antitubulin agents may not share a common pharmacophore). Instead, high throughput screening of small molecule libraries with structure diversity may be a good choice in the future discovery of antitubulin compounds.[260] The discovery of other mechanisms of taxoids, such as apoptosis and stimulation of the immune system, will prompt people to find new synergic use of taxoids with other drugs.

It is expected that our SAR and mechanistic knowledge will lead to the rational design of the next generation of taxoids with better properties in the future. New techniques including combinatorial chemistry, genomics, and proteomics will reshape the pharmaceutical industry in the future and accelerate the research and development of new drugs and, undoubtedly, benefit taxoid research.

## Notes

After this manuscript was prepared, an important article appeared that should be cited here.[261]

Based on analysis of electronic crystallography and NMR data for the bindings of Taxol and epothilone A to tubulin subunits, it was proposed that they did not share a common pharmacophore (similar binding mode and sites) as hypothesized for a long time, because they bind to their receptors uniquely and independently. Also, the T-shape conformation of Taxol binding to tubulin was supported from this study.

## ACKNOWLEDGMENTS

We thank Ms. Chun-yan Han and Xiao-yan Tian for their assistance in drawing strutures. Our taxoid research project is supported by the Foundation for the Author of National Excellent Doctoral Dissertation of P.R. China (Grant 199949).

## REFERENCES

1. Wani, M. C.; Taylor, H. L.; Wall, M. E.; Coggon, P.; McPhail, A. T. Plant antitumor agents. VI. The isolation and structure of taxol, a novel antileukemic and antitumor agent from *Taxus brevifolia*. *J. Am. Chem. Soc.*, **1971**, 93: 2325–2327.

2. Ojima, I.; Kuduk, S. D.; Chakravarty, S. Recent advances in the medicinal chemistry of taxoid anticancer agents. In: Maryanoff, B. E.; Reitz, A. B. editors. *Advances in Medicinal Chemsitry.* Stamford, CT: JAI Press Inc.; **1999**, 69–124.

3. Kingston, D. G. I. Taxol, a molecule for all seasons. *Chem. Commun.*, **2001**, 867–880.

4. Gueritte, F. General and recent aspects of the chemistry and structure-activity relationships of taxoids. *Curr. Pharm. Design*, **2001**, 7: 1229–1249.

5. Kingston, D. G. I.; Chaudhary, A. G.; Gunatilaka, A. A. L.; Middleton, M. L. Synthesis of taxol from baccatin III via an oxazoline intermediate. *Tetrahedron Lett.*, **1994**, 35: 4483–4484.

6. Magri, N. F.; Kingston, D. G. I.; Jitrangari, C.; Piccariello, T. Modified taxols. 3. Preparation and acylation of baccatin III. *J. Org. Chem.*, **1986**, 51: 3239–3242.

7. Samaranayake, G.; Neidigh, K. A.; Kingston, D. G. I. Modified taxols. 8. Deacylation and reacylation of baccatin III. *J. Nat. Prod.*, **1993**, 56: 884–898.

8. Denis, J.-N.; Greene, A. E.; Guénard, D.; Guéritte-Voegelein, F.; Mangatal, L.; Potier, P. A highly efficient, practical approach to natural taxol. *J. Am. Chem. Soc.*, **1988**, 110: 5917–5919.

9. Kant, J.; O'Keffe, W. S.; Chen, S.-H.; Farina, V.; Fairchild, C.; Johnston, K.; Kadow, J. F.; Long, B. H.; Vyas, D. A chemoselective approach to functionalize the C-10 position of 10-deacetylbaccatin III. Synthesis and biological properties of novel C-10 taxol analogues. *Tetrehedron Lett.*, **1994**, 35: 5543–5546.

10. Holton, R. A.; Zhang, Z.; Clarke, P. A.; Nadizadeh, H.; Procter, D. J. Selective protection of the C(7) and C(10) hydroxyl groups in 10-deacetyl baccatin III. *Tetrahedron Lett.*, **1998**, 39: 2883–2887.

11. Damen, E. W. P.; Braamer, L.; Scheeren, H. W. Lanthanide trifluoromethanesulfonate catalysed selective acylation of 10-deacetylbaccatin III. *Tetrahedron Lett.*, **1998**, 39: 6081–6082.

12. Lee, D.; Kim, K. C.; Kim, M. J. Selective enzymatic acylation of 10-deacetylbaccatin III. *Tetrahedron Lett.*, **1998**, 39: 9039–9042.

13. Guéritte-Voegelein, F.; Guénard, D.; Lavelle, F.; Le Goff, M.-T.; Mangatal, L.; Potier, P. Relationships between the structure of taxol analogs and their antimitotic activity. *J. Med. Chem.*, **1991**, 34: 992–998.

14. Swindell, C. S.; Krauss, N.; Horwitz, S. B.; Ringel, I. Biologically active taxol analogues with deleted A-ring side chain substitutents and variable C-2′ configurations. *J. Med. Chem.*, **1991**, 34: 1176–1184.

15. Kant, J.; Huang, S.; Wong, H.; Fairchild, C.; Vyas, D.; Farina, V. Studies toward structure-activity relationship of Taxol®: synthesis and cytotoxicity of Taxol® analogues with C-2′ modified pheynylisoserine side chains. *Bioorg. Med. Chem. Lett.*, **1993**, 3: 2471–2474.

16. Jayasinghe, L. R.; Datta, A.; Ali, S. M.; Zygmunt, J.; Vander Velde, D. G.; Georg, G. I. Structure-activity studies of antitumor taxanes: synthesis of novel C-13 side chain homologated taxol and taxotere analogs. *J. Med. Chem.*, **1994**, 37: 2981–2984.

17. Kant, J.; Schwartz, W. S.; Fairchild, C.; Gao, Q.; Huang, S.; Long, B. H.; Kadow, J. F.; Langley, D. R.; Farina, V.; Vyas, D. Diastereoselective addition of Grignard reagents to azetidine-1,3-dione: synthesis of novel taxol analogs. *Tetrahedron Lett.*, **1996**, 37: 6495–6498.

18. Denis, J.-N.; Fkyerat, A.; Gimber, Y.; Coutterez, C.; Mantellier, P.; Jost, C.; Greene, A. E. Docetaxel (Taxotere) derivatives: novel NbCl₃-based stereoselective approach to 2'-methylated docetaxel. *J. Chem. Soc. Perkin Trans. 1*, **1995**, 1811–1816.

19. Ojima, I.; Wang, T.; Delaloge, F. Extremely stereoselective alkylation of 3-silyloxy-β-lactams and its applications to the asymmetric syntheses of novel 2-alkylisoserines, their dipeptides, and taxoids. *Tetrahedron Lett.*, **1998**, 39: 3663–3666.

20. Battaglia, A.; Bernacki, R. J.; Bertucci, C.; Bombardelli, E.; Cimitan, S.; Ferlini, C.; Gontana, G.; Guerrini, A.; Riva, A. Synthesis and biological evaluation of 2'-methyl taxoids derived from baccatin III and 14β-OH-baccatin III 1,14-carbonate. *J. Med. Chem.*, **2003**, 46: 4822–4825.

21. Génisson, Y.; Massardier, C.; Gautier-Luneau, I.; Greene, A. E. Effective enantioselective approach to α-aminoalkylacrylic acid derivatives *via* a synthetic equivalent of an asymmetric Baylis-Hillman reaction: application to the synthesis of two C-2' hydroxymethyl analogs of docetaxel. *J. Chem. Soc. Perkin Trans. 1*, **1996**, 2869–2872.

22. Ojima, I.; Slater, J. C.; Michaud, E.; Kuduk, S. D.; Bounaud, P.-Y.; Vrignaud, P.; Bissery, M.-C.; Veith, J.; Pera, P.; Bernacki, R. J. Syntheses and structure-activity relationships of the second-generation antitumor taxoids: exceptional activity against drug-resistant cancer cells. *J. Med. Chem.*, **1996**, 39: 3889–3896.

23. Ojima, I.; Wang, T.; Miller, M. L.; Lin, S.; Borella, C. P.; Geng, X.; Bernacki, R. J. Synthesis and structure-activity relationships of new second-generation taxoids. *Bioorg. Med. Chem. Lett.*, **1999**, 9: 3423–3428.

24. Ojima, I.; Lin, S. Efficient asymmetric syntheses of β-lactams bearing a cyclopropane or an epoxide moiety and their application to the syntheses of novel isoserines and taxoids. *J. Org. Chem.*, **1998**, 63: 224–225.

25. Ojima, I.; Inoue, T.; Chadravarty, S. Enantiopure fluorine-containing taxoids: potent anticancer agents and versatile probes for biomedical problems. *J. Fluorine Chem.*, **1999**, 97: 3–10.

26. Ojima, I.; Slater, J. C.; Pera, P.; Veith, J. M.; Abouabdellah, A.; Bégué, J.-P.; Bernacki, R. J. Synthesis and biological activity of novel 3'-trifluoromethyl taxoids. *Bioorg. Med. Chem. Lett.*, **1997**, 7: 133–138.

27. Ojima, I.; Lin, S.; Slater, J. C.; Wang, T.; Pera, P.; Bernacki, R. J.; Ferlini, C.; Scambia, G. Syntheses and biological activity of C-3'-difluoromethyl-taxoids. *Bioorg. Med. Chem.*, **2000**, 8: 1619–1628.

28. Ojima, I. Use of fluorine in the medicinal chemistry and chemical biology of bioactive compounds—A case study on fluorinated taxane anticancer agents. *Chem. Bio. Chem.*, **2004**, 5: 628–635.

29. Liu, C.; Tamm, M.; Notzel, M. W.; de Meijere, A.; Schilling, J. K.; Kingston, D. G. I. Synthesis and bioactivities of paclitaxel analogs with a cyclopropanted side-chain. *Tetrahedron Lett.*, **2003**, 44: 2049–2052.

30. Li, L.; Thomas, S. A.; Klein, L. L.; Yeung, C. M.; Maring, C. J.; Grampovnik, D. J.; Lartey, P. A.; Plattner, J. J. Synthesis and biological evaluation of C-3'-modified analogs of 9(R)-dihydrotaxol. *J. Med. Chem.*, **1994**, 37: 2655–2663.

31. Lucatelli, C.; Viton, F.; Gimbert, Y.; Greene, A. E. Synthesis of C-3' methyl taxotere (docetaxel). *J. Org. Chem.*, **2002**, 67: 9468–9470.

32. Roh, E. J.; Kim, D.; Lee, C. O.; Choi, S. U.; Song, C. E. Structure-activity relationship study at the 3'-N-position of paclitaxel: synthesis and biological evaluation of 3'-N-acyl-paclitaxel analogues. *Bioorg. Med. Chem.*, **2002**, 10: 3145–3151.

33. Roh, E. J.; Kim, D.; Choi, J. Y.; Lee, B.-S.; Lee, C. O.; Song, C. E. Synthesis, biological activity and receptor-based 3-D QSAR study of 3'-N-substituted-3'-N- debenzoylpa-clitaxel analogues. *Bioorg. Med. Chem.*, **2002**, 10: 3135–3143.

34. Ali, S. M.; Hoemann, M. Z.; Aube, J.; Georg, G. I.; Mitscher, L. A. Butitaxel analogues: synthesis and structure-activity relationships. *J. Med. Chem.*, **1997**, 40: 236–241.

35. Ali, S. M.; Hoemann, M. Z.; Aube, J.; Mitscher, L. A.; Georg, G. I.; McCall, R.; Jayasinghe, L. R. Novel cytotoxic 3'-(tert-bytyl)-3-phenyl analogues of paclitaxel and docetaxel. *J. Med. Chem.*, **1995**, 38: 3821–3828.

36. Chen, S.-H.; Xue, M.; Huang, S.; Long, B. H.; Fairchild, C. A.; Rose, W. C.; Kadow, J. F.; Vyas, D. Structure-activity relationships study at the 3'-N postion of paclitaxel-part 1: synthesis and biological evaluation of the 3'-(t)-butyllaminocarbonyloxy bearing pacli-taxel analogues. *Bioorg. Med. Chem. Lett.*, **1997**, 7: 3057–3062.

37. Xue, M.; Long, B. H.; Fairchild, C.; Johnston, K.; Rose, W. C.; Kadow, J. F.; Vyas, D. M.; Chen, S.-H. Structure-activity relationship study at the 3'-N position of paclitaxel. Part 2: synthesis and biological evaluation of 3'-N-thiourea- and 3'-thiocarbamate-bearing paclitaxel analogues. *Bioorg. Med. Chem. Lett.*, **2000**, 10: 1327–1331.

38. Xiao, X. Y.; Parandoosh, Z.; Nova, M. P. Design and synthesis of a taxoid library using radiofrequency encoded combinatorial chemistry. *J. Org. Chem.*, **1997**, 62: 6029–6033.

39. Bhat, L.; Liu, Y.; Victory, S. F.; Hime, R. H.; Georg, G. I. Synthesis and evaluation of paclitaxel C7 derviatives: solution phase synthesis of combinatorial libraries. *Bioorg. Med. Chem. Lett.*, **1998**, 8: 3181–3136.

40. Liu, Y. B.; Ali, S. M.; Boge, T. C.; Georg, G. I.; Victory, S.; Zygmunt, J.; Marquez, R. T.; Himes, R. H. A systematic SAR study of C10 modified paclitaxel analogues using a combinatorial approach. *Combinatorial Chem. High Throughput Screening*, **2002**, 5: 39–48.

41. Jagtap, P. G.; Baloglu, E.; Barron, D. M.; Bane, S.; Kingston, D. G. I. Design and synthesis of a combinatorial chemistry library of 7-acyl, 10-acyl, and 7,10-diacyl analogues of paclitaxel (Taxol) using solid phase synthesis. *J. Nat. Prod.*, **2002**, 65: 1136–1142.

42. Pandey, R. C.; Yankov, L. K.; Poulev, A.; Nair, R.; Caccamese, S. Synthesis and separation of potential anticancer active dihalocephalomannine diastereomers from extracts of *Taxus yunnanensis*. *J. Nat. Prod.*, **1998**, 61: 57–63.

43. Zheng, Q. Y.; Murray, C. K.; Daughenbaugh, R. J. Cephalomannine epoxide, its analogues and a method for preparing the same. U.S. patent 5,892,063, April 6, **1999**.

44. Barboni, L.; Lambertucci, C.; Ballini, R.; Appendino, G.; Bombardelli, E. Synthesis of a conformationally restricted analogue of paclitaxel. *Tetrahedron Lett.*, **1998**, 39: 7177–7180.

45. Barboni, L.; Lambertucci, C.; Appendino, G.; Vander Velde, D. G.; Himes, R. H.; Bombardelli, E.; Wang, M.; Synder, J. P. Synthesis and NMR-driven conformational analysis of taxol analogues conformationally constrained on the C13 side chain. *J. Med. Chem.*, **2001**, 44: 1576–1587.

46. Chen, S.-H.; Farina, V.; Vyas, D. M.; Doyle, T. W.; Long, B. H.; Fairchild, C. Synthesis and biological evaluation of C-13 amide-linked paclitaxl (taxol) analogues. *J. Org. Chem.*, **1996**, 61: 2065–2070.

47. Hoemann, M. Z.; Vander Velde, D. G.; Aubé, J.; Georg, G. I.; Jayasinghe, L. R. Synthesis of 13-epi-taxol via a transannular delivery of a borohydride reagent. *J. Org. Chem.*, **1995**, 60: 2918–2921.

48. Kelly, R. C.; Wicnienski, N. A.; Gebhard, I.; Qualls, S. J.; Han, F.; Dobrowolski, P. J.; Nidy, E. G.; Johnson, R. A. 12,13-isobaccatin III. Taxane enol esters (12,13-isotaxanes). *J. Am. Chem. Soc.*, **1996**, 118: 919–920.

49. Samaranayake, G.; Magri, N. F.; Jitrangsri, C.; Kingston, D. G. I. Modified taxols. 5. Reaction of taxol with electrophilic reagents and preparation of a rearranged taxol derivative with tubulin assembly activity. *J. Org. Chem.*, **1991**, 56: 5115–5119.

50. Chordia, M. D.; Kingston, D. G. I.; Hamel, E.; Lin, C. M.; Long, B. H.; Fairchild, C. A.; Johnston, K. A.; Rose, W. C. Synthesis and biological activity of A-nor-paclitaxel analogues. *Bioorg. Med. Chem.*, **1997**, 5: 941–947.

51. Chen, S.-H.; Fairchild, C.; Mamber, S. W.; Farina, V. Taxol structure-activity relationships: synthesis and biological evaluation of 10-deoxytaxol. *J. Org. Chem.*, **1993**, 58: 2927–2928.

52. Harriman, G. C. B.; Jalluri, R. K.; Grunewald, G. L.; Vander Velde, D. G.; Georg, G. I. The chemistry of the taxane diterpene: stereoselective synthesis of 10-deacetoxy-11,12-epoxypaclitaxel. *Tetrahedron Lett.*, **1994**, 36: 8909–8912.

53. Marder, R.; Bricard, L.; Dubois, J.; Guénard, D.; Guéritte-Voegelein, F. Taxoids: 11,12-dihydro-4-deacetyldocetaxel. *Tetrahedron Lett.*, **1996**, 37: 1777–1780.

54. Ojima, I.; Fenoglio, I.; Park, Y. H.; Sun, C.-M.; Appendino, G.; Pera, P.; Bernacki, R. J. Synthesis and structure-activity relationships of novel nor-seco analogues of taxol and taxotere. *J. Org. Chem.*, **1994**, 59: 515–517.

55. Ojima, I.; Lin, S.; Chakravarty, S.; Fenoglio, I.; Park, Y. H.; Sun, C. M.; Appendino, G.; Pera, P.; Veith, J. M.; Bernacki, R. J. Syntheses and structure-activity relationships of novel nor-seco taxoids. *J. Org. Chem.*, **1998**, 63: 1637–1645.

56. Appendino, G.; Belloro, E.; Grosso, E. D.; Minassi, A.; Bombardelli, E. Synthesis and evaluation of 14-nor-A-secotaxoids. *Eur. J. Org. Chem.*, **2002**, 277–283.

57. Wang, M.; Xia, X.; Kim, Y.; Hwang, D.; Jansen, J. M.; Botta, M.; Liotta, D. C.; Snyder, J. P. A unified and quantitative receptor model for the microtubule binding of paclitaxel and epothilone. *Org. Lett.*, **1999**, 1: 43–46.

58. Nicoletti, M. I.; Colombo, T.; Rossi, C.; Monardo, C.; Stura, S.; Zucchetti, M.; Riva, A.; Morazzoni, P.; Donati, M. B.; Bombardelli, E.; D'Incalci, M.; Giavazzi, R. IDN5109, a taxane with oral bioavilabity and antitutmor acitivity. *Cancer Res.*, **2000**, 60: 842–846.

59. Huang, G. Y.; Guo, J. Y.; Liang, X. T. Studies on structure modificaton and structure-activity relationship of new taxoids derived from sinenxan A. *Acta Pharm. Sin.*, **1998**, 33: 576–586.

60. Yin, D.; Liu, R.; Wang, D.; Guo, J.; Liang, X.; Sekiguchi, Y.; Kameo, K. Synthesis of analogues of paclitaxel with 14β-side chain from sinenxan A. *J. Chin. Pharm. Sci.*, **1999**, 8: 191–196.

61. Liu, R. W.; Yin, D. L.; Wang, D. H.; Li, C.; Guo, J. Y.; Liang, X. T. Synthesis and structure-activity relationships of novel 14b-side chain taxol derivatives. *Acta Pharm. Sin.*, **1998**, 33: 910–918.

62. Chaudhary, A. G.; Chordia, M. D.; Kingston, D. G. I. A novel benzoyl group migration: synthesis and biological evaluation of 1-benzoyl-2-des(benzoyloxy)paclitaxel. *J. Org. Chem.*, **1995**, 60: 3260–3262.

63. Chen, S.-H.; Huang, S.; Gao, Q.; Golik, J.; Farina, V. The chemistry of taxanes: skeletal rearrangements of baccatin derivatives via radical intermediates. *J. Org. Chem.*, **1994**, 59: 1475–1484.

64. Kingston, D. G. I.; Chordia, M. D.; Jagtap, P. G.; Liang, J.; Shen, Y.-C.; Long, B. H.; Fairchild, C. R.; Johnston, K. A. Synthesis and biological evaluation of 1-deoxypaclitaxel analogues. *J. Org. Chem.*, **1999**, 64: 1814–1822.

65. Baloglu, E.; Hoch, J. M.; Chatterjee, S. K.; Ravindra, R.; Bane, S.; Kingston, D. G. I. Synthesis and biological evaluation of C-3′NH/C-10 and C-2/C-10 modified paclitaxel analogues. *Bioorg. Med. Chem.*, **2003**, 11: 1557–1568.

66. Datta, A.; Vander Velde, D. G.; Georg, G. I.; Himes, R. H. Syntheses of novel C-9 and C-10 modified bioactive taxanes. *Tetrahedron Lett.*, **1995**, 36: 1985–1988.

67. Walker, M. A.; Johnson, T. D.; Huang, S.; Vyas, D. M.; Kadwo, J. F. Synthesis of a novel C-10 spiro-epoxide of paclitaxel. *Bioorg. Med. Chem. Lett.*, **2001**, 11: 1683–1685.

68. Chaudhary, A. G.; Kingston, D. G. I. Synthesis of 10-deacetoxytaxol and 10-deoxytaxotere. *Tetrahedron Lett.*, **1993**, 34: 4921–4924.

69. Chen, S.-H.; Wei, J.-M.; Vyas, D. M.; Doyle, T. W.; Farina, V. Facile synthesis of 7,10-dideoxy taxol and 7-epi-10-deoxy taxol. *Tetrahedron Lett.*, **1993**, 34: 6845–6848.

70. Holton, R. A.; Somoza, C.; Chai, K.-B. A simple synthesis of 10-deacetoxytaxol derivatives. *Tetrahedron Lett.*, **1994**, 35: 1665–1668.

71. Georg, G. I.; Cheruvallath, Z. S. Samarium diiodide-mediated deoxygenation of taxol: a one-step synthesis of 10-deacetoxytaxol. *J. Org. Chem.*, **1994**, 59: 4015–4018.

72. Georg, G. I.; Harriman, G. C. B.; Vander Velde, D. G.; Boge, T. C.; Cheruvallath, Z. S.; Datta, A.; Hepperle, M.; Park, H.; Himes, R. H.; Jayasinghe, L. Medicinal chemistry of paclitaxel. Chemistry, structure-activity relationships, and conformational analysis. In Georg, G. I.; Chen, T. T.; Ojima, I.; Vyas, D. M., editors. *Advances in the Chemistry of Taxane and Taxoid Anticancer Agents*. Washington, DC: American Chemical Society, **1994**, 217–232.

73. Pulicani, J.-P.; Jean-Dominique, B.; Bouchard, H.; Commerçon, A. Elcetorchemical reduction of taxoids: selective preparation of 9-dihydro-, 10-deoxy- and 10-deacetoxy-taxoids. *Tetrahedron Lett.*, **1994**, 35: 4999–5002.

74. Klein, L. L. Synthesis of 9-dihydrotaxol: a novel bioactive taxane. *Tetrahedron Lett.*, **1993**, 34: 2047–2050.

75. Klein, L. L.; Li, C. M.; Yeung, C. M.; Maring, C. J.; Thomas, S. A.; Grampovnik, D. J.; Plattner, J. J. Chemistry and antitumor activity of 9(R)-dihydrotaxanes. In Georg, G. I.; Chen, T. T.; Ojima, I.; Vyas, D. M. editors. *Advances in the Chemistry of Taxane and Taxoid Anticancer Agents*. Washington, DC: American Chemical Society, **1994**, 276–287.

76. Georg, G. I.; Cheruvallath, Z. S.; Vander Velde, D. G.; Himes, R. H. Stereoselective synthesis of 9β-hydroxytaxanes via reduction with samarium diiodide. *Tetrahedron Lett.*, **1995**, 36: 1783–1786.

77. Ishiyama, M.; Iimura, S.; Yoshino, T.; Chiba, J.; Uoto, K.; Terasawa, H.; Soga, T. New highly active taxoids from 9β-dihydrobaccatin-9,10-acetals. Part 2. *Bioorg. Med. Chem. Lett.*, **2002**, 12: 2815–2820.

78. Ishiyama, T.; Iimura, S.; Ohsuki, S.; Uoto, K.; Tersawa, H.; Soga, T. New highly active taxoids from 9β-dihydrobaccatin-9,10-acetals. *Bioorg. Med. Chem. Lett.*, **2002**, 12: 1083–1086.

79. Uoto, K.; Akenoshita, H.; Yoshino, T.; Hirota, Y.; Anido, S.; Mitsui, I.; Tersawa, H.; Soga, T. Synthesis and evaluation of water-soluble non-produrg analogues of docetaxel bearing *sec*-Aminoethyl group at the C-10 position. *Chem. Pharm. Bull.*, **1998**, 46(5): 770–776.

80. Iimura, S.; Uoto, K.; Ohsuki, S.; Chiba, J.; Yoshino, T.; Iwahana, M.; Jimbo, T.; Terasawa, H.; Soga, T. Orally active docetaxel analogue: synthesis of 10-deoxy-10-*C*-morpholinoehtyl deocetaxel analogues. *Bioorg. Med. Chem. Lett.*, **2001**, 11: 407–410.

81. Takeda, Y.; Yoshino, T.; Uoto, K.; Chiba, J.; Ishiyama, T.; Iwahana, M.; Jimbo, T.; Tanaka, N.; Terasawa, H.; Soga, T. New highly active taxoids from 9β-dihydrobaccatin-9,10-acetals. Part 3. *Bioorg. Med. Chem. Lett.*, **2003**, 13: 185–190.

82. Cheng, Q.; Oritani, T.; Horiguchi, T. The synthesis and biological activity of 9- and 2′-cAMP 7-deoxypaclitaxel analogues from 5-cinnamoyltriacetyltaxicin-I. *Tetrahedron*, **2000**, 56: 1667–1679.

83. Cheng, Q.; Oritani, T.; Horiguchi, T.; Yamada, T.; Mong, Y. Synthesis and biological evaluation of novel 9-functional heterocyclic coupled 7-deoxy-9-dihydropaclitaxel analogues. *Bioorg. Med. Chem. Lett.*, **2000**, 10: 517–521.

84. Klein, L. L.; Yeung, C. M.; Li, L.; Planttner, J. J. Synthesis of 9-deoxytaxane analogues. *Tetrahedron Lett.*, **1994**, 35: 4707–4710.

85. Klein, L. L.; Maring, C. J.; Li, L.; Yueng, C. M.; Thomas, S. A.; Grampovnik, D. J.; Planttner, J. J.; Henry, R. F. Synthesis of ring B-rearranged taxane analogues. *J. Org. Chem.*, **1994**, 59: 2370–2373.

86. Appendino, G.; Jakupovic, J.; Varese, M.; Belloro, E.; Danieli, B.; Bombardelli, E. Synthesis of 7,9-nitrogen-substituted paclitaxel derivatives. *Tetradedron Lett.*, **1996**, 37: 7837–7840.

87. Margraff, R.; Bézard, D.; Bourzat, J. D.; Commerçon, A. Synthesis of 19-hydroxy docetaxel from a novel baccatin. *Bioorg. Med. Chem. Lett.*, **1994**, 4: 233–236.

88. Chaudhary, A. G.; Gharpure, M. M.; Rimoldi, J. M.; Chordia, M. D.; Gunatilaka, A. A. L.; Kingston, D. G. I.; Grover, S.; Lin, C. M.; Hamel, E. Unexpectedly facile hydrolysis of the 2-benzoate group of taxol and syntheses of analogues with increased activity. *J. Am. Chem. Soc.*, **1994**, 116: 4097–4098.

89. Datta, A.; Jayasinghe, L. R.; Georg, G. I. Intenal nucleophile assisted selective deesterification studies on baccatin III. Synthesis of 2-debenzoyl and 4-deacetylbaccatin III analogues. *J. Org. Chem.*, **1994**, 59: 4689–4690.

90. Pulicanik, J.-P.; Bézard, D.; Bourzat, J.-D.; Bouchard, H.; Zucco, M.; Deprez, D.; Commerçon, A. Direct access to 2-debenzoyl taxoids by electrochemistry, synthesis of 2-modified docetaxel analogues. *Tetrahedron Lett.*, **1994**, 52: 9717–9720.

91. Nicolaou, K. C.; Renaud, J.; Nantermet, P. G.; Couladouros, E. A.; Guy, R. K.; Wrasidlo, W. Chemical synthesis and biological evaluation of C-2 taxoids. *J. Am. Chem. Soc.*, **1995**, 117: 2409–2420.

92. Kingston, D. G. I.; Chaudhary, A. G.; Chordia, M. D.; Gharpure, M.; Gunatilaka, A. A. L.; Higgs, P. I.; Rimoldi, J. M.; Samala, L.; Jagtap, P. G. Synthesis and biological evaluation of 2-acyl analogues of paclitaxel (Taxol). *J. Med. Chem.*, **1998**, 41: 3715–3726.

93. Chordia, M. D.; Yuan, H.; Jagtap, P. G.; Kadow, J. F.; Long, B. H.; Fairchild, C. R.; Johnston, K. A.; Kingston, D. G. I. Synthesis and bioactiviy of 2, 4-diacyl analogues. *Bioorg. Med. Chem.*, **2001**, 9: 171–178.

94. Georg, G. I., Harriman, G. C. B.; Hepperle, M.; Clowers, J. S.; Vander Velde, D. G.; Himes, R. H. Synthesis, conformational analysis, and biological evaluation of hetero-aromatic taxanes. *J. Org. Chem.*, **1996**, 61: 2664–2676.

95. Boge, T. C.; Himes, R. H.; Vander Velde, D. G.; Georg, G. I. The effect of the aromatic rings of taxol on biological activity and solution conformation: synthesis and evaluation of saturated taxol and taxotere. *J. Med. Chem.*, **1994**, 37: 3337–3343.

96. Ojima, I.; Kuduk, S. D.; Pera, P.; Veith, J. M.; Bernacki, R. J. Synthesis and structure-activity relationships of nonaromatic taxoids: Effects of alkyl and alkenyl ester groups on cytotoxicity. *J. Med. Chem.*, **1997**, 40: 279–285.

97. Chen, S.-H.; Farina, V.; Vyas, D. M.; Doyle, T. W. Synthesis of a paclitaxel isomer: C-2-acetoxy- C-4-benzoate paclitaxel. *Bioorg. Med. Chem. Lett.*, **1998**, 8: 2227–2230.

98. Fang, W.-S.; Fang, Q.-C.; Liang, X.-T. Synthesis of the 2α-benzoylamido analog of docetaxel. *Tetrahedron Lett.*, **2001**, 42: 1331–1333.

99. Fang, W.-S.; Liu, Y.; Liu, H.-Y.; Xu, S.-F.; Wang, L.; Fang, Q.-C. Synthesis and cytotoxicities of 2α-amido docetaxel analogues. *Bioorg. Med. Chem. Lett.*, **2002**, 12: 1543–1546.

100. Fang, W.-S.; Liu, Y.; Fang, Q.-C. First synthesis of 2α-PhS analogue of docetaxel. *Chin. Chem. Lett.*, **2002**, 13: 708–710.

101. Chen, S.-H.; Wei, J.-M.; Farina, V. Taxol structure-activity relationships: synthesis and biological evaluation of 2-deoxytaxol. *Tetrahderon Lett.*, **1993**, 34: 3205–3206.

102. Chordia, M. D.; Kingston, D. G. I. Synthesis and biological evaluation of 2-epi-Paclitaxel. *J. Org. Chem.*, **1996**, 61: 799–801.

103. Fang, W.-S.; Fang, Q.-C.; Liang, X.-T. Reinvestigation to the C-7 epimerization of paclitaxel and related taxoids under basic conditions. *Synth. Commun.*, **1997**, 27: 2305–2310.

104. Ojima, I.; Bounaud, P.-Y.; Ahern, D. G.. New photoaffinity analogues of paclitaxel. *Bioorg. Med. Chem. Lett.*, **1999**, 9: 1189–1194.

105. Sambaiah, T.; King, K.-Y.; Tsay, S.-C.; Mei, N.-W.; Hakimclahi, S.; Lai, Y.-K.; Lieu, C.-H.; Hwu, J. R. Synthesis and immunofluorescence assay of a new biotinylated paclitaxel. *Eur. J. Med. Chem.*, **2002**, 37: 349–353.

106. Georg, G. I.; Harriman, G. C. B.; Himes, R. H.; Mejillano, M. R. Taxol photoaffinity label: 7-(*p*-azidobenoyl) taxol synthesis and biological evaluation. *Bioorg. Med. Chem. Lett.*, **1993**, 2: 735–738.

107. Maththew, A. E.; Mejillano, M. R.; Nath, J. P.; Himes, R. H.; Stella, V. J. Synthesis and evaluation of some water-soluble prodrugs and derivatives of taxol with antitumor activity. *J. Med. Chem.*, **1992**, 35: 145–151.

108. Jimenez-Barbero, J.; Souto, A. A.; Abal, M.; Barasoain, I.; Evangelio, J. A.; Acuna, A. U.; Andreu, J. M.; Amat-Duerri, F. Effect of 2′-OH acetylation on the bioactivity and conformation of 7-*O*-[*N*-(4′-fluoresceincarbonyl)-L-ananyl]taxol. A NMR-fluorescence microscopy study. *Bioorg. Med. Chem.*, **1998**, 6: 1857–1863.

109. Souto, A. A.; Acuna, U.; Andreu, J. M.; Barasoain, I.; Abal, M.; Amat-Guerri, F. New fluorescent water-soluble taxol derivatives. *Angew. Chem. Int. Ed. Engl.*, **1995**, 34: 2710–2712.

110. Holton, R. A.; Chai, K.-B. C7 acyloxy and hydrido taxane derivatives and pharmaceutical compositions containing them. U.S. patent 6,462,208, October 8, **2002**.

111. Golik, J.; Wong, H. S. L.; Chen, S.-H.; Doyle, T. W.; Wright, J. J. K.; Knipe, J.; Rose, W. C.; Casazza, A. M.; Vays, D. M. Synthesis and antitumor evaluation of paclitaxel phosphonooxymethyl ethers: A novel class of water soluble paclitaxel prodrugs. *Bioorg. Med. Chem. Lett.*, **1996**, 6: 1837–1842.

112. Alstadt, T. J.; Fairchild, C. R.; Golik, J.; Johnston, K. A.; Kadow, J. F.; Lee, F. Y.; Long, B. H.; Rose, W. C.; Vyas, D. M.; Wong, H.; Wu, M.-J.; Wittman, M. D. Synthesis and antitumor activity of novel C-7 paclitaxel ethers: discovery of BMS-184476. *J. Med. Chem.*, **2001**, 44: 4577–4583.

113. Nikolakakis, A.; Haidara, K.; Sauriol, F.; Mamer, O.; Zamir, L. O. Semi-synthesis of an O-glycosylated docetaxel analouge. *Bioorg. Med. Chem.*, **2003**, 11: 1551–1556.

114. Mastalerz, H.; Zhang, G.; Kadow, J.; Fairchild, C.; Long, B.; Vyas, D. M. Synthesis of 7β-sulfur analogues of paclitaxel utilizing a novel epimerization of the 7α-thiol group. *Org. Lett.*, **2001**, 3: 1613–1615.

115. Chen, S.-H.; Huang, S.; Farina, V. On the reaction of taxol with DAST. *Tetrahedron Lett.*, **1994**, 35: 41–44.

116. Chen, S.-H.; Huang, S.; Wei, J.; Farina, V. Serendipitous synthesis of a cyclopropane-containing taxol analog via anchimeric participation of an unexpected angular methyl group. *J. Org. Chem.*, **1993**, 58: 4520–4521.

117. Johnson, R. A.; Nidy, E. G.; Dobrowolski, P. J.; Gebhard, I.; Qualls, S. J.; Wicnienski, N. A.; Kelly, R. C. Taxol chemistry. 7-O-triflate as precursors to olefins and cyclopropanes. *Tetrahedron Lett.*, **1994**, 35: 7893–7896.

118. Chen, S.-H.; Kant, J.; Mamber, S. W.; Roth, G. P.; Wei, J.-W.; Marshall, D.; Vyas, D. M.; Farina, V. Taxol structure-activity relationships: synthesis and biological evaluation of taxol analogues modified at C-7. *Bioorg. Med. Chem. Lett.*, **1994**, 4: 2223–2228.

119. Chaudhary, A. G.; Rimoldi, J. M.; Kinsgton, D. G. I. Modified taxols. 10. Preparation of 7-deoxytaxol, a highly bioactive taxol derivative, and interconversion of taxol and 7-epi-taxol. *J. Org. Chem.*, **1993**, 58: 3798–3799.

120. Chen, S.-H.; Huang, S.; Kant, J.; Fairchild, C.; Wei, J.; Farina, V. Synthesis of 7-deoxy- and 7,10-dideoxytaxol via radical intermediates. *J. Org. Chem.*, **1993**, 58: 5028–5029.

121. Saičič, R. N.; Matovič, R. An efficient semisynthesis of 7-deoxypaclitaxel from taxine. *J. Chem. Soc. Perkin Trans.*, **2000**, 1: 59–65.

122. Takeda, Y.; Yoshino, T.; Uoto, K.; Terasawa, H.; Soga, T. A new method for synthesis of 7-deoxytaxane analogues by hydrogenation of $\Delta^{6,7}$-taxane derivatives. *Chem. Pharm. Bull.*, **2002**, 50: 1398–1400.

123. Altstadt, T. J.; Gao, Q.; Wittman, M. D.; Kadow, J. F.; Vyas, D. M. Crystallographic determination of the stereochemsitry of C-6,7 epoxy paclitaxel. *Tetrahedron Lett.*, **1998**, 39: 4965–4966.

124. Yuan, H.; Kingston, D. G. I. Synthesis of 6-hydroxypaclitaxel, the major human metabloite of paclitaxel. *Tetrahedron Lett.*, **1998**, 36: 4967–4970.

125. Wittman, M. D.; Altstadt, T. J.; Fairchild, C.; Hansel, S.; Johnston, K.; Kadow, J. F.; Long, B. H.; Rose, W. C.; Vyas, D. M.; Wu, M.-J.; Zoeckler, M. E. Synthesis of metabolically blocked paclitaxel analogues. *Bioorg. Med. Chem. Lett.*, **2001**, 11: 809–810.

126. Yuan, H.; Fairchild, C. R.; Liang, X.; Kingston, D. G. I. Synthesis and biological activity of C-6 and C-7 modified paclitaxels. *Tetrahedron*, **2000**, 56: 6407–6414.

127. Wittman, M. D.; Alstadt, T. J.; Kadow, J. F.; Vyas, D. M.; Johnston, K.; Fairchild, C.; Long, B. Stereospecific synthesis of 7-deoxy-6-hydroxy paclitaxel. *Tetrahedron Lett.*, **1999**, 40: 4943–4946.

128. Chen, S.-H.; Kadow, J. F.; Farina, V. First synthesis of novel paclitaxel (Taxol) analogues modified at the C4-position. *J. Org. Chem.*, **1994**, 59: 6156–6158.

129. Chen, S.-H. First syntheses of C-4 methyl ether paclitaxel analogues and the unexpected reactivity of 4-deacetyl-4-methyl ether baccatin III. *Tetrahedron Lett.*, **1996**, 37: 3935.

130. Georg, G. I.; Ali, S. M.; Boge, T. C.; Datta, A.; Falborg, L.; Himes, R. H. Selective C-2 and C-4 deacylation and acylation of taxol: the first synthesis of a C-4 substituted taxol analogue. *Tetrahedron Lett.*, **1994**, 35: 8931–8934.

131. Chen, S.-H.; Wei, J.-M.; Long, B. H.; Fairchild, C. A.; Carboni, J.; Mamber, S. W.; Rose, W. C.; Johnston, K.; Casazza, A. M.; Kadow, J. F.; Farina, V.; Vyas, D. M.; Doyle, T. W. Novel C-4 paclitaxel (Taxol) analogues: potent antitumor agents. *Bioorg. Med. Chem. Lett.*, **1995**, 5: 2741–2746.

132. Chen, S.-H.; Fairchild, C.; Long, B. H. Synthesis and biological evaluation of novel C-4 aziridine-bearing paclitaxel (Taxol) analogues. *J. Med. Chem.*, **1995**, 38: 2263–2267.

133. Yuan, H.; Kingston, D. G. I.; Sackett, D. L.; Hamel, E. Synthesis and biological activity of a novel C4-C6 bridged paclitaxel analogue. *Tetrahedron*, **1999**, 55: 9707–9716.

134. Liang, X.; Kingston, D. G. I.; Long, B. H.; Fairchild, C. A.; Johnston, K. A. Paclitaxel analogs modified in ring C: synthesis and biological evaluation. *Tetrahedron* **1997**, 53: 3441–3456.

135. Pulicani, J.-P.; Bouchard, H.; Bourzat, J.-D.; Commercon, A. Preparation of 7-modified doecetaxel analogs using electrochemistry. *Tetrahedron Lett.*, **1994**, 35: 9709-9712.

136. Marder-Karsenti, R.; Dubois, J.; Bricard, L.; Guénard, D.; Guéritte-Voegelein, F. Synthesis and biological evaluation of D-ring-modified taxanes: 5(20)-azadocetaxel analogues. *J. Org. Chem.*, **1997**, 62: 6631.

137. Gunatilaka, A. A. L.; Ramdayal, F. D.; Sarragiotto, M. H.; Kingston, D. G. I.; Sackett, D. L.; Hamel, E. Synthesis and biological evaluation of novel paclitaxel (Taxol) D-ring modified analogues. *J. Org. Chem.*, **1999**, 64: 2694–2703.

138. Cheng, Q.; Kiyota, H.; Yamaguchi, M.; Horiguchi, T.; Oritani, T. Synthesis and biological evaluation of 4-deacetoxy-1,7-didioxy azetidine paclitaxel analogues. *Bioorg. Med. Chem. Lett.*, **2003**, 13: 1075–1077.

139. Merckle, L.; Dubois, J.; Place, E.; Thoret, S.; Gueritte, F.; Guenard, D.; Poupat, C.; Ahand, A.; Potier, P. Semisynthesis of D-ring modified taxoids: novel thia derivatives of docetaxel. *J. Org. Chem.*, **2001**, 66: 5058–5065.

140. Dubois, J.; Thoret, S.; Gueritte, F.; Guenard, D. Synthesis of 5(20)deoxydocetaxel, a new active docetaxel analogue. *Tetrahedron Lett.*, **2000**, 41: 3331–3334.

141. Chordia, M. D.; Chaudhary, A. G.; Kingston, D. G. I.; Jiang, Y. Q.; Hamel, E. Synthesis and biological evaluation of 4-deacetoxypaclitaxel. *Tetrahedron Lett.*, **1994**, 35: 6843–6846.

142. Barboni, L.; Datta, A.; Dutta, D.; Georg, G. I.; Vander Velde, D. G.; Himes, R. H.; Wang, M.; Snyder, J. P. Novel D-seco paclitaxel analogues: synthesis, biological evaluation, and model testing. *J. Org. Chem.*, **2001**, 66: 3321–3329.

143. Beusker, P. H.; Veldhuis, H.; Brinkhorst, J.; Hetterscheid, D. G. H.; Feichter, N.; Bugaut, A.; Scheeren, H. W. Semisynthesis of 7-deoxypaclitaxel derivatives devoid of an oxetane D-ring, starting from taxine B. *Eur. J. Org. Chem.*, **2003**, 689–705.

144. Ojima, I.; Chakravarty, S.; Inoue, T.; Lin, S.; He, L.; Horwitz, S. B.; Kuduk, S. D.; Danishefsky, S. J. A common pharmacophore for cytotoxic natural products that stablize microtubules. *Proc. Natl. Acad. Sci. USA*, **1999**, 96: 4256–4261.

145. Ojima, I.; Lin, S.; Inouue, T.; Miller, M. L.; Borella, C. P.; Geng, X.; Walsh, J. J. Macrocyclic formation of ring-closing metathesis. Application to the syntheses of novel macrocyclic taxoids. *J. Am. Chem. Soc.*, **2000**, 122: 5343–5353.

146. Boge, T. C.; Wu, Z.-J.; Himes, R. H. Vander, Velde, D. G.; Georg, G. I. Conformationally restricted paclitaxel analogues: macrocyclic mimics of the "hydrophobic collapse" conformation. *Bioorg. Med. Chem. Lett.*, **1999**, 9: 3047–3052.

147. Ojima, I.; Geng, X.; Lin, S.; Pera, P.; Bernacki, R. Design, synthesis and biological activity of novel C2-C3′ N-linked macrocyclic taxoids. *Bioorg. Med. Chem. Lett.*, **2002**, 12: 349–352.

148. Querolle, O.; Dubois, J.; Thoret, S.; Dupont, C.; Guéritte, F.; Guénard, D. Synthesis of novel C-2,C3′-N linked macrocyclic taxoids with variable ring size. *Eur. J. Org. Chem.*, **2003**, 542–550.

149. Querolle, O.; Dubois, J.; Thoret, S.; Roussi, F.; Montiel-Smith, S.; Guéritte, F.; Guénard, D. Synthesis of novel macrocyclic docetaxel analogues. Influence of their macrocyclic ring size on tubulin activity. *J. Med. Chem.*, **2003**, 46: 3623–3630.

150. Metaferia, B. B.; Hoch, J.; Glass, T. E.; Bane, S. L.; Chatterjee, S. K.; Snyder, J. P.; Lakdawala, A.; Cornett, B.; Kingston, D. G. I. Synthesis and biological evaluation of novel macrocyclic paclitaxel analogues. *Org. Lett.*, **2001**, 3: 2461–2464.

151. Ganesh, T.; Guza, R. C.; Bane, S.; Ravindra, R.; Shanker, N.; Lakdawala, A. S.; Snyder, J. P.; Kingston, D. G. I. The bioactive taxol conformation of β-tubulin: experimental evidence from highly active constrained analogues. *Proc. Natl. Acad. Sci. USA*, **2004**, 101: 10006–10011.

152. Shi, Q.; Wang, H.-K.; Bastow, K.; Tachibana, Y.; Chen, K.; Lee, F.-Y.; Lee, K.-H. Antitumor agents 210. Synthesis and evaluation of taxoid-epipodophyllotoxin conjugates as novel cytotoxic agents. *Bioorg. Med. Chem. Lett.*, **2001**, 9: 2999–3004.

153. Ohtsu, H.; Nakanishi, Y.; Bastow, K. F.; Lee, F.-Y.; Lee, K.-H. Antitumor agents 216. Synthesis and evaluation of paclitaxel-camptothecin conjugates as novel cytotoxic agents. *Bioorg. Med. Chem.*, **2003**, 11: 1851–1857.

154. Horiguchi, T.; Oritani, T.; Kiyota, H. Synthesis and antitumor activity of 2-(m-substituted-benzoyl)baccatin III analogues from taxinine. *Tetrahedron*, **2003**, 59: 1529–1538.

155. Wu, J. H.; Zamir, L. O. A rational design of bioactive taxanes with side chains situated elsewhere than on C-13. *Anti-Cancer Drug Des.*, **2000**, 15: 73-.

156. Nikolakakis, A.; Wu, J. H.; Bastist, G..; Sauriol, F.; Mamer, O.; Zamir, L. O. Design and syntheses of putative bioactive taxanes. *Bioorg. Med. Chem.*, **2002**, 10: 2387–2395.

157. Soto, J.; Mascareñas, J. L.; Castedo, L. A paclitaxel analogue with a 2(3-20)Abeotaxane skeleton: synthesis and biological evaluation. *Bioorg. Med. Chem. Lett.*, **1998**, 8: 273.

158. Murugkar, A.; Padhye, S.; Guha-Roy, S.; Wagh, U. Metal complexes of Taxol precursor 1. Synthesis, characterization and antitumor activity of the copper complex of 10-deacetylbaccatin thiosemicarbazone. *Inorg. Chem. Commun.*, **1999**, 2: 545–548.

159. Jordan, M. A.; Wilson, L. Microtubules and actin filaments-dynamic targets for cancer chemotherapy. *Curr. Opin. Cell. Biol.*, **1998**, 10: 123–130.

160. Nogales, E.; Wolf, S. G.; Downing, K. H. Structure of the β-tubulin dimer by electron crystallography. *Nature*, **1998**, 391: 199–203.

161. Rao, S.; He, L.; Chadravarty, S.; Ojima, I.; Orr, G. A.; Horwitz, S. B. Characterization of the Taxol binding site on the microtubule: identification of Arg282 in β-tubulin as the site of photoincorporation of a 7-benzophenone analogue of Taxol. *J. Biol. Chem.*, **1999**, 31: 37990–37994.

162. Li, Y.; Poliks, B.; Cegelski, L.; Poliks, M.; Gryczynski, Z.; Piszczek, G.; Jagtap, P. G.; Studelska, D. R.; Kingston, D. G. I.; Schaefer, J., Bane, S. Conformation of microtubule-bound palictaxel determined by fluorescence spectroscopy and REDOR NMR. *Biochemistry*, **2000**, 29: 281–291.

163. Chatterjee, S. K.; Barron, D. M.; Vos, S.; Bane, S. Baccatin III induces assemble of purified tubulin into long microtubules. *Biochemistry*, **2001**, 40: 6964–6970.

164. Andreu, J. M.; Barasoain, I. The interaction of baccatin III with Taxol binding site of microtubules determined by a homogeneous assay with fluorescent taxoid. *Biochemistry*, **2001**, 40: 11975–11984.

165. He, L.; Jagtap, P. G.; Kingston, D. G. I.; Shen, H.-J.; Orr, G. A.; Horwitz, S. B. A common pharmacophore for taxol and the epothilones based on the biological activity of a taxane molecule lacking a C-13 side chain. *Biochemisty*, **2000**, 39: 3972–3978.

166. Li, Y.; Edsall, R., Jr.; Jagtap, P. G.; Kingston, D. G. I.; Bane, S. Equilibrium studies of a fluorescent palictaxel derivative binding to microtubules. *Biochemistry*, **2000**, 39: 616–623.

167. Diaz, J. F.; Strobe, R.; Engelborghs, Y.; Souto, A. A.; Andreu, J. M. Molecular recognition of Taxol by microtubules. *J. Biol. Chem.*, **2000**, 275: 26265–26276.

168. Nogales, E.; Whittaker, M.; Milligan, R. A.; Downing, K. H. High-resolution model of the microtubule. *Cell*, **1999**, 96: 79–88.

169. Lillo, M. P.; Cañadas, O.; Dale, R. E.; Acuña, A. U. Location and properties of the taxol binding center in microtubules: a picosecond laser study with fluorescent taxoids. *Biochemsitry*, **2002**, 41: 12436–12449.

170. Lowe, J.; Li, H.; Downing, K. H.; Nogales, E. Refined structure of β-tubulin at 3.5A resolution. *J. Mol. Biol.*, **2001**, 313: 1045–1057.

171. Grob-Meurer, P.; Kasparian, J.; Wade, R. H. Microtubule structure at improved resolution. *Biochemistry*, **2001**, 40: 8000–8008.

172. Banerjee, A.; Kasmala, L. T. Differential assembly kinetics of α-tubulin isoforms in the presence of paclitaxel. *Biochem. Biophys. Res. Commun.*, **1998**, 245: 349–351.

173. Giannakakou, P.; Gussio, R.; Nogales, E.; Downing, K. H.; Zaharevitz, D.; Bollbuck, B.; Poy, G.; Sackett, D.; Nicolaou, K. C.; Fojo, T. A common pharmacophore for epothilone and taxanes: molecular basis for drug resistance conferred by tubulin mutations in human cancer cells. *Proc. Natl. Acad. Sci. USA*, **2000**, 97: 2904–2909.

174. Snyder, J. P.; Nettles, J. H.; Cornett, B.; Downing, K. H.; Nogales, E. The binding conformation of Taxol in β-tubulin: A model based on electron crystallographic density. *Proc. Natl. Acad. Sci. USA*, **2001**, 98: 5312–5316.

175. Ojima, I.; Kuduk, S. D.; Chakravarty, S.; Ourevitch, M.; Begue, J. P. A novel approach to the study of solution sturctures and dynamic behavior of paclitaxel and docetaxel using fluorine-containing analogues as probes. *J. Am. Chem. Soc.*, **1997**, 119: 5519–5527.

176. Snyder, J. P.; Nevins, N.; Cicero, D. O.; Jansen, J. The conformations of Taxol in chloroform. *J. Am. Chem. Soc.*, **2000**, 122: 724–725.

177. Milanesio, M.; Ugliengo, P.; Viterbo, D.; Appendino, G. *Ab initio* conformational study of the phenylisoserine side chain of paclitaxel. *J. Med. Chem.*, **1999**, 42: 291–299.

178. Wang, M.; Cornett, B.; Nettle, J.; Liotta, D. C.; Snyder, J. P. The oxetane ring in Taxol. *J. Org. Chem.*, **2000**, 65: 1059–1068.

179. Boge, T. C.; Heperle, M.; Vander Velde, D. G.; Gunn, C. W.; Grunerwald, G. L.; Georg, G. I. The oxetane conformational lock of paclitaxel: structural analysis of D-secopaclitaxel. *Bioorg. Med. Chem. Lett.*, **1999**, 9: 3041–3046.

180. Ambudkar, S. V.; Dey, S.; Hrycyna, C. A.; Ramachandra, M.; Pastan, I.; Gottesman, M. M. Biochemical, cellular, and pharmacological aspects of the multidrug transporter. *Annu. Rev. Pharmacol. Toxicol.*, **1999**, 39: 361–398.

181. Distefano, M.; Scambia, G.; Ferlini, C.; Gallo, D.; De, Vincenzo, R.; Filippini, P.; Riva, A.; Bombardelli, E.; Mancuso, S. Antitumor activity of paclitaxel (Taxol) analogues on MDR-posititive human cancer cells. *Anticancer Drug Des.*, **1998**, 13: 489–499.

182. Shionoya, M.; Jimbo, T.; Kitagawa, M.; Soga, T.; Tohgo, A. DJ-927, a novel oral taxane, overcomes P-glycoprotein-medicated multidrug resistance in vitro and in vivo. *Cancer Sci.*, **2003**, 94: 459–466.

183. Appendino, G.; Belloro, E.; Jakupovic, S.; Danieli, B.; Jakupovic, J.; Bombardelli, E. Synthesis of paclitaxel (docetaxel) / 2-deacetaoxytaxinine J dimers. *Tetrahedron*, **1999**, 55: 6567–6576.

184. Ojima, I.; Bounaud, P.-Y.; Takeuchi, C.; Pera, P.; Bernacki, R. J. New taxanes as highly efficient reversal agents for multidrug resistance in cancer cells. *Bioorg. Med. Chem. Lett.*, **1998**, 8: 189–194.

185. Kobayashi, J.; Hosoyama, H.; Wang, X.-X.; Shigemori, H.; Koiso, Y.; Iwasaki, S.; Sasaki, T.; Naito, M.; Tsuruo, T. Effects of taxoids from *Taxus cuspidata* on microtubule depolymerization and vincristin accumulation in MDR cells. *Bioorg. Med. Chem. Lett.*, **1997**, 7: 393–398.

186. Kobayashi, J.; Hosoyama, H.; Wang, X.-X.; Shigemori, H.; Koiso, Y.; Sudo, Y.; Tsuruo, T. Modulation of multidrug resistance by taxuspine C and other taxoids from Japanese yew. *Bioorg. Med. Chem. Lett.*, **1998**, 8: 1555–1558.

187. Hosoyama, H.; Shigemori, H.; Tomida, A.; Tsuruo, T.; Kobayashi, J. Modulation of multidrug resistance in tumor cells by taxinine derivatives. *Bioorg. Med. Chem. Lett.*, **1999**, 9: 389–392.

188. Sako, M.; Suzuki, H.; Hirota, K. Syntheses of taxuspine C derivatives as functional inhibitors of p-glycoprotein, and ATP-associated cell-memberane transporter. *Chem. Pharm. Bull.*, **1998**, 46: 1135–1139.

189. Sako, M.; Suzuki, H.; Yamamoto, N.; Hirota, K. Highly increased cellular accumulation of vincristine, a useful hydrophobic antitumor-drug, in multidrug-resistant solid cancer cells induced by a simply reduced taxinine. *Bioorg. Med. Chem. Lett.*, **1999**, 9: 3403–3406.

190. Kosugi, K.; Saki, J-I.; Zhang, S.; Watanabe, Y.; Sasaki, H.; Suzuki, T.; Hagiwara, H.; Hirata, N.; Hirose, K; Ando, M.; Tomida, A.; Tsuruo, T. Neutral taxoids from *Taxus cuspidata* as modulators of multidrug-resistant tumor cells. *Phytochem.*, **2000**, 54: 839–845.

191. Kobayashi, J.; Shigemori, H. Bioactive taxoids from the Japanese yew *Taxus cuspidata*. *Med. Res. Rev.*, **2002**, 22: 305–328.

192. Morihira, K.; Nishimori, T.; Kusama, H.; Horiguchi, Y.; Kuwajima, I.; Tsuruo, T. Synthesis of C-ring aromatic taxoids and evaluation of their multi-drug resistance reversing activity. *Bioorg. Med. Chem. Lett.*, **1998**, 8: 2973–2976.

193. Morihira, K.; Nishimori, T.; Kusama, H.; Horiguchi, Y.; Kuwajima, I.; Tsuruo, T. Synthesis and evaluation of artificial taxoids with antitumor and multi-drug resistance reversing activities. *Bioorg. Med. Chem. Lett.*, **1998**, 9: 2977–2982.

194. Chibale, K.; Ojima, I.; Haupt, H.; Geng, X.; Pera, P.; Bernacki, R. J. Modulation of human mammary cell sensitivity to paclitaxel by new quinoline sulfonamides. *Bioorg. Med. Chem. Lett.*, **2001**, 11: 2457–2460.

195. Gruol, D. J.; Bernd, J.; Phippard, A. E.; Ojima, I.; Bernacki, R. J. The use of a novel taxane-base p-glycoprotein inhibitor to identifiy mutations that alter the interaction of the protein with paclitaxel. *Mol. Pharmacol.*, **2001**, 60: 104–113.

196. Wu, Q.; Bounaud, P.-Y.; Kuduk, S. D.; Yang, C.-H.; Ojima, I.; Horwitz, S. B.; Orr, G. A. Identification of the domains of photoincorporation of the 3′- and 7-benzophenone analogues of taxol in the carboxyl-terminal half of murine mdr1b p-glycoprotein. *Biochemistry*, **1998**, 37: 11272–11279.

197. Burkhart, C. A.; Kavallaris, M.; Horwitz, S. B. The role of β-tubulin isotypes in resistance to antimitotic drugs. *Biochim. Biophys. Acta.*, **2001**, 1471: O1–O9.

198. Sangrajrang, S.; Fellous, A. Taxol resistance. *Chemotherapy*, **2000**, 46: 327–334.

199. Verdier-Pinard, P.; Wang, F.; Martello, L.; Burd, B.; Orr, G. A.; Horwitz, S. B. Analysis of tubulin isotypes and mutations from taxol-resistant cells by combined isoelectrofocusing and mass spectroscopy. *Biochemistry*, **2002**, 42: 5349–5357.

200. Kavallaris, M.; Burkhart, C. A.; Horwitz, S. B. Antisense oligonucleotides to class III beta-tubulin sensitize drug-resistant cells to Taxol. *Br. J. Cancer*, **1999**, 80: 1020–1025.

201. Verrillils, N. M.; Flemming, C. L.; Liu, M.; Ivery, M. T.; Cobon, G. S.; Norris, M. D.; Haber, M.; Kavallaris, M. Microtubule alterations and mutations induced by desoxyepothilone B: implications for drug-target interactions. *Chem. Biol.*, **2003**, 10: 597–607.

202. Gonzalez-Garay, M. L.; Chang, L.; Blade, K.; Menick, D. R.; Cabral, F. A β-tubulin leucine cluster involved in microtubule assembly and paclitaxel resistance. *J. Biol. Chem.*, **1999**, 274: 23875–23882.

203. Giannakakou, P.; Gussio, R.; Nogales, E.; Downing, K. H.; Zaharevitz, D.; Bollbuck, B.; Poy, G.; Sackett, D.; Nicolaou, K. C.; Fojo, T. A common pharmacophore for epothilone and taxanes: molecular basis for drug resistance conferred by tubulin mutations in human cancer cells. *Proc. Natl. Acad. Sci. USA*, **2000**, 97: 2904–2909.

204. Monzo, M.; Rosell, R.; Sanchez, J. J.; Lee, J. S.; O'Brate, A.; Gonzalez-Larriba, J. L.; Alberola, V.; Lorenzo, J. C.; Nunez, L.; Ro, J. Y.; Martin, C. Paclitaxel resistance in non-small-cell lung cancer associated with beta-tubulin gene mutations. *J. Clin. Oncol.*, **1999**, 17: 1786–1793.

205. Goncalves, A.; Braguer, D.; Kamath, K.; Martello, L.; Briand, C.; Horwitz, S.; Wilson, L.; Jordan, M. A. Resistance to Taxol in lung cancer cells associated with increased microtubule dynamics. *Proc. Natl. Acad. Sci. USA*, **2001**, 98: 11737–11741.

206. Kan, P.; Chen, Z. B.; Lee, C. J.; Chu, I. M. Development of nonionic surfactant/ phospolipid o/w emulsion as a paclitaxel delivery system. *J. Controlled Release*, **1999**, 58: 271–278.

207. Crosasso, P.; Ceruti, M.; Brusa, P.; Arpicco, S.; Dosio, F.; Cattel, L. Preparation, characterization and properties of sterically stabilized paclitaxel-containing liposomes. *J. Controlled Release*, **2000**, 63: 19–30.

208. Ceruti, M.; Crosasso, P.; Brusa, P.; Arpicco, S.; Dosio, F.; Cattel, L. Preparation, characterization and properties in vitro and in vivo of liposomes containing water-soluble prodrugs of paclitaxel. *J. Controlled Release*, **2000**, 63: 141–153.

209. Paradis, R.; Page, M. New active paclitaxel amino acids derivatives with improved water solubility. *Anticancer Res.*, **1998**, 18: 2711–2716.

210. Li, C.; Yu, D.-F.; Newman, R. A.; Cabral, F.; Stephens, L. C.; Milas, L.; Wallace, S. Complete regression of well-established tumors using a novel water-soluble poly(L-glutamic acid)-paclitaxel conjugate. *Cancer Res.*, **1998**, 58: 2404–2409.

211. Li, C.; Price, J. E.; Milas, L.; Hunter, N. R.; Ke, S.; Yu, D.-F.; Charnsangavej, C.; Wallace, S. Antitumor activity of poly(L-glutamic acid)-paclitaxel on syngeneic and xenografted tumors. *Clin. Cancer Res.*, **1999**, 5: 891–897.

212. Damen, E. W. P.; Wiegerinck, P. H. G.; Braamer, L.; Sperling, D.; de Vos, D.; Scheeren, H. W. Paclitaxel esters of malic acid as prodrugs with improved water solubility. *Bioorg. Med. Chem.*, **2000**, 8: 427–432.

213. Niethammer, A.; Gaedicke, G.; Lode, H. N.; Wrasidlo, W. Synthesis and preclinical characterization of a paclitaxel prodrug with improved antitumor activity and water solubility. *Bioconjugate Chem.*, **2001**, 12: 414–420.

214. Seligson, A. L.; Terry, R. C.; Brssi, J. C. Dougalss, J. G., III; Sovak, M. A new prodrug of paclitaxel: synthesis of protaxel. *Anti-Cancer Drugs*, **2001**, 12: 305–313.

215. Hwu, J. R.; Tsay, S.-C.; Sambaiah, T.; Lai, Y.-K.; Lieu, C.-H.; Hakimelahi, G. H.; King, K.-Y. Design and synthesis of paclitaxel-containing aminoester phosphate and phosphoamidate. *Mendeleev Commun.*, **2001**, 216.

216. Takahashi, T.; Tsukamoto, H.; Yamaka, H. Design and synthesis of a water-soluble taxol analogue: taxol-sialyl conjugate. *Bioorg. Med. Chem. Lett.*, **1998**, 8: 113–116.

217. Greenwald, R. B.; Pendri, A.; Bolikal, D.; Gilbert, C. W. Highly water soluble taxol derivatives: 2′-polyethyleneglycol esters as potential prodrugs. *Bioorg. Med. Chem. Lett.*, **1994**, 4: 2465–2470.

218. Greenwald, R. B.; Gilbert, C. W.; Pendri, A.; Conover, C. D.; Xia, J.; Martinez, A. Drug delivery systems: water soluble taxol 2′-poly(ethylene glycol) ester prodrugs—design and in vivo effectiveness. *J. Med. Chem.*, **1996**, 39: 424–431.

219. Feng, X.; Yuan, Y.-J.; Wu, J.-C. Synthesis and evaluation of water-soluble paclitaxel prodrugs. *Bioorg. Med. Chem. Lett.*, **2002**, 12: 3301–3303.

220. Paradis, R.; Page, M. New active paclitaxel glucuronide derivative with improved water solubility. *Int. J. Oncol.*, **1998**, 12: 391–394.

221. Pendri, A.; Conover, C. D.; Greenwald, R. B. Antitumor activity of paclitaxel-2′-glycinate conjugated to poly(ethylene glycol): a water-soluble prodrug. *Anti-Cancer Drug Des.*, **1998**, 13: 387–395.

222. Dosio, F.; Arpicco, S.; Brusa, P.; Stella, B.; Cattel, L. Poly(ethylene glycol)-human serum albumin-paclitaxel conjugates: preparation, characterization and pharmacokinetics. *J. Controlled Release*, **2001**; 76: 107–117.

223. Sugahara, S.-I.; Kajiki, M.; Kuriyama, H.; Kobayashi, T.-R. Palictaxel delivery systems: the use of amino acid linkers in the conjugation of paclitaxel with carboxymethyldextran to create prodrugs. *Biol. Pharm. Bull.*, **2002**, 25: 632–641.

224. Rodrigues, P. C. A.; Scheuermann, K.; Stockmar, C.; Maier, G.; Fiebig, H. H.; Unger, C.; Mülhaupt, R.; Krazt, F. Synthesis and in vitro efficacy of acid-sensitive poly(ethylene glycol) palictaxel conjugates. *Bioorg. Med. Chem. Lett.*, **2003**, 13: 355–360.

225. Wrasidlo, W.; Gaedicke, G.; Guy, R. K.; Renaud, J.; Pitsinos, E.; Nicolaou, K. C.; Reisfeld, R. A.; Lode, H. N. A novel 2'-(N-methylpyridinium acetate) prodrug of paclitaxel induces superior antitumor responses in preclinical cancer models. *Bioconjugate Chem.*, **2002**, 13: 1093–1099.

226. Vrudhula, V. M.; MacMaster, J. F.; Li, Z.; Kerr, D. E.; Senter, P. D. Reductively activated disulfide prodrugs of paclitaxel. *Bioorg. Med. Chem. Lett.*, **2002**, 12: 3591–3594.

227. Ali, S.; Ahmad, I.; Peters, A.; Masters, G.; Minchey, S.; Janoff, A.; Mayhew, E. Hydrolyzable hydrophobic taxances: synthesis and anti-cancer activities. *Anti-Cancer Drugs*, **2001**, 12: 117–128.

228. Ali, S.; Minchey, S.; Janoff, A. S.; Mayhew, E. A differential scanning calorimetry study of phosphocholines mixed with paclitaxel and its bromoacylated taxanes. *Biophys. J.*, **2000**, 78: 246–256.

229. Rodrigues, M. L.; Carter, P.; Wirth, C.; Mullins, S.; Lee, A.; Balckburn, B. K. Synthesis and β-lactamase-mediated activation of a cephalosporin-taxol prodrug. *Chem. Biol.*, **1995**, 2: 223–227.

230. Vrudhula, V. M.; Kerr, D. E.; Siemers, N. O.; Dubowchik, G. M.; Senter, P. D. Cephalosporin prodrugs of paclitaxel for immunologically specific activation by L-49-sFv-β-lactamase fusion protein. *Bioorg. Med. Chem.*, **2003**, 13: 539–542.

231. De Bont, D. B. A.; Leenders, R. G. G.; Haisma, H. J.; Van der Meulen-Muileman, I.; Scheeren, H. W. Synthesis and biological activity of β-glucuronyl carbamate-based prodrugs of paclitaxel as potential candidates for ADEPT. *Bioorg. Med. Chem.*, **1997**, 5: 405–414.

232. Schmidt, F.; Ungureanu, I.; Duval, R.; Pompon, A.; Monneret, C. Cancer therapy: a paclitaxel prodrug for ADEPT (antibody-directed enzyme prodrug therapy). *Eur. J. Org. Chem.*, **2001**, 2129–2134.

233. Guillemard, V.; Saragovi, H. U. Taxane-antibody conjugates afford potent cytotoxicity, enhanced solubility and tumor target selectivity. *Cancer Res.*, **2001**, 61: 694–699.

234. Safavy, A.; Raisch, K. P.; Khazaeli, M. B.; Buchsbaum, D. J.; Bonner, J. A. Paclitaxel derivative for targeted therapy of cancer: Toward the development of smart taxanes. *J. Med. Chem.*, **1999**, 42: 4919–4924.

235. Safavy, A.; Bonner, J. A.; Waksal, H. W.; Buchsbaum, D. J.; Gillespie, G. Y.; Khazaeli, M. B.; Arani, R.; Chen, D.-T.; Carpenter, M.; Raisch, K. P. Synthesis and biological evaluation of paclitaxel-C225 conjugate as a model for targeted drug delivery. *Bioconjugate Chem.*, **2003**, 14: 302–310.

236. Ojima, I.; Geng, X.; Wu, X.; Qu, C.; Borella, C. P.; Xie, H.; Wilhelm, S. D.; Leece, B. A.; Bartle, L. M.; Goldmacher, V. S.; Chari, R. V. J. Tumor-specific novel taxoid-monoclonal antibody conjugates. *J. Med. Chem.*, **2002**, 45: 5620–5623.

237. Luo, Y.; Prestwich, G. D. Synthesis and selective cytotoxicity of a hyaluronic acid-antitumor bioconjugate. *Bioconjugate Chem.*, **1999**, 10: 755–763.

238. Lee, W. L.; Fuchs, P. L. Reduction of azides to primary amines in substrates bearing labile ester functionality. Synthesis of a PEG-solubilized, "Y"-shaped iminodiacetic acid reagent for preparation of folate-tethered drugs. *Org. Lett.*, **1999**, 1: 179–181.

239. Lee, W. L.; Lu, J. Y.; Low, P. S.; Fuchs, P. L. Synthesis and evaluation of taxol-folic acid conjugates as targeted antineoplastics. *Bioorg. Med. Chem. Lett.*, **2002**, 10: 2397–2414.

240. Dubowchik, G. M.; Mosure, K.; Knipe, J. O.; Firestone, R. A. Cathepsin B-sensitive dipeptide prodrugs. 2. Models of anticancer drugs paclitaxel (TaxolR), Mitomycin C and doxorubicin. *Bioorg. Med. Chem. Lett.*, **1998**, 8: 3347–3352.

241. De Groot, F. M. H.; Van Berkom, L. W. A.; Scheeren, H. W. Synthesis and biological evaluation of 2'-carbamte-linked and 2-carbonate-linked prodrugs of paclitaxel: Selective activation by the tumor-associated protease plasmin. *J. Med. Chem.*, **2000**, 43: 3093–3102.

242. Damen, E. W. P.; Nevalainen, T. J.; van den Bergh, T. J. M.; De Groot, F. M. H.; Scheeren, H. W. Synthesis of novel paclitaxel prodrugs for bioreductive activation in hypoxic tumour tissue. *Bioorg. Med. Chem.*, **2002**, 10: 71–77.

243. Liu, C.; Strobl, J. S.; Bane, S.; Schilling, J. K.; McCracken, M.; Chatterjeee, S. K.; Rahim-Bata, R.; Kingston, D. G. I. Design, synthesis, and bioactivities of steroid-linked taxol analogues as potential targeted drugs for prostate and breast cancer. *J Nat. Prod.*, **2004**, 67: 152–159.

244. Yeung, T. K.; Germond, C.; Chen, X.; Wang, Z. The mode of action of Taxol: apoptosis at low concentration and necrosis at high concentration. *Biochem. Biophys. Res. Commun.*, **1999**, 263: 398–404.

245. Srivastava, R. K.; Mi, Q.-S.; Hardwick, J. M.; Longo, D. L. Deletion of the loop region of bcl-2 completely blocks paclitaxel-induced apoptosis. *Proc. Natl. Acad. Sci. USA*, **1999**, 98: 3775–3780.

246. Ling, Y.-H.; Tornos, C.; Perez-Soler, R. Phosphorylation of Bcl-2 is a marker of M phase events and not a determinant of apoptosis. *J. Biol. Chem.*, **1998**, 273: 18984–18991.

247. Fan, W. Possible mechanisms of paclitaxel-induced apoptosis. *Biochem. Pharmacol.*, **1999**, 57: 1215–1221.

248. Balgosklonny, M. V. Unwinding the loop of bcl-2 phosphorylation. *Leukemia*, **2001**, 15: 869–874.

249. Ruvolo, P. P.; Deng, X.; May, W. S. Phosphorylation of bcl2 and regulation of apoptosis. *Leukemia*, **2001**, 15: 515–522.

250. Blagosklonny, M. V.; Fojo, T. Molecular effects of paclitaxel: myths and reality. *Int. J. Cancer*, **1999**, 83: 151–156.

251. Byrd, C. A.; Bornmann, W.; Erdjument-Bromage, H.; Tempst, P.; Pavletich, N.; Rosen, N.; Nathan, C. F.; Ding, A. Heat shock protein 90 mediates macrophage activation by Taxol and bacterial lipopolysaccharide. *Proc. Natl. Acad. Sci. USA*, **1999**, 96: 5645–5650.

252. Kawasaki, K.; Akashi, S.; Shimazu, R.; Yoshida, T.; Miyake, K.; Nishijima, M. Mouse toll-like receptor 4-MD-2 complex mediates lipopolysaccharide-mimetic signal transduction by Taxol. *J. Biol. Chem.*, **2000**, 275: 2251–2254.

253. Nakano, M.; Tominaga, K.; Saito, S.; Kirikae, F.; Lin, S.; Fumero, C. L.; Ojima, I.; Kirikae, T. Lipopolysaccharide- and paclitaxel (Taxol)-induced tolerance in murine peritoneal macrophages. *J. Endotoxin Res.*, **1999**, 5: 102–106.

254. Kirikae, T.; Ojima, I.; Ma, Z.; Kirikae, F.; Hirai, Y.; Nakano, M. Structural significance of the benzoyl group at the C-3'-N position of paclitaxel for nitric oxide and tumor necrosis factor production by murine macrophage. *Biochem. Biophys. Res. Commun.*, **1998**, 245: 298–704.

255. Kirikae, T.; Ojima, I.; Fuero-Oderda, C.; Lin, S.; Kirikae, F.; Hashimoto, M.; Nakano, M. Structural significance of the acyl group at C-10 position and the A ring of the taxane core of paclitaxel for inducing nitric oxide and tumor necrosis factor production by murine macrophages. *FEBS Lett.*, **2000**, 478: 221–226.

256. Lin, C.-E.; Garvey, D. S.; Janero, D. R.; Letts, J. L.; Marek, P.; Richardson, M. S.; Serbryanik, D.; Shumway, M. J.; Tam, S. W.; Trocha, A. M.; Young, D. V. Combination of paclitaxel and nitric oxide as a novel treatment for the reduction of restenosis. *J. Med. Chem.*, **2004**, 47: 2276–2282.

257. Altmann, K.-H. Microtubule-stabilizing agents: a growing class of important anticancer drugs. *Curr. Opin. Chem. Biol.*, **2001**, 5: 424–431.

258. Klar, U.; Graf, H.; Schenk, O.; Rohr, B.; Schulz, H. New synthetic inhibitors of microtubule depolymerization. *Bioorg. Med. Chem. Lett.*, **1998**, 8: 1397–1402.

259. Shintani, Y.; Tanaka, T.; Nozaki, Y. GS-164, a small synthetic compound, stimulates tubulin polymerization by a similar mechanism to that of Taxol. *Cancer Chemother. Pharmacol.*, **1997**, 40: 513–520.

260. Haggart, S. J.; Mayer, T. U.; Miyamoto, D. T.; Fathi, R.; King, R. W.; Mitchison, T. J.; Schreiber, S. L. Dissecting cellular processes using small molecules: identification of colchicine-like, taxol-like and other small molecules that perturb mitosis. *Chem Biol.*, **2000**, 7: 275–286.

261. Nettles, J. H.; Li, H.; Cornett, B.; Krahn, J. M.; Snyder, J. P.; Downing, K. H. The binding mode of epothilone A on α,β-tubulin by electron crystallography. *Science*, **2004**, 305: 866–869.

# 4

# THE OVERVIEW OF STUDIES ON HUPERZINE A: A NATURAL DRUG FOR THE TREATMENT OF ALZHEIMER'S DISEASE

Da-Yuan Zhu, Chang-Heng Tan, and Yi-Ming Li

*State Key Laboratory of Drug Research, Shanghai Institute of Materia Medica, Shanghai Institutes for Biological Sciences, Chinese Academy of Sciences, Shanghai, China*

## 4.1 INTRODUCTION

### 4.1.1 Powerful AChEI Originated From Traditional Chinese Medicine

Traditional Chinese medicine (TCM) has a long history of serving people, which tends to raise the natural defenses of the organism instead of trying to restore its natural functions. The accumulated clinical experience inspired the search for new drugs in modern times. Huperzine A (HA, **1**) is one successful example of this continuum.

HA is a natural-occurring alkaloid that was isolated from Chinese medicinal herbs, *Qian Ceng Ta* [*Huperzia serrata* (Thunb.) Trev. = *Lycopodium serratum*] and its related genera by Chinese scientists in the early 1980s.[1,2] Pharmacological studies in vitro and in vivo demonstrated that HA was a potent, selective, and reversible acetylcholinesterase inhibitor (AChEI), which crosses the blood-brain barrier smoothly and shows high specificity for AChE with a prolonged biological half-life.[3–5] It has been approved as a drug for the treatment of Alzheimer's disease (AD) in China. Because the isolation and use of HA was released without patent protection, it is sold in the United States as a dietary supplement. Several review articles for the research progress in the chemistry, pharmacology, structural biology, and clinical trials of HA have been published within the last 10 years.[4–10] This

---

*Medicinal Chemistry of Bioactive Natural Products* Edited by Xiao-Tian Liang and Wei-Shuo Fang
Copyright © 2006 John Wiley & Sons, Inc.

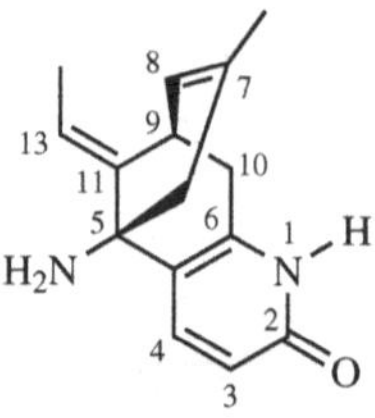

Huperzine A (**1**)
(5R, 9R, 11E)-5-amino-11-ethylidene-
5,6,9,10-tetrahydro-7-methyl-5,9-
methanocycloocta [*b*]pyridin-2[1*H*]-one

chapter represents a comprehensive documentation of the overview of studies on HA up to January 2004.

### 4.1.2  Alzheimer's Disease

AD is a progressive, degenerative disease of the brain. The disease is the most common form of dementia affecting elderly people, with a mean duration of around 8.5 years between the onset of clinical symptoms and death. The incidence of AD increases with age, even in the oldest age groups: from 0.5% at 65, it rises to nearly 8% at 85 years of age. Some 12 million persons have AD, and by 2025, that number is expected to increase to 22 million.

Neuropathologically, AD is characterized by (1) parenchymal amyloid deposits or neuritic plaques; (2) intraneuronal deposits of neurofibrillary tangles; (3) cerebral amyloid angiopathy, and (4) synaptic loss.[11] Current treatment for AD in most countries consists in the administration of AChEIs to increase the amount of acetylcholine (ACh) at the neuronal synaptic cleft by inhibiting AChE, based on the finding that ACh is dramatically low in the brains of AD patients. AChE is an enzyme that breaks down ACh, a neurotransmitter in the brain that is required for normal brain activity and is critical in the process of forming memories.

To date, four AChEIs, Cognex (tacrine), Aricept (donepezil or E2020), Exelon (rivastigmine), and Reminyl (galanthamine hydrobromide) currently are approved as prescription drugs by the United States to treat the symptoms of mild-to-moderate AD. However, the clinical usefulness of AChEIs has been limited by their short half-lives and excessive side effects caused by activation of peripheral cholinergic systems, as well as by hepatotoxicity, which is the most frequent and important side effect of tacrine therapy.[12-14]

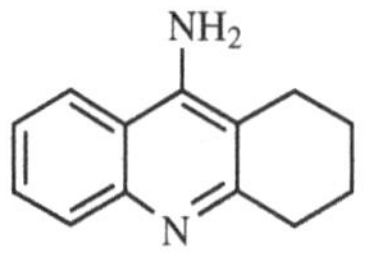

Tacrine

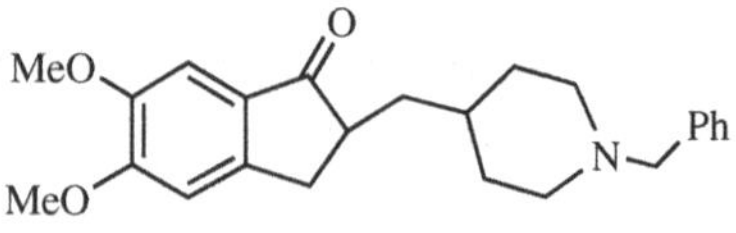

Donepezil

Galanthamine                    Rivastigmine

## 4.2  PROFILES OF HA

### 4.2.1  Discovery of HA

*H. serrata* and its related genera have been used as folk herbs for the treatment of memory disorder and schizophrenia in the east of China. Phytochemical studies disclosed that these plants contained mainly serratene-type triterpenes[15,16] and *Lycopodium* alkaloids.[17–24] In the early 1970s, Chinese scientists reported that the total alkaloids of *H. serrata* could relax the striated muscle and alleviate the symptom of myasthenia gravis on the animal model. Biodirected assay caused the phenolic alkaloids fraction to be spotlighted[25] and the following chemical component isolation resulted in the finding of HA.[1]

### 4.2.2  Physical Appearance of HA

HA is a rigid three-ring system molecule that consisted of a tetrahydroquinolinone, three-carbon bridge ring, exocyclic ethylidene, and primary amino group. Its empirical formula is $C_{15}H_{18}N_2O$, and its molecular weight is 242. The compound is optically active and in the plant is present only in its ($-$)-enantiomer. Its structure and stereochemistry were elucidated as (5$R$, 9$R$, 11$E$)-5-amino-11-ethylidene-5,6,9,10-tetrahydro-7-methyl-5,9-methanocycloocta [*b*]pyridin-2[1*H*]-one on the basis of nuclear magnetic resonance (NMR), infrared (IR), ultraviolet (UV), and circular dichroism (CD) data and chemical transformations by Liu et al.,[1] which was confirmed by x-ray crystallographic analysis.[26] Two similar reported molecules, selagine[27] and isoselagine,[28] were reexamined to be the same with HA by Sun et al.[29] and Ayer and Trifonov,[23] respectively.

From the viewpoint of biogenesis, HA belonged to the lycodine type of *Lycopodium* alkaloids with the opening of ring-C and a carbon atom lost. *Lycopodium* alkaloid is a class of alkaloids that possesses a unique ring system isolated from the family of Huperziaceae and Lycopodiaceae of Lycopodiales order. They have a common formula $C_{16}N$ or $C_{16}N_2$ with three or four rings, which may be subclassed into four types, namely, lycopodine, lycodine, fawcettimine, and miscellaneous, on the basis of carbon skeleton and probable biogenetic pathway.[23] More than 200 *Lycopodium* alkaloids have been isolated and identified so far. Surprisingly, they had no anti-ChE effect or far less than HA.[24]

## 4.3 PLANT RESOURCES

HA has been obtained from *H. serrata,*[1] *Phlegmariurus fordii* (Baker) Ching,[2] *L. selago,*[27] *L. serratum* (Thunb.) var. *Longipetiolatum Spring,*[28] and *Lycopodium varium.*[30] These clubmoss plants belong to the family of Huperziaceae or Lyco-podiaceae of Lycopodiales order from the viewpoint of plant taxonomy. We conducted a large-scale plant investigation for the HA source in China.[31] A total of 67 species, 11 varieties, and 2 forma were collected from 19 provinces in China, and the sources of the *Huperzia, Phlegmariurus,* and *Lycopodium* species were also investigated in those provinces. The examination of the chemical consti-tuent demonstrated that HA was mainly present in the Huperziaceae family (con-sisting of genus *Huperzia* and genus *Phlegmariurus*) (Table 4-1).[32] This result was in agreement with Chu's investigation.[33] In general, *H. serrata* is the most impor-tant source of HA from the term of plant quantities, distribution, and HA content. Ma et al. measured the HA content-collected season curve of *H. serrata* (Figure 4-1), suggesting that autumn is the best time for collecting HA.[32]

**TABLE 4-1. HA Detected by Thin-Layer Chromatography in Whole Plant Extract of *Huperzia* and Related Genera**

| Species | Collected Region | HA |
|---|---|---|
| **Family Huperziaceae** | | |
| Genus *Huperzia* | | |
| 1 *H. serrata* (Thunb.) Trev. | *Dongkou,* Hunan | + |
| 2 *H. serrata* f. *longipetiolata* (Spring) Ching | *Jinpin,* Yunnan | + |
| 3 *H. serrata* f. *intermedia* (Nakai) Ching | *Yanbi,* Yunnan | + |
| 4 *H. crispata* (Ching) Ching | *Nanchuan,* Sichuan | + |
| 5 *H. austrosinica* Ching | *Xingyi,* Guangdong | + |
| 6 *H. herteriana* (Kumm.) Sen et Sen | *Gongshan,* Yunnan | − |
| 7 *H. emeiensis* Ching et Kung | *Emei,* Sichuan | − |
| 8 *H. delavayi* (Christ et Herter) Ching | *Cangshan,* Yunnan | − |
| 9 *H. sutchueniana* (Herter) Ching | *Nanchuan,* Sichuan | − |
| 10 *H. selago* (L.) Brenh. ex Schrank | *Gongliu,* Xinjiang | + |
| 11 *H. chinensis* (Christ) Ching | *Taibaishan,* Shanxi | + |
| 12 *H. bucawangensis* Ching | *Jinpin,* Yunnan | + |
| 13 *H. selago var. appressa* (Desv.) Ching | *Gongshan,* Yunnan | + |
| 14 *H. ovatifolia* Ching | *Jinpin,* Yunnan | + |
| 15 *H. whangshanensis* Ching | *Huangshan,* Anhui | + |
| 16 *H. tibetica* (Ching) Ching | *Gongshan,* Yunnan | + |
| 17 *H. liangshanica* Ching et H. S. Kung | *Liangshan,* Sichuan | + |
| 18 *H. kunmingensis* Ching | *Kunming,* Yunnan | + |
| 19 *H. laipoensis* Ching | *Leipo,* Sichuan | + |
| 20 *H. nanchuanensis* Ching et H. S. Kung | *Nanchuan,* Sichuan | − |
| 21 *H. obscure-denticulata* Ching | *Nanchuan,* Sichuan | − |
| Genus *Phlegmariurus* | | |
| 22 *Ph. squarrosus* (Forest.) Love et | *Mengla,* Yunnan | − |
| 23 *Ph. fordii* (Baker) Ching | *Xingyi,* Guangdong | + |

**TABLE 4-1** (*Continued*)

| Species | Collected Region | HA |
|---|---|---|
| 24 *Ph. phlegmaria* (L.) Holub | *Mengla*, Yunnan | − |
| 25 *Ph. guangdongensis* Ching | *Xingyi*, Guangdong | − |
| 26 *Ph. cancellatus* (Spring) Ching | *Gongshan*, Yunnan | − |
| 27 *Ph. yunnanensis* Ching | *Gongshan*, Yunnan | + |
| 28 *Ph. carinatus* (Desv.) Ching | *Jinpin*, Yunnan | + |
| 29 *Ph. henryi* (Baker) Ching | *Mengla*, Yunnan | + |
| **Family Lycopodiaceae** | | |
| Genus Palhinhaea | | |
| 30 P. cernua (L.) A. Franco et Vasc. | *Jinpin*, Yunnan | − |
| Genus *Diphasiastrum* | | |
| 31 *D. complanatum* (L.) Holub | *Dali*, Yuannan | − |
| 32 *D. alpinum* (L.) Holub | *Changbaishan*, Jilin | − |
| Genus *Lycopodiastrum* | | |
| 33 *L. casuarinoides* (Spring) Holub | *Jinpin*, Yunnan | − |
| Genus Lycopodiella | | |
| 34 *L. inundata* (L.) Holub | *Lianchen*, Fujian | + |
| Genus *Lycopodium* | | |
| 35 *L. obscurum* L | *Changbaishan*, Jilin | − |
| 36 *L. annotinum* L. | *Emei*, Sichuan | − |
| 37 *L. japonicum* Thunb. | *Linan*, Zhejiang | − |

*Source*: Data from Ref. 32
*Note*: + —detectable; − —undetectable

*H. serrata* is a clubmoss fern that is distributed worldwide and grows in the forest, shrubbery, and roadside in the region at an altitude of 300–2700 m. It has a long growing period (8–15 years) and low HA content (in our experience, 1 ton dried whole plant yielded 70–90 g HA). So, the source of HA has been an obstacle to large-scale application of HA and its active derivates. Many scientists have tried to resolve this problem with the methods of synthesis, tissue culture, genetic engineer, or plant cultivation in the last decade. However, they did not successfully create a new HA source to replace of the natural plant thus far.

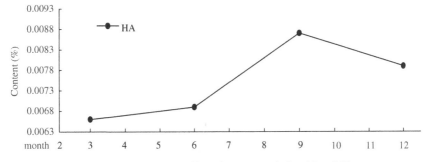

**Figure 4-1.** The HA content-collected seasons relationship of *H. serrata*.

**TABLE 4-2. Anticholinesterase Effects of ChEIs in Vitro**

| ChEI | IC$_{50}$ ($\mu$M) | | Ki* (nM) |
|------|-------------------|---|----------|
|      | AChE (rat cortex) | BuChE (rat serum) |  |
| HA | 0.082 | 74.43 | 24.9 |
| Galanthamine | 1.995 | 12.59 | 210.0 |
| Donepezil | 0.010 | 5.01 | 12.5 |
| Tacrine | 0.093 | 0.074 | 105.0 |

*Assayed with erythrocyte membrane AchE. Data from Ref. 35.

## 4.4  PHARMACOLOGY

### 4.4.1  Effects on Cholinesterase Activity

The cholinesterase (ChE) inhibition activity of HA has been evaluated in vitro and in vivo by Tang et al. using spectrophotometric methods[34] with slight modifications.[3,35] The concentration of the inhibitor yielding 50% inhibition of enzyme activity (IC$_{50}$) of HA on AChE and butyrylcholinesterase (BuChE) compared with other ChE inhibitors is listed in Table 4-2. HA initiated AChE (rat cortex) inhibition at 10 nM. The anti-AChE activity of HA was more effective than that of tacrine and galanthamine, but less than that of donepezil. The pattern of inhibition is of the mixed competitive type. In contrast, HA inhibited BuChE at a much higher concentration than donepezil, whereas tacrine was more potent toward BuChE. HA has the highest specificity for AChE. The $K_i$ values (inhibition constants, in nM) revealed that HA was more potent than tacrine and galanthamine, but about twofold less potent than donepezil (Table 4-2).

Compared with AChE in animals such as horse and rat, HA is a weaker inhibitor of human serum BuChE. This selectivity for AChE as opposed to BuChE (similar to that of galanthamine) may suggest a better side-effects profile.[36] However, a stronger inhibition of BuChE could be important in the later stage of AD[37] and could offer more protection over amyloid β-peptide (Aβ) plaque deposition.[38] In contrast to isoflurophate, the AChE activity did not decrease with the prolongation of incubation with HA in vitro, and the AChE activity returned to 94% of the control after being washed five times, which indicates a reversible inhibitory action.[3]

Significant inhibition of AChE activity was demonstrated in the cortex, hippocampus, striatum, medial septum, medulla oblongata, cerebellum, and hypothalamus of rats that were sacrificed 30 min after the administration of HA at several dose levels compared with saline control.[39-41]

After administration of oral HA at doses of 0.12–0.5 mg/kg, a clear, dose-dependent inhibition of AChE was demonstrated in the brains of rats.[39,40] In contrast to the AChE inhibition in vitro, the relative inhibitory effect of oral HA over AChE was found to be about 24- and 180-fold, on an equimolar basis, more potent than donepezil and tacrine, respectively. In rats, HA injected intraperitoneally (i.p.) exhibited similar efficacy of AChE inhibition as demonstrated after oral administration, whereas i.p. administration of tacrine and donepezil showed greater inhibition on both

**TABLE 4-3. Anti-ChE Activities of Oral HA, Donepezil, and Tacrine in Rats**

| ChEI | Dose mg/kg ($\mu$mol/kg) | AChE Inhibition (%) ($n = 6$) | | | BuChE Inhibition (%) Serum ($n = 3$) |
|---|---|---|---|---|---|
| | | Cortex | Hippocampus | Striatum | |
| HA | 0.36 (1.5) | $20 \pm 6^a$ | $17 \pm 3^a$ | $18 \pm 4^a$ | $18 \pm 10$ |
| | 0.24 (1.0) | $16 \pm 6^a$ | $15 \pm 3^a$ | $16 \pm 8^a$ | $16 \pm 14$ |
| | 0.12 (0.5) | $10 \pm 6^a$ | $8 \pm 7$ | $13 \pm 10^b$ | $7 \pm 12$ |
| Donepezil | 6.66 (16) | $18 \pm 6^a$ | $12 \pm 5^a$ | $12 \pm 8^b$ | $33 \pm 7^a$ |
| | 5.00 (12) | $11 \pm 6^a$ | $10 \pm 4^a$ | $10 \pm 6^b$ | $22 \pm 1^a$ |
| | 3.33 (8) | $9 \pm 11$ | $6 \pm 8$ | $8 \pm 6$ | $8 \pm 10$ |
| Tacrine | 28.2 (120) | $20 \pm 6^a$ | $11 \pm 10^b$ | $11 \pm 10^b$ | $52 \pm 5^a$ |
| | 21.1 (90) | $8 \pm 6^a$ | $9 \pm 6$ | $8 \pm 41^a$ | $40 \pm 20^b$ |
| | 14.1 (60) | $7 \pm 7$ | $2 \pm 2$ | $2 \pm 5$ | $24 \pm 17$ |

$^aP < 0.01$: $^bP < 0.05$ vs. saline group. Values expressed as percent inhibition (vs. saline control) $\pm$ standard deviation. Data from Ref. 5.

AChE activity and serum BuChE.[42] The inhibitory action of HA on brain AChE was less than that of donepezil after the intraventricular injection but more effective than that of tacrine.[39] Maximal AChE inhibition in rat cortex and whole brain was reached at 30–60 min and maintained for 360 min after oral administration of 0.36 mg/kg HA.[40–42] The oral administration of HA produced greater AChE inhibition compared with donepezil and tacrine, which indicated that it has greater bioavailability and more easily penetrates the blood-brain barrier (Table 4-3). Repeated doses of HA showed no significant decline in AChE inhibition as compared with that of a single dose, which demonstrates that no tolerance to HA occurred.[43]

### 4.4.2   Effects on Learning and Memory

HA has been found to be an effective cognition enhancer in a broad range of animal cognitive models by Tang et al.[35,41,44–55] and in clinical trials (see Section 4.4.3).[56] The effects of HA on nucleus basalis magnocellularis lesion-induced spatial working memory impairment were tested by means of a delayed-nonmatch-to-sample radial arm maze task. Unilateral nucleus basalis magnocellularis lesion by kainic acid impaired the rat's ability to perform this task. This working memory impairment could be ameliorated by HA.[52] HA ameliorates the impaired memory naturally occurring or induced by scopolamine in aged rats. The Morris water maze was used to investigate the effects of HA on acquisition and memory impairments. During 7-day acquisition trials, aged rats took longer latency to find the platform. HA at a dosage of 0.1–0.4 mg/kg subcutaneously (s.c.) could significantly reduce the latency or reverse the memory deficits induced by scopolamine.[54]

The effects of HA on the disruption of spatial memory induced by the muscarinic antagonist scopolamine and ($\gamma$-amino-$n$-butyric acid (GABA)) against muscimol in the passive avoidance task was tested in chicks. The avoidance rate was evaluated as memory retention. Both scopolamine (100 ng) and muscimol (50 ng), injected intracranially 5 minutes before training, resulted in a decreased avoidance rate. HA (25 ng), injected intracranially 15 minutes before training, reversed memory deficits at 30 min after training and persisted at least 1 hour. The improving effects exhibited a bell-shaped dose-response curve. The results indicated that HA improved the process of memory formation not only by acting as a highly potent and selective AChEI but also by antagonizing effects mediated via the GABA$_A$ receptors.[55]

Reserpine [0.1 mg/kg intramuscularly (i.m.)] or yohimbine (0.01 mg/kg i.m.) induces significant impairments in the monkey's ability to perform the delayed response task. HA at a dosage of 0.01 mg/kg i.m. for the yohimbine-treated monkeys markedly improved the memory impairments. The effects exhibited an inverted U-shaped dose-response pattern. The data suggest that HA may improve working memory via an adrenergic mechanism.[57]

It was reported that subchronic administration of HA did not induce deleterious effects on spatial memory in guinea pigs.[58] A systematic comparison of tolerances between the mixed acetyl-butyryl-ChE inhibitors and the selective AChEIs such as HA indicated that they showed a remarkably similar profile of behavioral symptoms associated with overdosing in rats.[59]

### 4.4.3 Effects on the Protection of Neuronal Cells

Another interesting property of HA pharmacology relates to a broad range of protective actions. It has been studied that HA could protect neuronal cells against nerve gas poisoning,[60] against glutamate toxicity,[61] against Aβ toxicity,[62–65] and against neuronal cells apoptosis induced by hypoxic-ischemic (HI) or oxidative stress.[66–68]

#### 4.4.3.1 Nerve Gas Poisoning
HA has been tested as a prophylactic drug against soman and other nerve gas poisoning with an excellent outcome.[69] It works by protecting cortical AChE from soman inhibition and by preventing subsequent seizures. This prophylactic use makes HA a potential protective agent against chemical weapons. It has been demonstrated that rats can be protected against low doses of soman with pretreatment with only HA, and without typical cholinergic side effects.[69] This protection was confirmed in a study with primates, where HA was compared with pyridostigmine: The cumulative dose of soman needed to produce convulsions and epileptic activity was 1.55-fold higher in the animals who received HA compared with the group of primates pretreated with pyridostigmine.[60] The same study demonstrated that HA selectively inhibited red cell AChE activity, whereas pyridostigmine also inhibited plasma BuChE. Thus, the superior protection offered by HA appears to be related both to the selectivity of HA for red cell AChE, preserving the scavenger capacity of plasma BuChEs for organophosphate (OP) agents, and to the protection conferred on cerebral AChE.[60]

#### 4.4.3.2 Glutamate Toxicity
HA also protects primary neuronal cell culture and animals from glutamate toxicity. Glutamate activates N-methyl-D-aspartate (NMDA) receptors and increases the flux of calcium ions into the neurons,[70] whereas calcium at toxic levels can kill the cells.[71]

Pretreatment of primary neuronal cells with HA reduced glutamate- and OP-induced toxicity and decreased neuronal death.[61] The consequence of excitatory amino-acid-induced overstimulation has been implicated in a variety of acute and chronic neurodegenerative disorders, including Parkinson's disease, dementia, neuroleptic drug-induced side effects, spasticity, ischemic brain damage, epilepsy, anxiogenesis, traumatic brain injury, AD, OP-induced seizures, and neuronal cell death.[72] Other ChE inhibitors available, such as donepezil, physostigmine, and tacrine, also exhibit an antagonist effect on the NMDA receptor in addition to their inhibitory effect on AChE.[73] A comparative study demonstrated that HA is the most powerful in protecting mature neurons, followed by donepezil, physostigmine, and tacrine.[61] In this research, HA was particularly effective in protecting more mature neurons against neurotoxicity because of the presence of more functional NMDA receptors in mature neurons.

In addition to the loss of cholinergic function in patients with AD, glutamatergic and GABAergic neurotransmitter systems may also be compromised.[74] Thus, HA, with its ability to attenuate glutamate-mediated toxicity, may treat dementia as a preventive agent by slowing or blocking the pathogenesis of AD at an early stage.[70]

### 4.4.3.3 Oxidative Stress

Increased oxidative stress, which results from free radical damage to cellular function, can be involved in the events leading to AD, and it is also connected to lesions called tangles and plaques. Plaques are caused by the deposition of Aβ and observed in the brains of AD patients.[75,76] HA and tacrine were compared for their ability to protect against Aβ-induced cell lesion, level of lipid peroxidation, and antioxidant enzyme activities in rat PC12 and primary cultured cortical neurons.[62,63] After pretreatment of both cells with HA or tacrine (0.1–10 mM) before Aβ exposure, the survival of the cells was significantly elevated. Wang et al. found that both drugs are similarly protective against Aβ toxicity, which results in a reduction of cell survival and glutathione peroxidase and catalase activity, and both increase the production of malondialdehyde and superoxide dismutase. Administration of HA reduced the apoptosis (programmed cell death) that normally followed β-amyloid injection.[67] Prevention in the expression of apoptosis-related proteins and limitation in the extent of apoptosis in widespread regions of the brain were also seen. Wang et al. suggested that these actions may reflect a regulation of expression of apoptosis-related genes.

### 4.4.3.4 Hypoxic-Ischemic Brain Injury

It has been suggested that by having effects in the cholinergic system and on the oxygen-free radical system and energy metabolism, HA may be useful for the treatment of vascular dementia.[77] The protective effect of HA on an HI brain injury was investigated in neonatal rats in which a combination of common carotid artery ligation and exposure to a hypoxic environment caused great brain damage.[68] HA administrated daily to neonatal rats, at the dose of 0.1 mg/kg i.p. for 5 weeks after HI injury, produced significant protection from damage after HI injury and on behavior (decreased escape latency in water maze) and neuropathology (less extensive brain injury). Consequently, Wang et al. concluded that HA might be effective in the treatment of HI encephalopathy in neonates. Similar protection was obtained by administering subchronical oral doses of HA (0.1 mg/kg, twice daily for 14 days) after 5 min of global ischemia in gerbils.[78]

### 4.4.4 Toxicology

Toxicological studies conducted in different animal species indicated less severe undesirable side effects associated with cholinergic activation for HA than for other AChEIs such as physostigmine and tacrine.[42,79] In mice, the $LD_{50}$ doses were 4.6 mg per os (p.o.), 3.0 mg s.c., 1.8 mg i.p., and 0.63 mg i.v. Histopathological examinations showed no changes in liver, kidney, heart, lung, and brain after administration of HA for 180 days, in dogs (0.6 mg/kg i.m.) and in rats (1.5 mg/kg p.o.). No mutagenicity was found in rats, and no teratogenic effect was found in mice or rabbits.[80]

### 4.4.5 Effects on Miscellaneous Targets

It was reported that HA inhibited nitric oxide production from rat C6 and human BT325 glioma cells.[81] The actions of HA on the fast transient potassium current and the sustained potassium current were investigated in acutely dissociated rat hippocampus neurons by Li et al.[82] HA reversibly inhibited the transient potassium current, being voltage independent and insensitive to atropine. In fact, the inhibition on the fast transient potassium current might form a potential toxic effect of HA in AD treatment. In this context, HA seems safer than tacrine, as the latter was much more potent in the inhibition of the transient potassium current. The results suggested that HA may act as a blocker at the external mouth of the A channel.[82,83]

Human studies have confirmed the analgesic action of AChEIs, such as physostigmine and neostigmine. The antinociceptive effect of HA was also investigated in the mouse hot plate and abdominal constriction tests by Galeotti et al.[84] The results showed that HA could produce the dose-dependent antinociception in mice, without impairing motor coordination, by potentiating endogenous cholinergic activity. HA is endowed by muscarinic antinociceptive properties mediated by the activation of the central $M_1$ muscarinic receptor. So HA and other AChEIs could be employed as analgesic for the relief of painful human conditions.[84]

In conclusion, as an AChEI, HA possesses different pharmacological actions other than hydrolysis of synaptic ACh. HA has direct actions on targets other than AChE. These noncholinergic roles of HA could also be important in AD treatment. The therapeutic effects of HA are probably based on a multitarget mechanism.

## 4.5 CLINICAL TRIALS

Scores of clinical studies with HA have been reported thus far. Favorable efficacy of HA was demonstrated in the treatment of more than 1000 patients suffering from age-related memory dysfunction or dementia in China. An early study conducted on 100 patients with probable AD oral HA (0.15–0.25 mg, t.i.d.) showed significant improvement in all rating scores evaluated by the Buschke Selective Reminding task. An inverted U-shaped dose response curve for memory improvement was observed.[48,85,86] The most frequently occurring side effects with HA were related to its cholinergic property. The incidence of adverse events such as dizziness, nausea, and diarrhea with HA 0.2 mg was comparable with that observed with placebo control. No liver and kidney toxicity was detected.[87,88]

In early study, 99% of 128 patients with myasthenia gravis showed controlled or improved clinical manifestations of the disease. The duration of action of HA lasted $7 \pm 6$ h, and side effects were minimal compared with neostigmine.[89]

In the United States, the safety and efficacy of HA were evaluated in 26 patients meeting the DSM IV-R and the NINCDS-ADRDA criteria for uncomplicated AD and possible or probable AD.[90] This study (office-based) lasted 3 months and was open label. Other therapies, including tacrine, donepezil, and *G. biloba* were continued. An oral dose of 50 µg HA was given twice a day to 22 patients, and the 4 other patients received a dose of 100 µg twice daily. A mean dementia baseline score of 22.6 was measured with the Mini-Mental State Examination (MMSE). The changes in this score, for the 50 µg group and for the 100 µg group, respectively, were 0.5 and 1.5 points at 1 month, 1.2 and 1.8 points at 2 months, and 1.1 and 1.0 points at 3 months. Despite the small number of patients, the authors observed dose-related improvements with higher MMSE scores at higher dosage and no serious side effects.

Sun et al. reported that HA enhanced the memory and learning performance of adolescent students.[56] With a double-blind and matched-pair method, 34 pairs of junior middle-school students complaining of memory inadequacy were divided into two groups. The memory quotient of the students receiving HA was higher than those of the placebo group, and the scores on Chinese language lessons in the treated group were also elevated markedly. They also finished a test in AD patients.[87] Sixty AD patients were divided into two groups taking HA ($4 \times 50$ µg p.o., b.i.d., for 60 days) in capsules and tablets, respectively. There were significant differences on all psychological evaluations between "before" and "after" the 60 days trials for the two groups. No severe side effects except moderate-to-mild nausea were observed. HA can reduce the pathological changes of the oxygen-free radicals in plasma and erythrocytes of AD patients as well.

A double-blind trial of HA on cognitive deterioration in 314 cases of benign senescent forgetfulness, vascular dementia, and AD was reported by Ma et al.[91,92] The first clinical trial was conducted by the double-blind method on 120 patients of age-associated memory impairment with a memory quotient <100. The dosage was 0.03 mg i.m., b.i.d., for 14–15 days. The effective rates were 68.3% and 26.4%, respectively, in the two groups. The second trial was conducted on 88 patients of age-associated memory impairment. The dosage was 0.1 mg HA p.o., q.i.d., for 14–15 days. The effective rates for the treated and control groups were 68.2% and 34.1%, respectively. No significant side effects were observed except for gastric discomfort, dizziness, insomnia, and mild excitement.

Another placebo-controlled, double-blind, randomized trial of HA in treatment of mild-to-moderate AD has been evaluated by Zhang et al.[93] Overall, 202 patients aged between 50 and 80 years enrolled from 15 centers nationwide in China were randomly divided into a HA treatment group ($n = 100$ p.o., 400 µg/day for 12 weeks) and a placebo group ($n = 102$) to undergo 12 weeks of testing. There was a significant difference between the two groups at 6 weeks, indicating that

HA improved the condition of the patients from Week 6. In comparison with the baseline data, the HA group improved significantly the cognitive function, activity of daily life (ADL), noncognitive disorders, and overall clinical efficacy. Mild and transient adverse events (edema of bilateral ankles and insomnia) were observed in 3% of HA-treated patients.

Chang et al. surveyed the effect of HA on promoting verbal recall in middle-aged and elderly patients with dysmnesia of varying severities and disclosed that HA has a fair effect on improving the ability of verbal recall, retention, and repetition in patients with mild and moderate dysmnesia.[94] The randomized double-blind crossing medication method was used for 50 middle-aged and elderly patients with dysmnesia. They were given each placebo or HA 100 µg p.o., b.i.d., for 2 weeks. Multiple selective verbal reminding tests were conducted before and after the medication. In patients with mild and moderate dysmnesia, values of $\Sigma$ recall ($\Sigma R$), long-term retrievel (LTR), random LTR, presentation, reminded recall and pass number, long-term storage, consistent LTR, and unreminded recall of HA group were markedly increased in contrast to those of the placebo group. Furthermore, HA was found to have no evident effect on most patients with severe dysmnesia. No severe adverse reactions and inhibition of blood ChE activity were encountered during the treatment with HA.

## 4.6  SYNTHESIS OF HA AND ITS ANALOGS

The multifaceted bioactivities of HA and its scarcity in nature have provided the impetus for renewed interests in the synthesis of this target molecule. Moreover, the structure activity relationship (SAR) of HA has been extensively studied.

### 4.6.1  Synthesis of Racemic HA

Total synthesis of racemic HA was first accomplished independently by both Qian and Ji[96] and Xia and Kozikowski[96] in 1989 (Schemes 4-1 and 4-2). Almost the same synthetic strategy was adopted starting from the β-keto ester **2** or **2a**. Qian and Ji prepared **2a** using traditional methods with 17.1% yield from ethyl aceto-acetate. The three-carbon bridge ring was constructed through tandem Michael-adlol reaction with methacrolein. Hereinafter, a MsOH elimination reaction to form an endocyclic double bond, a Wittig reaction for exocyclic double bond, and a Curtius rearrangement and deprotection reaction were successively conducted. The low yield of *E*-product of Wittig reaction is the important deficiency. Xia and Kozikowski synthesized **2a** using another route. In contrast to Qian and Ji's, it is longer and required expensive reagents such as PhSeCl and Pd(OH)$_2$. The remaining steps to *rac*-HA is similar to that of Qian and Ji's, with the exception of using PhSH/AIBN to enhance the *E/Z* product ratio from 10:90 to 90:10.

In 1990, Xia and Kozikowski improved the preparation of key intermediate **4** from **3** (Scheme 4-3).[97] They used an efficient one-pot, three-component process to prepare 2-pyridone **6** from a carbonyl compound, ammonia, and methyl propiolate, which enhanced the yield of **2a** and avoided those expensive reagents. In 1993,

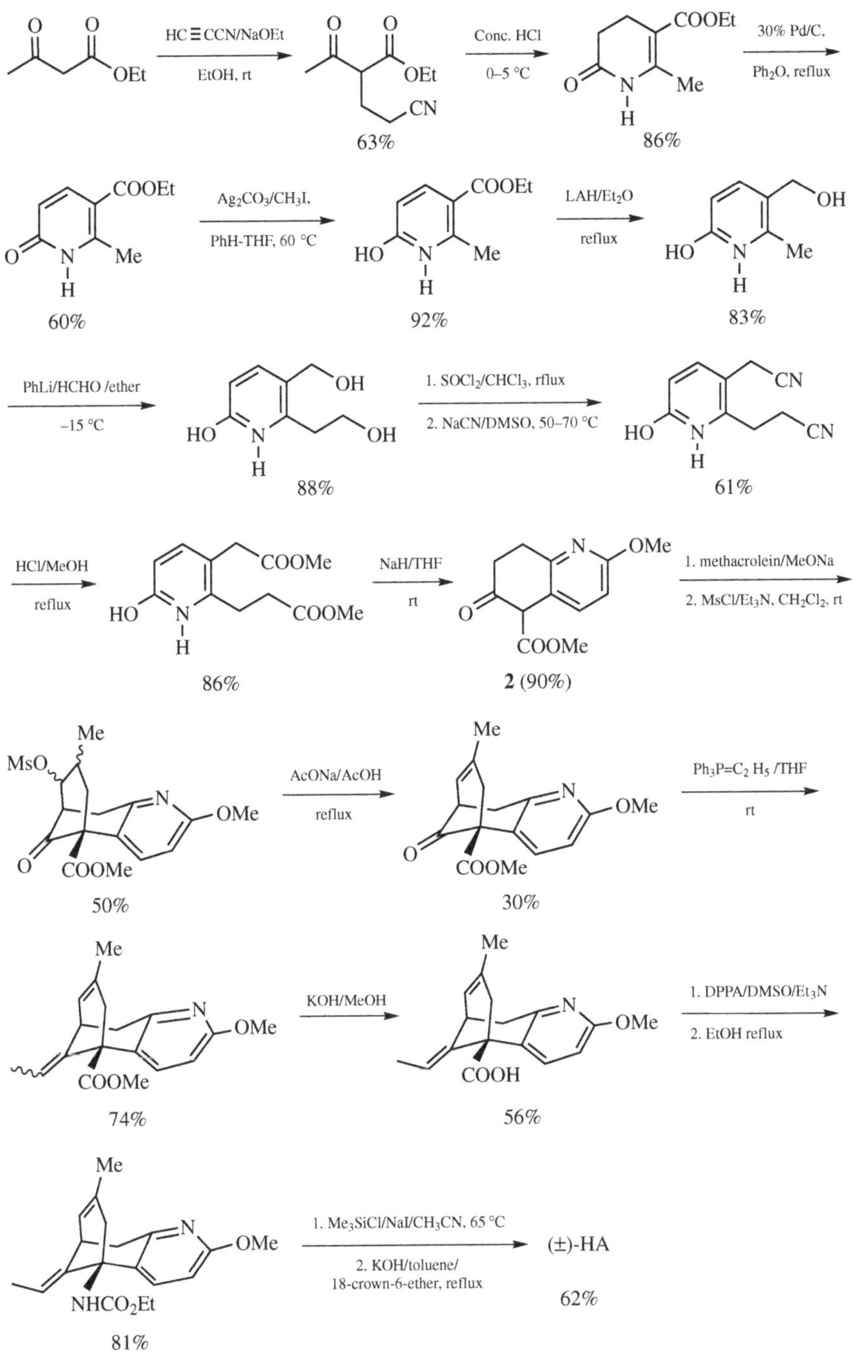

**Scheme 4-1.** Synthetic route to *rac*-HA by Qian and Ji.

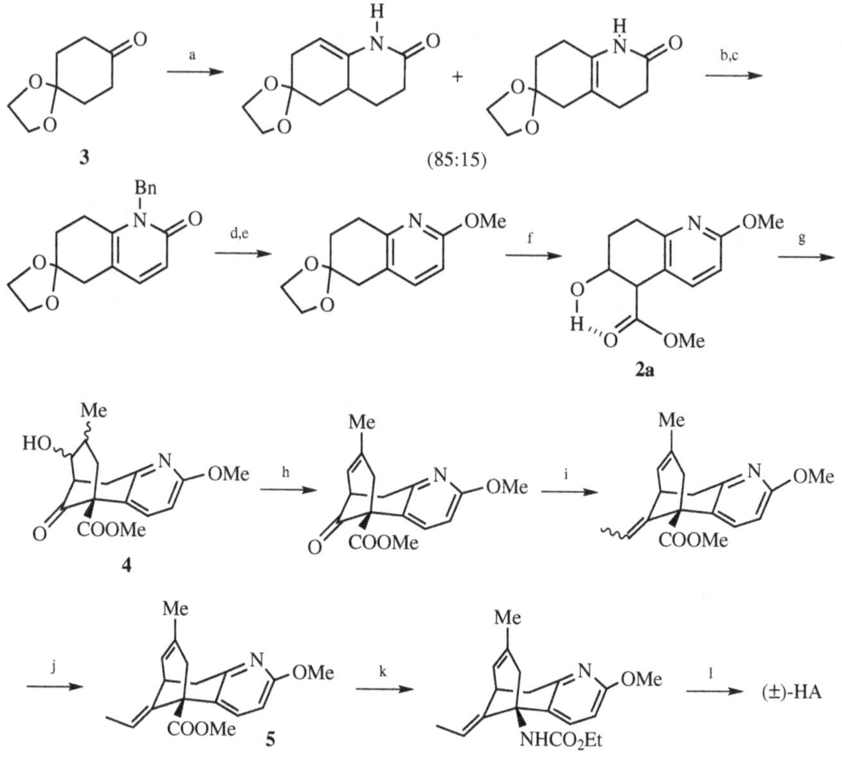

**Reagents:** a). Pyrrolidine, PhH, p-TsOH (catalyst), reflux; acrylamide, dioxane, reflux; H₂O, dioxane, reflux (70% overall); b). KH, BnCl, THF, rt (100%); c) LDA, PhSeCl, THF, –78 °C; NaIO₄; Et₃N, MeOH, Reflux (80%); d) H₂, Pd(OH)₂/C, HOAc, rt (80%); e) Ag₂CO₃, MeI, CHCl₃, rt (92%); f) 5% HCl, acetone, reflux (85%); KH, (MeO)₂CO, reflux (87%); g) methacrolein, tetramethylguanidine, CH₂Cl₂, rt (93%);
h) MsCl, Et₃N, DMAP, CH₂Cl₂ (96%); NaOAc, HOAc, 110 °C, 24 h (50%);
i) Ph₃P=CHCH₃, THF, 0 °C to rt (73%); j) PhSH, AIBN, 170 °C, 24h (100%);
k) 20% NaOH, THF, MeOH, reflux, 2 days (78% based on E ester); SOCl₂, toluene, 80 °C, 2h; NaN₃, 80 °C; MeOH, reflux (80% overall); l) TMSI, CHCl₃, reflux (92%).

**Scheme 4-2.** Synthetic approaches to (±)-HA by Xia and Kozikowski.

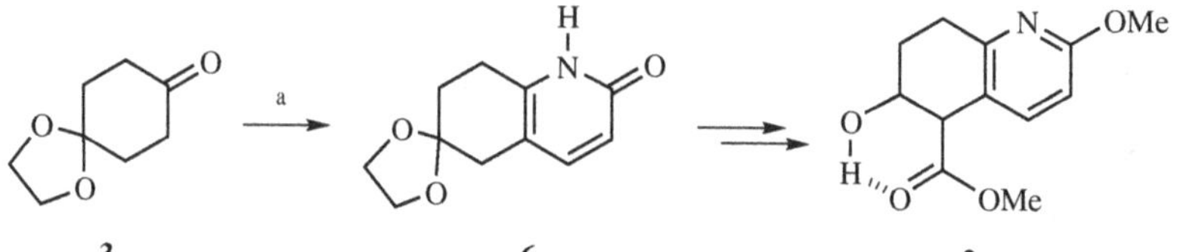

Reagents and conditions: a). methyl propiolate, NH₃, MeOH, 100 °C, 10 h (70%)

**Scheme 4-3.** One-pot process to prepare **6**.

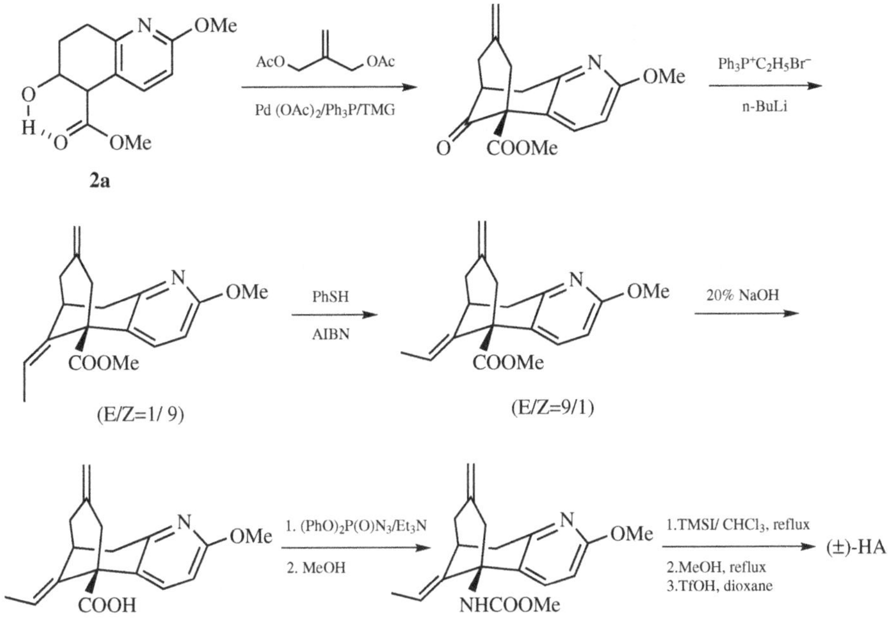

**Scheme 4-4.** Pd-catalyzed route to (±)-HA.

Xia and Kozikowski developed a palladium-catalyzed bicycloannulation route (Scheme 4-4) to racemic HA from **2a** in 40% overall yield.[98] The three-carbon bridge was more efficiently introduced by Pd-catalyzed alkylation of **2a** with 2-methylenepropane-1,3-diyl diacetate on both sides of the ketone carbonyl. Compared with (−)-HA ($IC_{50}$ for AChE inhibition 0.047 μM), the racemate exhibited an $IC_{50}$ of 0.073 μM, which is, within error, as expected if the unnatural enantiomer is inactive.

Camps et al. developed a route to (±)-HA from a keto capamate **7**, which was obtained in 22% overall yield from 1,4-cyclohexanedione monoethylene ketal (**3**) (Scheme 4-5).[99–101] This new approach to racemic HA features the elaboration of the pyridone moiety of IIA in a late stage. In this way, we can access the different heterocyclic analogs instead of the pyridone moiety in HA. However, the total yield of HA was not markedly improved, compared with that of other approaches, and the purification of the isomers proved to be a tedious and difficult task.

### 4.6.2  Synthesis of Optically Pure (−)-HA

For the AChE inhibition effect of natural (−)-HA being 38-fold more than its enantiomer (+)-HA, to synthesize natural (−)-HA attracted widespread attention. Yamada et al. first reported the route to optically pure (−)-HA in 1991.[102] On the basis of the route established to (±)-HA, Yamada et al. chose to introduce absolute stereochemistry at the stage of the Michael-aldol reaction, which creates the bridging ring of HA. As shown in Scheme 4-6, **2a** was transesterified with

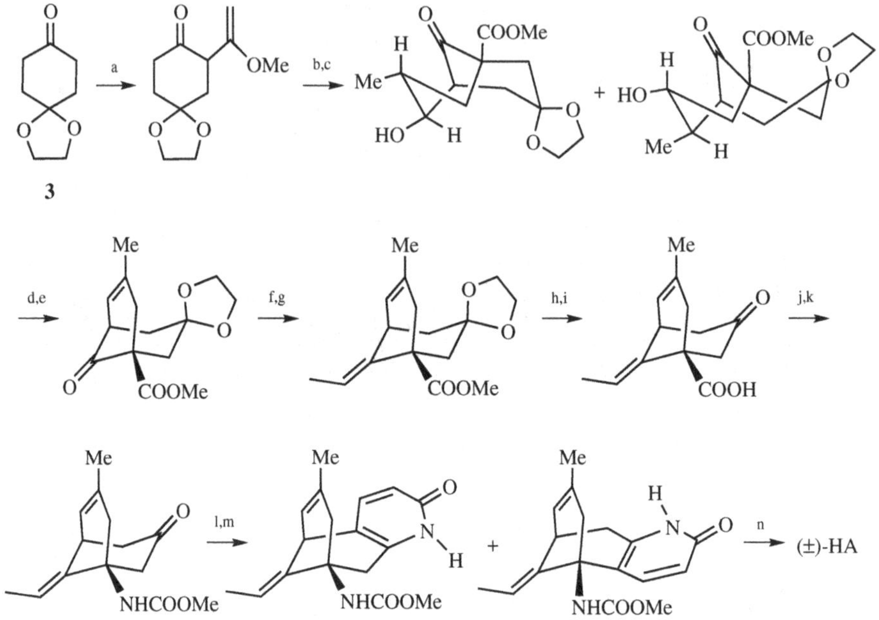

**Reagents:** a).Me$_2$CO$_3$, NaH/KH, THF; b) α-methylacrolein; c) TMG, CH$_2$Cl$_2$ or DBU, MeCN;
d) p-tolyl chlorothionoformate, Py; e) pyrolysis; f) ethyltriphenylphosphonium bromide, n-BuLi, THF;
g) thiophenol, AIBN, toluene; h) 20% NaOH, H$_2$O/THF/MeOH;
i) 2N HCl, dioxane; j) (PhO)$_2$P(O)N$_3$, Et$_3$N, chlorobenzene; k)MeOH;
l) pyrrolidine, molecular sieves, PhH; m) propiolamide;
n) n-PrSLi, HMPA; o) TMSI, CHCl$_3$; p) MeOH.

**Scheme 4-5.** Camps et al. approach to preparing racemic HA.

(−)-8-phenylmenthol and **8** reacted with methacrolein in the presence of tetramethylguanidine at room temperature (r.t.) over 2 days. A 90% yield of mixture **9** was isolated. Mixture **9** was transferred olefin **10** employing conditions identical with those reported previously.

Chen and Yang reported an approach to optical intermediate (5S, 9R)-**4**, which could be conveniently transformed optical (−)-HA via steps similar to Qian and Ji and Xia and Kozikowski's approaches to (±)-HA.[103] **2a** reacted with methacrolein in the presence of 0.1 equivalence quinine at r.t. over 10 days to obtain isomer (5S, 9R)-**4** (Scheme 4-7).

Kaneko et al. reported the preparation of the key intermediate (+)-**12** of (−)-HA via the asymmetric Pd-catalyzed bicycloannulation of the β-keto ester **2** with 2-methylene-1,3-propanediol diacetate **11** (Scheme 4-8).[104] The chiral ferrocenylphosphine ligand **13** gave 64% ee enantioselectivity.

Illuminated by these promising results, several new chiral ferrocenylphosphine ligands were thus prepared.[105,106] The enantioselectivity of the bicycloannulation

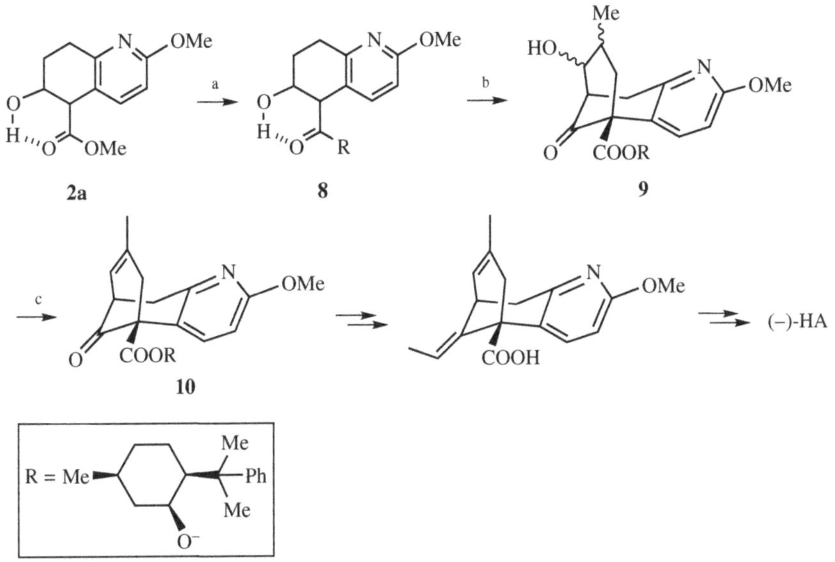

**Reagents and conditions:** a) RH, PhH, reflux, 3 days (91%); b).methacrolein, TMG, CH₂Cl₂, rt (90%); c). MsCl, Et₃N, DMAP, CH₂Cl₂, rt; NaOAc, HOAc, 110 °C.

**Scheme 4-6.** Route to optical pure (−)-HA by Yamada et al.

was evidently improved with ligand **14** to afford **12** in 81% ee. It was obvious that fine-tuning of the size of the *N*-substituent of the ligand with an appropriate chain length had a dramatic effect on the enantioselectivity of the reaction. Enantioselectivity of 90.3% ee for **12** was achieved with (*R*,*S*)-ferrocenylphosphine ligand **15** possessing a cyclopentyl group at the nitrogen. With the most efficient chiral ligand

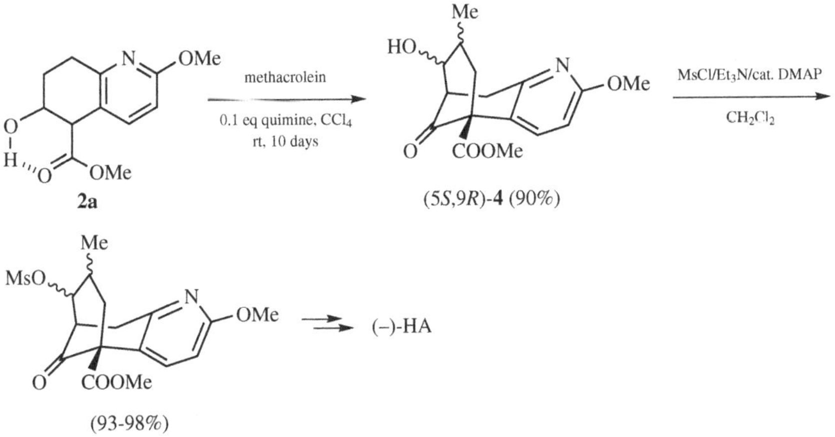

**Scheme 4-7.** The route to (5*S*,9*R*)-**4** by Chen and Yang.

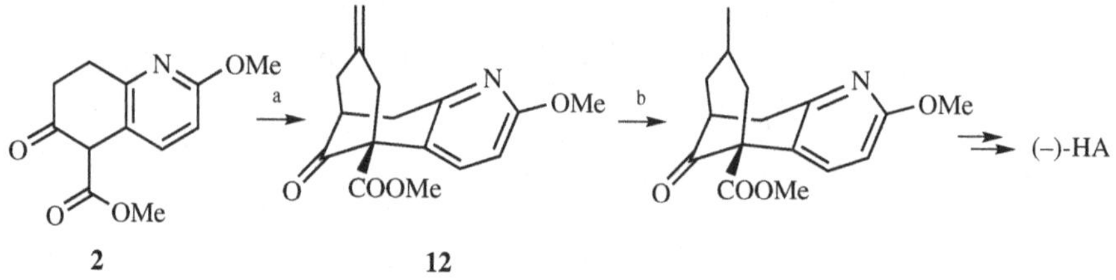

**2**                    **12**                    (–)-HA

**Reagents:** a) **11**, (η³-allyl)Pdcl, TMG,
chiral ligand, toluene; b) triflic acid, dioxane.

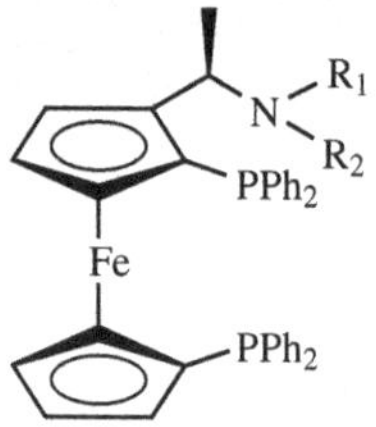

**13**: R₁=Me, R₂=(CH₂)₄OH
**14**: R₁=Et, R₂=(CH₂)₄OH
**15**: R₁=cyclopentyl, R₂=(CH₂)₅OH

**Scheme 4-8.** The preparation of (+)-**12** by Kaneko et al.

**15** in hand, the chiral nonracemic product **12** was obtained in the desired configuration for the synthesis of natural (–)-HA.[106]

Lee et al. developed a new method for the construction of the skeleton of HA via
the Mn(III)-mediated oxidative radical cyclization of allylic compound **16** derived
from **2**. The thermodynamically unstable *exo* double bond product **17** could be
easily isomerized to the *endo* olefin **18** by treating it with triflic acid as reported
in the literature (Scheme 4-9).[107]

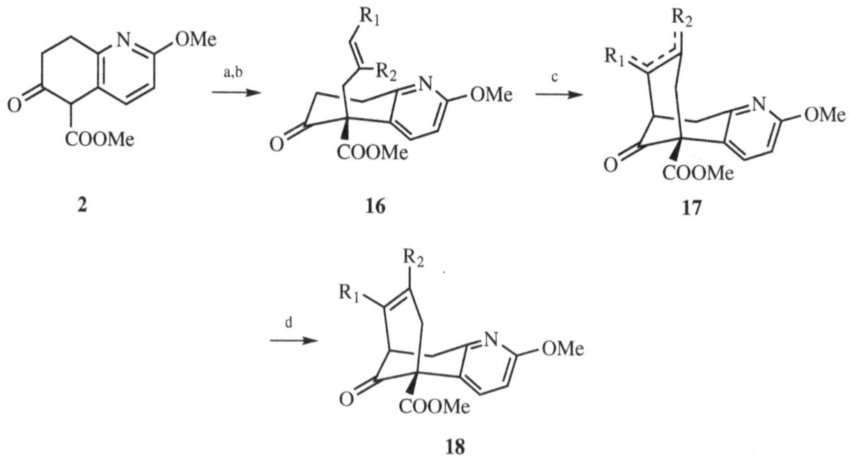

**2**                    **16**                    **17**

**18**

**Reagents**: NaH/DMF; b) allylic bromides; c) Mn(OAc)₃, Cu(OAc)₂, AcOH; d) triflic acid

**Scheme 4-9.** The new method constructing the skeleton of HA by Lee et al.

### 4.6.3 Studies on the Structure–Activity Relationship

#### 4.6.3.1 Synthetic (±)-HA Analogs

The powerful bioactivities of HA attracted the attention of scientists for its SAR. Series of HA analogs were generated by adding, omitting, or modifying substituents of HA by Kozikowski et al.,[98,108–113] He et al.,[114,115] Kaneko et al.,[116] Zeng et al.,[117] Zhou and Zhu,[118] and Hogenauer et al.[119,120] These HA analogs are listed in Figure 4-2, which were modified on the quinolinone ring (**19a–e, 20–22**), exocyclic ethylidene (**23a–i, 24a–d**), primary amino (**25a–k**), three-carbon bridge ring (**26a–f, 27a–f**), and multiplicate moieties (**28a–h**). AChE IC$_{50}$ value tests on many HA analogs disclosed that they were far less active than racemic HA. Furthermore, although C-10 axial methyl (**27b**), C-10 dimethyl (**27a**), and (−)-C-10

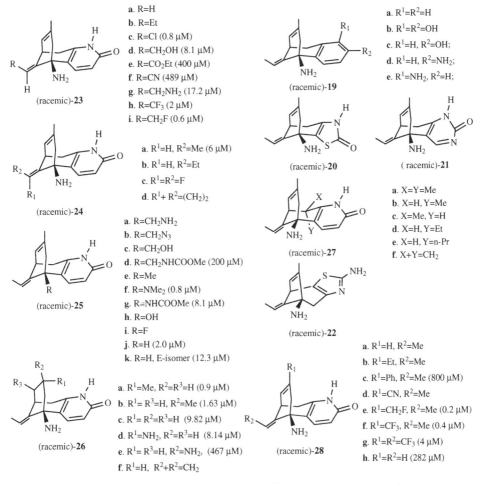

**Figure 4-2.** Racemic HA analogs (part IC$_{50}$ values in parentheses).

spirocyclopropyl (**27f**) analogs of HA have been found to have comparable or somewhat more potent anti-AChE activities than (±)-HA, the preparation of these analogs are laborious, and the costs will be even far more expensive than natural HA.

### 4.6.3.2  Derivatives From Natural HA

Due to the rigid configuration of HA, the structural modification for natural HA is focused on the pyridone ring and the primary amino group.[121,122] As shown in Figure 4-3, the reduction of two pairs of double bonds and the adding of substituents on the pyridone ring resulted in far lower anti-AChE activity. However, structural modification of the primary amino group had encouraging results, especially a few of Schiff base derivatives of HA.[123] Taking into account the chemical unstability of Schiff base derivatives, we designed and prepared ZT-1, which possessed an aromatic ring with a Cl-atom to attenuate the electron cloud density of the N=C group, and an intramolecular H-bond is formed through a six-numbered ring. ZT-1 has the longer duration time, lower toxicity, and better bioavailability compared with HA. Now, phase II of the clinical trials of ZT-1 is underway; for details, see Section 4.8.

### 4.6.3.3  Simplified Analogs of HA

As a promising lead compound, scientists have been interested in the chemical modification of HA in the search for new analogs that may possess higher activity, longer duration of action, less toxicity, and could be prepared by simpler and efficient approaches as compared with HA.

Several types of HA-simplified analogs have been designed and synthesized (Figure 4-4). All of them possess the supposed pharmacophore moiety of HA. 5-Substituted aminomethyl-2(1H)-pyridones **29**;[109,114,124,125] 5-substituted amino-5,6,7,8-tetrahydroquinolinones **30**, **31**, and **32**;[110,114,115,126] and 5-substituented aminoquinolinones 5-substituted aminoquinolinones **33**,[127] and simplified analogs without the pyridone ring **34**;[128] **35** and **36**[129] and **37**[127] were prepared and their anti-AChE were tested. Camps et al. prepared some HA analogs **38**, **39a–d**, **40a–c**, which kept the tetrahydroquinolinone and three-carbon bridge ring of HA.[101,130,131] These analogs were found to be inactive or less active in the inhibition of AChE. It means that the conformational constraints, hydrophobic binding, and steric and electrostatic fields provided by the unsaturated bridge and fused pyridone ring in HA must be involved in its AChE inhibitory activity.

### 4.6.3.4  Hybrids of HA With Other AChEIs

In 1998, Badia et al. reported that 17 polycyclic HA analogs combined the 4-aminoquinoline moiety of tacrine with the bridged carbobicyclic moiety of HA.[132] Hybrid compounds **41a–c** showed AChE inhibition activities approximately −2.5, 2, and 4 times higher than tacrine, respectively, but no direct comparison with HA.

The hybrid analogs **42** ($n = 4$–10, 12) comprising tacrine and a moiety of HA were all more potent AChEIs than tacrine and HA. When $n = 10$, **42** displayed the

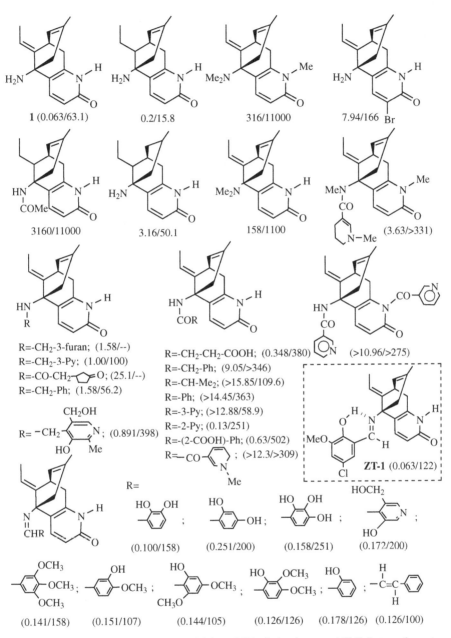

**Figure 4-3.** IC$_{50}$ (μM) of anti-ChE activities of HA derivatives on AChE (rat erythrocyte membrane) and BuChE (rat serum) over a concentration range from 1 nM to 10 nM. Data from Refs. 121 and 122.

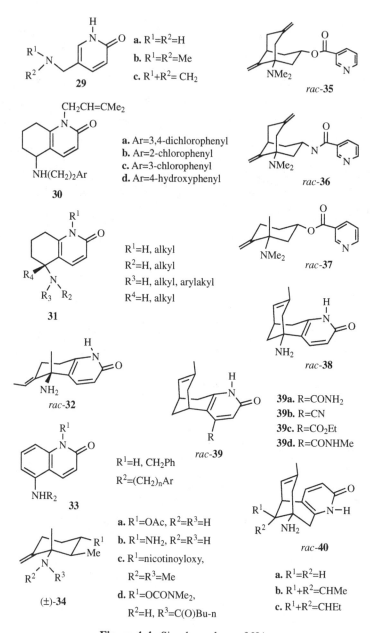

**Figure 4-4.** Simple analogs of HA.

highest potency against AChE, being 13-fold more potent than HA, but the selectivity far less than HA.[133]

Inspired by the bivalency or dimer strategy and the example of *bis*-tacrine,[134] Carlier et al. designed and prepared several alkylene-linked dimers of 5-amino-5,6,7,8-tetrahydroquinolin-2-one, bis(*n*)-hupyridones **43a–b** and **44** (see Figure 4-5).[135] The mixtures of *rac*- and *meso*-diastereomers **43a** showed dramatically enhanced anti-AChE potency. The highest potency was observed at $n = 12$, and the $IC_{50}$ value is 159 nM with the control of $(-)$-HA equal to 115 nM. (*rac, meso*)-**43b** were also optimized at $n = 12$, but they were less potent than (*rac, meso*)-**43a**. (*rac, meso*)-**44** ($n = 12$) was 31-fold less potent than (*rac, meso*)-**43a**.

Jin et al. designed and synthesized a series of *bis*-$(-)$-HA (**45**) with various lengths of the alkylene tethers.[136,137] Pharmacological tests found these dimers were less potent than HA. Zeng et al. reported on a HA-E2020 combined compound **46**,[138] which was a mixture of four stereoisomers. The $IC_{50}$ value of **46** ($>190 \ \mu M$) was much higher than that of $(-)$-HA ($0.082 \ \mu M$).

Based on the anti-AChE activity of huperzine B (HB, **47**),[139] a natural homologue of HA, Rajendran et al. prepared **48a–c**, hybrid analogs of HA and HB.[140,141] In comparison with $(\pm)$-HA, **48a–c** are approximately 10–100 times less active in AChE inhibition.

Badia et al. designed and synthesized more than 30 huprines, which are the tacrine-HA hybrids of the 4-aminoquinoline moiety of tacrine combined with the bridged carbobicyclic moiety, without the ethylidene substituents, of HA.[132] Pharmacological studies of these compounds demonstrated that they are a novel class of potent and selective AChEIs. 3-Chloro-substituted huprines **49a** and **49b** are the

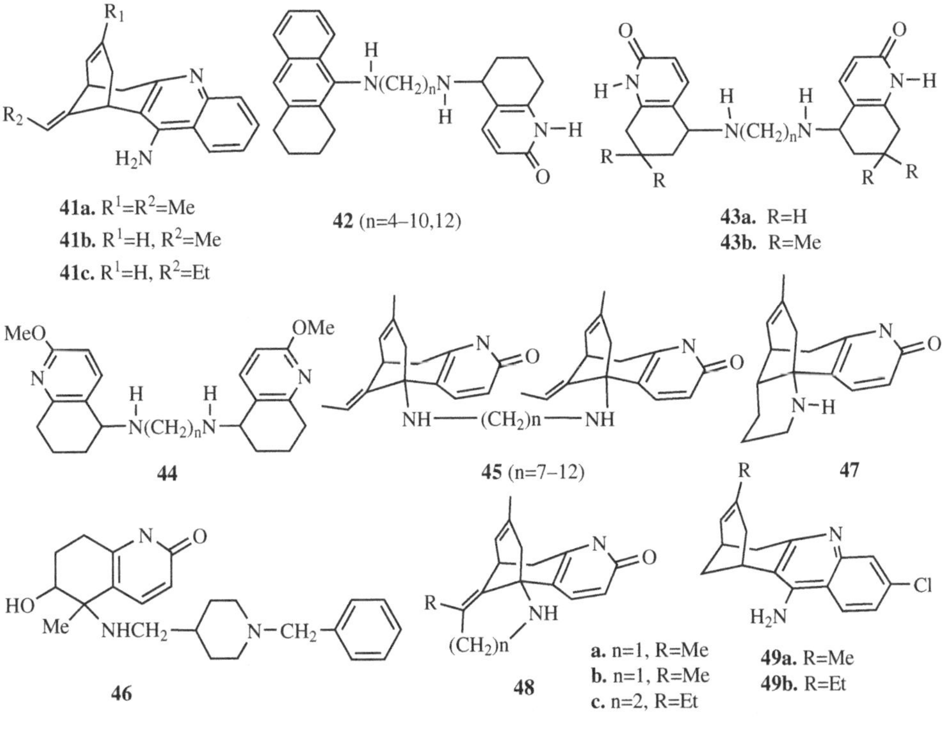

**41a.** $R^1=R^2=Me$
**41b.** $R^1=H, R^2=Me$
**41c.** $R^1=H, R^2=Et$

**42** (n=4–10,12)

**43a.** R=H
**43b.** R=Me

**44**

**45** (n=7–12)

**47**

**46**

**48**
  **a.** n=1, R=Me
  **b.** n=1, R=Me
  **c.** n=2, R=Et

**49a.** R=Me
**49b.** R=Et

**Figure 4-5.** Dimer hybrids of HA with other AChEIs.

most potent and selective human AChEIs among them. They showed high inhibitory activity toward human AChE (*h*AChE) with IC$_{50}$ values of 0.318 and 0.323 nM, respectively.

Additional studies on **49a** and **49b** have shown that both compounds act as the tight-binding and reversible AChEIs. They cross the blood-brain barrier and bind to the *h*AChE with a $K_i$ value of around 30 µM, which is one of the highest affinities reported in the literature. The affinity of both compounds for *h*AChE is 180-fold higher than that of HA.[142,143]

## 4.7 STRUCTURAL BIOLOGY

### 4.7.1 Interaction Between HA and AChE

The crystal structure of the complex of *Torpedo californica* AChE (*Tc*AChE) with natural HA at 2.5 Å resolution was conducted by Raves et al.[144] The result showed an unexpected orientation for the inhibitor, with surprisingly few strong direct interactions with protein residues to explain its high affinity. HA was found to be bound with aromatic residues in the active site gorge of *Tc*AChE, which localizes between tryptophan at position 86 (Trp86) and tyrosine at position 337 (Tyr337) in the enzyme.[9] Only one strong hydrogen bond is formed between the pyridine oxygen of HA and Tyr130. The ring nitrogen hydrogen binds to the protein through a water molecule. The hydrogen-bonding network is formed between the –NH$_3^+$ group and the protein through several waters (Figure 4-6a). The perfected orientation of HA within the active site makes the ethylidene methyl group form a cation-like (or termed the C–H$\cdots\pi$ hydrogen bond) interaction with Phe330 (Figure 4-6b). The

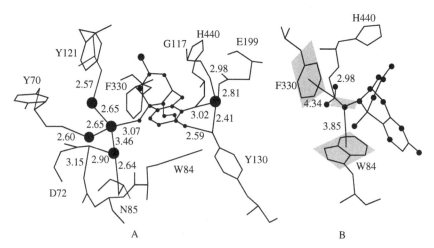

**Figure 4-6.** The interactions of (−)-HA with the active site of *Tc*AChE. (a) The hydrogen bonding networks. The water molecules are represented by red balls. (b) The C–H...π hydrogen bonds. The distances are given in ångströms. Copy from Ref. 10.

formation of the AChE–HA complex is rapid, and the dissociation is slow.[145] This complex has been studied with kinetic, computer-aided docking and x-ray crystallography approaches.

The three-dimensional (3-D) computer image of AChE–HA binding generated in the Raves et al. study revealed how the HA blocks the enzyme by sliding smoothly into the active site of AChE where ACh is broken down, and how it latches onto this site via many subtle chemical links. It was also demonstrated that HA can form an extra hydrogen bond with Tyr 337 within the choline site that exists only in mammalian AChE, but not in *Torpedo* enzyme and BuChE.[146,147] The stronger inhibitory property of HA for mammalian AChE than for the other two enzymes may rely on this particular interaction.

Xu et al. investigated how HA enter and leave the binding gorge of AChE with steered molecular dynamics (SMD)[148] simulations.[149] The analysis of the force required along the pathway shows that it is easier for HA to bind to the active site of AChE than to disassociate from it, which interprets at the atomic level the previous experimental result that the unbinding process of HA is much slower than its binding process to AChE. The direct hydrogen bonds, water bridges, and hydrophobic interactions were analyzed during two SMD simulations. The break of the direct hydrogen bond needs a great pulling force. The steric hindrance of the bottleneck might be the most important factor for producing the maximal rupture force for HA to leave the binding site, but it has little effect on the binding process of HA with AChE. Residue Asp72 forms a lot of water bridges, with HA leaving and entering the AChE binding gorge, acting as a clamp to take out HA from or put HA into the active site. The flip of the peptide bond between Gly117 and Gly118 has been detected during both the conventional MD and the SMD simulations. The simulation results indicate that this flip phenomenon could be an intrinsic property of AChE, and the Gly117-Gly118 peptide bond in both HA bound and unbound AChE structures tends to adopt the native enzyme structure. Finally, in a vacuum, the rupture force is increased up to 1500 pN, whereas in a water solution, the greatest rupture force is about 800 pN, which means water molecules in the binding gorge act as a lubricant to facilitate HA entering or leaving the binding gorge.

### 4.7.2  Structure-Based HA Analog Design

The x-ray structure of complexes of *Tc*AChE with HA and other AChE inhibitors displayed that these noncovalent inhibitors vary greatly in their structures and bind to different sites of the enzyme, offering many different starting points for future drug design. To rationalize the structural requirements of AChE inhibitors, Kaur and Zhang attempted to derive a coherent AChE-inhibitor recognition pattern based on literature data of molecular modeling and quantitative SAR analyses.[150] It is concluded that hydrophobicity and the presence of an ionizable nitrogen are the prerequisites for the inhibitors to interact with AChE. It is also recognized that water molecules play a crucial role in defining these different 3-D positions.

To date, more than 30 structures of the ligand-AChE complexes have been determined by x-ray crystallography (http://www.rcsb.org/pdb/index.html). Great efforts for designing potent novel inhibitors have been undertaken based on the available 3-D structures of the inhibitor-AChE complexes.[132–135,142,143,151–156]

Considering that the bridgehead amino group of HA is not part of a direct interaction with *Tc*AChE, Hogenauer et al. prepared 5-desamino HA (**25j**) and revealed that it had 100-fold less activity than HA, which indicates that the amino functionality is necessary for biological activity.[119]

The x-ray crystal structures of tacrine-*Tc*AChE[157] and (−)-HA-*Tc*AChE[144] complexes indicate that the binding sites for tacrine and (−)-HA within *Tc*AChE are adjacent and partially overlapped[158] Camps et al.[155] designed a series of huprines (**38–41** and **49**), which combine the probable pharmacophores of (−)-HA and tacrine. The synthesis and bioassay of these hybrids have been reviewed in Section 4-6. Of the series, huprine X **49a** showed the highest potency, which inhibited *h*AChE with an inhibition constant, $K_i$, of 26 pM, being about 180-fold more potent than (−)-HA and 1200-fold more potent than tacrine.[151] To explain the SARs at a more quantitative level, Camps et al. performed a molecular modeling study on a series of huprines.[152] The predicted free energy values are in general agreement with the inhibitory activity data of these inhibitors, and the modeling results rationalized the binding modes of these compounds to AChE. The crystal structure of **49a** complexing with *Tc*AChE was determined by Dvir et al. at 2.1 Å resolution.[158] In general, huprine X binds to the anionic site and hinders access to the esteratic site. Its aromatic portion occupies the same binding site as tacrine, stacking between the aromatic rings of Trp84 and Phe330, whereas the carbobicyclic unit occupies the same binding pocket as (−)-HA. Its chlorine substituent was found to lie in a hydrophobic pocket interacting with the rings of the aromatic residues Trp432 and Phe330 and with the methyl groups of Met436 and Ile439.

The complexes of *Tc*AChE with such bisquatrnary ligands as decamethonium [DECA, $Me_3N^+(CH_2)_{10}N^+Me_3$][157] and BW284C51[159] led to the assignment of Trp279 as the major element of a second, remote binding site, which is near the top of the active-site gorge, named the peripheral "anionic" site, about 14 Å apart from the active site. These structural assignments promoted the development of bivalent AChE inhibitors **42–46** capable of binding sites simultaneously to improve drug potency and selectivity.[133–135,156] The first example of bivalent inhibitors is the heptylene-linked tacrine dimer, *bis*(7)-tacrine, designed and synthesized by Pang et al. on the basis of computational studies.[134] *Bis*(7)-tacrine showed significantly higher potency and selectivity for inhibition of rat AChE than did monomeric tacrine.[134,156] Later on, Carlier et al. designed and synthesized the bivalent inhibitors **42** that are composed of a key fragment of HA and an intact tacrine unit.[133] The most active compound in this series is 13-fold more potent than HA and 25-fold more potent than tacrine; however, their selectivity is lower than HA. Also, Carlier et al.[135] designed and synthesized a series of dimers **43** of hupridone, an easily synthesized but pharmacologically inactive fragment of (−)-HA on the basis of docking modeling. Although these HA-like dimers are not as potent as *bis*(7)-tacrine or the tacrine–HA fragment heterodimer on rat AChE, being only 2-fold more potent than (−)-HA for the most active compound, they are superior to the latter dimers in terms of selectivity for AChE.[160]

In 2002, Dvir et al. determined the crystal structures of the complexes of HB (**47**) and (+)-HA with *Tc*AChE at 2.10 and 2.35 Å resolution, respectively.[161] The dissociation constants of (+)-HA, (−)-HA, and HB were reported with the values of 4.30, 0.18, and 0.33 μM, respectively. All three constants interact with the "anionic" subsite of the active site, primarily through π–π stacking and through Van Der Waals or C–H...π interactions with Trp84 and Phe330. Because their *R*-pyridone moieties are responsible for their key interactions with the active site via hydrogen bonding, and possibly via C–H...π interactions, all three maintain similar positions and orientations with respect to it. The carbonyl oxygens of all three seem to repel the carbonyl oxygen of Gly117, which thus causes the peptide bond between Gly117 and Gly118 to undergo a peptide flip. As a consequence, the position of the main chain nitrogen of Gly118 in the "oxyanion" hole in the native enzyme becomes occupied by the carbonyl of Gly117. Furthermore, the flipped conformation is stabilized by hydrogen bonding of Gly117*O* to Gly119*N* and Ala201*N*, the other two functional elements of the three-pronged "oxyanion hole" characteristic of ChEs. All three inhibitors thus would be expected to abolish hydrolysis of all ester substrates, whether charged or neutral.

Wong et al. determined the crystal structures of two bis-hupyridones, (*S,S*)-**43a** (*n* = 10) and (*S,S*)-**43a** (*n* = 12), the potent dual-site inhibitors of AChE.[160] The structures revealed that one hupyridone unit bound to the "anionic" subsite of the active-site, as observed for the *Tc*AChE-(−)-HA complex, and the second hupyridone unit was located near Trp279 in the "peripheral" anionic site at the top of the gorge. Both (*S,S*)-**43a** (*n* = 10) and (*S,S*)-**43a** (*n* = 12) fit the active-site gorge. The results confirm that the increased affinity of the dimeric HA analogs for AChE is conferred by binding to the two "anionic" sites of the enzyme. The structures provided a good explanation for the inhibition data showing that (*S,S*)-**43a** (*n* = 10) binds to *Tc*AChE about 6–7- and >170-fold more tightly than (*S,S*)-**43a** (*n* = 12) and (−)-HA, respectively. In comparison with the crystal structure of mouse AChE, Kozikowski et al.[110] rationalized the lower binding affinity of (*S,S*)-**43a** (*n* = 10) and (*S,S*)-**43a** (*n* = 12) for rat AChE, which shows that (*S,S*)-**43a** (*n* = 12) binds about three- and two-fold more tightly than (*S,S*)-**43a** (*n* = 10) and (−)-HA, respectively.

## 4.8   ZT-1: NEW GENERATION OF HA AChE

ZT-1 is a Schiff base derivative from natural HA, and its chemical name is [5*R*-(5α,9β,11*E*)]-5-{[(5-chloro-2-hydroxy-3-methoxyphenyl)methylene]amino}-11-ethylidene-5,6,9,10-tetrahydro-7-methyl-5,9-methanocycloocta[b]pyridin-2(1*H*)-one. ZT-1 is a prodrug and is transformed nonenzymatically into the active compound HA. In aqueous solution, ZT-1 is rapidly degraded into HA and 5-Cl-*o*-vanillin by hydrolysis. Now, ZT-1 is being developed as a drug candidate for the treatment of AD by Debiopharm S.A. of Switzerland.[a]

---

[a]Updated information is available for ZT-1. Please visit http://www.debio.com

### 4.8.1 Pharmacology

#### 4.8.1.1 ChE Inhibition

In vitro studies have demonstrated that ZT-1 is a potent and selective AChEI in rat cortex homogenate and in red blood cell AChE from different species (rat, bovine, and human). In contrast, ZT-1 presents a much weaker inhibitory activity on rat BuChE. In vitro, the AChE inhibitory effect of ZT-1 and HA is in the same range and slightly stronger than tacrine.

In vivo, a marked dose-dependent inhibition of AChE present in whole brain and in different brain regions (cortex, hippocampus, and striatum) was observed in rats after intragastric (i.g.) administration of ZT-1 (0.2–0.8 mg/kg) and was similar to HA (0.1–0.5 mg/kg), donepezil (3.4–6.7 mg/kg), and tacrine (14.1–28.1 mg/kg). The inhibition of serum BuChE was weaker with ZT-1 and HA than with donepezil and tacrine.

Maximal AChE inhibition in rat whole brain was reached 1 hour after i.g. administration of ZT-1 (0.4 mg/kg), HA (0.4 mg/kg), donepezil (6.7 mg/kg), and tacrine (28.1 mg/kg). Significant inhibition in cortex AChE was observed between 0.5 and 3 hours with all four AChEIs. Significant inhibition was still present for ZT-1 and HA at 6 hours, but not for donepezil and tacrine. Peak inhibition of serum BuChE was almost comparable for ZT-1 (28%) and HA (34%), slightly more pronounced for donepezil (39%), and clearly greater for tacrine (65%). The BuChE activity returned to near control level 6 hours postadministration for ZT-1 or HA, whereas it was still partially inhibited for donepezil and especially tacrine.

ZT-1 induced a dose-dependent elevation of ACh in the cortex of conscious rats. Compared with donepezil, ZT-1 was found to be 20-fold more potent at increasing cortical ACh levels; it showed a longer duration of action and fewer side effects.

#### 4.8.1.2 Cognitive Enhancing in Animal

The effects of ZT-1, HA, and donepezil on the scopolamine-induced memory deficits in rats were compared in the radial maze test. The memory deficits were significantly reversed by ZT-1 (0.1–0.3 mg/kg i.p. or i.g.), HA (0.1–0.2 i.p. or 0.2–0.4 mg/kg i.g.) and donepezil (0.3–0.6 mg/kg i.p. or 0.6–0.9 mg/kg i.g.), respectively. The dose-response curve of each compound was bell-shaped, with the maximum improvements at 0.2 mg/kg i.g. and i.p. for ZT-1 and HA and at 0.6 mg i.p. and 0.9 mg i.g. with donepezil.

A study in monkeys showed that ZT-1 (1.5–300 μg/kg) administered i.m. reversed the memory deficits induced by scopolamine (30 μg/kg i.m.) in young adult monkeys ($n = 4$) as well as in aged monkeys.

### 4.8.2 Toxicology

The ZT-1 toxicology investigations consisted of acute toxicity studies in rats and mice (oral and s.c. routes), 4-week repeated-dose toxicity studies in rats and dogs (oral and s.c. routes) with a 2-week recovery period, and toxicokinetic assessment, in vitro genotoxicity tests, and safety pharmacology. In these investigations, the observed effects were exaggerated pharmacological cholinergic effects characteristic of ChEIs, and most of these effects were no more present at the end of the recovery phase. Moreover, ZT-1 was not mutagenic in two in vitro tests (Ames test and mouse lymphoma test).

### 4.8.3   Pharmacokinetics

#### 4.8.3.1   In Animal
The pharmacokinetics of ZT-1 (i.v., p.o., and s.c.) was investigated in dogs. ZT-1 was rapidly absorbed after oral and s.c. administration, with a $T_{max}$ of 1.5 and 0.5 h, respectively. Whatever the route of administration, ZT-1 was rapidly hydrolysed into its active metabolite, HA, which showed a $T_{max}$ between 0.5 and 3 h. Terminal half-life was 0.6–1.4 h for ZT-1 and about 3 h for HA.

Peak concentrations ($C_{max}$) were 151, 10, and 33 nM for ZT-1 and 83, 55, and 228 nM for HA after an administration of the i.v. (0.2 mg/kg of ZT-1), oral (0.6 mg/kg), and s.c. (0.6 mg/kg) routes, respectively. Oral and s.c. bioavailability of ZT-1 was estimated as 8% and 26%, respectively. With respect to HA, its exposure represented 43% (oral) and 98% (s.c.) as compared with the i.v. administration.

In addition, for ZT-1 doses of 1–5 mg/kg administered orally and s.c. to dogs, drug exposure was proportional to dose, and a slight-to-moderate accumulation of ZT-1 and HA was observed after 4 weeks of daily ZT-1 administration.

#### 4.8.3.2   In Humans
After single escalating oral doses (1–3 mg), ZT-1 was rapidly absorbed and transformed into its active metabolite HA. Peak plasma concentrations were generally attained around 2 to 3 hours and around 1.3 to 2.5 hours postdosing, respectively. The terminal half-lives of ZT-1 and HA were fairly consistent across all dose levels and were typically about 4 and 20 hours, respectively. The mean residence times (MRTs) of ZT-1 and HA were about 7 and 27 hours, respectively. No ZT-1 was recovered in urine, and up to about 33% of the administered dose of ZT-1 was excreted as HA within 120 hours.

There was no statistical gender difference in pharmacokinetics, except for $C_{max}$ and AUC of both compounds that were sometimes increased in female subjects, which probably reflects the differences in body composition and weight.

Drug exposure was not proportional to dose, with ZT-1 and HA systemic availability increasing more than dose-proportionally in the 1- to 3-mg dose range, which suggests a nonlinearity at the level of absorption/first-pass metabolism or distribution. In addition, food delayed the peak concentrations of ZT-1 and HA and increased the systemic exposure to HA by 50–80%.

After the administration of repeated oral doses (0.5 mg, 1 mg, or 1.5 mg) over 14 days, steady-state plasma concentrations of HA were achieved within 3–5 days, which shows about a 2-fold accumulation compared with concentrations found after administration of single doses. No accumulation was observed for ZT-1. Drug exposure was fairly proportional to dose in the 1–1.5-mg dose range. The terminal half-lives of ZT-1 and HA were fairly similar across all dose levels (about 5 h and 20 h, respectively), and they were consistent with values measured after single administration of ZT-1.

### 4.8.4   Clinical Trials

Four clinical phase I studies involving 72 healthy elderly subjects have been performed and have showed that administration of ZT-1 in humans seems to be safe and well tolerated. The incidence of possibly drug-related adverse events, in particular nervous system and gastrointestinal symptoms, was similar to placebo for doses up to 1.5 mg. An international multicenter phase II trial for dose finding and efficacy assessment, in mild-to-moderate AD patients, is underway in 28 hospitals in Europe.

The initial assessment of the potential of ZT-1 to improve cognition showed that it could antagonize the cognitive impairment induced by scopolamine in healthy elderly volunteers. The study was conducted according to a randomized, placebo and positive-controlled, double-blind and crossover design. Donepezil was a positive internal control.

Overall, ZT-1 reduced the cognitive impairments produced by scopolamine on tasks measuring attention, working memory, episodic secondary memory, and eye–hand coordination. These findings suggest that ZT-1 may be an effective symptomatic treatment for the cognitive deficits associated with AD.

### ABBREVIATIONS

| | |
|---|---|
| ACh | acetylcholine |
| AChE | acetylcholinesterase |
| AChEI | acetylcholinesterase inhibitor |
| AD | alzheimer's disease |
| Aβ | amyloid β-peptide |
| b.i.d. | bis in die (= twice a day) |
| BuChE | butyrylcholinesterase |
| ChE | cholinesterase |
| ECG | electrocardiogram |
| GABA | γ-amino-n-butyric acid |
| HA | huperzine A |
| *h*AChE | human AChE |
| HB | huperzine B |
| HI | hypoxic-ischemic |
| *H. serrata* | *Huperzia serrata* (Thunb.) Trev. |
| i.g. | intragastric |
| i.m. | intramuscularly |
| i.p. | intraperitoneally |
| i.v. | intravenously |
| $IC_{50}$ | 50% inhibitory concentration |
| LTR | long-term retrievel |
| MMSE | Mini-Mental State Examination |
| NMDA | *N*-methyl-D-aspartate |

| OP | organophosphate |
| p.o. | Per os |
| q.i.d. | quarter in die ($=$ four times a day) |
| r.t. | room temperature |
| SAR | structure activity relationship |
| s.c. | subcutaneously |
| SMD | steered molecular dynamics |
| t.i.d. | ter in die ($=$ three times a day) |
| TCM | traditional Chinese medicine |
| *Tc*AChE | *Torpedo californica* AChE |

## REFERENCES

1. Liu, J. S.; Zhu, Y. L.; Yu, C. M.; Zhou, Y. Z.; Han, Y. Y.; Wu, F. W.; Qi, B. F. The structures of huperzines A and B, two new alkaloids exhibiting marked anticholinesterase activity. *Can. J. Chem.*, **1986**, 64: 837–839.

2. Xu, Z. L.; Chu, B. M.; Luan, X. H.; Wu, W. Z.; Cai, D. G. Structural identification of Fordine. *Med. J. Chinese PLA* (in Chinese), **1985**, 10(4): 263–264.

3. Wang, Y. E.; Yue, D. X.; Tang, X. C. Anticholinesterase activity of huperzine A. *Acta Pharmacol. Sin.*, **1986**, 7(2): 110–113.

4. Tang, X. C.; He, X. C.; Bai, D. L. Huperzine A: a novel acetylcholinesterase inhibitor. *Drugs Future*, **1999**, 24(6): 647–663.

5. Tang, X. C.; Han, Y. F. Pharmacological profile of huperzine A, a novel acetylcholinesterase inhibitor from Chinese herb. *CNS Drug Rev.*, **1999**, 5(3): 281–300.

6. Kozikowski, A. P.; Tuckmantel W. Chemistry, pharmacology, and clinical efficacy of chinese nootropic agent huperzine A. *Acc. Chem. Res.*, **1999**, 32: 641–650.

7. Bai, D. L.; Tang, X. C.; He, X. C. Huperzine A, potential therapeutic agent for treatment of Alzheimer's disease. *Curr. Med. Chem.*, **2000**, 7: 355–374.

8. Zeng, F. X.; Jiang, H. L.; Yang, Y. S.; Chen, K. X.; Ji, R. Y. Progress in synthesis and structural modification of huperzine A. *Prog. Chem.*, **2000**, 12(1): 63–76.

9. Zangara, A. The psychopharmacology of huperzine A: an alkaloid with cognitive enhancing and neuroprotective properties of interest in the treatment of Alzheimer's disease. *Pharmarcol. Biochem. Behav.*, **2003**, 75(3): 675–686.

10. Jiang, H. L.; Bai, D. L.; Luo, X. M. Progress in clinic, pharmacological, chemical and structural biological studies of huperzine A: a drug of traditional Chinese medicine origin for the treatment of Alzheimer's disease. *Curr. Med. Chem.*, **2003**, 10: 2231–2252.

11. Ghiso, J.; Frangione, B. Amyloidosis and Alzheimer's disease. *Adv. Drug Deliver Rev.*, **2002**, 54: 1539–1551.

12. Farlow, M.; Gracon, S. I.; Hershey, L. A.; Lewis, K. W.; Sadowsky, C. H.; DolanUreno, J. A controlled trial of tacrine in Alzheimer's disease. *JAMA*, **1992**, 268: 2523–2529.

13. Knapp, M. J.; Knopman, D. S.; Solomon, P. R.; Pendlebury, W. W.; Davis, C. S.; Gracon, S. I. A 30-week randomized controlled trial of high-dose tacrine in patients with Alzheimer's disease. *JAMA*, **1994**, 271: 985–991.

14. Rogers, S. L.; Farlow, M. R.; Doody, R. S.; Mohs, R.; Friedhoff, L.T. A 24-week, double-blind, placebo-controlled trial of donepezil in patients with Alzheimer's disease. *Neurology*, **1998**, 50(1): 136–145.

15. Zhou, H.; Jiang, S. H.; Tan, C. H.; Wang, B. D.; Zhu, D. Y. New epoxyserratanes from *Huperzia serrata*. *Planta Med.*, **2003**, 69(1): 91–94.

16. Tong, X. T.; Tan, C. H.; Zhou, H.; Jiang, S. H.; Ma, X. Q.; Zhu, D. Y. Triterpenoids constituents of *Huperzia miyoshiana*. *Chinese J. Chem.*, **2003**, 21(10): 1364–1368; Zhou, H.; Tan, C. H.; Jiang, S. H.; Zhu, D. Y. Serratenetype triterpenoids from *Huperzia serrata*. *J. Nat. Prod.*, **2003**, 66: 1328–1332.

17. Manske, R. H. F. In Manske, R. H. F., editor. *The Alkaloids*, vol. 5. New York: Academic Press; **1955**, 295–300.

18. Manske, R. H. F. In Manske, R. H. F., editor. *The Alkaloids*, vol. 7. New York: Academic Press; **1960**, 505–507.

19. MacLean, D. B. In Manske, R. H. F., editor. *The Alkaloids*, vol. 10. New York: Academic Press; **1968**, 305–382.

20. MacLean, D. B. In Manske, R. H. F., editor. *The Alkaloids*, vol. 14. New York: Academic Press; **1973**, 348–406.

21. Blumenkopf, T. A.; Heathcock, C. H. In Pelletier, S. W., *The Alkaloids*, vol. 3. New York: Academic Press; **1983**, 185–240.

22. Ayer, W. A. The *Lycopodium* alkaloids. *Nat. Prod. Rep.*, **1991**, 8: 455–463.

23. Ayer, W. A.; Trifonov, L. S. In Cordell, G. A.; Brossi, A., editors. *The Alkaloids: Chemistry and Pharmacology*, vol. 45. New York: Academic Press; **1994**, 233–266.

24. Tan, C. H.; Zhu, D. Y. The progress in the research of *Lycopodium* alkaloids. *Chin. J. Nat. Med.* (in Chinese), **2003**, 1: 1–7.

25. Yu, C. M.; Shen, W. Z.; Han, J. W.; Chen, Y. C.; Zhu, Y. L. Studies on the alkaloids of Chinese medicine, *Lycopodium serratum*. *Acta Pharm. Sin.*, **1982**, 17(10): 795–797.

26. Geib, S. J.; Tuckmantel, W.; Kozikowski, A. P. Huperzine A, a potent acetylcholines-terase inhibitor of use in the treatment of Alzheimer's disease. *Acta Crystallogr. C*, **1991**, 47: 824–827.

27. Valenta, Z.; Yoshimura. H.; Rogers, E. F.; Ternbah, M.; Wiesner, K. The structure of selagine. *Tetrahedron Lett.*, **1960**, 1(31): 26–33.

28. Chen, C. H.; Lee, S. S. Studies on the bioactive *Lycopodium* alkaloid—the structure and activity of isoselagine from *Lycopodium serratum var. longipetiolatum*. *J. Taiwan Pharm. Assoc.*, **1984**, 36: 1–7.

29. Sun, C. M.; Ho, L. K.; Sun, M. L. Revised stereochemistry of so-called isoselagine and spectroscopic analysis of 6α-hydroxylycopodine from *Lycopodium serratum var. long-ipetiolatum*. *Planta Med.*, **1993**, 59: 467–471.

30. Ainge, G. D.; Lorimer, S. D.; Gerard, P. J.; Ruf, L. D. Insecticidal activity of Huperzine A from the New Zealand clubmoss, *Lycopodium varium*. *J. Agr. Food Chem.*, **2002**, 50(3): 491–494.

31. Ma, X. Q. Ph.D thesis. Shanghai Institute of Materia Medica, Chinese Academy of Sciences, 1997.

32. Ma, X. Q.; Jiang, S. H.; Zhu, D. Y. Alkaloid patterns in huperzia and some related genera of Lycopodiaceae Sensu Lato occuring in China and their contribution to classification. *Biochem. Syst. Ecol.*, **1998**, 26: 723–727.

33. Chu, B. M.; Li, J. Applied high performed thin-layer scanning method to detect the content of fordine of 14 *Lycopodium* plants. *Trad. Chin. Herbs*, **1986**, 17(8): 109–110.

34. Ellman, G. L.; Courtney, K. D.; Andre, V.; Featherstone, R. M. A new and rapid colorimetric determination of acetylcholinesterase activity. *Biochem. Pharmacol.*, **1961**, 7: 88–95.

35. Cheng, D. H. Ren, H.; Tang, X. C. Huperzine A, a novel promising acetylcholinesterase inhibitor. *Neuroreport*, **1996**, 8: 97–101.

36. Scott, L. J.; Gao, K. L. Galantamine, a review of its use in Alzheimer's disease. *Drugs*, **2000**, 60(5): 1095–1122.

37. Ballard, C. G. Advances in the treatment of Alzheimer's disease: benefits of dual cholinesterase inhibition. *Eur. Neurol.*, **2002**, 47: 64–70.

38. Guillozet, A.; Smiley, J. F.; Mash, D. C.; Mesulam, M. M. Butyrylcholinesterase in the life cycle of amyloid plaques. *Ann. Neurol.*, **1997**, 42: 909–918.

39. Cheng, D. H.; Tang, X. C. Comparative studies of huperzine A, E2020, and tacrine on behavior and cholinesterase activities. *Pharmacol. Biochem. Behav.*, **1998**, 60: 377–386.

40. Tang, X. C.; De Sarno, P.; Sugaya, K.; Giacobini, E. Effect of huperzine A, a new cholinesterase inhibitor, on the central cholinergic system of the rat. *J. Neurosci. Res.*, **1989**, 24: 276–285.

41. Tang, X. C.; Kindel, G. H.; Kozikowski, A. P.; Hanin, I. Comparison of the effects of natural and synthetic huperzine A on rat brain cholinergic function *in vitro* and *in vivo*. *J. Enthnopharmacol.*, **1994**, 44: 147–155.

42. Wang, H.; Tang, X. C. Anticholinesterase effects of huperzine A, E2020, and tacrine in rats. *Acta Pharmacol. Sin.*, **1998**, 19: 27–30.

43. Laganiere, S.; Corey, J.; Tang, X. C.; Wulfert, E.; Hanin, I. Acute and chronic studies with the anticholinesterase huperzine A: effect on central nervous system cholinergic parameters. *Neuropharmacology*, **1991**, 30: 763–768.

44. Han, Y. F.; Tang, X. C. Preclinical and clinical progress with huperzine A: a novel acetylcholinesterase inhibitor. In Becker, R.; Giacobini, E., editors. *Alzheimer Disease: From Molecular Biology to Therapy*. Boston, MA: Birkhäuser; **1996**, 245–250.

45. Liu, J.; Zhang, H. Y.; Tang, X. C.; Wang, B.; He, X. C.; Bai, D. L. Effects of synthetic (−)-huperzine A on cholinesterase activities and mouse water maze performance. *Acta Pharmacol. Sin.*, **1998**, 19: 413–416.

46. Lu, W. H.; Shou, J.; Tang, X. C. Improving effect of huperzine A on discrimination performance in aged rats and adult rats with experimental cognitive impairment. *Acta Pharmacol. Sin.*, **1988**, 9: 11–15.

47. Tang, X. C.; Han, Y. F.; Chen, X. P.; Zhu, X. D. Effects of huperzine A on learning and retrieval process of discrimination performance in rats. *Acta Pharmacol. Sin.*, **1986**, 7: 501–511.

48. Tang, X. C.; Xiong, Z. Q.; Qian, B. C.; Zhou, Z. F.; Zhang, C. C. Cognitive improvement by oral huperzine A: a novel acetylcholinesterase inhibitor. In Giacobini, E.; Becker, R., editors. *Alzheimer Therapy: Therapeutic Strategies*. Boston, MA: Birkhäuser; **1994**, 113–119.

49. Wang, T.; Tang, X. C. Reversal of scopolamine-induced deficits in radial maze performance by (−)-huperzine A: comparison with E2020 and tacrine. *Eur. J. Pharmacol.*, **1998**, 349: 137–142.

50. Xiong, Z. Q.; Tang, X. C. Effects of huperzine A, a novel acetylcholinesterase inhibitor, on radial maze performance in rats. *Pharmacol. Biochem. Behav.*, **1995**, 51: 415–419.

51. Xiong, Z. Q.; Han, Y. F.; Tang, X. C. Huperzine A ameliorates the spatial working memory impairments induced by AF64A. *Neuroreport*, **1995**, 6: 2221–2224.

52. Xiong, Z. Q.; Cheng, D. H.; Tang, X. C. Effects of huperzine A on nucleus basalis magnocellularis lesion-induced spatial working memory deficit. *Acta Pharmacol. Sin.*, **1998**, 19: 128–132.

53. Ye, J. W.; Cai, J. X.; Wang, L. M.; Tang, X. C. Improving effects of huperzine A on spatial working memory in aged monkeys and young adult monkeys with experimental cognitive impairment. *J. Pharmacol. Exp. Ther.*, **1999**, 288: 814–819.

54. Ye, J. W.; Shang, Y. Z.; Wang, Z. M.; Tang, X. C. Huperzine A ameliorates the impaired memory of aged rat in the Morris water maze performance. *Acta Pharmacol. Sin.*, **2000**, 21(1): 65–69.

55. Gao, Y.; Tang, X. C.; Guan, L. C.; Kuang, P. Z. Huperzine A reverses scopolamine-and muscimol-induced memory deficits in chick. *Acta Pharmacol. Sin.*, **2000**, 21: 1169–1173.

56. Sun, Q. Q.; Xu, S. S.; Pan, J. L.; Guo, H. M.; Cao, W. Q. Huperzine-A capsules enhance memory and learning performance in 34 pairs of matched adolescent students. *Acta Pharmacol. Sin.*, **1999**, 20(7): 601–603.

57. Ou, L. Y.; Tang, X. C.; Cai, J. X. Effect of huperzine A on working memory in reserpine-or yohimbine-treated monkeys. *Eur. J. Pharmacol.*, **2001**, 433(2–3): 151–156.

58. Filliat, P.; Foquin, A.; Lallement, G. Effects of chronic administration of huperzine A on memory in guinea pigs. *Drug Chem. Toxicol.*, **2002**, 25(1): 9–24.

59. Schmidt, B. H.; Van Der Staay, F. J. *Int. J. Geriatr. Psychopharmacol.*, **1998**, 1: 134.

60. Lallement, G.; Baille, V.; Baubichon, D.; Carpentier, P.; Collombet, J. M.; Filliat, P.; Foquin, A.; Four, E.; Masqueliez, C.; Testylier, G.; Tonduli, L.; Dorandeu, F. Review of the value of huperzine as pretreatment of organophosphate poisoning. *Neurotoxicology*, **2002**, 23(1): 1–5.

61. Ved, H. S.; Koenig, M. L.; Dave, J. R.; Doctor, B. P. Huperzine A, a potential therapeutic agent for dementia, reduces neuronal cell death caused by glutamate. *Neuroreport*, **1997**, 8(4): 963–968.

62. Xiao, X. Q.; Zhang, H. Y.; Tang, X. C. Huperzine A attenuates amyloid β-peptide fragment 25-35-induced apoptosis in rat cortical neurons via inhibiting reactive oxygen species formation and caspase-3 activation. *J. Neurosci. Res.*, **2002**, 67(1): 30–36.

63. Xiao, X. Q.; Wang, R.; Tang, X. C. Huperzine A and tacrine attenuate β-amyloid peptide-induced oxidative injury. *J. Neurosci. Res.*, **2000**, 61(5): 564–569.

64. Xiao, X. Q.; Wang, R.; Han, Y. F.; Tang, X. C. Protective effects of huperzine A on β-amyloid (25–35) induced oxidative injury in rat pheochromocytoma cells. *Neurosci. Lett.*, **2000**, 286(3): 155–158.

65. Wang, R.; Zhang, H. Y.; Tang, X. C. Huperzine A attenuates cognitive dysfunction and neuronal degeneration caused by β-amyloid protein-(1–40) in rat. *Eur. J. Pharmacol.*, **2001**, 421(3): 149–156.

66. Xiao, X. Q.; Yang, J. W.; Tang, X. C. Huperzine A protects rat pheochromocytoma cells against hydrogen peroxide-induced injury. *Neurosci. Lett.*, **1999**, 275(2): 73–76.

67. Wang, R.; Xiao, X. Q.; Tang, X. C. Huperzine A attenuates hydrogen peroxide-induced apoptosis by regulating expression of apoptosis-related genes in rat PC12 cells. *Neuroreport*, **2001**, 12(12): 2629–2634.

68. Wang, L. S.; Zhou, J.; Shao, X. M.; Tang, X. C. Huperzine A attenuates cognitive deficits and brain injury in neonatal rats after hypoxia-ischemia. *Brain Res.*, **2002**, 949(1–2): 162–170.

69. Grunwald, J.; Raveh, L.; Doctor, B. P.; Ashani, Y. Huperzine A as a pretreatment candidate drug against nerve agent toxicity. *Life Sci.*, **1994**, 54: 991–997.

70. Gordon, R. K.; Nigam, S. V.; Weitz, J. A.; Dave, J. R.; Doctor, B. P.; Ved, H. S. The NMDA receptor ion channel: a site for binding of Huperzine A. *J. Appl. Toxicol.*, **2001**, 21(Suppl 1): S47–S51.

71. Sattler, R.; Tymianski, M. Molecular mechanisms of calcium-dependent excitotoxicity. *J. Mol. Med.*, **2000**, 78(1): 3–13.

72. Choi, D. W. Calcium and excitotoxic neuronal injury. *Ann. N Y Acad. Sci.*, **1994**, 747: 162–171.

73. Wang, X. D.; Zhang, J. M.; Yang, H. H.; Hu, G. Y. Modulation of NMDA receptor by huperzine A in rat cerebral cortex. *Acta Pharmacol. Sin.*, **1999**, 20: 31–35.

74. Vajda, F. J. Neuroprotection and neurodegenerative disease. *J. Clin. Neurosci.*, **2002**, 9(1): 4–8.

75. Perry, E. K.; Tomlinson, B. E.; Blessed, G.; Bergman, K.; Gibson, P. H.; Perry, R. H. Correlation of cholinergic abnormalities with senile plaques and mental scores. *Br. Med. J.*, **1978**, 2: 1457–1459.

76. Selkoe, D. J.; Abraham, C. R.; Podlisny, M. B.; Duffy, L. K. Isolation of low-molecular-weight proteins from amyloid plaque fibers in Alzheimer's disease. *J. Neurochem.*, **1986**, 46(6): 1820–1834.

77. Wang, L. M.; Han, Y. F.; Tang, X. C. Huperzine A improves cognitive deficits caused by chronic cerebral hypoperfusion in rats. *Eur. J. Pharmacol.*, **2002**, 398(1): 65–72.

78. Zhou, J.; Fu, Y.; Tang, X. C. Huperzine A protects rat pheochromocytoma cells against oxygen–glucose deprivation. *Neuroreport*, **2001**, 12: 2073–2077.

79. Yan, X. F.; Lu, W. H.; Lou, W. J.; Tang, X. C. Effects of huperzine A and B on skeletal muscle and electroencephalogram. *Acta Pharmacol. Sin.*, **1987**, 8: 117–123.

80. Tu, Z. H.; Wu, M. Y. Mutagenicity and comutagenicity of three nootropics: huperzine A, aniracetam and piracetam. *New Drugs Clin. Remedies*, **1990**, 9: 65–68.

81. Zhao, H. W.; Li, X. Y. Ginkgolide A, B, and huperzine A inhibit nitric oxide production from rat C6 and human BT325 glioma cells. *Acta Pharmacol. Sin.*, **1999**, 20(10): 941–943.

82. Li, Y.; Hu, G. Y. Huperzine A, a nootropic agent, inhibits fast transient potassium current in rat dissociated hippocampal neurons. *Neurosci. Lett.*, **2002**, 324(1): 25–28.

83. Li, Y.; Hu, G. Y. Huperzine A inhibits the sustained potassium current in rat dissociated hippocampal neurons. *Neurosci. Lett.*, **2002**, 329(2): 153–156.

84. Galeotti, N.; Ghelardini, C.; Mannelli, L. D.; Bartolini, A. Antinociceptive profile of the natural cholinesterase inhibitor huperzine A. *Drug Develop. Res.*, **2001**, 54(1): 19–26.

85. Zhang, C. L. Therapeutic effects of huperzine A on the aged with memory impairment. *New Drugs Clin. Remedies*, **1986**, 5: 260–262.

86. Zhang, C. L.; Wang, G. Z. Effects of huperzine A tablet on memory. *New Drugs Clin. Remedies*, **1990**, 9: 339–341.

87. Xu, S. S.; Cai, Z. Y.; Qu, Z. W.; Yang, R. M.; Cai, Y. L.; Wang, G. Q.; Su, X. Q.; Zhong, X. S.; Cheng, R. Y.; Xu, W. A.; Li, J. X.; Feng, B. Huperzine A in capsules and

tablets for treating patients with Alzheimer disease. *Acta Pharmacol. Sin.*, **1999**, 20(6): 486–490.

88. Xu, S. S.; Gao, Z. X.; Weng, Z.; Du, Z. M.; Xu, W. A.; Yang, J. S.; Zhang, M. L.; Tong, Z. H.; Fang, Y. S.; Chai, X. S. et al. Efficacy of tablet huperzine-A on memory, cognition and behavior in Alzheimer's disease. *Acta Pharmacol. Sin.*, **1995**, 16: 391–395.

89. Cheng, Y. S.; Lu, C. Z.; Ying, Z. L.; Ni, W. Y.; Zhang, C. L.; Sang, G. W. 128 cases of myasthenia gravis treated with huperzine A. *New Drugs Clin. Remedies*, **1986**, 5: 197–199.

90. Mazurek, A. An open label trial of huperzine A in the treatment of Alzheimer's disease. *Altern. Ther.*, **1999**, 5(2): 97–98.

91. Ma, Y. X.; Zhu, Y.; Gu, Y. D.; Yu, Z. Y.; Yu, S. M.; Ye, Y. Z. Double-blind trial of huperzine-A on cognitive deterioration in 314 cases of benign senescent forgetfulness, vascular dementia, and Alzheimer's disease. *Ann. NY Acad. Sci.*, **1998**, 854: 506–507.

92. Ma, Y. X.; Zhu, Y.; Gu, Y. D.; Yu, Z. Y.; Yu, S. M.; Ye, Y. Z. Double-blind trail of huperzine-A (HUP) on cognitive impairment and dementia in 314 cases. *N-S Arch. Pharmacol.*, **1998**, 358(Suppl. 1): P35194.

93. Zhang, Z. X.; Wang, X. D.; Chen, Q. T.; Shu, L.; Wang, J. Z.; Shan, G. L. Clinical efficacy and safety of huperzine A in treatment of mild to moderate Alzheimer disease, a placebo-controlled, double-blind, randomized trial. *Chinese Med. J. Peking*, **2002**, 82(14): 941–944.

94. Chang, S. Y.; Chen, S. M.; Cao, Q. L.; Liu, P; Wang, F. G.; Wang, Z. X. A clinical study of the effect of huperzine A to improve the ability of verbal recall, retention and repetition in middle-aged and elderly patients with dysmnesia. *Herald Med.* (in Chinese), **2002**, 21(5): 263–265.

95. Qian, L. G.; Ji, R. Y. A total synthesis of (±)Huperzine A. *Tetrahedron Lett.*, **1989**, 30: 2089–2090.

96. Xia, Y.; Kozikowski, A. P. A practical synthesis of the Chinese "nootropic" agent huperzine A: a possible lead in the treatment of Alzheimer's disease. *J. Am. Chem. Soc.*, **1989**, 111: 4116–4117.

97. Kozikwoski, A. P.; Reddy, E. R.; Miller, C. P. A simplified route to a key intermediate in the synthesis of the Chinese noortropic agent huperzine A. *J. Chem. Soc. Perkin. Trans. 1*, **1990**, 195–197.

98. Campiani, G.; Sun, L. Q.; Kozikowski, A. P.; Aagaard, P.; McKinney, M. A palladium-catalyzed route to huperzine A and its analogues and their anticholinesrterase activity. *J. Org. Chem.*, **1993**, 58: 7660–7669.

99. Kozikowski, A. P.; Campiani, G.; Tuckmantel, W. An approach to open-chain and modified heterocyclic-analogs of the acetylcholinesterase inhibitor huperzine-A through a bicyclo[3.3.1]nonane intermediate. *Heterocycles*, **1994**, 39(1): 101–116.

100. Camps, P.; Contreras, J.; Fontbardia, M.; Solans, X. Improved synthesis of methyl 7,7-ethylenedioxy-3-methyl-9-oxobicyclo[3.3.1]non-3-ene-1-carboxylate intermediate for the synthesis of huperzine A analogues. *Synthetic Commun.*, **1996**, 26(1): 9–18.

101. Camps, P.; Contreras, J.; El Achab, R.; Morral, J.; Munoz-Torrero, D.; Font-Bardia, M.; Solans, X.; Badia, A.; Vivas, N. M. New syntheses of *rac*-huperzine A and its *rac*-7-ethyl-derivative. Evaluation of several huperzine A analogues as acetylcholinesterase inhibitors. *Tetrahedron*, **2000**, 56(26): 4541–4553.

102. Yamada, F.; Kozikowski, A. P.; Reddy, E. R.; Pang, Y. P.; Miller, J. H.; Mckinney, M. A route to optically pure (−)-huperzine-A—molecular modeling and *in vitro* pharmacology. *J. Am. Chem. Soc.*, **1991**, 113(12): 4695–4696.

103. Chen, W. P.; Yang, F. Q. Asymmetric total synthesis of optically active huperzine A. *Chinese J. Med. Chem.*, **1995**, 5(1): 10–17.

104. Kaneko, S.; Yoshino, T.; Katoh, T.; Terashima, S. A novel enantioselective synthesis of the key intermediate of (−)-huperzine A employing asymmetric palladium-catalyzed bicycloannulation. *Tetrahedron-Asymmetry*, **1997**, 8(6): 829–832.

105. He, X. C.; Wang, B.; Bai, D. L. Studies on asymmetric synthesis of huperzine A-1. Palladium-catalyzed asymmetric bicycloannulation of 5,6,7,8-tetrahydro-2-methoxy-6-oxo-5-quinolinecarboxylic esters. *Tetrahedron Lett.*, **1998**, 39(5–6): 411–414.

106. He, X. C.; Wang, B.; Yu, G. L.; Bai, D. L. Studies on the asymmetric synthesis of huperzine A. Part 2: highly enantioselective palladium-catalyzed bicycloannulation of the beta-keto-ester using new chiral ferrocenylphosphine ligands. *Tetrahedron-Asymmetry*, **2001**, 12(23): 3213–3216.

107. Lee, I. Y. C.; Jung, M. H.; Lee, H. W.; Yang, J. Y. Synthesis of huperzine intermediates via Mn(III)-mediated radical cyclization. *Tetrahedron Lett.*, **2002**, 43(13): 2407–2409.

108. Kozikowski, A. P. Synthetic chemistry, neurotransmission and second messengers. *J. Heterocyclic Chem.*, **1990**, 27: 97–105.

109. Kozikowski, A. P.; Xia, Y.; Reddy, E. R.; Tuckmantel, W.; Hanin, I.; Tang, X. C. Synthesis of huperzine A and its analogues and their anticholinesterase activity. *J. Org. Chem.*, **1991**, 56(15): 4636–4645.

110. Kozikowski, A. P.; Miller, C. P.; Yamada, F.; Pang, Y. P.; Miller, J. H.; McKinney, M.; Ball, R. G. Delineating the pharmacophoric elements of huperzine A: importance of the unsaturated three-carbon bridge to its AChE inhibitory activity. *J. Med. Chem.*, **1991**, 34(12): 3399–3402.

111. Kozikowski, A. P.; Prakash, K. R. C.; Saxena, A.; Doctor, B. P. Synthesis and biological activity of an optically pure 10-spirocyclopropyl analog of huperzine A. *Chem. Commum.*, **1998**, 12: 1287–1288.

112. Campiani, G.; Kozikowski, A. P.; Wang, S.; Ming, L.; Nacci, V.; Saxena, A.; Doctor, B. P. Synthesis and anticholinesterase activity of huperzine A analogues containing phenol and catechol replacements for the pyridone ring. *Bioorg. Med. Chem. Lett.*, **1998**, 8(11): 1413–1418.

113. Kozikowski, A. P.; Campiani, G., Nacci, V., Sega, A., Saxena, A., Doctor, B. P. An approach to modified heterocyclic analogues of huperzine A and isohuperzine A. Synthesis of the pyrimidone and pyrazole analogues, and their anticholinesterase activity. *J. Chem. Soc. Perkin Trans. 1*, **1996**, 11: 1287–1297.

114. He, X. C.; Wang, Z. Y.; Li, Y. L.; Xu, Z. R.; Bai, D. L. Synthesis of analogues of huperzine A. *Chinese Chem. Lett.*, **1993**, 4: 597–600.

115. He, X. C.; Yu, G. L.; Bai, D. L. Studies on analogues of huperzine A for treatment of senile dementia. VI. Asymmetric total synthesis of 14-norhuperzine A and its inhibitory activity of acetylcholinesterase. *Acta Pharm. Sin.*, **2003**, 38(5): 346–349.

116. Kaneko, S.; Nakajima, N.; Shikano, M.; Katoh, T.; Terashima, S. Synthesis and acetylcholinesterase inhibitory activity of fluorinated analogues of huperzine A. *Bioorg. Med. Chem. Lett.*, **1996**, 6: 1927–1930.

117. Zeng, F. X.; Jiang, H. L.; Tang, X. C.; Chen, K. X.; Ji, R. Y. Synthesis and acetylcholinesterase inhibitory activity of (±)-14-fluorohuperzine A. *Bioorg. Med. Chem. Lett.*, **1998**, 8(13): 1661–1664.

118. Zhou, G. C.; Zhu, D. Y. Synthesis of 5-substituted analogues of huperzine A. *Bioorg. Med. Chem. Lett.*, **2000**, 10(18): 2055–2057.

119. Hogenauer, K.; Baumann, K.; Mulzer, J. Synthesis of (±)-desamino huperzine A. *Tetrahedron Lett.*, **2000**, 41(48): 9229–9232.

120. Hogenauer, K.; Baumann, K.; Enz, A.; Mulzer, J. Synthesis and acetylcholinesterase inhibition of 5-desamino huperzine A derivatives. *Bioorg. Med. Chem. Lett.*, **2001**, 11(19): 2627–2630.

121. Tang, X. C.; Xu, H.; Feng, J.; Zhou, T. X.; Liu, J. S. Effect of cholinesterase inhibition *in vitro* by huperzine A analogs. *Acta Pharmacol. Sin.*, **1994**, 15(2): 107–110.

122. Zhu, D. Y.; Tang, X. C., et al. Huperzine A derivatives, their preparation and their use. *European Patent*, 0 806 416 B1,

123. Xiong, Z. Q.; Tang, X. C.; Lin, J. L.; Zhu, D. Y. Effects of isovanihuperzine A on cholinesterase and scopolamine-induced memory impairment. *Acta Pharmacol. Sin.*, **1995**, 16(1): 21–25.

124. Kozikowski, A. P.; Thiels, E.; Tang, X. C.; Hanin. In Maryanoff, B. E.; Maryanoff, C. A., editors. *Advance in Medicinal Chemistry.*, vol. 1. Greenwich, CT: JAI Press; **1992**, 175–205.

125. He, X. C.; Wang, Z. Y.; Li, Y. L.; Bai, D. L. Studies on analogues of huperzine A for the treatment od Alzheimer's disease. I synthesis of 5-substituted aminomethyl-2(1*H*)-pyridones and 5-substituted amino-5,6,7,8-tetrahydroquinolones. *Chinese J. Med. Chem.*, **1994**, 4: 257–262.

126. Fink, D. M.; Bores, G. M.; Effland, R. C.; Huger, F. P.; Kurys, B. E.; Rush, D. K.; Selk, D. E. Synthesis and evaluation of 5-amino-5,6,7,8-tetrahydroquinolinones as potential agents for the treatment of Alzheimer's disease. *J. Med. Chem.*, **1995**, 38(18): 3645–3651.

127. Wu, B. G. Ph. D thesis, Shanghai Institute of Materia Medica, Chinese Academy of Sciences, 1996.

128. Wu, B. G.; Zhen, W. P.; Bo, Y. X.; He, X. C.; Bai, D. L. Synthesis of analogues of huperzine A-3. The preparation of some single ring analogues of huperzine A. *Chinese Chem. Lett.*, **1995**, 6(3): 193–196.

129. Wu, B. G.; Bai, D. L. Synthesis of huperzine A analogues. *Chinese Pharm. J.*, **1995**, 30(Suppl.): 63–66.

130. Camps, P.; Contreras, J.; Morral, J.; Munoztorrero, D.; Fontbardia, M.; Solans, X. Synthesis of an 11-unsubstituted analogue of (±)-huperzine A. *Tetrahedron*, **1999**, 55(28): 8481–8496.

131. Camps, P.; Munoztorrero, D.; Simon, M. Easy access to 4-substituted (±)-huperzine A analogues. *Synthetic Commun.*, **2001**, 31(22): 3507–3516.

132. Badia, A.; Banos, J. E.; Camps, P.; Contreras, J.; Gorbig, D. M.; Munoztorrero, D.; Simon, M.; Vivas, N. M. Synthesis and evaluation of tacrine-huperzine A hybrids as acetylcholinesterase inhibitors of potential interest for the treatment of Alzheimer's disease. *Bioorg. Med. Chem.*, **1998**, 6(4): 427–440.

133. Carlier, P. R.; Du, D. M.; Han, Y. F.; Liu, J.; Pang, Y. P. Potent, easily synthesized huperzine A-tacrine hybrid acetylcholinesterase inhibitors. *Bioorg. Med. Chem. Lett.*, **1999**, 9(16): 2335–2338.

134. Pang, Y. P.; Quiram, P.; Jelacic, T.; Hong, F.; Brimijoin, S. Highly potent, selective, and low cost bis-tetrahydroaminacrine inhibitors of acetylcholinesterase. *J. Biol. Chem.*, **1996**, 271: 23646–23649.

135. Carlier, P. R.; Du, D. M.; Han, Y. F.; Liu, J.; Perola, E.; Williams, I. D.; Pang, Y. P. Dimerization of an inactive fragment of huperzine A produces a drug with twice the potency of the natural product. *Angew. Chem. Int. Ed. Engl.*, **2000**, 39(10): 1775–1777.

136. Jin, G. Y.; He, X. C.; Zhang, H. Y.; Bai, D. L. Synthesis of alkylene-linked dieters of (−)-huperzine A. *Chinese Chem. Lett.*, 2002, 13(1): 23–26.

137. Jin, G. Y.; Luo, X. M.; He, X. C.; Jiang, H. L.; Zhang, H. Y.; Bai, D. L. Synthesis and docking studies of alkylene-linked dimers of (−)-huperzine A. *Arzneimittel-Forsch*, **2003**; 53(11): 753–757.

138. Zeng, F. X.; Jiang, H. L.; Yang, Y. S.; Liu, D. X.; Zhang, H. Y.; Chen, K. X.; Ji, R. Y. Synthesis and pharmacological study of huperzine A-E2020 combined compound. *Acta Chim. Sin.*, **2000**, 58(5): 580–587.

139. Liu, J.; Zhang, H. Y.; Wang, L. M.; Tang, X. C. Inhibitory effects of huperzine B on cholinesterase activity in mice. *Acta Pharmacol. Sin.*, **1999**, 20(2): 141–145.

140. Rajendran, V.; Rong, S. B.; Saxena, A.; Doctor, B. P.; Kozikowski, A. P. Synthesis of a hybrid analog of the acetylcholinesterase inhibitors huperzine A and huperzine B. *Tetrahedron Lett.*, **2001**, 42(32): 5359–5361.

141. Rajendran, V.; Saxena, A.; Doctor, B. P.; Kozikowski, A. P. Synthesis of more potent analogues of the acetylcholinesterase inhibitor, huperzine B. *Bioorg. Med. Chem. Lett.*, **2002**, 12(11): 1521–1523.

142. Camps, P.; Contreras, J.; Font-Bardia, M.; Morral, J.; Munoz-Torrero, D.; Solans, X. Enantioselective synthesis of tacrine-huperzine A hybrids. Preparative chiral MPLC separation of their racemic mixtures and absolute configuration assignments by x-ray diffraction analysis. *Tetrahedron-Asymmetry*, **1998**, 9(5): 835–849.

143. Camps, P.; El Achab, R.; Gorbig, D. M.; Morral, J.; Munoz-Torrero, D.; Badia, A.; Banos, J. E.; Vivas, N. M.; Barril, X.; Orozco, M.; Luque, F. J. Synthesis in vitro pharmacology and molecular modeling of very potent tacrine-huperzine A hybrids as acetylcholinesterase inhibitors of potential interest for the treatment of Alzheimer's disease. *J. Med. Chem.*, **1999**, 42(17): 3227–3242.

144. Raves, M. L.; Harel, M.; Pang, Y. P.; Silman, I.; Kozikowski, A. P.; Sussman, J. L. Structure of acetylcholinesterase complexed with the nootropic alkaloid, (−)-huperzine A. *Nat. Struct. Biol.*, **1997**, 4(1): 57–63.

145. Ashani, Y.; Peggins, J. O.; Doctor, B. P. Mechanism of inhibition of cholinesterases by huperzine-A. *Biochem. Biophys. Res. Commun.*, **1992**, 184(2): 719–726.

146. Saxena, A.; Qian, N. F.; Kovach, I. M.; Kozikowski, A. P.; Pang, Y. P.; Vellom, D. C.; Radic, Z.; Quinn, D.; Taylor, P.; Doctor, B. P. Identification of amino-acid-residues involved in the binding of huperzine-A to cholinesterases. *Protein Sci.*, **1994**, 10: 1770–1778.

147. Sussman, J. L.; Harel, M.; Frolow, F.; Oefner, C.; Goldman, A.; Toker, L.; Silman, I. Atomic structure of acetylcholinesterase from *Torpedo californica*: a prototypic acetylcholine-binding protein. *Science*, **1991**, 253: 872–879.

148. Isralewitz, B.; Gao, M.; Schulten, K. Steered molecular dynamics and mechanical functions of proteins. *Curr. Opin. Struc. Biol.*, **2001**, 11: 224–230.

149. Xu, Y. C.; Shen, J. H.; Luo, X. M.; Silman, I.; Sussman, J. L.; Chen, K. X.; Jiang, H. L. How does huperzine A enter and leave the binding gorge of acetylcholinesterase? Steered molecular dynamics simulations. *J. Am. Chem. Soc.*, **2003**, 125(37): 11340–11349.

150. Kaur, J.; Zhang, M. Q. Molecular modelling and QSAR of reversible acetylcholinesterase inhibitors. *Curr. Med. Chem.*, **2000**, 7(3): 273–294.

151. Camps, P.; Cusack, B.; Mallender, W. D.; El Achab, R.; Morral, J.; Munoz-Torrero, D.; Rosenberry, T. L. Huprine X is a novel high-affinity inhibitor of acetylcholinesterase that is of interest for treatment of Alzheimer's disease. *Mol. Pharmacol.*, **2000**, 57(2): 409–417.

152. Camps, P.; El Achab, R.; Morral, J.; Munoz-Torrero, D.; Badia, A.; Banos, J. E.; Vivas, N. M.; Barril, X.; Orozco, M.; Luque, F. J. New tacrine-huperzine A hybrids (huprines): highly potent tight-binding acetylcholinesterase inhibitors of interest for the treatment of Alzheimer's disease. *J. Med. Chem.*, **2000**, 43(24): 4657–4666.

153. Camps, P.; Gomez, E.; Munoz-Torrero, D.; Arno, M. On the regioselectivity of the Friedlander reaction leading to huprines: stereospecific acid-promoted isomerization of syn-huprines to their anti-regioisomers. *Tetrahedron-Asymmetry*, **2001**, 12(20): 2909–2914.

154. Ros, E.; Aleu, J.; De Aranda, I. G.; Munoz-Torrero, D.; Camps, P.; Badia, A.; Marsal, J.; Solsona, C. The pharmacology of novel acetylcholinesterase inhibitors (±)-huprines Y and X on the *Torpedo* electric organ. *Eur. J. Pharmacol.*, **2001**, 421(2): 77–84.

155. Camps, P.; Gomez, E.; Munoz-Torrero, D.; Badia, A.; Vivas, N. M.; Barril, X.; Orozco, M.; Luque, F. J. Synthesis in vitro pharmacology and molecular modeling of syn-huprines as acetylcholinesterase inhibitors. *J. Med. Chem.*, **2001**, 44(26): 4733–4736.

156. Carlier, P. R.; Han, Y. F.; Chow, E. S. H.; Li, C. P. L.; Wang, H.; Lieu, T. X.; Wong, H. S.; Pang, Y. P. Evaluation of short-tether bis-THA AChE inhibitors. A further test of the dual binding site hypothesis. *Bioorg. Med. Chem.*, **1999**, 7(2): 351–357.

157. Harel, M.; Schalk, I.; Ehret-Sabatier, L.; Bouet, F.; Goeldner, M.; Hirth, C.; Axelsen, P. H.; Silman, I.; Sussman, J. L. Quaternary ligand binding to aromatic residues in the active-site gorge of acetylcholinesterase. *P. Natl. Acad. Sci. USA*, **1993**, 90(19): 9031–9035.

158. Dvir, H.; Wong, D. M.; Harel, M.; Barril, X.; Orozco, M.; Luque, F. J.; Munoz-Torrero, D.; Camps, P.; Rosenberry, T. L.; Silman, I.; Sussman, J. L. 3D structure of *Torpedo californica* acetylcholinesterase complexed with huprine X at 2.1 angstrom resolution: kinetic and molecular dynamic correlates. *Biochemistry*, **2002**, 41(9): 2970–2981.

159. Felder, C. E.; Harel, M.; Silman, I.; Sussman, J. L. Structure of a complex of the potent and specific inhibitor BW284C51 with *Torpedo californica* acetylcholinesterase. *Acta Crystallogr. D Biol. Crystallogr.*, **2002**, 58: 1765–1771.

160. Wong, D. M.; Greenblatt, H. M.; Dvir, H.; Carlier, P. R.; Han, Y. F.; Pang, Y. P.; Silman, I.; Sussman, J. L. Acetylcholinesterase complexed with bivalent ligands related to huperzine A: experimental evidence for species-dependent protein-ligand complementarity. *J. Am. Chem. Soc.*, **2003**, 125(2): 363–373.

161. Dvir, H.; Jiang, H. L.; Wong, D. M.; Harel, M.; Chetrit, M.; He, X. C.; Jin, G. Y.; Yu, G. L.; Tang, X. C.; Silman, I.; Bai, D. L.; Sussman, J. L. X-ray structures of *Torpedo californica* acetylcholinesterase complexed with (+)-huperzine A and (−)-huperzine B: structural evidence for an active site rearrangement. *Biochemistry*, **2002**, 41(35): 10810–10818.

# 5

# QINGHAOSU (ARTEMISININ)—A FANTASTIC ANTIMALARIAL DRUG FROM A TRADITIONAL CHINESE MEDICINE

YING LI

*Shanghai Institute of Materia Medica, Shanghai Institutes for Biological Sciences, Chinese Academy of Sciences, Shanghai, China*

HAO HUANG AND YU-LIN WU

*State Key Laboratory of Bioorganic and Natural Products Chemistry, Shanghai Institute of Organic Chemistry, Chinese Academy of Sciences, Shanghai, China*

## 5.1 INTRODUCTION

Qinghaosu (**1**, artemisinin), a composition of the traditional Chinese medicine qinghao (*Artemisia annua* Linnaeus, composites), is a special sesquiterpene with a unique 1,2,4-trioxane segment and has excellent antimalarial activity, especially for multi-drug-resistant parasites. Qinghaosu and its derivatives have been recognized as a new generation of powerful antimalarial drug for combating the most popular infectious disease, malaria, worldwide. Artemether and artesunate, two qinghaosu derivatives and Coartem, were approved by the Chinese authority and collected in the "Essential Medicine List" by the World Health Organization. These qinghaosu medicines have been successfully applied to remedy several million malaria-suffering patients since their advent. Meanwhile, over 1000 research papers and dozens of reviews[1–16] have been published to record the rapid progress of

*Medicinal Chemistry of Bioactive Natural Products* Edited by Xiao-Tian Liang and Wei-Shuo Fang
Copyright © 2006 John Wiley & Sons, Inc.

qinghaosu research from the different disciplines of botany, chemistry, pharmacology, and clinic medicine during the last two decades. Undoubtedly, the discovery of qinghaosu is one of the most important achievements for the natural products chemistry during the last two decades of the twentieth century. It may be recognized as a milestone in the progress of natural products chemistry in comparison with quinine in the nineteenth century. Qinghaosu is also a star molecule just like Taxol in about the same period. For the modernization of traditional Chinese medicine, qinghaosu is also one of the most successful examples. This chapter intends to describe its structure determination, reaction, and synthesis. The congeneric natural products in qinghao are also mentioned. This chapter will review the progress in the search for derivatives of qinghaosu, the chemical biology study, and the exploration on the action mode. Because of the limitation of the volume and the massive publications on this subject, this chapter will preferentially, rather than comprehensively, introduce the progress achieved in China.

## 5.2   QINGHAOSU AND QINGHAO
(*ARTEMISIA ANNUA* L. COMPOSITES)

### 5.2.1   Discovery and Structure Determination of Qinghaosu

In the 1960s, drug-resistant malarial parasites developed and spread rapidly in Southeast Asia and Africa; therefore, existing antimalarial drugs, such as quinine, chloroquine, and pyrimethamine-sulphadoxine became less efficient. The introduction of a new generation of antimalarial drug was much anticipated by the 100,000,000 patients worldwide. Now malaria is no longer a serious infectious disease in China, but back then, Chinese people, especially those who lived in the southern provinces, faced a critical situation. In 1967, a program involving several hundred Chinese scientists nationwide was launched to take on this challenge. A part of this program, called "Program 523," endeavored to explore the traditional Chinese medicine and herb. More than 1000 samples from different herbs have been studied by the modern methods, and isolation of the active principles is monitored with antimalarial screening in animal models. Several active principles, such as yingzhaosu A (**2**, yingzhaosu) from yingzhao (*Artabotrys hexapetalu* (LF) Bhand),[17] agrimols (**3**) from xianhecao (*Agrimonia pilosa* L.),[18,19] robustanol (**4**) from dayean (*Eucalyptus robusta* Sm),[20] protopine (**5**) from nantianzhu (*Nandina domestica* T.),[21] bruceine D and E (**6** and **7**) from yadanzi (*Brucea javanica* (L) Merr),[22] and anluosu (**8**) from lingshuianluo [*Polyalthia nemoralis* A.(DC)],[23] have been identified. Total synthesis and structure modification of some principles, such as yingzhaosu, agrimols, and febrifugine (**9**), which was isolated[24] from the Chinese traditional antimalarial medicine changshan (*Dichroa febrifuga* Lour.) in 1948, were also identified thereafter (Structure 5-1). Unfortunately these natural products and their synthetic derivatives were hardly available, insufficiently active, or too toxic. However, just as the Chinese old proverb says, "Heavens never fail the people working with heart and soul," the great promise for the new generation of antimalarial medicine did rise from an obscure weed, the old Chinese traditional

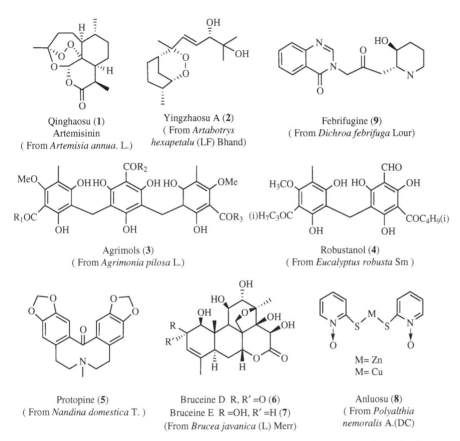

**Structure 5-1.** Some antimalarial natural products from traditional Chinese herbs.

herb qinghao in the 1970s after the cooperative hard work of several research groups from several provinces.

Qinghao has been used as a traditional medicine for at least 2000 years in China. The earliest written record in silk so far discovered is the *Recipes for 52 Kinds of Diseases*, which was unearthed from the Mawangdui Tomb of the West Han Dynasty (168 BC) in Changsha, Hunan Province. Figure 5-1 shows the weed qinghao in bloom with a yellow flower and the Chinese characters qinghao taken from this unearthed piece of silk. In this record, qinghao was used for the treatment of hemorrhoids. The first record of qinghao for the treatment of malaria (fever) was described in *The Handbook of Prescriptions for Emergency Treatments* written by Ge Hong (281–340 AD) (Figure 5-2). Since then, a series of Chinese medicine books, including the most famous book, *Compendium of Medical Herbs* (Bencao Gangmu) by Li Shizhen in 1596, described the application of qinghao for fever remedy. Practically qinghao have been widely used for the treatment of fever and other diseases, especially in the countryside. Therefore, the phytochemical groups of Program 523 paid special attention to this herb, and in the early 1970s, three

**Figure 5-1.** Qinghao in bloom with yellow flower.

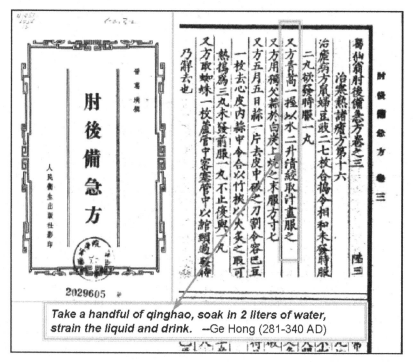

*Take a handful of qinghao, soak in 2 liters of water, strain the liquid and drink.* —Ge Hong (281-340 AD)

**Figure 5-2.** A page describing the treatment of malaria with qinghao from *The Handbook of Prescriptions for Emergency Treatments.*

groups in Beijing, Yunnan, and Shandong, almost simultaneously extracted the active fraction from it with diethyl ether, petroleum ether, or acetone, respectively, by monitoring with antimalarial screening in vivo. Afterward a colorless needle crystal was obtained and proved to be effective in the preliminary clinic trial. The herb "qinghao" available from the store of traditional Chinese medicine is a general name, and it consists of several species of weeds. Although it was clear that among the herb qinghao only huanghuahao (blooming with a yellow flower in Figure 5-1) is rich in this antimalarial principle, this principle was still called qinghao-su (su means principle in Chinese) instead of huanghuahao-su (the principle of huanghuahao) according to the customary terminology. In the phytotaxonomy, qinghao is the *A. annua* L. composite, so qinghaosu is also called artemisinin or rarely arteannuin.

The structure determination[25,26] of qinghaosu was performed by a joint research group that consisted of researchers from the Institute of Chinese Materia Medica and the Shanghai Institute of Organic Chemistry during the mid-1970s. It was proposed that this compound seemed to be a sesquiterpene from [1]H nuclear magnetic resonance (NMR), [13]C NMR, HRMS, and elemental analysis; hence, the molecular weight was 282, and the molecular formula was $C_{15}H_{22}O_5$. However, it was not so easy to assign its structure. The major difficulty was in arranging these five oxygen atoms in this 15-carbon molecule skeleton, which has only one proton attached at the carbon bearing oxygen (5.68 in singlet) that appeared in the [1]H NMR spectrum. In early 1975, the peroxide structure of yingzhaosu (**2**)[27] inspired researchers that qinghaosu might also be a peroxide compound. The hypothesis was confirmed by simple qualitative analysis (NaI-AcOH) and quantitative analysis (PPh₃) soon afterward. It was also revealed that the fragment 250 in the mass spectrum came from loss of a molecular oxygen from qinghaosu instead of loss of a methanol as believed before. Referring to the structure of arteannuin B (artemisinin B **10**)[28,29] isolated also from *A. annua* before and after the physical data, three structures (**11–13**) were possible, and structure **13** was preferable because of the existence of some peroxy lactones in the literature at that time (Structure 5-2). The real structure and the relative configuration were at last proved by x-ray crystal analysis. Finally the absolute configuration was obtained by abnormal diffraction x-ray crystal analysis.[30] Therefore, qinghaosu has really an unprecedented unique structure with an inter-peroxyl ketal–acetal–lactone consisting of a rare OO–C–O–C–O–C=O segment, and until now, no such structure has been found in other natural products.

**Structure 5-2**

### 5.2.2   The Phytochemistry of Qinghao and Other Natural Products From Qinghao

After the discovery of qinghaosu (**1**) from qinghao (*A. annua* L.) in the early 1970s, *A. annua* became one of the most extensively investigated plants thereafter. As the intriguing chemotaxonomic marker in *A. annua*, qinghaosu (**1**) attracted intense efforts initially devoted to the establishment of the highest content of qinghaosu in *A. annua* and possibly other *Artemisia* plants. Hence, studies on the time course of the levels of qinghaosu (**1**), its biosynthetic precursors, and the biosynthetically related sesquiterpenes were conducted by several research groups around the world.[31–35] According to the results, qinghaosu (**1**) is identified in all *A. annua* plants from different geographical origins, whereas its content is varied drastically with its growing area and stages of plant development. Qinghaosu (**1**) is present in the leaves and flowers of *A. annua* in ~0.01–1.1% of dry weight.[31,32,34] Production of qinghaosu (**1**) from *A. annua* rarely exceeds 1.0% of the dry weight, with the highest content just before flowering.

Apart from *A. annua*, qinghaosu (**1**) was detected in only one other *Artemisia* species: *Artemisia apiacea*.[32] But the abundance was too low (0.08%) to justify an isolation on a technical scale.

Since the discovery of qinghaosu, systematic phytochemical studies on *A. annua* have been also conducted. Different *A. annua* materials including the leaves, stems/flowers, roots, and seeds as well as the endophytes inside *A. annua* have been employed for phytochemical investigations. Up to the time of this writing, more than 150 natural products were reported to belong to different chemical structure types. Herein, we try to give a summary of these secondary metabolites isolated from *A. annua* to date.

### 5.2.2.1   Terpenoids from A. annua

As mentioned, qinghaosu chemically belongs to the cadinane sesquiterpene; therefore, the other sesquiterpene components in *A. annua* have been given preferential attention. From indigenous *A. annua* L., continuous phytochemical studies by Chinese researches in the early 1980s led to the excavation of another ten sesquiterpenes including deoxy-artemisinin (**14**),[36] artemisinin D (**15**),[37] artemisinin F (**16**),[38] artemisinin E (**17**),[37] artemisinin A (**18**),[36,37] epoxyarteannuinic acid (**19**),[39] artemisinic acid (**20**),[40,41] artemisinic acid methyl ester (**21**),[42] artemisinol (**22**),[42] and arteannuin B (**10**).[36] Among them, arteannuin B (**10**) was reported in the early 1970s.[28,29] They are all closely related to the amorphene series of sesquiterpene characterized by the presence of a *cis*-decalin skeleton with the isopropyl group *trans* to the hydrogen on the ring juncture. From a biogenetic viewpoint, artemisinic acid (**20**) or its 11,13-dihydro analoge, dihydro-artemisinic acid (**23**), which was isolated later from *A. annua*, are late precursors in the biogenesis of qinghaosu (**1**).[43] The two compounds **20** and **23** were first reported by Chinese researchers,[40,43] and procedures for their isolation were also reported in the early 1980s elsewhere.[44] In the late 1980s, another procedure for the isolation of artemisinic acid (**20**) was described by Roth and Acton.[45] By 1991, 16 closely related sesquiterpenes had been isolated from the aerial part of *A. annua* and briefly summarized by Zaman and Sharma.[4] Four additional sesquiterpenes include the β-epoxy isomer of arteannuin B (**24**),[46] 6-*epi*-deoxyarteannuin B (**25**),[47,48] 11,13-dehydro-qinghaosu artemisitene (**26**),[49] and 6,7-dehydro-artemisinic acid (**27**).[47] Qinghaosu (**1**) can be classified as a cadinane sesquiterpene oxygenated at the 12-position; the other 15 cadinanae sesquiterpenes share this common structural feature (see Table 5-1, Nos. 1–16). A new sesquiterpenes called artemisinin G (**28**) was purified from *A. annua* shortly after the Zaman and Sharma brief review by Chinese scientists in 1992.[50] Because the authors found that compound **28** was a decomposition product of qinghaosu by heating at 190 °C for 10 min, or in refluxing xylene for 22 h, the possibility that artemisinin G (**28**) could be an artifact during the isolation procedure was eliminated through a heating experiment with qinghaosu that mimicked the isolation process. A new member of the unusual cadinanolide series of sesquiterpenes, annulide (**29**), was described in 1993 with its structure and relative configuration determined by nuclear magnetic resonance (NMR) results, although the sample was limitedly purified.[51] Meanwhile the closely related structure isoannulide (**30**) was also described in the same article with poor purity; the complete and unambiguous NMR spectral assignments for isoannulide (**30**) were presented in a later article by the same author.[52] Three other structural relatives of artemisinin B (**10**), **31**,[53] **32**,[54] and 6α-hydroxyisoannulide (**33**),[55] were recorded in 1987, 1992, and 1994, respectively. Known as an acid hydrolysis product of artemisinin B (arteannuin B, **10**) based on an earlier synthetic research, compound **33** was checked by thin-layer chromatography (TLC) analysis of the crude extract and proved to be the endogenetic natural product from *A. annua* other than an artifact during purification. A bisnor-sesquiterpene, norannuic acid (**34**) was reported in 1993,[56] and three new cadinane sesquitepenes (**35–37**) were isolated and reported in 1994, both by Ahmad and Misra.[57] Compound **37** is of interest in that a 3-isobutyryl group was discovered for the first time in cadinane sesquiterpenes from *A. annua*. Brown described compound **38** (a pair of isomers), **39** in 1994,[58] and Sy and Brown described another new cadinane sesquitepene (**40**) in addition to a new eudesmane sesquiterpene (**41**) from the aerial parts of *A. annua* in 1998.[59] Compound **40** is unique in that it is oxygenated at the 7-position rather than at the 12-position found in most other cadinanes isolated from this species. Compound **41** is a 5α-hydroxyeudesmane incorporating an allylic tertiary hydroxide group evidenced by the results of HREIMS and 1D/2D-NMR, which is also a 5-hydroxy derivative of *trans*-β-selinene previously identified from *A. annua* growing in the United Kingdom.[51] Seven new sesquiterpenes, including a peroxylactone arteannuin H (**42**) and arteannuin I – M (**43–48**), were isolated by the same research group in 1998.[60] Meanwhile they also proposed that these compounds were biogenetically related to dihydroartemisinic acid (**23**) via some intermediate allylic hydroperoxide such as compound **49**, which was eventually isolated as a natural product in 1999 and proved not to be an artifact of isolation.[61] A reinvestigation of *A. annua* gave a novel cadinane diol, arteannuin O (**50**); its structure was established by two-dimensional (2D) NMR and x-ray crystallography.[62] Synthesis of arteannuin O (**50**) from dihydro-*epi*-deoxyarteannuin B (**22**) led the authors to

**TABLE 5-1. Sesquiterpenes Isolated From *A. annua* (1972–2004)**

| No. | No. of Comp. | Trivial Name(s) | Plant Part | Reference |
|-----|------|-----------------|------------|-----------|
| 1 | 1 | Qinghaosu, artemisinin, arteannuin | Aerial part | 25,26 |
| 2 | 10 | Artemisinin B, arteannuin B | Aerial part | 28,29,36 |
| 3 | 14 | Deoxyartemisinin, deoxyarteannuin, Qinghaosu III | Aerial part | 36 |
| 4 | 15 | Artemisinin D, arteannuin D, Qinghaosu IV | Aerial part | 37 |
| 5 | 16 | Artemisinin F | Aerial part | 38 |
| 6 | 17 | Artemisinin E, Qinghaosu V, arteannuin E | Aerial part | 37 |
| 7 | 18 | Artemisinin A, arteannuin A, Qinghaosu I | Aerial part | 36,37 |
| 8 | 19 | Epoxyarteannuinic acid | Aerial part/Seeds | 39 |
| 9 | 20 | Artemisinic acid, Qinghao acid, Arteannuinic acid | Aerial part | 40,41 |
| 10 | 21 | Artemisinic acid methyl ester | Aerial part | 42 |
| 11 | 22 | Artemisinol | Aerial part | 42 |
| 12 | 23 | Dihydro-artemisinic acid | Aerial part | 43 |
| 13 | 24 | Artemisinin C, Arteannuin C | Aerial part | 46 |
| 14 | 25 | (+)-deoxyisoartemisinin B, *epi*-deoxyarteannuin B | Aerial part | 47,48 |
| 15 | 26 | Artemisitene | Aerial part | 49 |
| 16 | 27 | 6,7-dehydroartemisinic acid | Aerial part | 47 |
| 17 | 28 | Artemisinin G | Leaf | 50 |
| 18 | 29 | Annulide | Aerial part | 51 |
| 19 | 30 | Isoannulide | Aerial part | 51,52 |
| 20 | 31 | Dihydroarteannuin B | | 53 |
| 21 | 32 | Dihydro-*epi*-deoxyarteannuin B | Aerial part | 54 |
| 22 | 33 | 6α-hydroxyisoannulide | Aerial part | 55 |
| 23 | 34 | Norannuic acid | Aerial part | 56 |
| 24 | 35 | Cadin-4,7(11)-dien-12-al | Aerial part | 57 |
| 25 | 36 | Cadin-4(15),11-dien-9-one | Aerial part | 57 |
| 26 | 37 | 3-isobutyryl cadin-4-en-11-ol | Aerial part | 57 |
| 27 | 38 | | Aerial part | 58 |
| 28 | 39 | | Aerial part | 58 |
| 29 | 40 | | Leaf | 59 |
| 30 | 41 | | Leaf | 59 |
| 31 | 42 | Arteannuin H | Leaf | 60 |
| 32 | 43 | Arteannuin I | Leaf | 60 |
| 33 | 44 | Arteannuin J | Leaf | 60 |
| 34 | 45 | Arteannuin N | Leaf | 60 |
| 35 | 46 | Arteannuin K | Leaf | 60,61 |
| 36 | 47 | Arteannuin L | Leaf | 60,61 |
| 37 | 48 | Arteannuin M | Leaf | 60,61 |
| 38 | 49 | Dihydroartemisinic acid hydroperoxide | | 61 |
| 39 | 50 | Arteannuin O | Leaf | 62 |
| 40 | 51 | Deoxyarteannuin B | Aerial part | 52 |

**TABLE 5-1.**  (*Continued*)

| No. | No. of Comp. | Trivial Name(s) | Plant Part | Reference |
|-----|------|------|------|------|
| 41 | 52 | Dihydro-deoxyarteannuin B | Aerial part | 52 |
| 42 | 53 | | Seeds | 62 |
| 43 | 54 | | Seeds | 62 |
| 44 | 55 | | Seeds | 62 |
| 45 | 56 | | Seeds | 62 |
| 45 | 57 | | Seeds | 62 |
| 46 | 58 | | Seeds | 62 |
| 47 | 59 | | Seeds | 62 |
| 48 | 60 | | Seeds | 62 |
| 49 | 61 | | Seeds | 62 |
| 50 | 62 | | Seeds | 62 |
| 51 | 63 | | Seeds | 62 |
| 52 | 64 | | Seeds | 62 |
| 53 | 65 | | Seeds | 62 |
| 54 | 66 | Nortaylorione | Essential oil* | 63 |
| 55 | 67 | 3α,5β-dihydroxy-4α,11-epoxybisnor cadinane | Aerial part | 64 |
| 56 | 68 | Abscisic acid | Aerial part | 65 |
| 57 | 69 | Abscisic acid, methyl ester | Aerial part | 65 |

*The new nor-sesquiterpene was detected and identified in the essential oil extract from *A. annua*; the structure was indirectly elucidated and verified by GC, GC/MS, and synthesis.

propose a structure revision of the stereochemistry claimed for the 5-OH group in arteannuins K (**46**), L (**47**) and M (**48**) as shown in their strutures.[62] Two amorphane sesquiterpenes, deoxyarteannuin B (**51**) and dihydro-deoxyarteannuin B (**52**), were introduced to the sesquiterpene family and isolated from the aerial parts of *A. annua* in 2001.[52] Recently, the first phytochemical investigation of natural products from the seeds of *A. annua* was conducted by Sy et al., which led to the discovery of 14 new sesquiterpenes (**53–65**).[63] The structures of these compounds were elucidated mainly from the results of 2D NMR spectroscopic studies including HSQC, HMBC, [1]H-[1]H COSY, and NOESY. (+)-Nortaylorione (**66**), a nor-sesquiterpene, was described as a new natural product from *A. annua* by Marsaioli et al.[64] The structure elucidation including its relative and absolute configuration of compound **66** was not based on the real isolation from essential oil extract but on organic synthesis. The new bisnor cadinane sesquiterpene **67** was then isolated from *A. annua*.[65] In addition, two sesquiterpene plant hormones, abscisic acid (**68**) and its methyl ester (**69**), were found from Indian-grown *A. annua* (Table 5-1, Structure 5-3).[66]

Besides sesquiterpenes, several mono- (**70–75**), di- (**76**), and triterpenoids (**77–79**) have been obtained from *A. annua*, accounting for some ten compounds (**70–79**).[63,67–70] Several common triterpenes also found in *A. annua* are α-amyrenone, α-amyrin, β-amyrin, taraxasterone, oleanolic acid, and baurenol (Structure 5-4).[57]

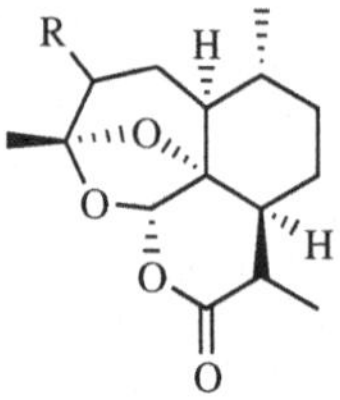

**14** R=H
**15** R=α-OH

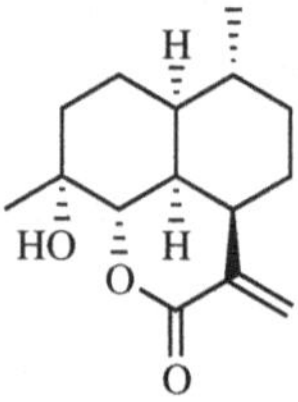

**16**

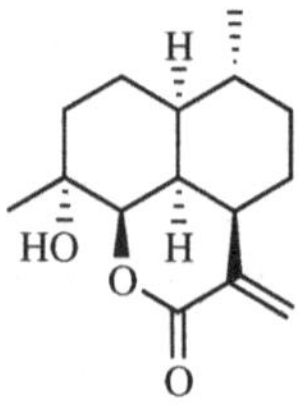

**17**

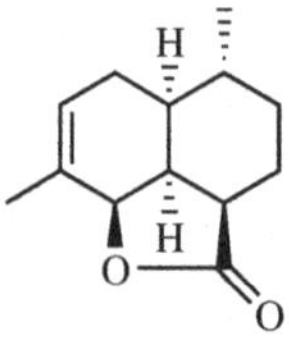

**18**

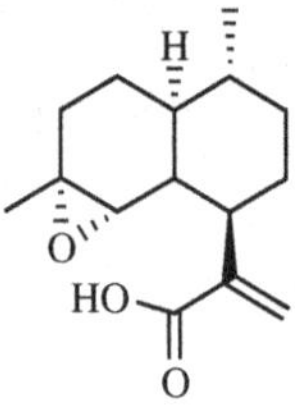

**19**

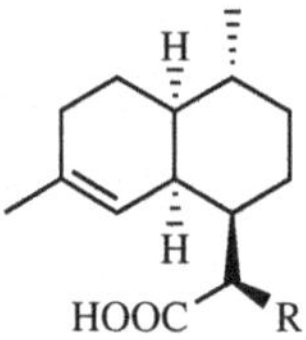

**20** R =CH$_2$
**23** R = CH$_3$

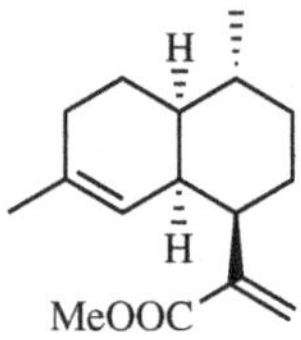

**21**

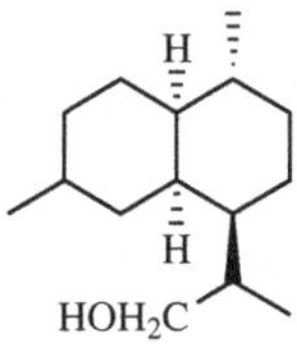

**22**

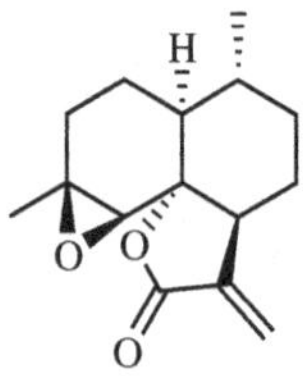

**24**

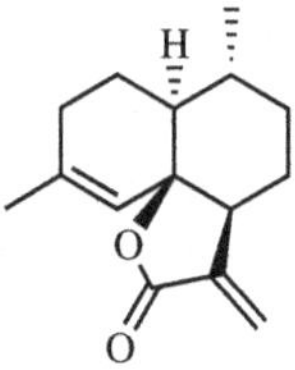

**25**

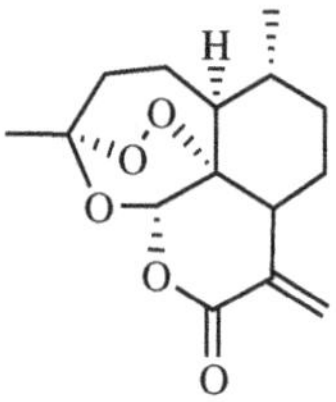

**26**

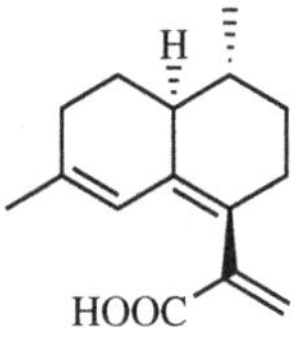

**27**

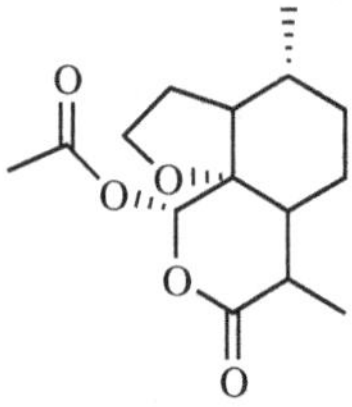

**28**

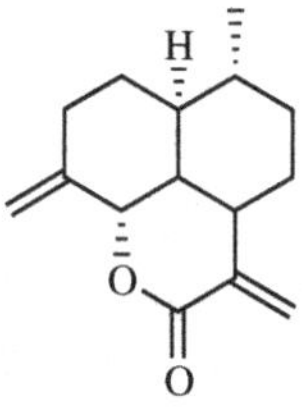

**29**

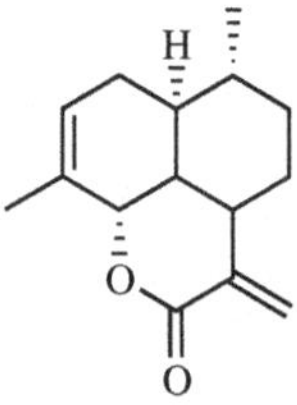

**30**

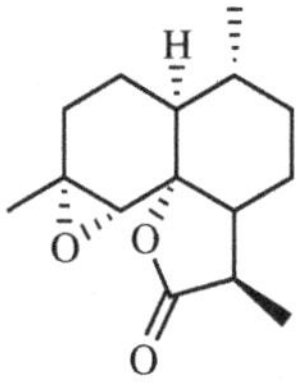

**31**

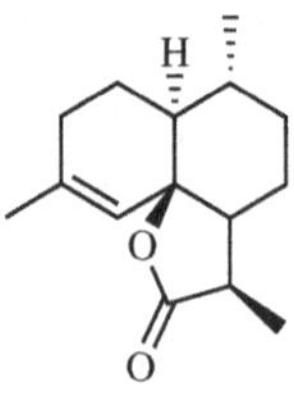

**32**

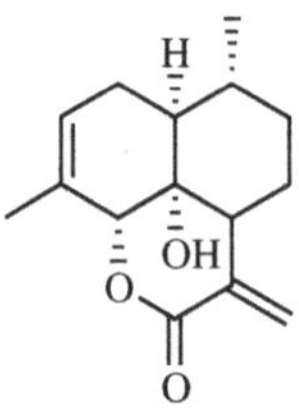

**33**

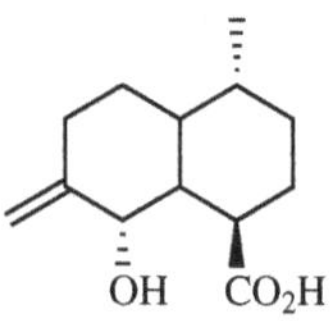

**34**

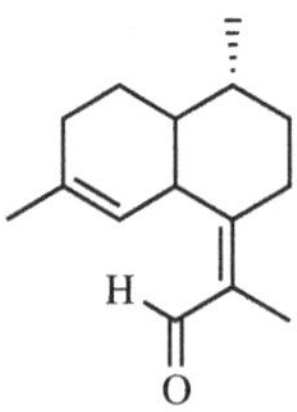

**35**

**Structure 5-3.** Structures of sesquiterpenes isolated from *A. annua* (1972–2004).

**Structure 5-3.** (*Continued*)

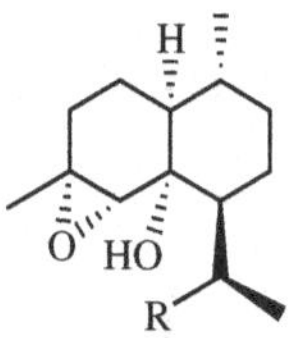

56 R=CO$_2$H
57 R=CH$_2$OH

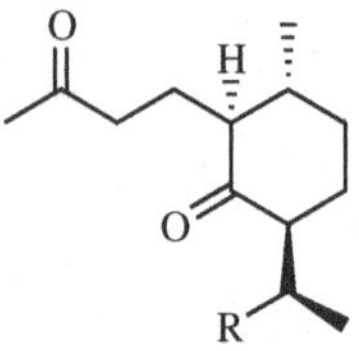

58 R=CH$_2$O-CHO
59 R=CO$_2$H

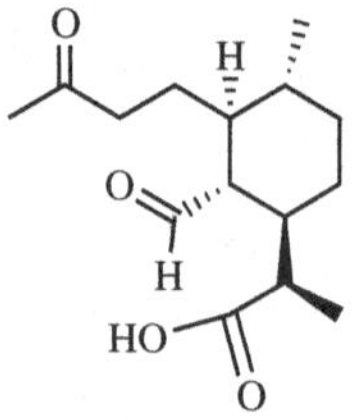

60

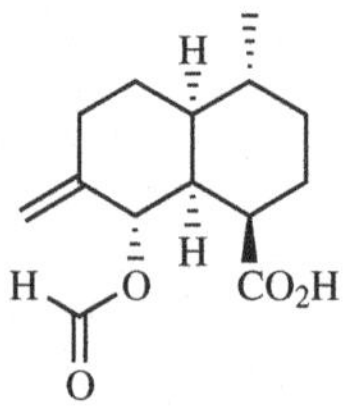

61

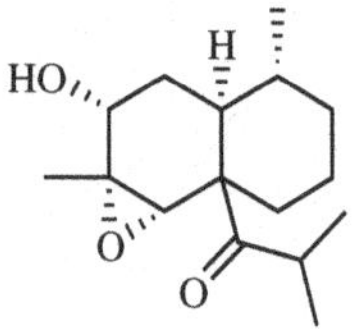

62

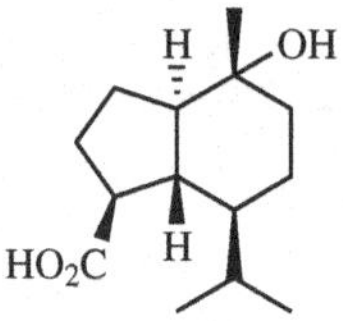

63

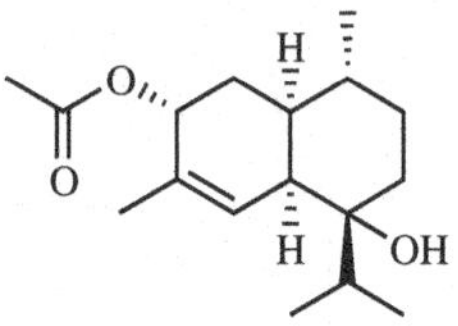

64

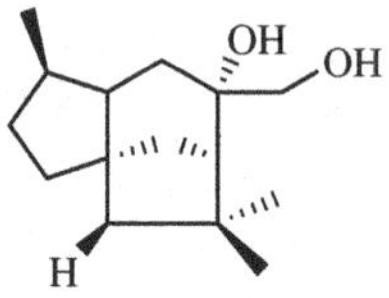

65

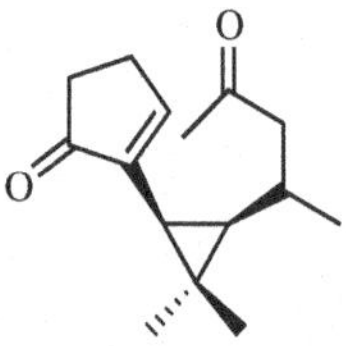

66

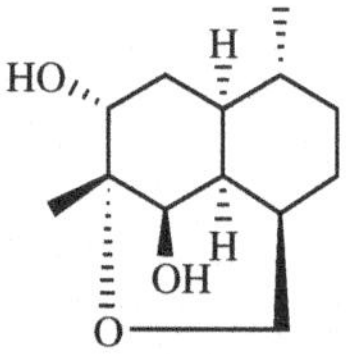

67

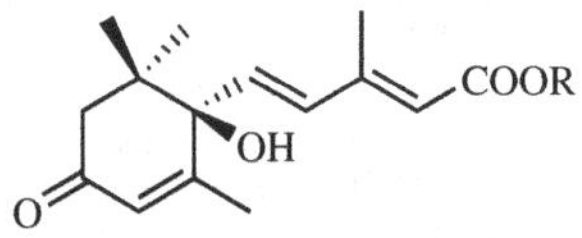

68 R=H
69 R=CH$_3$

**Structure 5-3.** (*Continued*)

### 5.2.2.2 Essential Oils From A. annua

Essential oil from *A. annua* is another active research interest as it could be potentially used in perfume, cosmetics, and aromatherapy. Depending on its geographical origin, the oil yield in *A. annua* ranges from 0.02% to 0.49% on a fresh weight basis and from 0.04% to 1.9% on a dry weight basis.[71] The major components in the oil were reported to be artemisia ketone (**80**), isoartemisia ketone (**81**), 1,8-cineole (**82**), and camphor (**83**) (Structure 5-5). GC/MS was employed to analyze the chemical composition in the essential oil; more than 70 constituents have been identified. For more detailed information on the oil composition of essential oil from *A. annua*, the readers are referred to Refs. 65, 66 and 72–81.

### 5.2.2.3 Flavonoids and Coumarins From A. annua

Up to today, 46 flavonoids have been isolated from *A. annua*.[66,79,80,82] They are as follows: apigenin, artemetin, astragalin, axillarin, casticin, chrysoeriol, chrysoplenetin, chrysosplenol, chrysosplenol D, 3'-methoxy, chrysosplenol, cirsilineol, cirsiliol, cirsimaritin, cynaroside, eupatorin, 2',4',5-trihydroxy-5',6,7-trimethoxy flavone, 3',5,7,8-tetrahydroxy-3,4'-dimethoxy flavone, 3,3',5-trihydroxy-4',6,7-trimethoxy flavone, 3,5-dihydroxy-3',4',6,7-tetramethoxy flavone, 4,5,5'-trihydroxy-3,5,6,7-tetramethoxy flavone, 5-hydroxy-3,4',6,7-tetramethoxy flavone, 5-hydroxy-3,4',

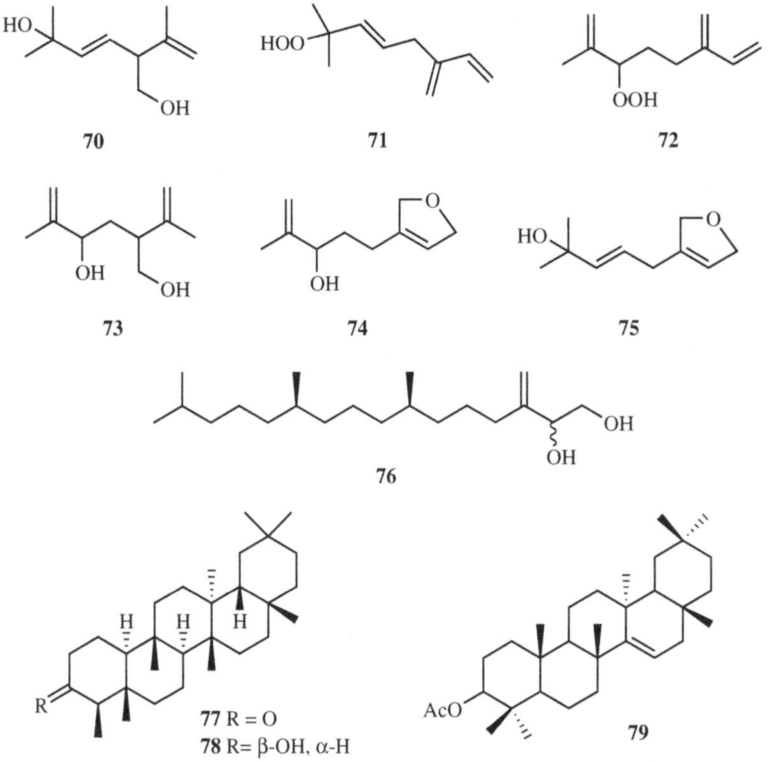

**Structure 5-4**

6,7-tetramethoxy flavone, isokaempferide, kaempferol, kaempferol-6-methoxy-3-*O*-β-D-glucoside, luteolin, luteolin-7-methyl ether, pachypodol, patuletin, patuletin-3-*O*-β-D-glucoside, penduletin, quercetagetin-3′,4′,6,7-tetramethyl ether, quercetagetin-3,4′-dimethyl ether, quercetagetin-3,6-dimethyl ether, quercetagetin-4,6′,7-trimethyl ether, quercetagetin-4′-methyl ether, quercetin, quercetin-3′-*O*-β-D-glucoside, quercetin-3-methyl ether, quercimeritrin, isoquercitrin, retusin, rhamnentin, rutin, and tamarixetin.

About seven commonly occurring coumarins were found in *A. annua*, namely scopoletin, scopolin, aesculetin, 6,8-dimethoxy-7-hydroxy coumarin, 5,6-dimethoxy-7-hydroxy coumarin, tomentin, and coumarin.[66,82–84]

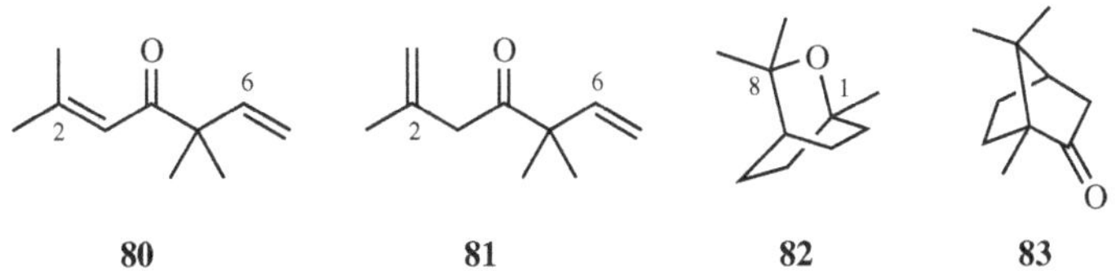

**Structure 5-5**

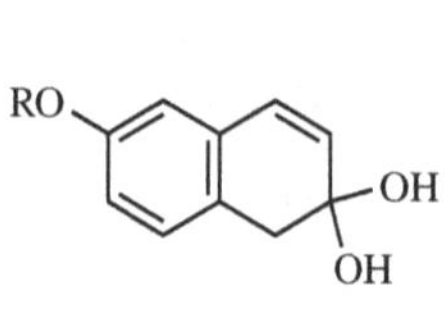

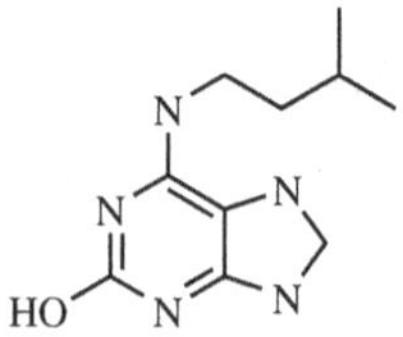

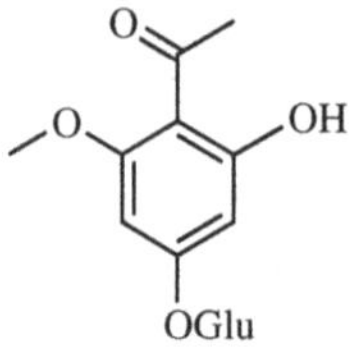

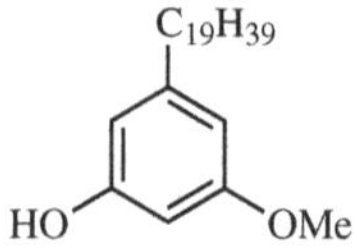

84 R=CH₃
85 R=H

86    87    88

**Structure 5-6**

### 5.2.2.4 Miscellaneous Components and Natural Products From Endophytes in A. annua

Two new chromene derivatives have been isolated from the aerial parts of *A. annua*. Their structures were resolved through normal NMR spectra as 2,2-dihydroxy-6-methoxychromene (**84**) and 2,2,6-trihydroxychromene (**85**).[83]

A novel cytokinin, 6-(3′-methylbutylamino)-2-hydroxy-7,8-dihydropurine (**86**), was obtained from a methanolic extract of the aerial part of Indian-grown *A. annua*.[85]

Two phenolic compounds have been described as new natural products from *A. annua*. The water-soluble part of an ethanol extract of the aerial parts afforded annphenone (**87**), a phenolic acetophenone. Column chromatography followed by high-performance liquid chromatography of an Et₂O extract of the aerial parts yielded the new compound 5-nonadecylresorcinol-3-*O*-methyl ether (**88**); its structure was deduced largely from NMR spectroscopy and confirmed by chemical synthesis (Structure 5-6).[86,87]

A new highly unstable polyacetylene (**89**) as well as the known polyacetylene ponticaepoxide (**90**) were obtained after repeated chromatographic purification of the crude petroleum ether extract of *A. annua*.[88] The new polyacetylene was called annuadiepoxide (**89**).

A new lipid constituent methyl-11,12,15-trihydroxy-13(14)-octadecanoate (**91**) was recently isolated from the leaves of *A. annua*.[65] A lipophilic fraction of *A. annua* was found to contain aurantiamide acetate, a dipeptide (**92**) (Structure 5-7).[89]

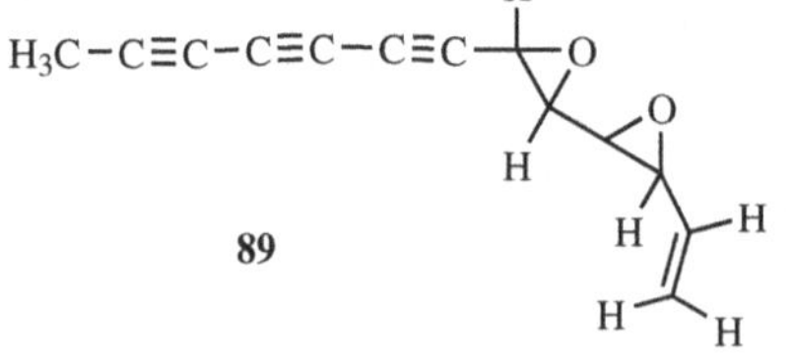

89 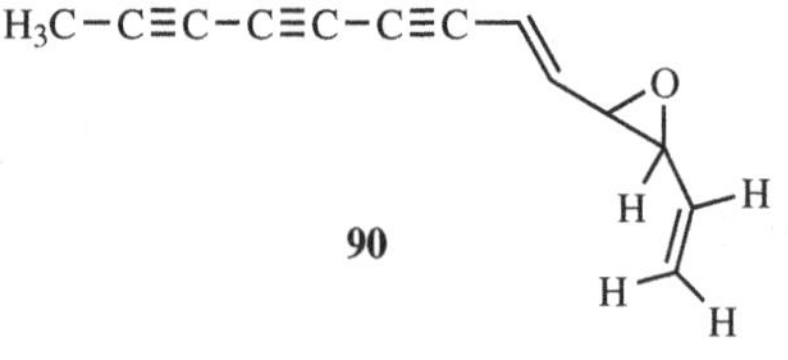    90

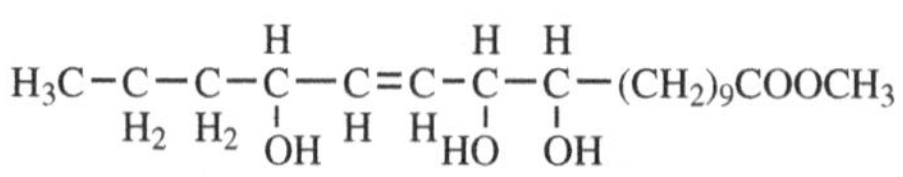

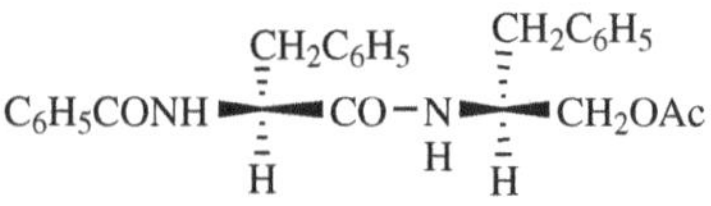

91    92

**Structure 5-7**

93                    94 R=COCH$_3$                        96                    97
                      95 R=COCH$_2$C$_6$H$_5$

**Structure 5-8**

It is interesting to note that the well-known *A. annua* is seldom attacked by any phytopathogenic fungi, which could be partially associated with the presence of endophytes.[90] Two endophytic fungi in *A. annua* have been phytochemically explored. From *Colletotrichum* sp., an endophyte isolated from inside the stem of *A. annua*, 11 chemical constituents were isolated including three new antimicrobial metabolites.[91] Several known steroids were recorded as stigmasterol,[44,92] ergosterol,[92] 3β,5α,6β-trihydroxyergosta-7,22-diene, 3β-hydroxy-ergosta-5-ene, 3-oxo-ergosta-4,6,8(14),22-tetraene, 3β-hydroxy-5α,8α-epidioxy-ergosta-6,22-diene, 3β-hydroxy-5α,8α-epidioxy-ergosta-6,9(11),22-triene, 3-oxo-ergosta-4-ene, and plant hormone indole-3-acetic acid. The chemical structures of the three new metabolites were elucidated by a combination of spectroscopic methods (infrared, MS, and NMR) as 6-isoprenyliindole-3-carboxylic acid (**93**), 3β,5α-dihydroxy-6β-acetoxy-ergosta-7,22-diene (**94**), and 3β,5α-dihydroxy-6β-phenylacetyloxy-ergosta-7,22-diene (**95**), respectively. Two new metabolites with novel carbon skeleton, leptosphaerone (**96**) and leptosphaeric acid (**97**), were discovered from the AcOEt extract of endophytic fungus *Leptosphaeria* so. IV 403 (Structure 5-8).[93,94]

In summary, ~150 secondary metabolites, besides those common compositions of the plant, have been found and isolated from *A. annua*, including compounds isolated from endophytes inside *A. annua*. To date, all phytochemical studies in relation to *A. annua* have led to ~60 sesquiterpenes; ~16 mono-, di-, and triterpenes; 46 flavonoids; 7 coumarins; 9 miscellaneous components; and 13 endophyte-produced natural products.

## 5.3  REACTION OF QINGHAOSU

Study on the reactions of qinghaosu is heuristic, not only for determination of its structure, but also especially for understanding chemical properties and hence for the modification and utilization of the qinghaosu molecule to develop a new medicinal application. Among them, reduction and acidic degradation have been paid the most attention and have received more practical application.

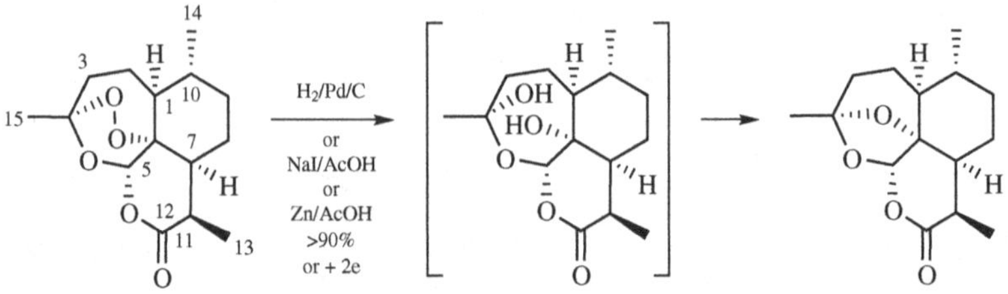

**Scheme 5-1**

### 5.3.1 Reduction of Qinghaosu

The tandem peroxy, ketal, acetal, and lactone groups in the qinghaosu molecule are all reducible under different reaction conditions, but the two sides, peroxy group or lactone, will be reduced at first. The peroxy group can be reduced by hydrogenation in the presence of palladium/charcoal to afford a dihydroxy intermediate that in turn will convert into a stable product, deoxyqinghaosu (**14**), under stand or by treatment with catalytic amount of acid.[26] **14** also can be obtained by reduction with zinc dust–acetic acid.[95] The inactivity of deoxyqinghaosu indicates that the peroxy group is a principle segment for antimalaria.[96,97] Recently, it was found that the above-mentioned qualitative analysis for the peroxy group converts qinghaosu to deoxyqinghaosu too, but the conversion is only 27%. Bromide, unlike iodide, could not reduce qinghaosu under the same reaction condition. The reaction of triphenyl phosphine with qinghaosu used for the quantitative analysis of peroxy group is complicated; however, deoxyqinghaosu still can be separated from the product mixture in 23% yield.[98] The electrochemical reduction of qinghaosu has been reported recently from several research groups, and a two-electron unreversible reduction process was observed.[99–104] Deoxyqinghaosu (**14**) was found to be the only product when we repeated this slow electrochemical reduction (Scheme 5-1).[98] Another important reaction of the peroxy group is the single-electron reduction with ferrous ion, copper(I) ion, and so on which is related to the antimalarial mechanism and will be discussed in a later section of this chapter in detail.

The lactone of qinghaosu could be reduced with mild hydride-reducing agents, such as sodium borohydride, potassium borohydride, and lithium borohydride to lactol, dihydroqinghaosu or reduced qinghaosu (**98**) in over 90% yield.[26] It is a novel reduction, because normal lactone could not be reduced with sodium borohydride under the same reaction condition (0–5°C, in methanol). It was surprising to find that the lactone was reduced, but that the peroxy group survived. However, the lactone of deoxyqinghaosu (**14**) resisted reduction with sodium borohydride and could only be reduced with isobutyl aluminium hydride to the lactol, deoxydihydroqinghaosu (**99**),[105] which was identified with the product from hydrogenation of **98** (Scheme 5-2). These results show that the peroxy group assists the reduction of lactone with sodium borohydride to a lactol, but not to the over reduction product alcohol. No clear explanation for this reduction process exists. The easy availability of

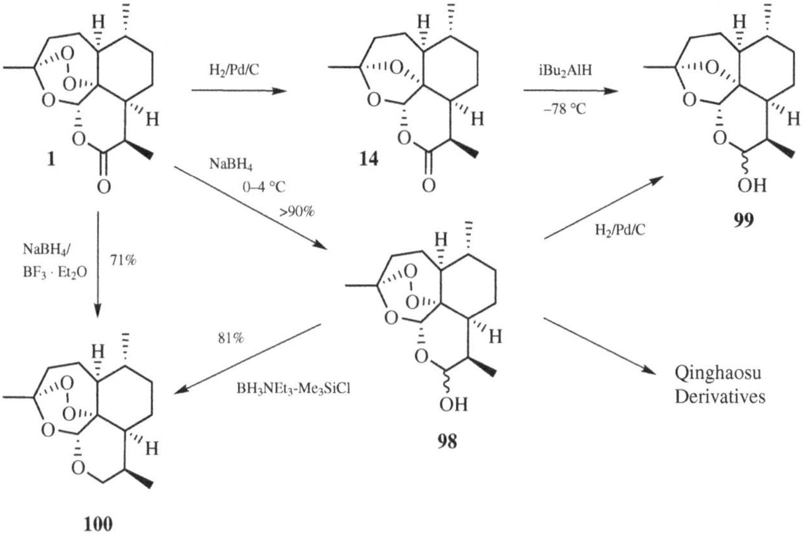

**Scheme 5-2**

dihydroqinghaosu makes the derivation of qinghaosu possible; a detailed discussion of this takes place in Section 5.5.

Qinghaosu can be reduced even more with sodium borohydride in the presence of boron trifluoride to deoxoqinghaosu (**100**).[106] **100** can also be obtained by reduction of **98** with BH$_3$NEt$_3$ and Me$_3$SiCl in DME.[107] The more powerful reducing agent lithium aluminium hydride reduces not only lactone and peroxy group, but also acetal and ketal, to yield the exhaustively reduced product **101** and partially reduced products (Scheme 5-3).[108,109]

### 5.3.2 Acidic Degradation of Qinghaosu

Treatment of qinghaosu in a mixture of glacial acetic acid and concentrated sulfuric acid (10:1) at room temperature yields a mixture of one carbon less products, among which several ketone-lactone or α,β-unsaturated ketone can be isolated.[110] X-ray crystal analysis of the major component **103** shows that its C-7 configuration is inverted in comparison with that of qinghaosu.[111] An intermediate **102** for the formation of these products has been proposed (Scheme 5-4).

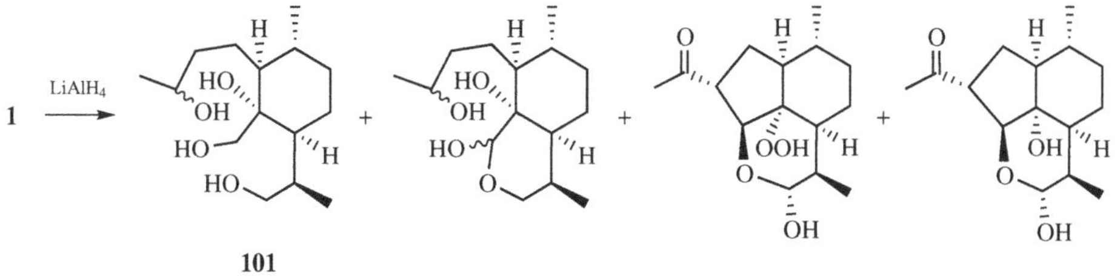

**Scheme 5-3**

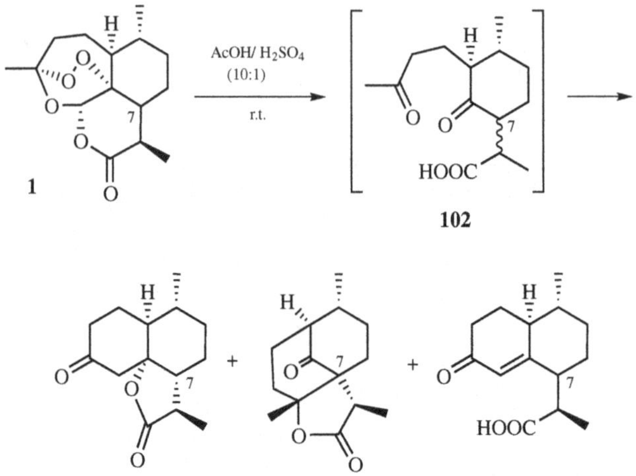


1 102

103
and its epimers and its epimers and its epimers

**Scheme 5-4**

In a continued investigation, it is found that refluxing of a solution of qinghaosu and a catalytic amount of acid in methanol affords a mixture of methyl esters **104**, the treatment of which with glacial acetic acid and concentrated sulfuric acid (10:1) at 0–5°C gives in turn a C-7 configuration reserved diketone ester **105** and minor recovered qinghaosu. The overall yield based on the recovered starting material can be over 90%. Intermediate **104** can be purified and identified and can be ring-closed to qinghaosu, **104** may be reduced to deoxyqinghaosu (**14**) or peroxy-reserved lactone **106** (Scheme 5-5).[112] Diketone ester is a useful relay intermediate for the

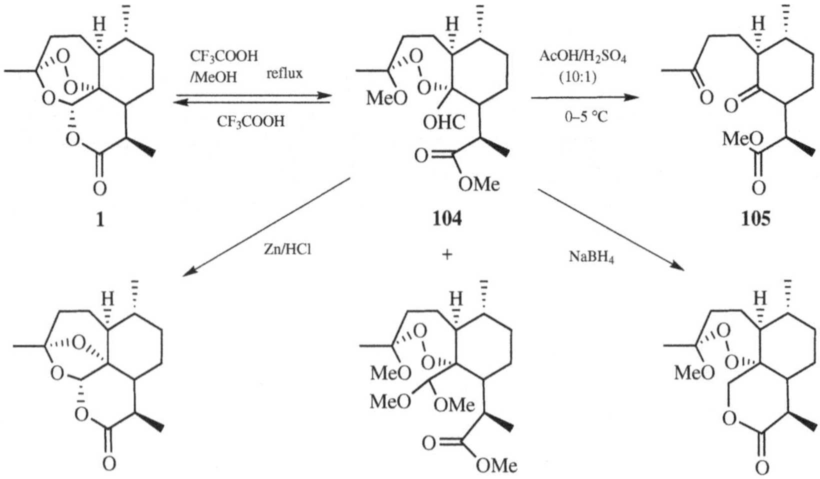


1 104 105

14 106

**Scheme 5-5**

synthesis of qinghaosu and its derivatives, and it will be mentioned in the following sections. Somewhat late, Imakura et al. also reported that treatment of qinghaosu in methanol or ethanol with TsOH or 14% hydrochloric acid afforded the methyl or ethyl ester of **104** and **105**, but in low yields.[113]

### 5.3.3 Miscellaneous Chemical Reaction

#### 5.3.3.1 Degradation in an Alkaline Medium
Qinghaosu is unstable in an alkaline medium; it may rapidly decompose in potassium carbonate–methanol–water at room temperature to complicate the products, among which an octa-hydro-indene **107** can be isolated in 10% yield.[26] It may be necessary to pay attention to this property when handling the qinghaosu sample.

#### 5.3.3.2 Pyrolysis
Qinghaosu is a stable compound in comparison with common peroxides; no decomposition is observed even at its melting point at 156–157 °C. However, pyrolysis takes place at 190 °C for 10 min, which provides a product mixture, from which compounds **108** (4%), **28** (12%), and **15** (10%) can be separated.[114,115] It is interesting that the latter two compounds are also the compositions of qinghao (*A. annua*),[37,50] the metabolites of qinghaosu in vivo,[116,117] and the reaction products of qinghaosu and ferrous ion (*vide infra*) (Scheme 5-6).

### 5.3.4 Biotransformation

Microbial transformation study can serve as a model for the study of qinghaosu metabolism in the mammalian and can unshed the new pathway to qinghaosu derivatives. Therefore, several research groups have endeavored to transform qinghaosu with different microbes and have found that the hydroxy group can be introduced in some inactive carbon positions of qinghaosu.

It was reported in 1989 that qinghaosu can be transferred to deoxyqinghaosu (**14**) by *Nocardia corallinaz* and 3α-hydroxy-deoxyqinghaosu (**15**) by *Penicilliam*

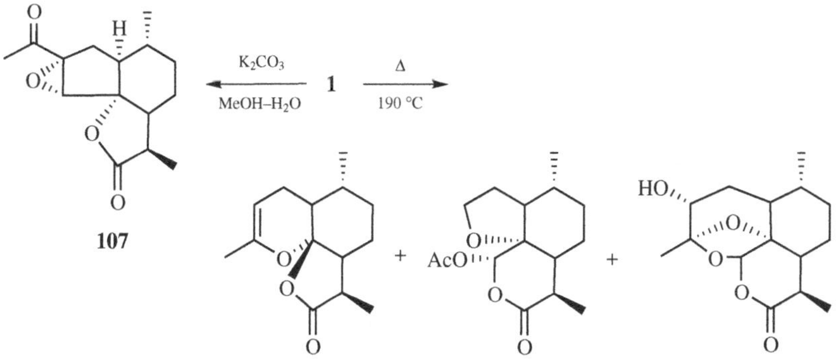

**Scheme 5-6**

**Structure 5-9**

*chrysogenum* in low yield.[118] However, in 2002, Zhan et al. obtained the biotransfer product 9β-hydroxy-qinghaosu (**109**), 3β-hydroxy-qinghaosu (**110**), **14** and **15** with *Mucor polymorphosporus*, and **15** and 1α-hydroxyqinghaosu (**111**) with *Aspergillus niger*.[119] In another report, 10β-hydroxyqinghaosu (**112**) was obtained with *Cunninghamella echinulata* (Structure 5-9).[120] It is worth noticing that compound **15** might be the reaction product of qinghaosu with iron ion–reducing agents in the incubating medium.

In the meantime, there were several reports about the microbial transformation of artemether and arteether, from which several 1α, 9α, 9β, 14-hydroxy derivatives and the products derived probably from the reaction with iron ion–reducing agents in the incubating medium were identified.[121–124]

## 5.4   CHEMICAL SYNTHESIS AND BIOSYNTHESIS OF QINGHAOSU

### 5.4.1   Partial Synthesis and Total Synthesis of Qinghaosu

The outstanding antimalarial activity and the unique structure of qinghaosu have attracted great attention in the research area of organic synthesis. Over nine synthetic routes for qinghaosu have appeared in the literature. In general, the starting materials for these synthetic routes are usually from optical monoterpene or sesquiterpene and the key step introducing peroxy group, except one, is based on the photooxidation. The photooxidation can be performed with singlet oxygen proceeding a [2 + 2] addition with enol ether or a biomimetic manner. These syntheses are outlined in Scheme 5-7. In the earliest two synthetic routes (**A**)[125–128] and (**B**),[129] enol methyl ether **117** is the common substrate of the key step photooxidation. (−)-Isopulegol (**115**) is the starting material in route (**B**), whereas in the route (**A**), artemisinic acid (**20**, arteannuinic acid, qinghao acid) is the starting material to synthesize **117** through dihydro-artmisinic acid (**23**) to finish the partial synthesis and then **23** and **20** are synthesized from the commercially available citronellal (**114**) to finish the total synthesis. In route (**C**), a degradation product **105** serves as the relay intermediate to conveniently prepare the key intermediate **118**.[130] Both route (**D**)[131] and route (**E**)[132,133] published in 1990 and 2003, respectively, use

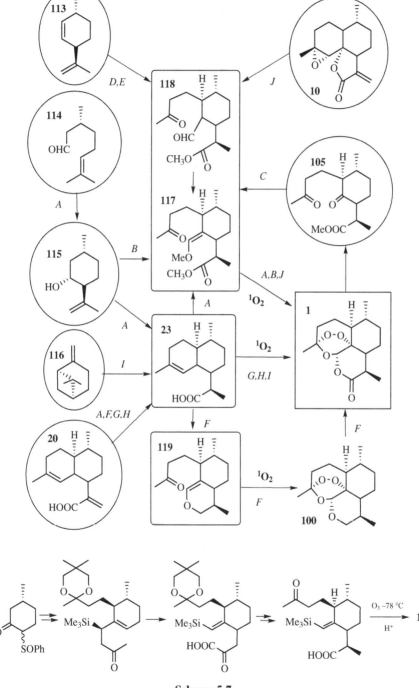

**Scheme 5-7**

(+)- isolimonene (113) as the common starting material. The intramolecular Diels–Alder reaction or iodolactonization followed by Michael addition and Wittig reaction is, respectively, taken as the key steps to afford the key intermediates 23 and 117, and hence, the formal synthesis of qinghaosu occurs. Starting also from dihydro-artmisinic acid (23) as route (A), but using cyclo-enol ether 119 as the substrate of photooxidation, qinghaosu (1) can be selectively synthesized in better overall yield in route (F).[134,135] Another advantage in this route is that the more active compound deoxoqinghaosu (100) can be also synthesized as the last intermediate. The so-called biomimetic synthesis of 1 from 23 occurs in a manner of direct photooxidation in route (G)[136,137] and route (H).[138] The 1993 route (I)[139] reports the synthesis of 23 from α-pinene (116) and then to qinghaosu (1) according to route (G). The synthesis of qinghaosu (1) from another composition 10 in A. annua was reported in the 1992, 1998 route (J), among which the key step is the deoxygenation followed by ozonization to give the ketal protected intermediate 117 (Scheme 5-7).[140,141]

The exceptional synthesis of qinghaosu (1) is the utilization of the ozonization of vinylsilane to build the peroxy group instead of the photooxidation methods (Scheme 5-7).[142,143]

All of these total or partial syntheses have found academic interest. However, the yield of the key step for the formation of the peroxy group is always not so satisfied. Therefore, new methodology for the synthesis of 1 is still expected.

### 5.4.2  Biogenetic Synthesis of Qinghaosu

Since the 1980s, several laboratories have paid attention to the biogenetic synthesis of qinghaosu. From the biogenesis point of view, qinghaosu as a sesquiterpene seems to be synthesized from isoprene and in turn from mavelonic acid lactone (MVA, 120), wherefore three laboratories have found that qinghaosu can be synthesized indeed from MVA or isopentenyl pyrophosphate (IPP, 121), as they were incubated with the homogenate prepared from the leaves of qinghao (A. annua).[144–147] Based on the consideration that there are some plausible biogenetic relationships among three major compositions 1, 20, and 10 in qinghao, Huang et al. and Wang et al. succeeded in the incorporation of MVA into 20 in qinghao and then realized the biotransformation of 20 into 1 and 10 in the homogenate.[144,145] These experiments confirm the previous proposal that 20 is the biosynthesis precursor of 1. Their additional studies on the biotransformation of 20 also find that epoxide 19 is the precursor of 10, but not the precursor of 1, and 10 is not the precursor of 1. The other qinghao component 23 also can be biogenetically transferred into 1, but it is not clear whether artemisinic acid (20) converts to 1 through dihydroartemisinic acid (23) or artemisitene (26) (Scheme 5-8).[148]

Since then, several laboratories also report in their research that artemisinic acid (20) is confirmed to be the biogenetic intermediate of qinghaosu, and dihydroartemisinic acid (23) can also be biotransferred to qinghaosu.[149] Furthermore, on incubation with the cell-free extracts of qinghao leaves, 20 can also be transferred into 10 and in turn 10 as well as dihydro-10 can be transferred to qinghaosu.

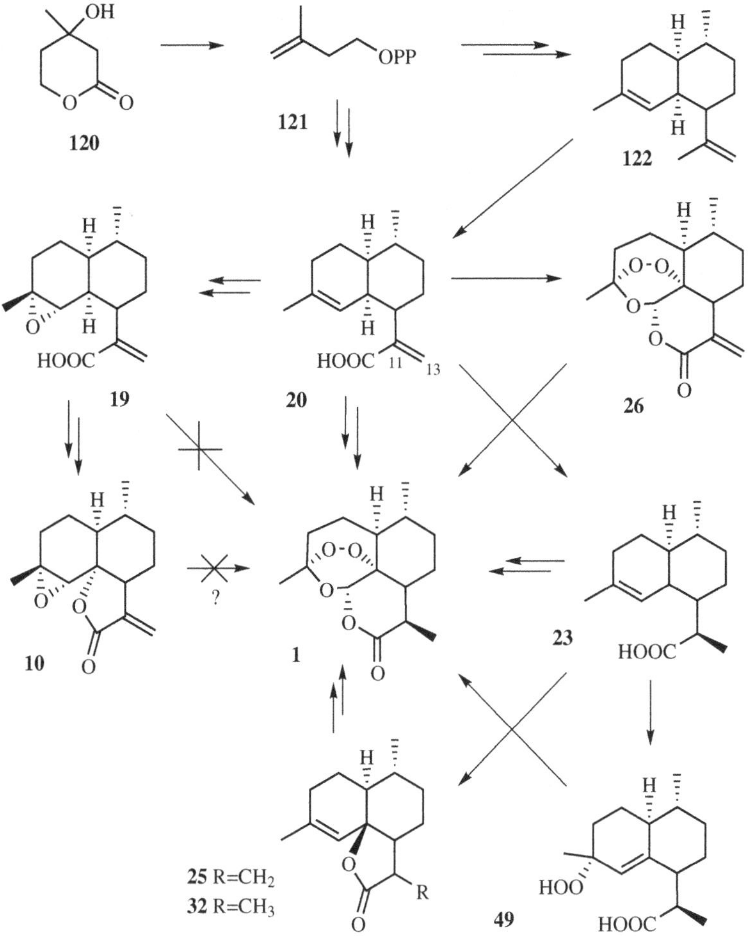

**Scheme 5-8**

Therefore, they conclude that **10** is the biosynthesis precursor of qinghaosu, but it has not been mentioned whether **10** is the necessary intermediate for the biosynthesis of **1**.[150–152] It is unclear whether the different conclusions about the role played by **10** come from the different experimental conditions, the isotopic-labeled precursor in the homogenate[148] versus unlabeled precursor in the cell-free system.[151] The detail process of conversion of **20** or **10** into **1** has not been understood yet, although Wang et al. has suggested that deoxyisoartemisinin B (**25**) and dihydro-deoxyisoartemisinin B (**32**) are the intermediate in the conversion of **20** and **23** to **1**.[153] It has been shown that photo-irradiation will accelerate this conversion. However, it is still uncertain whether some enzymes catalyze a photooxidation in the biosynthesis of **1**. Recently Brown and Sy have performed an in vivo transformation with 15-labeled **23** by feeding them via the root to intact *A. annua* and have concluded that allylic hydroperoxide (49) are the key intermediate in the conversion of 23 to 1 and its other congeners, and this conversion may not need to invoke the participation of enzymes.[154] This biogenetic synthesis has been employed to prepare isotopic-labeled qinghaosu.[155] Since the 1990s, several attempts have been made to enhance qinghaosu production in the cell and tissue culture by omittance or addition of medium components, precursor feedings, and modulating the biosynthesis route,[149,156] but at this time, the biosynthesis approach is still an academic method to provide this antimalarial drug.

On the other hand, the progress in the production of qinghaosu is also made from the selection and breeding of high-yielding cultivars.[157,158] In this respect, hybrid lines containing up to 1.4% qinghaosu on a dry leaves basis have been obtained by selection and crossing, in wild populations, of genotypes with high qinghaosu concentration.

The genetic engineering of qinghao (A. annua) has also been paid great attention recently; some preliminary results about the early stage of qinghaosu biosynthesis have been reported. For example, amorpha-4,11-diene synthase, an enzyme responsible for the cyclization of farnesyl diphosphate into ring sesquiterpene, has been expressed in Escherichia coli and production of amorpha-4,11-diene (122) was identified.[159,160]

## 5.5   DERIVATIVES AND ANTIMALARIAL ACTIVITY

Early pharmacological and clinical studies showed that qinghaosu possessed fast action, low toxicity, and high activity on both drug-resistant and drug-sensitive malaria, even if the severe patients suffering from cerebral malaria could rapidly recover after nasal feeding of qinghaosu. However, the high rate of parasite recrudescence was observed. There was great need for improvement on the inconvenient administration and the high recrudescence rate.

It has been noted that qinghaosu has a special structure bearing peroxy group and rare –O–C–O–C–O–C–O segment, which is different from that of all known antimalarial drugs. In the primary chemical structure-activity study,[96,97] the function of the peroxy group for antimalarial activity was first examined. The negative result of deoxyartemisinin for the antimalarial activity against Plasmodium berghei in mice showed that the peroxy group was essential. Soon afterward, it was found that some other simple peroxides including monoterpene ascaridol had no activity. These facts demonstrated that the peroxy group was an essential, but not a sufficient, factor.

When dihydroartemisinin was found to be more active than qinghaosu and the introduction of the hydroxy group into the molecular nucleus could not improve its solubility in water, three types of dihydroartemisinin derivatives were synthesized and evaluated in China (Scheme 5-9).[96,97,161]

The first 25 compounds (in oil solution) were tested in the mice-infected chloroquine-resistant P. berghei by administration of intramuscular injection.[162] Most of these derivatives showed more activity than qinghaosu and dihydroartemisinin.

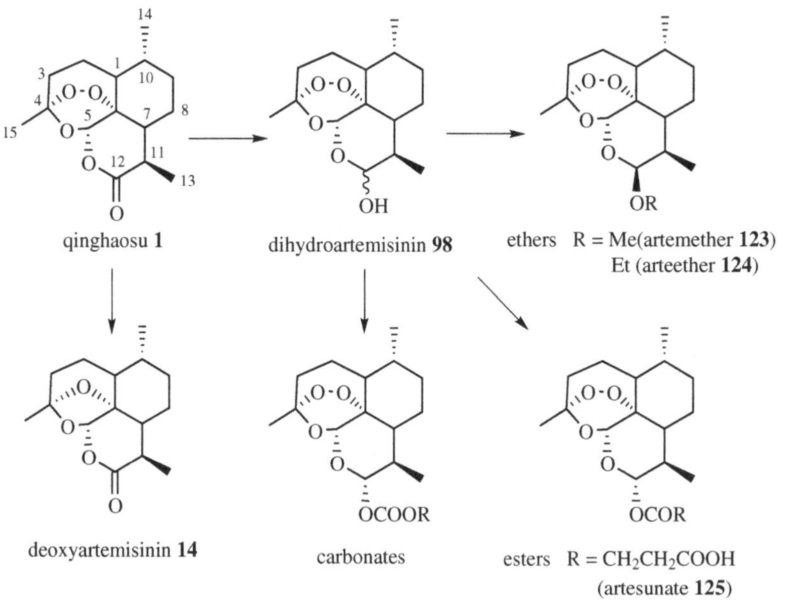

**Scheme 5-9**

Oil-soluble artemether and water-soluble sodium artesunate were developed and approved as new antimalarial drugs by the Chinese authorities in 1987. After 1992, dihydroartemisinin, Coartem (a combination of artemether and benflumetol), and Artekin (a combination of dihydroartemisinin and piperaquine) were also marketed as new antimalarial drugs. Since then, over 10 million malaria patients on a global scale have been cured after administration of these drugs. As a result, artemether, artesunate, and Coartem were added by the World Health Organization to the ninth, eleventh, and twelfth *Essential Medicine List* respectively.

When artemether and sodium artesunate were successfully used by intramuscular or intravenous administration for treatment of severe malarial patients, their shortcomings, such as short half-life and instability of aqueous solution of sodium artesunate, were cognized. Hence, qinghaosu derivatives and analogs numbering in the thousands were synthesized and evaluated by many research groups worldwide.

### 5.5.1 Modification on C-12 of Qinghaosu

From the view of chemistry, the C-12 position of dihydroartemisin is similar to the C-1 position of carbohydrates. So, these C-12 derivatives may be divided into three types: *O*-glycosides, *N*-glycosides, and *C*-glycosides using a similar term in carbohydrate chemistry.

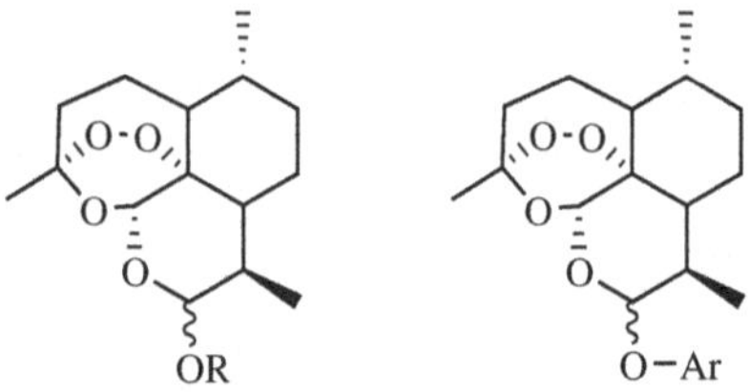

126                    127

**Structure 5-10**

### 5.5.1.1   O-glycosides (126 and 127)

The ethers and esters of dihydroartemisinin noted above may be considered as its *O*-glycosides (**126**) (Structure 5-10).

Because of the high content (0.6–1.1%) of qinghaosu in *A. annua L.* planted in some regions and an efficient process of extraction, a large quantity of qinghaosu is available in China. Based on the early work, it was known that little changes of C-12 substituents always lead to a great difference in the antimalarial activity; hence, more dihydroartemisinin derivatives were synthesized. A series of ethers and esters in which C-12 substituents contained halogen (especially, F), nitrogen, sulfur atoms, and others were prepared.[163] Afterward, several mono- and polyfluorinated artemisinin derivatives were reported by Pu et al.[164] Recently, trifluoromethyl analoges of dihydroartemisinin, artemether, artesunate, and other analogs (Structure 5-11) were synthesized by French groups.[165–167] The presence of the CF$_3$ group at C-12 of artemisinin clearly increased the chemical stability under simulated stomach acid conditions.

Venugopalan et al. also prepared some ethers and thioethers of dihydroartemisinin.[168] Haynes et al. synthesized many new C-12 esters and ethers of dihydro- artemisinin.[169] Compared with similar work performed 20 years ago, they successfully prepared some β-aromatic esters of dihydroartemisinin by means of Schmit and Mitsunobu procedures.

Another type of *O*-glycosides (**127**), 12-β aryl ethers of dihydroartemisinin, was synthesized by a reaction of acetyl dihydroartemisinin or trifluoroacetyl dihydroartemisinin with various substituted phenols in the presence of trifluoroacetic acid. Most of these compounds were proved to have better antimalarial activity against *P. berghei* in mice than qinghaosu, but less activity than artemether. Unexpectedly,

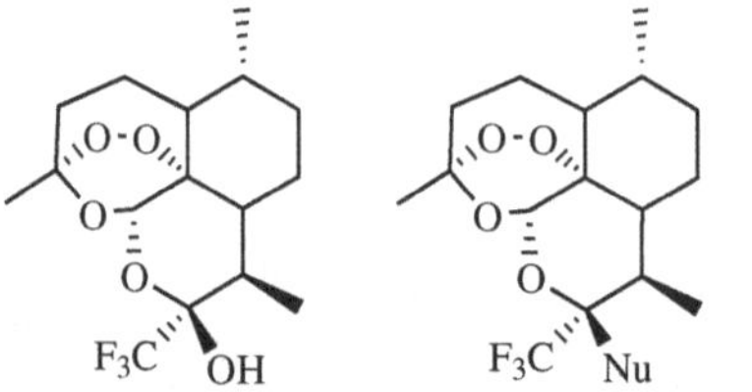

Nu = OCH$_2$CF$_3$, OCOCH$_2$CH$_2$COOH,
OOH, OMe, OEt, OCH$_2$CH=CH$_2$
OCH$_2$CH$_2$OH, NHC$_6$H$_4$OMe(4),
OCH$_2$C$_6$H$_4$ COOMe(4)

**Structure 5-11**

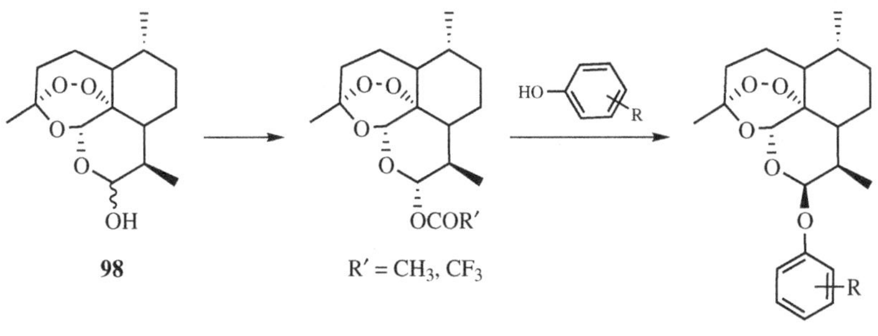

**Scheme 5-10**

some compounds showed much higher cytotoxicity (against KB, HCT-8, A2780 cell lines) than artemether (Scheme 5-10).[170]

O'Neill et al. also synthesized this kind of derivative by the reaction of dihydroartemisinin and phenols under the TMSOTf and AgClO$_4$ promotion at $-78\,°C$. The p-trifluoromethyl derivative was selected for in vivo biological evaluation, preliminary metabolism and mechanism of action studies (Scheme 5-11).[171]

### 5.5.1.2 N-glycosides
Dihydroartemisinin or trifluoroacetyl dihydroartemisinin reacted with aromatic amines or heterocycles, such as triazole and benzotriazole, in the presence of acidic catalyst to afford its N-glycosides, which were more active in vivo than qinghaosu (Scheme 5-12).[172,173]

### 5.5.1.3 C-glycosides
Because the C-glycosides could not be converted into dihydroartemisinin **98**, and their in vivo half-life might be significantly longer than the O-glycosides of **98**, many C-glycosides **129** have been synthesized. At the beginning, Jung and Haynes groups prepared several 12-alkyl-deoxoartemisinin from qinghao acid in five to seven steps.[174–176] However, the scarcity of qinghaosu acid, the low overall yield, and the production of both the 12α-isomer and the 12β-isomer indicated the need for another approach. Since then, Ziffer et al. employed **98** or its acetate **128** as the starting material, which reacted with allyltrimethylsilane in the presence of an acid catalyst (boron trifluoride etherate or titanium tetrachloride) to prepare

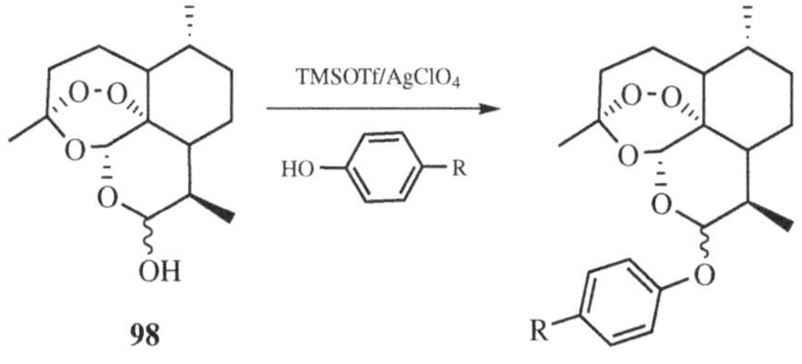

**Scheme 5-11**

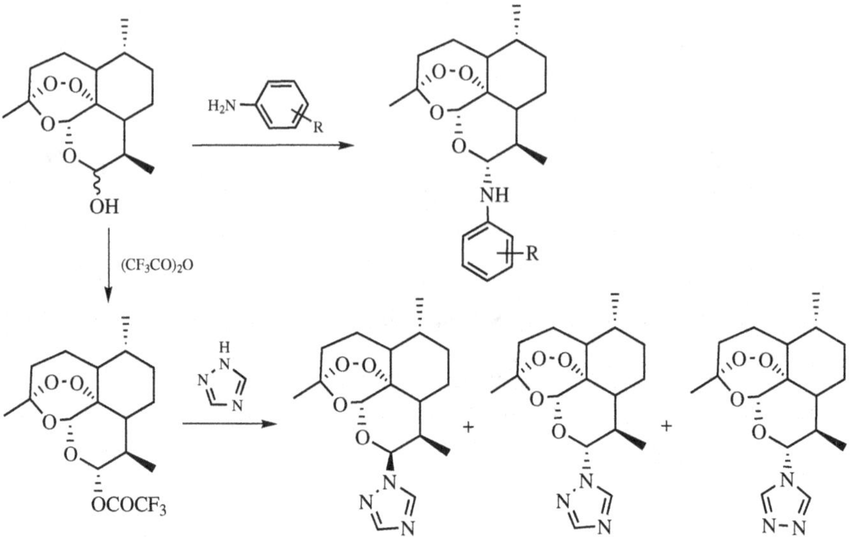

Scheme 5-12

12β-allyldeoxoartemisinin **129** (R = CH$_2$CH=CH$_2$) and related compounds in high stereoselectivity (Scheme 5-13).[177,178] The most active compound 12β-*n*-propyldeoxo-artemisinin, **129** (R = CH$_2$CH$_2$CH$_3$), proved to be as active and toxic as arteether.

Recently they synthesized more 12-alkyldeoxoartemisinins (Scheme 5-13) according to this method. Compound **129** [R = β-CH$_2$CH(OH)Et, β-CH$_2$CH(OH)-Bu(*t*)] showed five to seven times greater activity in vitro than qinghaosu.[179]

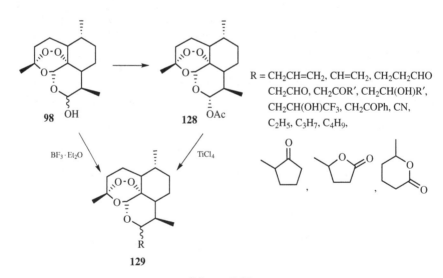

R = CH$_2$CH=CH$_2$, CH=CH$_2$, CH$_2$CH$_2$CHO
CH$_2$CHO, CH$_2$COR′, CH$_2$CH(OH)R′,
CH$_2$CH(OH)CF$_3$, CH$_2$COPh, CN,
C$_2$H$_5$, C$_3$H$_7$, C$_4$H$_9$,

Scheme 5-13

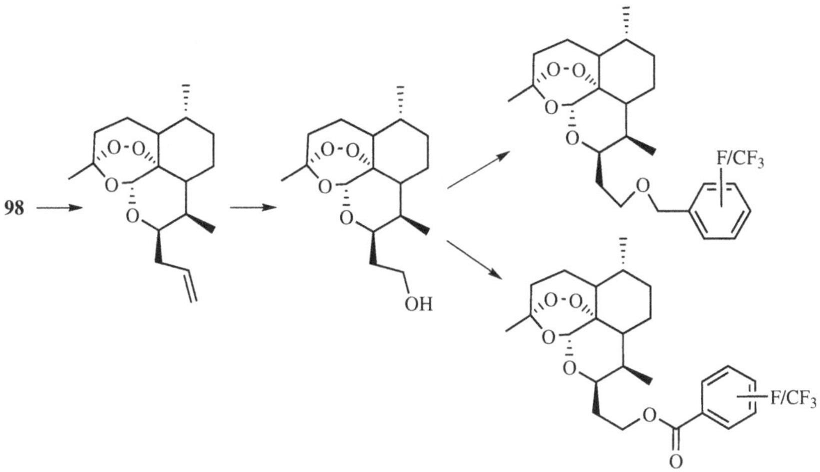

**Scheme 5-14**

O'Neill et al. synthesized 12-C ethanol of deoxoartemisinin and its ethers and esters (Scheme 5-14). The selected derivatives were generally less potent than the dihydroartemisinin in vivo test.[180]

Posner et al. synthesized some C-glycosides (Scheme 5-15) by using 12-F-deoxoartemisinin as the intermediate.[181,182] These compounds had high antimalarial potencies in vitro against *Plasmodium falciparum*. Some were active in vivo, but less than arteether.

At the beginning of the 1990s, dihydroartemisinin acetate **128**, as an electrophilic reagent, was reacted with aromatic substrates in the presence of boron

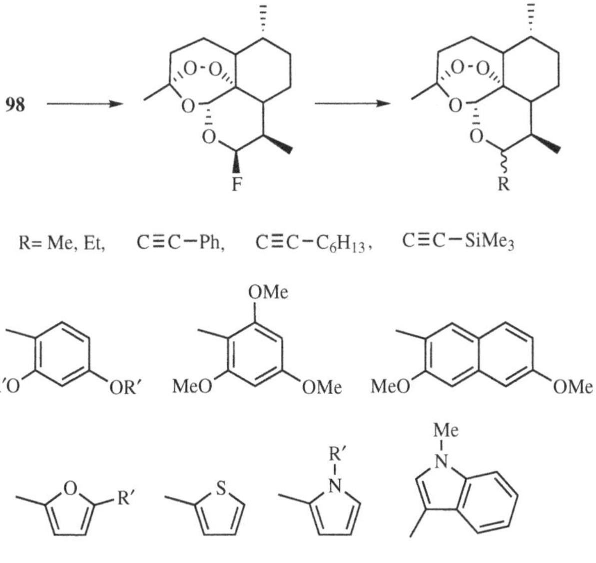

R= Me, Et,    C≡C−Ph,    C≡C−C$_6$H$_{13}$,    C≡C−SiMe$_3$

**Scheme 5-15**

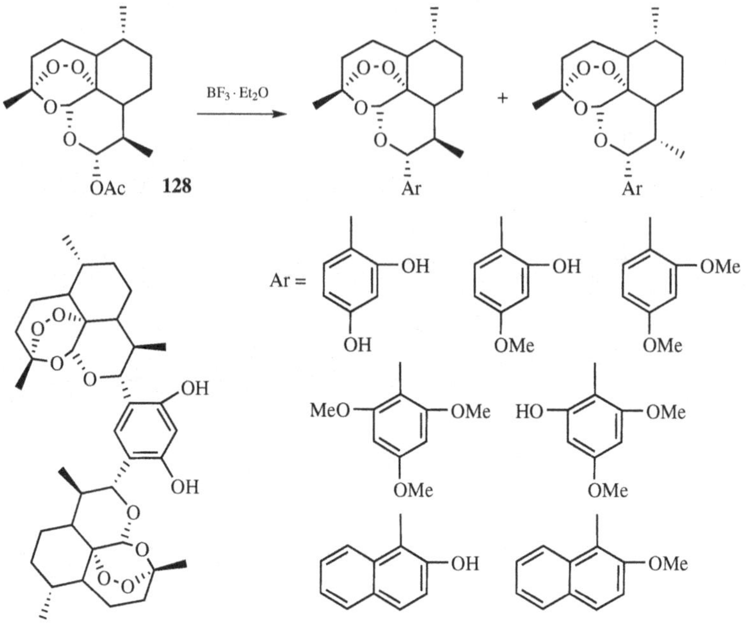

**Scheme 5-16**

trifluoride-etherate to yield 12-α aryl derivatives and a byproduct 11-β epimers (Scheme 5-16). Their antimalarial activity was higher than qinghaosu, and some compounds were even higher than artemether. Also, some compounds showed other bioactivities.[183–187]

Recently, a new approach to the synthesis of 12 C-glycosides (Scheme 5-17) was also reported.[188]

### 5.5.2  Water-Soluble Qinghaosu Derivatives

Sodium artesunate is the first water-soluble qinghaosu derivative and used for treatment of the severe malaria patients by intravenous or intramuscular administration. However, the aqueous solution is unstable, and its hydrolysis product,

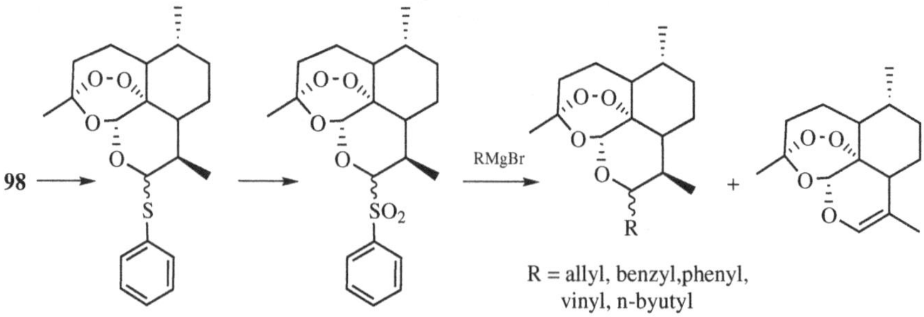

R = allyl, benzyl, phenyl, vinyl, n-byutyl

**Scheme 5-17**

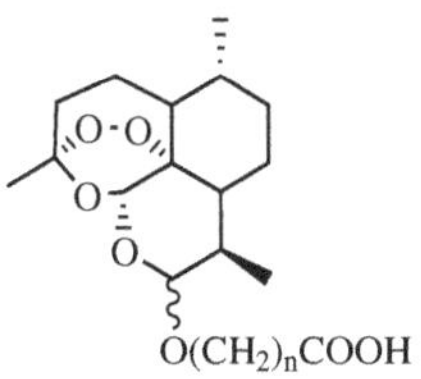

**130**

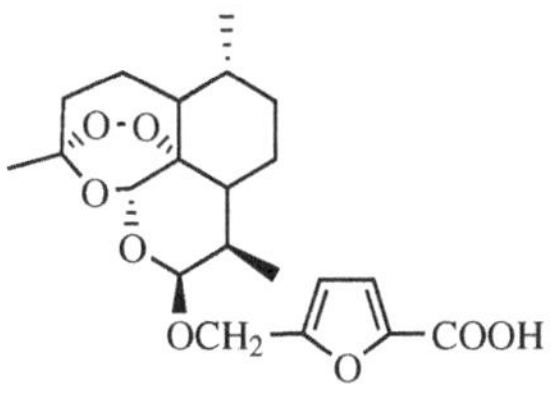

**131**

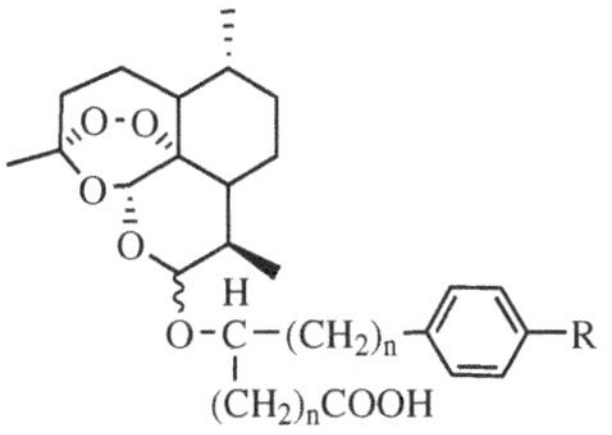

**132**

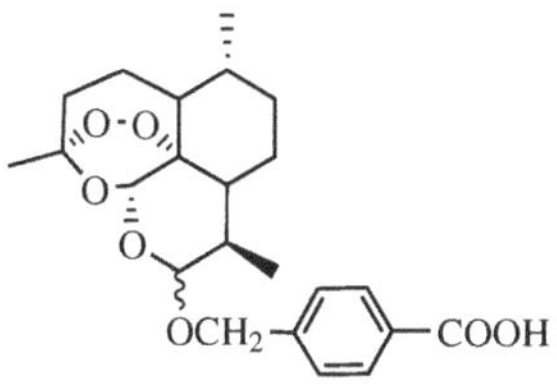

Artelinic acid **133**

**Structure 5-12**

dihydroartemisinin, quickly subsides. Hence, the synthesis of stable, water-soluble qinghaosu derivatives is an important research program.

### 5.5.2.1 Qinghaosu Derivatives Containing the Carboxyl Group

Because artemether has greater stability than artesunate, it was supposed that the replacement of the ester linkage by the ether linkage in the artesunate molecule

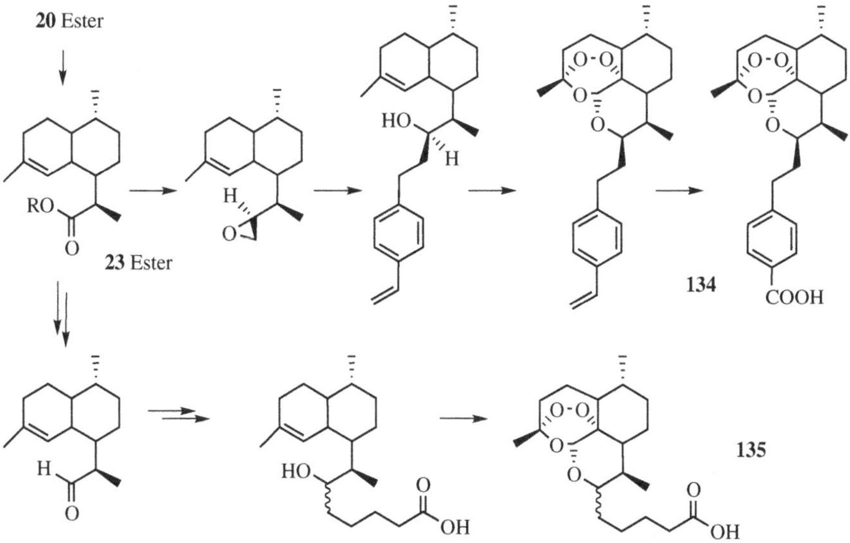

**Scheme 5-18**

O(CH$_2$)nNR$_1$R$_2$          OH
                              O(CH$_2$)nNR$_1$R$_2$          OCH$_2$CH(OH)CH$_2$NR$_1$R$_2$

OCH$_2$— OH / CH$_2$NR$_2$          OCO— OH / CH$_2$NR$_2$

**Structure 5-13**

would enable the derivative to be more stable. In fact, the sodium salts of compounds **130, 131** are much less active than sodium artesunate and their solubility in water is still poor.[189] Lin et al. prepared compound **132, 133** and found the sodium salt of artelinic acid (**133**) to be stable in aqueous solution.[190,191] However, no report about its clinical trial was publish (Structure 5-12).

To search for stable, water-soluble dihydroartemisinin derivatives with higher efficacy and longer plasma half-life than artesunate and artelinic acid, deoxoartelinic acid **134** was prepared (Scheme 5-18) and tested in vitro and in vivo.[192] It was reported that **134** showed superior antimalarial activity and was more stable in simulated stomach acid than arteether. In 1992, Haynes et al. already reported on the synthesis of 5-carba-4-deoxoartesunic acid (**135**) from artemisinic acid (**20**) in a similar way, but they did not mention its activity at that time.[176]

### 5.5.2.2 Qinghaosu Derivatives Containing the Basic Substituent

In view of the known basic antimalarial drugs (such as chloroquine, quinine) that are being used as salts for injection, it was proposed that introducing an amino group into the qinghaosu molecule may lead to water-soluble derivatives. Thus, five types of basic qinghaosu derivatives were synthesized (Structure 5-13).[189,193–195]

These basic compounds combined with organic acid (such as oxalic acid and maleic acid) to yield the corresponding salts. Generally, they had good water-solubility and stability. Some compounds showed much more activity against *P. berghei* in mice than artesunate. However, their efficacies were less than that of artesunate against *P. knowlesi* in rhesus monkeys.[189,195] In addition, more qinghaosu derivatives containing the amino group were also reported (Structure 5-14).[196–199]

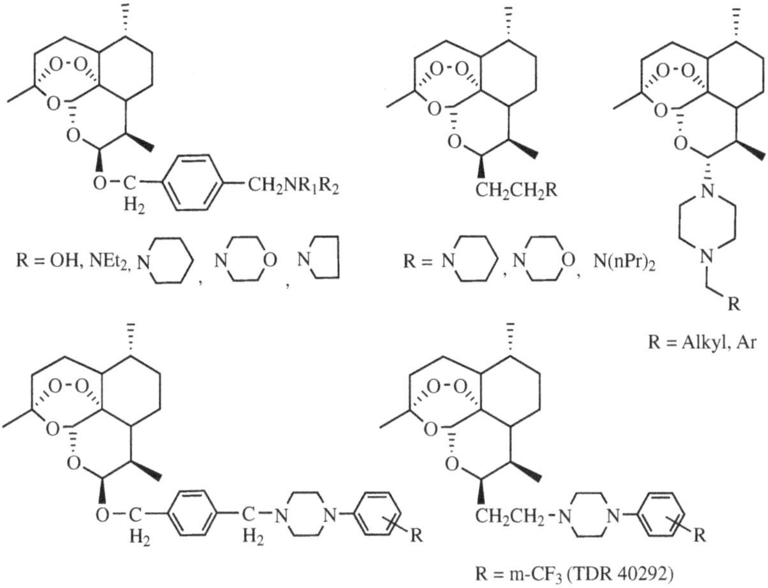

**Structure 5-14**

### 5.5.3 Modification on C-11 or/and C-12

Some C-11 substituted qinghaosu and its derivatives were prepared and tested against *P. berghei* in mice.[200,201] Their lower antimalarial activity may be attributed to the introduction of 11-α substituent (Structure 5-15).[202–205]

### 5.5.4 Modification on C-4 or/and C-12

The 4-methyl is located near the peroxy group, so the modification on C-4 may offer the important information about the SAR. Some compounds (Structure 5-16) were therefore synthesized.[206,207] These compounds showed more activity than qinghaosu. It was noteworthy that deoxoartemisinin was also more active than qinghaosu in vitro and in vivo.[208,209]

Avery et al. have prepared a lot of 4- alkyl-, 4-(arylalkayl)-, and 4-(carboxyalkyl)-qinghaosu by their method with the reaction of vinylsilane and ozone as the

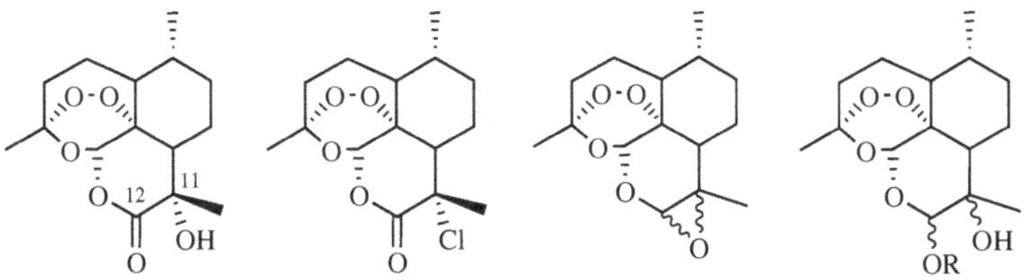

**Structure 5-15**

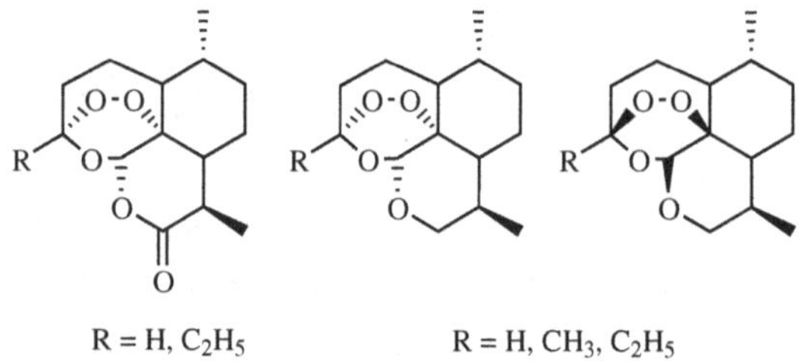

R = H, C$_2$H$_5$                R = H, CH$_3$, C$_2$H$_5$

**Structure 5-16**

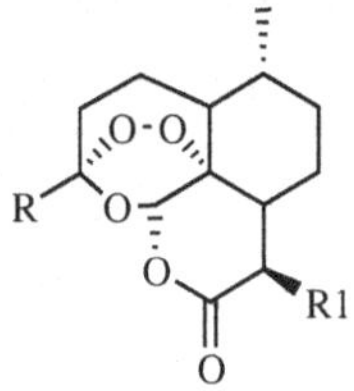

**Structure 5-17**

key step. Some of their derivatives were more active than qinghaosu in vitro (Structure 5-17).[210,211]

### 5.5.5 Modification on C-3 or/and C-13

A series of qinghaosu analogs of C-3 or/and C-13 modification were prepared from artemisinic acid by Lee et al. (Scheme 5-19).[212] Among these analogs, only 13-nitromethyl qinghaosu had antimalarial activity comparable with qinghaosu.

### 5.5.6 Modification on C-13

Under non- or base-catalyzed conditions, artemisitene reacted with triazole, benzotriazole, or benzimidazole to yield a series of Michael addition products (Scheme 5-20). All of these compounds had antimalarial activity in vivo.[213]

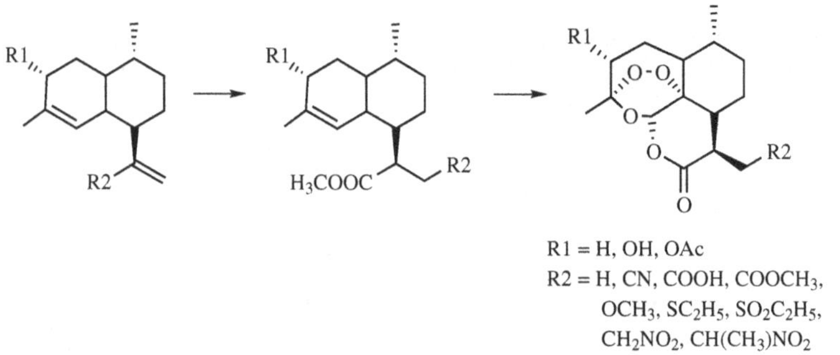

R1 = H, OH, OAc
R2 = H, CN, COOH, COOCH$_3$,
     OCH$_3$, SC$_2$H$_5$, SO$_2$C$_2$H$_5$,
     CH$_2$NO$_2$, CH(CH$_3$)NO$_2$

**Scheme 5-19**

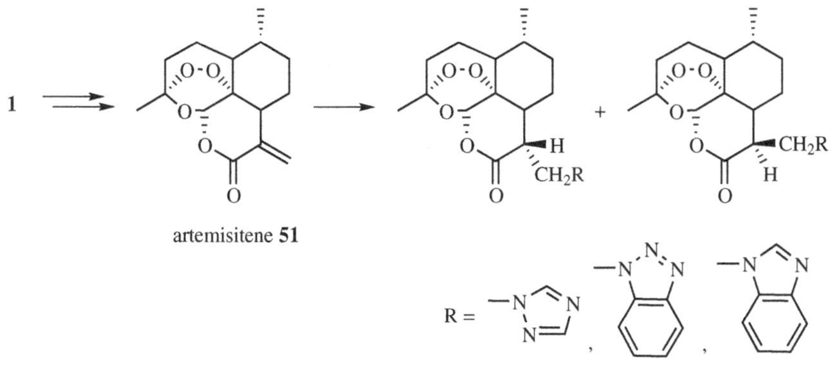

**Scheme 5-20**

Ma et al. synthesized another type of C-13 derivative by the acid-catalyzed Michael addition of artemisitene.[214]

### 5.5.7 Modification on C-11 and C-12

Jung et al. prepared 11-substituted deoxoartemisinin **136** from artemisinic acid (**20**) using photooxidation as the key step (Scheme 5-21).[215] Compound **136** (R = CH$_2$OH) was more active than qinghaosu and artesunate in vitro.

### 5.5.8 Azaartemisinin

Torok et al. reported that the reaction of artemisinin and methanolic ammonia or primary alkyl-and heteroaromatic amines yielded azaartemisinin or N-substituted azaartemisinin (**137**) and N-substituted azadesoxyartemisinin (**138**) as byproducts (Scheme 5-22). Some N-substituted azaartemisinin had good antimalarial activity, such as compound **137** (R = CH$_2$CHO), which was 26 times more active in vitro and 4 times more active in vivo than artemisinin.[216]

More N-alkyl derivatives **140** were prepared by means of Michael additions to azaartemisinin (**139**) (Scheme 5-23).[217]

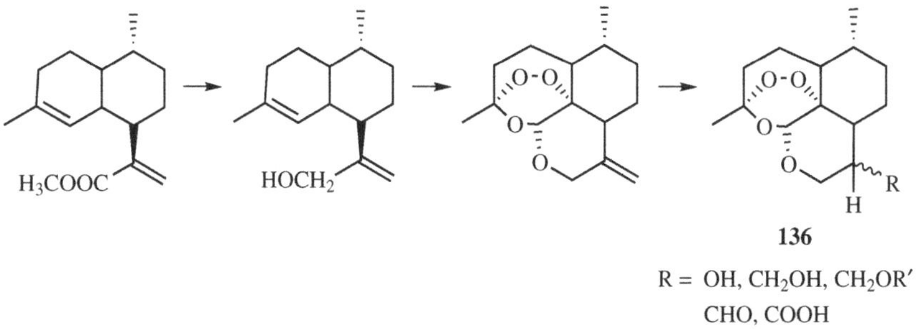

**136**

R = OH, CH$_2$OH, CH$_2$OR'
CHO, COOH

**Scheme 5-21**

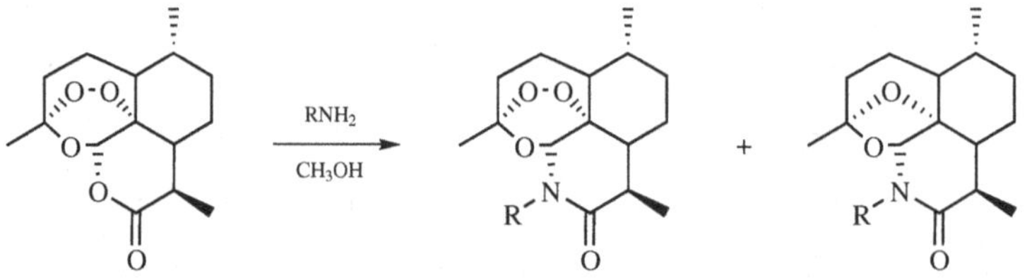

**137**        **138**

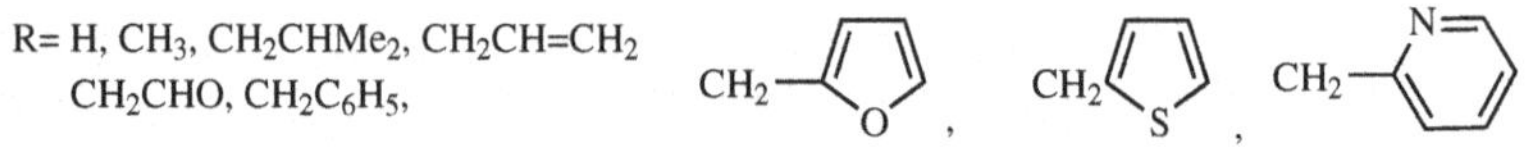

R= H, CH$_3$, CH$_2$CHMe$_2$, CH$_2$CH=CH$_2$
CH$_2$CHO, CH$_2$C$_6$H$_5$,

CH$_2$—⟨furan⟩ ,   CH$_2$—⟨thiophene⟩ ,   CH$_2$—⟨pyridine⟩

**Scheme 5-22**

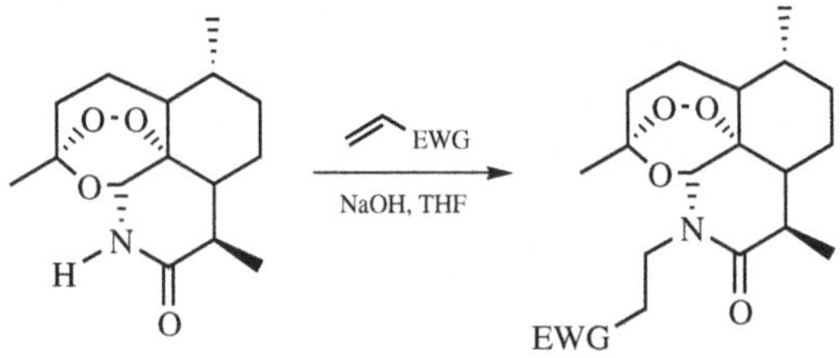

**139**        **140**

EWG = COOC$_2$H$_5$, CN, COCH$_3$, SO$_3$C$_6$H$_5$, SO$_2$C$_6$H$_5$, SOC$_6$H$_5$

**Scheme 5-23**

### 5.5.9  Carbaartemisinin

To inspect the effect of the segment of O–O–C–O–C–O–C=O in the artemisinin molecule, carbaartemisinin **141** and its analogs **142–144** (Structure 5-18) were synthesized and evaluated. These compounds displayed much lower antimalarial activity in vitro than artemisinin.[218]

### 5.5.10  Steroidal Qinghaosu Derivatives

Some research groups synthesized steroidal qinghaosu derivatives in which the qinghaosu nucleus, trioxane, or tetraoxane combined with a steroidal skeleton in

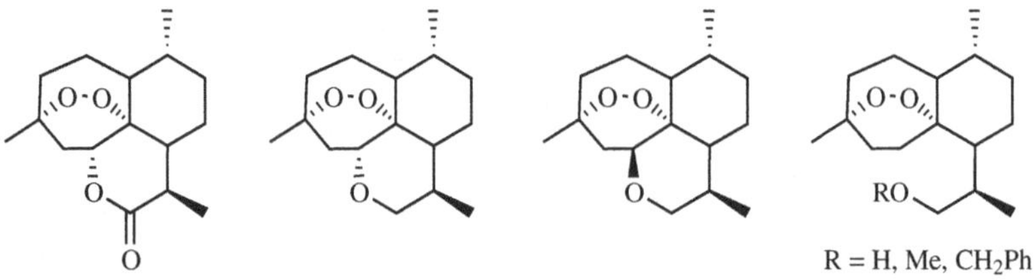

R = H, Me, CH$_2$Ph

**141**        **142**        **143**        **144**

**Structure 5-18**

**Structure 5-19**

different styles (Structure 5–19).[194,219–221] These compounds showed antimalarial activity. For the first time, mixed steroidal tetraoxanes (Structure 5-20) were screened against *Mycobacterium tberculosis* with minimum inhibitory concentrations as low as 4.73 µM against the H37Rv strain.[222]

## 5.5.11 Dimers and Trimers

In medical research, coupling two active centers in one molecule is a common strategy to enhance the activity. So the dimers of qinghaosu (Structure 5-21) were also synthesized.[13,223,224] Most of these compounds were more active than qinghaosu, but less active than artemether.

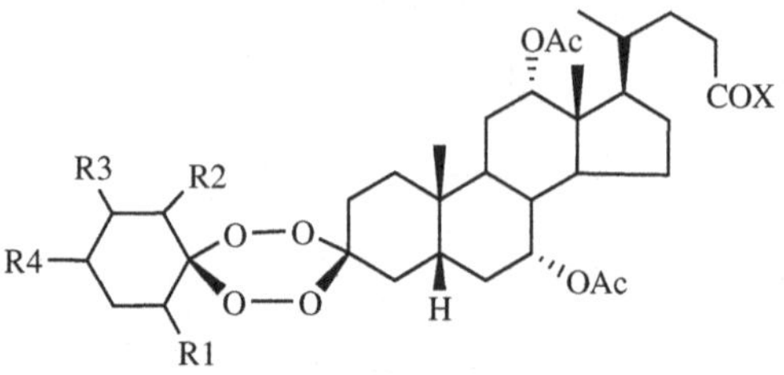

X = OH, OMe, NHR,   R = H, Me, Et, nP
R1, R2, R3, R4 = H, Me, Et

**Structure 5-20**

Posner et al. reported that some new types of qinghaosu dimers (Structure 5-22) had antimalarial, antiproliferative, and antitumor activities. Compound **149** was 50 times more potent than the parent drug artemisinin and about 15 times more potent than the clinically used acetal artemether. Dimers **145–150** were especially potent and selective at inhibiting the growth of some human cancer cell lines.[225-227]

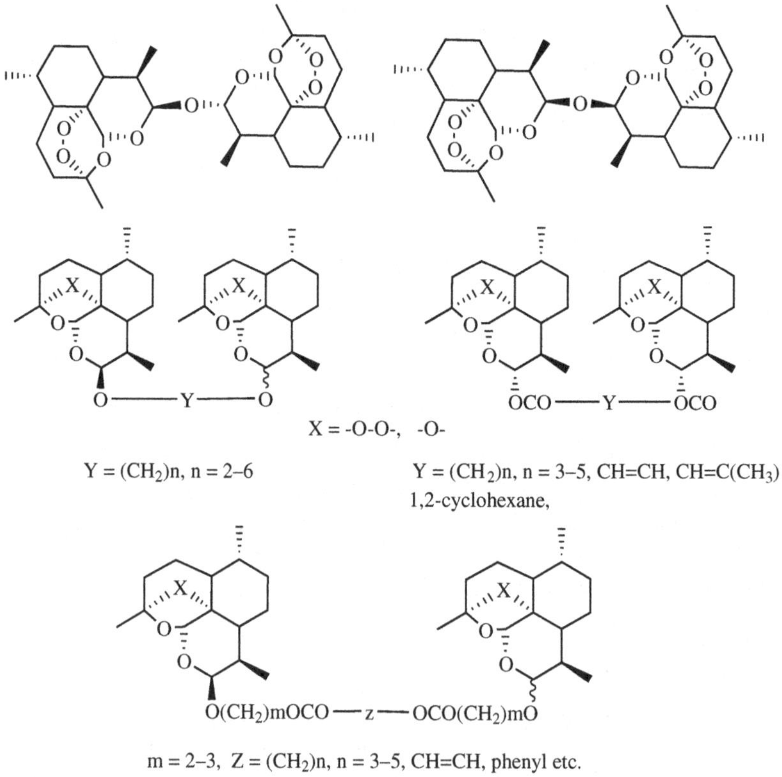

X = -O-O-, -O-

Y = (CH₂)n, n = 2–6

Y = (CH₂)n, n = 3–5, CH=CH, CH=C(CH₃)
1,2-cyclohexane,

m = 2–3, Z = (CH₂)n, n = 3–5, CH=CH, phenyl etc.

**Structure 5-21**

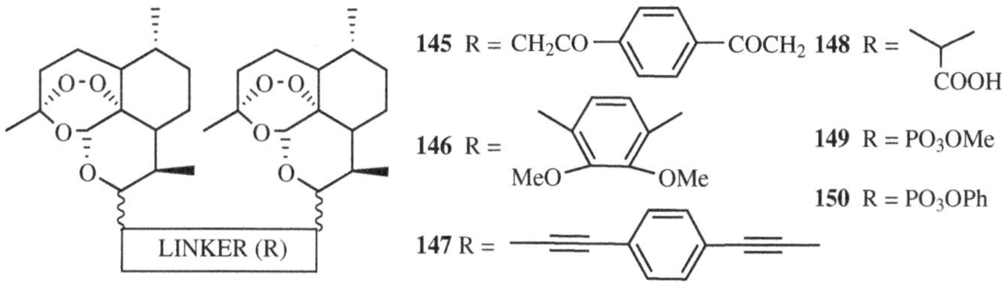

145 R = CH$_2$CO— benzene —COCH$_2$   148 R = isopropyl-COOH

146 R = (dimethyl dimethoxy benzene) MeO   OMe

147 R = —≡— benzene —≡—

149 R = PO$_3$OMe

150 R = PO$_3$OPh

LINKER (R)

**Structure 5-22**

In compound **151**, a carbon chain connected with two qinghaosu nuclei was also synthesized by metathesis, but its activity has not been measured yet (Scheme 5-24).[228]

Also, some amide-linked dimers, sulfide-linked dimers, sulfone-linked dimers, and trimers were synthesized by Jung et al. These compounds showed potent and selective inhibition on the growth of certain human cancer cell lines (Structure 5-23). In particular, trimer **152** was comparable with that of clinically used anticancer drugs.[229]

### 5.5.12   1,2,4-Trioxanes and 1,2,4,5-Tetraoxanes

Since the 1990s, many research studies have demonstrated that 1,2,4-trioxanes and 1,2,4,5-tetraoxanes are important qinghaosu analogs. Some compounds are promising because of their high antimalarial activity and easy preparation. Some reviews about these peroxides have been published.[9,13,230–232]

## 5.6.   PHARMACOLOGY AND CHEMICAL BIOLOGY OF QINGHAOSU AND ITS DERIVATIVES

### 5.6.1   Bioactivities of Qinghaosu Derivatives and Analogs

#### 5.6.1.1   Antimalarial Activity[233, 234]

Since artemisinin antimalarial drugs were developed by Chinese scientists in the 1980s, over 10 million patients infected with falciparum malaria including

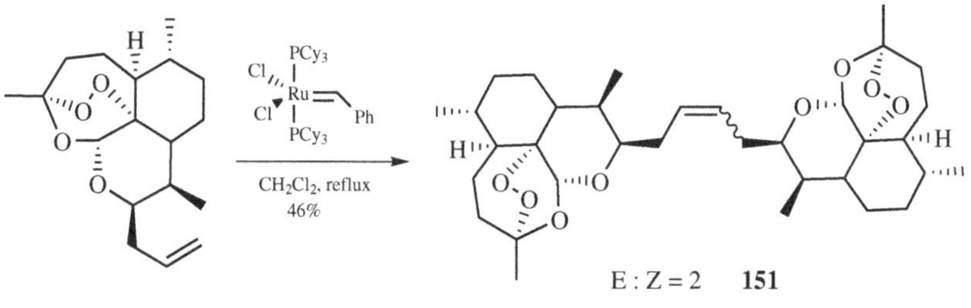

$E : Z = 2$   **151**

**Scheme 5-24**

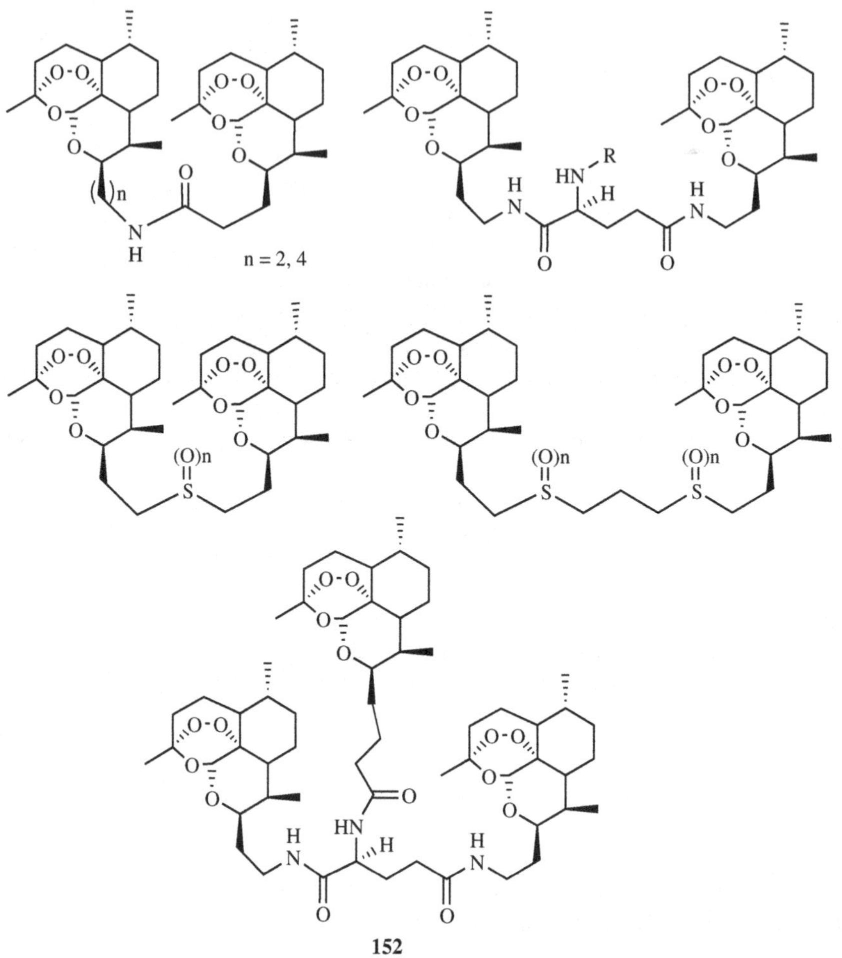

**152**

**Structure 5-23**

multidrug-resistant *P. falciparum* in all areas of the world were cured. These drugs derived from the natural *A. annua* L. have many advantages: quick reduction of fevers, fast clearing parasites in blood (90% of malaria patients recovered within 48 hours), and no significant side effects. Although the neurotoxicity was found in animals after high doses of certain compounds, no related clinical toxicity has been observed in humans. Similarly, high doses of artemisinins may induce foetal resorption (no mutagenic or teratogenic) in experimental animals; however, the monitor for hundreds of women in severe malaria or uncomplicated malaria in pregnancy showed artemisinins to be safe for these mothers and babies.

### 5.6.1.2 *Against Other Parasites*
Many experimental and clinical studies performed in China reveal that artemisinin, artemether, and artesunate are not only the potent antimalarial drugs but also the

useful agents for other diseases, especially as an antiparasitic agent, such as against *Schistosoma japonicum*, *Clonorchis Sinensis*, *Theileria annulatan*, and *Toxoplasma gondii*. In the 1970s, artemether and artesunate were confirmed to be more active than artemisinin in both animal models.[235–237] They strongly killed the immature worms living in mice, but praziquantel could not act. Their prevention of the development of the mature female worms was also proved in other animal models (rat, rabbit, and dog).[238–240] Since 1993, artemether and artesunate were studied in randomized, double-blind, placebo-controlled trials in China[241–249] and approved as the prevention drugs for schistosomiasis by the Chinese authorities in 1996. Afterward, these drugs showed similar activity against *S. mansoni* and *S. haematobium* in the laboratory studies and clinical trials in other countries.[250–253]

### 5.6.1.3 Antitumor Activity

Some components of *A. annua* L., such as qinghaosu (**1**), artemisinin B (**10**), artemisinic acid (**20**), artemisitene (**26**), flavnoids, and other terpenoids, showed antitumor activities at varying concentrations against L-1210, P-388, A-549, HT-29, MCF-7, and KB in vitro.[69,254,255]

In the assay of cytotoxicity of qinghaosu and related compounds against Ehrlich ascites tumor cells, qinghaosu, artemether, arteether, and artesunate exhibited cytotoxicity (IC$_{50}$ 12.2~29.8 $\mu$M), artemisitene **51** was more active (IC$_{50}$ 6.8 $\mu$M), and the dimer of dihydroartemisinin was the most potent (IC$_{50}$ 1.4 $\mu$M).[256–258]

The antitumor effect of artesunate was tested in vitro and in vivo in China.[259–261] It possessed cytotoxicity for six cell lines (IC$_{50}$ 1~100 $\mu$g/mL) and antitumor effects on human nasopharyngeal cancer (CNE2, SUNE-1) and human liver cancer (BEL-7402) in nude mice. Recently artesunate has been analyzed for its antitumor activity against 55 cell lines.[262] It was most active against leukemia and colon cancer cell lines. It is notable that no CEM leukemia sublines, which are resistant to either doxorubicin, vincristine, methotrexate, or hydroxyurea, showed cross resistance to artesunate.

It was found that dihydroartemisinin can selectively kill cancer cells in the presence of holotransferrin, which can increase intracellular iron concentrations, and normal breast cells (HTB 125) and lymphocytes had nonsignificant changes. It seems the mechanisms of anticancer action and of antimalarial activity are similar.[263–265]

The antitumor activity of *N*-glycoside, *O*-glycoside, and some dimers have been mentioned. Recently, compound **153** was found to have antitumor activity. Additional study discovered the most active compound **154** in this series ( R = *p*-Br, IC$_{50}$ = 11 nM, and 27 nM against P 388 and A 549 cell lines, respectively), but its deoxy analog **155** is inactive.[266,267] Compound **156** yielded by coupling that the cyanoarylmethyl group with artesunate can not show higher antitumor activity than **154**.[268] Flow cytometry data showed that these compounds caused an accumulation of L1210 and P388 cells in the G1-phase of the cell cycle and apoptosis in the P388 cells (Structure 5-24).[266-268]

More studies on antitumor action of qinghaosu analogs were reported.[269,270]

**Structure 5-24**

### 5.6.1.4  *Immunosuppression*

Generally, antimalarial drugs possess immunosuppressive action and are often used by physicians for the treatment of dermatoses, such as chloroquine and hydroxychloroquine for lupus erythematosus and multiple solar dermatitis. The immunopharmacological action of qinghaosu and its derivatives has been studied for a long time in China. The experimental results suggested that qinghaosu, artesunate, and artemether had both immunosuppressive and immunostimulating activities.[271]

Qinghao extraction and qinghaosu were smoothly tried to treat systemic lupus erythematosus (SLE) patients in the 1980s.[272] Because of the high immunosuppressive action of sodium artesunate on the SLE mice model, 56 patients with lupus erythematosus (DLE 16, SCLE 10, SLE 30) were treated by intravenous sodium artesunate (60 mg, once a day, 15 days a course, two to four courses), with an effect rate of 94%, 90%, and 80%, respectively.[273]

### 5.6.2  Early Biologically Morphologic Observation of the Antimalarial Action of Qinghaosu

The life cycle of the parasite in both mosquitoes and humans is complex. When an infected mosquito bites, sporozoites are injected into the blood stream of the human victim and then travel to liver tissue where they invade the parenchymal cell. During development and multiplication in the liver, which is known as the preerythrocytic stage, the host is asymptomatic. After 1 or 2 weeks, merozoites are released from the liver and the parasites take up residence in the red blood cells (erythrocytic stage). The parasite feeds on the protein portion of hemoglobin, and hemozoin, a waste product, accumulates in the host cell cytoplasm. After the parasite undergoes nuclear divisions, the red blood cell bursts and merozoites, parasites waste, and cell debris are released that cause he body temperature to rise (malarial fever). The newly released merozoites invade other red blood cells. After the circulate repeats several times, a few merozoites become differentiated into male and female gametocytes. When a mosquito takes the blood, the gametocytes begin sexual reproduction in its digestive track.

Qinghaosu can be used by physicians for the treatment for chloroquine-resistant malaria; it must have a different mode of action from that of chloroquine. Early study by Chinese scientists demonstrated that artemisinin drugs had a direct parasiticidal action against *P. falciparum* in the erythrocytic stage both in vitro and in vivo.[274] Also, the morphologic changes were observed under the electron microscope.[274] Qinghaosu drugs were added to the media, and samples were taken at definite intervals for electromicroscopic examination. The injuries of membrane structures of the parasite included swelling of the limiting membrane and the nuclear membrane, and, formation of the autophagic vacuole. It was also found that some free radical scavengers, such as vitamin E, would reduce the efficiency of qinghaosu. However, the inherent reason for these observed phenomena have not been acknowledged.

### 5.6.3   The Free Radical Reaction of Qinghaosu and Its Derivatives With Fe(II)

Qinghaosu is a sesquiterpene molecule containing carbon, hydrogen, oxygen, and no nitrogen atoms and can used by physicians for the treatment of multidrug-resistant strains of *P. falciparum*. It is obvious that its antimalarial mechanism is different from previous alkaloidal antimalarial drugs such as quinine and chloroquine. Since the discovery of **1**, what will be its action mode on the molecular level is a widely interesting question, although it is a difficult task. Actually until now, the action mode of quinine and other synthetic alkaloidal antimalarial drugs has not been so clearly understood.[275]

Qinghaosu acts parasite at its intra-erythrocytic asexual stage. At this stage, the parasite takes hemoglobin as its nutritional resource, digests hemoglobin, and leaves free heme, which is then polymerized to parasite safety poly-heme (hemozoin). Two other points should be mentioned: Over 95% iron in the human body exists as heme in the red blood cell, and the peroxide segment of **1** and its derivatives is essentially responsible for its activity.

Being aware of the DNA cleavage with the Fenton reagent[276,277] and the above-mentioned situation of qinghaosu-parasite-red blood cell, this laboratory has studied the reaction of qinghaosu and its derivatives with ferrous ion in aqueous acetonitrile since the early 1990s. At first, the reaction of **1** and ferrous sulfate (1:1 in mole) was run in $H_2O$–$CH_3CN$ (1:1 in volume, pH 4) at room temperature. It was interesting to find that the two major products were tetrahydrofuran compound **28** and 3-hydroxy deoxyqinghaosu **15**, which have been identified as the natural products of qinghao, pyrolysis products, and the metabolites of qinghaosu in vivo mentioned above. After careful chromatography, a miner product epoxide **157** was identified. In addition, acetylation of the remaining high-polarity products yielded the acetyl tetrahydrofuran compound **158**. Based on the analysis of these products, a reaction mechanism of an oxygen-centered free radical followed by single- electron rearrangement was suggested in 1995–1996 (Scheme 5-25).[278]

Since then, several qinghaosu derivatives have been treated with ferrous sulfate in the same reaction condition. Except for some hydrolysis products, similar derivatives of tetrahydrofuran compound **28** and 3-hydroxy deoxyqinghaosu **15** were

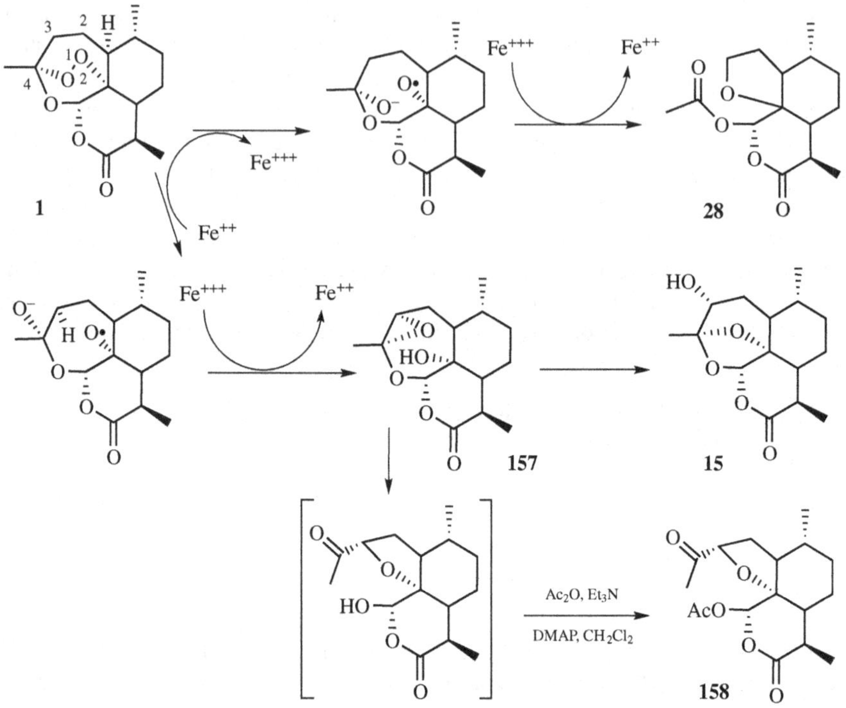

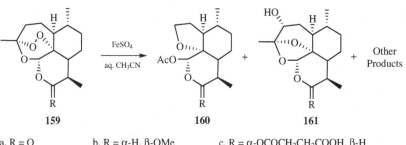

**Scheme 5-25**

also isolated as the two major products, but in somewhat different ratio (Scheme 5-26, Table 5-2). Usually these derivatives with higher antimalarial activity produced a higher ratio of tetrahydrofuran compounds (**160s**) than that of **160a** (**28**) from **1**. However, it is hard to say whether a correlation exists between this reaction and activity at this stage.

a, R = O          b, R = α-H, β-OMe          c, R = α-OCOCH$_2$CH$_2$COOH, β-H

d, R = α-H, β-OCH$_2$Ph     e, R = OH, H     f, R = α-OCOPh, β-H

**Scheme 5-26**

**TABLE 5-2. The Results of Cleavage of 1 and Its Derivatives With FeSO₄ in Aqueous CH₃CN**

| Entry | Compound | | Products (yields) | | |
|---|---|---|---|---|---|
| 1 | **159a (1)** | **160a (28)** (25%) | **161a (15)** (67%) | **Others** (<10%) |
| 2 | **159b (123)** | **160b** (37%) | **161b** (45%) | **161e** (4%, α+β) |
| 3 | **159c (125)** | **160c** (45%) | **161c** (23%) | **161e** (25%) |
| 4 | **159d** | **160d\*** (39%) | **161d** (56%) | |
| 5 | **159e (98)** | **160e** (46%) | **161e** (25%) | |
| 6 | **159f** | **160f** (59%) | **161f** (25%) | |

*Note*: Hydrolysis of **160d** led to a dialdehyde.

During this project, an electron spin resonance (ESR) signal of secondary carbon-centered free radical (**163**) was detected in the reaction of **1** and equivalent ferrous sulfate in aqueous acetonitrile with MNP as a trapping agent.[279] In the same year, Butler et al. detected the ESR signals of both primary (**162**) and secondary free radicals (**163**) with DMPO and DBNBS as trapping agents.[280] Figures 5-3 and 5-4 show the ESR spectra with MNP and DBNBS as trapping agent, respectively.

Based on these new evidences and the results published from other laboratories, the reaction mechanism of **1** and ferrous ion was revised so that this reaction proceeded through short-lived oxygen-centered free radicals and then

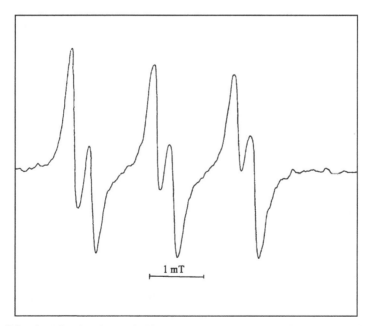

**Figure 5-3.** The ESR signal recorded in a run in aq. CH₃CN with qinghaosu as substrate in the presence of 1 equiv. of FeSO₄ with MNP as trapping agent.

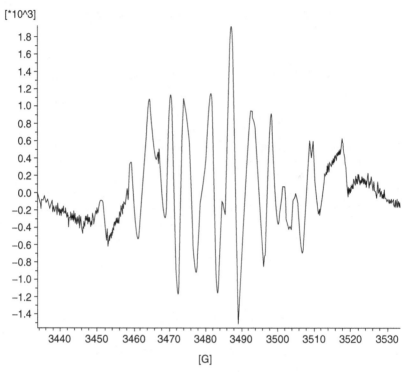

**Figure 5-4.** The ESR signal recorded in a run in aq. $CH_3CN$ with qinghaosu as substrate in the presence of 1 equiv. of $FeSO_4$ with DBNBS as trapping agent.

carbon-centered free radicals.[279] There were two kinds of carbon-centered free radicals: a primary C-centered free radical **162** and a secondary one **163**. Both carbon-centered free radicals were then confirmed by the isolation of their hydrogen-abstracted products **164** and **14**. The proposed mechanism is concisely shown in Scheme 5-27. Thus, tetrahydrofuran compound **28** is derived from a primary C-centered free radical **162** and 3-hydroxy deoxyqinghaosu (**15**) from **163** via **157**. This free radical mechanism may also explain why these similar products were obtained from the reaction of derivatives of **1**.

It has been found that less than 1 equivalent of ferrous ion can also decompose qinghaosu to its free radical degradation products, although for a longer reaction time. However, in the presence of excess of other reducing agents, such as mercaptans, cysteine, and ascorbic acid, qinghaosu can be degraded by as little as $10^{-3}$ equivalents of ferrous sulfate to give compound **28** and **15**. In the presence of 2 equivalents of cysteine, besides **28** and **15,** compound **164, 14**, and after treatment with acetic anhydride, compound **165** also could be separated. **164** and **14** were derived from abstracting a proton of the proposed free radical **162** and **163**, respectively. However, compound **165** was supposed to be derived from **166**, the adduct of primary free radical **162** and cysteine (Scheme 5-28).[281] This deduction was then

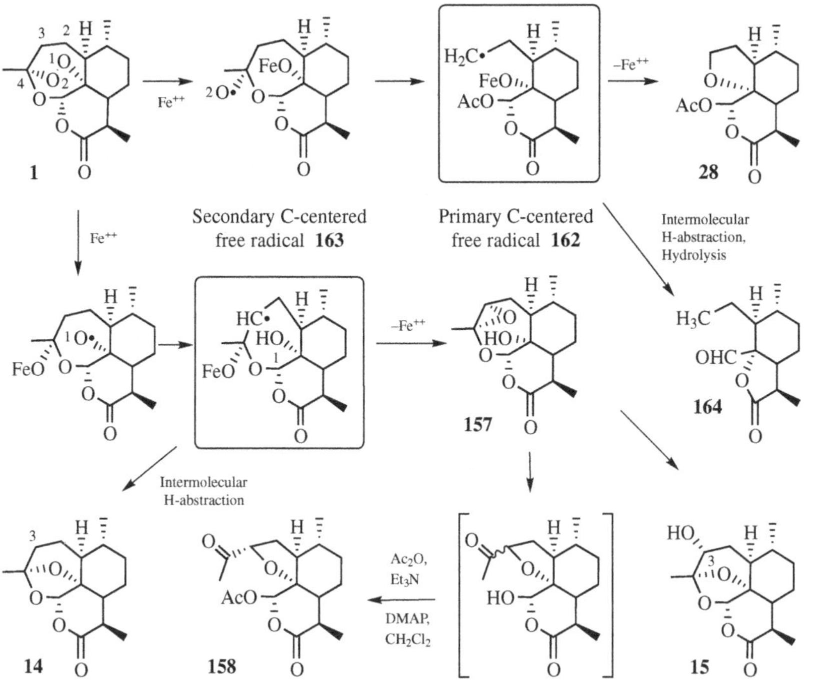

**Scheme 5-27**

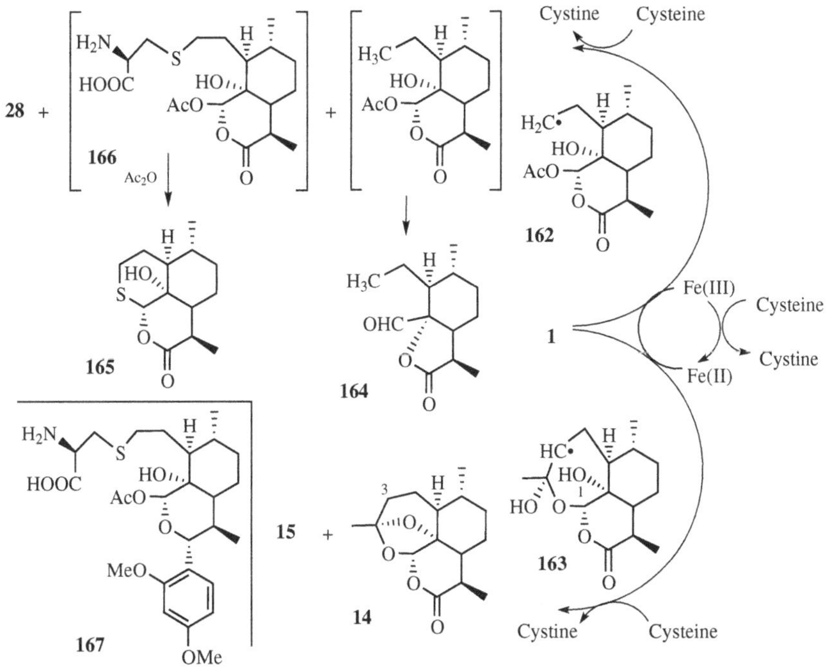

**Scheme 5-28**

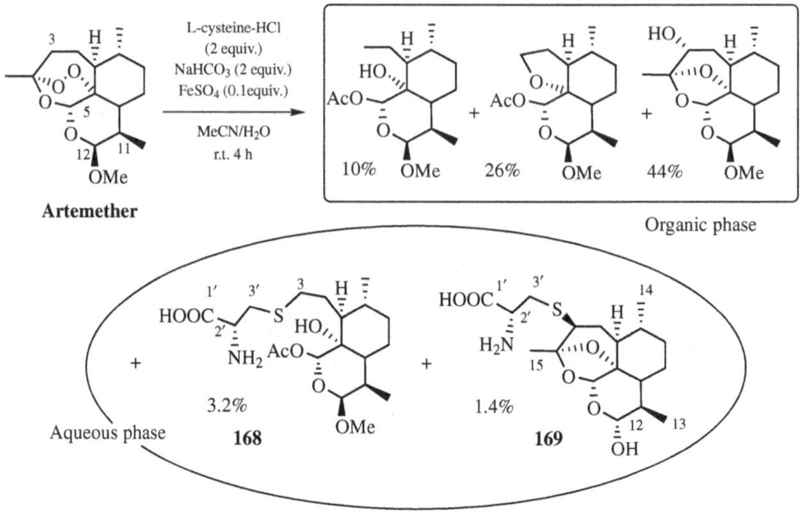

**Scheme 5-29**

confirmed by the separation of an adduct **167** of cysteine and a derivative of qinghaosu from their Fe catalyzed reaction mixture (Scheme 5-28).[186] Recently the degradation reaction with artemether and a catalytic amount of Fe(II/III) in the presence of cysteine was also performed, which gives not only the adduct (**168**) of the primary radical, but also the adduct (**169**) of the secondary radical for the first time (Scheme 5-29).[282]

In the 1990s, several laboratories engaged in the study on the reaction of ferrous ion and qinghaosu compounds and proposed that it was a free radical reaction.[283–287] Posner et al. have proposed that a high-valent iron-oxo was also intermediated during this reaction, but this viewpoint has not been generally accepted.[283] At the same time, Robert et al. also identified the adduct of radical **162** and tetraphenylporphyrin or heme and hence confirmed the intermediacy of the carbon-centered free radical.[287] In line with these data, it can be concluded that the reaction of qinghaosu and its derivatives with ferrous ion is definitely a free radical reaction through a short-lived O-radical-anion and subsequent primary and secondary C-centered radical.

### 5.6.4 Antimalarial Activity and the Free Radical Reaction of Qinghaosu and Its Derivatives

With the clarification of the C-centered free radical's participated mechanism of the reaction of **1** and ferrous ion, it is then interesting to note whether this free radical mechanism is related to its antimalarial activity. Recently, for the study of the mode of action, several stable and UV-detectable C-12 aromatic substituted derivatives of **1** were synthesized. Using the usual Lewis acid as the catalyst, the Friedel–Crafts alkylation gave the desired product **170** or **172** and 11-methyl epimer **171** or **173** as

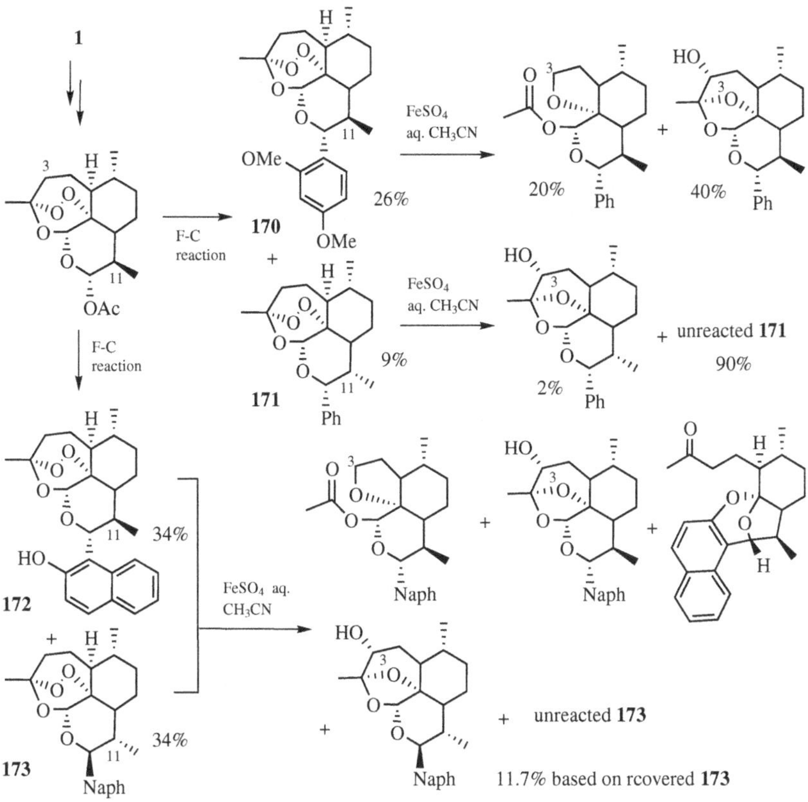

**Scheme 5-30**

the byproduct. These products were separated and subjected to the bioassay and chemical reaction with ferrous ion, respectively. It is interesting to find that these derivatives with normal configuration at C-11 showed higher bioactivity and higher chemical reactivity in the reaction with ferrous ion. However, their C-11 epimers were obviously less active for malaria and almost inert to the reaction with ferrous ion (Scheme 5-30, Table 5-3).[186,187]

**TABLE 5-3. The $ED_{50}$ and $ED_{90}$ Values Against**
***P. berghei* K173 Strain (administered orally to mice as suspensions in Tween 80)**

| Compound | $ED_{50}$ mg kg$^{-1}$ | $ED_{90}$ mg kg$^{-1}$ |
|---|---|---|
| Artemether | 1 | 3.1 |
| **170** | 1.27 | 5.27 |
| **171** | 4.18 | 76.27 |
| **172** | 0.58 | 1.73 |
| **173** | 7.08 | 60.99 |

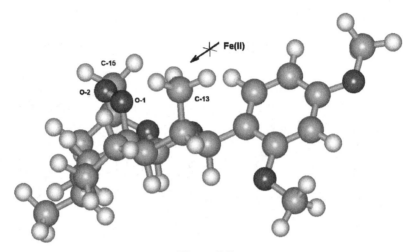

**Figure 5-5**

The 11α-epimers **171** and **173** are much less reactive than their corresponding 11β-epimers. The unfavorable influence from the 11α-substituents can also be found in other examples.[200] This lower reactivity may be attributed to the steric hindrance around the O-1 atom in 11α-epimers, which blocks the way for Fe(II) to attack O-1 (Figure 5-5). These experimental results show that the cleavage of peroxide with Fe(II) and then the formation of C-centered free radical, especially primary C-centered radical, is essential for the antimalarial activity.

Posner et al. have synthesized the 3-methyl-derivatives of qinghaosu and found that the antimalarial activity of 3β-methyl derivative in vitro was about the same as qinghaosu; however, the activity of the 3α-methyl- or 3,3-dimethy-derivative was at least two orders less than that of qinghaosu. This difference was supposed to come from the availability of the C-3 free radical for the later derivatives, but it was not mentioned whether these bio-inactive compounds were also inert in the reaction with ferrous ion.[288]

### 5.6.5   Interaction of Biomolecules with Carbon-Centered Free Radical

The C-centered free radical is the active species, but what target will be attacked by these radicals derived from qinghaosu and its derivatives in the biosystem is still a puzzle. This topic is an interesting one in the qinghaosu research area.

#### 5.6.5.1   Interaction With DNA

As mentioned, DNA could be cleaved with the Fenton reagent, and then we can take into consideration whether the qinghaosu–ferrous ion could also cleave DNA, although there are two different kinds of free radicals: the oxygen radical from the Fenton reaction and the carbon free radical from the reaction of the qinghaosu–ferrous ion. It is makes sense that the malaria parasite is mainly living in the red blood cell; however, the mature red blood cell has no nucleus, but the

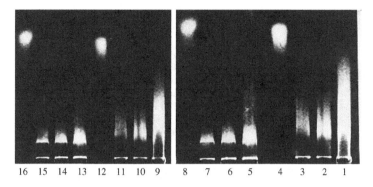

|  | 16 | 15 | 14 | 13 | 12 | 11 | 10 | 9 | 8 | 7 | 6 | 5 | 4 | 3 | 2 | 1 |  |
|---|---|---|---|---|---|---|---|---|---|---|---|---|---|---|---|---|---|
| FeSO4 | + |  | + |  | + |  | + |  | + |  | + |  | + |  |  |  | FeSO4 |
| Artemether | + | + |  |  | + | + |  |  | + | + |  |  | + | + |  |  | QHS |
| CT-DNA |  |  |  | + | + | + | + |  |  |  |  |  | + | + | + | + | CT-DNA |
| Salmon DNA | + | + | + | + |  |  |  |  | + | + | + | + |  |  |  |  | Salmon-DNA |

**Figure 5-6.** Agarose gel illustrates the cleavage reactions of calf thymus DNA and salmon DNA by qinghaosu or artemether and ferrous ion at 37 °C for 12 hours in a phosphate buffer solution.

parasite does. If qinghaosu could permeate the membrane, reach the nucleus of the parasite, and react with Fe(II), it would be possible that the interaction with DNA takes place. Therefore, this probability may explain why qinghaosu is only toxic to the parasite, but not to the normal red blood cell.

Considering that the pH in blood serum is about 7.35–7.45 and that the ferrous ion will precipitate above pH 7, the DNA damage experiments with qinghaosu and stoichiometric ferrous ion were performed in aqueous acetonitrile (1:1), at 37 °C and at pH 6.5 adjusted with a phosphate buffer. It was interesting to find that the cleavage of calf thymus, salmon, and a supercoiled DNA pUC 18 was observed yielding a DNA fragment with about 100 base-pair (the marks not shown in the figures) (Figures 5-6 and 5-7).[278,289] The phosphate buffer is important; otherwise, the calf thymus DNA was totally cleaved, when the concentration of $FeSO_4$ was 200–2000 µM even in the absence of qinghaosu. Changing forms and no change was only observed as the concentration of $FeSO_4$ was 10–25 µM and below 10 µM for the supercoiled DNA in the absence of a phosphate buffer and qinghaosu.

It was known that qinghaosu and its derivatives could react with a catalytic amount of ferrous or ferric ions in the presence of excess reduced agents including cysteine or glutathione (GSH) to afford the free radical reaction products (*vide ante*). When the concentration of pUC18 and qinghaosu in aqueous acetonitril was 0.040 µg/µL and 2 mM, the combination of 10 µM of $FeSO_4$ and 400 µM of L-cysteine/$NaHCO_3$ would cleave the supercoiled DNA pUC18 totally. The absence of either qinghaosu, $FeSO_4$, or L-cysteine/$NaHCO_3$ would cause the DNA to not be cleaved. The results were similar, if the calf thymus DNA was used instead of pUC18 as the substrate, but a fragment of 100 BP could be detected

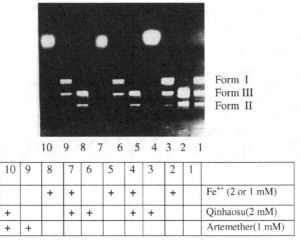

|    | 10 | 9 | 8 | 7 | 6 | 5 | 4 | 3 | 2 | 1 |                  |
|----|----|---|---|---|---|---|---|---|---|---|------------------|
|    |    |   | + | + |   | + | + |   | + |   | $Fe^{++}$ (2 or 1 mM) |
|    | +  |   |   | + | + |   | + | + |   |   | Qinhaosu(2 mM)   |
|    | +  | + |   |   |   |   |   |   |   |   | Artemether(1 mM) |

**Figure 5-7.** Agarose gel illustrates the cleavage reactions of DNA pUC18 by qinghaosu or artemether and ferrous ion at 37 °C for 12 hours in a phosphate buffer solution.

as the cleavage product (Figure 5-8). The situation was similar, as reduced GSH instead of cysteine was used as the reduced agent.[290]

These experimental results confirm that these qinghaosu-produced free radicals can cleave the DNA. At this stage, it is hard to say that the preconditions for the

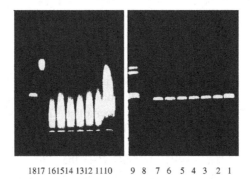

|                            | 18 | 17 | 16 | 15 | 14 | 13 | 12 | 11 | 10 | 9 | 8 | 7 | 6 | 5 | 4 | 3 | 2 | 1 |
|----------------------------|----|----|----|----|----|----|----|----|----|---|---|---|---|---|---|---|---|---|
| QHS (2 mM)                 |    | +  |    | +  | +  |    | +  | +  |    |   | + |   | + | + |   |   | + |   |
| $FeSO_4$ (10 µM)           |    | +  | +  |    | +  |    | +  |    |    |   | + | + |   | + | + |   |   |   |
| L-cys/NaHCO$_3$ (400 µM)   |    | +  | +  | +  |    | +  |    |    |    |   | + | + | + |   | + |   |   |   |

**Figure 5-8.** Agarose gel illustrates the cleavage reactions of DNA pUC18 (40 ng/µL, lines 1–8) or calf thymus DNA (1 µg/µL, lines 10–17) by qinghaosu, catalytic amount of ferrous ion, and cysteine at 37 °C for 12 hours. Lines 9 and 18 Mark DNA (pUC18 DNA +Hinf I, 65, 75, 214, 396, 517, 1419 BP from top sequentially).

parasite DNA damage: (Qinghaosu could permeate the membrane, reach the nucleus of parasite, and react with Fe(II)) could be fulfilled, but it is possible that the DNA damage may be responsible for the toxicity of qinghaosu and its derivatives on the tumor cell line, just as in the case of some other antitumor compounds, such as enediyne compounds. The positions of DNA attacked by the qinghaosu free radical are being studied, and it was observed that deoxygunosine (dG), perhaps deoxyadenosine (dA), in some cases, was attacked.

### 5.6.5.2   Interaction With Amino Acid, Peptide, and Protein

As early as the 1980s, it was indicated that some proteins such as cytochrome oxidase in the membranes and mitochondria were a target for the action of qinghaosu.[291] Meshnick et al. have performed a series of experiments about the interaction between qinghaosu and proteins in the presence of heme and have concluded that the binding between qinghaosu and albumin probably involves thiol and amino groups via both iron-dependent and -independent reactions. However, they have not isolated and confirmed such covalent adducts and have presented the question, "how does protein alkylation lead to parasite death?"[8,292–294]

Recent studies on the chemistry of a digestive vacuole (pH 5.0–5.4) within *P. falciparum* have revealed a defined metabolic pathway for the degradation of hemoglobin. *Plasmodium* has a limited capacity for *de novo* amino acid synthesis, so hemoglobin proteolysis may be essential for its survival. However, hemoglobin degradation alone seems insufficient for the parasite's metabolic needs because it is a poor source of methionine, cysteine, glutamine, and glutamate and contains no isoleucine. On the other hand, as pointed out by Fracis et al., several experiments show that cysteine protease has a key role in the hemoglobin degradation pathway; it has even been hypothesized that the plasmepsins generate hemoglobin fragments that cannot be further catabolized without cysteine protease action.[295]

On the other hand, malaria parasite-infected red blood cells have a high concentration of the reduced GSH, the main reducing agent in physiological systems.[296] It was also reported that excess GSH in a parasite may be responsible for protecting the parasite from the toxicity of heme.[297,298] In general, GSH takes part in many biological functions, including the detoxification of cytosolic hydrogen peroxide and organic peroxides and then protects cells from being damaged by oxidative stress. Therefore, depletion of GSH or inhibiting GSH reductase in a parasite cell will induce oxidative stress and then kill these cells.[299]

Accordingly, as the first step, the interaction of qinghaosu and cysteine in the presence of a catalytic amount of Fe(II/III) was studied. From the reaction mixture, a water-soluble compound was isolated. This compound could be visualized with ninhydrin on TLC, and it showed a formula of $C_{16}H_{27}NO_6S \cdot H_2O$. Treatment of this compound with acetic anhydride yielded a cyclic thioether **165**, which in turn undoubtedly showed the formation of adduct **166** of **1** and cysteine through a σ bond between C-3 and sulfur.[281] A stable adduct **167** of cysteine and **170** was then isolated in 33% yield with the same reaction protocol.[185] As mentioned, both adducts of cysteine with primary and secondary free radical derived from artemether were also identified recently, albeit in low yield (Structure 5-25).[282] More

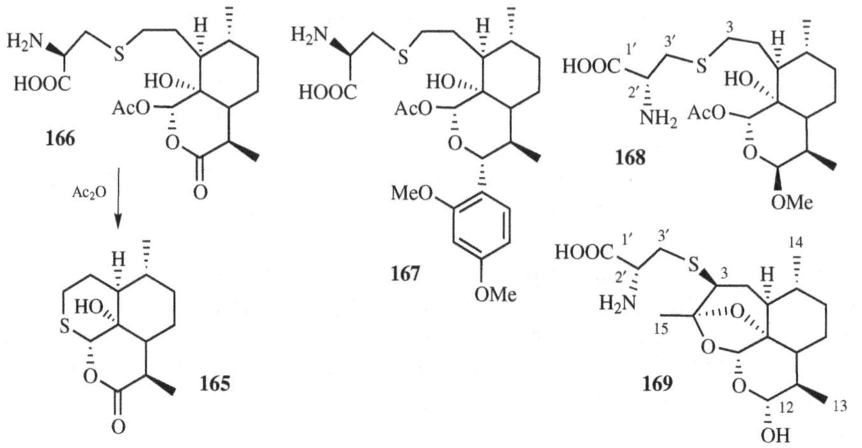

**Structure 5-25**

recently, using heme as the Fe(III), resource adduct **167** still could be identified.[300] These results confidentially show that free radicals derived from **1** and its peroxide derivatives can attack cysteine, but at this stage, it is still not known whether these adducts might be the inhibitor of cysteine protease and/or other enzymes.

The successful identification of cysteine adducts encouraged us to study the reaction of **170** and GSH-cat Fe(III). After careful isolation, an adduct **174**, similar to **167**, was obtained in 1% yield from the aqueous layer, which was easily rearranged to compound **175** in acidic medium. Thereafter, a similar adduct **176** between GSH and the primary C-radical derived from qinghaosu was isolated and structurally confirmed by NMR and other spectroscopy (Structure 5-26).[185]

As mentioned, GSH plays an important role in protecting the parasite from oxidative stress. Therefore, these results are significant for understanding the action mode of qinghaosu and its derivatives. The formation of the GSH–qinghaosu adduct reduces the amount of GSH, and the adduct might also cause inhibition

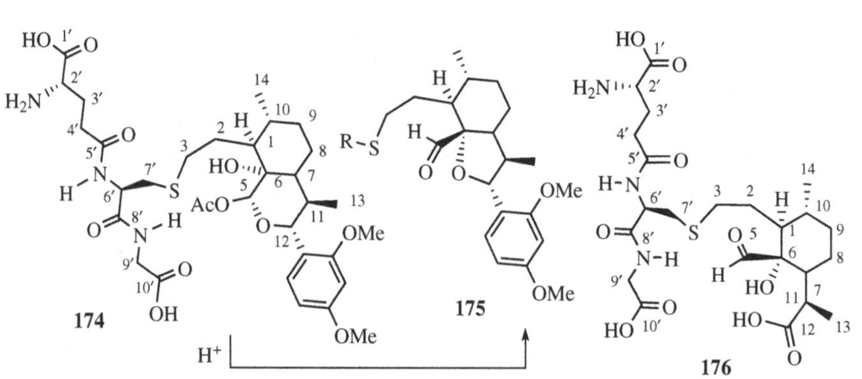

**Structure 5-26**

of the GSH reductase and other enzymes in the parasite. On the other hand, it is instructive that those C-centered radicals derived from **1** and its derivatives could attack not only free cysteine and cysteine residue in peptide, but also probably in cysteine-contained proteins. The formation of the covalent bond adduct between parasite proteins and **1** and its derivatives mentioned in the literature is mostly possible. In 2003, Eckstein-Ludwig et al. reported that the malarial calcium-dependent ATPase (PfATP6) might be the molecular target for qinghaosu.[301] PfATP6 is located in the sarco/endoplasmic reticulum situated outside the food vacuole of the parasite. This inhibition of PfATP6 was iron-dependent; that is, the free radical derived from the qinghaosu–iron ion was the active species. However, they did not have definitive evidence of whether the free radicals from qinghaosu bind to PfATP6 in several sites. In this respect, our experimental result that an adduct may be formed between the cysteine residue and the free radicals from qinghaosu is inspired to understand Eckstein-Ludwig et al.'s discovery.

### 5.6.5.3  Interaction With Heme

Recently, Robert et al. reviewed their mechanistic study on qinghaosu derivatives.[14,287] They identified some adducts of heme and the primary C-centered free radical **162** through the *meso*-position from the reaction of qinghaosu with Fe(III)-heme and 2,3-dimethylhydroquinone in methylene chloride[302] or Fe(III)-heme and GSH in dimethyl sulfoxide.[303] Structure 5-27 shows the major qinghaosu–heme adduct after demetallation. However, they did not clearly point how these adducts were related with the action mode of the antimalarial or the inhibition of the hemozoin formation. Also, in 2002, a heme–qinghaosu named hemart was reported and showed that hemart stalls all mechanisms of heme polymerization, which results in the death of the malarial parasite.[304] However, their hemart was synthesized just by mixing equivalent heme and qinghaosu in dimethyl acetamide (DMA) at 37 °C for 24 hours. It is not clear that hermart is a complex or a covalent adduct. The primary C-centered free radical **162** is only formed by reaction of qinghaosu and heme in the presence of reducing agents, but not in the absence of the reductants. In 2003, Haynes et al. pointed out that qinghaosu antimalarials do not inhibit hemozoin formation and hence ruled out the role of the qinghaosu–heme adduct as an inhibitor.[305]

**Structure 5-27**

### 5.6.6    Another Point of View and Summary

In the early 1980s, Wu and Ji performed the Hansch analysis of antimalarial activity and the distribution coefficient between oil and water of qinghaosu derivatives and found that the more lipophilic, the more active the derivative.[306] It is understandable that the lipophilic property of the qinghaosu derivative is related to its permeating ability across the membrane of the cell, but it is just the first step for the mode of action. Recently, several laboratories have performed the cyclic voltammetry study of qinghaosu and its derivatives. A correlation of the activities of qinghaosu derivatives with their reduction potentials was reported.[104] However, it was also indicated that the electrochemical reduction of qinghaosu and its derivatives was a two-electron transfer,[100] and it produced deoxyqinghaosu and its derivatives, but not free radical reduction products, which were confirmed by isolation and identification of the electrochemical reduction product.[98] The major metabolites of qinghaosu in vivo are the same as free radical reduction products, so it is confusing if there is a correlation between the electrochemical reduction and the cleavage of peroxy bridge of **1** in vivo.

Recently, Jefford has concluded that the killing of the parasite by alkylation with a carbon radical is a logical and convincing sequel, but the death of the parasite also may occur by oxygen atom transfer or by the action of an oxy-electrophilic species.[285] However, Haynes has argued totally against the C-centered radical proposal based on the consideration of redox chemistry and SARs.[307] In their recent article, Haynes et al. conclude that the antimalarial activity does not correlate with the chemical reactivity of qinghaosu derivatives against Fe(II) based on an experimental observation of a special C-12 nitrogen-substituted derivative.[308] Another totally different viewpoint was presented in 2001 that the antimalarial activity of **1** may be caused by (1) the interaction of the intact compound without chemical reaction, (2) the chemical reaction of **1** and/or its degradation products with the parasite biomolecules, or (3) the oxygen free radical occurring during redox reaction.[309] Their consideration is based on an undefined experimental result that dihydroqinghaosu is the product from the reaction of **1** and ferrous sulfate in an aqueous buffer. The formation of dihydroqinghaosu determined only by TLC has no precedent, and the reduction of lactone to lactol with ferrous ion is also unbelievable in chemistry.

In a recent account, O'Neill and Posner again mentioned that high-valent iron–oxo-intermediate is important for high antimalarial activity, but this opinion is not widely accepted.[16]

In summary, the mode of antimalarial action of **1** has been a noticeable research object since its discovery. Several theories have been postulated until now, however, their further development has still been difficult as mentioned by Wu.[310] Right now, from several experimental results and proposed theories, the postulation that the carbon-centered free radical is the key active species killing the parasite would most likely be convincing. The additional key point is what might be the genuine target attacked by these catalytic amount of radicals and what might be destroying the whole biosystem of the parasite.

## 5.7. CONCLUSION

Since the discovery of qinghaosu three decades ago, qinghaosu and its derivatives, as well as other corresponding peroxide compounds, have received great attention. A new generation of antimalarial drug was born from traditional Chinese medicine. It is expected that more medicinal applications will also be produced from this natural gift. New ground in chemistry and biology has also been broken during this exploration. With the continuing endeavor around the world, it is expected that more achievements will appear in the qinghaosu research area.

## REFERENCES

1. Qinghaosu Antimalarial Coordinating Research Group. Antimalaria studies on qinghaosu. *Chin. Med. J.*, **1979**, 92(12): 811–816.

2. Klayman, D. L. Qinghaosu (artemisinin), an antimalarial drug from China. *Science*, **1985**, 228: 1049–1055.

3. Luo, X.-D.; Shen, C.-C. The chemistry, pharmacology and clinical applications of qinghaosu (artemisinin) and its derivatives. *Med. Res. Rev.*, **1987**, 7(1): 29–52.

4. Zaman, S.-S.; Sharma, R.-P. Some aspects of the chemistry and biological activity of artemisinin and related antimalarials. *Heterocycles*, **1991**, 32(8): 1593–1637.

5. Butler, A. R.; Wu, Y.-L. Artemisinin (Qinghaosu): a new type of antimalarial drug. *Chem. Soc. Rev.*, **1992**, 85–90.

6. Hien, T. T.; White, N. J. Qinghaosu. *Lancet*, **1993**, 341: 603–608.

7. Wu, Y.-L.; Li, Y. Study on the chemistry of qinghaosu (artemisinin). *Med. Chem. Res.*, **1995**, 5(8): 569–586.

8. Meshnick, S. R.; Taylor, T. E.; Kamchonwongpaisan, S. Artemisinin and the antimalarial endoperoxides: from herbal remedy to targeted chemotherapy. *Microbiological Rev.*, **1996**, 60(2): 301–315.

9. Jefford, C. W. Peroxidic antimalarials. In Testa, B.; Meyer, A. U., editors. *Advances in Drug Research*, vol. 29. New York: Academic Press; **1997**, 271–320.

10. Ziffer, H.; Highet, R. J.; Klayman, D. L. Artemisinin: an endoperoxidic antimalarial from *Artemisia annua*, L. *Progr. Chem. Nat. Prod.*, **1997**, 72: 121–214.

11. Li, Y.; Wu, Y.-L. How Chinese scientists discovered qinghaosu (artemisinin) and developed its derivatives? What are the future perspectives? *Medecine Tropicale*, **1998**, 58(3S): 9–12.

12. Avery, M. A.; Alvimgaston, M.; Woolfrey, J. R. Synthesis and structure-activity - relationships of peroxidic antimalarials based on artemisinin. *Advanc. Medicinal Chem.*, **1999**, 4: 125–217.

13. Bhattacharya, A. K.; Sharma, R. P. Recent developments on the chemistry and biological-activity of artemisinin and related antimalarials—an update. *Heterocycles*, **1999**, 51(7): 1681–1745.

14. Robert, A.; Dechycabaret, O.; Cazelles, J.; Benoitvical, F.; Meunier, B. Recent advances in malaria chemotherapy. *J. Chin. Chem. Soc.*, **2002**, 49(3): 301–310.

15. Li, Y.; Wu, Y. L. An over 4-millennium story behind qinghaosu (artemisinin)—a fantastic antimalarial-drug from a traditional Chinese herb. *Curr. Med. Chem.*, **2003**, 10(21): 2197–2230.

16. O'Neill, P. M.; Posner, G. H. A medicinal chemistry perspective on artemisinin and related endoperoxides. *J. Med. Chem.*, **2004**, 47(12): 2945–2964.

17. Liang, X. T.; Yu, D. Q.; Wu, W. L.; Deng, H. C. The structure of yingzhaosu A. *Huaxue Xuebao (Acta Chimica Sinica)*, **1979**, 37: 215–230.

18. Shanghai Institute of Materia Medica, Shanghai 14th Pharmaceutical Plant. Studies on the active principles of Shian-Ho-Tsao 1. The isolation, structure and synthesis of Agrimol C. *Huaxue Xuebao*, **1975**, 33: 23–33.

19. Chen, C. L.; Zhu, D. Y.; Wang, H. C. Huang, B. S.; Qin, G. W. The structures of Agrimol A, B, and D. *Huaxue Xuebao*, **1978**, 36: 35–41.

20. Qin, G. W.; Chen, Z. X.; Wang, H. C.; Qian, M. K. The structure and synthesis of robustanol A. *Huaxue Xuebao*, **1981**, 39: 83–89.

21. Zhao, Y.; Zheng, J. J.; Huang, S. Y.; Li, X. J.; Lin, Q. Y.; Zhang, Q. X. Experimental studies on antimalarial action of protopine derivatives. *Chin. Pharmaceu. Bull.*, **1981**, 16: 327–330.

22. Lin, L. Z.; Zhang, J. S.; Chen, Z. L.; Xu, R. S. Studies on chemical constituents of *Brucea javanica* (L.) Merr.1. Isolation and identification of bruceaketolic acid and other four quassinoids. *Huaxue Xuebao*, **1982**, 40: 73–78.

23. Han, G. Y.; Xu, B. X.; Wang, X. P.; Liu, M. Z.; Xu, X. Y.; Meng, L. N.; Chen, Z. L.; Zhu, D. Y. Study on the active principle of *Polyalthia nemoralis* 1. The isolation and identification of natural zinc compound. *Huaxue Xuebao*, **1981**, 39: 433.

24. Chou, T. Q.; Fu, F. Y.; Kao, Y. S. Antimalarial constituents of Chinese drug, Ch'ang shan, *Dichroa Febrifuga Lour. J. Am. Chem. Soc.*, **1948**, 70: 1765–1767.

25. Coordinating Research Group for the Structure of Artemisinin. A new type of sesquiterpene lactone—artemisinin. *Kexue Tongbao (Chin. Sci. Bull.)*, **1977**, 22: 142.

26. Liu, J.-M.; Ni, M.-Y.; Fan, Y.-F.; Tu, Y.-Y.; Wu, Z.-H.; Wu, Y.-L.; Chou, W.-S. Structure and reactions of arteannuin. *Huaxue Xuebao*, **1979**, 37(2): 129–141.

27. Xu, X.-X.; Zhu, J.; Huang, D.-Z.; Zhou, W.-S. Total synthesis of (+)-yingzhaosu. *Tetrahedron Lett.*, **1991**, 32: 5785–5788.

28. Jeremic, D.; Jokic, A.; Behbud, A.; Stefanovic, M. A new type of sesqiterpene lactone isolated from *artemisia annua* L. arteannuin B *Tetrahedron Lett.*, **1973**, 32: 3039–3042.

29. Uskokovic, M. R.; Williams, T. H.; Blount, J. F. The structure and absolute configuration of arteannuin B. *Helv. Chim. Acta*, **1974**, 57(3): 600–602.

30. Qinghaosu Antimalarial Coordinating Research Group, Institute of Biophysics, CAS. The crystal structure and absolute configuration of qinghaosu. *Scientia Sinica*, **1979**, 11: 1114–1128.

31. Singh, A.; Vishwakarma, R. A.; Husain, A. Evaluation of *Artemisia annua* strains for higher artemisinin production. *Planta Medica*, **1988**, 54(5): 475–476.

32. Leirsh, R.; Soicke, H.; Stehr, C.; Tüllner, H. U. Formation of artemisinin in *Artemisia annua*, during one vegetation period. *Planta Medica*, **1986**, 52: 387–390.

33. Li, G.-D.; Zhou, Q.; Zhao, C.-W.; Zhai, F.-Y.; Huang, L.-F. Situation in research of artemisinin drugs. *Chin. Pharm. J.*, **1998**, 33(7): 385–389.

34. Wallaart, T. E.; van Uden, W.; Lubberink, H. G. M.; Woerdengag, H. J.; Pras, N.; Quax, W. J. Isolation and identification of dihydroartemisinic acid from *Artemisia-annua* and its possible role in the biosynthesis of artemisinin. *J. Nat. Prod.*, **1999**, 62(3): 430–433.

35. van Agtmael, M. A.; Eggelte, T. A.; van Boxtel, C. J. Artemisinin drugs in the treatment of malaria: from medicinal herb to registered medication. *Trends Pharmacol. Sci.*, **1999**, 20(5): 199–204.

36. Tu, Y.-Y.; Ni, M.-Y.; Zhong, Y.-R.; Li, L.-N.; Cui, S.-L.; Zhang, M.-Q.; Wang, X.-Z.; Liang, X.-T. Studies on the chemical constituents of *Artemisia annua*, L. *Acta Pharmaceutica Sinica*, **1981**, 16(5): 366–370.

37. Tu, Y.-Y.; Ni, M.-Y.; Zhong, Y.-R.; Li, L.-N.; Cui, S.-L.; Zhang, M.-Q.; Wang, X.-Z.; Liang, X.-T. Studies on the constituents of *Artemisia annua*. Part II. *Planta Medica*, **1982**, 44: 143–145.

38. Zhu, D.-Y; Deng, D.-A.; Zhang, S.-G; Xu, R.-S. Structure of artemisilactone. *Acta Chim. Sinica*, **1984**, 42(9): 937–939.

39. Wu, Z.-H.; Wang, Y.-Y. Structure and synthesis of arteannuin and related compounds. XI. Identification of epoxyarteannuinic acid. *Acta Chim. Sinica*, **1984**, 42: 596.

40. Deng, D.-A.; Zhu, D.-Y.; Gao, Y.-L.; Dai, J.-Y.; Xu, R.-S. Studies on the structure of artemisic acid. *Kexue Tongbao*, **1981**, 19: 1209–1211.

41. Tu, Y.-Y.; Ni, M.-Y.; Chung, Y.-Y.; Li, L.-N. Chemical constituents in *Artemisia annua*, L. and the derivatives of artemisinine. *Zhong Yao T'ong Bao (Bull. Chin. Nat. Med.)*, **1981**, 6(2): 31.

42. Zhu, D.-Y.; Zhang, S.-G.; Liu, B.-N.; Fan, G.-J.; Liu, J.; Xu, R.-S. Study on antibacterial constituents of Qing Hao (*Artemisia annua* L.). *Zhongcaoyao (Chin. Trad, Herb Drug)*, **1982**, 13(2): 54.

43. Huang, J.-J.; Xia, Z.-Q.; Wu, L.-F. Constituents of *Artemisia annua* L. I. Isolation and identification of 11R-(−)-dihydroarteannuic acid. *Acta Chim. Sin.*, **1987**, 45: 609–612.

44. Tian, Y.; Wei, Z.-Y.; Wu, Z.-H. Studies on the chemical constituents of Qing Hao (*Artemisia annua*), a traditional Chinese herb. *Zhongcaoyao*, **1982**, 13: 249–251.

45. Roth, R. J.; Acton, N. Isolation of arteannuic acid from *Artemisia annua*. *Planta Medica*, **1987**, 53: 501–502.

46. Misra, L. N. Arteannuin C, a sesquiterpene lactone from *Artemisia annua*. *Phytochemistry*, **1986**, 25(12): 2892–2893.

47. El-Feraly, F.-S.; Al-Meshal, I. A.; Khalifa, S. I. epi-Deoxyarteannuin B and 6,7-dehydroartemisinic acid from *Artemisia annua*. *J. Nat. Prod.*, **1989**, 52: 196–198.

48. Roth, R. J.; Acton, N. Isolation of epi-deoxyarteannuin B from *Artemisia annua*. *Planta Med.*, **1987**, 53(6): 576.

49. Acton, N.; Klayman, D. L. Artemisitene, a new sesquiterpene lactone endoperoxide from *Artemisia annua*. *Planta Med.*, **1985**, 5: 441–442.

50. Wei, Z. X.; Pan, J. P.; Li, Y. Artemisinin-G—a sesquiterpene from *Artemisia annua*. *Planta Med.*, **1992**, 58(3): 300.

51. Brown, G. D. Annulide, a sesquiterpene lactone from *Artemisia-Annua*. *Phytochemistry*, **1993**, 32(2): 391–393.

52. Sy, L. K.; Brown, G. D. Deoxyarteannuin-B, dihydro-deoxyarteannuin-B and tans-5-hydroxy- 2-isopropenyl-5-methylhex-3-en-1-ol from *Artemisia anuua*. *Phytochemistry*, **2001**, 58(8): 1159–1166.

53. Elmarkaby, S. A. Microbial transformation studies on arteannuin B. *J. Nat. Prod.*, **1987**, 50(5): 903–909.

54. Brown, G. D. Two new compounds from *Artemisia annua*. *J. Nat. Prod.*, **1992**, 55(12): 1756–1760.

55. Pathak, A. K.; Jain, D. C.; Bhakuni, R. S.; Chaudhuri, P. K.; Sharma, R. P. Deepoxidation of arteannuin-B with chlorotrimethylsilane and sodium-iodide. *J. Nat. Prod.*, **1994**, 57(12): 1708–1710.

56. Misra, L. N.; Ahmad, A.; Thakur, R. S.; Jakupovic, J. Bisnor-cadinanes from *Artemisia annua* and definitive C-13 NMR assignments of beta-arteether. *Phytochemistry*, **1993**, 33(6): 1461–1464.

57. Ahmad, A.; Misra, L. N. Terpenoids from *Artemisia annua* and constituents of its essential oil. *Phytochemistry*, **1994**, 37(1): 183–186.

58. Brown, G. D. Cadinanes from *Artemisia annua* that may be intermediates in the biosynthesis of artemisinin. *Phytochemistry*, **1994**, 36(3): 637–641.

59. Sy, L. K.; Brown, G. D. Three sesquiterpenes from *Artemisia annua*. *Phytochemistry*, **1998**, 48(7): 1207–1211.

60. Sy, L. K.; Brown, G. D.; Haynes, R. A novel endoperoxide and related sesquiterpenes from *Artemisia annua* which are possibly derived from allylic hydroperoxides. *Tetrahedron*, **1998**, 54(17): 4345–4356.

61. Wallaart, T. E.; Pras, N.; Quax, W. J. Isolation and identification of dihydroartemisinic acid hydroperoxide from *Artemisia annua*—a novel biosynthetic precursor of artemisinin. *J. Nat. Prod.*, **1999**, 62(8): 1160–1162.

62. Sy, L. K.; Cheung, K. K.; Zhu, N. Y.; Brown, G. D. Structure elucidation of Arteannuin-O, a novel cadinane diol from *Artemisia annua* and the synthesis of Arteannuin-K, Arteannuin-L, Arteannuin-M and Arteannuin-O. *Tetrahedron*, **2001**, 57(40): 8481–8493.

63. Brown, G. D.; Liang, G. Y.; Sy, L. K. Terpenoids from the seeds of *Artemisia annua*. *Phytochemistry*, **2003**, 64(1): 303–323.

64. Deoliveira, C. M.; Ferracini, V. L.; Foglio, M. A.; Demeijere, A.; Marsaioli, A. J. Detection, synthesis and absolute-configuration of (+)-nortaylorione, a new terpene from *Artemisia annua*. *Tetrahedron Asymmetry*, **1997**, 8(11): 1833–1839.

65. Tewari, A.; Bhakuni, R. S. Terpenoid and lipid constituents from *Artemisia annua*. *Indian J. Chem. Sect. B*, **2003**, 42(7): 1782–1785.

66. Bhakuni, R. S.; Jain, D. C.; Sharma, R. P.; Kumar, S. Secondary metabolites of *Artemisia annua* and their biological activity. *Curr. Sci.*, **2001**, 80(1): 35–48.

67. Ruecker, G.; Mayer, R.; Manns, D. α- and β-Myrcene hydroperoxide from *Artemisia annua*. *J. Nat. Prod.*, **1987**, 50(2): 287–289.

68. Brown, G. D. Phytene-1,2-diol from *Artemisia annua*. *Phytochemistry*, **1994**, 36(6): 1553–1554.

69. Zheng, G.-Q. Cytotoxic terpenoids and flavonoids from *Artemisia annua*. *Planta Medica*, **1994**, 60: 54–57.

70. Agrawal, P. K.; Bishnoi, V. Studies on Indian medical plants. 42. Sterol and taraxastane derivatives from *Artemisia annua* and a rational approach based upon C-13 NMR for the identification of skeletal type of amorphane sesquiterpenoids. *Indian J. Chem.*, **1996**, 35B: 86–88.

71. Bagchi, G. D.; Haider, F.; Dwivedi, P. D.; Singh, A.; Naqvi, A. A. Essential oil constituents of *Artemisia annua* during different growth periods at monsoon

conditions of subtropical north Indian plains. *J. Essent. Oil Res.*, **2003**, 15(4): 248–250.

72. Takemoto, T.; Nakajima, T. Study on the essential oil of *Artemisia annua* L. I. Isolation of a new ester compound. *Yakugaku Zasshi*, **1957**, 77: 1307–1309.

73. Takemoto, T.; Nakajima, T. Study on the essential oil of *Artemisia annua* L. II. Structure of *l*-β-artemisia alcohol. *ibid* 1310–1313.

74. Takemoto, T.; Nakajima, T. Study on the essential oil of *Artemisia annua* L. III. Discussion on the structure of artemisia ketone. *ibid* 1339–1344.

75. Takemoto, T.; Nakajima, T. Study on the essential oil of *Artemisia annua* L. IV. Study on the essential oil of *Artemisia-Annua* L. I. On dihydroartemisia ketone. *ibid* 1344–1347.

76. Nakajima, T. Study on the essential oil of *Artemisia annua* L. V. *ibid*, **1960**, 82: 1323–1326.

77. Tellez, M. R.; Canel, C.; Rimando, A. M.; Duke, S. O. Differential accumulation of isoprenoids in glanded and glandless *Artemisia annua* L. *Phytochemistry*, **1999**, 52: 1035–1040.

78. Zou, Y.-H.; Shi, J.-D.; Shi, H.-T. Analysis of volatile components from *Artemisia annua* Linn. *J. Instrument. Anal. (Fenxi Ceshi Xuebao)*, **1999**, 18(1): 55–57.

79. Jain, N.; Srivastava, S. K.; Aggarwal, K. K.; Kumar, S.; Syamasundar, K. V. Essential oil composition of *Artemisia annua* L. 'Asha' from the plains of northern India. *J. Essent. Oil Res.*, **2002**, 14(4): 305–307.

80. Rasooli, I.; Rezaee, M. B.; Moosavi, M. L.; Jaimand, K. Microbial sensitivity to and chemical properties of the essential oil of *Artemisia annua* L. *J. Essent. Oil Res.*, **2003**, 15(1): 59–62.

81. Juteau, F.; Masotti, V.; Bessiere, J. M.; Dherbomez, M.; Viano, J. Antibacterial and antioxidant activities of *Artemisia annua* essential oil. *Fitoterapia*, **2002**, 73(6): 532–535.

82. Liu, H.-M.; Li, G.-L.; Wu, H.-Z. Studies on the chemical constituents of *Artemisia annua* L. *Acta Pharma Sin*, **1981**, 16(1): 65–67.

83. Yang, S.-L.; Roberts, M. F.; O'Neill, M. J.; Bucar, F.; Phillipson, J. D. Flavonoids and chromenes from *Artemisia annua*. *Phytochemistry*, **1995**, 38(1): 255–257.

84. Yang, S.-L.; Roberts, M. F.; Phillipson, J. D. Methoxylated flavones and coumarins from *Artemisia annua*. *Phytochemistry*, **1989**, 28(5): 1509–1511.

85. Shukla, A. A.; Farooqi, H. A.; Shukla, Y. N. A new adenine derivative from *Artemisia annua*. *J. Indian Chem. Soc.*, **1997**, 74: 59.

86. Brown, G. D. Two new compounds from *Artemisia annua*. *J. Nat. Prod.*, **1992**, 55(12): 1756–1760.

87. Singh, A. K.; Pathak, V.; Agrawal, P. K. Annphenone, a phenolic acetophenone from *Artemisia annua*. *Phytochemistry*, **1997**, 44(1): 555–557.

88. Manns, D.; Hartmann, R. Annuadiepoxide, a new polyacetylene from the aerial parts of *Artemisia annua*. *J. Nat. Prod.*, **1992**, 55(1): 29–32.

89. Tu, Y.-Y.; Yin, J.-P.; Ji, L.; Huang, M.-M.; Liang, X.-T. Chemical constituents of sweet wormword (*Artemisia annua*) (III). *Zhongcaoyao*, **1985**, 16(5): 200–201.

90. Liu, C.-H.; Zou, W.-X.; Lu, H.; Tan, R.-X. Antifungal activity of *Artemisia annua* endophyte cultures against phytopathogenic fungi. *J. Biotechnology*, **2001**, 88(3): 277–282.

91. Lu, H.; Zou, W.-X.; Meng, J. C.; Hu, J.; Tan, R.-X. New bioactive metabolites produced by *Colletotrichum* sp., an endophytic fungus in *Artemisia annua*. *Plant Sci.*, **2000**, 151(1): 67–73.

92. Tu, Y.-Y.; Zhu, Q.-C.; Shen, X. Constituents of young *Artemisia annua*. *Zhongyao Tongbao*, **1985**, 10(9): 419–420.

93. Liu, J.-Y.; Liu, C.-H.; Zou, W.-X.; Tian, X.; Tan, R.-X. Leptosphaerone, a metabolite with a novel skeleton from *Leptosphaeria-Sp.* Iv403, an endophytic fungus in *Artemisia-Annua*. *Helv. Chim. Acta*, **2002**, 85: 2664–2667.

94. Liu, J.-Y.; Liu, C.-H.; Zou, W.-X.; Tan, R.-X. Leptosphaeric acid, a metabolite with a novel carbon skeleton from *Leptosphaeria Sp.* Iv403, an endophytic fungus in *Artemisia-annua*. *Helv. Chim. Acta*, **2003**, 86: 657–660.

95. Li, Y.; Yu, P.-L.; Chen, Y.-X.; Zhang J.-L.; Wu, Y.-L. Studies on analogs of qinghaosu—some acidic degradations of qinghaosu. *Kexue Tongbao (Engl.)*, **1986**, 31(35): 1038–1040.

96. Li, Y.; Yu, P.-L.; Chen, Y.-X.; Li, L.-Q.; Gai, Y.-Z.; Wang, D.-S.; Zheng, Y.-P. Synthesis of some derivatives of artemisinin. *Kexue Tongbao*, **1979**, 24(14): 667–669.

97. Li, Y.; Yu, P.-L.; Chen, Y.-X.; Li, L.-Q.; Gai, Y.-Z.; Wang, D.-S.; Zheng, Y.-P. Studies on analogs of artemisinin I. The synthesis of ethers, carboxylic esters and carbonates of dihydroartemisinin. *Yaoxue Xuebao (Acta Pharm. Sin.)*, **1981**, 16(6): 429–439.

98. Wu, W.-M.; Wu, Y.-L. Chemical and electro-chemical reduction of qinghaosu (artemisinin). *J. Chem. Soc. Perkin Trans. I*, **2000**, 24: 4279–4283.

99. Chen, Y.; He, C.-X.; Zhu S.-M.; Chen, H.-Y. Electrocatalytic reduction of artemether by hemin. *J. Electrochem. Soc.*, **1997**, 144(6): 1891–1894.

100. Chen, Y.; Zhu, S.-M.; Chen H.-Y.; Li, Y. Study on the electrochemical behaviors of artemisinin and its derivatives. I. reduction of artemisinin at Hg electrode. *Huaxue Xuebao*, **1997**, 55: 921–925.

101. Chen, Y.; Zhu, S.-M.; Chen, H.-Y., Study on the electrochemical behaviors of artemisinin (qinghaosu) and its derivatives II. Reduction mechanism of artemisinin in the presence of hemin. *Huaxue Xuebao*, **1998**, 56: 925–929.

102. Chen, Y.; Zheng, J. M.; Zhu, S.-M.; Chen, H.-Y. Evidence for hemin inducing the cleavage of peroxide bond of artemisinin (qinghaosu) — cyclic voltammetry and in-situ FT IR spectroelectrochemical studies on the reduction-mechanism of artemisinin in the presence of hemin. *Electrochemica Acta*, **1999**, 44(14): 2345–2350.

103. Zhang, F.; Gosser, D. K. Jr.,; Meshnick, S. R. Hemin-catalyzed decomposition of artemisinin (qinghaosu). *Biochem. Pharmacol.*, **1992**, 43(8): 1805–1809.

104. Jiang, H.-L.; Chen, K.-X.; Tang, Y.; Chen, J.-Z.; Li, Y.; Wang, Q.-M.; Ji, R.-Y.; Zhuang, Q.-K. Theoretical and cyclic voltammetry studies on antimalarial mechanism of artemisinin (Qinghaosu) derivatives. *Indian J. Chem.*, **1997**, 36B: 154–160.

105. Chen, Z.-R. Shanghai Institute of Organic Chemistry, Chinese Academy of Sciences, Unpublished data.

106. Jung, M.; Li, X.; Bostos, D. A.; ElSohly, H. N.; McChesney, J. D. A short and stereospecific synthesis of (+)-deoxoartemisinin and (−)-deoxodesoxyartemisinin. *Tetrahedron Lett.*, **1989**, 30(44): 5973–5976.

107. Rong, Y.-J.; Ye, B.; Zhang, C.; Wu, Y.-L. An efficient synthesis of deoxoqinghaosu from dihydro-qinghaosu. *Chin. Chem. Lett.*, **1993**, 4(10): 859–860.

108. Wu, Y.-L.; Zhang, J.-L., Reduction of qinghaosu with lithium aluminium hydride. *Youji Huaxue (Chin. J. Org. Chem.)*, **1986**, 6: 153–156.

109. Sy, L.-K.; Hui, S.-M.; Cheung, K.-K.; Brown, G.-D. A rearranged hydroperoxide from the reduction of artemisinin. *Tetrahedron*, **1997**, 53(22): 7493–7500.

110. Zhou, W.-S.; Wen, Y.-C. Studies on structure and synthesis of arteannuin and related compound VI. The structure of arteannuin degradation products. *Acta Chim. Sinica*, **1984**, 42(5): 455–459.

111. Gu, Y.-X. Determination of crystal structure of norsesquiterpenoid lactone. *Acta Physica Sinica*, **1982**, 31(7): 963–968.

112. Li, Y.; Yu, P.-L.; Chen, Y.-X.; Zhang, J.-L.; Wu, Y.-L. Studies on analogs of qinghaosu, some acidic degradations of qinghaosu. *Kexue Tongbao (Eng.)*, **1986**, 31(15): 1038–1040.

113. Imakura, Y; Hachiya, K.; Ikemoto, T.; Yamashita, S.; Kihara, M.; Kobayashi, S.; Shingu, T.; Mihous, W. K.; Lee, K. H. Acid degradation products of qinghaosu and their structure-activity relationships. *Heterocycles*, **1990**, 31(6): 1011–1016.

114. Lin, A.-J.; Klayman, D. L.; Hoch, J. M.; Silverton, J. V.; George, C. F. Thermal rearrangement and decomposition products of artemisinin (qinghaosu). *J. Org. Chem.*, **1985**, 50(23): 4504–4508.

115. Luo, X.-D.; Yeh, H.-J. C.; Brossi, A. The chemistry of drugs. VI. Thermal decomposition of qinghaosu. *Heterocycles*, **1985**, 23(4): 881–887.

116. Zhu, D.-Y.; Huang, B.-S.; Chen, Z.-L.; Yin, M.-L.; Yang, Y.-M.; Dai, M.-L.; Wang, B.-D.; Huang, Z.-H. Isolation and identification of the metabolite of artemisinin in human. *Zhongguo Yaoli Xuebao (Acta Pharma. Sin.)*, **1983**, 4(3): 194–197.

117. Lee, I.-S.; Hufford, C. D. Metabolism of antimalarial sesquiterpene lactones. *Pharmac. Ther.*, **1990**, 48(3): 345–355.

118. Lee, I.-S.; El-Sohly, H. N.; Croom, E. M., Jr.; Hufford, C. D. Microbial metabolism studies of the antimalarial sesquiterpene artemisinin. *J. Nat. Prod.*, **1989**, 52(2): 337–341.

119. Zhan, J. X.; Zhang, Y. X.; Guo, H. Z.; Han, J.; Ning, L. L.; Guo, D. A. microbial-metabolism of artemisinin by *Mucor-Polymorphosporus* and *Aspergillus-Niger. J. Nat. Prod.*, **2002**, 65(11): 1693–1695.

120. Zhan, J. X.; Guo, H. Z.; Dai, J. G.; Zhang, Y. X.; Guo, D. Microbial transformations of artemisinin by *Cunninghamella-Echinulata* and *Aspergillus-Niger. Tetrahedron Lett.*, **2002**, 43(25): 4519–4521.

121. Hu, Y.-L.; Highet, R. J.; Marion, D.; Ziffer, H. Microbial hydroxylation of a dihydroartemisinin derivative. *J. Chem. Soc. Chem. Commun.*, **1991**, 1176–1177

122. Hu, Y. L.; Ziffer, H.; Li, G.; Yeh, H. J. C. Microbial oxidation of the antimalarial drug arteether. *Bioorg. Chem.*, **1992**, 20(2): 148–154.

123. Hufford, C. D.; Khalifa, S. I.; Orabi, K. Y.; Wiggers, F. T.; Kumar, R.; Rogers, R. D.; Campana, C. F. 1α-Hydroxyarteether, a new microbial transformation product. *J. Nat. Prod.*, **1995**, 58(5): 751–755.

124. Abourashed, E. A.; Hufford, C. D. Microbial transformation of artemether. *J. Nat. Prod.*, **1996**, 59(3): 251–253.

125. Xu, X.-X.; Zhu, J.; Huang, D.-Z.; Zhou, W.-S. Studies on structure and synthesis of arteannuin and related compound X. The stereocontrolled synthesis of arteannuin and deoxyarteannuin from arteannuic acid. *Huaxue Xuebao*, **1983**, 41(6): 574–576.

126. Xu, X.-X.; Zhu, J.; Huang, D.-Z.; Zhou, W.-S. Total synthesis of arteannuin and deoxyarteannuin. *Tetrahedron*, **1986**, 42(3): 819–828.

127. Zhou, W.-S. Total synthesis of arteannuin (qinghaosu) and related compounds. *Pure Appl. Chem.*, **1986**, 58(5): 817–824.

128. Xu, X.-X.; Zhu, J.; Huang, D.-Z.; Zhou, W.-S. Studies on the structure and synthesis of arteannuin and related compound XVII. The stereocontrolled total synthesis of methyl dihydroarteannuate—the total synthesis of arteannuin. *Huaxue Xuebao*, **1984**, 42: 940–942.

129. Schmid, G.; Hofheinz, W. Total synthesis of qinghaosu. *J. Am. Chem. Soc.*, **1983**, 105(3): 624–625.

130. Wu, Y.-L.; Zhang, J.-L.; Li, J.-C. Studies on the synthesis of qinghaosu and its analogs–Reconstruction of qinghaosu from its degradation product. *Huaxue Xuebao*, **1985**, 43: 901–903.

131. Ravindranathan, T.; Kumar, M. A.; Menon, R. B.; Hiremath, S. V. Stereoselective synthesis of artemisinin. *Tetrahedron Lett.*, **1990**, 31(5): 755–758.

132. Yadav, J. S.; Babu, R. S.; Sabitha, G. Stereoselective total synthesis of (+)-artemisinin. *Tetrahedron Lett.*, **2003**, 44: 387–389.

133. Yadav, J. S.; Babu, R. S.; Sabitha, G. Total synthesis of (+)-artemisinin. *ARKIVOC*, **2003**, 125–139.

134. Ye, B.; Wu, Y.-L. Syntheses of carba-analogues of qinghaosu. *Tetrahedron*, **1989**, 45(23): 7287–7290.

135. Ye, B.; Wu, Y.-L. An efficient synthesis of qinghaosu and deoxoqinghaosu from arteanuic acid. *J. Chem. Soc. Chem. Commun.*, **1990**, 726–727.

136. Roth, R. J.; Acton, N. A simple conversion of artemisinic acid into artemisinin. *J. Nat. Prod.*, **1989**, 52(5): 1183–1185.

137. Acton, N.; Roth, R. J. On the conversion of dihydroartemisinic acid into artemisinin. *J. Org. Chem.*, **1992**, 57(13): 3610–3614.

138. Haynes, R. K.; Vonwiller, S. C. Catalysed oxygenation of allylic hydroperoxides derived from Qinghao (artemisinic) acid. Conversion of qinghao acid into dehydroqinghaosu (artemisitene) and qinghaosu (artemisinin). *J. Chem. Soc. Chem. Commun.*, **1990**, 451–453.

139. Liu, H.-J.; Yeh, W.-L.; Chew, S. Y. A total synthesis of the antimalarial natural product (+)-qinghaosu. *Tetrahedron Lett.*, **1993**, 34(28): 4435–4438.

140. Lansbury, P. T.; Nowak, D. M. An efficient partial synthesis of (+)-artemisinin and (+)-deoxo-artemisinin. *Tetrahedron Lett.*, **1992**, 33(8): 1029–1032.

141. Nowak, D. M.; Lansbury, P. T. Synthesis of (+)-artemisinin and (+)-deoxoartemisinin from arteannuin B and arteannuin acid. *Tetrahedron*, **1998**, 54(3/4): 319–336.

142. Avery, M. A.; Jennings-White, C.; Chong, W. K. M. The total synthesis of (+)- artemisinin and (+)-9-desmethylartemisinin. *Tetrahedron Lett.*, **1987**, 28(40): 4629–4632.

143. Avery, M. A.; Chong, W. K. M.; Jennings-White, C. Stereoselective total synthesis of (+)-artemisinin, the antimalarial constituent of *Artemisia annua* L. *J. Am. Chem. Soc.*, **1992**, 114(3): 974–979.

144. Huang, J.-J.; Zhou, F.-Y.; Wu, L.-F.; Zhen, G.-H. Studies on the biosynthesis of arteannuin I. The biosynthesis of arteannuinic acid in *Artemisia annua* L. *Acta Chim. Sin. Engl. Ed.*, **1988**, 6: 383–385.

145. Wang, Y.; Xia, Z.-Q.; Zhou, F.-Y.; Wu, Y.-L.; Huang, J.-J.; Wang, Z.-Z. Studies on the biosynthesis of arteannuin III. Arteannuic acid as a key intermediate in the biosyntheses of arteannuin and arteannuin B. *Acta Chim. Sin. Engl. Ed.*, **1988**, 6: 386–387.

146. Akhila, A.; Thakur, R. S.; Popli, S. P. Biosynthesis of artemisinin in *Artemisia annua*. *Phytochemistry*, **1987**, 26(7): 1927–1930.

147. Kudakasseril, G. J.; Lam, L.; Staba, E. J. Effect of sterol inhibitors on the incorporation of $^{14}$C-isopentenyl pyrophosphate into artemisinin by a cell-free system from *Artemisia annua* tissue cultures and plants. *Planta Medica*, **1987**, 53: 280–284.

148. Wang, Y.; Xia, Z.-Q.; Zhou, F.-Y.; Wu, Y.-L.; Huang, J.-J.; Wang, Z.-Z. Studies on the biosynthesis of arteannuin. IV. The biosynthesis of arteannuin and arteannuin B by the leaf homogenate of *Artemisia annua* L. *Chin. J. Chem.*, **1993**, 11(5): 457–463.

149. Abdin, M. Z.; Israr, M.; Rehman, R. U.; Jain, S. K. Artemisinin, a novel antimalarial drug: Biochemical and molecular approaches for enhanced production. *Planta Medica*, **2003**, 69(4): 289–299.

150. Sangwan, R. S.; Agarwal, K.; Luthra, R.; Thakur, R. S.; Singhsangwan, N. Biotransformation of arteannuic acid into arteannuin-B and artemisinin in *Artemisia annua*. *Phytochemistry*, **1993**, 34(5): 1301–1302.

151. Nair, M. S. R.; Basile, D. V. Bioconversion of arteannuin-B to artemisinin. *J. Nat. Prod.*, **1993**, 56(9): 1559–1566.

152. Bharel, S.; Gulati, A.; Abdin, M. Z.; Srivastava, P. S.; Vishwakarma, R. A.; Jain, S. K. Enzymatic synthesis of artemisinin from natural and synthetic precursors. *J. Nat. Prod.*, **1998**, 61(5): 633–636.

153. Wang, Y.; Shen, Z. W.; Xia, Z. Q.; Zhou, F. Y. Studies on the biosynthesis of arteannuin. 5. The role of 6-epi-deoxyarteannuin-B in arteannuin biosynthesis. *Chinese J. Chem.*, **1993**, 11(5): 476–478.

154. Brown, G. D.; Sy, L. K. In-vivo transformations of dihydroartemisinic acid in *Artemisia annua* plants. *Tetrahedron*, **2004**, 60(5): 1139–1159.

155. Li, Y.; Yang, Z.-X.; Chen, Y.-X.; Zhang, X. Synthesis of [15-$^{14}$C] labeled artemisinin. *Yaoxue Xuebao*, **1994**, 29(9): 713–716.

156. Liu, C. Z.; Wang, Y. C.; Guo, C.; Ouyang, F.; Ye, H. C.; Li, G. F. Production of artemisinin by shoot cultures of *Artemisia annua* L. in a modified inner-loop mist bioreactor. *Plant Sci.*, **1998**, 135(2): 211–217.

157. Delabays, N.; Simonnet, X.; Gaudin, M. The genetics of artemisinin content in *Artemisia annua* L. and the breeding of high yielding cultivars. *Curr. Med. Chem.*, **2001**, 8(15): 1795–1801.

158. Chen, D. H.; Ye, H. C.; Li, G. F.; Liu, Y. Advances in molecular biology of plant isoprenoid metabolic pathway. *Acta Botan. Sin.*, **2000**, 42(6): 551–558.

159. Chang, Y. J.; Song, S. H.; Park, S. H.; Kim, S. U. Amorpha-4,11-diene synthase of *Artemisia annua*: cDNA isolation and bacterial expression of a terpene synthase involved in artemisinin biosynthesis. *Arch. Biochem. Biophys.*, **2000**, 383(2): 178–184.

160. Martin, V. J. J.; Pitera, D. J.; Withers, S. T.; Newman, J. D.; Keasling, J. D. Engineering a mevalonate pathway in *Escherichia coli* for production of terpenoids. *Nature Biotechnol.*, **2003**, 21(7): 796–802.

161. Liu, X. Study on derivatives of artemisinin *Chin. Pharmaceu. Bull.*, **1980**, 15: 183.

162. Gu, H. M.; Lu, B. F.; Qu, Z. Q. Antimalarial activities of 25 derivatives of artemisinin against chloroquine-resistant *Plasmodium berghei. Acta Pharmacol. Sin.*, **1980**, 1: 48–50.

163. Yu, P. L.; Chen, Y. X.; Li, Y.; Ji, R. Y. Studies on analogs of artemisinin. IV. Synthesis of derivatives of qinghaosu containing halogen, nitrogen and sulfur heteroatoms. *Chin. Pharmaceu. Bull.*, **1980**, 15: 44; *Acta Pharmaceu. Sin.*, **1985**, 20: 357–365.

164. Pu, Y. M.; Torok, D. S.; Ziffer, H.; Pan, X.-Q.; Meshnick, S. R. Synthesis and antimalarial activities of several fluorinated artemisinin derivatives. *J. Med. Chem.*, **1995**, 38: 4120–4124.

165. Abouabdellah, A.; Bégué, J.-P.; Bonnet-Delpon, D.; Gantier, J. C.; Truong Thi Thanh, N.; Truong, D. T. Synthesis and in vivo antimalarial activity of 12 α-trifluoromethylhydroartemisinin. *Bioorg. Med. Chem. Lett.*, **1996**, 6: 2717–2720.

166. Nga, T. T. T.; Ménage, C.; Bégué, J.-P.; Bonnet-Delpon, D.; Gantier, J.-C.; Pradines, B.; Doury, J.-C.; Thac, T. Synthesis and antimalarial activities of fluoroalkyl derivatives of dihydroartemisinin. *J. Med. Chem.*, **1998**, 41: 4101–4108.

167. Magueur, G.; Crousse, B.; Charneau, S.; Grellier, P.; Bégué, J.-P.; Bonnet-Delpon, D. Fluoroartemisinin: Trifluoromethyl analogues of artemether and artesunate. *J. Med. Chem.*, **2004**, 47: 2694–2699.

168. Venugopalan, B.; Karnik, P. J.; Bapat, C. P.; Chatterjee, D. K.; Iyer, N.; Lepcha, D. Antimalarial activity of new ethers and thioethers of dihydroartemisinin. *Eur. J. Med. Chem.*, **1995**, 30: 697–706.

169. Haynes, R. K.; Chan, H. W.; Cheung, M. K.;. Lam, W. L.; Soo, M. K.; Tsang, H. W.; Voerste, A.; Williams, I. D. C-10 Ester and ether derivatives of dihydroartemisinin-10α artesunate, preparation of authentic 10-β artesunate, and of other ester and ether derivatives bearing potential aromatic intercalating groups at C-10. *Eur. J. Org. Chem.*, **2002**, 1: 113–132.

170. Liang, J.; Li, Y. Synthesis of the aryl ether derivatives of artemisinin. *Chinese J. Med. Chem.*, **1996**, 6: 22–25.

171. O'Neill, P. M.; Miller, A.; Bishop, L. P. D.; Hindley, S.; Maggs, J. L.; Ward, S. A.; Roberts, S. M.; Scheinmann, F.; Stachulski, A. V.; Posner, G. H.; Park, B. K. Synthesis, antimalarial activity, biomimetic iron(II) chemistry, and in-vivo metabolism of novel, potent C-10-phenoxy derivatives of dihydroartemisinin. *J. Med. Chem.*, **2001**, 44: 58–68.

172. Yang, Y. H.; Li, Y.; Shi, Y. L.; Yang, J. D.; Wu, B. A. Artemisinin derivatives with 12-aniline substitution: Synthesis and antimalarial activity. *Bioorg. Med. Chem. Lett.*, **1995**, 5: 1791–1794.

173. Li, Y.; Liao, X. B. CN Patent: ZL 99,124,012.

174. Jung, M.; Bustos, D. A.; El-Sohly, H. N.; McChesney, J. D. A concise and stereoselective synthesis of (+)-12-*n*-butyldeoxoartemisinin. *Synlett.*, **1990**, 743–744.

175. Jung, M.; Yu, D. Y.; Bustos, D.; El-Sohly, H. N.; McChesney, J. D. A concise synthesis of 12-(3′-hydroxy-*n*-propyl)-deoxoartemisinin. *Bioorg. Med. Chem. Lett.*, **1991**, 1: 741–744.

176. Haynes, R. K.; Vonwiller, S. C. Effecient preparation of novel qinghaosu(artemisinin) derivatives: Conversion of qinghaosu (artemisinic) acid into deoxoqinghaosu derivatives and 5-carba-4-deoxoartesunic acid. *Synlett.*, **1992**, 481–483.

177. Pu, Y. M.; Ziffer, H. Synthesis and antimalarial activities of 12β-allyldeoxoartemisinin and its derivatives. *J. Med. Chem.*, **1995**, 38: 613–616.

178. Ma, J. Y.; Katz, E.; Ziffer, H. A new synthetic route to 10-beta-alkyl deoxoartemisinins. *Tetrahedron Lett.*, **1999**, 40: 8543–8545.

179. Ma, J. Y.; Katz, E.; Kyle, D. E.; Ziffer, H. Syntheses and antimalarial activities of 10-substituted deoxoartemisinins. *J. Med. Chem.*, **2000**, 43: 4228–4232.

180. O'Neill, P. M.; Searle, N. L.; Kan, K. W.; Storr, R. C.; Maggs, J. L.; Ward, S. A.; Raynes, K.; Park, B. K. Novel, potent, semisynthetic antimalarial carba analogs of the first-generation 1,2,4-trioxane artemether. *J. Med. Chem.*, **1999**, 42: 5487–5493.

181. Woo, S. H.; Parker, M. H.; Ploypradith, P.; Nosten, F.; Posner, G. H. Direct conversion of pyranose anomeric OH–F–R in the artemisinin family of antimalarial trioxanes. *Tetrahedron Lett.*, **1998**, 39: 1533–1536.

182. Posner, G. H.; Parker, M. H.; Northrop, J.; Elias, J. S.; Ploypradith, P.; Xie, S. J.; Shapiro, T. A. Orally active, hydrolytically stable, semisynthetic, antimalarial trioxanes in the artemisinin family. *J. Med. Chem.*, **1999**, 42: 300–304.

183. Li, Y.; Yang, Y. H.; Liang, J.; Shan, F.; Wu, G. S. CN Patent ZL 94,113,982.4.

184. Li, Y.; Shan, F.; Xiao, S. H.; Xu, H. H. CN Patent ZL 98,110,770.2.

185. Wang, D.-Y.; Wu, Y.-L. A possible antimalarial action mode of qinghaosu (artemisinin) series compounds - alkylation of reduced glutathione by C-centered primary radicals produced from antimalarial compound qinghaosu and 12-(2,4-dimethoxyphenyl)-12-deoxoqinghaosu. *Chem. Commun.*, **2000**, 2193–2194.

186. Wang, D.-Y.; Wu, Y.-L.; Wu, Y. K.; Liang, J.; Li, Y. Further evidence for the participation of primary carbon-centered free-radicals in the antimalarial action of the qinghaosu (artemisinin) series of compounds. *J. Chem. Soc. Perkin Trans. 1*, **2001**, 605–609.

187. Wang, D.-Y.; Wu, Y. K.; Wu, Y.-L.; Li, Y.; Shan, F. Synthesis, iron(II)-induced cleavage and *in vivo* antimalarial efficacy of 10-(2-hydroxy-1-naphthyl)-deoxoqinghaosu (-deoxoartemisinin). *J. Chem. Soc. Perkin Trans. 1*, **1999**, 1827–1831.

188. Lee, S.; Oh, S. A simple synthesis of C-10 substituted deoxoartemisinin and 9-epi-deoxoartemisinin with various organozinc reagents. *Tetrahedron Lett.*, **2002**, 43: 2891–2894.

189. Li, Y.; Zhu, Y. M.; Jiang, H. J.; Pan, J. P.; Wu, G. S.; Wu, J. M.; Shi, Y. L.; Yang, J. D.; Wu, B. A. Synthesis and antimalarial activity of artemisinin derivatives containing an amino group. *J. Med. Chem.*, **2000**, 43: 1635–1640.

190. Lin, A. J.; Klayman, D. L.; Milhous, W. K. Antimalarial activity of new water-soluble dihydroartemisinin derivatives. *J. Med. Chem.*, **1987**, 30: 2147–2150.

191. Lin, A. J.; Miller, R. E. Antimalarial activity of new dihydroartemisinin derivatives. 6. a-alkylbenzylic ethers. *J. Med. Chem.*, **1995**, 38: 764–770.

192. Jung, M.; Lee, K.; Kendrick, H.; Robinson, B. L.; Croft, L. Synthesis, stability, and antimalarial activity of new hydrolytically stable and water-soluble (+)-deoxoartelinic acid. *J. Med. Chem.*, **2002**, 45: 4940–4944.

193. Li, Y.; Wu, S. X.; Jiang, H. J.; Zhu, M. Y.; Zhu, Y. M. CN Patent ZL 89,109,562.4.

194. Li, Y.; Jiang, H. J.; Pan, J. P.; Cao, B. J.; Chen, Y. X.; Yu, P. L.; Wu, G. S.; Zhang, H. CN Patent ZL 93,112,454.9.

195. Li, Y.; Yang, Z. S.; Zhang, H.; Cao, B. J.; Wang, F. D.; Zhang, Y.; Shi, Y. L.; Yang, J. D.; Wu, B. A. Artemisinin derivatives bearing Mannich base group—synthesis and antimalarial activity. *Bio. Med. Chem.*, **2003**, 11: 4363–4368.

196. Kamchonwongpaisan, S.; Paitayatat, S.; Thebtaranonth, Y.; Wilairat, P.; Yuthavong, Y. Mechanism-based development of new antimalarials: Synthesis of derivatives of artemisinin attached to iron chelators. *J. Med. Chem.*, **1995**, 38: 2311–2316.

197. O'Neill, P. M.; Bishop, L. P.; Storr, R. C.; Hawley, S. R.; Maggs, J. L.; Ward, S. A.; Park, B. K. Mechanism-based design of parasite-targeted artemisinin derivatives: Synthesis

and antimalarial activity of benzylamino and alkylamino ether analogues of artemisinin. *J. Med. Chem.*, **1996**, 39: 4511–4514.

198. Haynes, R. K. Artemisinin and derivatives: The future for malaria treatment? *Curr. Opin. Infect. Dis.*, **2001**, 14: 719–726.

199. Hindley, S.; Ward, S. A.; Storr, R. C.; Searle, N. L.; Bray, P. G. et al. Mechanism-based design of parasite-targeted artemisinin derivatives: Synthesis and antimalarialactivity of new diamine containing analogues. *J. Med. Chem.*, **2002**, 45: 1052–1063.

200. Li, Y.; Lu, D. W.; Zhang, W. B. Synthesis of 11-α hydroxy– and-α chloroartemisinin. *Chin. Chem. Lett.*, **1993**, 4: 99–100.

201. Hufford, C. D.; Khalifa, S. I. Preparation and characterization of new C-11 oxygenated artemisinin derivatives. *J. Nat. Prod.*, **1993**, 56: 62–66.

202. Yagen, B.; Pu, Y. M.; Yeh, H. J. C.; Ziffer, H. Tandem silica gel-catalysed rearrangements and subsequent Baeyer–Villiger reactions of artemisinin derivatives. *J. Chem. Soc. Perkin Trans. 1.*, **1994**, 843–846.

203. Pu, Y. M.; Yagen, B.; Ziffer, H. Stereoselective oxidations of a β-methylglycal, anhydrodihydroartemisinin. *Tetrahedron Lett.*, **1994**, 35: 2129–2132.

204. Lu, D. W.; Li, Y. Synthesis and configurational determination of 11-chloroartemisinin. *Chem. J. Chin. Univ.*, **1995**, 16: 237–238.

205. Li, Y.; Zhang, H. B.; Ye, Y. P. Synthesis of esters of dihydroartemisinin and 11,12-dihydroxyartemisinin. *Chin. J. Med. Chem.*, **1995**, 5: 127–130.

206. Rong, Y.-J.; Wu, Y.-L. Synthesis of C-4-substituted qinghaosu analogues. *J. Chem. Soc. Perkin Trans. 1*, **1993**, 2147–2148.

207. Ye, B.; Zhang, C.; Wu, Y.-L. Synthetic studies on 15-nor-qinhaosu. *Chin. Chem. Lett.* (Eng.), **1993**, 4: 569–572.

208. Jung, M.; Li, X.; Bustos, D. A.; El-Sohly, H. N.; McChesney, J. D.; Milhous, W. K. Synthesis and antimalarial activity of (+)-deoxoartemisinin. *J. Med. Chem.*, **1990**, 33: 1516–1518.

209. Ye, B.; Wu, Y.-L.; Li, G.-F.; Jiao, X.-Q. Antimalarial activity of deoxoqinghaosu. *Acta Pharmaceu. Sin.*, **1991**, 26: 228–230.

210. Avery, M. A.; Mehrotra, S.; Johnson, T. L.; Bonk, J. D.; Vroman, J. A.; Miller, R. Structure-activity relationships of the antimalarial agent artemisinin. 5. Analogs of 10-deoxoartemisinin substituted at C-3 and C-9. *J. Med. Chem.*, **1996**, 39: 4149–4155.

211. Avery, M. A.; Mehrotra, S.; Bonk, J. D.; Vroman, J. A.; Goins, D. K.; Miller, R. Structure-activity relationships of the antimalarial agent artemisinin. 4. Effect of substitution at C-3. *J. Med. Chem.*, **1996**, 39: 2900–2906.

212. Han, J.; Lee, J. G.; Min, S. S.; Park, S. H.; Angerhofer, C. K.; Cordell, G. A.; Kim, S. U. Synthesis of new artemisinin analogs from artemisinic acid modified at C-3 and C-13 and their antimalarial activity. *J. Nat. Prod.*, **2001**, 64: 1201–1205.

213. Liao, X. B.; Han, J. Y.; Li, Y. Michael addition of artemisitene. *Tetrahedron Lett.*, **2001**, 42: 2843–2845.

214. Ma, J. Y.; Weiss, E.; Kyle, D. E.; Ziffer, H. Acid catalyzed michael additions to artemisitene. *Bioorg. Med. Chem. Lett.*, **2000**, 10: 1601–1603.

215. Jung, M.; Lee, K.; Jung, H. First synthesis of (+)-deoxoartemisitene and its novel C-11 derivatives. *Tetrahedron Lett.*, **2001**, 42: 3997–4000.

216. Torok, D. S.; Ziffer, H.; Meshnick, S. R.; Pan, X.-Q.; Ager, A. Syntheses and antimalarial activities of N-substituted 11-azaartemisinins. *J. Med. Chem.*, **1995**, 38: 5045–5050.

217. Mekonnen, B.; Ziffer, H. A new route to N-substituted 11-azazrtemisinins. *Tetrahedron Lett.*, **1997**, 38: 731–734.

218. Avery, M. A.; Fan, P.; Karle, J. M.; Bonk, J. D.; Miller, R.; Goins, D. K. Structure-activity relationships of the antimalarial agent artemisinin. 3. Total synthesis of (+)-13-carbaartemisinin and related tetra- and tricyclic structures. *J. Med. Chem.*, **1996**, 39: 1885–1897.

219. Rong, Y.-J.; Wu, Y.-L. Synthesis of steroidal 1,2,4-trioxane as potential antimalarial agent. *J. Chem. Soc. Perkin Trans.1.*, **1993**, 2149–2150.

220. Opsenica, D.; Pocsfalvi, G.; Juranic, Z.; Tinant, B.; Declercq, J. P.; Kyle, D. E.; Milhous, W. K.; Salaja, B. A. Cholic-acid derivatives as 1,2,4,5-tetraoxane carriers—structure and antimalarial and antiproliferative activity. *J. Med. Chem.*, **2000**, 43: 3274–3282.

221. Todorovic, N. M.; Stefanovic, M.; Tinant, B.; Declercq, J. P.; Makler, M. T.; Solaja, B. A. Steroidal geminal dihydroperoxides and 1,2,4,5-tetraoxanes: Structure determination and their antimalarial activity. *Steroids*, **1996**, 61: 688–696.

222. Solaja, B. A.; Terzic, N.; Pocsfalvi, G.; Gerena, L.; Tinant, B.; Opsenica, D.; Milhous, W. K. Mixed steroidal 1,2,4,5-tetraoxanes—antimalarial and antimycobacterial activity. *J. Med. Chem.*, **2002**, 45: 3331–3336.

223. Chen, Y. X.; Yu, P. L.; Li, Y.; Ji, R. Y. Studies on analogs of qinghaosu III. The synthesis of diacidesters and mono esters of dihydroqinghaosu. *Acta Pharmaceu. Sin.*, **1985**, 20: 105–111.

224. Chen, Y. X.; Yu, P. L.; Li, Y.; Ji, R. Y. Studies on analogs of qinghaosu VII. The synthesis of bis(dihydroqinghaosu) and bis(dihydrodeoxyqinghaosu). *Acta Pharmaceu. Sin.*, **1985**, 20: 470–473.

225. Posner G. H.; Ploypradith, P.; Parker, M. H.; O'Dowd, H. S.; Woo, H.; Northrop, J.; Krasavin, M.; Dolan, P.; Kensler, T. W.; Xie, S. J.; Shapiro, T. A. Antimalarial, antiproliferative and antitumor activities of artemisinin-derived, chemically robust, trioxane dimmers. *J. Med. Chem.*, **1999**, 42: 4275–4280.

226. Posner, G. H.; McRiner, A. J.; Paik, I. H.; Sur, S.; Borstnik, K. Xie, S.; Shapiro, T. A.; Alagbala, A.; Foster, B. Anticancer and antimalarial efficacy and safety of artemisinin-derived trioxane dimers in rodents of artemisinin-derived trioxane dimmers in rodents. *J Med. Chem.*, **2004**, 47: 1299–1301.

227. Jeyadevan, J. P.; Bray, P. G.; Chadwick, J.; Mercer, A. E.; Byrne, A.; Ward, S. A.; Park, B. K.; Williams, D. P.; Cosstick, R.; Davies, J.; Higson, A. P.; Irving, E.; Posner, G. H.; O'Neill, P. M. Antimalarial and antitumour evaluation of novel C-10 non-acetal dimers of 10β-(2-hydroxyethyl)-deoxoartemisinin. *J. Med. Chem.*, **2004**, 47: 1290–1298.

228. Wang, D.-Y.; Wu, Y.-L. to be published.

229. Jung, M.; Lee, S.; Ham, J.; Lee, K.; Kim, H.; Kim, S. K. Antitumor activity of novel deoxoartemisinin monomers, dimmers, and trimer. *J. Med. Chem.*, **2003**, 46: 987–994.

230. McCullough, K. J.; Nojima, M. Recent advances in the chemistry of cyclic peroxides. *Curr. Org. Chem.*, **2001**, 5: 601–636.

231. Dong, Y.-X. Synthesis and antimalarial activity of 1,2,4,5-tetraoxanes. *Mini Rev. Med. Chem.*, **2002**, 2: 113–123.

232. Tang, Y. Q,; Dong, Y. X.; Vennerstrom, J. L. Synthetic peroxides as antimalarials. *Med. Res. Rev.*, **2004**, 24: 425–448.

233. White, N. J.; Olliaro, P. Artemisinin and derivatives in the treatment of uncomplicated malaria. *Med. Trop.*, **1998**, 58(Suppl. 3): 54–56.

234. McIntosh, H. M.; Olliaro, P. Treatment of severe malaria with artemisinin derivatives, a systematic review of randomized controlled trials. *Med. Trop.*, **1998**, 58(Suppl. 3): 61–62.

235. Le, W. J.; Huang, G. F.; You, J. Q.; Xie, R. R.; Mei, J. Y. The experimental studies of artemisinin derivatives against Schistosoma japonicum in animals. *Chin. Pharmaceu. Bull.*, **1980**, 15: 182.

236. Le, W.-J.; You, J.-Q.; Yang, Y.-Q.; Mei, J.-Y.; Guo, H.-F.; Yang, H.-Z.; Zhang, C.-W. Studies on the efficacy of artemether in experimental schistosomiasis. *Acta Pharmaceu. Sin.*, **1982**, 17: 187–193.

237. Wu, L. J.; Xu, P. S.; Xuan, Y. X.; Li, S. W. The experimental studies of early treatment of artesunate against schistosomiasis. *Chin. J. Schistosom. Control*, **1995**, 7(3): 129–133.

238. Xiao, S.-H.; You, J.-Q.; Jiao, P.-Y.; Mei, J.-Y. Effect of early treatment of artemether against Schistosomiasis in mice. *Chin. J. Parasitol. Parasit. Dis.*, **1994**, 12: 7–12.

239. Xiao, S.-H.; You, J.-Q.; Mei, J.-Y.; Jiao, P.-Y.; Guo, H.-F.; Feng, J.-J.; Wang, S.-F.; Xie, R.-R. Early treatment of artemether and praziquantel in rabbits repeatedly infected with *Schistosoma japonicum* cercariae. *Chin. J. Parasitol. Parasit. Dis.*, **1994**, 12: 252–256.

240. Xiao, S. H.; Yang, Y. Q.; Zhang, C. W.; You, J. Q. Microscopic observations on livers of rabbits and dogs infected with *Schistosoma japonicum cercariae* and early treatment with artemether or praziquantel. *Acta Pharmacol. Sin.*, **1996**, 17: 167–170.

241. Wu, L. J.; Li, S. W.; Xian, Y. X.; Xu, P. S.; Liu, Z. D.; Hu, L. S.; Zhou, S. Y.; Qiu, Y. X.; Liu, Y. M. Field studies on preventive effect of 346 patients infected with *Schistosoma japonicum* by administration of artesunate. *Chin. J. Schistosom. Control*, **1995**, 7(6): 323–327.

242. Xiao, S.-H.; Shi, Z.-G.; Zhuo, S.-J.; Wang, C.-Z.; Zhang, Z.-G.; Chu, B.; Zheng, J.; Chen, M.-G. Field studies on preventive effect of artemether against infection with *Schistosoma japonicum*. *Chin. J. Parasitol. Parasit. Dis.*, **1995**, 13: 170–173.

243. Liu, Z.-D.; Hu, L.-S.; Liu, Y.-M.; Hu, G.-H.; Hu, F.; Qiu, Y.-X.; Gao, Z.-L.; Liu, H.-Y.; Li, J.-Y.; Su, L.-H.; Li, S.-W.; Wu, L.-J. Expanded experimental study on the prevention of Schistosomiasis Japonica by oral artesunate. *Chin. J. Parasitol. Dis. Cont.*, **1996**, 9: 37–39.

244. Tian, Z.-Y.; Xiao, S.-H.; Xiao, J.-W.; Zhou, Y.-C.; Liu, D.-S.; Zheng, J.; Chen, M.-G.; Qu, G.-S.; Zhang, X.-Y.; Yao, X.-M.; Zhang, X.-Z.; Zhang, D.-L.; Wang, G. X. Reduction of *Schistosoma Japonicum* infection in an endemic area in islet with erembankment after prophylaxis with oral artemether throughout the transmission season. *Chin. J. Parasitol. Parasit. Dis.*, **1997**, 15: 208–211.

245. Xu, M.-S.; Zhu, C.-G.; Wang, H.; Gao, F.-H.; Wu, Y.-X.; Chui, D.-Y.; Zhang X.-Z.; Ou, N. Study on preventive effect of artesunate against infection due to *Schistosoma japonicum* in an endemic marshland area. *Chin. J. Schistosom. Cont.*, **1997**, 9: 268–271.

246. Dai, Y. H.; Lu, G. Y.; Liu, Z. Y.; Zhang, D. P. The effect analysis of using artesunate on large-scale to prevent Schistosomaisis during fighting a flood in the Yangtze River. *Chin. J. Schistosom. Cont.*, **1999**, 11: 143–145.

247. Luo, R.-G.; Wang, X.-Y.; Li, Z.-H.; Yan, J.-L.; Wang, Y.-N.; Liu, Z.-D.; Hu, L.-S.; Wu, L.-J.; Li, S.-W. Clinical report of artesunate for the prevention of of *Schistosoma japonicum*. *J. Pract. Parasit. Dis.*, **1999**, 5: 48.

248. Xu, M. S.; Zhang, S. Q.; Li, S. W.; Wang, T. P. Field application of oral artesumate for preventing *Schistosoma japonicum* infection. *Chin. J. Parasitol. Parasit. Dis.*, **1999**, 17: 241–243.

249. Xiao, S.-H.; Catto, B. A. In vitro and in vivo studies of the effect of artemether on Schistosoma mansoni. *Antimicrob. Agents Chemother.*, **1989**, 33: 1557–1562.

250. Utzinger, J.; N'Goran, E. K.; N'Dri, A.; Lengeler, C.; Xiao, S. H.; Tanner, M. Oral artemether for prevention of *Schistosoma mansoni* infection: randomized controlled trial. *Lancet*, **2000**, 355(9212): 1320–1325.

251. Utzinger, J.; Xiao, S. H.; Keiser, J.; Chen, M. G.; Zheng, J.; Tanner, M. Current progress in the development and use of artemether for chemoprophylaxis of major human schistosome parasites. *Curr. Med. Chem.*, **2001**, 8: 1841–1860.

252. De Clercq, D.; Vercruysse, J.; Kongs, A.; Verle, P.; Dompnier, J. P.; Faye, P. C. Efficacy of artesunate and praziquantel in Schistosoma haematobium infected schoolchildren. *Acta Trop.*, **2002**, 82: 61–66.

253. Xiao, S. H.; Tanner, M.; N'Goran, E. K.; Utzinger, J.; Chollet, J.; Bergquist, R.; Minggang, C.; Zheng, J. Recent investigations of artemether, a novel agent for the prevention of schistosomiasis japonica, mansoni and haematobia. *Acta Trop.*, **2002**, 82: 175–181.

254. Jung, M.; ElSohly, H. N.; McCHesney, J. D. Artemisinic acid: A versatile Chiral synthon and bioprecursor to natural products. *Planta Medica*, **1990**, 56: 56.

255. Jung, M. Synthesis and cytotoxicity of novel artemisinin analogs. *Bioorg. Med. Chem. Lett.*, **1997**, 7(8): 1091–1094.

256. Woerdenbag, H. J.; Moskai, T. A.; Pras, N.; Malingre, T. M.; El-Feraly, F. S.; Kampinga, H. H.; Konings, A. W. T. Cytotoxicity of artemisinin-related endoperoxides to *Ehrlich ascites* tumor cells. *J. Nat. Prod.*, **1993**, 56(6): 849–856.

257. Beekman, A. C.; Barentsen, A. R. W.; Woerdenbag, H. J.; Uden, W. V.; Pras, N.; Konings, A. W. T.; El-Feraly, F. S.; Galal, A. M.; Wikström, H. Stereochemistry-dependent cytotoxicity of some artemisinin derivatives. *J. Nat. Prod.*, **1997**, 60(4): 325–330.

258. Beekman, A. C.; Wierenga, P. K.; Woerdenbag, H. J.; Uden, W. V.; Pras, N.; Konings, A. W. T.; El-Feraly, F. S.; Galal, A. M.; Wikström, H. Artemisinin-derived sesquiterpene lactones as potential antitumour compounds: Cytotoxic action against bone marrow and tumour cell. *Planta Med.*, **1998**, 64(7): 615.

259. Yang, X. P.; Pan, Q. C.; Liang, Y. G.; Zhang, Y. L. Study on antitumor effect of sodium artesunate. *Cancer (in Chinese)*, **1997**, 16(3): 186.

260. Zhang, X.; Yang, X. P.; Pan, Q. C. Studies on the antitumor effect and apoptosis induction in human liver cancer cell line (BEL-7402) by sodium artesunate. *Chin. Trad. Herb Drug*, **1998**, 29(7): 467–469.

261. Zhang, J. X.; Wang, S. X.; Zhang, F. G.; Zhang, Y. H.; Liu, A. H. Effect of sodium artesunate toward antiproliferative activity of human cancer cell (HeLa, SACC-83) *in vitro*. *Chin. Trad. Herb Drug*, **2001**, 32(4): 345–346.

262. Efferth, T.; Dunstan, H.; Sauerbrey, A.; Miyachi, H.; Chitambar, C. R. The anti-malarial artesunate is also active against cancer. *Int. J. Oncol.*, **2001**, 18(4): 767–773.

263. Lai, H.; Singh, N. P. Selective cancer cell cytotoxicity from exposure to dihydroartemisinin and holotransferrin. *Cancer Lett.*, **1995**, 91: 41–46.

264. Moore, J. C.; Lai, H.; Li, J. R.; Ren, R. L.; McDougall, J. A.; Singh, N. P.; Chou, C. K. Oral administration of dihydroartemisinin and ferrous sulfate retarded implanted fibrosarcoma growth in the rat. *Cancer Lett.*, **1995**, 98: 83–87.

265. Singh, N. P.; Lai, H. Selective toxicity of dihydroartemisinin and holotransferrin toward human breast cancer cells. *Life Sci.*, **2001**, 70(1): 49–56.

266. Li, Y.; Shan, F.; Wu, J. M.; Wu, G. S.; Ding, J.; Xiao, D.; Yang, W. Y.; Atassi, G.; Leonce, S.; Caignard, D. H.; Renard, P. Novel antitumor artemisinin derivatives targeting G1 phase of the cell-cycle. *Bioorg. Med. Chem. Lett.*, **2001**, 11: 5–8.

267. Li, Y.; Wu, J. M.; Shan, F.; Wu, G. S.; Ding, J.; Xiao, D.; Han, J. X.; Atassi, G.; Leonce, S.; Caignard, D. H.; Renard, P. Synthesis and cytotoxicity of dihydroartemisinin ethers containing cyanoaryl methyl group. *Bioorg. Med. Chem.*, **2003**, 11(6): 977–984.

268. Wu, J. M.; Shan, F.; Wu, G. S.; Li, Y.; Ding, J.; Xiao, D.; Han, J. X.; Atassi, G.; Leonce, S.; Caignard, D. H.; Renard, P. Synthesis and cytotoxicity of artemisinin derivatives containing cyanoarylmethyl group. *Euro. J. Med. Chem.*, **2002**, 36: 469–479.

269. Galal, A. M.; Ross, S. A.; ElSohly, M. A.; ElSohly, H. N.; El-Feraly, F. S.; Ahmed, M. S.; McPhail, T. Deoxyartemisinin derivatives from photooxygenation of anhydrodeoxydi-hydroartemisinin and their cytotoxic evaluation. *J. Nat. Prod.*, **2002**, 65: 184–188.

270. Sadava, D.; Phillips, T.; Lin, C.; Kane, S. E. Transferrin overcomes drug resistance to artemisinin in human small-cell lung carainoma cells. *Cancer Lett.*, **2002**, 179(2): 151–156.

271. Lin, P. Y.; Pan, J. Q.; Feng, Z. M.; Zhang, D.; Xiao, L. Y. Comparison of immuno-pharmacologic action of artemisinin and its derivatives. In Zhou, J. H.; Li, X. Y.; Rong, K. T., editors. *Progress in Immuno-Pharmacology*. Beijing: Chinese Science and Technique Press; **1993**, 325–337.

272. Zhuang, G. K.; Zou, M. X.; Xu, X.; Zhu, Y. Clinical study on Qinghao in treatment of discoid lupus erythematosus. *Nat. Med. J. China*, **1982**, 62: 365.

273. Yu, Q. B.; Gao, Y. X. Treatment of 56 lupus erythematosus patients with sodium artesunate. *Chin. J. Dermatol.*, **1997**, 30: 51–52.

274. China Cooperative Research Group on Qinghaosu and its Derivatives as Antimalarials. Antimalarial efficacy and mode of action of qinghaosu and its derivatives in experimental models. *J. Trad. Chin. Med.*, **1982**, 2(1): 17–24.

275. Frederich, M.; Dogne, J. M.; Angenot, L.; De Mol, P. New trends in anti-malarial agents. *Curr. Med. Chem.*, **2002**, 9(15): 1435–1456.

276. Imlay, J. A.; Chin, S. M.; Linn, S. Toxic DNA damage by hydrogen peroxide through the Fenton reduction *in vivo* and *in vitro*. *Science*, **1988**, 240(4852): 640–642.

277. Imlay, J. A.; Linn, S. DNA damage and oxygen radical toxicity. *Science*, **1988**, 240(4857): 1302–1309.

278. Wu, W.-M.; Yao, Z.-J.; Wu, Y.-L.; Jiang, K.; Wang, Y.-F.; Chen, H.-B.; Shan, F.; Li, Y. Ferrous ion induced cleavage of the peroxy bond in qinghaosu and its derivatives and the DNA damage associated with this process. *J. Chem. Soc. Chem. Commun.*, **1996**, 18: 2213–2214.

279. Wu, W.-M.; Wu, Y.-K.; Wu, Y.-L.; Yao, Z.-J.; Zhou, C.-M.; Li, Y.; Shan, F. A unified mechanism framework for the Fe(II)-induced cleavage of qinghaosu and derivatives/analogs. The first spin-trapping evidence for the earlier postulated secondary C-4 radical. *J. Am. Chem. Soc.*, **1998**, 120(14): 3316–3325.

280. Butler, A. R.; Gilbert, B. C.; Hulme, P.; Irvine, L. R.; Renton, L.; Whitwood, A. C. EPR evidence for the involvement of free radicals in the iron-catalysed decomposition of qinghaosu(artemisinin) and some derivatives: Antimalarial action of some polycyclic endoperoxides. *Free Rad. Res.*, **1998**, 28: 471–476.

281. Wu, Y.-K.; Yue, Z.-Y.; Wu, Y.-L. Interaction of qinghaosu (artemisinin) with cysteine sulfhydryl mediated by traces of nonheme iron. *Angew. Chem. Int. Ed.*, **1999**, 38(17): 2580–2582.

282. Wu, W. M.; Chen, Y. L.; Zhai, Z. L.; Xiao, S. H.; Wu, Y. L. Study on the mechanism of action of artemether against schistosomes—The identification of cysteine adducts of both carbon-centered free-radicals derived from artemether. *Bioorg. Med. Chem. Lett.*, **2003**, 13(10): 1645–1647.

283. Posner, G. H.; Park, S. B.; Gonzalez, L.; Wang, D.-S.; Cumming, J. N.; Klinedinst, D.; Shapiro, T. A.; Balci, M. D. Evidence for the importance of high-valent Fe=O and of a diketone in the molecular mechanism of action of antimalarial trioxane analogs of artemisinin. *J. Am. Chem. Soc.*, **1996**, 118(14): 3537–3538.

284. Cumming, J. N.; Posner, G. H. Antimalarial activity of artemisinin (qinghaosu) and related trioxanes: Mechanism(s) of action. *Adv. Pharmacol. (San Diego)*, **1997**, 37: 253–297.

285. Jefford, C. W. Why artemisinin and certain synthetic peroxides are potent antimalarials—implications for the mode of action. *Curr. Med. Chem.*, **2001**, 8(15): 1803–1826.

286. Olliaro, P. L.; Haynes, R. K.; Meunier, B.; Yuthavong, Y. Possible modes of action of the artemisinin-type compounds. *Trends Parasitol.*, **2001**, 17(3): 122–126.

287. Robert, A.; Dechycabaret, O.; Cazelles, J.; Meunier, B. From mechanistic studies on artemisinin derivatives to new modular antimalarial-drugs. *Account Chem. Res.*, **2002**, 35(3): 167–174.

288. Posner, G. H.; Oh, C. H.; Wang, D.; Gerena, L.; Milhous, W. K.; Meshnick, S. R.; Asawamahasadka, W. Mechanism-based design, synthesis, and *in vitro* antimalarial testing of new 4-methylated trioxanes structurally related to artemisinin: The importance of a carbon-centered radical for antimalarial activity. *J. Med. Chem.*, **1994**, 37(9): 1256–1258.

289. Wu, Y.-L.; Chen, H.-B.; Jiang, K.; Li, Y.; Shan, F.; Wang, D.-Y.; Wang, Y.-F.; Wu, W.-M.; Wu, Y.; Yao, Z.-J.; Yue, Z.-Y.; Zhou, C.-M. Interaction of biomolecules with qinghaosu (artemisinin) and its derivatives in the presence of ferrous ion—an exploration of antimalarial mechanism. *Pure Appl. Chem.*, **1999**, 71(6): 1139–1142.

290. Wu, W.-M.; Xu, Y.-Z.; Wu, Y.; Chen, H.-B.; Wu, Y.-L. to be published.

291. Zhao, Y.; Hanton, W. K.; Lee, K.-H. Antimalarial agents, 2. Artesunate, an inhibitor of cytochrome oxidase activity in *Plasmodium berghei*. *J. Nat. Prod.*, **1986**, 49(1): 139–142.

292. Yang, Y.-Z.; Asawamahasadka, W.; Meshnick, S. R. Alkylation of human albumin by the antimalarial artemisinin. *Biochem. Pharmacol.*, **1993**, 46(2): 336–339.

293. Yang, Y.-Z.; Little, B.; Meshnick, S. R. Alkylation of proteins by artemisinin: Effects of heme, PH, and drugs structure. *Biochem. Pharmacol.*, **1994**, 48(3): 569–573.

294. Meshnick, S. R. Artemisinin antimalarials: Mechanism of action and resistance. *Medecine Tropicale*, **1998**, 58(3S): 13–17.

295. Fracis, S. E.; Sullivan, D. J., Jr.; Goldberg, D. E. Hemoglobin metabolism in the malaria parasite *Plasmodium falciparum*. *Annu. Rev. Microbiol.*, **1997**, 51: 97–123.

296. Atamna, H.; Ginsburg, H. The malaria parasite supplies glutathione to its host cell—investigation of glutathione transport and metabolism in human erythrocytes infected with *Plasmodium falciparum*. *Eur. J. Biochem.*, **1997**, 250: 670–679.

297. Ginsburg, H.; Famin, O.; Zhang, J.; Krugliak, M. Inhibition of glutathione-dependent degradation of heme by chloroquine and amodiaquine as a possible basis for their antimalarial mode of action. *Biochem. Pharmacol.*, **1998**, 56(10): 1305–1313.

298. Famin, O.; Krugliak, M.; Ginsburg, H. Kinetics of inhibition of glutathione-mediated degradation of ferriprotoporphyrin IX by antimalarial drugs. *Biochem. Pharmacol.*, **1999**, 58(1): 59–68.

299. Schirmer, R. H.; Muller, J. G.; Krauth-Siegel, R. L. Disulfide-reductase inhibitors as chemotherapeutic agents: the design of drugs for trypanosomiasis and malaria. *Angew. Chem. Int. Ed. Engl.*, **1995**, 34: 141–154.

300. Wu, W.-M.; Chen, Y.-L.; Wu, Y.-L. to be published.

301. Eckstein-Ludwig, U.; Webb, R. J.; van Goethem, I. D. A.; East, J. M.; Lee, A. G.; Kimura, M.; O'Neill, P. M.; Bray, P. G.; Ward, S. A.; Krishna, S. Artemisinins target the SERCA of *Plasmodium falciparum. Nature*, **2003**, 424(6951): 957–961.

302. Robert, A.; Cazelles, J.; Meunier, B. Characterization of the alkylation product of heme by the antimalarial drug artemisinin. *Angew. Chem. Int. Ed.*, **2001**, 40(10): 1954–1957.

303. Robert, A.; Coppel, Y.; Meunier, B. Alkylation of heme by the antimalarial drug artemisinin. *Chem. Commun.*, **2002**, 5: 414–415.

304. Kannan, R.; Sahal, D.; Chauhan, V. S. Heme-artemisinin adducts are crucial mediators of the ability of artemisisnin to inhibit heme polymerization. *Chem. Biol.*, **2002**, 9(3): 321–332.

305. Haynes, R. K.; Monti, D.; Taramelli, D.; Basilico, N.; Parapini, S.; Olliaro, P. Artemisinin antimalarials do not inhibit hemozoin formation. *Antimicrob. Agents Chemother.*, **2003**, 47(3): 1175–1175.

306. Wu, J.-A.; Ji, R.-Y. A quantitative structure–activity study on artemisinin analogs. *Zhongguo Yaoli Xuebao*, **1982**, 3(1): 55–60.

307. Haynes, R. K. Artemisinin and derivatives—the future for malaria treatment. *Curr. Opin. Infect. Dis.*, **2001**, 14(6): 719–726; Haynes, R. K.; Ho, W.-Y.; Tsang, H. W. *Abstract of The Third International Symposium for Chinese Medicinal Chemists, Hong Kong*, Dec. 28–31, **2002**, 13.

308. Haynes, R. K.; Ho, W.-Y.; Chan, H.-W.; Fugmann, B.; Stetter, J.; Croft, S. L.; Vivas, L.; Peters, W.; Robinson, B. L. Highly antimalaria-active artemisinin derivatives: Biological activity does not correlate with chemical reactivivty. *Angew. Chem. Int. Ed.*, **2004**, 43: 1381–1385.

309. Sibmooh, N.; Udomsangpetch, R.; Kijjoa, A.; Chantharaksri, U.; Mankhetkorn, S. Redox reaction of artemisinin with ferrous and ferric ions in aqueous buffer. *Chem. Pharm. Bull. Tokyo*, **2001**, 49(12): 1541–1546.

310. Wu, Y. K. How might qinghaosu (artemisinin) and related-compounds kill the intraerythrocytic malaria parasite—a chemists view. *Account. Chem. Res.*, **2002**, 35(5): 255–259.

# 6

# PROGRESS OF STUDIES ON THE NATURAL CEMBRANOIDS FROM THE SOFT CORAL SPECIES OF *SARCOPHYTON* GENUS

YULIN LI, LIZENG PENG, AND TAO ZHANG

*State Key Laboratory of Applied Organic Chemistry, Institute of Organic Chemistry, Lanzhou University, Lanzhou, P. R. China*

## 6.1 INTRODUCTION

Many cembrane-type diterpenoids have been isolated from terrestrial and especially from marine sources in the past several decades.[1] The cembrane diterpenoids basic skeleton **1** is an isoprenoid 14-membered carbocyclic ring, substituted with an isopropyl residue at C-1 and three symmetrically disposed methyl groups at positions 4, 8, and 12. It turns out that they constitute the most widely distributed of all diterpene families, even though the first representative, cembrene (**2**), was only discovered in 1962 (Figure 6-1).[2,3] Many structural variants are now known from both the plant and the animal kingdoms, and they display a wide range of biological activities. Most known cembranes have come from pine trees and tobacco plants, but many others have come from marine sources. The soft corals have been a particularly rich source of cembrane and cembranolide lactone natural products. In recent years, there has been much interest in the metabolites of marine invertebrates such as soft corals, sponges, and sea cucumbers, and the chemistry and biology of these compounds in general have been reviewed.[4–8] Soft corals form a significant group of marine organisms occurring widely in the coral reefs all over the world. Of those, the corals of the genera *Cespitularia, Clavularia, Gersemia, Lobophytum, Nephthea,*

*Medicinal Chemistry of Bioactive Natural Products* Edited by Xiao-Tian Liang and Wei-Shuo Fang
Copyright © 2006 John Wiley & Sons, Inc.

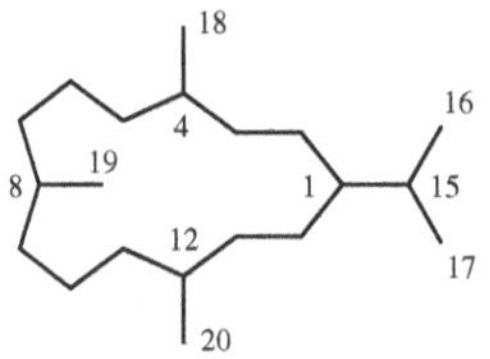

1 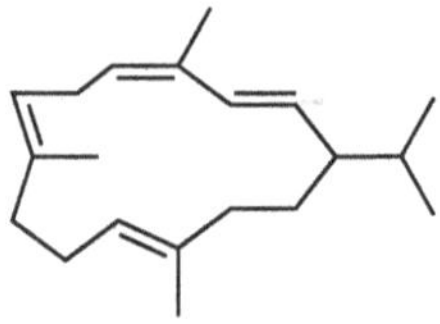                           Cembrene (2)

**Figure 6-1**

*Sarcophyton*, and *Sinularia* are the most prolific. This review mainly pertains to the cembrane-type constituents of the soft corals of the *Sarcophyton* genus.

## 6.2 CEMBRANE-TYPE CONSTITUENTS FROM THE *SARCOPHYTON* GENUS

*Sarcophyton* species contain diterpenes up to 10% of their dry weight, and this large quantity of secondary metabolites plays an important role in the survival of Octocorals with defensive, competitive, reproductive, and possibly pheromonal functions.[5] Soft corals lacking physical defense thus seem to be protected from predation by the presence of diterpene toxins in their tissue. Nearly 25 species of this genus occurring in different seawaters have been examined chemically so far, and more than 80 cembranoid diterpenes have been isolated from the *Sarcophyton* genus since the 1970s.

### 6.2.1   Sarcophytols from the *Sarcophyton* Genus

Up to 20 sarcophytols were isolated from the *Sarcophyton* genus with the name of sarcophytol from A to T (except sarcophytol L, including two sarcophytol T) (Figure 6-2).

Chemical examination of *Sarcophyton glaucum* collected at Ishigaki island, Okinawa Prefecture, resulted in the isolation of seven cembranoid diterpenes, namely sarcophytol A (**3**), sarcophytol A acetate (**4**), sarcophytol B (**5**), sarcophytonin A (**6**), and minor constituents sarcophytol C (**7**), D (**8**), and E (**9**).[9] These compounds were found to be susceptible to autooxidation while being purified. The structural determination of these compounds was made mainly based on proton and carbon nuclear magnetic resonance (NMR) spectral evidence and degradative studies by ozonolysis. X-ray crystallographic analysis for the two crystalline compounds, sarcophytol B (**5**) and D (**8**), has been reported. The total lipid extracts of *S. glaucum* comprise about 40% sarcophytol A (**3**), 5% each of sarcophytol A acetate (**4**) and sarcophytonin A (**6**), about 1% sarcophytol B (**5**), and minor amounts of sarcophytol C (**7**), D (**8**), and E (**9**).

Eight new cembranoids, sarcophytol G (**11**), sarcophytol H (**12**) and its acetate (**13**), sarcophytol I (**14**), sarcophytol J (**15**), sarcophytol M (**17**), sarcophytol N (**19**),

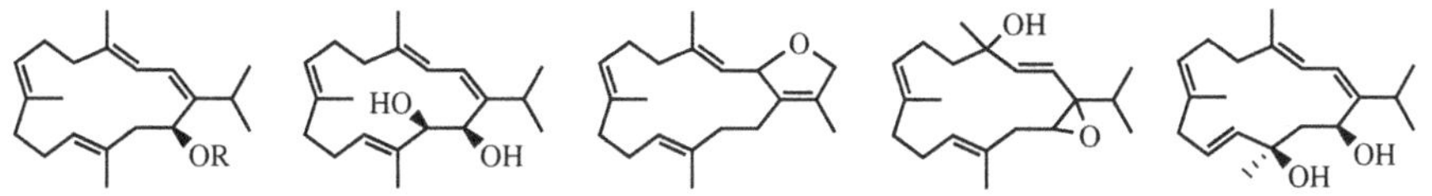

R=H, Sarcophytol-A (**3**)    Sarcophytol-B (**5**)    Sarcophytonin A (**6**)    Sarcophytol-C (**7**)    Sarcophytol-D (**8**)
R=Ac, Sarcophytol-A
    acetate (**4**)

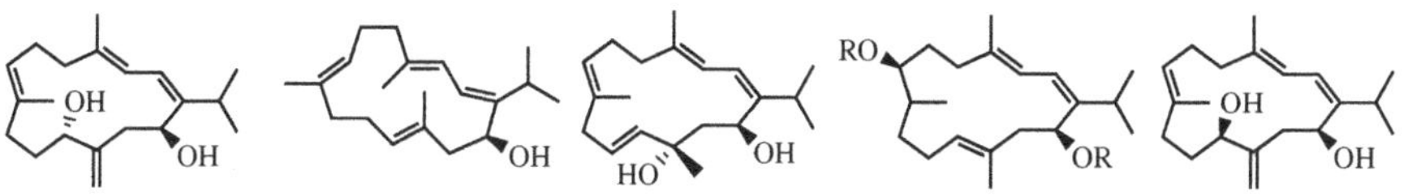

Sarcophytol-E (**9**)    Sarcophytol-F (**10**)    Sarcophytol-G (**11**)    R=H, Sarcophytol-H (**12**)    Sarcophytol-I (**14**)
                                                                            R=Ac, Sarcophytol-H
                                                                                acetate (**13**)

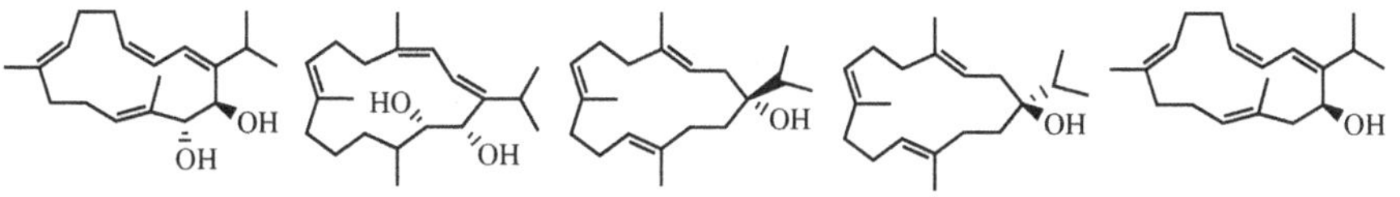

Sarcophytol-J (**15**)    Sarcophytol-K (**16**)    Sarcophytol-M (**17**)    Serratol (**18**)    Sarcophytol-N (**19**)

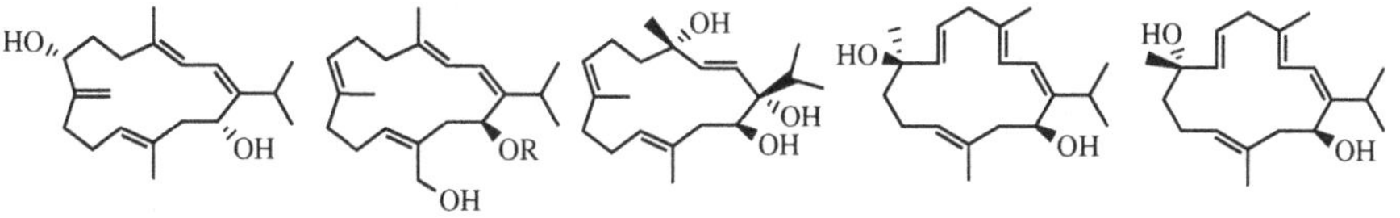

Sarcophytol-O (**20**)    Sarcophytol-P (**21**)    Sarcophytol-Q (**22**)    Sarcophytol-R (**23**)    Sarcophytol-S (**24**)

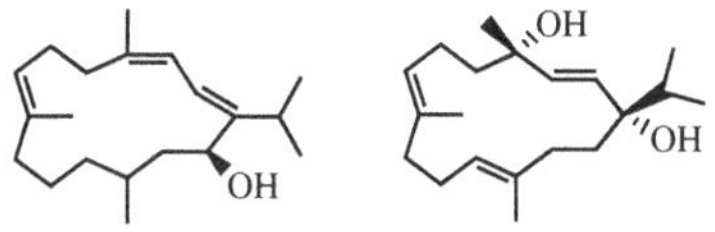

Sarcophytol-T (**25**)    Sarcophytol-T (**26**)
    in 1990    in 1998

**Figure 6-2**

and sarcophytol O (**20**), in addition to two known cembranoids, (−)-nephthenol
(**40**) and sinulariol D (**74**), were reported as being isolated from the soft coral
*S. glaucum*.[10] The structures were determined by spectroscopic data and chemical conversion; sarcophytol M (**17**) was found to be the enantiomer of the known
compound cembrenol, serratol (**18**). Six new cembranoids, sarcophytol F (**10**), K
(**16**), P (**21**), Q (**22**), R (**23**), and S (**24**) together with sarcophytol B (**5**), C (**7**), D
(**8**), E (**9**), G (**11**), H (**12**), I (**14**), J (**15**), M (**17**), N (**19**), and O (**20**) were reported
coming from a Japanese species.[11] Sarcophytol P (**21**) was shown to be the
20-hydroxy derivative of sarcophytol A (**3**); sarcophytol R (**23**) and S (**24**) were cor-
related with sarcophytol A (**3**) by conversion of its 7*R*, 8*R* and 7*S*, 8*S* epoxide

derivatives. Sarcophytol Q (**22**) was shown to be a 1, 4, 14-trihydroxycembranoid, and sarcophytol K (**16**) is a 13, 14-dihydroxy-cembranoid bearing a 1*E*, 3*Z*-diene moiety. Sarcophytol F is the 1*E* isomer of sarcophytol A (**3**).

A new cembranoid derivative sarcophytol T (**25**) in addition to sarcophytol A (**3**), sarcophytol N (**19**), and sarcophytol F (**10**) were reported from the same species.[12] Sarcophytol T (**25**) as well as its isomers sarcophytol N (**19**) and sarcophytol F (**10**) were converted by autooxidation to bicyclo [9.3.0] tetradecane derivatives, when they were kept in CHCl₃ solution at room temperature. Interestingly, König and Wright reported a second sarcophytol T (**26**) isolated from an Australian sample of *S. ehrenbergi* von Marenzeller (Alcyoniidae, Octocorallia) in 1998[13]; its structure was determined by spectroscopic data and confirmed by chemical correlation (Figure 6-2).

## 6.2.2  The Other Cembrane-Type Constituents from the *Sarcophyton* Genus

Chemical examination of *Sarcophytonbirkilandi* collected off Peloris Island, Australia, resulted in the isolation of five cembranoid diterpenes.[14] These diterpenes are cembrene C (**27**), 11,12-epoxycembrene C (**28**), (2*R*, 11*R*, 12*R*)-isosarcophytoxide (**29**), its epimer (2*S*, 11*R*, 12*R*)-isosarcophytoxide (**30**), and (−)-sarcophytoxide (**31**). The structure and absolute stereochemistry of **30** was determined by x-ray analysis. Sarcophytoxide (**31**) showed ichthyotoxic activity to *Gambusia affinis*. Two new cembranoid diterpenes, (−)-sarcophine (**32**) and (−)-sarcophytoxide (**31**), along with three known cembranoids (**33**, **34** and **35**) were reported as being isolated from an Australian species of *Sarcophyton crassocaule*.[15] Compounds **31** and **32**, which are released by *S. crassocaule* into the seawaters, were regarded as allelochemicals (Figure 6-3).[16] The crude extract of *S. crassocaule* was found to immediately arrest the forward mobility of rat cauda epidiglymal sperm without apparent suppression of their trial movement. Sarcophinone (**36**) was isolated from the Chinese species *Sarcophytondecaryi*, and its structure was determined by one-dimensional (1-D) NMR spectral data and confirmed by x-ray analysis.[17] The same species yielded (−)-sarcophine (**32**) along with five other lipid compounds.[18] The same species collected from the Pacific Ocean was reported to yield five cembranoid diterpenes, namely decaryiol (**37**), thunbergol (**38**), trochliophorol (**39**), nephthenol (**40**), and 3, 4-epoxynephthenol (**41**), of which decaryiol (**37**) was found to be a new one.[19] Their structures were determined on the basis of spectral data and by chemical correlations. Sarcophyroxide (**31**) and cembrene C (**27**), two cembranoid diterpenes were also isolated from an Australian species *Sarcophyton ethrenbergi* (Figure 6-3).[20]

Several researchers had chemically examined *S. glaucum* occurring in different seawaters. (+)-sarcophine (**42**), the first new epoxycembranoid, was isolated from this species.[21] It shows toxicity against *Gambusia affinis* and is toxic to mice, rats, and guinea pigs. It also exhibits strong antiacetylcholine action on the isolated guinea pig ileum and is a competitive inhibitor of cholinesterase in vitro.[22] Sarcoglaucol (**43**), a novel cembranoid, was reported from the same species.[23] Its structure and relative stereochemistry was determined by x-ray analysis. The same species

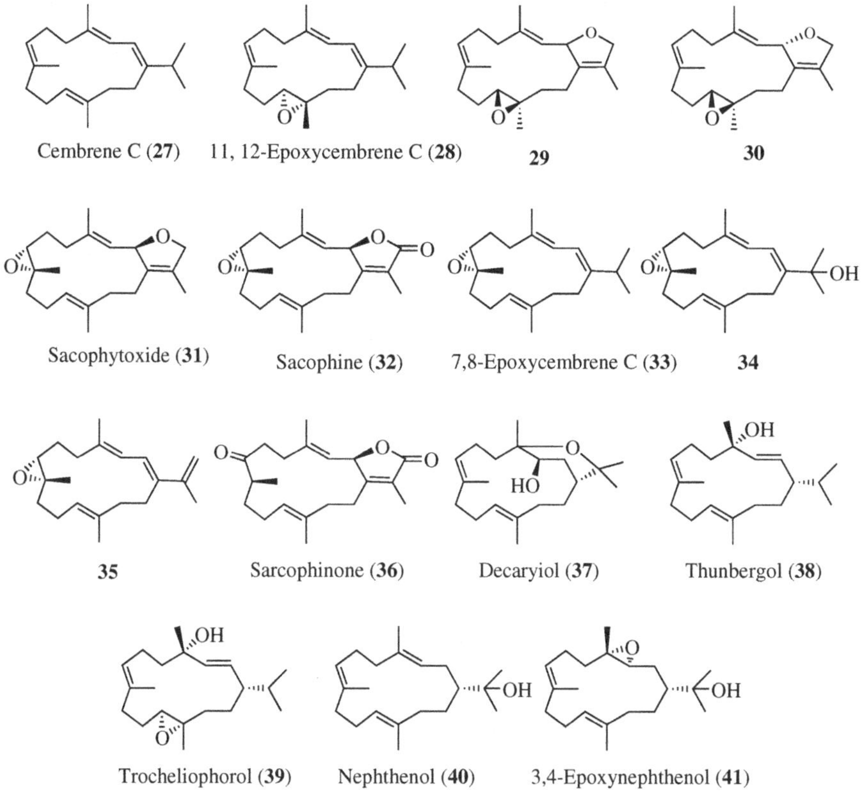

Cembrene C (**27**)    11, 12-Epoxycembrene C (**28**)    **29**    **30**

Sacophytoxide (**31**)    Sacophine (**32**)    7,8-Epoxycembrene C (**33**)    **34**

**35**    Sarcophinone (**36**)    Decaryiol (**37**)    Thunbergol (**38**)

Trocheliophorol (**39**)    Nephthenol (**40**)    3,4-Epoxynephthenol (**41**)

**Figure 6-3**

collected from the Pacific Ocean coast furnished four known cembranoids, Sarco-phytol A (**3**), sarcophytol A acetate (**4**), sarcophytol B (**5**), and sarcophytonin A (**6**).[9] Another Japanese species was reported to yield three cembranoid derivatives: 10-oxocembrene (**44**), 10-hydroxy cembrene (**45**), and 10-methoxycembrene (**46**).[24] Chemical examination of an unidentified species of the *Sarcophyton* genus (Primp museum specimen number FN 3544/000/01, collected at Nara Inlet, Hook Island, Queensland) resulted in the isolation of two cembranoid diterpenes. These diterpenes are methyl (1Z,2S,3E,11Z,13R)-2,16-epoxy-13-hydroxycembra-1(15), 3,7,11-tetraen-20-oate (**47**) and (1Z,2S,3E,11Z,13R)-2,16-epoxy-13-hydroxy-16-oxocembra-1 (**15**), 3, 7, 11-tetraen-20-oate (**48**), of which **48** was the compound responsible for the effect on the central nervous system demonstrated by the crude extract (Figure 6-4).[25]

A new epoxy-bridged cembranoid diterpene **49** possessing antifouling activity was reported from an unidentified Thai soft coral species of the *Sarcophyton* genus, whose structure was assigned by ¹H NMR and ¹³C NMR experiments.[26] Two new cembranoids, **50** and **51**, which showed cytotoxic activity on ehrlichacities tumor

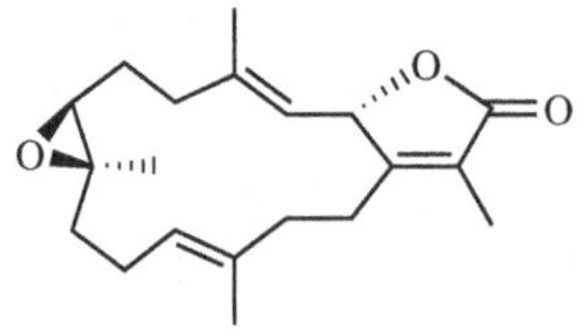

(+)-Sarcophine (42) 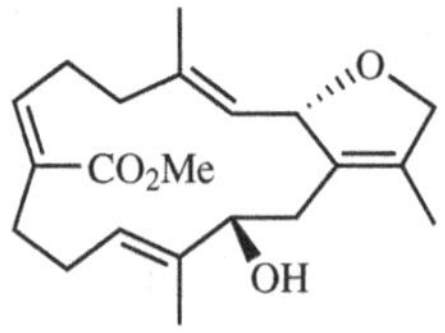          Sarcoglaucol (43) 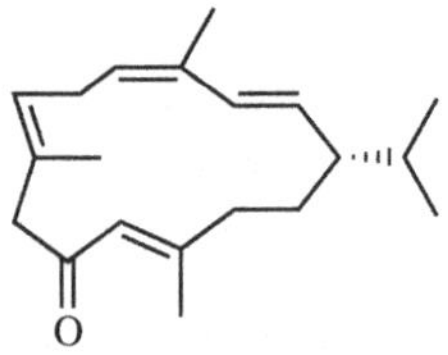          10-Oxocembrene (44)

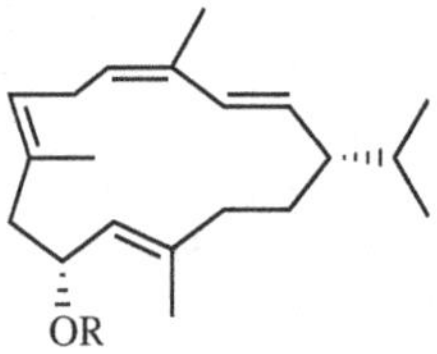

R=H, 10-Hydroxycembrene (45)          47          48
R=Me, 10-Methoxycembrene (46)

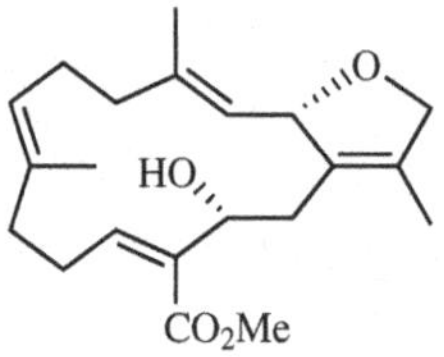

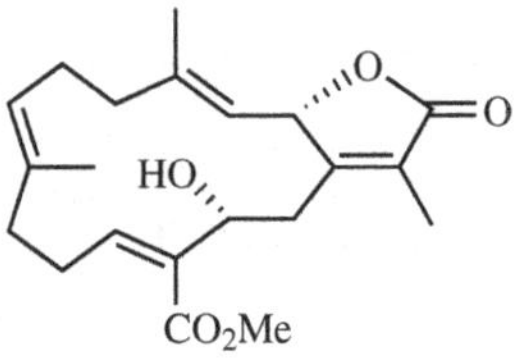

**Figure 6-4**

cells, along with eight known cembrane-type compounds (52–59), were isolated from the hexane extract of the *Sarcophyton trocheliophorum* collected from the Pacific Ocean, and their structures were identified by 1-D NMR spectral data.[27,28] Two cyto-toxic cembranoid diterpenes 30 and 60 were isolated from the Chinese species.[29] A new macrocyclic diterpenoid lactone, (−)-sartrochine (61), was reported from the same species, which showed cytotoxic activity against S180 cells and an antibiotic effect on Streptococus.[30] Its structure was determined by x-ray analysis, and the abso-lute configuration was assigned by CD method. 13-Hydroxysarcophine (62), cembrene C (27), 11, 12-epoxycembrene C (28), nephthenol (40), and (−)-sarcophine (32) were isolated from an Andaman species *S. trocheliophorum*.[31] Two new cembranoid diterpenes, 63 and 64, related to sarcophine were reported from an Australian species. Compound 64 showed convulsant activity (Figure 6-5).[25]

A new triketocembranoid, methylsaroate (65), was reported from a Japanese species.[32] A Okinawan species yielded cembrene C (27), sarcophytonin A (6), and deoxysarcophine (66) along with two lactonic cembranoids, sarcophytonin B (67) and sarcophytonin C (68).[33] (−)-Sarcophytoxide (31) was also isolated from *Sarcophyton pauciplicatum* in 1978.[34] A new bisepoxide 69 along with one known compound 65 were isolated from an Australian species of *Sarcophyton tortuosum*, whose structures were elucidated by 1-D and two-dimensional (2-D) NMR spectral data.[35] Greenland and Bowden reported on the isolation of 70 and 7,8-epoxycembrene C (33).[36] Coll et al. reported on a new cembranoid diterpene 71 from an Australian species.[37] A Red Sea species yielded nephthenol (40). Two new cembranoids, isosarcophytoxide C (68) and diepoxide 72, were isolated from the same species, and their structures were determined on the basis of their spectral data and chemical transformations.[38] A new dihydrofuran cembranoid, sar-cophytonin E (73), was isolated from another species (Figure 6-6).[39]

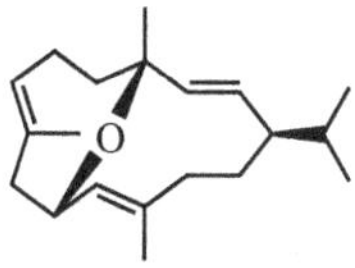

49

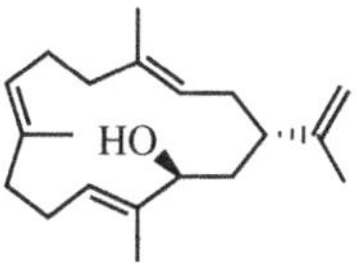

50

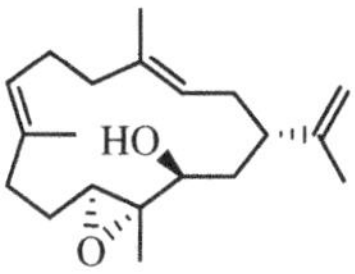

51

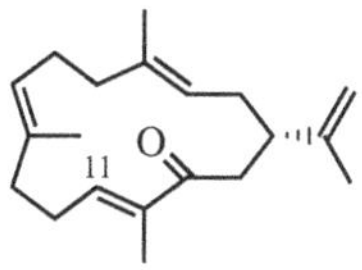

11Z, **52**
11E, **53**

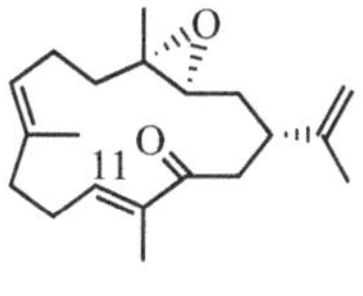

11Z, **54**
11E, **55**

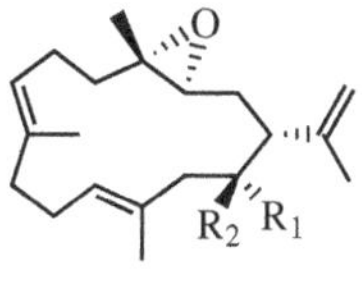

R₁=R₂=H, **56**
R₁=OH, R₂=H, **57**
R₁R₂=O, **58**

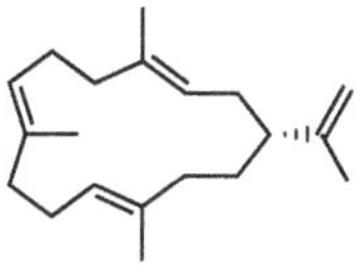

59

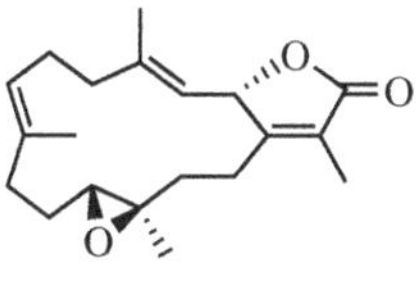

(+)-Isosarcophine (**60**)

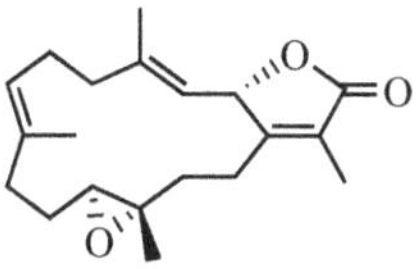

(−)-Sartrochine (**61**)

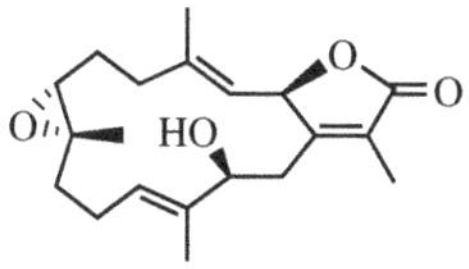

13-Hydroxysarcophine (**62**)

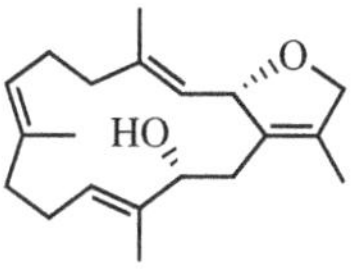

63

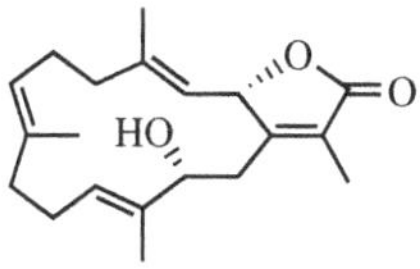

64

**Figure 6-5**

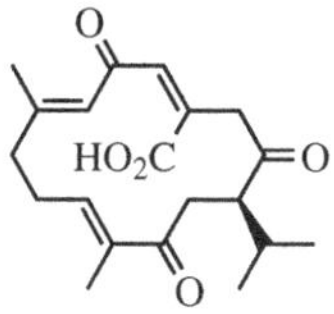

65

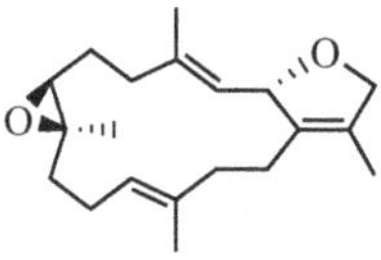

66

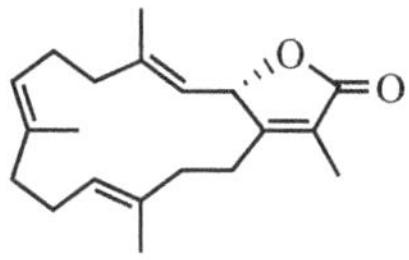

Sarcophytonin B (**67**)

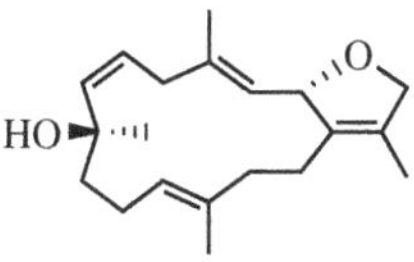

Sarcophytonin C (**68**)

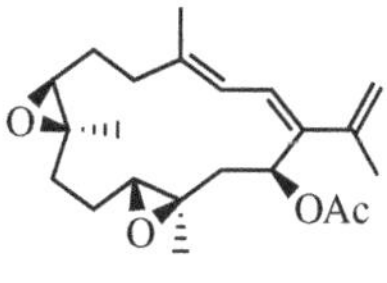

69

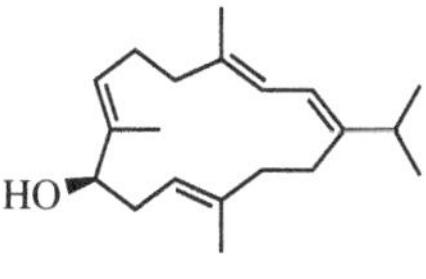

70

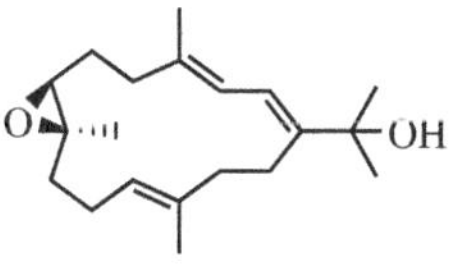

71

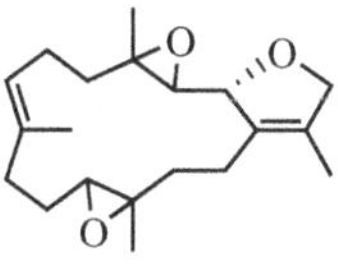

72

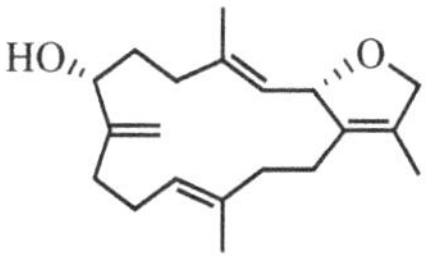

Sarcophytonin E (**73**)

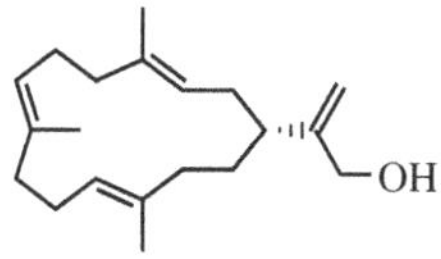

Sinulariol D (**74**)

**Figure 6-6**

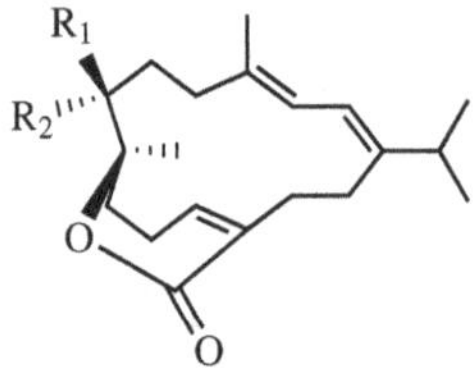

| No. | R₁ | R₂ |
|-----|-----|-----|
| 75 | H | OH |
| 76 | H | AcO |
| 77 | OH | H |
| 78 | =O | |

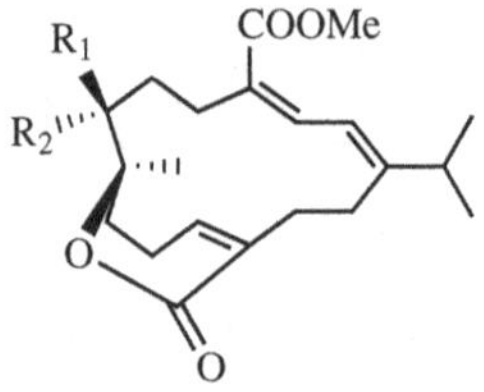

| No. | R₁ | R₂ |
|-----|-----|-----|
| 79 | H | OH |
| 80 | H | AcO |
| 81 | OH | H |
| 82 | =O | |

**Figure 6-7**

An unidentified Japanese species of the same genus yielded several cembranoids (**75–78**), all having seven-membered lactone systems. Their structures were elucidated based on spectral data and chemical correlations.[40] Several other seven-membered cembranoid lactones, (1*E*,3*E*,8*S*,11*Z*)-4,8-dimethyl-1-isopropyl-7-oxo-cyclotetradeca-1,3,11-triene and its 7*R* and 7*S*-hydroxyderivative (**79–82**), were isolated from an Australian species (Figure 6-7).[41]

Two novel irregular cembranoids possessing a 13-membered carbocyclic skeleton, sarcotol (**83**), and sarcotal acetata (**84**) were reported from another unidentified *Sarcophyton* species (Figure 6-8), whose structures were determined by spectroscopic and single-crystal x-ray analyses and chemical conversion.[42,43]

Two new cytotoxic tetraterpenes, methylsarcophytotate (**85**) and methylchlorosarcophytoate (**86**), were isolated from the species of Japanese seawater.[44] Another novel tetraterpenoid, methylisosartortuate (**88**), was reported being isolated from *S. tortuosum*; its structure was determined by spectral data and confirmed by x-ray studies.[45] The same species yielded one more unique tetracyclic tetraterpenoid, methylsartortuate (**89**), and its structure was confirmed by x-ray diffraction.[46] A new biscembranoid, methylneosartotuate (**87**), and a novel bisepoxidecembranoid **66** along with a known compound, methylsarcoate (**65**), were isolated from an Australian species of *S. tortuosum*; their structures were elucidated by 1-D and 2-D NMR spectral data.[35] A unique new tetraterpenoid **90** was also isolated from the Chinese species *S. tortuosum*, and its structure was established by comparison of its spectra with its derivatives (Figure 6-9).[47]

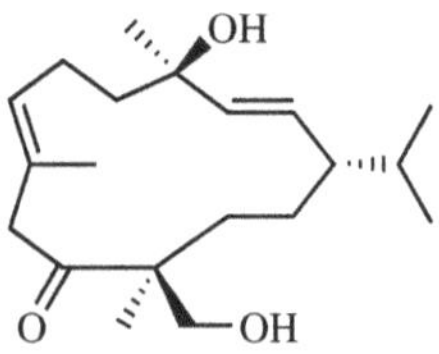

Sarcotol (**83**) 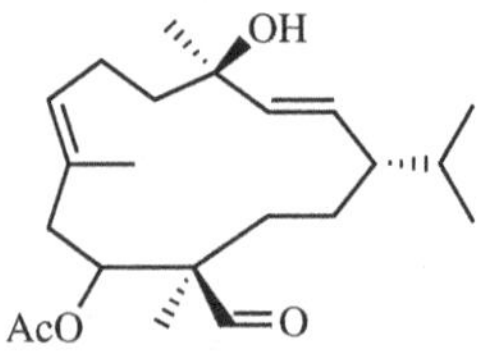          Sarcotal acetate (**84**)

**Figure 6-8**

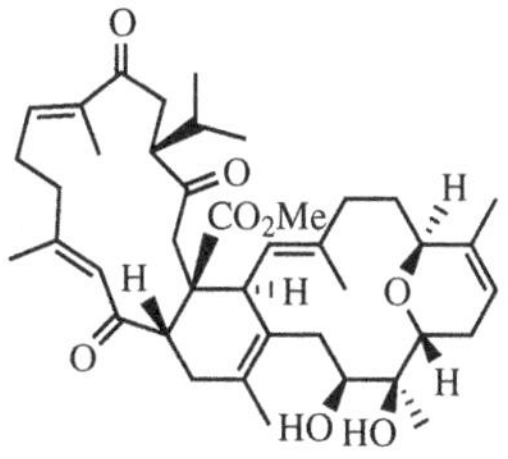

Methylsarcophytotate (**85**) 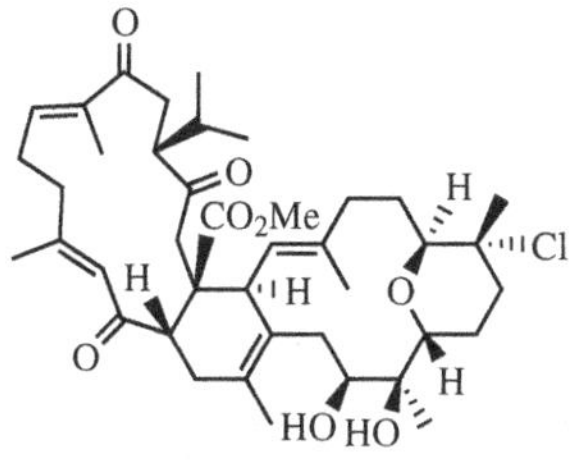     Methylchlorosarcophytotate (**86**) 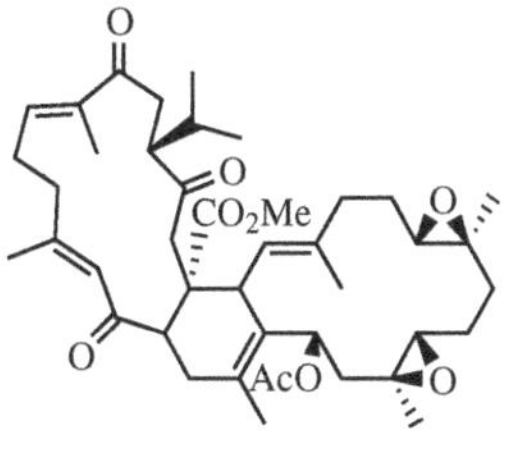     Methylneosartotutate (**87**)

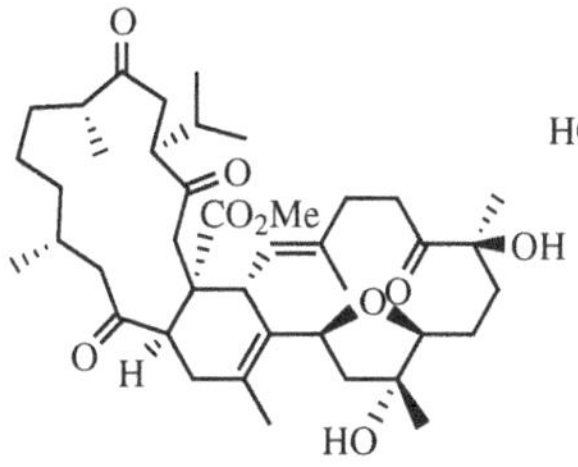

Methylisosartortuate (**88**) 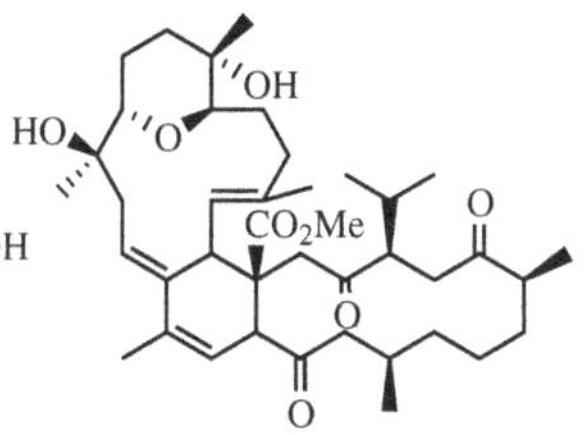          Methylsartortuate (**89**) 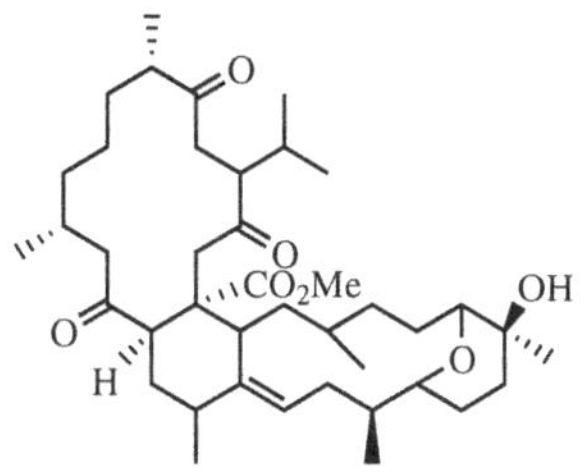               **90**

**Figure 6-9**

## 6.3   PHYSIOLOGICAL ACTION OF SARCOPHYTOL A AND SARCOPHYTOL B

The range of biological activity that has been recorded for 14-membered-ring diterpenoids is remarkably wide: They were found to function as insect trail pheromones, termite allomones, neurotoxins, cytotoxins, and anti-inflammatory and antimitotic agents.

Sarcophytol A (**3**) has been reported to exhibit antitumor activity and potent inhibitory activities against various classes of tumor promoters.[48] Both sarcophytol A (**3**) and sarcophytol B (**5**) have been shown to be active as a potent antitumor promoter in a two-stage mouse skin carcinogenesis model as well as an inhibitor of the hyperplasia of mouse skin; thus, the possibility exists that the compound inhibits chemical carcinogenesis. Furthermore with sarcophytol A (**3**), pancreatic carcinogenesis induced by *N*-nitrosobis(2-oxypropyl) amine is blocked by antipromotion and antiprogression processes in hamsters, and the growth of transplanted human pancreatic cancer cells in nude mice is suppressed. Moreover, it was demonstrated that **3** in diet inhibited spontaneous tumor development in organs such as the mamma, liver, and thymus; it also inhibits chemical carcinogenesis in the colon. In those experiments, sarcophytol A (**3**) did not show any toxic effect on animals. Recently, inhibition of tumor necrosis factor-alpha (TNF-$\alpha$) release from cells was proposed as a mechanism of the anticarcinogenesis of sarcophytol A (**3**).

Although comprehensive structure activity relationship studies have yet to be performed, some indications of the mode of inhibitory activity mediated by sarcophytol A (**3**) have been provided.[49] In particular, inhibition of the okadaic acid-induced pathway of tumor promotion by **3** has been associated with reduced expression of protein phosphatases, which in turn leads to an increase in protein phosphorylation, subsequent cell proliferation, and induction of TNF-α. Protein iso-prenylation was not reduced.[50] Several 12-*O*-tetradecanoylphorbol 13-acetate (TPA)-mediated responses are also inhibited by **3**, such as lens opacity, oxidative stress, and oxidant formation, in addition to decreasing infiltration of neutrophils, the level of myeloperoxidase, thymidine glycol, 8-hydroxyl-2-deoxyguanosine, and 5-hydroxymethyl-2-deoxyuridine.[51] It also alleviated TPA-induced inflammation and infiltration of phagocytes. Thus, a variety of mechanisms seem to contribute to the suppression of tumor promotion by **3**. Although human intervention trials with possible cancer chemopreventive agents are now being conducted in the United States, more studies to find new inhibitory and nontoxic agents are required.

## 6.4   TOTAL SYNTHESIS OF THE NATURAL CEMBRANOIDS

The wide-ranging biological activities exhibited by members of the cembranoid family of marine diterpenes, for example, deoxysarcophine, lophotoxin, sinularin, aseperdiol, and sarcophytol A, in combination with their diversity of structure and the lack of flexible synthetic methodology for the elaboration of 14-ring carbocycles, have made this family of natural products a particularly challenging area for the synthetic chemist. Methods that have been developed for macrocyclization in the cembranoid field include

(1) Ni[0]-Mediated intramolecular coupling of bis-allylic bromides[52]

(2) Intramolecular Wadsworth-Horner-Emmons olefinations[53–56]

(3) Sulphide/sulpho-stabilized carbanion alkylations[57,58]

(4) Cyanohydrin-stabilized carbanion alkylations[59–61]

(5) Intramolecular Friedel-Crafts acylations[62–64]

(6) Additions of allyl organometallics to aldehydes[65,66]

(7) [5, 5]-Sigmatropic ring expansion reactions[67–70]

(8) Ring contraction by stereoselective Wittig rearrangements[71,72]

(9) Intramolecular allylic radical cyclizations[73–76]

(10) Ti[0]-Iinduced intramolecular pinacol coupling and olefination-coupling of dicarbonyl compounds[77]

(11) Intramolecular Pd-catalyzed Stille cross-coupling between alkenyl-stannane and alkenylhalide[78,79]

The synthetic strategy of cembrane-type diterpenes include (1) efficient methods for the construction of a 14-membered carbocyclic ring, (2) functional group assembly, and (3) stereochemistry control as the key reactions. Because the cyclization reaction is often the key step in the total synthesis, each of these strategies will be described in Sections 6.4.1–6.4.4.

### 6.4.1 Total Synthesis of Sarcophytols

#### 6.4.1.1 Total Synthesis of Sarcophytol A

The first total synthesis of sarcophytol A (**3**), which was reported by Takayanagi et al. in 1990,[80,81] was achieved in a highly stereo- and enantioselective manner starting from $E,E$-farnesol (**91**); it included (1) a newly developed Z-selective ($Z : E \geq 35 : 1$) Horner–Emmons reaction with a phosphonate nitrile, (2) modified cyanohydrin macrocyclization, and (3) enantioselective (93% ee) reduction of macrocyclic ketone **100** as its key steps (Scheme 6-1).

Enantiomerically pure sarcophytol A (**3**) was obtained by a single recrystallization. An improved synthetic process appropriate for a large-scale preparation with geraniol as the starting material was also developed. It included a new ketal Claisen rearrangement with 2,2-dimethoxy-2,3-dimethylbatan-2-ol for α-ketol isoprene elongation, which was applied to the sequence for highly stereoselective ($E : Z \geq 99 : 1$) synthesis of trisubstituted γ, δ-unsaturated aldehydes, and acids (Scheme 6-2).[82]

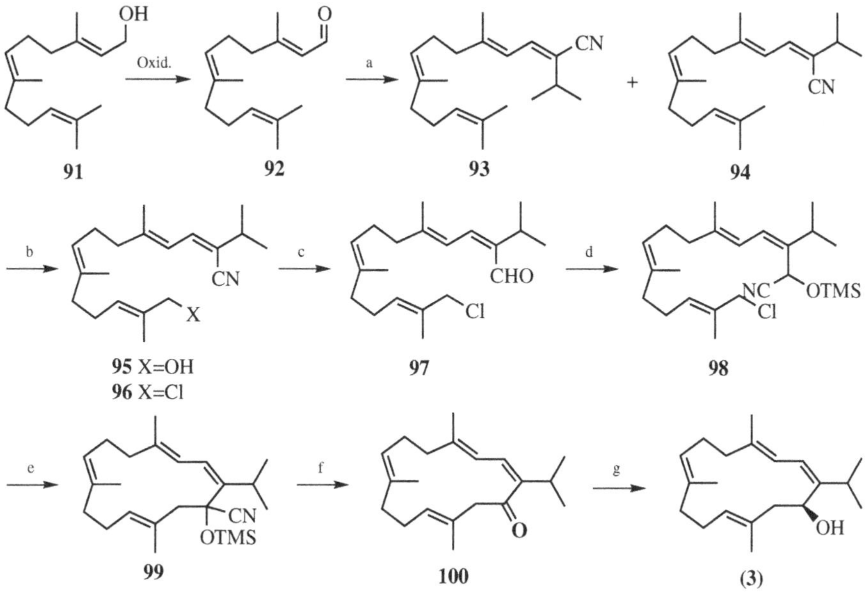

*Reagents and Conditions:* (a) $(EtO)_2POCH(CN)i$-Pr, Tol, $-78\,^\circ$C, 96%, (**94: 93** $\geq$ 35: 1); (b) i. 3 mol% $SeO_2$, 80% $t$-BuOOH, $CH_2Cl_2$, 52%; ii. $CCl_4$, $Ph_3P$, 93%; (c) DIBAL-H, $n$-hexane, $-78\,^\circ$C, then aq.AcOH, 79%; (d) TMSCN, cat. KCN/18-crown-6 ether complex; (e) LiN(TMS)$_2$, THF, $-55\,^\circ$C; (f) $n$-Bu$_4$N$^+$F$^-$ in 10% aqueous THF, 89% from **97**; (g) chirally modified LiAlH$_4$ reagents, 88%.

**Scheme 6-1**

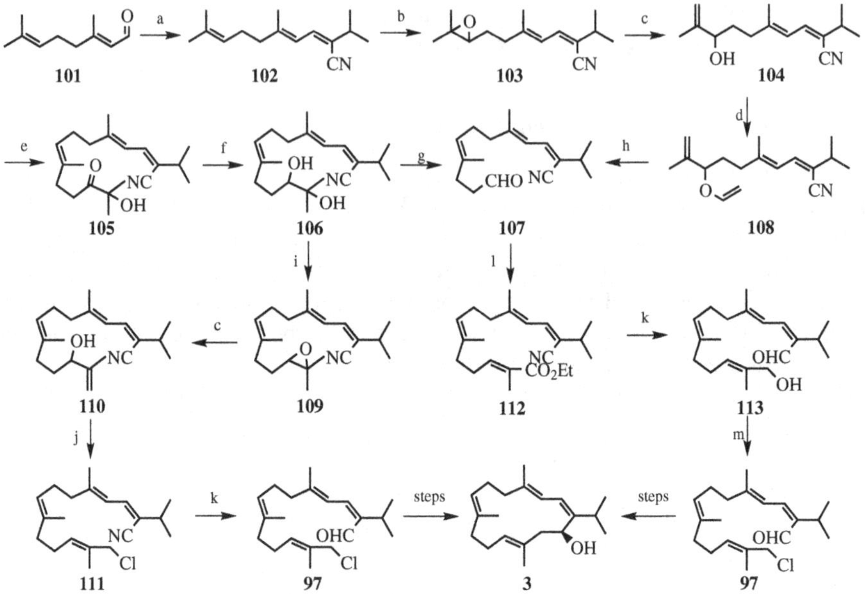

*Reagents and Conditions:* (a) KN(TMS)$_2$, (EtO)$_2$POCH(CN)$i$-Pr, toluene, $-78\,^{\circ}$C; (b) $m$-CPBA, CH$_2$Cl$_2$; (c) Al($i$-OPr)$_3$, toluene; (d) EtOCH=CH$_2$, Hg(AcO)$_2$; (e) 2,2-dimethoxy-2,3-dimethyl-butan-ol, 2 mol% of 2,4-dinitrophenol, 130$\,^{\circ}$C; (f) NaBH$_4$, MeOH; (g) NaIO$_4$, MeOH-H$_2$O; (h) toluene, 110$\,^{\circ}$C; (i) i. MsCl, Py; ii. K$_2$CO$_3$, MeOH; (j) SOCl$_2$, MeOH; (k) DIBAL-H toluene, $-78\,^{\circ}$C; (l) Ph$_3$P=C(CH$_3$)CO$_2$Et, CH$_2$Cl$_2$; (m) MsCl, LiCl, DMF.

**Scheme 6-2**

Macrocyclic ethers were obtained from a product of the ketal Claisen rearrangement via baker's yeast reduction. A [2,3]-Wittig rearrangement of these compounds afforded enantiomers of both sarcophytol A (**3**) and sarcophytol T (**25**) (in 1990) with high enantioselectivity (91–98% ee) in high yields. Surprisingly, almost-perfect reversal of chirality was observed between this geometrical isomer; that is, ($R$)-ether gave ($S$)-alcohol as (2$Z$, 4$E$) isomer, whereas it gave ($R$)-alcohol as (2$E$, 4$Z$)-isomer (Scheme 6-3).

Then Li et al. reported on a concise total synthesis of ($\pm$)-sarcophytol A (**3**) from the derivative of $E,E$-farnesol (**91**) by a low-valent titanium-mediated intramolecular McMurry olefination strategy (Scheme 6-4).[83]

### 6.4.1.2 Total Synthesis of Sarcophytol B

The first total synthesis of ($\pm$)-sarcophytol B (**5**) from $E,E$-farnesal (**92**), which was reported by McMurry et al. in 1989, used a low-temperature titanium-induced pinacol coupling reaction of 1,14-dialdehyde as the key step. They concluded that the natural sarcophytol B has the stereochemistry of a *trans* diol (Scheme 6-5).[84]

Li et al. reported on a concise total synthesis of ($\pm$)-sarcophytol B (**5**) from $E,E$-farnesol (**91**) by a low-valent titanium-mediated intramolecular McMurry

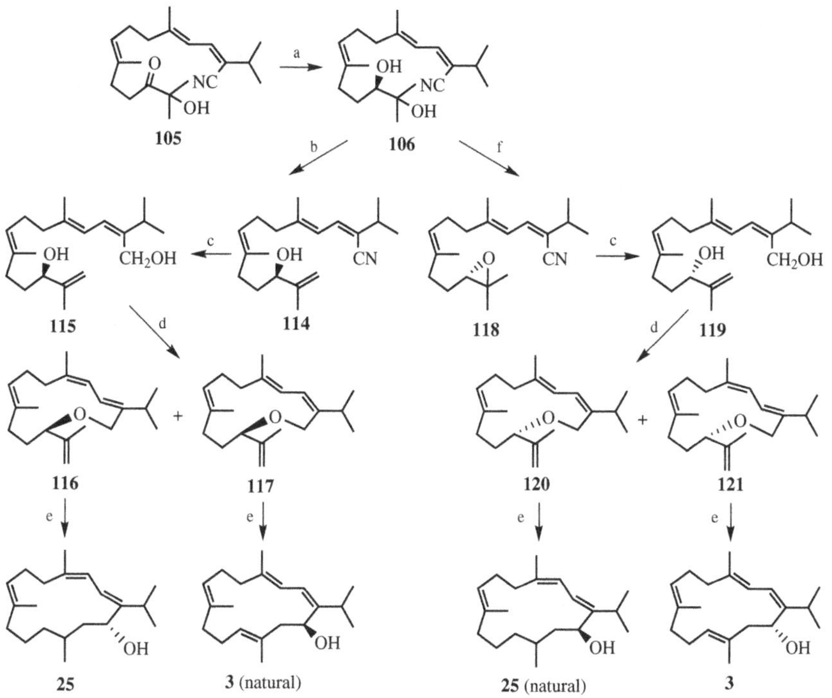

Reagents and Conditions: (a) Baker's yeast; (b) i. Ac$_2$O, Py; ii. POCl$_3$, Py; iii. LiOH, MeOH; (c) i. DIBAL-H; ii. (COOH)$_2$; iii. DIBAL-H; (d) i. Cl$_3$CCN, NaH; ii. p-TsOH, pentane; (e) t-BuLi, THF, $-78\,°C-0\,°C$; (f) i. MsCl, Py; ii. K$_2$CO$_3$.

**Scheme 6-3**

olefination strategy. In this article, they described the epimerzation of the C-14 center of the acetonide α to the keto carboxyl to afford the desired stereochemistry. Thus, the natural sarcophytol B (**5**) proved to be a cis diol (Scheme 6-6).[83]

### 6.4.1.3 Total Synthesis of (±)-Sarcophytol M

The closure of the macrocyclic ring by means of an intramolecular S$_N$2 reaction is a straightforward approach, and sulfur-stabilized carbanion alkylation has been successfully applied in the synthesis of 14-membered cembranoids. The synthesis of nephthenol (**40**) and cembrene A (**59**) is an example of this methodology. In Li and Yue's report, the total synthesis of (±)-sarcophytol M (**17**) was achieved from geraniol (**137**) through 12 steps and in 8.9% overall yield with an intramolecular nucleophilic addition of a sulfur-stabilized carbanion to a ketone as a key step.[85] This example is the first of the closure of a macrocyclic ring with the intramolecular nucleophilic addition of a sulfur-stabilized carbanion to a ketone (Scheme 6-7).

And in 1994, Yue and Li reported on an improved total synthesis of (±)-sarcophytol M (**17**), starting from E,E-farnesol (**91**) through seven steps and

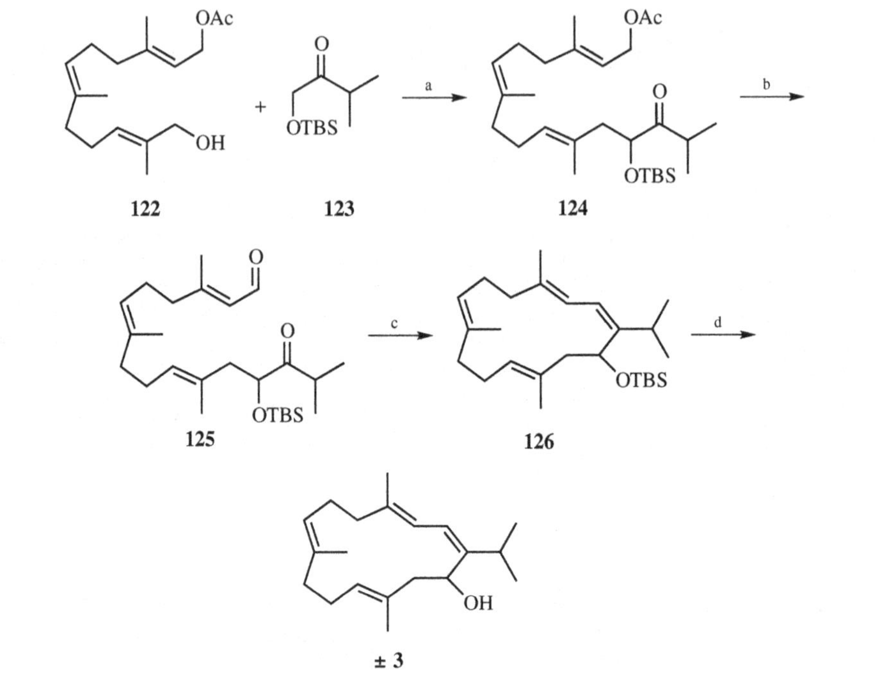

*Reagents and Conditions:* (a) i. **122**, Ph₃P, Imid., I₂, MeCN/Et₂O; ii. **123**, LDA, −78 °C; (b) i. K₂CO₃, MeOH; ii. MnO₂, CH₂Cl₂; (c) TiCl₄, Zn, THF, reflux, 25 h; (d) TBAF, THF.

**Scheme 6-4**

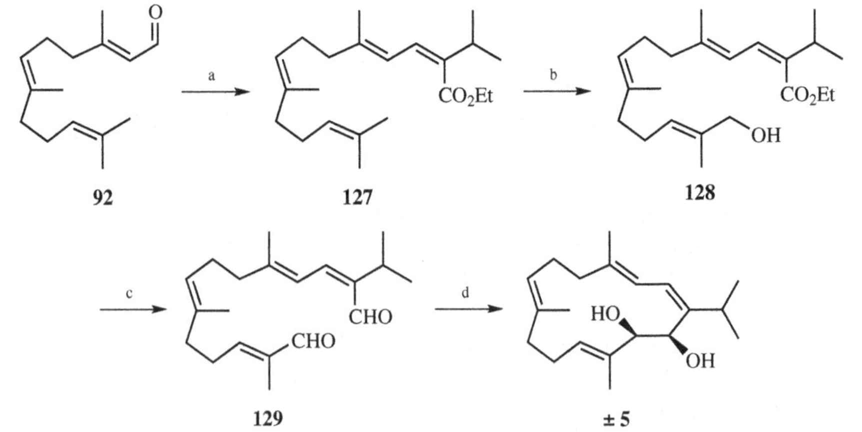

*Reagents and Conditions:* (a) *n*-BuLi, (EtO)₂POC(*i*-Pr)CO₂Et, THF, −78 °C, 60%; (b) SeO₂, *t*-BuOOH, CH₂Cl₂, 45%; (c) LiAlH₄, ether, then BaMnO₄, CH₂Cl₂, 70%; (d) TiCl₃ (DME)₂/Zn-Cu, DME, −40 °C, then H₂O, K₂CO₃, 46%.

**Scheme 6-5**

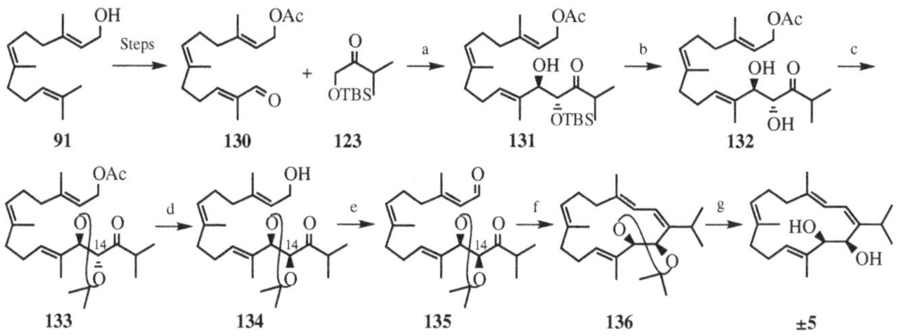

*Reagents and Conditions:* (a) LDA, −78 °C, 1 h, 84%; (b) TBAF, THF, 25 °C, 0.5 h; (c) Me$_2$C(OMe)$_2$, *p*-TsOH, DMF, 0.5 h; (d) K$_2$CO$_3$, MeOH, 2.5 h; (e) MnO$_2$, Et$_2$O, 4 h; (f) TiCl$_4$/Zn, THF, reflux, Py, 20 h, 58%; (g) 1N HCl-MeOH, 30 °C, 0.5 h, 64%.

**Scheme 6-6**

in 13.7% overall yield by means of an intramolecular nucleophilic addition of sulfur-stabilized carbanion to ketone as key steps (Scheme 6-8).[86]

### 6.4.2 Total Synthesis of Cembrene A and C

(*R*)-Cembrene A (**59**), isolated from termite, some plants, and soft coral successively, has been shown to be a highly active scent-trail pheromone.[87–90] Cembrene

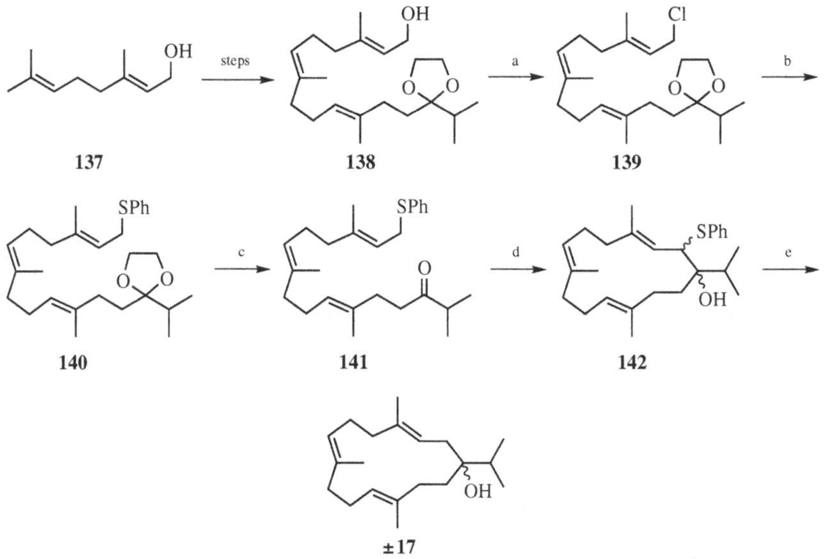

*Reagents and Conditions:* (a) CCl$_4$, PPh$_3$, reflux, 78%; (b) PhSNa, MeOH, 80%; (c) *p*-TsOH, acetone, 98%; (d) LDA, THF, −78 °C, 58%; (e) Li-Et-NH$_2$, −78 °C, 78%.

**Scheme 6-7**

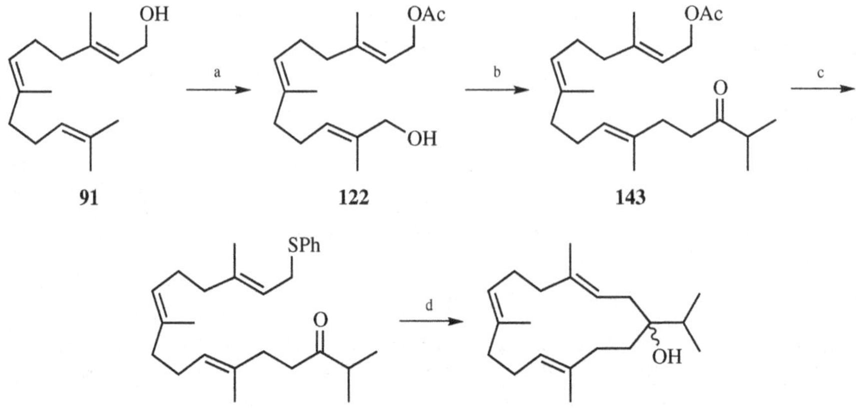

Reagents and Conditions: (a) i. Ac₂O, Py, DMAP, rt, 2 h, 98%; ii. SeO₂, t-BuOOH, salicyclic acid, rt, 64%; (b) methyl isopropyl ketone, THF, LDA, −78 °C, then iodide of **122**, 57%; (c) i. K₂CO₃, MeOH, rt, 30 min, 95%; ii. n-BuLi, TsCl, then PhSLi, 89%; (d) i. LDA, Dabco, THF, −78 °C, 58%; ii. Li-Et-NH₂, −78 °C, 78%.

**Scheme 6-8**

C (**27**) was first isolated as a component of oleoresin of *Pinus koraiensis* by Raldugin in 1971, but its structure was not confirmed.[90] In 1978, Vanderah et al. isolated it from the soft coral *Nephthea* spp. and established its structure as (E,E,E,E)-1,7,11-trimethyl-4-isopropyl-1,3,7,11-cyclotetradecatetraene. It was also found in another soft coral *Sarcophyton ehrenbergi* afterward. Cembrene C (**27**) was synthesized in a trace amount by isomerization of naturally occurring cembrene A (**59**) (Scheme 6-9).

In view of their biological activity and challenging structural features, some total syntheses of **59** and **27** have been reported.

### 6.4.2.1 Total Synthesis of Cembrene A

The closure of the macrocyclic ring by means of an intramolecular S$_N$2 reaction is a straightforward approach. For example, an intramolecular nucleophilic addition of sulfur-stabilized carbanion to epoxide was used in the synthesis of nephthenol (**40**) and cembrene A (**59**) from *trans,trans*-geranyllianlool (**144**) (Scheme 6-10).[91]

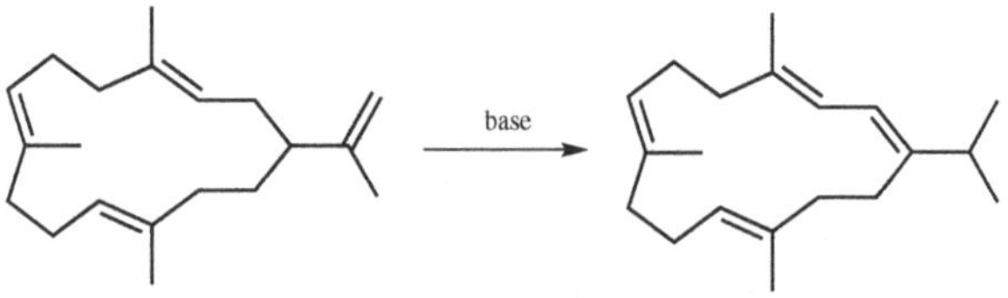

Cembrene A (**59**)          Cembrene C (**27**)

**Scheme 6-9**

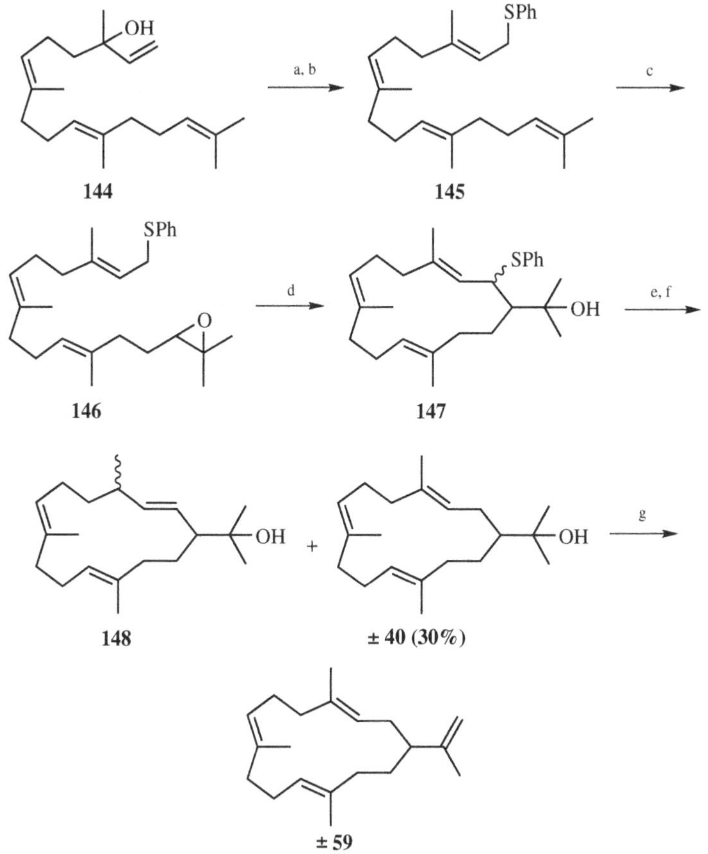

Reagents and Conditions: (a) PBr$_3$; (b) NaSPh; (c) NBS, aqueous THF, K$_2$CO$_3$, MeOH; (d) excess
n-BuLi, THF, DBU, −78 °C; (e) Li, EtNH$_2$; (f) preparative TLC, AgNO$_3$; (g) SOCl$_2$, Py.

**Scheme 6-10**

An alternative strategy for preparing cembrene A (59) was described. The key
step here is the regiospecific coupling of two functionalized geranyl units. The
advantage of this convergent approach is that it allows regioselective introduction
of functionalities before cyclization (Scheme 6-11).[92]

A classic method for preparing macrocyclic rings is ring expansion. Wender
et al. have used this technique to synthesize (−)-3Z-cembrene A (159) (Scheme
6-12).[67–70]

The first enantioselective total synthesis of R-(−)-cembrene A (59) and R-(−)-
nephthenol (40) were achieved by employing an intramolecular nucleophilic addition of sulfur-stabilized carbanion to asymmetric epoxide as the key step, starting
from L-serine (Scheme 6-13).[93]

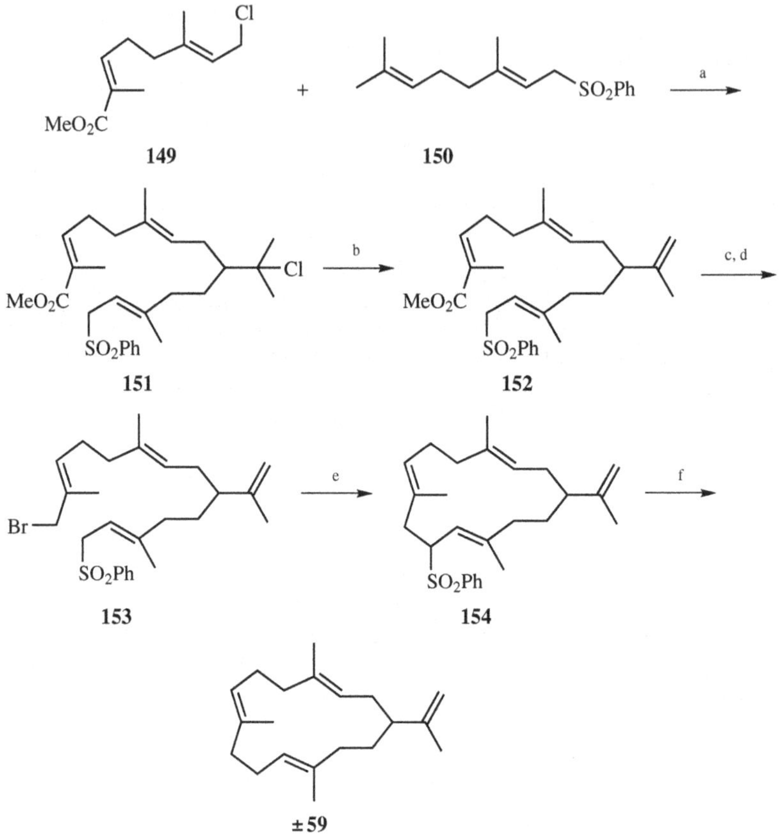

*Reagents and Conditions:* (a) SnCl$_4$, CH$_2$Cl$_2$, −94 °C, 3 h; (b) LiBr, Li$_2$CO$_3$, DMF, 100 °C; (c) LAH, −20 °C; (d) PPh$_3$, CBr$_4$, CH$_3$CN, room temperature; (e) LDA, THF, −78 °C; (f) Li, EtNH$_2$, −78 °C.

**Scheme 6-11**


(*S*)-cembrene A (**59**) has been synthesized by a new enantioselective intramolecular cyclization reaction of (*E, E, E*)-geranylgeranyl ether with an (*R*)-1,1′-binaphthyl-2-benzoxy-2′-oxy auxiliary as a chiral leaving group in the presence of tin(IV) chloride by Ishihara et al. in 2000. The desired product was obtained in 10% yield and 32% ee (Scheme 6-14).[94]

Yue et al. reported on the total synthesis of *R*-(−)-cembrene A by using the modified Wittig reaction and titanium-induced intramolecular carbonyl coupling as key steps; the overall yield is 29% starting from *E*-geranylacetone (**164**) and *R*-(+)-limonene (**168**) (Scheme 6-15).[95]

An alternative strategy to prepare (±)-cembrene A (**59**) by this group is described below. The key step here is a titanium-induced intramolecular carbonyl coupling reaction starting from *E*-geranylacetone and geraniol (**137**) (Scheme 6-16).[96]

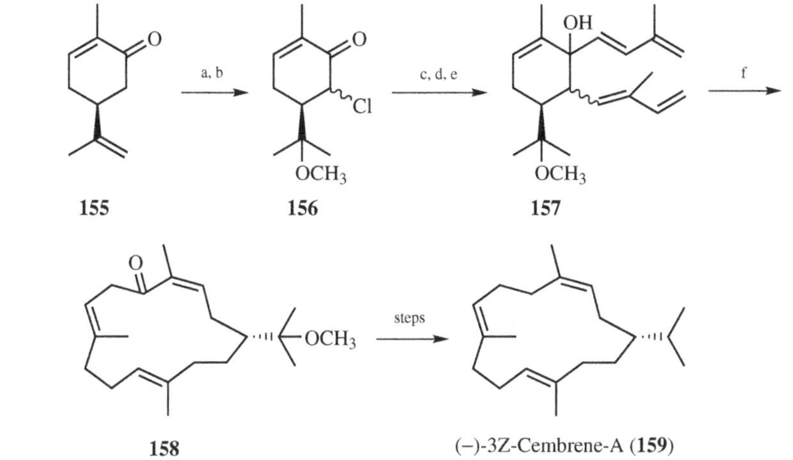

155    156    157

158    (−)-3Z-Cembrene-A (159)

*Reagents and Conditions:* (a) LDA; (b) F₃CSO₂Cl; (c) (*E*)-1-lithio-2-methyl-1, 3-butadiene; (d) lithium isopropenyl acetylide; (e) LAH; (f) KH, 18-crown-6, THF, rt.

**Scheme 6-12**

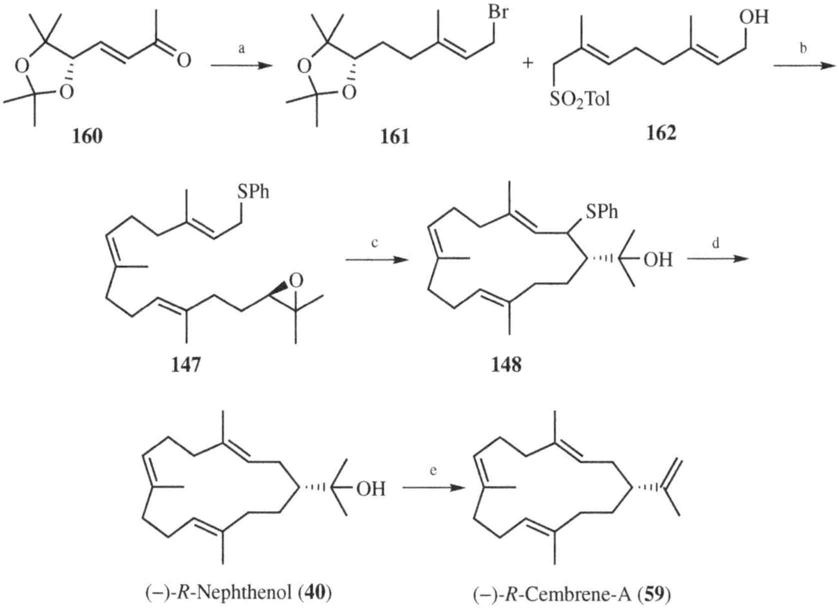

160    161    162

147    148

(−)-*R*-Nephthenol (40)    (−)-*R*-Cembrene-A (59)

*Reagents and Conditions:* (a) i. H₂, Pd/C; ii. (EtO)₂POCH₂CO₂Bu-*t*; iii. DIBAL-H; iv. HPLC; v. PBr₃; (b) i. *n*-BuLi; ii. Na(Hg); iii. PBr₃, PhSNa; iv. MsCl, K₂CO₃; (c) *n*-BuLi; (d) Na; (e) SOCl₂, Py.

**Scheme 6-13**

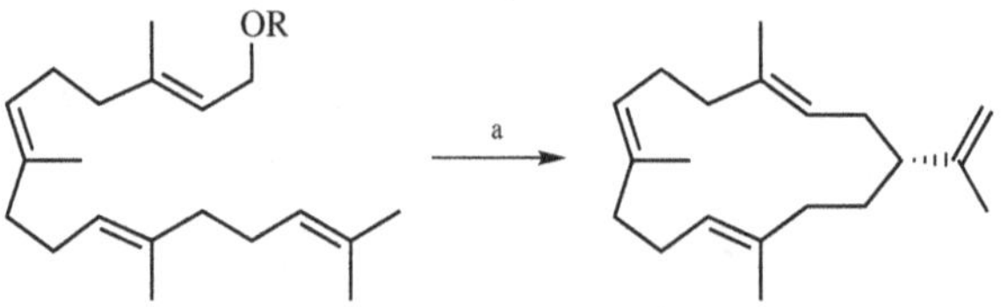

163                          (−)-R-Cembrene-A (59)

*Reagents and Conditions:* (a) SnCl$_4$, CH$_2$Cl$_2$, −78 °C, 60 h. R*=(R)-BINOL.

**Scheme 6-14**

### 6.4.2.2  Total Synthesis of Cembrene C

Li et al. reported on the total synthesis of cembrene C from geraniol (**137**) or *trans, trans*-farnesol (**91**) by using the titantium-induced intramolecular coupling of the dicarbonyl precursor keto enal **176** as the key macrocyclization step (Scheme 6-17).

From *E,E*-farnesol, cembrene C (**27**) was prepared via five steps in 21% overall yield by tiantium-induced macrocyclization (Scheme 6-18).[97]

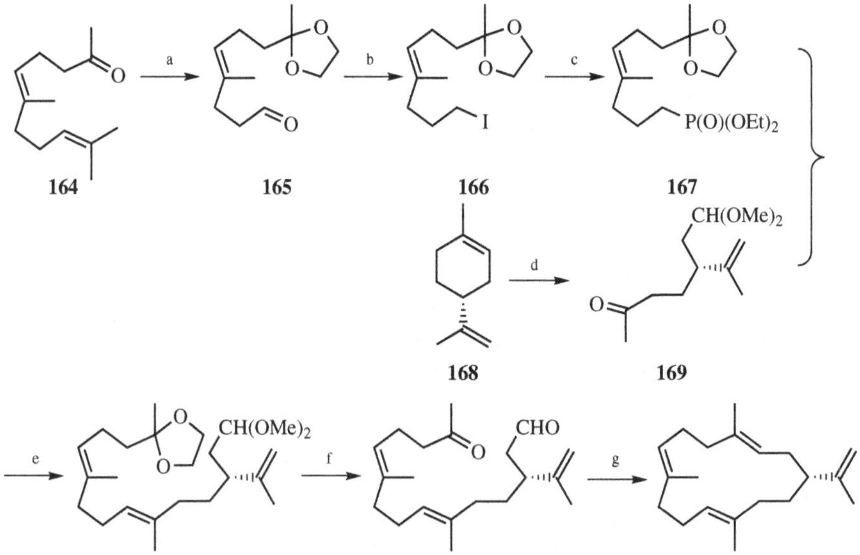

164          165          166          167

168          169

170                          171                          (−)-R-Cembrene-A (59)

*Reagents and Conditions:* (a) 1) ethylene glycol, *p*-TsOH, C$_6$H$_6$, reflux, 100%; 2) O$_3$, CH$_2$Cl$_2$, −78 °C, then rt, 78%; (b) 1) NaBH$_4$, MeOH, 0 °C, 8%; 2) Ph$_3$P, imidazole, I$_2$, 0 °C, 92%; (c) (EtO)$_3$P, reflux, 78%; (d) O$_3$, MeOH, −78 °C, then, Me$_2$S, *p*-TsOH, −78 °C ∼ rt, 73%; (e) *n*-BuLi, **167**, −78 °C, 0.5 h, then **169**, THF, reflux overnight, 58%; (f) i) *p*-TsOH, acetone, felux, 2 h; ii) CF$_3$CO$_2$H, CHCl$_3$, 0 °C, 96%; (g) TiCl$_4$, Zn, DME, reflux, 81%.

**Scheme 6-15**

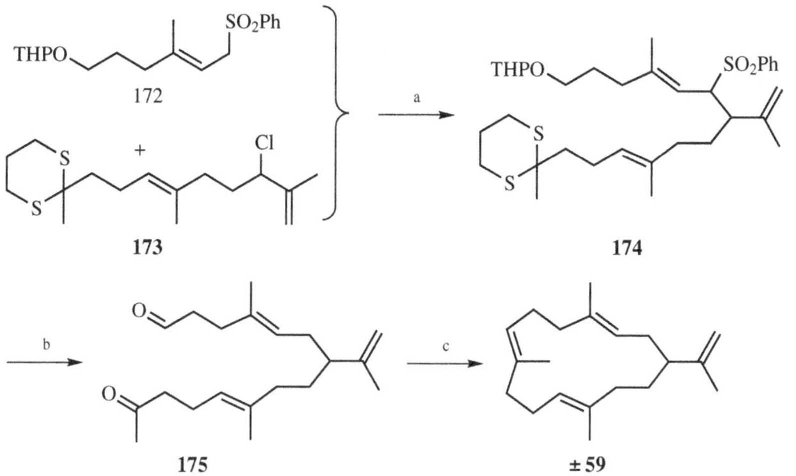

*Reagents and Conditions:* (a) 50% NaOH/TBAB, rt, 20 h; (b) i. PPTs/MeOH, ii. Na(Hg), Na$_2$HPO$_4$, MeOH; iii. HgCl$_2$, CaCO$_3$; iv. PCC, CH$_2$Cl$_2$; (c) TiCl$_3$/AlCl$_3$, Zn-Cu, DME, 36 h.

**Scheme 6-16**

An alternative strategy for preparing cembrene C (**27**) by this group is described below. The dicarbonyl precursor keto enal **176** was prepared from geraniol (Scheme 6-19).[98]

### 6.4.3 Total Synthesis of Several Natural Epoxy Cembrenoids

Naturally occurring cembranoid epoxides (cembranoxides) (Figure 6-10) have been found as chemical components of various tropical marine soft corals and represent a

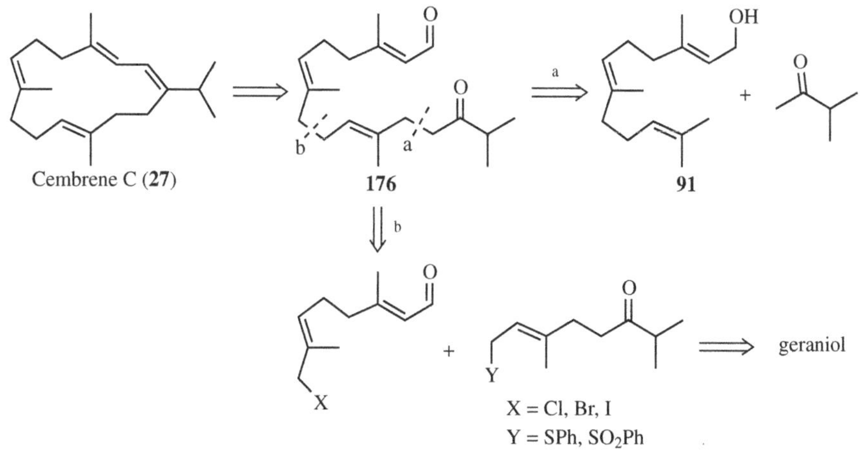

**Scheme 6-17**

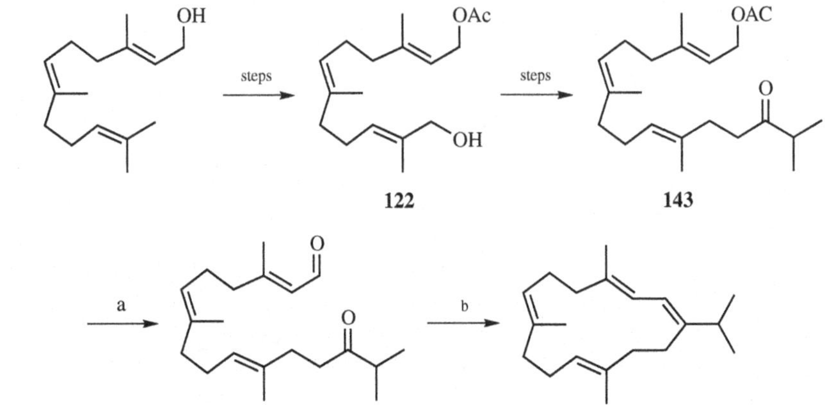

**122**  **143**

**176**  **27**

*Reagents and Conditions:* (a) i. $K_2CO_3$, MeOH; ii. PCC, $CH_2Cl_2$; (b) $TiCl_4/Zn$, THF, reflux, 25 h.

**Scheme 6-18**

class of oxidative metabolites of cembrene diterpenes. (+)-11,12-epoxycembrene C (**28**), a novel cembrane epoxide, was first isolated in 1978 by Bowden et al. from the Australian soft coral *Sinularia grayi*[99] and was subsequently found in various marine soft coral, that is, *Nephthea* spp.,[100] *Lobophytum* spp.,[101] *Eunicea* spp.,[102] *Sinularia* spp.,[103] and *Sarcophyton* spp. Its structure was determined by means of extensive spectroscopic techniques and chemical degradation. (−)-7, 8-epoxycembrene C (**33**) was first isolated in 1980 by Bowden et al. from the soft coral

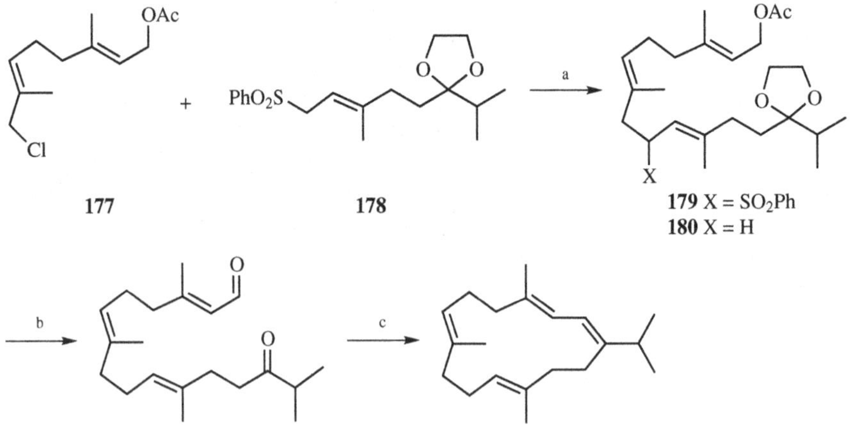

**177**  **178**  **179** X = SO$_2$Ph
**180** X = H

**176**  **27**

*Reagents and Conditions:* (a) n-BuLi, THF, −40 °C; (b) i. Na(Hg), Na$_2$HPO$_4$, MeOH; ii. K$_2$CO$_3$, MeOH; iii. p-TsOH, MeOH; iv. PCC, CH$_2$Cl$_2$; (c) TiCl$_4$/Zn, THF, reflux, 25 h.

**Scheme 6-19**

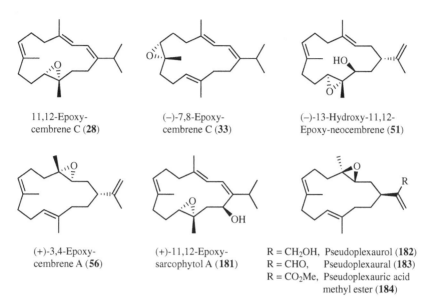

11,12-Epoxy-
cembrene C (**28**)

(–)-7,8-Epoxy-
cembrene C (**33**)

(–)-13-Hydroxy-11,12-
Epoxy-neocembrene (**51**)

(+)-3,4-Epoxy-
cembrene A (**56**)

(+)-11,12-Epoxy-
sarcophytol A (**181**)

R = CH₂OH,  Pseudoplexaurol (**182**)
R = CHO,     Pseudoplexaural (**183**)
R = CO₂Me,  Pseudoplexauric acid
                    methyl ester (**184**)

**Figure 6-10**

*Sarcophyton crassocaule* and then isolated from the soft coral *Eunicea* spp.[15] in 1993 by Shin and Fenical. Although its chemical structure was characterized spectroscopically,[27] the absolute configuration of the epoxide moiety remains undetermined so far. (−)-13-Hydroxy-11, 12-epoxyneocembrene (**51**), isolated in 1988 by Suleimenova et al. from the soft coral *Sarcophyton trocheliophorum*, has been shown to be an effective inductor of the release of labeled glucose from the lecithincholesterol liposomes and to exhibit cytostatic activities.[104] It was characterized spectroscopically. The absolute configuration of C-1 and C-13 has been defined to be (1*S*,13*S*) by chemical transformation, but the absolute configuration of the epoxide of **51** remains undetermined. (+)-3,4-Epoxycembrene A (**56**), a naturally occurring epoxy derivative of cembrane diterpene *R*-(−)-cembrene A (**59**), was first isolated in 1981 by Bowden et al. from the Australian soft coral *Sinularia facile*,[105] and it was subsequently found in various marine soft coral, that is, *Nephthea brassica* (Formosan),[28] and *Sarcophyton* spp. The absolute configuration of **56** was determined as (1*R*,3*S*,4*S*) by means of extensive spectroscopic techniques and chemical degradation. And **56** has been shown to be a potent cytotoxic contituent against A549, HT29, KB, and P388 cell lines by Duh et al.[106,107] (+)-11,12-Epoxysarcophytol A (**181**), an epoxy cembrane diterpene, was first isolated by Bowden et al. in 1983 from an Australian marine soft coral *Lobophytum* sp.[100] and characterized spectroscopically and chemically as (1*Z*,3*E*,7*E*)-14-hydroxyl-11,12-epoxy cembra-1,3,7-triene. The configuration of 14-hydroxyl was confirmed as *S* by a zinc-copper couple-mediated reductive elimination of the epoxide moiety leading to the formation of a known cembrane diterpenoid sarcophytol A (**3**), a possible biosynthesis precursor of **181**. However, the *absolute configuration* of

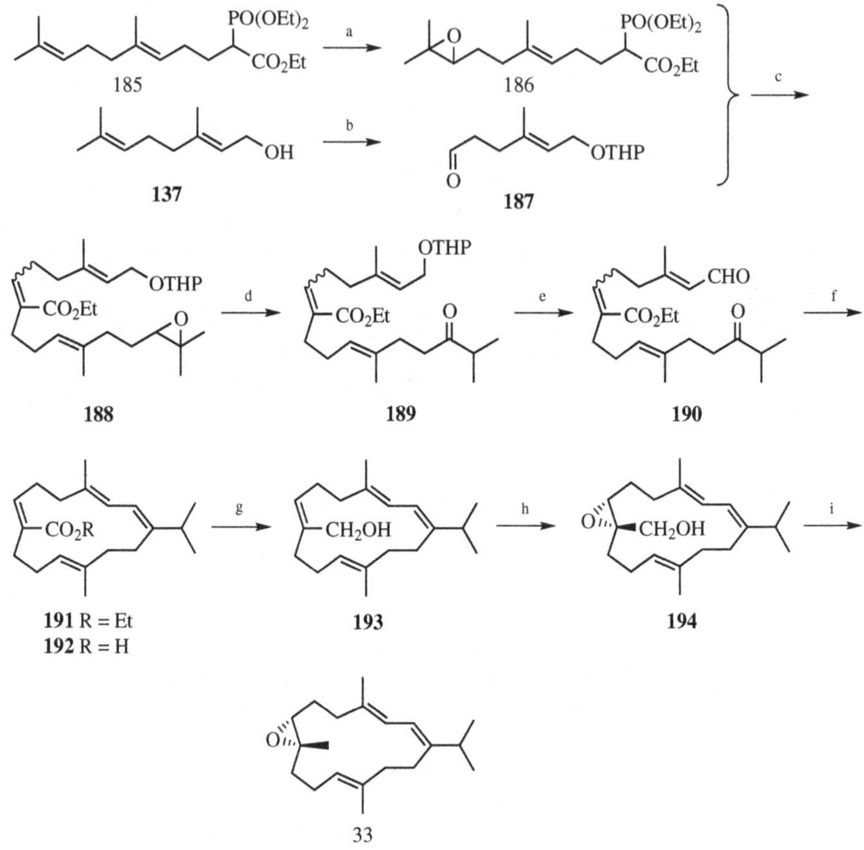

185 → 186 (a)   } c

137 → 187 (b)

188 → 189 (d) → 190 (e) → (f)

191 R = Et
192 R = H → 193 (g) → 194 (h) → (i)

33

*Reagents and Conditions:* (a) i. NBS, THF-water, 0 °C; ii. K$_2$CO$_3$, MeOH, 72%; (b) i. DHP, *p*-TsOH, CH$_2$Cl$_2$, rt; ii. O$_3$, Py, CH$_2$Cl$_2$, −78 °C, 61%; (c) LDA, −78 °C, 30 min then aldehyde **187**, −78 °C ∼ 23 °C., 61%; (d) LiClO$_4$, C$_6$H$_6$, reflux, 75%; (e) i. *p*-TsOH, MeOH, 23 °C; ii. MnO$_2$, *n*-Hexane, 23 °C, 71%; (f) TiCl$_4$, Zn, THF, reflux, 42%; (g) i. KOH, EtOH-H$_2$O, reflux, 90%; ii. LiAlH$_4$, Et$_2$O, 23 °C, 90%; (h) Ti(O$^i$Pr)$_4$, L-(+)-DET, *t*-BuOOH, 20 °C, 92%; (i) i. Ph$_3$P, I$_2$, imidazole, Py, Et$_2$O-CH$_3$CN (3:1), 0 °C; ii. NaBH$_3$CN, HMPA, THF, 60 °C, 80%.

**Scheme 6-20**

the epoxide of **181** remains undetermined and was assumed as (11*S*, 12*S*) by the authors. Pseudoplexaurol (**182**), a novel cembrenoid, was first isolated from the Caribbean gorgonian *Pseudoplexaura porosa* by Rodríguez and Martínez in 1993.[108] Although the chemical structure of **182** was characterized on the basis of spectral data and chemical degradation, the relative as well as the absolute configurations of three stereogenic centers (C-1, C-3, and C-4) remained to be assigned unambiguously. And **182** has been shown to exhibit significant cytotoxic antitumor activities against several human tumor cell lines. Pseudoplexaural (**183**) and pseudoplexauric acid methyl ester (**184**)[109] have also been found in other Caribbean marine species *Eunicea succinea* (sea wip) and gorgonian *Eunicea mammosa*, respectively, and their structures were determined by spectroscopic analysis and chemical correlation with natural pseudoplexaurol (**182**).

Stereoselective construction of the epoxide functionality in the macrocyclic cembrene skeleton comprises a challenging task for total synthesis. We report herein the completion of the total syntheses of the above cembranoxides in detail.

### 6.4.3.1 Total Synthesis of (−)-7,8-Epoxycembrene C

The first enantioselective total synthesis of (−)-7,8-epoxycembrene C (**33**) was achieved via a general approach by employing an intramolecular McMurry coupling and Sharpless asymmetric epoxidation as key steps from readily available starting material. The syntheses presented here verified the absolute stereochemistry assignment of the epoxy configuration of **33** as assumed (7R,8R) (Scheme 6-20).[110]

### 6.4.3.2 Total Synthesis of (+)-11,12-Epoxycembrene C

The first enantioselective total synthesis of (+)-11,12-epoxycembrene C (**28**) has been accomplished via a macro-olefination strategy by employing titanium-mediated McMurry coupling as a key step and the Sharpless asymmetric epoxidation for the introduction of chiral epoxide. Based on the enantioselective Sharpless asymmetric epoxidation, Li et al. assumed the configuration of natural **28** to be (11S,12S) (Scheme 6-21).[111]

### 6.4.3.3 Total Synthesis of 11,12-Epoxy-13-Hydroxycembrene A

11,12-Epoxy-13-hydroxycembrene A (**51**) can be prepared from 13-hydroxycembrene A (**50**) by epoxidation of **50** with VO(acac)$_2$-$t$-BuOOH in benzene (Scheme 6-22).[112] (±)-**50** was synthesized by Xing et al. in 1999 starting from geraniol (**137**) and **206** (Scheme 6-23).[113] Then Liu et al. reported on a more general and efficient enantioselective synthesis of (−)-**50** by employing an intramolecular McMurry coupling as the key step from readily available starting material S-(+)-carvone (**156**) (Scheme 6-24).[114]

### 6.4.3.4 Total Synthesis of 3,4-Epoxycembrene A

Starting from the chiral pool (R-(+)-limonene), the total synthesis of natural (+)-3,4-epoxycembrene A (**56**) has been achieved by Liu et al. with the low-valent titanium-mediated intramolecular pinacol coupling of the corresponding sec-keto aldehyde precursor **171** (Scheme 6-25). A more general and efficient enantioselective synthesis of (+)-3,4-epoxy-cembrene A (**56**) with a chiral pool protocol and Sharpless asymmetric epoxidation for the introduction of three chiral centers has also been reported by the same authors in 2001 (Scheme 6-26).[115]

### 6.4.3.5 Total Synthesis of (+)-11,12-Epoxysarcophytol A

A concise and efficient enantioselective synthesis of (+)-(11S,12S)-epoxysarcophytol-A (**181**) was accomplished for the first time in six simple operations from readily

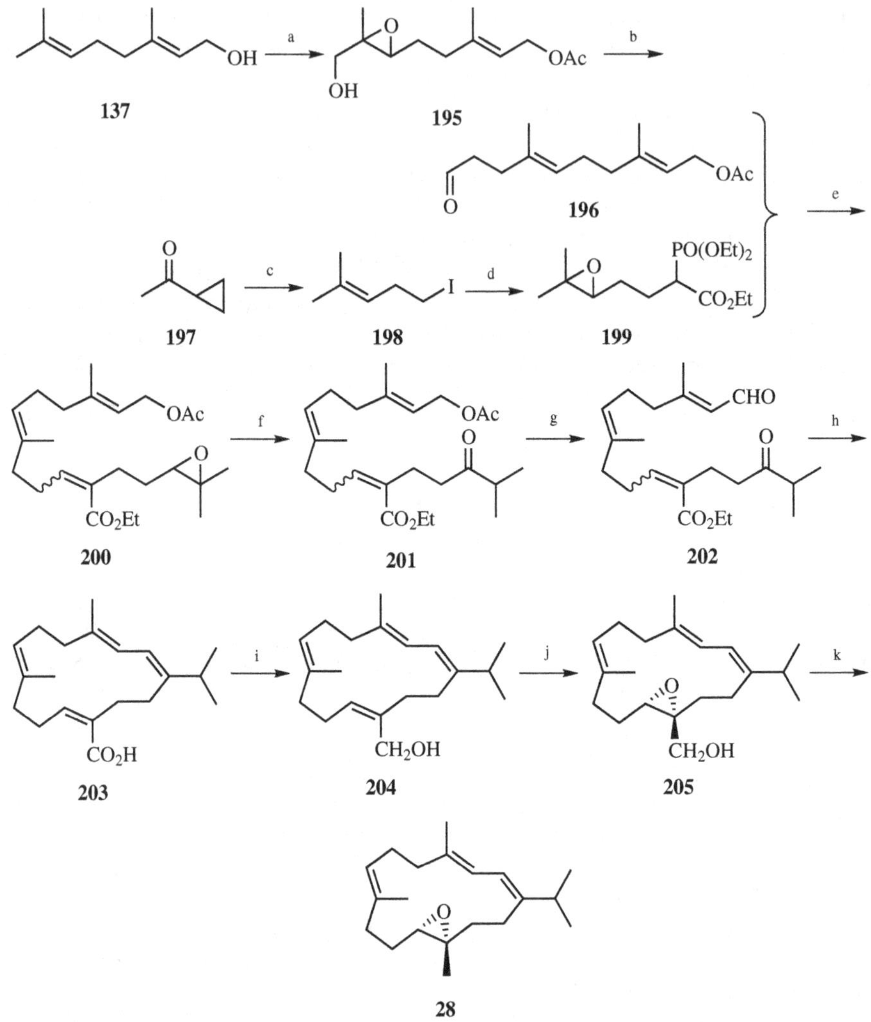

**28**

*Reagents and Conditions:* (a) i. Ac$_2$O, pyridine, rt, 2 h, 98%; ii. SeO$_2$, *t*-BuOOH, CH$_2$Cl$_2$, rt, 3 h, 73%; iii. VO (acac)$_2$, *t*-BuOOH, C$_6$H$_6$, reflux, 2 h, 90%; (b) i. Ph$_3$P, I$_2$, pyridine, Et$_2$O/CH$_3$CN (5/3), 0 °C, 1 h, then added 1*eq* H$_2$O, 38 °C, 6 h, 93%; ii. Hg(OAc)$_2$, ethyl vinyl ether, reflux, 83%; iii. sealed tube, 110 °C, 90%; (c) MeMgI, Et$_2$O, reflux, then 30% H$_2$SO$_4$, 0 °C, 1 h, 76%; (d) i. NaH, (EtO)$_2$P(O)CH$_2$CO$_2$Et, DMF, 60 °C, 88%; ii. *m*-CPBA, CH$_2$Cl$_2$, 0 °C, 93%; (e) LDA, −78 °C, 30 min then aldehyde **196**, −78 °C ∼ r.t., 73%; (f) LiClO$_4$, C$_6$H$_6$, reflux, 81%; (g) i. K$_2$CO$_3$, MeOH, r.t.; ii. MnO$_2$, *n*-Hexane, r.t., 72%; (h) i. TiCl$_4$, Zn, THF, reflux, 48%; ii. KOH, EtOH-H$_2$O, reflux, 88%; (i) LiAlH$_4$, Et$_2$O, r.t., 92%; (j) Ti(O*i*-Pr)$_4$, D-(−)-DET, *t*-BuOOH, −20 °C, 85%; (k) i. Ph$_3$P, I$_2$, imidazole, Py, Et$_2$O-CH$_3$CN (3:1), 0 °C, 90%; ii. NaBH$_3$CN, HMPA, THF, 60 °C, 90%.

**Scheme 6-21**

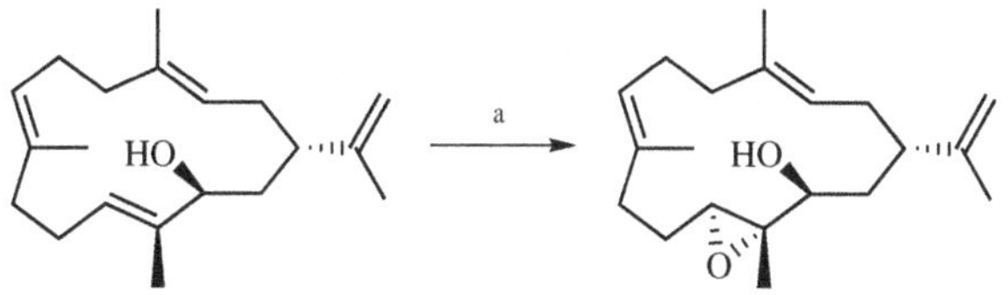

(−)-13-Hydroxy-
neocembrene (**50**)

(−)-13-Hydroxy-11,12-
Epoxy-neocembrene (**51**)

*Reagents and Conditions:* (a) VO(acac)$_2$, *t*-BuOOH, benzene, 23 °C, 2 h, 95%.

**Scheme 6-22**

available *trans, trans*-farnesol derivative **122** with an overall yield of 42%, by which the absolute stereochemistry of natural product **181** was assigned unambiguously.[116] The synthesis presented here features a combination of the highly enantioselective Sharpless epoxidation and CBS reduction for the assembly of three chiral centers and the macrocyclization of cyanohydrin silyl ether-derived carbanion alkylation to lead to the 14-membered cembrane cyclic skeleton (Scheme 6-27).

The spectroscopy data of synthetic **181**, as well as the specific rotation, were found to be identical with that reported for the natural product. The corresponding acetate of synthetic **181**, obtained by direct acylation, was identical with the natural product in all respects. Therefore, we could conclude that the absolute configuration of C-11, C-12 of 181 to be (11*S*,12*S*).

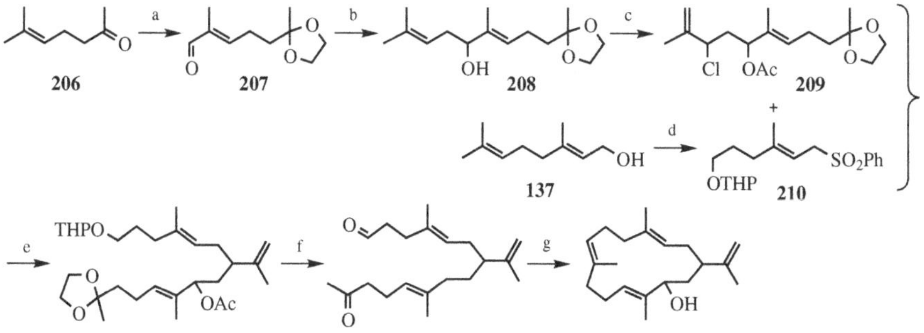

*Reagents and Conditions:* (a) i. HOCH$_2$CH$_2$OH, *p*-TsOH, C$_6$H$_6$, reflux, 92%; ii. SeO$_2$, *t*-BuOOH, CH$_2$Cl$_2$, rt, 64%; (b) i. (CH$_3$)$_2$C=CHCH$_2$SO$_2$Ph, LDA, THF, −78 °C, then added **207**; ii. Li, EtNH$_2$, THF; (c) i. Ac$_2$O, Py, 92%; ii. *t*-BuOCl, silica gel-H, 60%; (d) i. PBr$_3$, Py; 2) PhSO$_2$Na, 73%; ii. O$_3$, CH$_2$Cl$_2$, −78 °C, then Me$_2$S, 80%; iii. NaBH$_4$, MeOH, 0 °C, 85%; iv. DHP, *p*-TsOH, CH$_2$Cl$_2$, rt, 70%; (e) i. 50% NaOH, TBAB, 50%; ii. Li-EtNH$_2$, THF, 65%; iii. Ac$_2$O, Py, 95%; (f) i. *p*-TsOH, MeOH, 50 °C, 82%; ii. PCC, CH$_2$Cl$_2$, 90%; (g) i. TiCl$_4$, Zn-Cu, THF, reflux, 30 h, 38%; ii. K$_2$CO$_3$, MeOH, 77%.

**Scheme 6-23**

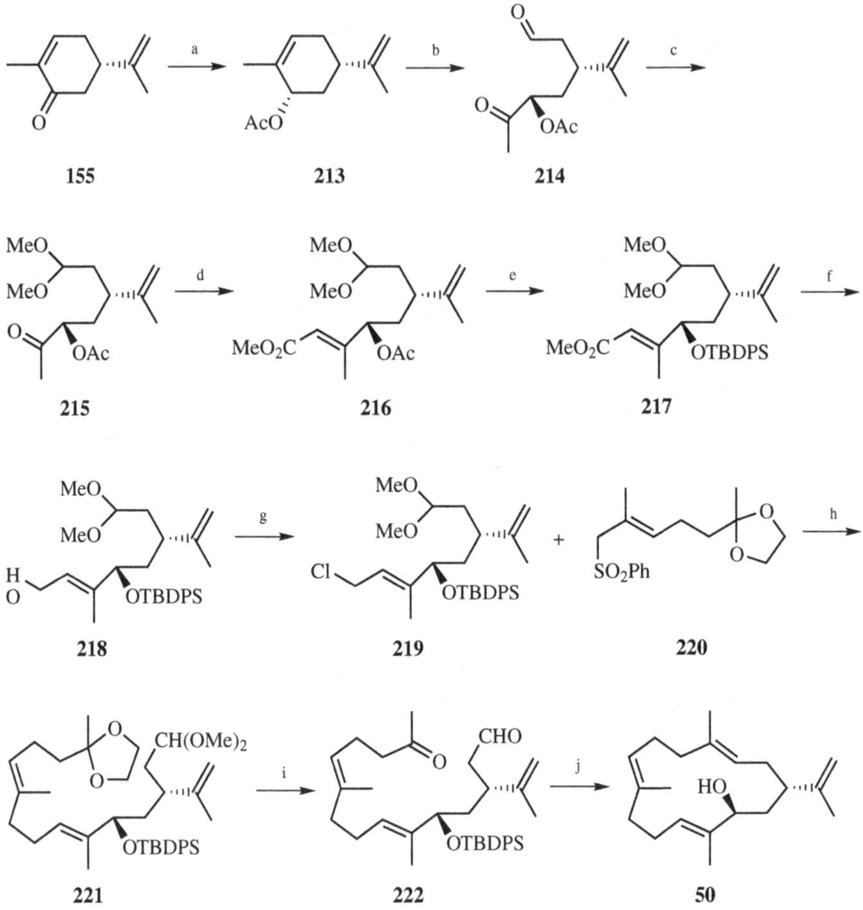
a
b
c
d
e
f
g
h
i
j
AcO
OAc
MeO
MeO₂C
OTBDPS
Cl
SO₂Ph
CH(OMe)₂
CHO
HO
155
213
214
215
216
217
218
219
220
221
222
50

*Reagents and Conditions:* (a) i. $LiAlH_4$, $Et_2O$, −78 °C, 2 h; ii. $Ac_2O$, Py, DMAP, 23 °C, 99%; (b) i. $O_3$, $CH_2Cl_2$-MeOH (3:2), −78 °C; 2. $Me_2S$, −78 °C ~ 23 °C, 24 h, 62%; (c) $NH_4Cl$, MeOH, 50 °C, 6 h, 88%; (d) $Ph_3P{=}CHCO_2Me$, $PhCO_2H$, toluene, reflux, 72 h, 82%; (e) i. $K_2CO_3$, MeOH, 23 °C, 1 h, 93%; ii. TBDPSCl, imidazole, DMF, 23 °C, 2 h, 91%; (f) DIBAL-H, $CH_2Cl_2$, −78 °C, 2 h, 87%; (g) NCS, $Ph_3P$, THF, 23 °C, 12 h, 92%; (h) i. *n*-BuLi, −78 °C, 30 min then added **219**, −78 °C ~ 23 °C, 4 h, 92%; ii. Na(Hg), $Na_2HPO_4$, MeOH, 23 °C, 6 h, 91%; (i) *p*-TsOH, acetone, 40 °C; 4 h, 89%; (j) i. $TiCl_4$, Zn, Py, DME, reflux, 10 h, 81%; ii. *n*-$Bu_4NF$, THF, 60 °C, 24 h, 85%.

**Scheme 6-24**

#### *6.4.3.6 Asymmetric Total Synthesis of Pseudoplexaurol and 14-Deoxycrassin*

14-Deoxycrassin (**239**), a novel crassin-like cembrenoid, was first isolated from the Caribbean gorgonian *Pseudoplexaura porosa* by Rodríguez and Martínez in 1993 along with pseudoplexaurol (**182**). Two more abundant cembranolides, crassin (**240**) and its acetate (**241**), with closely related structures and potent bioactivities were identified previously from the same species. Although the chemical structures of **182** and **239** were characterized on the basis of spectral data and chemical

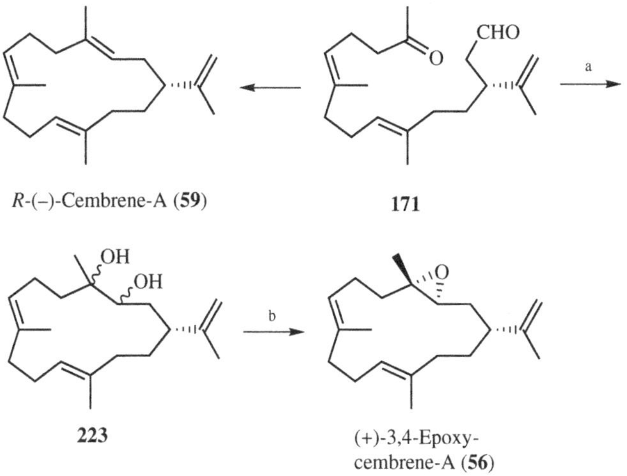

R-(–)-Cembrene-A (**59**)                    **171**

**223**                    (+)-3,4-Epoxy-
                         cembrene-A (**56**)

*Reagents and Conditions:* (a) TiCl$_4$, Zn, DME, 0 °C, 81%; (b) MsCl, Py then K$_2$CO$_3$, MeOH, 65%.

**Scheme 6-25**

degradation, the relative as well as the absolute configurations of three stereogenic centers (C-1, C-3, and C-4) were postulated to be as shown (Figure 6-11) by biogenetic correlation and remained to be assigned unambiguously. Both **182** and **239** have been shown to exhibit significant cytotoxic antitumor activities against several human tumor cell lines. Pseudoplexaural (**183**) and pseudoplexauric acid methyl ester (**184**) have also been found in other Caribbean marine species *Eunicea succinea* (sea wip) and gorgonian *Eunicea mammosa,* respectively, and their structures were determined by spectroscopic analysis and chemical correlation with natural pseudoplexaurol (**182**). Epoxy alcohol **182** was assumed to be a logical biosynthetic precursor of **239**, but attempts to synthesize lactone **239** from epoxy alcohol **182** via pseudoplexauric acid intermediate failed (Scheme 6-28).[117] A recent successful translactonization of euniolide to crassin (**236**), which was reported by Rodríguez et al.,[118] using aqueous alkaline hydrolysis under the assistance of sonication (Scheme 6-29) implies a possible conversion of **182** to **239** (Scheme 6-30). Synthetic studies on the closely related natural cembranolides bearing a 3, 4-epoxy and/or lactone functions, such as crassin (**240**), euniolide (**242**), and isolobophytolide (**243**), have been conducted over the past decades by several groups (Figure 6-11).[119–124]

Liu et al. have reported on the total synthesis of **182** and **239** to verify their structural relationship and to confirm the relative and absolute configurations.[125] The overall strategic plan involves (1) chemical conversion of epoxy alcohol **182** to lactone **239** via an intramolecular lactonization, (2) construction of the 14-membered cembrane ring with geometrically defined double bonds and substituents, (3) implementation of C-1 stereogenic center via a readily available chiral pool (i.e., *S*-limonene), and (4) introduction of chiral epoxide functionality via a Sharpless

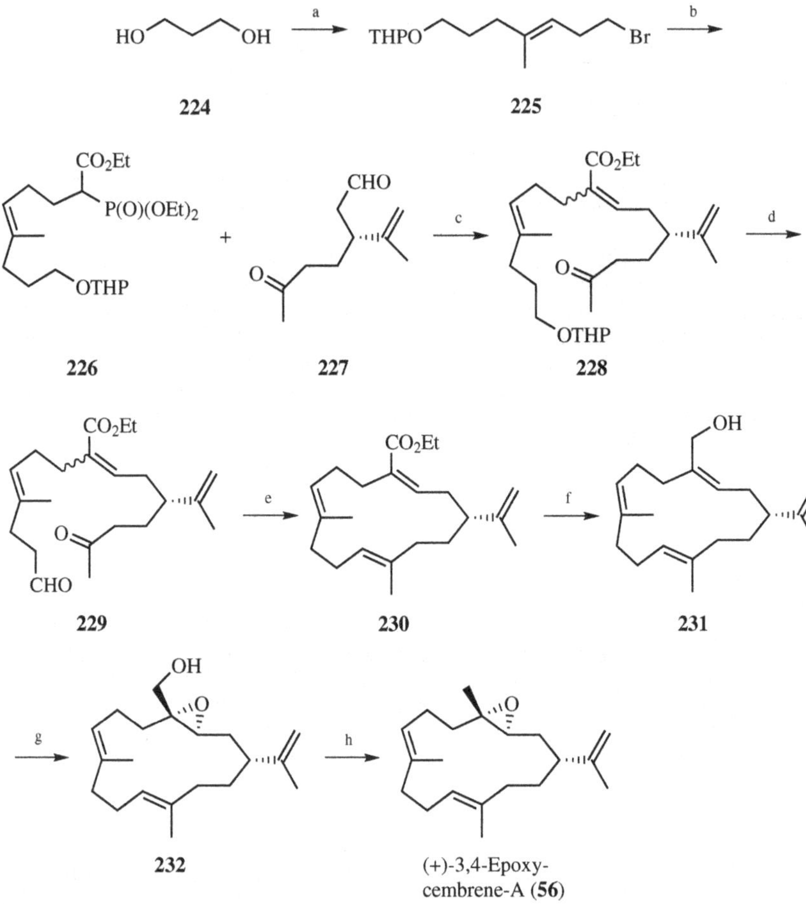

**Scheme 6-26**

*Reagents and Conditions:* (a) i. 48% HBr, C$_6$H$_6$, reflux, 12 h, 80%; ii. DHP, *p*-TsOH, CH$_2$Cl$_2$, rt, 2 h, 99%; iii. Li, THF, methyl cyclopropylketone, r.t., 1 h, 84%; iv. LiBr, TMSCl, CH$_2$Cl$_2$, r.t, 2 h, 79%; (b) NaH, (EtO)$_2$P(O)CH$_2$CO$_2$Et, DMF, 60 °C, 6 h, 79%; (c) *n*-BuLi, **226**,−78 °C, THF, 0.5 h, −78 °C ∼ −20 °C, 0.5 h, then added alehyde **227**, −40 °C, 2 h, 70%; (d) i. *p*-TsOH, MeOH, r.t., 2 h; ii. PCC, NaOAc, CH$_2$Cl$_2$, r.t., 2 h, 85%; (e) TiCl$_3$, Zn-Cu, DME, reflux, 6 h, 41%; (f) DIBAL-H, CH$_2$Cl$_2$, −78 °C, 1 h, 92%; (g) Ti(O$^i$Pr)$_4$, D-(−)-DET, *t*-BuOOH, CH$_2$Cl$_2$, −20 °C, 8 h, 90%; (h) i. Ph$_3$P, imidazole, Py, I$_2$, Et$_2$O−CH$_3$CN, 0 °C, 2 h; ii. NaBH$_3$CN, HMPA−THF, 60 °C, 24 h, 77%.

asymmetric epoxidation of corresponding macrocyclic allylic alcohol intermediate. The key reaction we employed for the effective construction of the 14-membered carbocyclic ring is the low-valent titanium-mediated macrocyclization, which is either an intramolecular pinacol-type or olefination coupling (McMurry coupling of dicarbonyl compounds) (Scheme 6-31).

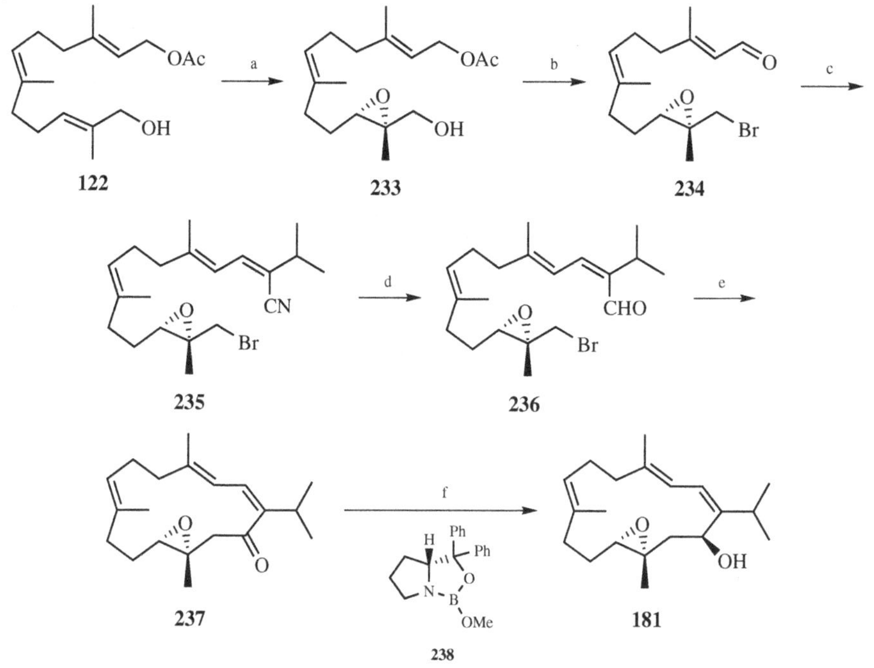

**Scheme 6-27**

*Reagents and Conditions:* (a) Ti(O-*i*-Pr)$_4$, L-(+)-DET, *t*-BuO$_2$H, CH$_2$Cl$_2$, $-40\,°C$, 95%; (b) i. MsCl, Et$_3$N, CH$_2$Cl$_2$, $-10\,°C \sim 0\,°C$; then LiBr, acetone, $50\,°C$, 82%; ii. K$_2$CO$_3$, MeOH, $23\,°C$; then MnO$_2$, Hexane, $23\,°C$, 91%; (c) LiN(SiMe$_3$)$_2$/(Me)$_2$CHCH(CN)P(O)(OEt)$_2$, toluene, $-78\,°C$; then **234**, 90%; (d) DIBAL-H, Hexane, $-78\,°C$; then 10% *aq.* oxalic acid, $0\,°C$, 88%; (e) i. Me$_3$SiCN, *cat.* KCN/18-C-6, THF; ii. LiN(SiMe$_3$)$_2$, THF, reflux; then TBAF, 85% from **236**; (f) BH$_3 \cdot$Me$_2$S, **238** (10 mol%), toluene, $0\,°C$, 88%.

Chemical conversion of pseudoplexaurol (**182**) to 14-deoxycrassin (**239**) was realized by a Lewis acid-catalyzed lactonization process (Scheme 6-32).

## 6.4.4  Total Synthesis of Cembranolides

Sarcophytonin B (**67**) was first isolated from *Sarcophyton* spp. by Kobayashi and Hirase[27] in 1990. *cis*- and *trans*-Cembranolides were isolated from *Sinularia mayi* by Uchio et al. (Scheme 6-33).[126] and from *Lobophytum michaelae* by Coll et al.[76] Cembranolides were assumed to be logical biosynthetic precursors of sarcophytonin B. Several syntheses of the cembranolides and sarcophytonin B (**67**) have been reported.

($\pm$)-*cis*-Cembranolide was synthesized via Cr(II)-mediated intramolecular macrocyclization of $\omega$-formyl-$\beta$-methoxycarbonylallyl halides, which were prepared from geranylgeraniol (Scheme 6-34).[127,128]

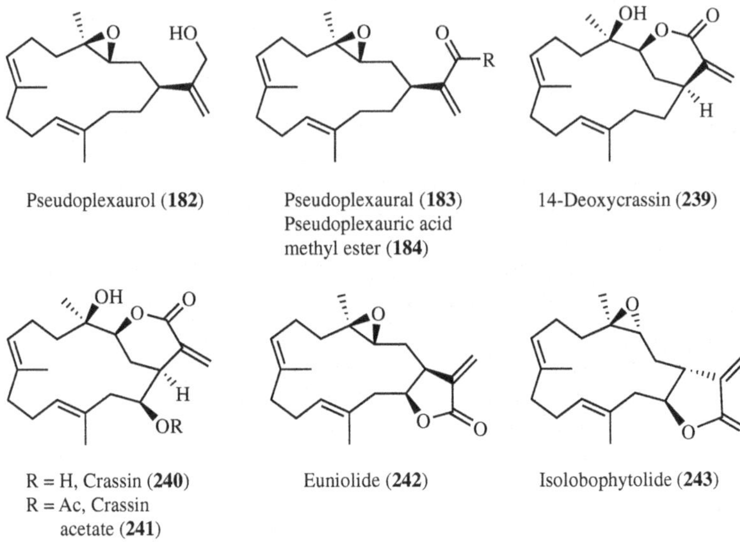

Pseudoplexaurol (**182**)

Pseudoplexaural (**183**)
Pseudoplexauric acid
methyl ester (**184**)

14-Deoxycrassin (**239**)

R = H, Crassin (**240**)
R = Ac, Crassin
      acetate (**241**)

Euniolide (**242**)

Isolobophytolide (**243**)

**Figure 6-11**

The first synthesis of naturally occurring cembranolides has been reported by Kodama et al. in 1982 detailed below (Scheme 6-35).[129]

(±)-*cis*-Cembranolide has also been prepared from **268** (Scheme 6-36). The base-catalyzed retro-aldol reaction of the tertiary alcohol derived from **269** produced ketone **270**. The alkylation of **270** with ethyl iodoacetate gave **271**. After

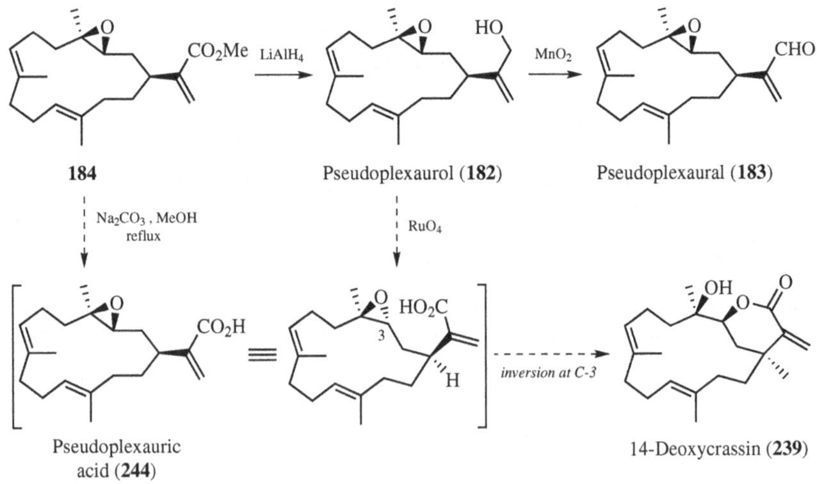

**184**

Pseudoplexaurol (**182**)

Pseudoplexaural (**183**)

Pseudoplexauric
acid (**244**)

14-Deoxycrassin (**239**)

Chemical correlation of pseudoplexaurol (**182**) with **183, 184** and attempted conversion to 14-deoxycrassin (**239**).

**Scheme 6-28**

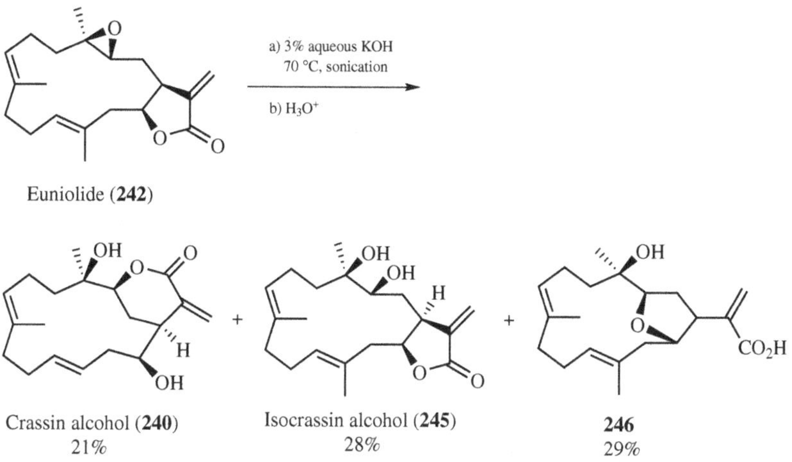

a) 3% aqueous KOH
   70 °C, sonication

b) H₃O⁺

Euniolide (**242**)

+

+

Crassin alcohol (**240**)
21%

Isocrassin alcohol (**245**)
28%

**246**
29%

Translactonization of euniolide to crassin alcohol (**240**).

**Scheme 6-29**

reduction with sodium borohydride, lactone **272** was generated. Introduction of the exomethylene group completed the synthesis of **256**.[62]

In 1987, Marshall and Crooks reported the addition of lithium dimethylcuprate to the conjugated cycloalkynone **276** to yield, after equilibration, a single *E*-enone **277**, which was converted via acidic hydrolysis, oxidation, and reduction to

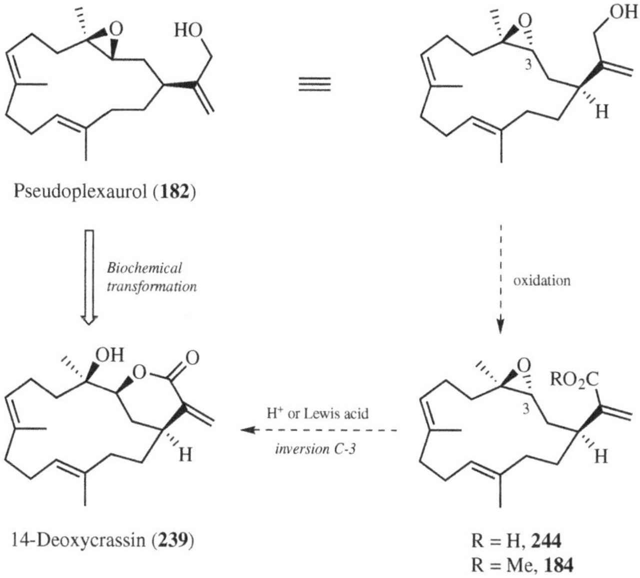

Pseudoplexaurol (**182**)

≡

Biochemical
transformation

oxidation

14-Deoxycrassin (**239**)

H⁺ or Lewis acid
- - - - - - - - -
*inversion C-3*

R = H, **244**
R = Me, **184**

Possible biogenetic relationship of 14-deoxycrassin (**239**) and pseudoplexaurol (**182**).

**Scheme 6-30**

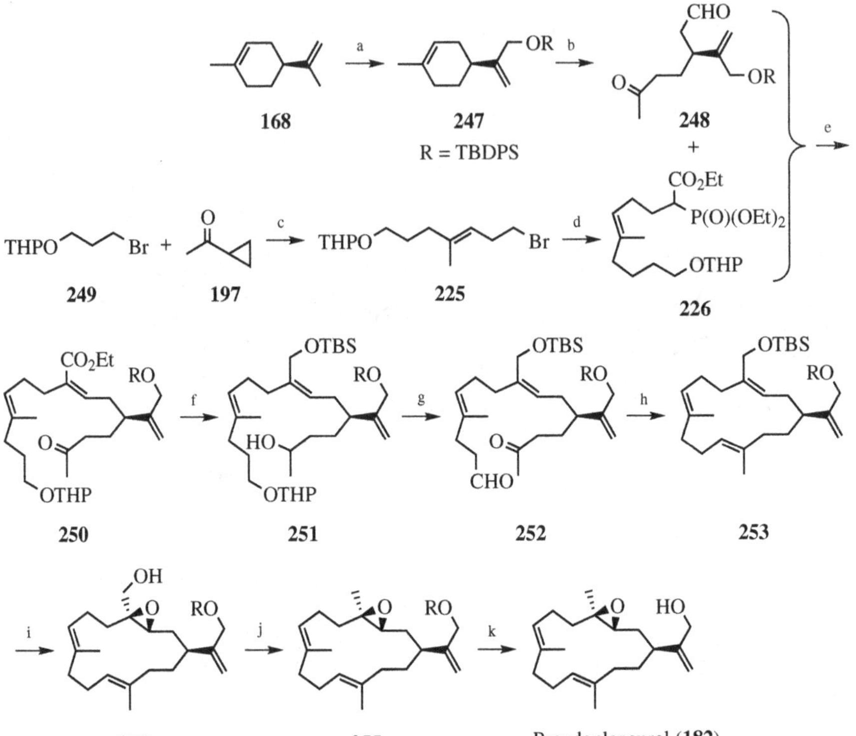


168 247 248 249 197 225 226 250 251 252 253 254 255 Pseudoplexaurol (182)

*Reagents and conditions:* (a) i. *n*-BuLi, TMEDA, THF, 23 °C, 24 hr, then $O_2$; ii. TBDPSCl, imidazole, DMF; (b) $O_3$, $CH_2Cl_2$, −78 °C, then $Me_2S$; (c) i. Li, THF, r.t., then cyclopropyl methyl ketone (**197**); ii. LiBr, TMSCl, $CH_2Cl_2$, r.t.; (d) $(EtO)_2P(O)CH_2CO_2Et$, NaH, DMF, 60 °C; (e) *n*-BuLi, THF, −78 °C; (f) i. DIBAL-H, $CH_2Cl_2$, −78 °C; ii. TBSCl, imidazole, DMF; (g) i. $MgBr_2$, $Et_2O$, r.t.; ii. Swern oxidation; (h). $TiCl_4$, Zn, Py, DME, reflux; (i) i. PPTs, EtOH, r.t.; ii. $Ti(O^iPr)_4$, *L*-(+)-DET, *t*-BuOOH, $CH_2Cl_2$, −20 °C; (j) i. $I_2$, $Ph_3P$, imidazole, $Et_2O$-$CH_3CN$, 0 °C; ii. $NaBH_3CN$, THF, HMPA, 60 °C; (k) TBAF, THF.

**Scheme 6-31**

the *cis*-lactone. α-Methylenation afforded the natural cembranolide **256** (Scheme 6-37).[66,130]

The homochiral hydroxyl enol ether **280**, which was secured through $BF_3$-promoted cyclization of alkoxy stannane **274**, was elaborated to the *cis*-cembranolide **256**. The synthesis confirmed the absolute stereochemistry of the natural cembranolide (Scheme 6-38).[131]

The first asymmetric total synthesis of (−)-*trans*-cembranolide (**295**) was reported by Taber and Song in 1997 (Scheme 6-39).[132] The key step in the synthesis is the diastereoselective Rh-mediated cyclization of the enantiomerically pure diazo ester **288** to the tetrahydrofuran **289**.

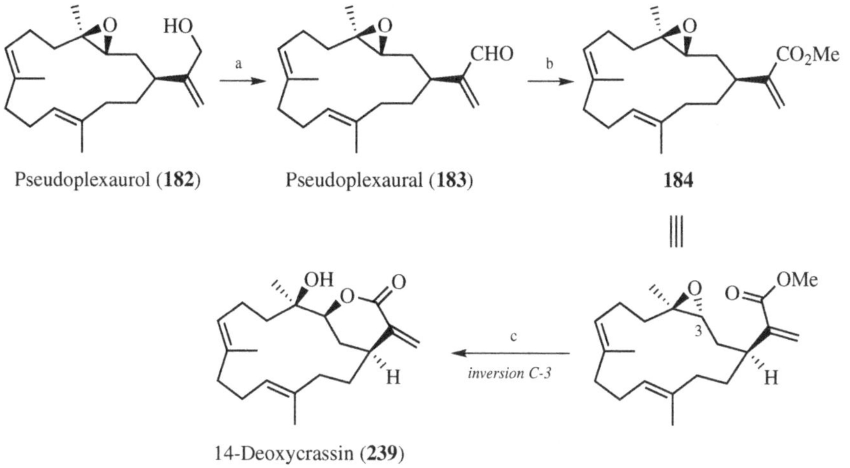
HO
O
a
CHO
b
CO2Me
Pseudoplexaurol (182)
Pseudoplexaural (183)
184
OH
O
O
H
c
inversion C-3
O
O
OMe
3
H
14-Deoxycrassin (239)

*Reagents and conditions:* (a) $MnO_2$, 92%; (b) $MnO_2$, NaCN, MeOH, HOAc, 58%; (c) $ZnCl_2$, $CH_2Cl_2$, 0 °C to 30 °C, 71%.

**Scheme 6-32**

## 6.5 STUDIES ON NOVEL MACROCYCLIZATION METHODS OF CEMBRANE-TYPE DITERPENOIDS

Although several synthetic strategies and methods for the construction of a 14-membered-ring system have appeared in literature over the past three decades and notable progress has been made in this field, the lack of a general method for the preparation of 14-membrened rings presents an ongoing challenge for total synthesis. A novel general and diversified strategy for the stereocontrolled total synthesis of cembrane-type diterpenoids is highly desirable.

### 6.5.1 A Stille Cyclization Approach to (±)-Isocembrene

Characteristic to some important cembrane diterpenoids, such as isocembrene, cembrene, cembrene C, and sarcophytol A, is a 1,3-diene unit in the macrocyclic skeleton. In continuation of our ongoing project on the total synthesis of cembrane-type diterpenoids, we intended to explore a novel macrocyclization method to

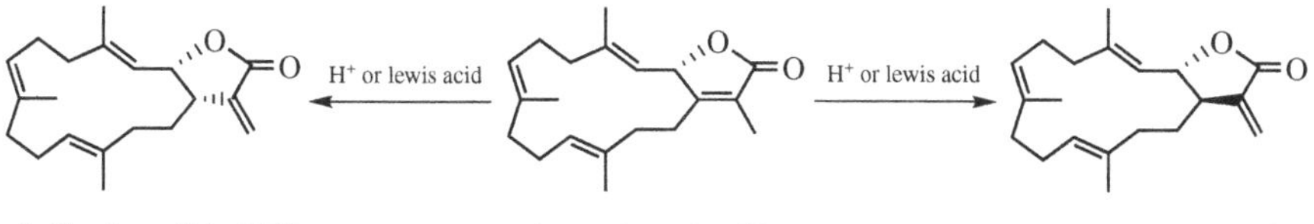
O
O
H+ or lewis acid
O
O
H+ or lewis acid
O
O

*cis*-Cembranolide (**256**) Sarcophytonin (**67**) *trans*-Cembranolide (**257**)

**Scheme 6-33**

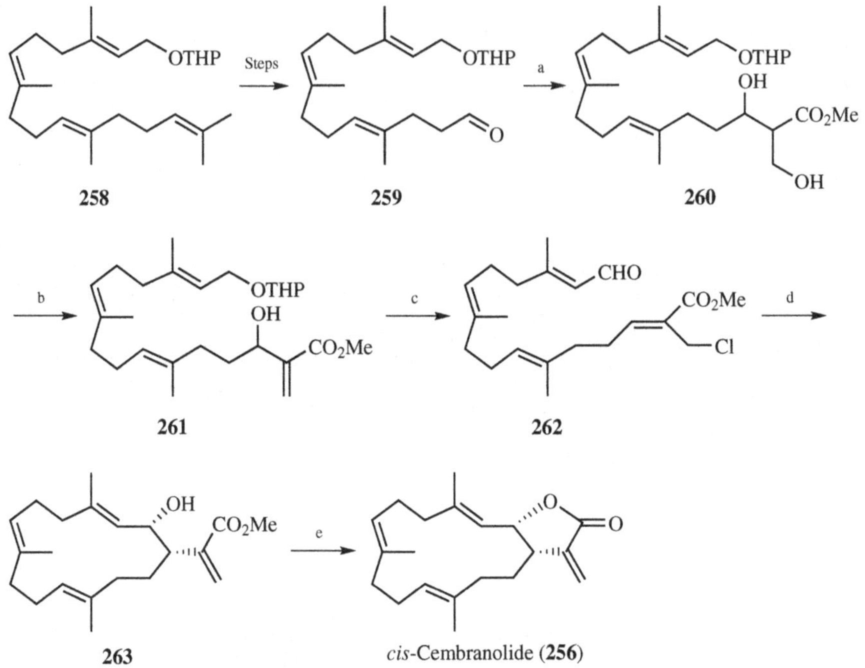

**258** → **259** → **260** → **261** → **262** → **263** → *cis*-Cembranolide (**256**)

*Reagents and Conditions:* (a) 1) LDA, HOCH₂CH₂CO₂Me, THF, −78 °C, 56%, conv. 70%; (b) i. TsCl, Py, 96%; ii. DBU, Et₂O, rt, 83%; (c) i. CCl₄, Oct₃P, rt, 71%; ii. PPTs, MeOH, 92%; iii. Swern oxidation, 89%; (d) CrCl₂, DMF, rt, 81%; (e) *p*-TsOH, C₆H₆, rt, 87%.

**Scheme 6-34**

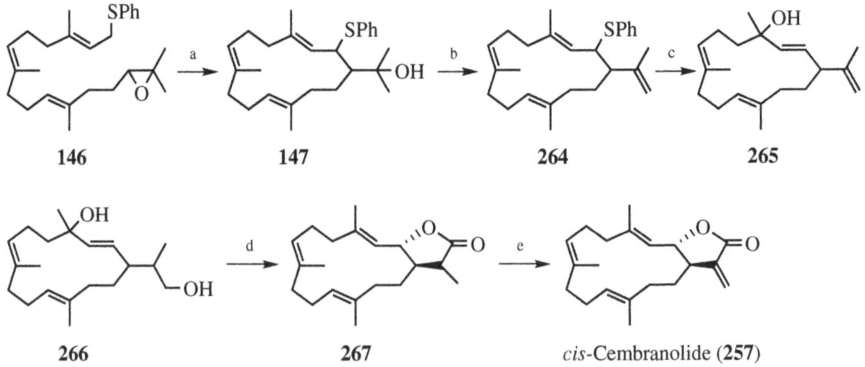

**146** → **147** → **264** → **265** → **266** → **267** → *cis*-Cembranolide (**257**)

*Reagents and Conditions:* (a) *n*-BuLi, Dabco, THF, −78 °C; (b) SOCl₂, Py; (c) i. NaIO₄, MeOH, 0 °C; ii. P(OMe)₃, MeOH, rt, 67%; (d) i. 9-BBN, THF, rt; ii. H₂O₂, NaOH, 0 °C, 81%; (e) PDC, DMF, rt, 37%; (f) i. PhSeCl, LDA, THF, −78 °C; ii. NaIO₄, MeOH, rt, 72%.

**Scheme 6-35**

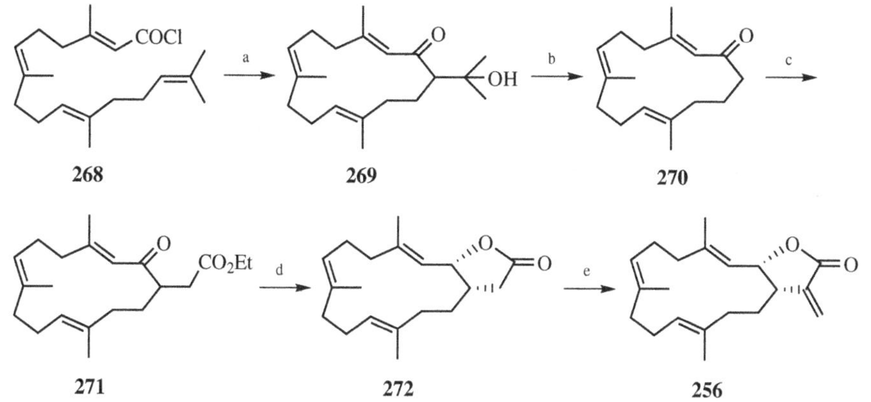

**268** → **269** → **270** → **271** → **272** → **256**

*Reagents and Conditions:* (a) SnCl$_4$, CH$_2$Cl$_2$, −78 °C; (b) i. ZnO, AcOH, 45 °C, 71%; ii. LiOH, *aq* dioxane, 75 °C, 50%; (c) LDA, THF-HMPA, −78 °C, ICH$_2$CO$_2$Et, 66%; (d) i. NaBH$_4$, EtOH, ii. PPTs, benene, reflux, 100%; (e) i. LDA, HCHO (g), −78 °C, 65%; ii. MsCl, Py, DMAP (cat), CH$_2$Cl$_2$; iii. DBU, benene, rt, overnight, 96%.

**Scheme 6-36**

construct the macrocyclic skeleton that bears a 1,3-diene unit. Intramolecular Pd-catalyzed cross-coupling between alkenyl-stannane and alkenyl halide (or triflate) functionality is now firmly established as an important methodology for the construction of unsaturated heterocycles and carbocycles.[133,134] This reaction is relatively insensitive to moisture and air and tolerant to a variety of functional groups on either coupling partner; furthermore, it is stereospecific and regio-selective.[135] Thus, this reaction seemed to be ideal for the synthesis of a variety of cembrane-type compounds (Figure 6-12).

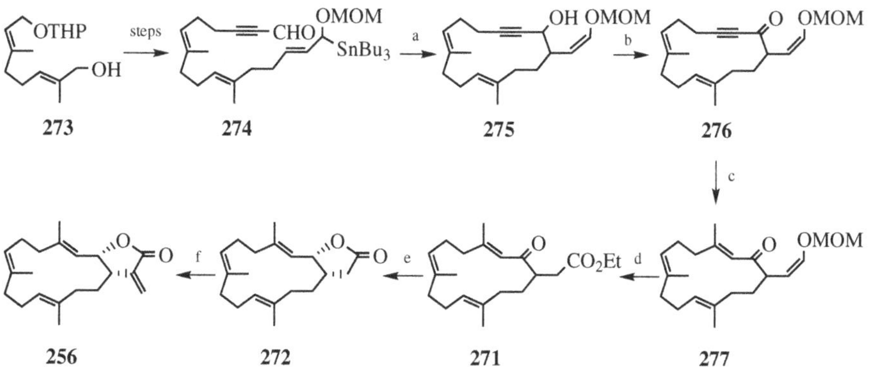

**273** → **274** → **275** → **276** → **277** → **271** → **272** → **256**

*Reagents and Conditions:* (a) BF$_3$·Et$_2$O, CH$_2$Cl$_2$, −78 °C, 88%; (b) Swern oxidation; (c) i. LiCuMe$_2$, THF; ii. LiSiPr; (d) i. HCl, H$_2$O; ii. PDC, DMF, rt; iii. CH$_2$N$_2$, Et$_2$O; (e) i. NaBH$_4$, EtOH; ii. PPTs, benzene, reflux; (f) i. LDA, CH$_2$O (g), THF, −78 °C; ii. MCDI, CuCl$_2$.

**Scheme 6-37**

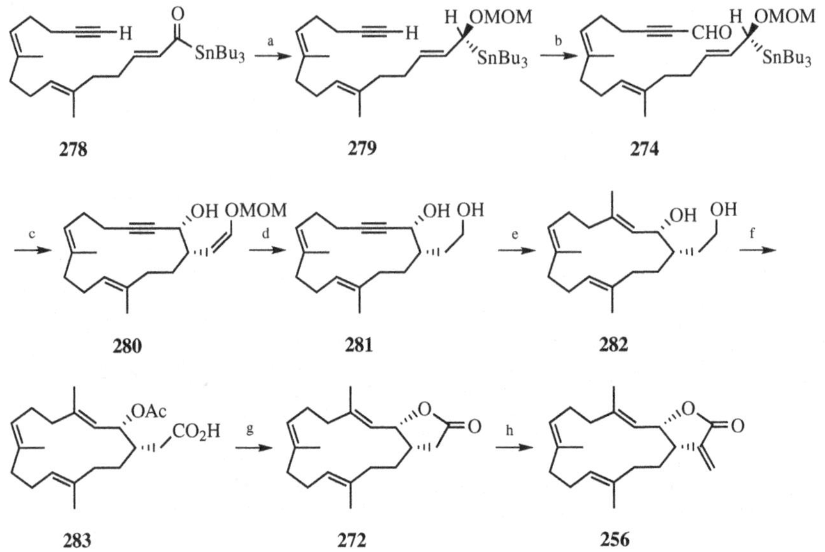

**278** → **279** → **274**

**280** → **281** → **282** →

**283** → **272** → **256**

*Reagents and Conditions:* (a) (R)-(+)-BINOL-H, then MOMCl, $^i$Pr$_2$NEt; (b) i. LDA, CH$_2$O; ii. BuOMgBr, (CH$_2$)$_5$CON=NCO(CH$_2$)$_5$; (c) BF$_3$·Et$_2$O, CH$_2$Cl$_2$, −78 °C, 88%; (d) i. 10% HCl, THF; ii. NaBH$_4$, EtOH, 72%; (e) i. Red-Al, THF, I$_2$, THF; ii. LiCuMe$_2$, THF, 65%; (f) i. TBSCl, Et$_3$N, CH$_2$Cl$_2$; ii. AcCl, Py, 0 °C; iii. TBAF, HOAc, THF; iv. PDC, DMF, 52%; (g) i. K$_2$CO$_3$, MeOH; ii. DCC, DMAP, CH$_2$Cl$_2$; (h) i. LDA, CH$_2$O (g), THF, −78 °C; ii. MsCl, Py; 3) DBU, benene, reflux.

### Scheme 6-38

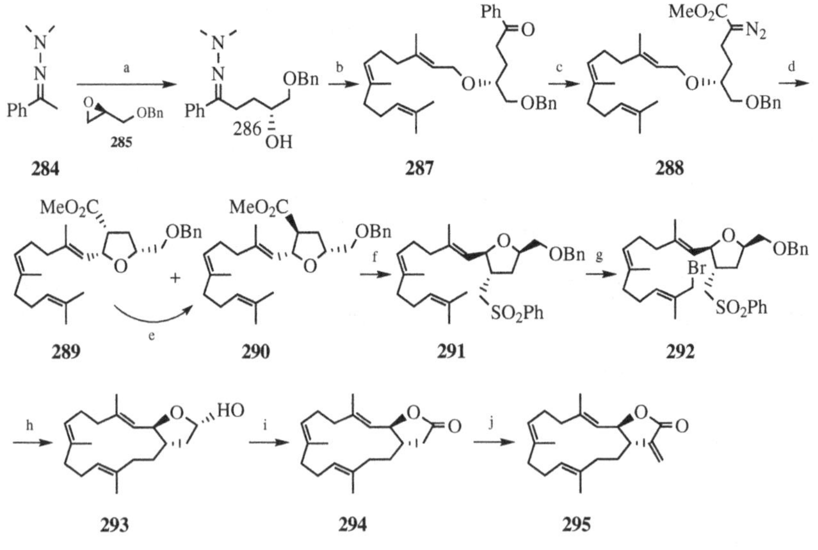

**284** → **285** / **286** → **287** → **288** →

**289** + **290** → **291** → **292** →

**293** → **294** → **295**

*Reagents and Conditions:* (a) LHMDS, then epoxide **286**; (b) i. NaH, farnesyl bromide; ii. *aq* HCl, THF, 74%; (c) NaH, (MeO)$_2$CO, DBU, PNBSA, 63%; (d) Rh(Oct)$_2$, 90%; (e) NaOMe, 88%; (f) i. LiAlH$_4$, 84%; ii. TsCl, Et$_3$N; 3) NaSO$_2$Ph, Bu$_4$NI, DMF, 84%; (g) i. SeO$_2$, t-BuOOH; ii. CBr$_4$, Ph$_3$P, 50%; (h) i. LDA, THF; ii. Na/NH$_3$, 61%; (i) PDC, Ac$_2$O, 58%; (j) i. CH(OMe)(NMe$_2$)$_2$; ii. DIBAL-H, 64%.

### Scheme 6-39

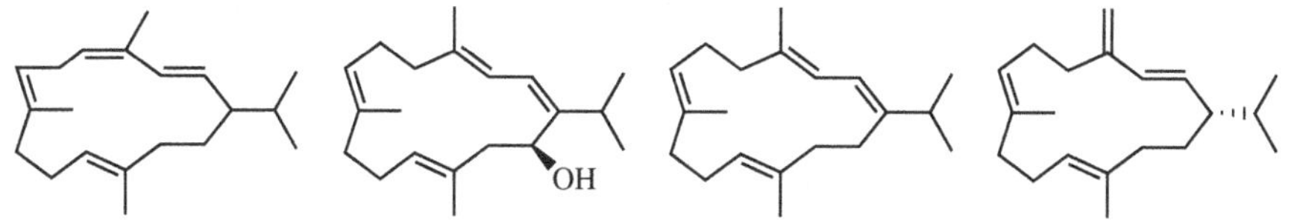

Cembrene (2)    Sarcophytol A (3)    Cembrene C (27)    Isocembrene (296)

**Figure 6-12**

Isocembrene (**296**), a simple cembrane diterpenoids, was first isolated by Kashtanova et al.[136] in 1968 from a Russia marine soft coral *Pinus sibirica* and characterized spectroscopically and chemically as (1*S*, 2*E*, 7*E*, 11*E*)-2, 4 (18), 7, 11-cembratetraene. Pattenden and Smithies accomplished the first total synthesis of (±)-**296** in 1996.[137] In continuation of our ongoing project on the development of novel macrocyclization methods for cembrane diterpenoids, we report herein the completion of the total synthesis of **296** using the Stille sp$^2$-sp$^2$ macrocyclization reaction as the key step to elaborating the stereodefined 1,3-diene unit in this compound. The synthetic strategy from available *E*-geranyl acetone outlined in Scheme 1 involves (1) chemoselective synthesis of *E*-alkenylstannanes from keto aldehydes using Bu$_3$SnCHI$_2$ in DMF, (2) chemoselective synthesis of vinyl triflate from ketonic carbonyl, and (3) macrocyclization of precursor **303** containing vinyl tin and vinyl triflate groups at the termini of the chain by the intramolecular Stille cross-coupling reaction (Scheme 6-40).

Thus, an efficient and convergent total synthesis of (±)-isocembrene has been accomplished in seven steps from the known geranylacetone through an intramolecular Stille cross-coupling reaction. The strategy is of great potential for the divergent synthesis of complex cembrane-type diterpenoids such as cembrene-C and sarcophytols.

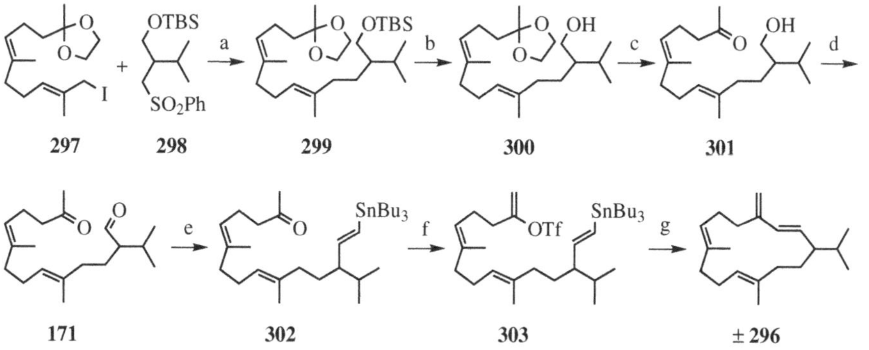

297    298    299    300    301

171    302    303    ± 296

*Reagents and Conditions:* (a) 1. *n*-BuLi, THF, −40 °C, 1.5 h, then **297**, −78 °C-rt, 1.5 h, 89%; 2. Na (Hg), Na$_2$HPO$_4$, MeOH, rt, 16 h, 84%; (b) TBAF, THF, rt, 30 min, 95%; (c) *p*-TsOH, acetone, rt, 4 h, 95%; (d)Swern oxidation, 92%; (e) Bu$_3$SnCHI$_2$, CrCl$_2$, DMF, rt, 5 h, 74%; (f) NaHMDS, PhNTf$_2$, THF, −78 °C, 2 h, 84%; (g) (Ph$_3$P)$_4$Pd, LiCl, THF, reflux, 6 h, 76%.

**Scheme 6-40**

## ACKNOWLEDGMENTS

We are grateful for financial support from the National Natural Science Foundation of China and the Special Research Grant for Doctoral Sites in Chinese Universities.

## REFERENCES

1. Tursch, B.; Braeckman, J. C.; Dolaze, D.; Kaisin, M. In Scheuer, P. J., editor. *Marine Natural Products: Chemical and Biological Perspectives*, vol. 2. New York: Academic Press; **1978**, 247.

2. Dauben, W. G.; Thiessen, W. E.; Resnich, P. R. *J. Am. Chem. Soc.*, **1962**, 84: 2015.

3. Kobayashi, H.; Akiyosh, S. *Bull. Chem. Soc. Japan.*, **1962**, 35: 1044.

4. Faulkner, D. J. *Nat. Prod. Rep.*, **1995**, 12: 223, and references therein.

5. Coll, J. C. *Chem. Rev.*, **1992**, 92: 613.

6. Coll, J. C. *Chem. Rev.*, **1993**, 93: 1693.

7. Anjaneyulu, A. S. R.; Rao, G. V. *J. Indian Chem. Soc.*, **1997**, 74: 272.

8. Tius, M. A. *Chem. Rev.*, **1988**, 88: 719.

9. Kobayashi, M.; Nakagawa, T.; Mrrsuhashi, H. *Chem. Pharm. Bull.*, **1979**, 27: 2382.

10. Kobayashi, M.; Osabe, K. *Chem. Pharm. Bull.*, **1989**, 37: 631.

11. Kobayashi, M.; Ishizaka, T.; Nakano, E. *Chem. Pharm. Bull.*, **1989**, 37: 2053.

12. Kobayashi, M.; Nakano, E. *J. Org. Chem.*, **1990**, 55: 1947.

13. König, G. M.; Wright, A. D. *J. Nat. Prod.*, **1998**, 61: 494.

14. Bowden, B. F.; Coll, J. C.; Heaton, A.; Konig, G.; Bruck, M. A.; Cramer, R. E.; Klein, D. M.; Scheuer, P. J. *J. Nat. Prod.*, **1987**, 50: 650.

15. Bowden, B. F.; Coll, J. C.; Mrrchell, S. J. *Aust. J. Chem.*, **1980**, 33: 879.

16. Coll, J. C.; Bowden, B. F.; Tapiolas, D. M.; Dunlop, W. C. *J. Exp. Mar. Biol. Ecol.*, **1982**, 60: 293.

17. Yan, Z.; Chen, C.; Zeng, L. *Redai Haiyang.*, **1985**, 4: 80.

18. Yan, Z.; Chen, C.; Zeng, L. *Redai Haiyang.*, **1984**, 3: 78.

19. Carmely, S.; Groweiss, A.; Kashman, Y. *J. Org. Chem.*, **1981**, 46: 4279.

20. Bowden, B. F.; Coll, J. C.; Hicks, W.; Kazlauskas, R.; Mitchell, S. J. *Aust. J. Chem.*, **1978**, 31: 2707.

21. Kashman, Y.; Zadock, E.; Neeman, I. *Tetrahedron.*, **1974**, 30: 3615.

22. Neeman, I.; Tishelson, L.; Kashman, Y. *Toxicon.*, **1974**, 12: 593.

23. Albericci, M.; Breakman, J. C.; Daloze, D.; Tursch, B.; Declerck, J. P.; Germain, G.; Meersche, M. V. *Bull. Soc. Chim. Belg.*, **1987**, 87: 487.

24. Uchio, Y.; Nitta, M.; Toyata, J.; Nakayama, M.; Nishizona, Y.; Iwagawa, T.; Hase, T. *Chem. Lett.*, **1983**, 1719.

25. Kaziauskas, R.; Biard-Lambard, J. A.; Murphy, P. T.; Wells, R. J. *Aust. J. Chem.*, **1982**, 35: 61.

26. Anthoni, U.; Bock, K.; Christophersen, C.; Duus, J. O.; Kjaer, E. B.; Nielsen, P. H. *Tetrahedron Lett.*, **1991**, 32: 2825.

27. Suleimenova, A. M.; Kalinovskii, A. I.; Raldugin, A. I.; Shevtsov, S. A.; Bagryanskaya, I. Y.; Gatilov, Y. V.; Kuznetsova, T. V.; Elyakov, G. B. *Khim. Prir. Soedin*, **1988**, 24: 535.

28. Ravi, B. N.; Faulkner, D. J. *J. Org. Chem.*, **1978**, 43: 2127.

29. Wu, Y. C.; Hsiech, P. W.; Duh, C. Y.; Wang, S. K.; Song, K.; Fang, L. *J. Chin. Chem. Soc.*, **1992**, 39: 355.

30. Su, J.; Zhong, Y.; Lou, G.; Li, X.; Zeng, L.; Huang, Y.; Hu, S. *Huaxue Xuebao* (in Chinese). **1994**, 52: 813.

31. Kumar, S. M. D. Ph.D. thesis, Andhra University. **1992**.

32. Ishitsuka, M. O.; Kusumi, T.; Kakisawa, H. *Tetrahedron Lett.*, **1991**, 32: 2917.

33. Kobayashi, M.; Hirase, T. *Chem. Pharm. Bull.*, **1990**, 38: 2442.

34. Groweiss, A.; Kashman, Y.; Vanderah, D. J.; Tursch, B.; Cornet, P.; Breakman, J. C.; Daloze, D. *Bull. Soc. Chim. Belg.*, **1978**, 87: 277.

35. Leong, P. A.; Bowden, B. F.; Carroll, A. R.; Coll, J. C.; Meehan, G. V. *J. Nat. Prod.*, **1993**, 56: 521.

36. Greenland, G. L.; Bowden, B. F. *Aust. J. Chem.*, **1994**, 47: 2013.

37. Coll, J. C.; Hawes, G. B.; Liyanage, N.; Oberhanshi, W.; Wells, R. J. *Aust. J. Chem.*, **1977**, 30: 1305.

38. Kashman, Y.; Bodner, M.; Loya, Y.; Benayahu, Y. *Isr. J. Chem.*, **1977**, 16: 1.

39. Kobayashi, M.; Hirase, T. *Chem. Pharm. Bull.*, **1991**, 39: 3055.

40. Uchio, Y.; Nitta, M.; Toyata, J.; Nakayama, M.; Nishizona, Y.; Iwagawa, T.; Hase, T. *Chem. Lett.*, **1983**, 22: 613.

41. Bowden, B. F.; Coll, J. C.; Wills, R. H. *Aust. J. Chem.*, **1982**, 35: 621.

42. Iwagawa, T.; Shibata, Y.; Okamura, H.; Nakatani, M.; Shiro, M. *Tetrahedron Lett.*, **1994**, 35: 8415.

43. Iwagawa, T.; Nakamura, S.; Shibata, Y.; Masuda, T.; Okamura, H.; Nakatani, M.; Shiro, M. *Tetrahedron*, **1995**, 51: 5291.

44. Kudumi, T.; Igari, M.; Ishitsuka, M. O.; Ichikawa, A.; Itazone, Y.; Nakayama, N.; Kakisawa, H. *J. Org. Chem.*, **1990**, 55: 6268.

45. Su, J.; Long, K.; Peng, T.; He, C.; Clardy, J. *J. Am. Chem. Soc.*, **1986**, 108: 177.

46. Su, J.; Long, K.; Peng, T.; Zheng, Q.; Lin, X. *Huaxue Xuebao* (in Chinese), **1985**, 43: 796.

47. Su, J.; Long, K.; Peng, T.; Zeng, T.; Zheng, Q.; Lin, X. *Sci. Sin. Ser. B.*, **1988**, 31: 1172.

48. Mitsubishi Kasei Corp. Japanese Patent 8,161,318.

49. Fujiki, H.; Sugimura, T. *Cancer Surveys*, **1983**, 2: 539.

50. Yamauchi, O.; Omori, M.; Ninomiya, M.; Okuno, M.; Moriwaki, H.; Suganuma, M.; Fujiki, H.; Muto, Y. *Jpn. J. Cancer Res.*, **1991**, 82: 1234.

51. Narisawa, T.; Takahashi, M.; Niwa, M.; Fukaura, Y.; Fujiki, H. *Cancer Res.*, **1989**, 49: 3287.

52. Crombie, L.; Kenne, G.; Pattenden, G. *J. Chem. Soc., Chem. Commun.*, **1976**, 66: 4785.

53. Tius, M. A.; Fang, A. H. *J. Am. Chem. Soc.*, **1986**, 108: 3323.

54. Marshall, J. A.; DeHoff, B. S. *Tetrahedron Lett.*, **1986**, 27: 4873.

55. Kodarna, M.; Shiobara, Y.; Sumitomo, H.; Fukuzumi, K.; Mimami, H.; Niyamoto, Y. *Tetrahedron Lett.*, **1986**, 27: 2157.

56. Marshall, J. A.; DeHoff, B. S. *Tetrahedron*, **1987**, 43: 4849.

57. Dauben, W. G.; Saugier, R. K.; Fleischhauer, I. *J. Org Chem.*, **1985**, 50: 3767.

58. Marshall, J. A.; Cleary, D. G. *J. Org. Chem.*, **1986**, 51: 858.

59. Takahashi, T.; Nemoto, H.; Tsuji, J.; Miura, I. *Tetrahedron Lett.*, **1983**, 24: 3485.

60. Takahashi, T.; Nagashima, T.; Tsuji, J. *Tetrahedron Lett.*, **1981**, 22: 1359.

61. Takahashi, T.; Kitamura, K.; Tsuji, J. *Tetrahedron Lett.*, **1983**, 24: 4695.

62. Kato, T., Suzuki, M.; Kobayashi, T.; Moore, B. P. *J. Org. Chem.*, **1980**, 45: 1126.

63. Kato, T.; Suzuki, M.; Nakazima, Y.; Shimizu, K.; Kitahara, Y. *Chem. Lett.*, **1977**, 16: 705.

64. Aoki, M.; Uyehara, T.; Kato, T. *Chem. Lett.*, **1984**, 24: 695.

65. Marshall, J. A.; DeHoff, B. S.; Crooks, S. L. *Tetrahedron Lett.*, **1987**, 28: 527.

66. Marshall, J. A.; Crooks, S. L. *Tetrahedron Lett.*, **1987**, 28: 5081.

67. Wender, P. A.; Sieburth, S. McN.; Petraitis, J. J.; Singh, S. K. *Tetrahedron*, **1981**, 37: 3967.

68. Wender, P. A.; Holt, D. A.; Sieburth, S. McN. *J. Am. Chem. Soc.*, **1983**, 105: 3348.

69. Wender, P A.; Sieburth, S. McN. *Tetrahedron Lett.*, **1981**, 22: 2471.

70. Wender, P A.; Ternasky, R. J.; Sieburth, S. McN. *Tetrahedron Lett.*, **1985**, 26: 4319.

71. Marshall, J. A.; Jenson, T. M.; DeHoff, B. S. *J. Org. Chem.*, **1987**, 52: 3860.

72. Kodama, M.; Yoshino, S., Yamaguchi, S. *Tetrahedron Lett.*, **1993**, 34: 845.

73. Poter, N. A.; Magnin, D. R.; Wright, B. T. *J. Am. Chem. Soc.*, **1986**, 108: 2787.

74. Cox, N. J. G.; Pattenden, G. *Tetrahedron Lett.*, **1989**, 5: 621.

75. Uchio, Y.; Eguchi, S.: Nakayama, M.; Hase, T. *Chem. Lett.*, **1982**, 277.

76. Coll, J. C.; Mitchell, S. J.; Stokie, G. J. *Aust. J. Chem.*, **1977**, 30: 1859.

77. McMurry, J. E.; Matz, J. R.; Kees, K. L.; Bock, P. A. *Tetrahedron Lett.*, **1982**, 23: 1777.

78. Peng, L. Z.; Zhang, F. Z.; Mei, T. S.; Zhang, T.; Li, Y. L. *Tetrahedron Lett.*, **2003**, 44: 5921.

79. Zhang, F. Z.; Peng, L. Z.; Mei, T. S.; Zhang, T.; Li, Y. L. *Synth. Commun.*, **2003**, 33: 3761.

80. Takayanagi, H.; Kitano, Y.; Morinaka, Y. *Tetrahedron Lett.*, **1990**, 31: 3317.

81. Takayanagi, H.; Kitano, Y.; Morinaka, Y. *J. Org. Chem.*, **1994**, 59: 2700.

82. Takayanagi, H.; Sugiyama, S.; Morinaka, Y. *J. Chem. Soc. Perkin Trans. I*, **1995**, 751.

83. Li, W. D. Z.; Li, Y.; Li, Y. L. *Tetrahedron Lett.*, **1999**, 40: 965.

84. McMurry, J. E.; Rico, J. G.; Shih, Y. N. *Tetrahedron Lett.*, **1989**, 30: 1173.

85. Li, Y. L.; Yue, X. J. *Tetrahedron Lett.*, **1993**, 34: 2799.

86. Yue, X. J.; Li, Y. L. *Bull. Soc. Chim. Belg.*, **1994**, 103: 35.

87. Moore, B. P. *Nature* (London), **1966**, 211: 746.

88. Birch, A. J.; Brown, M. W. V.; Corrie, J. E. T.; Moore, B. P. *J. Chem. Soc., Perkin Trans.I*, **1972**, 2653.

89. Patil, V. D.; Nayak, U. R.; Dev, S. *Tetrahedron*, **1973**, 29: 341.

90. Vanderah, D. J.; Rutledge, N.; Schmitz, F. J.; Ciereszko, L. S. *J. Org. Chem.*, **1978**, 43: 1614.

91. Kodama, M.; Matsuki, Y.; Ito, S. *Tetrahedron Lett.*, **1975**, 35: 3065.

92. Takayanagi, T.; Uyehara, Y.; Kato, T. *J. Chem. Soc. Chem. Commun.*, **1978**, 359.

93. Schwabe, R.; Farkas, I.; Pfander, H. *Helv. Chem. Acta*, **1988**, 71: 292.
94. Ishihara, K.; Nakamura, H.; Yamamoto, H. *Synlett.*, **2000**, 1245.
95. Yue, X. J.; Li, Y. L. *Synthesis*, **1996**, 6: 736.
96. Li, W. D.; Li, Y.; Li, Y. L. *Science in China (Series B)*, **1993**, 36: 1161.
97. Li, Y. Ph.D. thesis, Lanzhou University, **1993**.
98. Mao, J. M.; Li, Y.; Hou, Z. J.; Li, Y. L.; Liang, X. T. *Science in China (Series B)*, **1992**, 35: 257.
99. Poet, S. E.; Ravi, B. N. *Aust. J. Chem.*, **1982**, 35: 77.
100. Bowden, B. F.; Coll, J. C.; Tapiolas, D. M. *Aust. J. Chem.*, **1983**, 36: 2289.
101. Subrahmanyam, C.; Rao, C. V.; Anjaneyulu, V.; Satyanarayana, P.; Rao, P. V. S.; Ward, R. S.; Pelter, A. *Tetrahedron*, **1992**, 48: 3111.
102. Shin, J.; Fenical, W. *Tetrahedron*, **1993**, 49: 515.
103. Anjaneyulu, A. S. R.; Raju, K. V. S. *Indian J. Chem.*, **1995**, 34B: 463.
104. Kuznetsova, T. A.; Popov, A. M.; Agafonova, I. G. *Khim. Prir. Soedin.*, **1989**, 1: 137.
105. Bowden, B. F.; Coll, J. C.; Mitchell, S. J.; Kazlauskas, R. *Aust. J. Chem.*, **1981**, 34: 1551.
106. Blackman, A. J.; Bowden, B. F.; Coll, J. C.; Frick, B.; Mahendran, M.; Mitchell, S. J. *Aust. J. Chem.*, **1982**, 35: 1873.
107. Duh, C. Y.; Wang, S. K.; Weng, Y. L.; Chiang, M. Y.; Dai, C. F. *J. Nat. Prod.*, **1999**, 62: 1518.
108. Rodríguez, A. D.; Martínez, N. *Experientia*, **1993**, 49: 179.
109. Rodríguez, A. D.; Acosta, A. L. *J. Nat. Prod.*, **1997**, 60: 1134.
110. Liu, Z. S.; Li, W. D. Z.; Peng, L. Z.; Li, Y.; Li, Y. L. *J. Chem. Soc., Perkin Trans. 1*, **2000**, 4250.
111. Li, Y. L.; Liu, Z. S.; Lan, J.; Li, J.; Peng, L. Z.; Li, W. D. Z.; Li, Y. L.; Chan, A. S. C. *Tetrahedron Lett.*, **2000**, 41: 7465.
112. Zhang, T; Liu, Z. S.; Li, Y. L. *Synthesis*, **2001**, 393.
113. Xing, Y. C.; Cen, W.; Lan, J.; Li, Y.; Li, Y. L. *J. Chin. Chem. Soc.*, **1999**, 46: 595.
114. Liu, Z. S.; Zhang, T.; Li, Y. L. *Tetrahedron Lett.*, **2001**, 42: 275.
115. Liu, Z. S.; Li, W. D. Z.; Li, Y. L. *Tetrahedron: Asymmetry*, **2001**, 12: 95.
116. Lan, J.; Liu, Z. S.; Yuan, H.; Peng, L. Z.; Li, W. D.; Li, Y.; Li. Y. L.; Chan, A.-S. *Tetrahedron Lett.*, **2000**, 41: 2181.
117. Rodríguez, A. D.; Li, Y. X.; Dhasmana, H.; Barnes, C. L. *J. Nat. Prod.*, **1993**, 56: 1101.
118. Rodríguez, A. D.; Piña, I. C.; Acosta, A. L.; Ramírez, C.; Soto, J. J. *J. Org. Chem.*, **2001**, 66: 648.
119. Tius, M. A.; Reddy, N. K. *Tetrahedron Lett.*, **1991**, 32: 3605.
120. Dauben, W. G.; Wang, T. Z.; Stephens, R. W. *Tetrahedron Lett.*, **1990**, 31: 2393.
121. McMurry, J. E.; Dushin, R. G. *J. Am. Chem. Soc.*, **1990**, 112: 6942.
122. McMurry, J. E.; Dushin, R. G. *J. Am. Chem. Soc.*, **1989**, 111: 8928.
123. Marshall, J. A.; Andrews, R. C.; Lebioda, L. *J. Org. Chem.*, **1987**, 52: 2378.
124. Marshall, J. A.; Royce, R. D. Jr. *J. Org. Chem.*, **1982**, 47: 693.
125. Liu, Z. S.; Peng, L. Z.; Li, W. D. Z.; Li, Y. L. *Synlett.*, **2003**, 1977.
126. Uchio, Y.; Eguchi, S.; Nakayama, M.; Hase, T. *Chem. Lett.*, **1982**, 277.

127. Nishitani, K.; Konomi, T.; Mimaki, Y.; Tsunoda, T.; Yamakawa, K. *Heterocycles*, **1993**, 36: 1957.

128. Nishitani, K.; Konomi, T.; Okada, K.; Yamakawa, K. *Heterocycles*, **1994**, 37: 679.

129. Kodama, M.; Kakahashi, T.; Ito, S. *Tetrahedron Lett.*, **1982**, 23: 5175.

130. Marshall, J. A.; Crooks, S. L.; DeHoff, B. S. *J. Org. Chem.*, **1988**, 53: 1616.

131. Marshall, J. A.; Gung, W. Y. *Tetrahedron Lett.*, **1988**, 29: 3899.

132. Taber, D. F.; Song, Y. *J. Org. Chem.*, **1997**, 62: 6603.

133. Stille, J. K. *Angew Chem., Int. Ed. Engl.*, **1986**, 25: 508.

134. Farina, V.; Krishnamurthy, V.; Scott, W. J. *Org. React.*, **1997**, 50: 1.

135. Duncton, M. A. J.; Pattenden, G. *J. Chem. Soc., Perkin Trans. I*, **1999**, 1235 and references cited therein.

136. Kashtanova, N. K.; Lisina, A. I.; Pentegova, V. A. *Khim. Prir. Soedin.*, **1968**, 4: 52.

137. Pattenden, G.; Smithies, A. J. *J. Chem. Soc., Perkin Trans. I*, **1996**, 57.

# 7

# MEDICINAL CHEMISTRY OF GINKGOLIDES FROM *GINKGO BILOBA*

KRISTIAN STRØMGAARD

*Department of Medicinal Chemistry, The Danish University of Pharmaceutical Sciences, Copenhagen, Denmark*

## 7.1   INTRODUCTION

Ginkgolides were first isolated from bitter principles of *Ginkgo biloba* by Furukawa in 1932, but it was not until 1967 that Nakanishi and coworkers elucidated their highly complex structure. Later, interesting biological activities for ginkgolides were discovered: In 1985, it was demonstrated that ginkgolides are potent antagonists of the platelet-activating factor (PAF) receptor, and recently, it was shown that ginkgolides also antagonize the glycine receptor. Extensive structure activity relationship (SAR) studies have been carried out for the study of these receptors. Moreover, ginkgolides have served as an inspiration in studies ranging from total chemical synthesis, elucidation of the biosynthetic pathways, and unraveling of biological effects. The chemistry and biology of the terpene trilactones in general was recently reviewed.[1]

### 7.1.1   *Ginkgo biloba* Extract

The *Ginkgo* tree (*Ginkgo biloba* L.) (Figure 7-1), is the only surviving member of a family of trees, *Ginkgoaceae*, that appeared 170 million years ago, and it is often called a "living fossil."[2] It is distinct from all other living plants and is often categorized in its own division, Ginkgophyta. *G. biloba* is dioecious, namely, the male

*Medicinal Chemistry of Bioactive Natural Products* Edited by Xiao-Tian Liang and Wei-Shuo Fang
Copyright © 2006 John Wiley & Sons, Inc.

**Figure 7-1.** A *Ginkgo biloba* tree.

and female structures exist on separate trees. Ginkgo trees can grow over 35 m high, with the main stem up to 10 m in girth and can reach ages in excess of 1000 years.[3] The tree is characterized by fan-shaped leaves split in the middle, which served as inspiration for the name "biloba" meaning two-lobed (Figure 7-2). The shape of

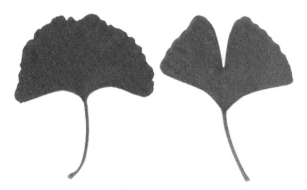

**Figure 7-2.** Leaves from *Ginkgo biloba*.

the leaves also inspired German poet, scientist, botanist, and philosopher Johann Wolfgang von Goethe (1749–1832), to write a poem called "Gin(k)go biloba," which is dedicated to his former lover Marianne von Willemer.[4]

The earliest records on the use of G. *biloba* as medicine dates back to 1505 AD,[5] where G. *biloba* treated aging members of the royal court for senility. Around 1965, leaf preparations of G. *biloba* were introduced to the Western world by Dr. Willmar Schwabe,[5] and together with Beaufour-Ipsen (now Ipsen), a standardized G. *biloba* extract called EGb 761 was developed.[6] Many G. *biloba* products have entered the market, and G. *biloba* extract is now among the best-selling herbal medications worldwide. G. *biloba* was originally grown throughout China and Korea, but it was introduced to Europe and North America in the eighteenth century. Today over 50 million G. *biloba* trees are grown, particularly in China, France, and South Carolina in the United States, producing approximately 8000 tons of dried leaves each year.[7]

EGb 761 is standardized with respect to the content of terpene trilactones (6%) and flavonoids (24%).[5] The terpene trilactones are the five ginkgolides and bilobalide, whereas the flavonoids are mainly flavonol-*O*-glycosides. EGb 761 contains many other components, including proanthocyanidins (prodelphinidins) and organic acids, particularly ginkgolic acids (anarcardic acids), which have allergenic properties; hence, the content in EGb 761 is limited to 5 ppm.[5] In studies of the pharmacological effects of G. *biloba*, particularly on effects in the central nervous system (CNS), EGb 761 has been widely used,[8–12] and the effects include improvement of cognition, antioxidant effects, increased cerebral blood flow and circulation, modification of neurotransmission, and protection against apoptosis.[11]

The most extensive clinical studies with EGb 761 have focused on alleviation of Alzheimer's disease (AD).[13–15] Two studies with a total of 549 AD patients showed that treatment with EGb 761 significantly slowed the loss of cognitive symptoms of dementia.[16,17] However, two recent clinical studies with more then 400 patients have cast doubt on the positive clinical effect of EGb 761.[18,19] Several meta-analyses have reviewed over 50 clinical studies using EGb 761 for treatment of dementia and cognitive functions associated with AD, and these studies conclude that EGb 761 have a small but significant effect on objective measures of cognitive function in AD, without significant adverse effects.[20] Currently, at least two major clinical trials are ongoing. The U.S. National Institutes of Health is sponsoring the Ginkgo evaluation of memory (GEM) study, which has enrolled more than 3000 older people recruited from four medical centers in the United States. In France, the pharmaceutical company Ipsen is sponsoring another clinical trial, the GuidAge study, which examines prevention of AD in patients over the age of 70 with memory impairment. The results from these two studies will obviously be of major importance in evaluating and determining the effects of GBE in relation to dementia. Therefore, EGb 761 might be beneficial in relieving symptoms of AD, although the reported effects are often small. However, in light of the current lack of treatment for AD patients,[21,22] EGb 761 could prove useful as an alternative to the currently available treatments.

It is not understood which components of EGb 761 are efficacious, but it is assumed that the major components flavonoids and terpene trilactones play important roles. However, when dealing with effects in the CNS, it is assumed that the flavonoids do not penetrate the blood-brain barrier (BBB),[10] whereas it has been suggested that terpene trilactones do penetrate the BBB.[23-25] However, in contrast to the wealth of studies that have been performed with EGb 761, far fewer studies have looked at the effect of the individual components of these extracts, in particular flavonoids and terpene trilactones.

### 7.1.2    Isolation and Structure Elucidation of Ginkgolides

The terpene trilactones, ginkgolides and bilobalide,[26] are the unique constituents of *G. biloba*, which are found exclusively in the *G. biloba* tree. The ginkgolides are diterpenes with a cage skeleton consisting of six five-membered rings, including three lactones, a tetrahydrofuran ring, and a spiro[4.4]nonane skeleton, and a characteristic *tert*-butyl group. The ginkgolides vary only in the number and positions of their hydroxyl groups. Bilobalide is also a terpene trilactones with a structure similar to the ginkgolides and is the major single component in EGb 761, comprising about 3% of the total extract, whereas the five ginkgolides take up another 3%.

bilobalide (BB)

|  | R$_1$ | R$_2$ | R$_3$ |
|---|---|---|---|
| ginkgolide A (GA) | H | H | OH |
| ginkgolide B (GB) | OH | H | OH |
| ginkgolide C (GC) | OH | OH | OH |
| ginkgolide J (GJ) | H | OH | OH |
| ginkgolide M (GM) | OH | OH | H |

The complex structures of ginkgolides were discovered in 1967 by pioneering work of Nakanishi et al. Since then, a wealth of studies involving ginkgolides have been published, including intriguing total syntheses pioneered by Corey, studies of their biosynthesis revealing a novel, general biosynthetic pathway, and preparation of several derivatives for SAR studies at the PAF receptor and, more recently, at the glycine (Gly) receptors.

The structural studies by Nakanishi et al. that led to the discovery of the structures of ginkgolides rank among the greatest achievements in natural products research, and the fascinating account of these studies has been given elsewhere.[27,28]

Briefly, ginkgolides were isolated from the root bark of *G. biloba*, and after extensive purification, multigram quantities of ginkgolides A, B, and C were achieved, whereas 200 mg of ginkgolide M was isolated. Extensive nuclear magnetic resonance studies, which included the pioneering work on application of the nuclear Overhauser effect (NOE), and numerous chemical reactions providing about 70 derivatives, was required for the determination of the truly complex structures of ginkgolides.[29–34] Around the same time, Okabe et al. determined the structures of GA, GB, and GC from leaves of *G. Biloba* with x-ray crystallography.[35,36] Later, another ginkgolide, ginkgolide J (GJ), was isolated from the leaves of *G. biloba*.[37]

The distinctive stability of the ginkgolide core was also encountered by Nakanishi et al. For example, treatment of GA with 50% NaOH at 160 °C for 30 minutes resulted in the loss of two carbons as oxalic acid to give a hemiacetal bisnor GA. Treatment of GA with sodium dichromate in concentrated sulfuric acid simply oxidized the hydroxyl lactone, which subsequently undergoes a photocyclization to give photodehydro-GA.

photodehydro-GA

Since the initial structural studies, a substantial effort has been made to simplify the isolation and quantification of ginkgolides, which was recently reviewed by van Beek.[38,39] The first step is extraction from the leaves, which in most cases has been accomplished with various water-containing solvent systems, thereby excluding apolar constituents and collecting the ginkgolides. A more simplified extraction method was recently published, in which the dried leaves are boiled in dilute hydrogen peroxide and extracted with ethyl acetate to give an off-white powder with a content of 60–70% ginkgolides and bilobalide.[40] Having the crude mixture of ging-golides (and bilobalide), it is a challenge to isolate the individual ginkgolides. A simple improvement in separation has been achieved by van Beek and Lelyveld with silica gel impregnated with sodium acetate, which disrupts the internal hydrogen bonds between the hydroxyl groups.[41]

Because ginkgolides lack common chromophores, ultraviolet detection is not suitable and other detection methods such as refractive index (RI), evaporation light scattering detection (ELSD), and mass spectrometry (MS) have been used. Several recent reports use high performance liquid chromatography (HPLC) MS to separate and quantify the content of ginkgolides and bilobalide, using either electrospray ionization (ESI)[42] or atmospheric pressure chemical ionization (APCI).[43,44] As an alternative to MS, ELSD has also been successfully applied to quantify ginkgolides and bilobalide in *G. biloba* extracts.[45,46]

### 7.1.3 Biosynthesis of Ginkgolides

Early attempts to reveal the biosynthetic route of ginkgolides, using [2-[14]C]-acetate and dl-[2-[14]C]-mevalonate, suggested the overall terpenoid origin of ginkgolides, and it was believed that ginkgolides were biosynthesized through the conventional mevalonate pathway.[47] However, only recently it was realized that the two conventional precursors, dimethylallyl pyrophosphate (DMAPP) and isopentenyl pyrophosphate (IPP), can participate in what is now known as the nonmevalonate pathway.

Independently, Rohmer and Arigoni et al. showed the existence of the nonmevalonate or deoxylulose phosphate pathway, in which pyruvate and glyceraldehyde 3-phosphate react to produce 2C-methyl-D-erythritol 2,4-cyclodiphosphate and ultimately DMAPP and IPP.[48–51] Actually, Arigoni et al. were studying the ginkgolide biosynthesis with a *G. biloba* embryo system and [13]C-labeled glucose, and they showed that ginkgolides were biosynthesized via the nonmevalonate pathway.[49]

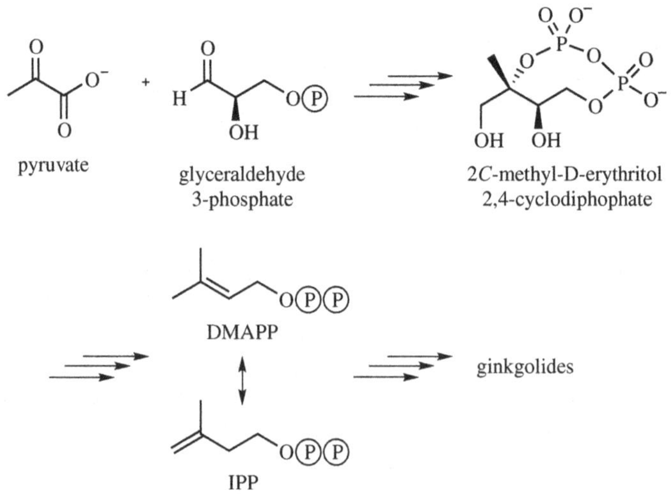

pyruvate      glyceraldehyde          2C-methyl-D-erythritol
              3-phosphate             2,4-cyclodiphophate

DMAPP

ginkgolides

IPP

Recently, a key player in the ginkgolide biosynthesis has been isolated and characterized: A cDNA that encodes *G. biloba* levopimaradiene synthase, a diterpene synthase involved in ginkgolide biosynthesis, was isolated and characterized.[52] Similarly, a new geranylgeranyl diphosphate synthase gene from *G. biloba*, was characterized.[53]

### 7.1.4  Chemistry of Ginkgolides

Since the discovery of the ginkgolides in 1967, numerous studies have explored the chemistry of these structures. The ginkgolide skeleton is made up of six highly oxygenated rings including 10–12 stereogenic centers, and a *tert*-butyl group, which provides an enormous challenge for total synthesis. Total syntheses of both GA[54] and GB[55] was achieved by Corey et al. in 1988, which rank among the greatest achievements in total synthesis of natural products. Recently, a novel synthesis of GB was been achieved by Crimmins et al.[56,57]

Several synthetic studies of ginkgolides have been accomplished by Weinges et al.[58–63] Most recently, they described an approach to the preparation of radiolabeled ginkgolides, in this case [$^{14}$C]-GA, although the actual radioligand was not synthesized.[64] In 2004, two radiolabeled analogs of ginkgolides were in fact prepared, including [$^3$H]GB[65] and GB labeled with [$^{18}$F] for positron emission tomography (PET) studies.[66] Thus, it is anticipated that these radiotracers will provide important information on biodistribution and binding sites of ginkgolides.

A few studies have looked at the conversion of GC into GB, which includes elimination of the 7-OH group present in GC. Weinges and Schick and Corey et al. have independently described similar four-step procedures, involving protection of 1-OH of GC followed by alcohol deoxygenation, and deprotection to provide GB.[67,68] A convenient, two-step procedure was described in a patent by Teng,[69] in which GC was reacted with triflic anhydride yielding exclusively 7-O-triflate-GC that was reduced with $Bu_4NBH_4$ to give GB. These studies revealed different reactivities of 1-, 7-, and 10-OH of ginkgolides. Generally, 1- and 10-OH are the most reactive OH-groups[68,70]: A bulky silyl group reacts preferentially at 1-OH, whereas all benzyl reagents react at the 10-OH, and triflic anhydride reacts exclusively at 7-OH. Similarly, it was recently shown that acetylation, which generally takes place at 10-OH under strongly acidic conditions takes place at 7-OH of GC.[70]

During the structural studies of ginkgolides, several analogs were synthesized. The so-called iso-ginkgolide derivatives, which stem from a translactonization of ring E, are of particular interest. In 2002, an x-ray crystallographic structure of an iso- ginkgolide derivative, 1,6,10-triacetate-isoGC, was obtained, and a mechanism for the formation of iso-derivatives that includes opening of ring E, stabilization of the intermediate by hydrogen bonding to 3-OH, and capture by acetic anhydride was suggested.[70]

## 7.2 GINKGOLIDES AND THE PAF RECEPTOR

The first direct interaction of ginkgolides with an important biological target was shown in 1985 when it was discovered that ginkgolides, particularly GB, are antagonists of the PAF receptor.[71–73]

The PAF receptor is a member of the G protein-coupled receptor (GPCR) family and has been identified in several cells and tissues, including those in the CNS.[74] PAF, the native phospholipid agonist of the PAF receptor, is involved in the modulation of various CNS and peripheral processes.[75] PAF receptor antagonists have been suggested as treatments for various inflammatory diseases, and they were pursued by several pharmaceutical companies as anti-inflammatory agents. Recently, the PAF receptor was suggested as a target for slowing the progressions of neurodegenerative diseases.[76] To date, however, no PAF receptor antagonist has been registered as a drug.

PAF (1-O-alkyl-2-acetyl-sn-glycero-3-phosphocholine) is involved in several events in the CNS, including modulation of long-term potentiation (LTP),[77–80] increase of intracellular $Ca^{2+}$,[81] and immediate early gene expression.[82–84] In LTP, PAF has been suggested as a retrograde messenger;[77] however, studies with PAF receptor knockout mice produced diverging results.[85,86] Recently, it was demonstrated that GB could inhibit hippocampal LTP, an effect mediated by PAF receptor antagonism, rather than by Gly receptor antagonism (see Section 7.3).[87]

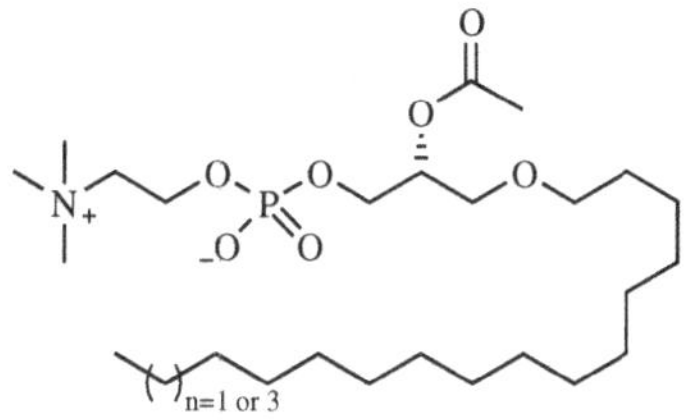

PAF

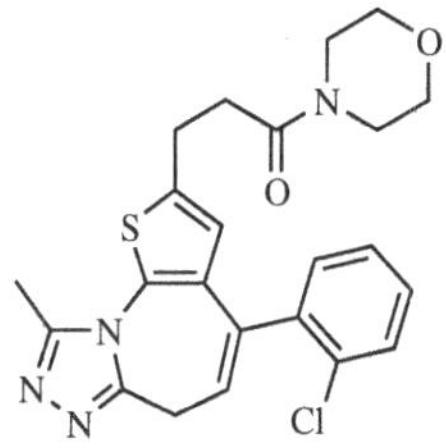

WEB 2086

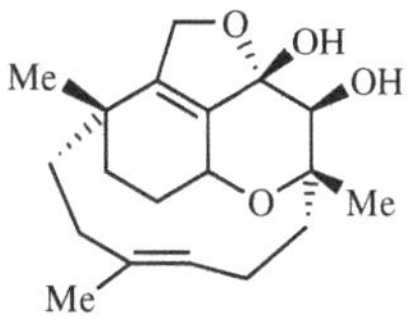

Phomactin A

A remarkable feature of PAF receptor antagonists is their structural diversity, ranging from WEB 2086, phomactin A to ginkgolides, all structurally different from PAF, but still competitive antagonists. Until recently, SAR studies of ginkgolides on the PAF receptor have focused on derivatives of GB, the most potent antagonist of the PAF receptor, and in all cases, the derivatives were evaluated for their ability to prevent PAF-induced aggregation of rabbit platelets.

In the initial description of the ability of ginkgolides to inhibit the PAF receptor, it was shown that GB was the most potent ginkgolide with an $IC_{50}$ of 1.88 µM, GA was slightly less potent, and GC was a weak antagonist (Table 7-1).[88] A few years later, methoxy and ethoxy analogs of GA, GB, and GC were prepared, in which the alkyl groups were introduced at C-1 or C-10 by reaction with diazoalkane to yield

**TABLE 7-1. Pharmacological Activity of Methoxy- and Ethoxy Analogs of GA, GB, and GC**

| Compounds | $R_1$ | $R_2$ | $R_3$ | $IC_{50}$ ($\mu$M) |
|---|---|---|---|---|
| GA | H | H | H | 0.74 |
| GB | OH | H | H | 0.25 |
| GC | OH | H | OH | 7.1 |
| 10-Me-GA | H | $CH_3$ | H | 13 |
| 10-Et-GA | H | $CH_3CH_2$ | H | 62 |
| 1-Me-GB | $OCH_3$ | H | H | 0.66 |
| 10-Me-GB | OH | $CH_3$ | H | 0.29 |
| 1-Et-GB | $OCH_3CH_2$ | H | H | 1.1 |
| 10-Et-GB | OH | $CH_3CH_2$ | H | 7.2 |
| 1-Me-GC | $OCH_3$ | H | OH | 4.2 |
| 10-Me-GC | OH | $CH_3$ | OH | 3.0 |
| 1-Et-GC | $OCH_3CH_2$ | H | OH | 8.5 |
| 10-Et-GC | OH | $CH_3CH_2$ | OH | 9.3 |

mixtures of 1- and 10-substituted analogs that were separated by column chromatography.[89] Interestingly, 1- and 10-methoxy analogs of GB were equipotent to GB, whereas the corresponding ethoxy analogs were less potent. The 10-substitued analogs of GA were significantly less potent than GA, whereas methyl analogs of GC were more potent and ethyl analogs were equipotent to GC (Table 7-1).

Corey and Gavai investigated various intermediates in the total syntheses of GA[54] and GB,[55] and they found that the lactone F was not essential for activity and could be replaced by other lipophilic groups,[90] whereas the unique *tert*-butyl group was critical for activity.[91] Villhauer and Anderson synthesized the CDE ring system of ginkgolides, which was ineffective as a PAF receptor antagonist.[92]

The most comprehensive SAR study on ginkgolides and PAF receptor was performed by Park et al., who synthesized more than 200 derivatives of GB, with particular focus on aromatic substituents at 10-OH.[93] These derivatives were generally synthesized by treatment of GB with a base and a benzyl halide derivative providing, in most cases, selective derivatization at 10-OH. Whereas GB had an $IC_{50}$ = 0.258 $\mu$M, most 10-*O*-benzylated derivatives were more potent, including 10-(3,5-dimethyl-2-pyridinyl)-methoxy-GB ($IC_{50}$ of 0.0245 $\mu$M) that was ten-fold more potent than GB. The same group also investigated elimination products of GB, as well as derivatives that were bridged between 1- and 10-OH, but these analogs were much less potent than GB.[94] GB derivatives, many of which were identical to those synthesized by Park et al. were prepared by Hu et al. with a slightly modified procedure and obtained benzylated GB derivatives that were more potent than GB.[95,96] Later, Hu et al. prepared various degradation and elimination products of GA and GB, as well as amide derivatives lacking rings C and D, but in all cases, decreased PAF receptor antagonism was observed.[97,98]

One goal of SAR studies is to put forward a pharmacophore model that can elucidate the activities of synthesized derivatives, as well as predict the activity of novel derivatives. A three-dimensional quantitative SAR study[99] was attempted for ginkgolides and the PAF receptor, using comparative molecular field analysis and 25 ginkgolide analogs, mainly those synthesized by Corey et al.[54,55,90] In agreement with the SAR studies just described, this pharmacophore model predicted that substituents in the 10-OH position of GB would improve activity.

Clarification of the interactions between ginkgolides and the PAF receptor on a molecular level can be carried out with photolabeling techniques; therefore, photoactivatable derivatives of GB and GC were recently prepared.[100] This study generated highly potent analogs, with benzophenone, trifluoromethyldiazirine, and tetrafluorophenyl azide groups in the 10-OH position of GB as the most active with $K_i$ values of 90–150 nM. These derivatives are promising tools for characterizing the PAF receptor–ginkgolide interaction. This study also provided the first evaluation of the interaction of ginkgolides with the cloned PAF receptor using a radioligand binding assay. In another recent study, the effect of acetate derivatives of ginkgolides was investigated, which showed that acetylation generally decreases antagonistic effects at the PAF receptor.[70] Recently, a study looked at the effect of substitutions at the 7-position of ginkgolides and showed that in contrast to previous reports, potent ginkgolide derivatives could be prepared by introducing chlorine; 7-chloro-GB, with a $K_i = 110$ nM, was eight-fold more potent than GB.[101]

In 2003, Krane et al. used a functional assay, microphysiometry, to evaluate the effects of ginkgolides on the human PAF receptor. GB and 10-*O*-benzyl-GB showed 81% and 93% (100-μM concentration) inhibition, respectively, of the PAF receptor response, the latter being equivalent to the effect of WEB2086. *G. biloba* extract mixtures were also tested, which indicated potential synergistic effects among the components in the extract.[102]

Thus, several ginkgolide derivatives that have been prepared and tested for their ability to antagonize the PAF receptor have resulted in a good understanding of the structural features required for inhibition. Clearly, a step forward in this understanding this interaction would come from photolabeling experiments or similar studies providing a more detailed insight. More investigation is required to determine the potential physiological effects in the CNS.

## 7.3   GINKGOLIDES AND GLYCINE RECEPTORS

Until 2002, the only specific target for ginkgolides was the PAF receptor, whose importance in CNS functions is not clear. Therefore, the recent finding that GB was a potent and selective antagonist of Gly receptors[103–105] provided an important target for additional studies on the neuromodulatory properties of ginkgolides.

The Gly receptors are found primarily in the spinal cord and brain stem but also in higher brain regions such as the hippocampus. They are, together with γ-aminobutyric acid (GABA$_A$) receptors, the main inhibitory receptors in the CNS.[106,107] Both Gly and GABA$_A$ receptors are ligand-gated ion channels that, together with nicotinic acetylcholine (nACh) and serotonin (5-HT$_3$) receptors, constitute a superfamily of membrane-bound receptors that mediate fast chemical synaptic transmission in the CNS.[108] Gly receptors share several structural similarities with these receptors, including a pentameric arrangement of subunits, each composed of four transmembrane domains (M1–M4) and an extracellular N-terminal 15-residue cysteine-loop motif.[109] Recently, there has been a renewed interest in ligands for Gly receptors, because modulators of the Gly receptor function could serve as muscle relaxants as well as sedative and analgesic agents.[110] A recent study suggests that α3 homomeric Gly receptors are important mediators of pain.[111]

glycine

strychnine                    picrotoxinin

The best known Gly receptor ligand is strychnine, which is a potent competitive antagonist and has been a highly valuable tool in the studies of Gly receptors. Another antagonist is the convulsant picrotoxin, which is a noncompetitive antagonist of both $GABA_A$ and Gly receptors and is believed to bind to the channel-forming region of the receptors. Electrophysiological studies showed that GB antagonizes Gly receptors in neocortical slices[105] and hippocampal cells,[103] and that the inhibition is noncompetitive, use-dependent, and probably voltage-dependent, which thus indicates that GB binds to the central pore of the ion channel. It was also shown that GC and GM were almost equipotent to GB, whereas GA and GJ were significantly less potent,[104,105] which suggests an important function of the 1-OH group present in GB, GC, and GM, but absent in GA and GJ. This assumption was corroborated by molecular modeling studies showing a striking structural similarity between picrotoxinin, the active component of picrotoxin, and ginkgolides;[105] the x-ray crystallographic structures of picrotoxinin and GB were overlaid showing a remarkable similarity in the three-dimensional structure, as indicated below. Thus, ginkgolides are highly useful pharmacological tools for studying the function and properties of Gly receptors.

picrotoxinin                    ginkgolide B

Although these studies showed that ginkgolides are selective antagonists of Gly receptors, a recent study looking at recombinant $\alpha_1\beta_2\gamma_{2L}$ $GABA_A$ receptors showed that ginkgolides are in fact moderately potent antagonists of these receptors, with $K_i$ values around 15 μM.[112] Another study suggested that GB might exhibit subunit selectivity at Gly receptors, as GB preferentially blocks chloride channels formed by heteromeric Gly receptors in hippocampal pyramidal neurons of rat.[113] The study, however, only showed a small difference, and studies on cloned Gly receptors are required to clarify this point.

Because the interaction of ginkgolides and Gly receptors was only discovered recently, the number of SAR studies carried out is also limited. In addition to that mentioned here, one study has looked at the five native ginkgolides, as well as at one synthetic ginkgolide (HE-196), showing that GB and GC are the most potent with $IC_{50}$ values of 0.273 and 0.267 µM, respectively, whereas GA and GJ had $IC_{50}$ values around 2 µM.[114] In the same study, the effects on glutamate-gated ion channels, such as $N$-methyl-D-aspartate and α-amino-3-hydroxy-5-methyl-4-isoxazole propionic acid receptors, was examined, and ginkgolides had essentially no effect on these receptors.[114]

To date, only one SAR study has been carried out in which 40 novel ginkgolide derivatives were tested for their ability to antagonize the Gly receptor.[115] GC served as the template, because GC is one of the most potent ginkgolides, and GC has the most hydroxyl groups, which thus allows more diverse functionalizations of these groups. All derivatives were substituted on one or more of the hydroxyl groups in GC, including rearranged iso-ginkgolides and substituents of varying size. The derivatives were investigated in a fluorescence-based membrane potential assay,[116] and somewhat surprisingly, all novel derivatives were significantly less potent than GC.[115] Thus, this result suggests important functions for the hydroxyl groups in ginkgolides.

The studies carried out so far clearly demonstrate that the effect on Gly receptors is distinct from the effect on PAF receptors, as SAR studies have shown that the structural requirements are different for the two receptors. The potential importance of the antagonism of Gly receptors also remains to be investigated.

## 7.4   VARIOUS EFFECTS OF GINKGOLIDES

The interactions of ginkgolides with the PAF and Gly receptors described above are the best-characterized interactions of ginkgolides with targets found in the CNS. Varying effects of ginkgolides have been observed using numerous assays with different tissues and conditions. No study has provided a clear-cut target for ginkgolides but instead has introduced many different pharmacological effects that may or may not be related to the targets described.

Ginkgolides have, however, modulating effects on several targets. One of the best characterized is the modulation of peripheral benzodiazepine (PB) receptors. The PB receptors are located mainly in peripheral tissues and glial cells in the brain and are distinct from the benzodiazepine site on $GABA_A$ receptors.[117,118] The function of PB receptors is not entirely clear, but they have been suggested to be involved in steroidogenesis, cell proliferation, and stress and anxiety disorders. The latter is supported by an increase in the number of PB receptors in specific brain regions in neurodegenerative disorders and after brain damage.[117] Initially it was shown that the ligand binding capacity of PB receptors was decreased along with a decrease in protein and mRNA expression after treatment with ginkgolides.[119,120] Recent studies have shown that the primary action of GB is the inhibition of PB receptor expression,[121] which is mediated through binding to a transcription factor, and it has been suggested that GB regulates excess glucocorticoid formation, through the PB receptor-controlled steroidogenesis.[122,123]

Numerous studies have shown that ginkgolides protect against various CNS incidents, such as ischemia and cerebrovascular and traumatic brain injury, as well as inflammation[11]; It is not understood how ginkgolides exert this effect, but it has been suggested that they interference with oxygen free radicals,[124] or reduction of glutamate-induced neurotoxicity[125,126] could play a role. Moreover, ginkgolides might reduce cytotoxic nitric oxide.[127] Finally, in vitro studies with ginkgolides have also shown cardioprotective effects.[124,128,129]

## 7.5 CONCLUSIONS AND OUTLOOK

The intriguing and complex structures of ginkgolides were discovered more than 35 years ago by a dazzling endeavor. Since then, many chemical and biological studies have been carried out including total syntheses of these complex natural products and elucidation of a novel, and general, biosynthetic pathway. The finding that ginkgolides are antagonists of the PAF receptor led to extensive SAR studies providing important information of the ligand–receptor interaction. Certainly, we will see even more detailed studies of this interaction in the future, when photoactivatable and radiolabeled ginkgolides are employed. The recent finding that ginkgolides are also antagonists of inhibitory ligand-gated ion channels, particularly glycine receptors, has opened up a new area for SAR studies as well as for detailed mechanistic studies. Moreover, the physiological implications of this effect, for example, for people taking *G. biloba* extract, are not yet known.

## ACKNOWLEDGMENT

Dr. Stanislav Jaracz is thanked for providing illustrations to Figures 7-1 and 7-2.

## REFERENCES

1. Strømgaard, K.; Nakanishi, K. Chemistry and biology of terpene trilactones from Ginkgo biloba. *Angew. Chem. Int. Ed.*, **2004**, 43: 1640–1658.
2. Major, R. T. The ginkgo, the most ancient living tree. *Science*, **1967**, 157: 1270–1273.
3. Jacobs, B. P.; Browner, W. S. Ginkgo biloba: A living fossil. *Am. J. Med.*, **2000**, 108: 341–342.
4. Goethe, J. W. *West-Ostlicher Divan*. Tübingen: Niemeyer, 1819.
5. Drieu, K.; Jaggy, H. History, development, and constituents of EGb 761. In *Medicinal and Aromatic Plants—Industrial Profiles, Vol. 12: Ginkgo biloba*. Amsterdam: Harwood Academic Publishers, **2000**, 267–277.

6. McKenna, D. J.; Jones, K.; Hughes, K. Efficacy, safety, and use of Ginkgo biloba in clinical and preclinical applications. *Alternat. Therap.*, **2001**, 7: 70–90.

7. Schmid, W. *Ginkgo* thrives. *Nature*, **1997**, 386: 755.

8. Darlington, C. L.; Smith, P. F.; Maclennan, K. The neuroprotective properties of ginkgo extracts. In *Medicinal and Aromatic Plants — Industrial Profiles, Vol. 12: Ginkgo biloba*, Amsterdam: Harwood Academic Publishers, **2000**, 331–344.

9. Di Renzo, G. Ginkgo biloba and the central nervous system. *Fitoterapia*, **2000**, 71: S43–S47.

10. DeFeudis, F. V.; Drieu, K. *Ginkgo biloba* extract (EGb 761) and CNS functions: basic studies and clinical applications. *Curr. Drug Targets*, **2000**, 1: 25–58.

11. Maclennan, K.; Darlington, C. L.; Smith, P. F. The CNS effects of Ginkgo biloba extracts and ginkgolide B. *Prog. Neurobiol.*, **2002**, 67: 235–257.

12. Luo, Y. Ginkgo biloba neuroprotection: therapeuic implications in Alzheimer's disease. *J. Alzheimers Dis.*, **2001**, 3: 401–407.

13. Zimmermann, M.; Colciaghi, F.; Cattabeni, F.; Di Luca, M. Ginkgo biloba extract: from molecular mechanisms to the treatment of Alzheimers disease. *Cell. Mol. Biol.*, **2002**, 48: 613–623.

14. Cohen-Salmon, C.; Venault, P.; Martin, B.; Raffalli-Sebille, M. J.; Barkats, M.; Clostre, F.; Pardon, M. C.; Christen, Y.; Chapouthier, G. Effects of *Ginkgo biloba* extract (EGb 761) on learning and possible actions on aging. *J. Physiol. (Paris)*, **1997**, 91: 291–300.

15. Warburton, D. M. *Ginkgo biloba* extract and cognitive decline. *Br. J. Clin. Pharmacol.*, **1993**, 36: 137.

16. Kanowski, S.; Hermann, W. M.; Stephan, K.; Wierich, W.; Horr, R. Proof of efficacy of *Ginkgo biloba* special extract EGb 761 in outpatients suffering from mild to moderate primary degenerative dementia of the Alzheimer type or multi-infarct dementia. *Pharmacopsychiatry*, **1996**, 29: 47–56.

17. Le Bars, P. L.; Katz, M. M.; Berman, N.; Itil, T. M.; Freedman, A. M.; Schatzberg, A. F. A placebo-controlled double-blind, randomized trial of an extract of Ginkgo biloba for dementia. North American EGb study group. *J. Am. Med. Assoc.*, **1997**, 278: 1327–1332.

18. Birks, J.; Evans, J. G. Ginkgo biloba for cognitive impairment and dementia (Cochrane Review). *The Cochrane Library*. Chichester, UK: John Wiley & Sons, **2004**.

19. Solomon, P. R.; Adams, F.; Silver, A.; Zimmer, J.; De Veaux, R. Ginkgo for memory enhancement. A randomized controlled trial. *J. Am. Med. Assoc.*, **2002**, 288: 835–840.

20. van Dongen, M.; van Rossum, E.; Kessels, A.; Knipschild, P. Ginkgo for elderly people with dementia and age-associated memory impairment: A randomized clinical trial. *J. Clin. Epidemiol.*, **2003**, 56: 367–376.

21. Wolfe, M. S. Therapeutical strategies for Alzheimer's diesease. *Nat. Rev. Drug Discovery*, **2002**, 1: 859–866.

22. Hardy, J.; Selkoe, D. J. The amyloid hypothesis of Alzheimer's disease: Progress and problems on the road to therapeutics. *Science*, **2002**, 297: 353–356.

23. Li, C. L.; Wong, Y. Y. The bioavailability of ginkgolides in Ginkgo biloba extracts. *Planta Med.*, **1997**, 63: 563–565.

24. Biber, A.; Koch, E. Bioavailability of ginkgolides and bilobalide from extracts of Ginkgo biloba using GC/MS. *Planta Med.*, **1999**, 65: 192–193.

25. Drago, F.; Floriddia, M. L.; Cro, M.; Giuffrida, S. Pharmacokinetics and bioavailability of Ginkgo biloba extract. *J. Ocul. Pharmacol. Ther.*, **2002**, 18: 197–202.

26. Hasler, A. Chemical constituents of *Ginkgo biloba*. In *Medicinal and Aromatic Plants— Industrial Profiles, Vol. 12: Ginkgo biloba*. Amsterdam: Harwood Academic Publishers, **2000**, 109–142.

27. Nakanishi, K. A personal account of the early ginkgolide structural studies. In *Medicinal and Aromatic Plants–Industrial Profiles, Vol. 12. Ginkgo biloba*. Amsterdam: Harwood Academic Publishers, **2000**, 143–150.

28. Nakanishi, K. Ginkgolides — isolation and structural studies carried out in the mid 1960's. In *Ginkgolides — Chemistry, Biology, Pharmacology*. Barcelona, Spain: J. R. Prous Science Publishers, **1988**, 27–36.

29. Maruyama, M.; Terahara, A.; Itagaki, Y.; Nakanishi, K. The ginkgolides I. Isolation and the characterization of the various groups. *Tetrahedron Lett.*, **1967**, 4: 299–303.

30. Maruyama, M.; Terahara, A.; Itagaki, Y.; Nakanishi, K. The ginkgolides II. Derivation of partial structures. *Tetrahedron Lett.*, **1967**, 4: 303–308.

31. Maruyama, M.; Terahara, A.; Nakadaira, Y.; Woods, M. C.; Nakanishi, K. The ginkgolides III. The structure of the ginkgolides. *Tetrahedron Lett.*, **1967**, 4: 309–313.

32. Maruyama, M.; Terahara, A.; Nakadaira, Y.; Woods, M. C.; Takagi, Y.; Nakanishi, K. The ginkgolides IV. Stereochemistry of the ginkgolides. *Tetrahedron Lett.*, **1967**, 4: 315–319.

33. Woods, M. C.; Miura, I.; Nakadaira, Y.; Terahara, A.; Maruyama, M.; Nakanishi, K. The ginkgolides V. Some aspects of their NMR spectra. *Tetrahedron Lett.*, **1967**, 4: 321–326.

34. Nakanishi, K. The ginkgolides. *Pure Appl. Chem.*, **1967**, 14: 89–113.

35. Okabe, K.; Yamada, K.; Yamamura, S.; Takada, S. Ginkgolides. *J. Chem. Soc. C*, **1967**, 2201–2206.

36. Sakabe, N.; Takada, S.; Okabe, K. The structure of ginkgolide A, a novel diterpenoid trilactone. *Chem. Commun.*, **1967**, 259–261.

37. Weinges, K.; Hepp, M.; Jaggy, H. Chemie der ginkgolide, II. Isolierung und struktur-aufklärung eines neuen ginkgolids. *Liebigs Ann. Chem.*, **1987**, 521–526.

38. van Beek, T. A. Chemical analysis of *Ginkgo* terpene trilactones. In: *Medicinal and Aromatic Plants—Industrial Profiles, Vol. 12. Gingko biloba*. Amsterdam: Harwood Academic Publishers, **2000**, 151–178.

39. van Beek, T. A. Chemical analysis of Ginkgo biloba leaves and extracts. *J. Chromatogr. A*, **2002**, 967: 21–55.

40. Lichtblau, D.; Berger, J. M.; Nakanishi, K. Efficient extraction of ginkgolides and bilobalide from Ginkgo biloba leaves. *J. Nat. Prod.*, **2002**, 65: 1501–1504.

41. van Beek, T. A.; Lelyveld, G. P. Preparative isolation and separation procedure for ginkgolides A, B, C, and J and bilobalide. *J. Nat. Prod.*, **1997**, 60: 735–738.

42. Mauri, P.; Migliazza, B.; Pietta, P. Liquid chromatography/electrospray mass spectrometry of bioactive terpenoids in Ginkgo biloba L. *J. Mass. Spec.*, **1999**, 34: 1361–1367.

43. Mauri, P.; Simonetti, P.; Gardana, C.; Minoggio, M.; Morazzoni, P.; Bombardelli, E.; Pietta, P. Liquid chromatography/atmospheric pressure chemical ionization mass spectrometry of terpene trilactones in plasma of volunteers dosed with *Ginkgo biloba* L. extracts. *Rapid Commun. Mass Spec.*, **2001**, 15: 929–934.

44. Jensen, A. G.; Ndjoko, K.; Wolfender, J. L.; Hostettmann, K.; Camponovo, F.; Soldati, F. Liquid chromatography-atmospheric pressure chemical ionisation/mass spectrometry: A rapid and selective method for the quantitative determination of ginkolides and bilobalide in Ginkgo leaf extracts and phytopharmaceuticals. *Phytochem. Anal.*, **2002**, 13: 31–38.

45. Camponovo, F.; Wolfender, J. L.; Maillard, M. P.; Potterat, O.; Hostettmann, K. Evaporative light scattering and thermospray mass spectrometry: Two alternative methods for detection and quantitative liquid chromatographic detrmination of ginkgolides and bilbalide in Ginkgo biloba leaf extracts an phytopharmaceuticals. *Phytochem. Anal.*, **1995**, 6: 141–148.

46. Ganzera, M.; Zhao, J.; Khan, I. A. Analysis of terpenelactones in *Ginkgo biloba* by high performance liquid chromatography and evaporative light scattering detection. *Chem. Pharm. Bull.*, **2001**, 49: 1170–1173.

47. Nakanishi, K.; Habaguchi, K. Biosynthesis of ginkgolide B, its diterpenoid nature, and origin of the *tert*-butyl group. *J. Am. Chem. Soc.*, **1971**, 93: 3546–3547.

48. Rohmer, M. A mevalonate-independent route to isopentyl diphopshate. In *Comprehensive Natural Products Chemistry*. Oxford, UK: Elsevier, **1999**, 45–67.

49. Schwarz, M.; Arigoni, D. Ginkgolide biosynthesis. In *Comprehensive Natural Products Chemistry*. Oxford, UK: Elsevier, **1999**, 367–400.

50. Rohdich, F.; Kis, K.; Bacher, A.; Eisenreich, W. The non-mevalonate pathway of isoprenoids: Genes, enzymes and intermediates. *Curr. Opin. Chem. Biol.*, **2001**, 5: 535–540.

51. Eisenreich, W.; Schwarz, M.; Cartayrade, A.; Arigoni, D.; Zenk, M. H.; Bacher, A. The deoxylulose phosphate pathway of terpenoid biosynthesis in plants and microorganisms. *Chem. Biol.*, **1998**, 5: R221–R233.

52. Schepmann, H. G.; Pang, J.; Matsuda, S. P. T. Cloning and characterization of *Ginkgo biloba* levopimaradiene synthase, which catalyzes the first comitted step in ginkgolide biosynthesis. *Arch. Biochem. Biophys.*, **2001**, 392: 263–269.

53. Liao, Z.; Chen, M.; Gong, Y.; Guo, L.; Tan, Q.; Feng, X.; Sun, X.; Tan, F.; Tang, K. A new geranylgeranyl diphosphate synthase gene from Ginkgo biloba, which intermediates the biosynthesis of the key precursor for ginkgolides. *DNA Sequence*, **2004**, 15: 153–158.

54. Corey, E. J.; Gosh, A. K. Total synthesis of ginkgolide A. *Tetrahedron Lett.*, **1988**, 29: 3205–3206.

55. Corey, E. J.; Kang, M. C.; Desai, M. C.; Ghosh, A. K.; Houpis, I. N. Total synthesis of (±)-ginkgolide B. *J. Am. Chem. Soc.*, **1988**, 110: 649–651.

56. Crimmins, M. T.; Pace, J. M.; Nantermet, P. G.; Kim-Meade, A. S.; Thomas, J. B.; Watterson, S. H.; Wagman, A. S. Total synthesis of (±)-ginkgolide B. *J. Am. Chem. Soc.*, **1999**, 121: 10249–10250.

57. Crimmins, M. T.; Pace, J. M.; Nantermet, P. G.; Kim-Meade, A. S.; Thomas, J. B.; Watterson, S. H.; Wagman, A. S. The total synthesis of (±)-ginkgolide B. *J. Am. Chem. Soc.*, **2000**, 122: 8453–8463.

58. Weinges, K.; Hepp, M.; Huber-Patz, U.; Rodewald, H.; Irngartinger, H. Chemie der ginkgolide, I. 10-Acetyl-1-methoxycarbonyl-2,3,14,15,16-pentanorginkgolid A, ein zwischenprodukt zur herstellung von bilobalid. *Liebigs Ann. Chem.*, **1986**, 1057–1066.

59. Weinges, K.; Rümmler, M.; Schick, H. Chemie der ginkgolide, VI. Herstellung von 1,10-dihydroxy- und 1,7,10-trihydroxyginkgolid aus 1,3,7,10-tetrahydroxyginkgolid. *Liebigs Ann. Chem.*, **1993**, 1023–1027.

60. Weinges, K.; Rümmler, M.; Schick, H.; Schilling, G. Chemistry of the ginkgolides, V. On the preparation of the ginkgolide skeleton. *Liebigs Ann. Chem.*, **1993**, 287–291.

61. Weinges, K.; Schick, H.; Irngartinger, H.; Oeser, T. Chemistry of ginkgolides, VIII. Synthesis and crystal structure of a dimeric 14,15,16-trinorginkgolide derivative. *Liebigs Ann. Chem.*, **1997**, 1607–1609.

62. Weinges, K.; Schick, H.; Irngartinger, H.; Oeser, T. Chemistry of ginkgolides, IX. Ozonolysis of (10R)-10-acetoxy-3,14-didehydroginkgolide. *Liebigs Ann. Chem.*, **1997**, 1755–1756.

63. Weinges, K.; Schick, H.; Pritzkow, H. Chemistry of ginkgolides, VII. Preparation and crystal structure analysis of a secondary ozonide from ginkgolide A. *Liebigs Ann. Chem.*, **1997**, 991–993.

64. Weinges, K.; Schick, H.; Irngartinger, H.; Oeser, T. Chemistry of ginkgolides. Part 10: Access to [14]C-labelled ginkgolide A. *Tetrahedron*, **2000**, 56: 3173–3176.

65. Strømgaard, K.; Suehiro, M.; Nakanishi, K. Preparation of a tritiated ginkgolide. *Bioorg. Med. Chem. Lett.*, **2004**, 14: 5673–5675.

66. Suehiro, M.; Simpson, N. R.; Van Heertum, R. L. Radiolabeling of ginkgolide B with [18]F. *J. Labelled Compd. Radiopharm.*, **2004**, 47: 485–491.

67. Weinges, K.; Schick, H. Chemie der ginkgolid, IV. Herstellung von ginkgolid B aus ginkgolid C. *Liebigs Ann. Chem.*, **1991**, 81–83.

68. Corey, E. J.; Rao, K. S.; Ghosh, A. K. Intramolecular and intermolecular hydroxyl reactivity differences in ginkgolides A, B and C and their chemical applications. *Tetrahedron Lett.*, **1992**, 33: 6955–6958.

69. Teng, B. P. (SCRAS) Preparation process of ginkgolide B from ginkgolide C. US Patent US5241084, 1993 August 31.

70. Jaracz, S.; Strømgaard, K.; Nakanishi, K. Ginkgolides: selective acetylations, translactonization, and biological evaluation. *J. Org. Chem.*, **2002**, 67: 4623–4626.

71. Braquet, P. Ginkgolides — chemistry, biology, pharmacology, and clinical perspectives. In *Ginkgolides — Chemistry, Biology, Pharmacology, and Clinical Perspectives*. Barcelona, Spain: J. R. Prous Science, **1988**, 794.

72. Braquet, P.; Esanu, A.; Buisine, E.; Hosford, D.; Broquet, C.; Koltai, M. Recent progress in ginkgolide research. *Med. Res. Rev.*, **1991**, 11: 295–355.

73. Braquet, P. The ginkgolides: potent platelet-activating factor antagonists isolated from Ginkgo biloba L.: chemistry, pharmacology and clinical applications. *Drugs Future*, **1987**, 12: 643–699.

74. Marcheselli, V. L.; Rossowska, M. J.; Domingo, M. T.; Braquet, P.; Bazan, N. G. Distinct platelet-activating factor binding sites in synaptic endings and in intracellular membranes of rat cerebral cortex. *J. Biol. Chem.*, **1990**, 265: 9140–9145.

75. Ishii, S.; Shimizu, T. Platelet-activating factor (PAF) receptor and genetically engineered PAF receptor mutant mice. *Prog. Lipid. Res.*, **2000**, 39: 41–82.

76. Singh, M.; Saraf, M. K. Platelet-activating factor: a new target site for the development of nootropic agents. *Drugs Future*, **2001**, 26: 883–888.

77. Kato, K.; Clark, G. D.; Bazan, N. G.; Zorumski, C. F. Platelet-activating factor as a potential retrograde messenger in CA1 hippocampal long-term potentiation. *Nature*, **1994**, 367: 175–179.

78. Kornecki, E.; Wieraszko, A.; Chan, J. C.; Ehrlich, Y. H. Platelet activating factor (PAF) in memory formation: Role as a retrograde messenger in long-term potentiation. *J. Lipid Mediators Cell Signal.*, **1996**, 14: 115–126.

79. Izquierdo, I.; Fin, C.; Schmitz, P. K.; Da Silva, R. C.; Jerusalinsky, D.; Quillfeldt, J. A.; Ferreira, M. B.; Medina, J. H.; Bazan, N. G. Memory enhancement by intrahippocampal, intraamygdala, or intraentorhinal infusion of platelet-activating factor measured in an inhibitory avoidance task. *Proc. Natl. Acad. Sci. USA*, **1995**, 92: 5047–5051.

80. Kato, K. Modulation of long-term potentiation in the CA1 area of rat hippocampus by platelet-activiting factor. *Adv. Exp. Med. Biol.*, **1999**, 469: 221–227.

81. Bito, H.; Nakamura, M.; Honda, Z.; Izumi, T.; Iwatsubo, T.; Seyama, Y.; Ogura, A.; Kudo, Y.; Shimizu, T. Platelet-activating factor (PAF) receptor in rat brain: PAF mobilizes intracelluar $Ca^{2+}$ in hippocampal neurones. *Neuron*, **1992**, 9: 285–294.

82. Bazan, N. G.; Allan, G. Platelet-activating factor in the modulation of excitatory amino acid neurotransmitter release and of gene expression. *J. Lipid Mediators Cell Signal.*, **1996**, 14: 321–330.

83. Dell' Albani, P.; Condorelli, D. F.; Mudo, G.; Amico, C.; Bindoni, M.; Belluardo, N. Platelet-activating factor and its methoxy-analogue ET-18-OCH₃ stimulate immediate early gene expression in rat astroglial cultures. *Neurochem. Int.*, **1993**, 22: 567–574.

84. Doucet, J. P.; Bazan, N. G. Excitable membranes, lipid messengers, and immediate-early genes. Alteration of signal transduction in neuromodulation and neurotrauma. *Mol. Neurobiol.*, **1992**, 6: 407–424.

85. Kobayashi, K.; Ishii, S.; Kume, K.; Takahashi, T.; Shimizu, T.; Manabe, T. Platelet-activating factor receptor is not required for long-term potentiation in the hippocampal CA1 region. *Eur. J. Neurosci.*, **1999**, 11: 1313–1316.

86. Chen, C.; Magee, J. C.; Marcheselli, V.; Hardy, M.; Bazan, N. G. Attenuated LTP in hippocampal dentate gyrus neurons of mice deficient in the PAF receptor. *J. Neurophysiol.*, **2001**, 85: 384–390.

87. Kondratskaya, E. L.; Pankratov, Y. V.; Lalo, U. V.; Chatterjee, S. S.; Krishtal, O. A. Inhibition of hippocampal LTP by ginkgolide B is mediated by its blocking action on PAF rather than glycine receptors. *Neurochem. Int.*, **2004**, 44: 171–177.

88. Braquet, P.; Spinnewyn, B.; Braquet, M.; Bourgain, R. H.; Taylor, J. E.; Etienne, A.; Drieu, K. BN 52021 and related compounds: a new series of highly specific PAF-acether antagonists isolated from *Ginkgo biloba* L. *Blood Vessels*, **1985**, 16: 558–572.

89. Braquet, P.; Esan, A. Ginkgolide derivatives. U.K. Patent 2,211,841, **1989**.

90. Corey, E. J.; Gavai, A. V. Simple analogs of ginkgolide-B which are highly-active antagonists of platelet activating factor. *Tetrahedron Lett.*, **1989**, 30: 6959–6962.

91. Corey, E. J.; Rao, K. S. Enantioselective total synthesis of ginkgolide derivatives lacking the tert-butyl group, an essential structural subunit for antagonism of platelet activating factor. *Tetrahedron Lett.*, **1991**, 32: 4623–4626.

92. Villhauer, E. B.; Anderson, R. C. Synthesis of the CDE ring system of the ginkgolides. *J. Org. Chem.*, **1987**, 52: 1186–1189.

93. Park, P. U.; Pyo, S.; Lee, S. K.; Sung, J. H.; Kwak, W. J.; Park, H. K.; Cho, Y. B.; Ryu, G. H.; Kim, T. S. (Sunkyong Industries). Ginkgolide derivatives. Int. Patent WO 9,518,131, **1995**.

94. Park, H. K.; Lee, S. K.; Park, P. U.; Kwan, W. J. (Sunkyong Industries). Ginkgolide derivatives and a process for preparing them. Int. Patent WO 9,306,107, **1993**.

95. Hu, L.; Chen, Z.; Cheng, X.; Xie, Y. Chemistry of ginkgolides: Structure-activity relationship as PAF antagonists. *Pure Appl. Chem.*, **1999**, 71: 1153–1156.

96. Hu, L.; Chen, Z.; Xie, Y.; Jiang, H.; Zhen, H. Alkyl and alkoxycarbonyl derivatives of ginkgolide B: Synthesis and biological evaluation of PAF inhibitory activity. *Bioorg. Med. Chem.*, **2000**, 8: 1515–1521.

97. Hu, L.; Chen, Z.; Xie, Y. Synthesis and biological activity of amide derivatives of ginkgolide A. *J. Asian Nat. Prod. Res.*, **2001**, 3: 219–227.

98. Hu, L.; Chen, Z.; Xie, Y.; Jiang, Y.; Zhen, H. New products from alkali fusion of ginkgolides A and B. *J. Asian Nat. Prod. Res.*, **2000**, 2: 103–110.

99. Chen, J.; Hu, L.; Jiang, H.; Gu, J.; Zhu, W.; Chen, Z.; Chen., K.; Ji, R. A 3D-QSAR study on ginkgolides and their analogues with comparative molecular field analysis. *Bioorg. Med. Chem. Lett.*, **1998**, 8: 1291–1296.

100. Strømgaard, K.; Saito, D. R.; Shindou, H.; Ishii, S.; Shimizu, T.; Nakanishi, K. Ginkgolide derivatives for photolabelling studies: preparation and pharmacological evaluation. *J. Med. Chem.*, **2002**, 45: 4038–4046.

101. Vogensen, S. B.; Strømgaard, K.; Shindou, H.; Jaracz, S.; Suehiro, M.; Ishii, S.; Shimizu, T.; Nakanishi, K. Preparation of 7-substituted ginkgolide derivatives: Potent platelet activating factor (PAF) receptor antagonists. *J. Med. Chem.*, **2003**, 46: 601–608.

102. Krane, S.; Kim, S. R.; Abrell, L. M.; Nakanishi, K. Microphysiometric measurement of PAF receptor responses to ginkgolides. *Helv. Chim. Acta*, **2003**, 86: 3776–3786.

103. Kondratskaya, E. L.; Lishko, P. V.; Chatterjee, S. S.; Krishtal, O. A. BN52021, a platelet factor antagonist, is a selective blocker of glycine-gated chloride channel. *Neurochem. Int.*, **2002**, 40: 647–653.

104. Kondratskaya, E. L.; Krishtal, O. A. Effects of Ginkgo biloba extract constitutents on glycine-activated strychnine-sensitive receptors in hippocampal pyramidal neurons of the rat. *Neurophysiology*, **2002**, 34: 155–157.

105. Ivic, L.; Sands, T. T. J.; Fishkin, N.; Nakanishi, K.; Kriegstein, A. R.; Strømgaard, K. Terpene trilactones from Ginkgo biloba are antagonists of cortical glycine and GABA$_A$ receptors. *J. Biol. Chem.*, **2003**, 278: 49279–49285.

106. Betz, H.; Harvey, R. J.; Schloss, P. Structures, diversity and pharmacology of glycine receptors and transporters. In *Pharmacology of GABA and Glycine Neurotransmission*. Berlin: Springer-Verlag, **2001**.

107. Betz, H.; Kuhse, J.; Schmieden, V.; Laube, B.; Kirsch, J.; Harvey, R. J. Structure and functions of inhibitory and excitatory glycine receptors. *Ann. NY Acad. Sci.*, **1999**, 868: 667–676.

108. Rajendra, S.; Lynch, J. W.; Schofield, P. R. The glycine receptor. *Pharmacol. Ther.*, **1997**, 73: 121–146.

109. Breitinger, H.-G.; Becker, C.-M. The inhibitory glycine receptor—simple views of a complicated channel. *ChemBioChem*, **2002**, 3: 1042–1052.

110. Laube, B.; Maksay, G.; Schemm, R.; Betz, H. Modulation of glycine receptor function: A novel approach for therapeutic intervention at inhibitory synapses? *Trends Pharmacol. Sci.*, **2002**, 23: 519–527.

111. Harvey, R. J.; Depner, U. B.; Wässle, H.; Ahmadi, S.; Heindl, C.; Reinold, H.; Smart, T. G.; Harvey, K.; Schütz, B.; Abo-Salem, O. M.; Zimmer, A.; Poisbeau, P.; Welzl, H.; Wolfer, D. P.; Betz, H.; Zeilhofer, H. U.; Müller, U. GlyR α3: An essential target for spinal PGE2-mediated inflammatory pain sensitization. *Science*, **2004**, 304: 884–887.

112. Huang, S. H.; Duke, R. K.; Chebib, M.; Sasaki, K.; Wada, K.; Johnston, G. A. R. Ginkgolides, diterpene trilactones of Ginkgo biloba, as antagonists at recombinant $\alpha_1\beta_2\gamma_{2L}$ GABA$_A$ receptors. *Eur. J. Pharmacol.*, **2004**, 494: 131–138.

113. Kondratskaya, E. L.; Fisyunov, A. I.; Chatterjee, S. S.; Krishtal, O. A. Ginkgolide B preferentially blocks chloride channels formed by heteromeric glycine receptors in hippocampal pyramidal neurons of rat. *Brain Res. Bull.*, **2004**, 63: 309–314.

114. Chatterjee, S. S.; Kondratskaya, E. L.; Krishtal, O. A. Structure-activity studies with Ginkgo biloba extract constituents as receptor-gated chloride channel blockers and modulators. *Pharmacopsychiatry*, **2003**, 36(Suppl. 1): S68–S77.

115. Jaracz, S.; Nakanishi, K.; Jensen, A. A.; Strømgaard, K. Ginkgolides and glycine receptors: A structure-activity relationship study. *Chem. Eur. J.*, **2004**, 10: 1507–1518.

116. Jensen, A. A.; Kristiansen, U. Functional characterisation of the human α1 glycine receptor in a fluorescence-based membrane potential assay. *Biochem. Pharmacol.*, **2004**, 67: 1789–1799.

117. Gavish, M.; Bachman, I.; Shoukrun, R.; Katz, Y.; Veenman, L.; Weisinger, G.; Weizman, A. Enigma of the peripheral benzodiazepine receptor. *Pharmacol. Rev.*, **1999**, 51: 629–650.

118. Veenman, L.; Gavish, M. Peripheral-type benzodiazepine receptors: their implication in brain disease. *Drug Develop. Res.*, **2000**, 50: 355–370.

119. Amri, H.; Ogwuegbu, S. O.; Boujrad, N.; Drieu, K.; Papadopoulus, V. *In vivo* regulation of the peripheral-type benzodiazepine receptor and glucocorticoid synthesis by the *Ginkgo biloba* extract EGb and isolated ginkgolides. *Endocrinology*, **1996**, 137: 5707–5718.

120. Amri, H.; Drieu, K.; Papadopoulos, V. Ex vivo regulation of adrenal cortical cell steroid and protein synthesis, in response to adrenocorticotropic hormon stimulation by the Ginkgo biloba extract EGb 761 and isolated ginkgolide B. *Endocrinology*, **1997**, 138: 5415–5426.

121. Drieu, K. (SCRAS). Ginkgolides for inhibition of membrane expression of benzodiazepine receptors. U.S. Patent 6,274,621, **2001**.

122. Amri, H.; Drieu, K.; Papadopoulos, V. Transcriptional suppression of adrenal cortical peripheral-type benzodiazepine receptor gene and inhibition of steroid synthesis by ginkgolide B. *Biochem. Pharmacol.*, **2003**, 65: 717–729.

123. Amri, H.; Drieu, K.; Papadopoulos, V. Use of ginkgolide B and a ginkgolide-activated response element to control gene transcription: example of the adrenocortical fperipheral-type benzodiazepine receptor. *Cell. Mol. Biol.*, **2002**, 48: 633–639.

124. Pietri, S.; Maurelli, E.; Drieu, K.; Culcasi, M. Cardioprotective and antioxidant effects of the terpenoid constituents of *Ginkgo biloba* extract. *J. Mol. Cell. Cardiol.*, **1997**, 29: 833–742.

125. Prehn, J. H.; Krieglstein, J. Platelet-activating factor antagonists reduce excitotoxic damage in cultured neurons from embryonic chick telencephalon and protect the rat hippocampus and neocortex from ischemic injury *in vivo. J. Neurosci. Res.*, **1993**, 34: 179–188.

126. Krieglstein, J. Neuroprotective properties of *Ginkgo biloba* constituents. *Phytother.*, **1994**, 15: 92–96.

127. Cheung, F.; Siow, Y. L.; Karmin, O. Inhibition by ginkgolides and bilobalide of the production of nitric oxide in marcophages (THP-1) but not in endothelial cells (HUVEC). *Biochem. Pharmacol.*, **2001**, 61: 503–510.

128. Liebgott, T.; Miollan, M.; Berchadsky, Y.; Drieu, K.; Culcasi, M.; Pietri, S. Complementary cardioprotective effects of flavonoid metabolites and terpenoid constituents of Ginkgo biloba extract (EGb 761) duringf ischemia and reperfusion. *Basic Res. Cardiol.*, **2000**, 95: 368–377.

129. Pietri, S.; Liebgott, T.; Finet, J.-P.; Culcasi, M.; Billottet, L.; Bernard-Henriet, C. Synthesis and biological studies of a new ginkgolide C derivative: evidence that the cardioprotective effect of ginkgolides is unrelated to PAF inhibition. *Drug. Develop. Res.*, **2001**, 54: 191–201.

# 8

# RECENT PROGRESS IN *CALOPHYLLUM* COUMARINS AS POTENT ANTI-HIV AGENTS

LIN WANG, TAO MA, AND GANG LIU

*Department of Medicinal Chemistry, Institute of Materia Medica, Chinese Academy of Medical Sciences & Peking Union Medical College, Beijing, China*

## 8.1  INTRODUCTION

Human immunodeficiency virus (HIV) is the cause of acquired immunodeficiency syndrome (AIDS). The chemotherapy of AIDS is considered to be one of the most challenged scientific projects, and it is one of the leading causes of morbidity and mortality in the world.[1] Although highly active antiretroviral therapy (HAART) using a combination of anti-HIV drugs has been highly effective in suppressing HIV load and decreasing mortality of AIDS patients, the emergence of the drug resistance among HIV and toxicity of the therapy have made the continued search for novel anti-HIV drugs necessary.[2,3] Currently, the U.S. Food and Drug Administration (FDA) approved HIV drugs including seven nucleoside HIV reverse transcriptase (HIV-RT) inhibitors (NRTIs), Abacavir (ABC), Didanosine (ddI), Emtricitabine (ETC), Lamivudine (3TC), Stavudine (d4T), Zalcitabine (ddC), and Zidovudine (AZT); one nucleotide RT inhibitor (NtRRTI) Tenofovir; three non-nucleoside RT inhibitors (NNRTIs), Delavirdine, Efavirenz, and Nevirapine; eight protease inhibitors (PIs), Amprenavir, Atazanavir, Fosamprenavir, Indinavir, Lopinavir, Nelfinavir, Ritonavir, and Saguinavir; and one fussion inhibitor, Enfuvirtide.[4] However, the rapid emergence of resistance toward currently used anti-HIV drugs is an important determinant in the eventual drug failure. New anti-HIV drug development strategies might overcome the virus-drug resistance problem by focusing on either novel targets or new compounds capable of suppressing HIV strains that are resistant to the currently used anti-HIV drugs. Therefore, medicinal chemists are interested in the development of novel anti-HIV agents that might be particularly effective in controlling strains of HIV that are resistant to the current NRTIs, NNRTIs, PIs, and so on.

Natural products have been important sources of new pharmacological active agents. Plant-derived products have led to the discovery of many clinically useful drugs for the treatment of human diseases such as antitumor and anti-infective drugs. Coumarins are well-known natural products displaying a broad range of biological activities.[5,6] Coumarin derivatives have been extensively used as therapeutic agents, active media for tunable dye lasers, optical bleaching agents, luminescent probes, and triplet sensitizers.[7] In 1992, Kashman et al.[8] first reported on a novel dipyranocoumain, calanolide A (**1**), isolated from the Malaysia tree *Calophyllum lanigerum*, and Patil et al.[9] reported on a related compound inophyllum B (**3**) from *Calophyllum inophyllum*. Both of them are representatives of a distinct class of NNRTI. Therefore, scientists are interested in the plant of genus *Calophyllum*. The chemical structures of naturally occurring coumarins isolated from *Calophyllum* species are summarized in Structure 8-1.

Numerous dipyranocoumarins have been isolated from different species of *Calophyllum*. According to their chemical skeleton, they have three heterocycle rings, ring B, C, and D, constructed from a phloroglucinol core A. These compounds fall into three basic structural types: (1) tetracycle dipyranocoumarins in which the D rings have a gem-dimethyl group, such as (+)-calanolide A (**1**), (+)-calanolide B (**2**), (+)-inophyllum B (**3**), and (+)-cordatolide A (**5**); (2) tetracyclic dipyranocoumarins with reversed D and C pyran rings; namely, the gem-dimethyl is present in ring C, as in the (+)-pseudocalanolide C (**21**) and (+)-pseudocalanolide D (**25**); and (3) tricyclic pyranocoumarins, for example, (+)-calanolide E (**26**) and (+)-cordatolide E (**27**), which contain a noncyclized equivalent of ring C of the calanolide tetracyclic structures. Individual members of the groups vary with respect to the C-4 substituent on the lactone ring (ring B) of the coumarin, where *n*-propyl (calanolides), phenyl (inophyllums), and methyl groups (cordatolides) may be encountered.

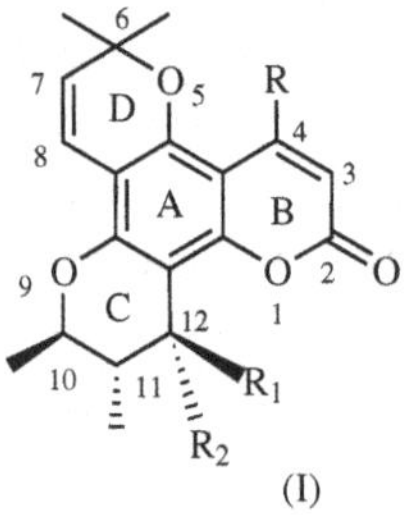

(I)

**1**  R=*n*-C$_3$H$_7$ R$_1$=OH R$_2$=H (+)-calanolide A[8]
**2**  R=*n*-C$_3$H$_7$ R$_2$=OH R$_1$=H (+)-calanolide B[8,10]

**Scheme 8-1.**  Chemical structures of naturally occurring *Calophyllum* coumarins.

**3** R=C$_6$H$_5$ R$_1$=OH R$_2$=H (+)-inophyllum B[9-11]
**4** R=C$_6$H$_5$ R$_2$=OH R$_1$=H (+)-inophyllum P[9,10]
**5** R=CH$_3$ R$_1$=OH R$_2$=H (+)-cordatolide A[12,13]
**6** R=CH$_3$ R$_2$=OH R$_1$=H (−)-cordatolide B[12,13]
**7** R=$n$-C$_3$H$_7$ R$_1$=OCH$_3$ R$_2$=H (+)-12-methoxy calanolide A[8]
**8** R=$n$-C$_3$H$_7$ R$_1$=OAc R$_2$=H (+)-12-acetoxy calanolide A[8]
**9** R=$n$-C$_3$H$_7$ R$_2$=OCH$_3$ R$_1$=H (+)-12-methoxy calanolide B[8]
**10** R=CH$_3$ R$_2$=OCH$_3$ R$_1$=H (+)-12-methoxy cordatolide B[13]

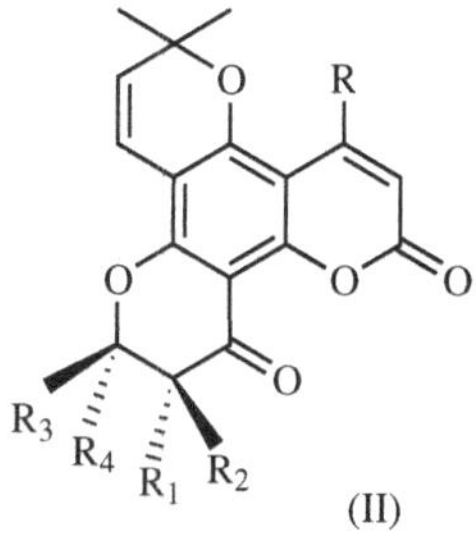

(II)

**11** R=C$_6$H$_5$ R$_1$=R$_3$=CH$_3$ R$_2$=R$_4$=H (+)-inophyllum C[9,10,11,14]
**12** R=C$_6$H$_5$ R$_2$=R$_3$=CH$_3$ R$_1$=R$_4$=H (+)-inophyllum E[9,10,11,14]
**13** R=C$_6$H$_5$ R$_2$=R$_4$=CH$_3$ R$_1$=R$_3$=H (−)-soulattrolone[15]

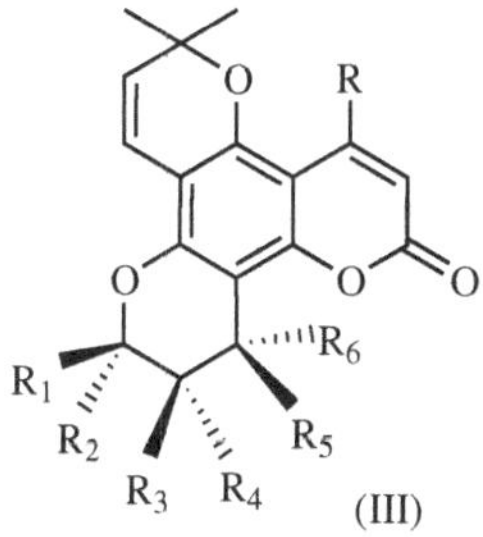

(III)

**14** R=$n$-C$_3$H$_7$, R$_2$=R$_4$=CH$_3$, R$_1$=R$_3$=R$_6$=H, R$_5$=OH (−)-calanolide F[16]
**15** R=$n$-C$_3$H$_7$, R$_2$=R$_3$=CH$_3$, R$_1$=R$_4$=R$_6$=H, R$_5$=OH (−)-calanolide B (cost-atolide)[17,18]
**16** R=C$_6$H$_5$, R$_2$=R$_3$=CH$_3$, R$_1$=R$_4$=H, R$_5$=OH (−)-inophyllum P (soulattro-lide)[19]
**17** R=C$_6$H$_5$, R$_1$=R$_3$=CH$_3$, R$_2$=R$_4$=R$_6$=H, R$_5$=OH (+)-inophyllum A[9,20]
**18** R=C$_6$H$_5$, R$_1$=R$_3$=CH$_3$, R$_2$=R$_4$=R$_5$=H, R$_6$=OH (+)-inophyllum D[9,11]

**Scheme 8-1.** (*Continued*)

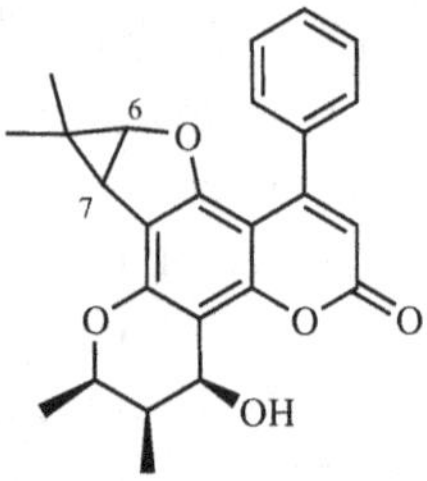

(IV)

**19** 6α,7α: (+)-inophyllum G-1[9]
**20** 6β,7β: (−)-inophyllum G-2[9]

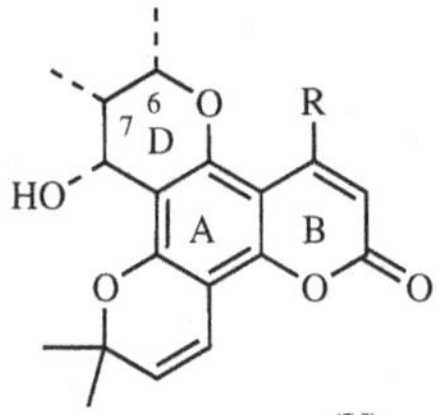

(V)

**21** R=$n$-C$_3$H$_7$ (+)-pseudo-calanolide C[8,16,21,22]
**22** R=CH$_3$ (+)-pseudo-cordatolide C[16]
**23** R=C$_6$H$_5$ 6,7-gem CH$_3$, *trans*-tomentolide A[23]

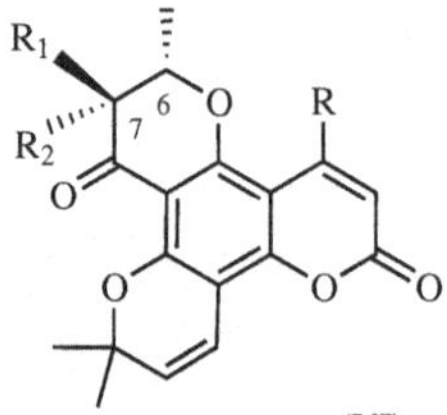

(VI)

**24** R=$n$-C$_3$H$_7$ R$_1$=CH$_3$ R$_2$=H tomentolide B[23]
**25** R=$n$-C$_3$H$_7$ R$_2$=CH$_3$ R$_1$=H (+)-pseudo-calanolide D[8,16,21]

Tricyclic pyranodihydrocoumarins

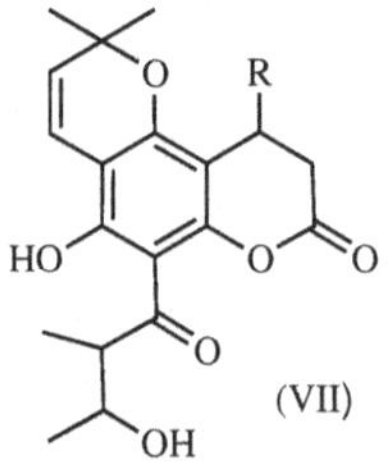

(VII)

**Scheme 8-1.** (*Continued*)

**26**   R=n-$C_3H_7$ (+)-calanolide E[24]
**27**   R=$CH_3$ (+)-cordatolide E[16]

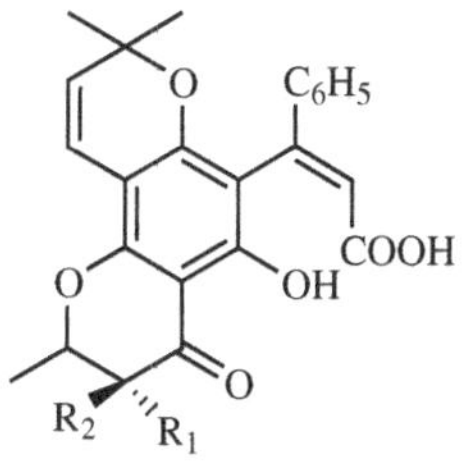

(VIII)

**28**   $R_1$=$CH_3$, $R_2$=H calophyllic acid[9]
**29**   $R_1$=H, $R_2$=$CH_3$ isocalophyllic acid[9]

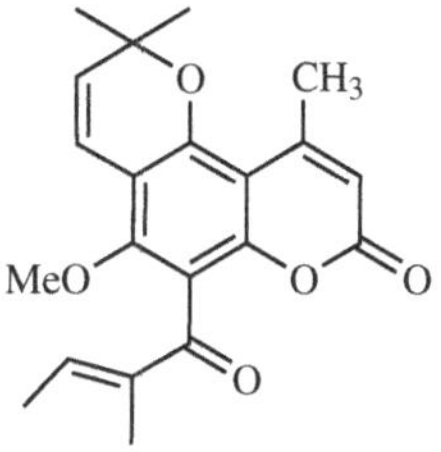

(IX)

**30**   oblongulide[12,13]

**Scheme 8-1.**   (*Continued*)

## 8.2   ANTI-HIV-1 ACTIVITY OF *CALOPHYLLUM* COUMARINS

### 8.2.1   Anti-HIV-1 Activity of Calanolides

#### *8.2.1.1   Anti-HIV-1 Activity of Calanolides in Cell Culture*
Flavin et al.[25] reported on the anti-HIV-1 activity of synthetic (±)-calanolide A and resolved (+)-calanolide A (**1**) and (−)-calanolide A [(−)-**1**] against both strains and clinical isolates of HIV-1 shown in Table 8-1. The cytotoxicity in the different cell lines examined of the (±)-**1**, (+)-**1**, and (−)-calanolide A was approximately the same levels. However, only (+)-**1** exhibited HIV-1 activity, which was similar to the data reported for the natural product [(+)-calanolide A, (**1**)],[8] and (−)-**1** was inactive. Both the AZT-resistant strain G910-6 and the pyridinone-resistant strain A17 were inhibited by (±)-**1** and (+)-**1**.[8,25,26] The (+)-**1** was more active than (±)-**1** against the AZT-resistant strain G910-6 with an $EC_{50}$ value of 0.027 and 0.108 μM, respectively. It was interesting that the activity of (±)-**1**

**TABLE 8-1. Anti-HIV-1 Activity of $(\pm)$-1, $(+)$-1, and $(-)$-Calanolide A $[(-)$-1]**

| Strain/cell | | $(\pm)$-1 | $(+)$-1 | $(-)$-1 | AZT[e] | DDC[e] |
|---|---|---|---|---|---|---|
| RF$_{11}$/CEM | EC$_{50}$ ($\mu$M) | 0.486 | 0.267 | | 0.023 | 0.189 |
| | IC$_{50}$ ($\mu$M) | 22.81 | 22.96 | 18.70 | 301.60 | 47.34 |
| | TI[d] | 47 | 86 | | 13 113 | 250 |
| III$_B$/MT2 | EC$_{50}$ ($\mu$M) | 0.108 | 0.053 | | 0.029 | 0.900 |
| | IC$_{50}$ ($\mu$M) | 6.86 | 14.80 | 7.31 | 51.64 | 83.80 |
| | TI[d] | 64 | 279 | | 1780 | 93 |
| H112–2[a]/ | EC$_{50}$ ($\mu$M) | 0.135 | 0.107 | | 0.037 | 1.562 |
| MT2 | IC$_{50}$ ($\mu$M) | 6.53 | 7.15 | 6.21 | 119.84 | 258.97 |
| | TI[d] | 48 | 67 | | 3236 | 166 |
| G910–6[b]/ | EC$_{50}$ ($\mu$M) | 0.108 | 0.027 | | | 0.994 |
| MT2 | IC$_{50}$ ($\mu$M) | 7.42 | 7.17 | 6.16 | 131.71 | 212.10 |
| | TI[d] | 69 | 266 | | | 213 |
| A17[c]/MT2 | EC$_{50}$ ($\mu$M) | 0.297 | 0.427 | | 0.014 | 0.331 |
| | IC$_{50}$ ($\mu$M) | 6.94 | 6.99 | 5.89 | 83.44 | 134.93 |
| | TI[d] | 23 | 16 | | 5960 | 408 |

[a]Pre-AZT treatment isolate. [b]Post-AZT treatment isolate. [c]Pyridinone-resistant isolate.
[d]TI = IC$_{50}$/EC$_{50}$. [e]AZT and DDC were positive control drugs.

(EC$_{50}$ = 0.297 $\mu$M) against the pyridinone-resistant strain A17 was more comparable with that of $(+)$-**1** (EC$_{50}$ = 0.427 $\mu$M).

It was discovered that the synthetic intermediate [12-oxo-$(\pm)$-calanolide A, (**3**)] exhibited anti-HIV-1 activity[27] with one less chiral center than calanolide A. This unique feature makes **3** an attractive candidate for drug development because it can be conveniently prepared. Therefore, $(\pm)$-**3**, $(+)$-**3**, and $(-)$-**3**, were prepared and evaluated for antiviral activities against HIV-1, simian immunodeficiency virus (SIV), and HIV-2 using CEM-SS cells infected in various laboratories and clinical isolates of viruses.[28] The antiviral results were summarized in Table 8-2 and indicated that these 12-oxo-calanolide A were active against both HIV-1 and SIV, but they were inactive against HIV-2. It was also worthy to note that compound **3** was the first reported calanolide analog capable of inhibiting SIV.

**TABLE 8-2. Antiviral Activity of $(\pm)$-, $(+)$-, and $(-)$-12-Oxo-calanolide A (3) Against Various Infected CEM-SS Cells**

| Virus | EC$_{50}$($\mu$M) | | | | |
|---|---|---|---|---|---|
| | $(\pm)$-**3** | $(+)$-**3** | $(-)$-**3** | DDC[a] | $(+)$-Calanolide A |
| HIV-1 (R$_F$) | 0.4 | 0.9 | 3.41 | 0.05 | 0.27 |
| HIV-1(III$_B$) | 0.51 | 1.0 | 1.88 | 0.02 | 0.17 |
| HIV-1(SK1) | 0.17 | 0.17 | 0.27 | 0.05 | 0.14 |
| SIV(Delta) | 1.24 | 1.66 | 6.12 | 0.19 | inactive |
| HIV-2(ROD) | 5.57 | 15.90 | ND[b] | 0.03 | inactive |

[a]DDC was used as a positive control drug. [b]ND: not determined.

Natural products have been and will be an important resource for new bioactive compounds. (+)-calanolide A (**1**) is the only natural product undergoing clinical trial for AIDS patients. A combination of the (+)-calanolide A and (−)-calanolide B with a variety of mechanistically diverse HIV inhibitory drugs such as AZT, 3TC, ddC, and ddI (NRTIs), nevirapin (NVP, NNRTI), indinavir (IDV), saquinavir (SQV), ritonavir (RTV), and nefinavir (NFV), protease inhibitors exhibited either additive or synergistic interactions. Furthermore, a combination of (+)-calanolide A with one NRTI and one PI yielded synergistic anti-HIV activity, whereas the combination of the (+)-calanolide A with one NRTI and one NNRTI showed an additive effect.[29] (+)-calanolide A, (−)-calanolide B (costatolide), and dihydrocostatolide were evaluated in different cell lines (CEM-SS, H9, MT2, AA5, V937, etc.) infected with both laboratory-derived and clinical strains of HIV-1, HIV-2, and SIV.[30] The anti-HIV-1 activity ($EC_{50}$) of these compounds ranged from 0.08 to 0.5, 0.06 to 1.4, and 0.1 to 8.8 μM for (+)-calanolide A, (−)-calanolide B, and dihydrocostatolide, respectively. However, no compound exhibited activity against HIV-2 or SIV. Furthermore, calanolides were active against a wide range of HIV-strains in various cell lines. The three calanolide stereoisomers showed their antiviral activities for resistance to a variety of NNRTIs as shown in Table 8-3.

The calanolides were active against viral isolates with the Y181C amino acid mutation in the HIV-1 RT, which is a commonly observed mutation in both laboratory and clinical viral isolates, and it was associated with high-level resistance to most other NNRTIs, such as nevirapine, pyridinone, E-BPTU, UC38, and diphenyl-sulfone. All tested calanolides exhibited no antiviral activity against NNRTI-related mutations, L100I, K103N, and Y188H. A unique and important feature of calanolides distinguishes it from the current marketed and developmental NNRTI. Additionally, (+)-calanolide A possessed anti-*Mycobacterium tuberculosis* (TB).[31] No anti-HIV agent, either in development or approved by the FDA, that exhibits this therapeutic capability. However, AIDS patients are more susceptible to TB compared with the healthy people. It makes (+)-calanolide A a valuable therapeutic agent to AIDS patients. A preclinical program of antituberculosis properties of (+)-calanolide A has been investigated by Sarawake MediChem Pharmaceuticals.

### 8.2.1.2 Anti-HIV-1 Activity of Calanolides in Hollow Fiber Mouse
Evaluation of (+)-calanolide A (**1**) in a hollow fiber culture-based in a SCID mouse assay of antiviral efficacy indicated that (+)-calanolide A exhibited significant anti-HIV-1 activity after oral or parenteral administration on a once-daily (200 mg/kg/dose) or twice-daily (150 mg/kg/dose) treatment. Furthermore, a synergistic effect was observed in the combination of (+)-calanolide A and AZT.[32]

### 8.2.2 Anti-HIV-1 Activity of Inophyllums

An inophyllum series was isolated from the Malaysia tree, *Calophyllum inophyllum* Linn (bioassay-guided procedure) (+)-inophyllum B (**3**), P (**4**), C (**11**), E (**12**), A (**17**), D (**18**), inophyllum G-1 (**19**), G-2 (**20**), calophyllic acid (**28**), and isocalophyllic acid (**29**). The (+)-inophyllum P (**4**), an enantiomer of soulattrolide (**16**) and as

TABLE 8-3. Activities of Calanolides Against Viruses Resistant to HIV-1-Specific Inhibitors

| Drug(s) to Which Isolates are Resistant (mutation) | EC50 (μM)[a] | | | | |
|---|---|---|---|---|---|
| | (+)-Calanolide A | (−)-Calanolide B | (−)-Dihydrocalanolide B | Nevirapine[a] | AZT[a] |
| IIIB (control) | 0.1 | 0.2 | 0.2 | 0.01 | 0.05 |
| Oxathiin carboxanilide (L100I) | >27 | >270 | >20 | 0.1 | 0.04 |
| UC10-costatolide (K103N) | >27 | >270 | >20 | ND[b] | 0.003 |
| Thiazolobenzimidazole (V108I) | 24.0 | 4.4 | 3.5 | 0.3 | 0.04 |
| TIBO-R82150 (A98G-V108I) | 22.0 | 1.6 | 5.1 | 0.6 | 0.05 |
| Calanolide A (T139I) | >27 | 4.5 | >20 | 0.01 | 0.01 |
| Diphenylsulfone (Y181C) | 0.08 | 0.08 | <0.01 | 5.9 | 0.01 |
| Nevirapine (Y181C) | <0.01 | <0.01 | 0.09 | >38 | 0.03 |
| Pyridinone (Y181C-L103N) | 0.12 | 0.8 | 0.8 | >38 | 0.01 |
| E-BPTU(Y181C) | 0.1 | <0.08 | <0.06 | 1.9 | 0.03 |
| UC38(Y181C) | 0.2 | <0.03 | 0.1 | 1.9 | 0.01 |
| 3TC(M184V) | 0.3 | 1.3 | 1.0 | 0.01 | 0.02 |
| Costatolide (Y188H) | >27 | >27 | >27 | ND | 0.004 |
| HEPT (P236L) | 0.6 | 1.1 | 0.2 | 0.02 | 0.01 |

[a]Nevirapine (NNRTI) and AZT (NRTI) are the positive control compounds.
[b]ND: not determined.

the C12-epimer of (+)-inophyllum B (**3**), and inophyllums G-1 (**19**) and G-2 (**20**) were novel compounds.[9] All 11 natural products were evaluated for their inhibitory activity against HIV-1 RT. Among the most potent active compounds, inophyllum B (**3**), and inophyllum P (**4**), inhibited HIV-1 RT with $IC_{50}$ values that were 38 nM and 130 nM, respectively, and both were active against HIV-1 in cell culture with $EC_{50}$ values that were 1.4 μM and 1.6 μM, respectively. However, both inophyllum B (**3**) and inophyllum P (**4**) possessed the *trans*-$C_{10}$,$C_{11}$-dimethyl chromanone ring, which is different from the stereochemistry of the hydroxy group at the $C_{12}$ position. The stereochemistry of the hydroxy group at the $C_{12}$ position was not critical for antiviral activity, but the subsituents of carboxyl group at the $C_{12}$ position lowered the activity significantly. For example, inophyllum C (**11**) and Inophyllum E (**12**) were much less active, whereas the closely related (+)-inophyllum D (**18**) and (+)-inophyllum A (**17**) possessing a *cis*-$C_{10}$, $C_{11}$ dimethylchromanol ring were less active. Furthermore, (−)-inophyllum P[33] (soulattrolide, **16**), a close structural resemblence to (−)-calanolide B (**15**), was found to be a potent inhibitor of HIV-1 RT with an $IC_{50}$ of 0.34 μM when HIV-RT was assessed for RNA-dependent DNA polymerase (RDDP). No appreciable activity was observed toward HIV-2 RT and avian myeloblastosis virus reverse transcriptase (AMVRT). The absolute configuration of (−)-inophyllum P as (10S, 11R, 12S) and other structurally related HIV-1 inhibitors, including (+)-inophyllum B (**3**), (+)-inophyllum P (**4**), (+)-calanolide A (**1**), (+)-calanolide B (**2**), (−)-calanolide B (**15**), (+)-cordatolide A (**5**), and (+)-cordatolide B (**6**) were also assigned.[34]

### 8.2.3  Anti-HIV-1 Activity of Cordatolides

(+)-Cordatolide A (**5**) and (−)-cordatolide B (**6**) were first isolated from *Calophyllum Cordato-oblongum*[12,13] for biological evaluation and found to mediate significant inhibition of HIV-1 RT with $IC_{50}$ values of 12.3 and 19.0 μM, respectively.[35] (+)-Cordatolide A (V) with the 12β-hydroxy configuration was somewhat more active than the corresponding 12α-hydroxy Cordatolide B (**6**). However, (+)-cordatolide A has the same stereochemistry as (+)-calanolide A (**1**) and (+)-inophyllum B (**3**), and the activity with inhibition of HIV-1 RT was much lower than calanolide A (**1**) and inophyllum B (**3**), but there is no report on the anti-HIV-1 activity of cordatolide A and B in the cell line. It seems to be that the substituents located at the $C_4$ position are important for antiviral activity because the propyl group at $C_4$ in calanolide A (**1**) or the phenyl group in inophyllum B (**3**) exhibited the highest activity, which indicates that the bulky substituents at the $C_4$ position may be necessary for the antiviral activity.

### 8.3  PHARMACOLOGY OF CALANOLIDES

### 8.3.1  Pharmacology of (+)-Calanolide A

(+)-Calanolide A (**1**) inhibited HIV-1 in cell culture but was inactive against HIV-2 or SIV. In the viral life-cycle investigations, it was demonstrated that calanolide A acts early in the infection process, similar to the NRTI such as zalcitabine. In enzyme inhibition assays, calanolide A potently and selectively inhibited recombinant HIV-1 RT but not cellular DNA polymerases or HIV-2 RT. In vitro antiviral studies have indicated that protective activity of (+)-calanolide A against a wide variety of HIV-1 isolates including synctium-inducing and non-synctium-inducing viruses and both T-cell tropic and monocycle-macrophage tropic viruses.[36]

Kinetic analysis demonstrated that (+)-calanolide A inhibited HIV-1 RT by a complex mechanism involving two possible binding sites; one of the calanolide binding sites is near both the pyrophosphate binding site and the active site of the RT enzyme.[37,38]

Evaluation of the safety and pharmacokinetics of a single dose of (+)-calanolide A has been accomplished in 47 healthy HIV uninfected adult volunteers. Pharmacokinetic parameters were highly variable. The half-life ($t_{1/2}$) was about 21 hours in those receiving an 800-mg dose, which suggests that (+)-calanolide A may be suitable for once-daily dosing. Both mean maximum plasma concentration ($C_{max}$) and area under the concentration curve (AUC) values increased with the dose. However, women seemed to have a higher plasma drug level and a longer elimination half-life than men, because this difference may be caused by the differences in their body weight.[36]

### 8.3.2  Clinical Trial of (+)-Calanolide A

The clinical development of calanolide A commenced in 1997. According to the phase IA clinical studies, it was indicated that (+)-calanolide A was safe and well tolerated over a range of different single- and multiple-dose regimens in 94 healthy HIV uninfected adult volunteers.[39] A phase IB study to evaluate calanolide A with different twice-a-day regimens for 14 days in 32 HIV-positive patients with no previous antiretroviral therapy. The phase IB studies showed a trend in viral reduction as the dosing was increased in AIDS patients. However, there was no existence of the emergency of viral mutants over the period of study.[40,41] It was possible to indicate that calanolide A could delay the onset of drug-resistant viral strains. The most frequently reported clinical adverse events were nausea, dyspepsia, and headache. No rashes have been observed. Sarawake MediChem Pharmaceuticals Inc. plans to conduct a phase II/III trial in 2004.

### 8.4  PREPARATION OF *CALOPHYLLUM* COUMARINS

### 8.4.1  Total Synthesis of Racemic *Calophyllum* Coumarins

#### *8.4.1.1  Total Synthesis of Racemic Calanolides*
Chenara et al.[42] was the first research group to establish the racemic calanolide A (±**1**) in 1993 (Scheme 8-1) and the related (±)-calanolide C and D via a five-step synthesis with 15% overall yield starting from phlorglucinol and then constructed the coumarin (A, B rings) followed by the chromanone ring (C) using a Lewis

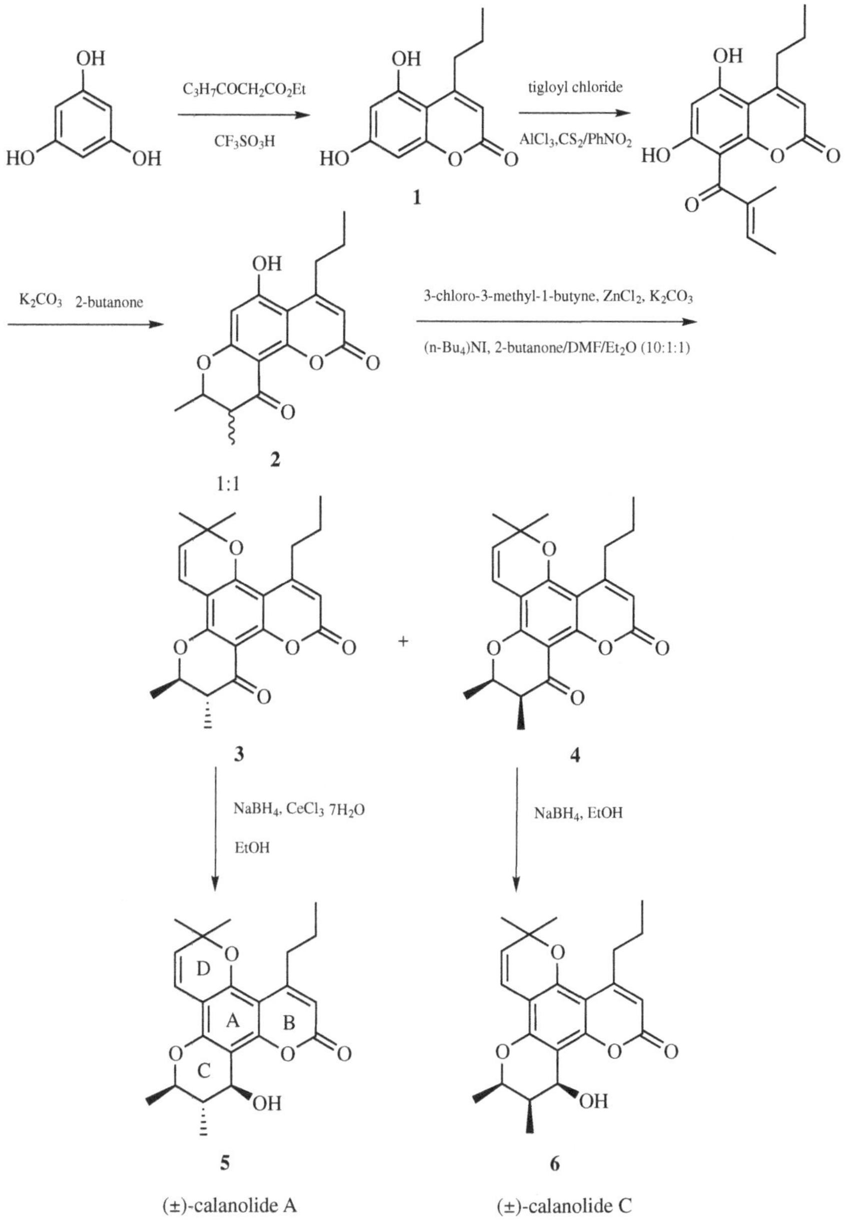

**Scheme 8-1**

acid-promoted Claisen rearrangement. The chromene ring was built last. Finally, Luche reduction of the chromanone provided racemic calanolide A (±**1**).

One year later, Kucherenko et al.[43] reported on a novel method to synthesize racemic calanolide A (Scheme 8-2).

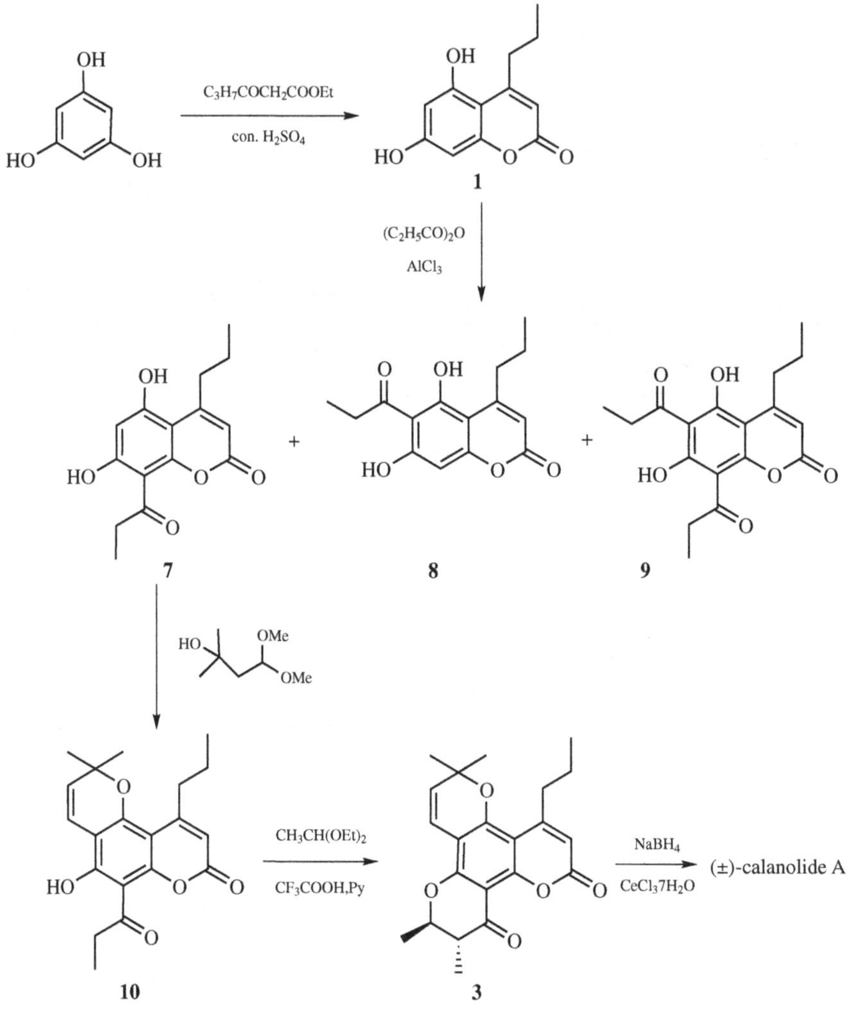

**Scheme 8-2**

This novel approach for synthesis of calanolide A (**1**) was somewhat different from the first one reported.[42] Use of 8-propionyl-5,7-dihydroxy-4-propylcoumarin (**7**) constructed the chromene ring (ring D) with 4,4-dimethoxy-2-methyl butan-2-ol. The chromanone ring (ring C) was built last by treatment of chromene (**10**) with acetoaldehyde diethyl acetal in the presence of trifluoroacetic acid and pyridine to give racemic chromanone (**3**) with the desired stereochemical arrangement. Luche reduction of the chromanone (**3**) with $NaBH_4/CeCl_3$ at $-10\ °C$ was highly selective in giving ($\pm$)-calanolide A in 90% yield with 90% purity, which was purified by preparative high-performance liquid chromatography (HPLC). However, the total yield of (+)-calanolide A was around only 5–6.5%, which was less than other synthetic approaches. Because propionation of 5,7-dihydroxy-4-propyl coumarin with propionic anhydride offered a mixture of an 8-position propionated product (**7**) along with 6-acylated (**8**) and *bis*-acylated compound (**9**), which was separated from **8** and **9** by silica gel chromatography, and the yield of the 8-acylated compound **7** was 30%.

In our research,[44,45] we prepared racemic calanolide A under the same strategy. calanolide A has three different heterocyclic rings, B, C, and D, which are constructed from the phloroglucinol core (A). We used phloroglucinol as the starting material and then constructed the coumarin scaffold, followed by the chromanone ring (ring C) as a key intermediate (**2**). The chromene ring was built last (Scheme 8-3). This method is different from the case of chromene reported before.[42] The Friedel–Crafts acylation and the ring closure of coumarin in a one-step reaction using tigloyl chloride in the presence of $AlCl_3$ formed a key intermediate (**2**) directly with 66% yield. The chromene ring was then introduced using pyridine-catalyzed condensation of 1,1-diethoxy-3-methyl-2-butene with the compound (**2**) in the presence of toluene. The reaction proceeded readily to give chromene (±)-**3** and (±)-**4** in the ratio of 1.75 : 1 in 68% yield. Luche reduction of the ketone (±)-**3** using $NaBH_4/CeCl_3$ at 0~5 °C produced (±)-calanolide A in 63% yield. Its spectral

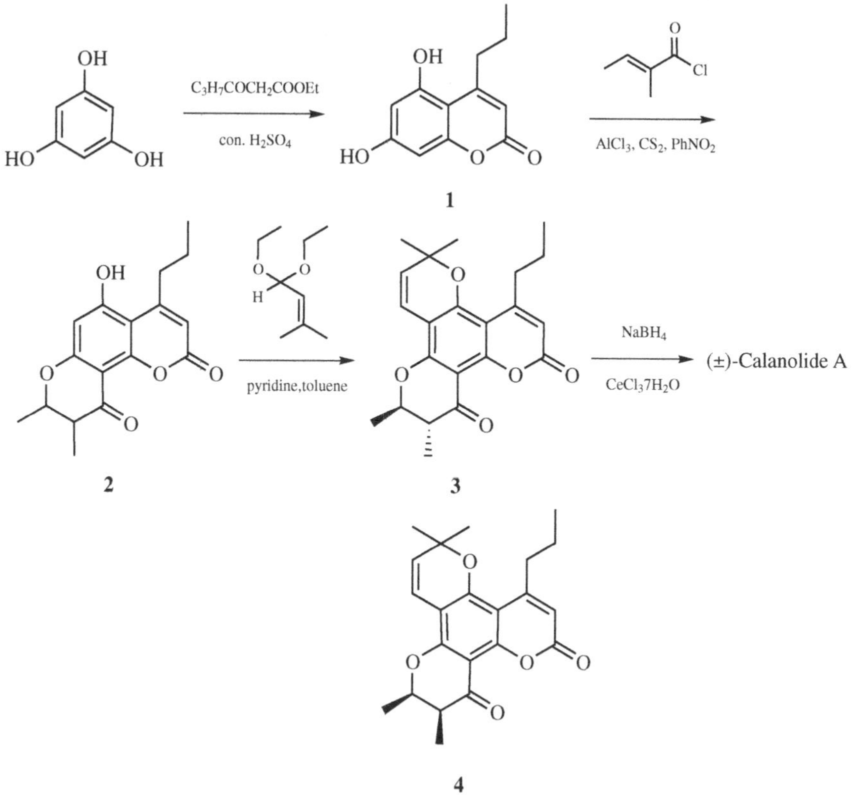

**Scheme 8-3**

data $^1$H nuclear magnetic resonance (NMR), infrared (IR), and mass spectrometry (MS) were identical to the data reported for the natural product (+)-calanolide A. The four-step synthesis of racemic calanolide A was accomplished with 16% overall yield. This four-step procedure to synthesize calanolide A (1) is not only shorter, but also the reagents are cheaper and commercially available.

### 8.4.1.2  Total Synthesis of Racemic Inophyllum

(+)-Inophyllum B (3), which is isolated from the *C. inophyllum* in 1972, has a similar skeleton of (±)-calanolide A with a phenyl group at the C$_4$ position instead of the n-propyl group.

The synthetic procedure of (±)-inophyllum B was accomplished similar to that of (±)-calanolide A as shown in Scheme 8-4.[46] The reaction of phloroglucinol with 4-phenyl-acetoacetate in the presence of hydrogen chloride afforded 5,7-dihydroxy-4-phenyl coumarin using the Pechman reaction. Then acylation of the coumarin using tigloyl chloride in the presence of AlCl$_3$ in PhNO$_2$ and CS$_2$ gave 8-position acylated coumarin (11) in 70% yield. This result was different from the case of

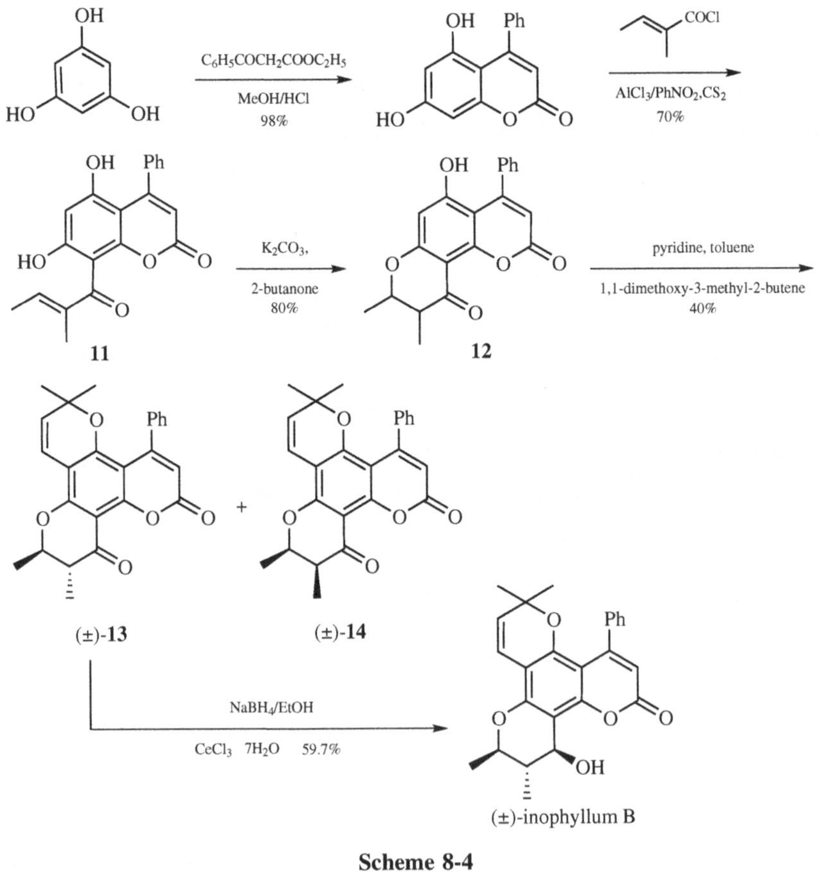

**Scheme 8-4**

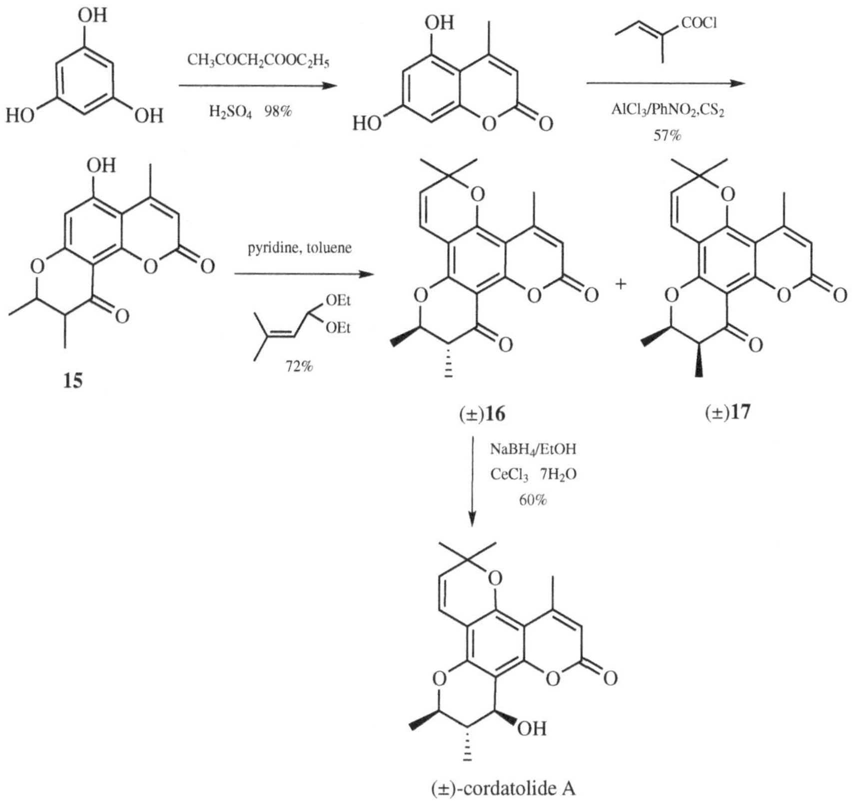

OH / HO / OH

CH$_3$COCH$_2$COOC$_2$H$_5$

H$_2$SO$_4$ 98%

OH / HO

COCl

AlCl$_3$/PhNO$_2$,CS$_2$
57%

OH

15

pyridine, toluene

OEt
OEt

72%

(±)16 + (±)17

NaBH$_4$/EtOH
CeCl$_3$ 7H$_2$O
60%

OH

(±)-cordatolide A

**Scheme 8-5**

acylation with cyclization directly for synthesis of (±)-calanolide A (Scheme 8-3) and (±)-cordatolide B (Scheme 8-5). Ring closure of the 8-tigloyl coumarin (11) in the presence of K$_2$CO$_3$ and 2-butanone gave a key intermediate racemic chromanone (12) in 80% yield. The chromene ring was then introduced by the pyridine-catalyzed condensation of 1,1-diethoxy-3-methyl-2-butene with the chromanone (12) to give dipyranocoumarins (13) and its stereoisomer (14) with a ratio of 2 : 1 in 60% yield. The desired stereo product (13) was separated by column chromatography in 40% yield and then reduced using NaBH$_4$/CeCl$_3$ · 7H$_2$O to afford (±)-inophyllum B by Luche reduction and finally purified by column chromatography. The struture was identified by $^1$H NMR, IR, MS, and elemental analysis. The five-step synthesis of (±)-inophyllum B in an overall yield of about 13.1% provided a useful method for further medicinal investigation of the dipyranocomarin class.

### 8.4.1.3  Total Synthesis of Racemic Cordatolides

(+)-Cordatolide A (5), isolated from the light petrol extract of the leaves of *Calophyllum Cordato-Oblangum* in 1985, has a similar structure of (+)-calanolide A with a methyl group at 4-position instead of the *n*-propyl group. It is also a dipyranocoumarin. The anti-HIV activity of (+)-cordatolide A was lower than that of (+)-calanolide A and (+)-inophyllum B.

The synthetic procedure of (±)-cordatolide A[47] was carried out similar to that of (±)-calanolide A. We used phloroglucinol as a starting material and then constructed the 5,7-dihydroxy-4-methyl coumarin by Pechman reaction in the presence of concentrated sulfuric acid in 98% yield. The acylation and ring closure of the coumarin in a one-step reaction using tigloyl chloride in the presence of AlCl$_3$ formed a key intermediate **15** in 57% yield. The chromene ring was then constructed by the same procedure mentioned above to obtain chromanone of (±)**16** and its stereoisomer (±)**17** with a ratio of 1.5 : 1 in 72% yield. (±)**16** was separated by column chromatography in 50% yield. Finally, Luche reaction reduced the chromanone (±)-**16** to obtain (±)-cordatolide A in 60% yield as shown in Scheme 8-5. The spectral data including $^1$H NMR, IR, and MS of (±)-cordatolide A were identical to those of natural product (+)-cordatolide A. This four-step synthesis of (±)-cordatolide A was accomplished in about 24% overall yield.

### 8.4.1.4   General Procedure of the Total Synthesis of Racemic Calophyllum *Coumarins*

Palmer and Josephs[48] reported on a general method for preparation of (±)-calanolide A, (±)-inophyllum B, and (±)-cordatolide A as shown in Scheme 8-6.

Starting from 8-(2-methylbutyryl) coumarin (**18a–c**), which was prepared by phloroglucinol through two steps (Scheme 8-6), treatment of **18a–c** with 3-methyl-3-hydroxybutyraldehyde in pyridine, followed by methylation with methyl iodide, gave **19a–c**. The conversion of the 2-methyl butyryl side chain in **19a–c** into the (*E*)-2-methylbut-2-enoyl (tigloyl) group to form coumarins **22a–c** by a four-step hydrobromiration-bromination-double dehydrobromination (**20,21**) sequence. Dimethylation of 8-acylcoumarins (**21**) with magnesium iodide in the presence of ether afforded the 7-hydroxy coumarins (**22**). Treatment of **22** with triethylamine gave the desired *trans*-and *cis*-dipyranocoumarins (**23**) and (**24**) that were separated by chromatography. Finally, reduction of **23** with NaBH$_4$/potassium biphthalate reduction gave (±)-calanolide A, (±)-cordatolide B, and (±)-inophyllum B (**25a–c**), respectively. This general method starting either from 8-(2-methylbutyryl)-5, 7-dihydroxycoumarins (**18a–c**) via 9-step sequence or from phloroglucinol via 11-step sequence for the synthesis of the racemic *Calophyllum* coumarins.

### 8.4.2   Preparation of Optically Active Calophyllum **Coumarins**

### 8.4.2.1   Synthesis of Optically Active Calanolide A (1) and Calanolide B (2)

(+)-Calanolide A (**1**) and (−)-calanolide B (**15**) have been shown to be potent inhibitors of HIV-1 RT. Synthesis of both optically active calanolide A and calanolide B have been accomplished by Deshpande et al. (Scheme 8-7).[49,50] They reported on the first enantioselective total synthesis of (+)-calanolide A (**1**) and (+)-calanolide B (**2**) and their (−)enantiomers, (−)-calanolide A (a new compound) and (−)-calanolide B (**15**), using a process that generated all three contiguous chiral centers (the (−)-calanolide B is also known as costatolide), respectively.

Formylation of 5,7-dihydroxy-4-(*n*-propyl)-coumarin (**1**) provided on 8-formylated product (**26**). Treatment of the compound **26** with 3-chloro-3-methyl-1-butyne to introduce regioselectively the chromene **27** because the phenolic hydroxyl group at the $C_7$ position was less accessible for formylated substitution because of a presumed hydrogen-bonding interaction. To construct the enantiomerically pure *trans*-2,3-dimethyl chroman-4-ol system, Deshpande et al.[49,50] used organoborone

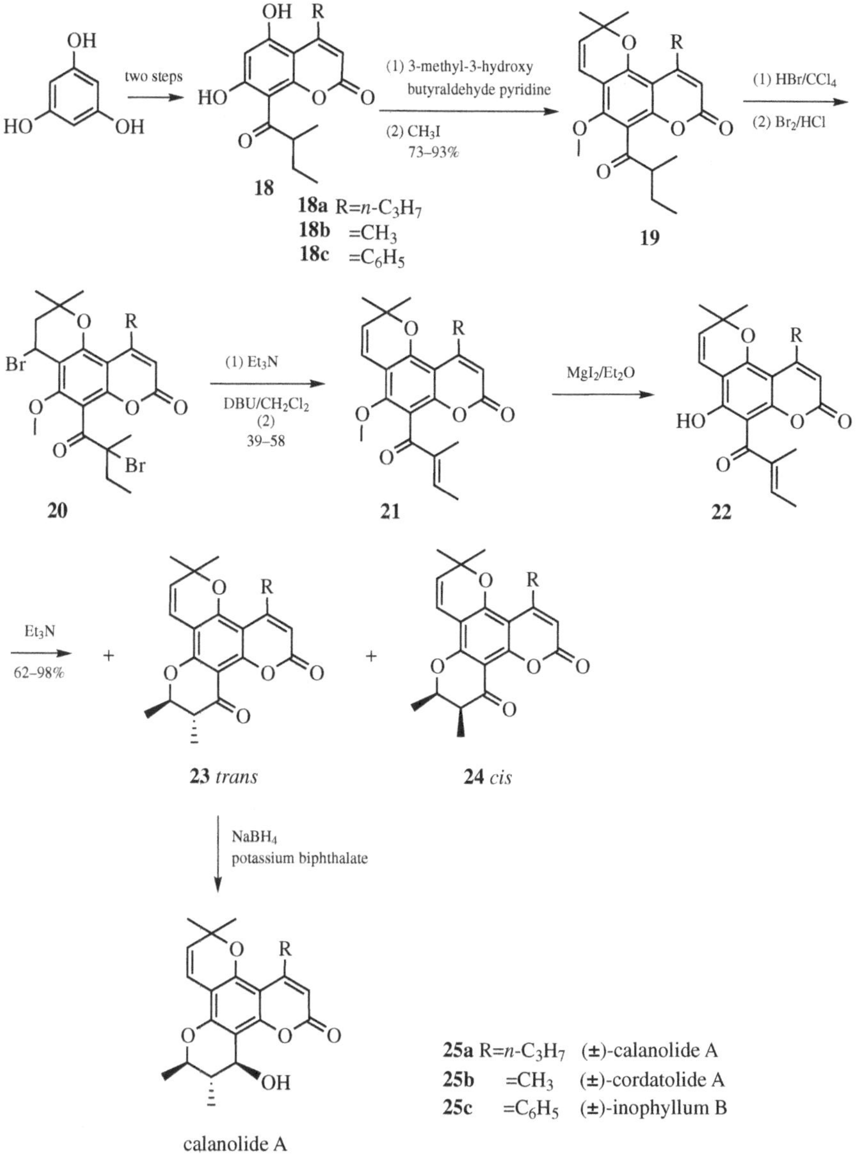


**25a** R=$n$-$C_3H_7$ (±)-calanolide A
**25b** =$CH_3$ (±)-cordatolide A
**25c** =$C_6H_5$ (±)-inophyllum B

**Scheme 8-6**

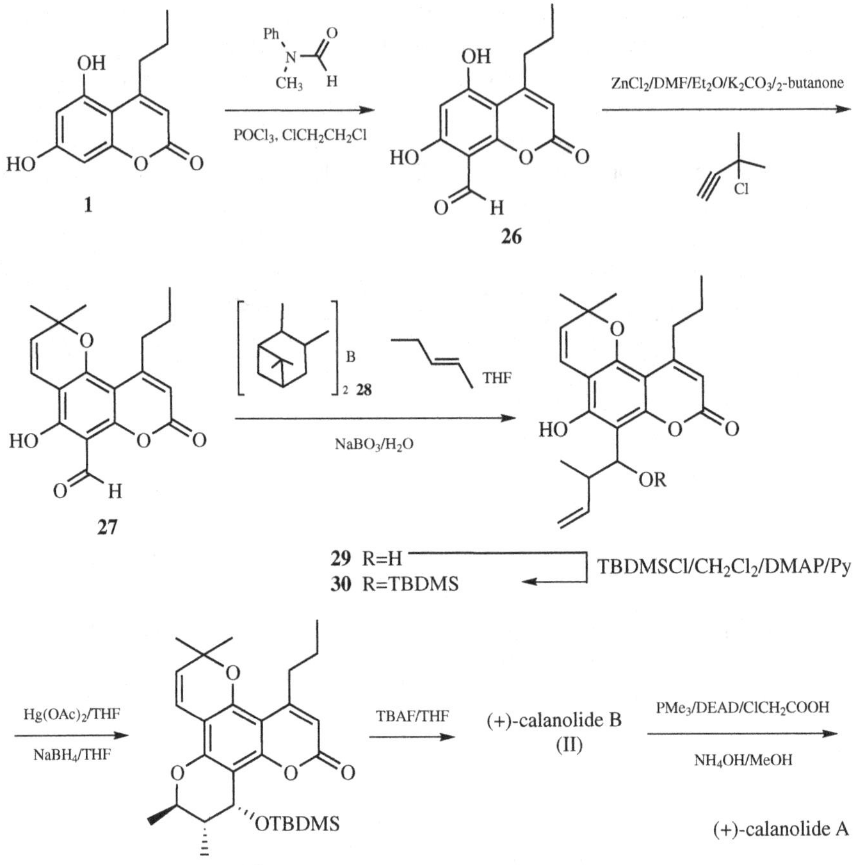

**Scheme 8-7**

reagent $(+)$-$(E)$-crotyldiisopinocampheylborane (**28**), which was prepared according to the procedure of Brown and Bhat. Treatment of the aldehyde (**27**) with the organoborone reagent (**28**) gave *threo* β-methyl homoallylic alcohol **29**, $[\alpha]_D^{20} = +78°$ with 66% yield. Acylation of **29** with TBDMSCl to give the monosillyated product **30** ($[\alpha]_D^{20} = +26°$) with 95% yield caused by the phenolic hydroxyl group was less active than the secondary homoallylic alcohol, which lead to preferred monosillylation. Cyclization of the compound **30** was carried out with mercury (II) acetate in THF, and the resulting intermediate was reduced with an excess of sodium borohydride to give silyl-protected $(+)$-calanolide B (**2**) in 85% yield ($[\alpha]_D^{20} = -41.5°$). Finally, deprotection of the silyl group with tetra-butylammonium fluoride gave $(+)$-calanolide B (**2**) in 86% yield $[\alpha]_D^{20} = +44°$. The $^1$H and $^{13}$C NMR spectra were identical with those reported for the corresponding natural products. The conversion of $(+)$-calanolide B into $(+)$-calanolide A in 80% yield ($[\alpha]_D^{20} = +66°$) was accomplished with a modified Mitsunobo reaction. The total yields of this seven-step process for synthesizing $(+)$-calanolide B and eight-step for (+)-calanolide A was 22.3% and 18.1%, respectively (from the starting material 5,7-dihydrox-4- (n-propyl)- coumarin). The entire procedure shown in Scheme 8-7 was repeated just using organoborone reagent (−)-(E)-crotyldiisopino-campheyl-borane (**28a**) instead of its enantiomer **28** to give (−)-calanolide B (**15**) (costatolide, $[\alpha]_D^{20} = -45°$). Finally, treatment with (−)-calanolide B using modified Mitsunobu conditions gave unknown (−)-calanolide A (1a) $[\alpha]_D^{20} = -66°$.

### 8.4.2.2 Synthesis of (+)-Inophyllum B (3) and (+)-Inophyllum P (4) Using (−)-Quinine Catalyzed Intramolecular oxo-Michael Addition[51]

Ishikawa et al. reported[52,53] that (−)-quinine-catalyzed asymmetric intramolecular oxo-Michael addition (IMA) of 7-hydroxy-8-tigloylcoumarin gave *cis*-2,3-dimethyl-4-chromanone systems with high enantioselectivity and moderate diastereoselectivity, especially when chlorobenzene was used as a solvent.[54] Therefore, total synthesis of (+)-calanolide A (**1**) was achieved by application of the (−)-quinine-catalyzed asymmetric IMA.[51] However, the synthetic route starting from 1,3,5- trimethoxybenzene was too long (13 steps with 3.5% overall yield) to practice. Finally, the authors improved and shortened the original synthetic route by application of MgI$_2$-assisted demethylation.

5,7-Dimethoxy coumarin (**32**) was prepared from the treatment of phloroglucinol with 4-phenylacetoacetate in the presence of triflic acid by Pechman reaction and then methylation with dimethylsulfate (Me$_2$SO$_4$). The Friedel–Crafts acylation of the resulting 5,7-dimethoxy coumarin (**32**) with tigloyl chloride and regioselective demethylation at 7-position with MgI$_2$ in the presence of potassium carbonate gave 7-hydroxy-5-methoxy-4-phenyl-8-tigloylcoumarin (**33**) in 40% overall yield (Scheme 8-8). The reaction of **33** with a catalytic amount (10 mol%) of (−)-quinine in chlorobenzene gave *cis*-(+)-chromanone **34** and the *trans*-(−)-chromanone **35** in a 3 : 1 ratio (*cis* : *trans*) in 93% overall yield. The enantiopurities of **34** and **35** was accomplished by chiral HPLC analysis. The reflux of *cis*-(+)-**34** in benzene with MgI$_2$ in the presence of K$_2$CO$_3$ for 14 days gave the desired *trans*-chromanone *trans*-**37** and *cis*-**36** in 42% and 37% yields, respectively. Refluxing *trans*-**37** with senecinaldehyde in the presence of phenylboronic acid, propionic acid, and toluene gave the expected cyclized products *trans*-(+)-**38**, (+)-inophyllum C (**11**) in 51% yield along with an isomerized *cis*-(+)-**38**, (+)-inophyllum E (**12**) in 23% yield. Reduction of *trans*-(+)-**38** with LiAlH$_4$ at −20 °C gave (+)-inophyllum B (**3**) in 91% yield and (+)-inophyllum P (**4**) in 9% yield. The synthetic procedure comprised eight steps from the starting material phloroglucinol.

### 8.4.2.3 Total Synthesis of (+)-Calanolide A (1) Using (−)-Quinine Catalyzed IMA[51]

The total synthesis of (+)-calanolide A (**1**) was carried out by application of the synthetic method for (+)-inophyllum B (**3**) using the same strategy.

The synthesis of 5,7-dimethoxy coumarin **39** started from phloroglucinol through coumarin construction by Pechman reaction followed by methylation via two steps. The resulting 8-tigloylated coumarin **40** (Scheme 8-9) was obtained by treatment of the coumarin (**39**) with tigloyl chloride according to the preparation method described in Scheme 8-8.

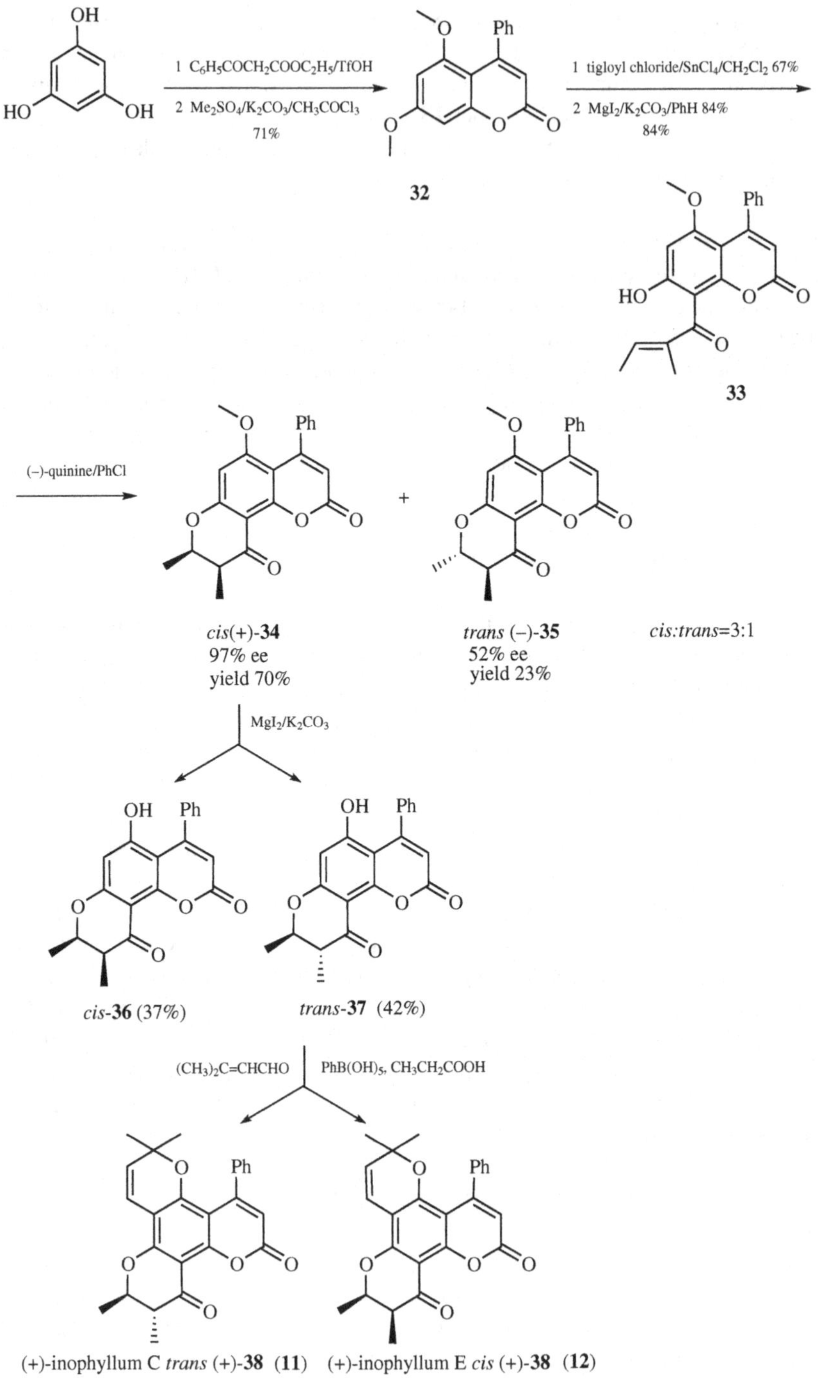
OH
HO
OH
1 $C_6H_5COCH_2COOC_2H_5$/TfOH
2 $Me_2SO_4/K_2CO_3/CH_3COCl_3$
71%
Ph
32
1 tigloyl chloride/$SnCl_4/CH_2Cl_2$ 67%
2 $MgI_2/K_2CO_3$/PhH 84%
84%
33
(–)-quinine/PhCl
*cis*(+)-**34**
97% ee
yield 70%
*trans* (–)-**35**
52% ee
yield 23%
*cis:trans*=3:1
$MgI_2/K_2CO_3$
*cis*-**36** (37%)
*trans*-**37** (42%)
$(CH_3)_2C=CHCHO$
$PhB(OH)_5$, $CH_3CH_2COOH$
(+)-inophyllum C *trans* (+)-**38** (**11**)
(+)-inophyllum E *cis* (+)-**38** (**12**)

**Scheme 8-8**

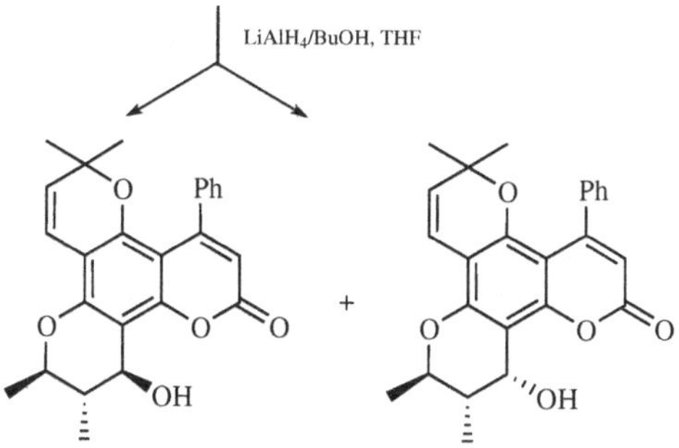

(+)-inophyllum B (**3**)        (+)-inophyllum P (**4**)
2, 91%                                9%

**Scheme 8-8**    (*Continued*)

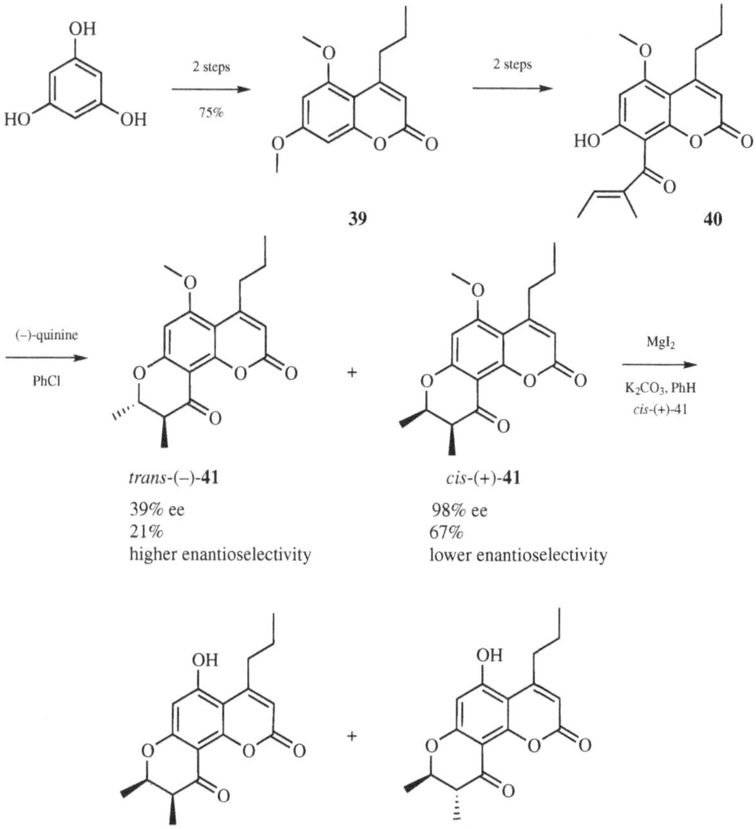

*trans*-(−)-**41**          *cis*-(+)-**41**

39% ee                    98% ee
21%                         67%
higher enantioselectivity    lower enantioselectivity

*cis*-**42**                        *trans*-(**42**)   major product

diastereomeric mixture
without separation

**Scheme 8-9**

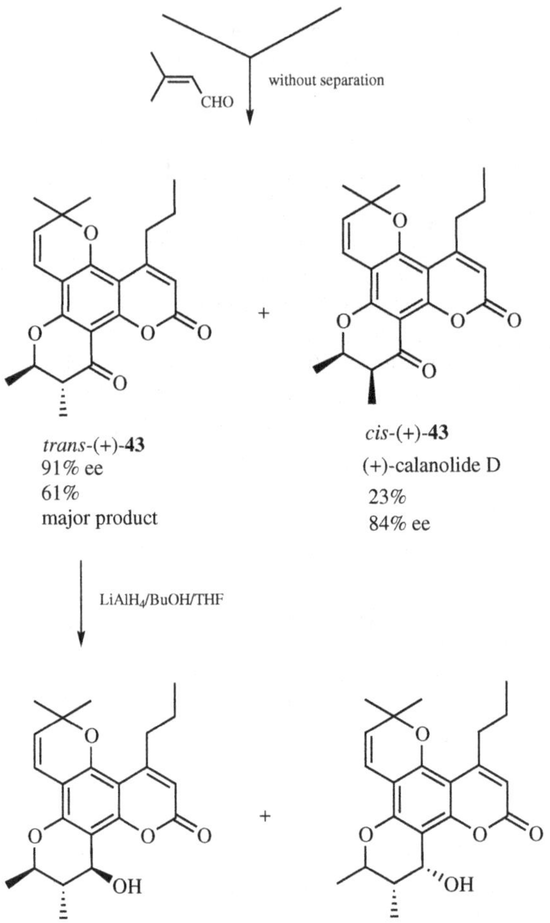


*trans*-(+)-**43**
91% ee
61%
major product

*cis*-(+)-**43**
(+)-calanolide D
23%
84% ee

(+)-calanolide A (**1**)
88%

(+)-calanolide B (**2**)

**Scheme 8-9** (*Continued*)

The (−)-quinine-catalyzed IMA of **40** gave *cis*-(+)-methoxy chromanone coumarin (*cis*-**41**) in 67% yield with 98% ee and its *trans*-**41** in 21% yield with lower eantioselectivity (39% ee). The $MgI_2$-assisted isomerization of *cis*-(+)-**41** accompanied by demethylation followed by the reaction of the formed diastereomeric mixture of chromanone **42** (without separation) with senecioyl aldehyde in the presence of phenyl boronic acid under the same condition to construct the 2,2-dimethyl chromanone ring reported in the synthesis of (+)-inophyllum B (**3**). The yield of *trans*-(+)-**43** in 61% with 91% ee and its *cis*-isomer (+)-**43** corresponded to the (+)-calanolide D as a minor product in 23% yield with 84% ee. At last, the synthesis of (+)-calanolide A (**1**) was accomplished by hydride reduction of (+)-**43** with

LiAlH$_4$ with 88% yield, and (+)-calanolide B was obtained as a minor reduction product with 12% yield.

The overall reaction for synthesizing (+)-calanolide B and (+)-calanolide A from the starting material phloroglucinol in eight steps results in 5% and 10% yields, respectively.

Although a long reaction time was needed in some reaction steps, this strategy would be applicable to a large-scale experiment under simple operation.

### 8.4.2.4 Synthesis of (−)-Calanolide A and (−)-Calanolide B Using a Catalytic Enantioselective Approach

Trost and Toste[55] developed a new route to synthesize (−)-calanolide A and B starting from compound **44**, which was available from phloroglucinol in one step (Scheme 8-10). Clemensen reduction of the ketone **44** gave the corresponding phenol **45**, which was coupled with alkyne **46** under the recently developed

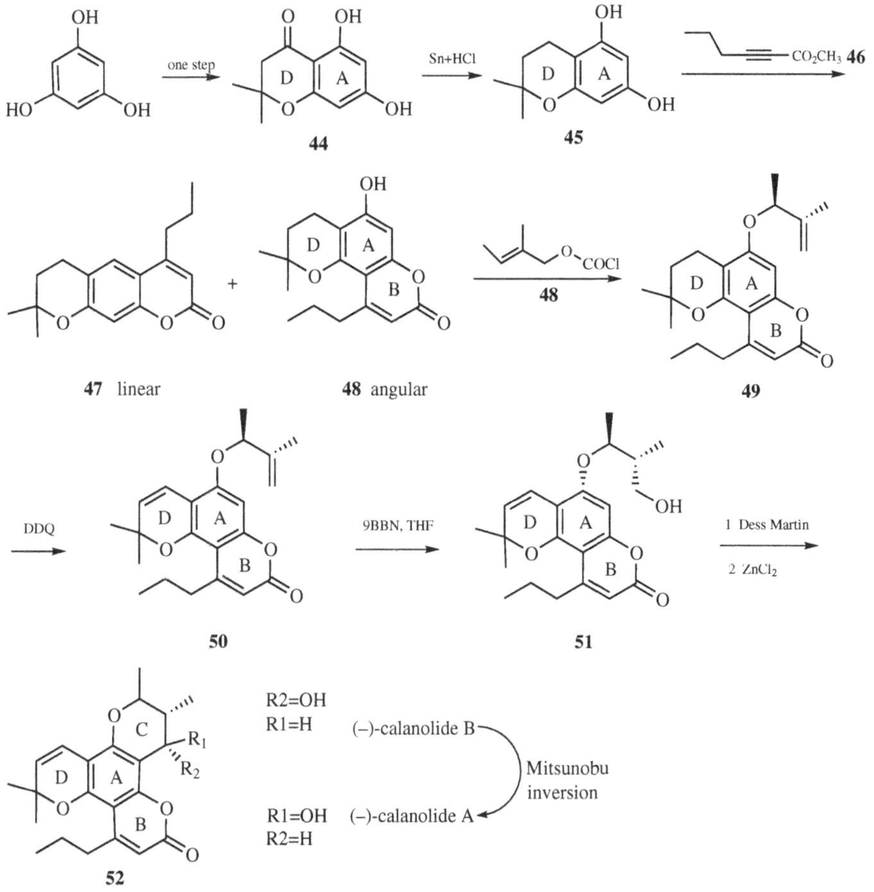

**Scheme 8-10**

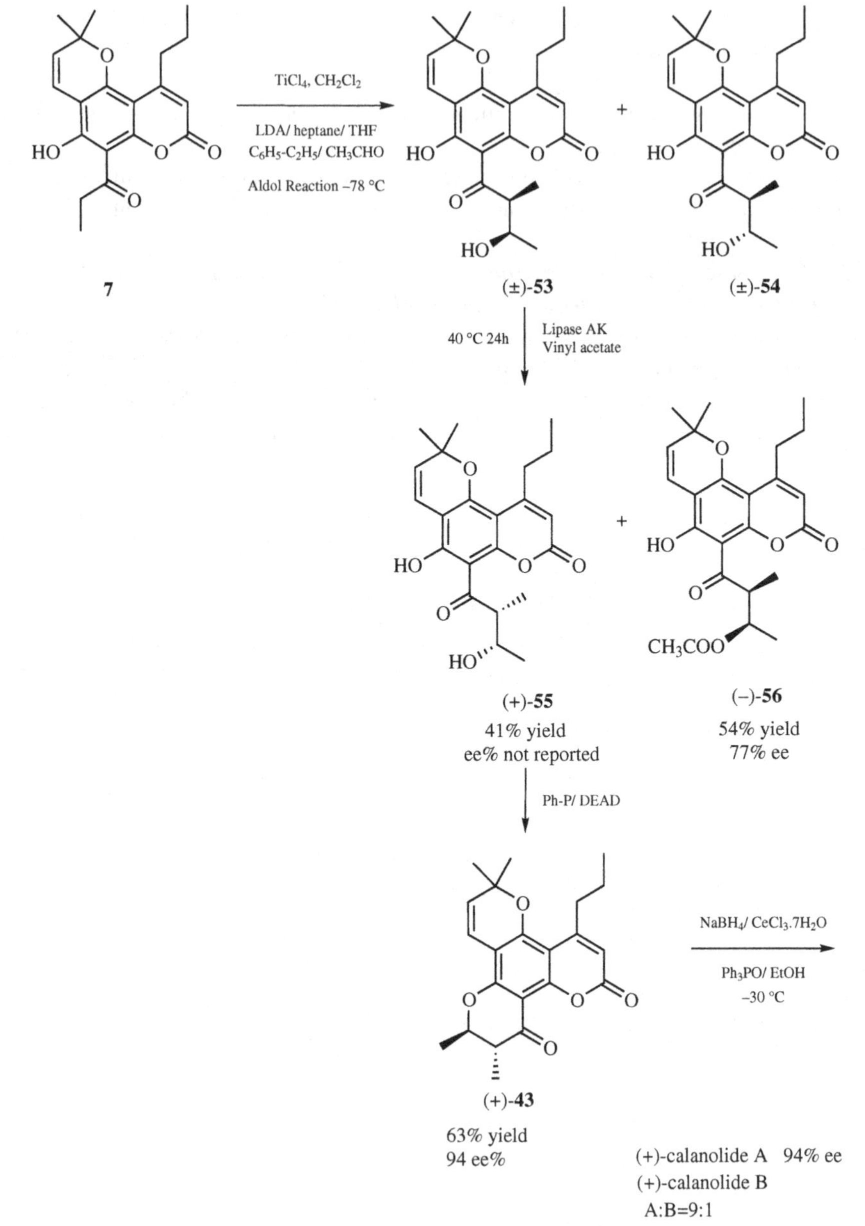

**Scheme 8-11**

palladium-catalyzed coumarin formation reaction. The resulting coumarins angular **48** and linear **47** were a mixture of regioisomers with a ratio of 4 : 1. The palladium-catalyzed reaction with coumarin (**48**) and tigloyl methyl carbonate in the presence of anthracene-based ligand in THF afforded **49**, the regiomer with a ratio 91 : 9 in 90% ee. Oxidization of **49** with DDQ obtained chromene **50** (ring D), which was chemoselectively hydroborated with 9-BBN to give the alcohol (**51**) with a 93 : 7 diastereomeric ratio for the newly formed stereocenter. The resulting chromene (**51**) was then readily oxidized with Dess Martin periodinane, and the formed aldehye was diastereoselectively cyclized with $ZnCl_2$ to obtain the thermodynamically less stable (−)-calanolide B (10 : 1 diastereoselectivity), which was then converted into (−)-calanolide A by a Mitsunobo inversion. This strategy includes asymmetric regioselective O-alkylation followed by electrophilic aromatic substitution to construct carbon–carbon constituents, an effective approach for enantioselective chromene and chromanone synthesis.

### 8.4.2.5    Synthesis of (+)-Calanolide A From Enzyme-Catalyzed Optical Resolution of Compound (+)-53

Synthesis of (+) calanolide A (Scheme 8-11)[56] was achieved by enzyme catalyzed resolution of the aldol products (±)-**53**. Compound **7** with acetaldehyde by aldol reaction in the presence of LDA/TiCl₄ stereoselectively produced a mixture of (±)-**53** and (±)-**54** (94% yield), the ratio of which was 96 : 4. (±)-**53** was then resolved by lipase AK-catalyzed acylation reaction in the presence of *tert*-butyl methyl ether and vinyl acetate at 40 °C to obtain 41% yield of (+)-**55** and 54% yield of the acetate (−)-**56**. Mitsunobu cyclization of (+)-**55** in the presence of triphenylphosphine and dielthyl azodicarboxylate afforded 63% yield of (+)-**43** with 94% ee as determined by chiral HPLC. Luche reaction on (+)-**43** with $CeCl_3 \cdot 7H_2O$ and triphenyl phosphine oxide and NaBH₄ in the presence of ethanol at −30 °C gave the crude product. It was purified by column chromatography on silica gel to give 78% yield of a mixture containing 90% of (+)-calanolide A and 10% (+)-calanolide B, which were further separated by HPLC.

### 8.4.2.6    Resolution of Synthetic (±)-Calanolide A by Chiral HPLC

The synthetic (±)-calanolide A was resolved into its enantiomers, (+)-calanolide A (**1**) and (−)-calanolide A, by using a semipreparative chiral HPLC column packed with amylose carbamate eluting with hexane/ethnol (95 : 5).[25] The ultraviolet detection was set at a wavelength of 254 nm. (+)-calanolide A and its enantiomer (−)-calanolide A were collected, and their chemical structures were identified based on their optical rotations and spectroscopic data, as compared with the corresponding natural and synthesis compounds.

## 8.5   STRUCTURE MODIFICATION OF CALANOLIDES

Four research groups were engaged in the structural modification of calanolides for the structure activity relationship studies. Galinis et al.[57] have focused on a catalytic reduction of the $\Delta^{7,8}$-olefinic bond in (+)-calanolide A (**1**) and (−)-calanolide B (**15**), such as chemical modifications of $C_{12}$ hydroxy group in (−)-calanolide B (**15**) because **15** was not only active against HIV-1 with a potency similar to that of (+)-calanolide A but also produced a selectively high yield (≥15%) isolated from the latex of *Calophyllum teysmanni* var. *inophyllolide*.[17] A total of 14 analogs were synthesized and tested by the NCI primary anti-HIV evaluation using XTT assay, whereas no compounds exhibited anti-HIV activity superior to the two natural lead products (**1** and **15**). Some structure activity requirements were apparent from the relative anti-HIV potencies of the various analogs. Reduction of the $\Delta^{7,8}$-olefinic bond indicated only a slight change in potency, and a hetero atom is required at $C_{12}$, for activity such as $(-)$-12-oxo-calanolide B and $(+)$-12-oxo-calanolide A still showed antiviral activity.

Zembower et al.[27] prepared a series of racemic structural analogs of $(\pm)$-calanolide A focused on the modification of the *trans*-10,11-dimethyldihydropyran-12-ol ring (ring C) and evaluated their anti-HIV activity using a CEM-SS cytoprotection assay. Introduction of the extra methyl group into either the $C_{10}$ or $C_{11}$ position resulted in only two chiral centers with lowering of the activity relative to $(\pm)$-calanolide A. Substitution of the $C_{10}$-methyl group with an isopropyl chain led to loss of activity. Analogs containing a *cis*-relationship between the $C_{10}$ and $C_{11}$ alkyl groups were completely devoid of the potency except for the synthetic intermediate in which the $C_{12}$-hydroxyl group in the ketone form possessed activity less potent than that of calanolide A with either the $C_{10}$-$C_{11}$-*cis*- or $C_{10}$-$C_{11}$-*trans*-relationship. It is interesting that these two ketones represent the first derivatives in the calanolide series to exhibit anti-HIV activity while not containing a $C_{12}$-hydroxyl group and including a *cis*-relationship between the $C_{10}$- and $C_{11}$-methyl groups.

Some structural modifications of racemic calanolide A have been studied by us.[58] Removal of $C_{11}$-methyl or methyl groups at both $C_{11}$ and $C_6$ positions maintains the basic stereochemistry of $(\pm)$-calanolide A as well as evaluates their anti-HIV activity. $(\pm)$-$C_{11}$-demethyl calanolide A and $(\pm)$-calanolide A exhibited similar potencies, whereas the 6,6,11-demethyl calanolide A also had diminished activity. The primary biological indicated that two methyl groups in the $C_6$ position might be necessary for antiviral activity.

Recently Sharma et al.[59,60] envisaged that replacement of the coumarin ring oxygen atom (at position $C_1$) of calanolide A with the nitrogen atom lead to dipyranogsuinolinone with a calanolide skeleton, namely, aza-calanolide A. The $EC_{50}$ value of $(\pm)$-aza-calanolide A (0.12 μM) is much lower than that of the natural product $(\pm)$-calanolide A (0.27 μM). The $IC_{50}$ value for $(\pm)$-aza-calanolide A and $(+)$-calanolide A are 15 μM and 23 μM, respectively. Thus, these compounds were useful as an anti-HIV agent and were found to possess a better therapeutic index than calanolide A.

## 8.6    CONCLUSION

$(+)$-Calanolide A (**1**), $(-)$-calanolide B (**15**), and $(+)$-inophyllum B (**3**) are the most potent candidates of anti-HIV-1 agents among *Calophyllum* coumarins. $(+)$-calanolide A (**1**) is the only naturally occurring anti-HIV agent to be at an advanced stage of a phase II clinical trial, but it has been obtained in relatively low yield isolated from *Calophyllum lanigerum*. Several synthetic routes for preparation of racemic calanolide A, cordatolide A, and inophyllum B were established, respectively, and enantioselective synthesis of optically active (+)-calanolide A and (+)-inophyllum B were also accomplished.

Some structural modifications of (+)-calanolide A or (−)-calanolide B have been carried out. However, no compounds showed anti-HIV-1 activity superior to the two lead compounds (1 and 15). Hopefully, calanolide A continues to have promise as an anti-HIV drug in the near future.

## REFERENCES

1. Piot, P.; Bartos, M.; Ghys, P. D.; Walker, N.; Schwartlander, B. The global impact of HIV/AIDS. *Nature*, **2001**, 410: 968–973.

2. Menendez, A. L. Target HIV: Antiretroviral therapy and development of drug resistance. *Trends Pharmacol. Sci.*, **2002**, 23: 381–388.

3. Richman D. D. HIV themotherapy. *Nature*, **2001**, 410: 995–1001.

4. http://www.aidsinfo.nih.gov/drugs/.

5. Vlietinck, A. J.; Bruyne, T. D.; Apers, S.; Pieters, L. A. Plant-derived leading compounds for chemotherapy of human immunodeficiency virus (HIV) infection. *Planta Medica*, **1998**, 64: 97–109.

6. Yang, S. S.; Cragg, G. M.; Newman, D. J.; Bader, J. P. Natural product-based anti-HIV drug discovery and development facilitated by the NCI developmental therapeutics program. *J. Nat. Prod.*, **2001**, 64: 265–277.

7. Song, A. M.; Zhang, J. H.; Lam, K. S. Synthesis and reaction of 7-fluoro-4-methyl-6-nitro-2-oxo- 2H–1-benzopyran-3-carboxylic acid: A novel scaffold for combinatorial synthesis of coumarins. *J. Com. Chem.*, **2004**, 6(1): 112–120.

8. Kashman, Y.; Gustafson, K. R.; Fuller, R. W.; Cardellina, J. H. II.; McMahon, J. B.; Currens, M. J.; Buckheit, R. W.; Jr.; Hughes, S. H.; Cragg, G. M.; Boyd, M. R. HIV inhibitory natural products. Part 7. The calanolides, a novel HIV-inhibitory class of coumarin derivatives from the tropical rainforest tree, *Calophyllum lanigerum*. *J. Med. Chem.*, **1992**, 35: 2735–2743.

9. Patil, A. D.; Freyer, A. J.; Eggleston, D. S.; Haltiwanger, R. C.; Bean, M. F.; Taylor, P. B.; Caranfa, M. J.; Breen, A. L.; Bartus, H. R. The inophyllums, novel inhibitors of HIV-1 reverse transcriptase isolated from the Malaysian tree, *Calophyllum inophyllum* Linn. *J. Med. Chem.*, **1993**, 36: 4131–4138.

10. Spino, C.; Dodier, M.; Sotheeswaran, S. Anti-HIV coumarins from *Calophyllum* seed oil. *Bioorg. Med. Chem. Lett.*, **1998**, 8: 3475–3478.

11. Kawazu, K.; Ohigashi, H.; Mitsui, T. The piscicidal constituents of *Calophyllum Inophyllum* Linn. *Tetrahedron Lett.*, **1968**, 19: 2383–2385.

12. Dharmaratne, H. R. W.; Sotheeswaran, S.; Balasubramaniam, S.; Waight, E. S. Triterpenoids and coumarins from the leaves of *Calophyllum Cordato-Oblongum*. *Phytochemistry*, **1985**, 24(7): 1553–1556.

13. Dharmaratne, H. R. W.; Sajeevani, J. R. D. M.; Marasinghe, G. P. K.; Ekanayache, E. M. H. G. S. Distribution of pyranocoumarins in *Calophyllum Cordato-Oblongum*. *Phytochemistry*, **1998**, 49(4): 995–998.

14. Cao, S. G.; Sim, K. Y.; Pfreira, J.; Goh, S. H. Coumarins from *Calophyllum teysmannii*. *Phytochemistry*, **1998**, 47(5): 773–777.

15. Gustafson, K. R.; Bokesch, H. R.; Fuller, R. W.; Cardellina, J. H.; Kadushin, M. R.; Soejarto, D. D.; Boyd, M. R. Calanone, a novel coumarin from *Calophyllum teysmannii*. *Tetrahedron Lett.*, **1994**, 35(34): 5821–5824.

16. McKee, T. C.; Fuller, R. W.; Covington, C. D.; Cardellina, J. H.; Gulakowski, R.; Krepps, B. L.; McMahon, J. B.; Boyd, M. R. New pyranocoumarins isolated from *Calophyllum lanigerum* and *Calophyllum teysmannii*. *J. Nat. Prod.*, **1996**, 59: 754–758.

17. Fuller, R. W.; Bokesch, H. R.; Gustafson, K. R.; McKee, T. C.; Cardellina, J. H.; McMahon, J. B.; Cragg, G. M.; Soejarto, D. D.; Boyd, M. R. HIV-inhibitory coumarins from latex of the tropical rainforest tree Calophyllum teysmannii var. inophylloide. *Bioorg. Med. Chem. Lett.*, **1994**, 4: 1961–1964.

18. Stout, G. H.; Stevens, K. L. The structure of Costatolide. *J. Org. Chem.*, **1964**, 29: 3604–3609.

19. Bandara, B. M. R.; Dharmaratne, H. R. W.; Sotheeswaran, S.; Balasubramaniam, S. Two chemically distinct groups of *Calophyllum* species from Sri Lanka. *Phytochemistry*, **1986**, 25(2): 425–428.

20. Gunasekera, S. P.; Jayatilake, G. S.; Selliah, S. S.; Sultanbawa, M. U. S. Chemical investigation of ceylonese plants. Part 27. Extractives of *Calophyllum cuneifolium* Thw. and *Calophyllum soulattri* Burm. f. (Guttiferae). *J. Chem. Soc., Perkin Tran. 1*, **1977**, 13: 1505–1511.

21. McKee, T. C.; Cardellina, J. H.; Dreyer, G. B.; Boyd, M. R. The pseudocalanolides: structure revision of calanolide C and D. *J. Nat. Prod.*, **1995**, 58: 916–920.

22. Dharmaratne, H. R. W.; Sotheeswaran, S.; Balasubramaniam, S. Triterpenes and neoflavonoids of *Calophyllum lankaensis* and *Calophyllum thwaitesii*. *Phytochemistry*, **1984**, 23(11): 2601–2603.

23. Nigam, S. K.; Mitra, C. R. Constituents of *Calophyllum Tomentosum* and *Calophyllum Apetalum* nuts: Structure of a new 4-alkyl- and of two new 4-phenyl-coumarins. *Tetrahedron Lett.*, **1967**, 28: 2633–2636.

24. McKee, T. C.; Covington, C. D.; Fuller, R. W.; Bokesch, H. R.; Young, S.; Cardellina, J. H., II.; Kadushin, M. R.; Soejarto, D. D.; Stevens, P. F.; Cragg, G. M.; Boyd, M. R. Pyranocoumarins from tropical species of the genus *Calophyllum*: A chemotaxonomic study of extracts in the National Cancer Institute collection. *J. Nat. Prod.*, **1998**, 61: 1252–1256.

25. Flavin, M. T.; Rizzo, J. D.; Khilevich, A.; Kucherenko, A.; Sheinkman, A. K.; Vilaychack, V.; Lin, L.; Chen, W.; Mata, E. Synthesis, chromatographic resolution, and anti-human immunodeficiency virus activity of ($\pm$)-calanolide A and its enantiomers. *J. Med. Chem.*, **1996**, 39: 1303–1313.

26. Buckheit, R. W.; Jr.; Fliakas-Boltz, V.; Decker, W. D.; Roberson, J. B.; Stup, T. L.; Pyle, C. A.; White, E. L.; McMahon, J. B.; Currens, M. J.; Boyd, M. R.; Bader, J. P. Comparative anti-HIV evaluation of diverse HIV-1-specific reverse transcriptase inhibitor-resistant virus isolates demonstrates the existence of distinct phenotypic subgroups. *Antiviral Res.*, **1995**, 26: 117–132.

27. Zembower, D. E.; Liao, S.; Flavin, M. T.; Xu, Z. Q.; Stup, T. L.; Buckheit, R. W.; Jr.; Khilevich, A.; Mar, A. A.; Sheinkman, A. K. Structrul analogues of the calanolide anti-HIV agents. Modification of the *trans*-10,11-dimethyl-dihydropyran-12-ol ring (ring C). *J. Med. Chem.*, **1997**, 40: 1005.

28. Xu, Z. Q.; Buckheit, R. W.; Jr.; Stup, T. L.; Flavin, M. T.; Khilevich, A.; Rizzo, J. D.; Lin, L.; Zembower, D. E. In vitro anti-human immunodeficiency virus (HIV) activity of the chromanone derivative, 12-oxocalanolide A. A novel NNRTI. *Bioorg. Med. Chem. Lett.*, **1998**, 8: 2179–2184.

29. Buchheit, R. W; Jr.; Russell, J. D.; Xu, Z. Q.; Flavin, M. Anti-HIV-1 activity of calanolides used in combination with other mechanistically diverse inhibitors of HIV-1 replication. *Antiviral Chem. Chemother.*, **2000**, 11: 321–327.

30. Buchheit, R. W; Jr.; White, E. L.; Eliakas-Boltz, V.; Russell, J.; Stup, T. L.; Kinjerski, T. L.; Osterling, M. C.; Weigand, A.; Bader, J. P. Unique anti-human immunodeficiency virus activities of the nonnucleoside reverse transcriptase inhibitors calanolide A, costatolide, and dihydrocostatolide. *Antimicrob. Agents Chemother.*, **1999**, 43: 1827–1834.

31. Xu, Z. Q.; Lin, Y. M.; Flavin, M. T. U.S. Patent Application No. 09/417,672/1999; U.S. Patent 6,268,393 B1, 2001

32. Xu, Z. Q.; Hollingshead, M. G.; Borgel, S.; Elder, C.; Khilevich, A.; Flavin, M. T. In vivo anti-HIV activity of (+)-calanolide A in the hollow fiber mouse model. *Bioorg. Med. Chem. Lett.*, **1999**, 9: 133–138.

33. Pengsuparp, T.; Serit, M.; Hughes, S. H.; Soejarto, D. D.; Pezzuto, J. M. Specific inhibition of human immunodeficiency virus type 1 reverse transcriptase mediated by soulattrolide, a coumarin isolated from the latex of *Calophyllum teysmannii*. *J. Nat. Prod.*, **1996**, 59: 839–842.

34. Shi, X. W.; Attygalle, A. B.; Liwo, A.; Hao, M. H.; Meinwald, J. Absolute stereochemistry of Soulattrolide and its analogues. *J. Org. Chem.*, **1998**, 63: 1233–1238.

35. Dharmaratne, H. R. W.; Wanigasekera, W. M. A. P.; Mata-Greenwood, E.; Pezzuto, J. M. Inhibition of human immunodeficiency virus type 1 reverse transcriptase activity by cordatolides isolated from *Calophyllum cordato-oblongum*. *Planta Med.*, **1998**, 64: 460–461.

36. Creagh, T.; Ruckle, J. L.; Tolbert, D. T.; Giltner, J.; Eiznhamer, D. A.; Dutt, B.; Flavin, M. T.; Xu, Z. Q. Safety and pharmacokinetics of single doses of (+)-calanolide, A.: A novel, naturally occurring nonnucleoside reverse transcriptase inhibitor, in health, human immunodeficiency virus-negative human subjects. *Antimicrob. Agents Chemother.*, **2001**, 45: 1379–1386.

37. Currens, M. J.; Gulakowski, R. J.; Mariner, J. M.; Moran, R. A.; Buckheit, R. W.; Jr.; Gustafson, K. R.; McMahon, J. B.; Boyd, M. R. Antiviral activity and mechanism of action of calanolide A against the human immunodeficiency virus type-1. *J. Pharmacol. Exp. Ther.*, **1996**, 279: 645–651.

38. Currens, M. J.; Mariner, J. M.; McMahon, J. B.; Boyd, M. R. Kinetic analysis of inhibition of human immunodeficiency virus type-1 reverse transcriptase by calanolide A. *J. Pharmaol. Exp. Ther.*, **1996**, 279: 652–661.

39. Buckle, J.; Giltner, J.; Creagh, T.; Dutta, B.; Tolbert, D.; Xu, Z. Q. Clinical safety and pharmacokinetics of (+)-calanolide A, a naturally occurring, NNRTI, in normal healthy and HIV-infected volunteers. 6th Conf. on Retroviruses and Opportunistic Infections, **1999**, abstract 606.

40. Bipul, D.; Renslow, S.; Roger, A.; Gerald, P.; Richard, P.; Timothy, C.; Xu, Z. Q. Preliminary results of a phse IB study on (+)-calanolide A, a naturally occurring novel NNTRI in treatment naïve HIV-1 infected patients. 5th International Congress on AIDS in Asia and the Pacific, **1999**, abstract 113.

41. Sarawak Medichem Pharmaceuticals. A phase 1B dose-range study to evaluate the safety, pharmacokinetics, and effects of (+)-calanolide A on surrogate marker in HIV-positive patients with no previous antiretroviral therapy. Protocol FDA 297A, **1998**.

42. Chenera, B.; West, M. L.; Finkelstein, J. A.; Dreyer, G. B. Total synthesis of (±)-calanolide A, a non-nucleoside inhibitor of HIV-1 reverse transcriptase. *J. Org. Chem.*, **1993**, 58: 5605.

43. Kucherenko, A.; Flavin, M. T.; Boulanger, W. A.; Khilevich, A.; Shone, R. L.; Rizzo, J. D.; Sheinkmann, A. K.; Xu, Z. Q. Novel approach for synthesis of (±)-calanolide A and its anti-HIV activity. *Tetrahedron Lett.*, **1995**, 36(31): 5475–5478.

44. Zhou, C. M.; Wang, L.; Zhang, M. L.; Zhao, Z. Z.; Zhang, X. Q.; Chen, X. H.; Chen, H. S. Total synthesis of (±)-calanolide A and its anti-HIV activity. *Chin. Chem. Lett.*, **1998**, 9: 433–434.

45. Zhou, C. M.; Wang, L.; Zhang, M. L.; Zhao, Z. Z.; Zhang, X. Q.; Chen, X. H.; Chen, H. S. Synthesis and anti-HIV activity of (±)-calanolide A and its analogues. *Acta Pharmaceutica Sinica*, **1999**, 34: 673–678.

46. Gao, Q.; Wang, L.; Liang, X. T. Total synthesis of (±)-inophyllum B. *Chin. Chem. Lett.*, **2002**, 13: 714–716.

47. Gao, Q.; Wang, L.; Liang, X. T. Total synthesis of (±)-cordatolide A and its anti-HIV acitivity. *Chin. Chem. Lett.*, **1999**, 10: 653–656.

48. Palmer, C. J.; Josephs, J. L. Synthesis of the *Calophyllum* coumarins. *Tetrahedron Lett.*, **1994**, 35: 5363–5366.

49. Deshpande, P. P.; Tagliaferri, F.; Victory, S. F.; Yan, S.; Baker, D. C. Synthesis of optically active calanolide A and B. *J. Org. Chem.*, **1995**, 60(10): 2964–2965.

50. Deshpande, P. P.; Baker, D. C. A simple approach to the synthesis of the chiral substituted chroman ring of *calophyllum* coumarins. *Synthesis*, **1995**, 630–632.

51. Sekino, E.; Kumamoto, T.; Tanaka, T.; Ikeda, T.; Ishikawa, T. Concise synthesis of anti-HIV-1 active (+)-inophyllum B and (+)-calanolide A by application of (−)-quinine-catalyzed intramolecular oxo-Michael addition. *J. Org. Chem.*, **2004**, 69: 2760–2769.

52. Ishikawa, T.; Oku, Y.; Tanaka, T.; Kumanoto, T. An approach to anti-HIV-1 active *calophyllum* coumarin synthesis: An enantioselective construction of 2,3-dimethyl-4-chromanone ring by quinine-assisted intramolecular Michael-type addition. *Tetrahedron Lett.*, **1999**, 40: 3777–3780.

53. Tanaka, T.; Kumanoto, T.; Ishikawa, T. Solvent effects on stereoselectivity in 2,3-dimethyl-4- chromanone cyclization by quinine-catalyzed asymmetric intromolecular oxo-Michael addition. *Tetrahedron-Asymmetry*, **2000**, 11: 4633–4637.

54. Tanaka, T.; Kumanoto, T.; Ishikawa, T. Enantioselective total synthesis of anti HIV-1 active (+)-calanolide A through a quinine-catalyzed asymmetric intromolecular oxo-Michael addition. *Tetrahedron Lett.*, **2000**, 41: 10229–10232.

55. Trost, B. M.; Toste, F. D. A catalytic enantioselective approach to chromans and chromanols. A total synthesis of (−)-calanolide A and B and the Vitamin E nucleus. *J. Am. Chem. Soc.*, **1998**, 120: 9074–9075.

56. Khilevich, A.; Mar, A.; Flavin, M. T.; Rizzo, J. D.; Lin, L.; Dzekhtser, S.; Brankovic, D.; Zhang, H. P.; Chen, W.; Liao, S. Y. Synthesis of (+)-calanolide, A.: An anti-HIV agent, via enzyme-catalyzed resolution of the aldol products. *Tetrahedron-Asymmetry*, **1996**, 7: 3315–3326.

57. Galinis, D. L.; Fuller, R. W.; McKee, T. C.; Cardellina, J. H.; Gulakowski, R. J.; McMahon, J. B.; Boyd, M. R. Structure-activity modifications of the HIV-1 inhibitors (+)-calanolide A and (−)-calanolide B. *J. Med. Chem.*, **1996**, 39: 4507–4510.

58. Zhou, C. M.; Wang, L.; Zhao, Z. Z. Synthesis of 11-demethyl and 6,6-demethyl calanolide A. *Chin. Chem. Lett.*, **1997**, 8: 859–860.

59. Sharma, G. V. M.; Ilangovan, A.; Narayanan, V. L.; Gurjar, M. K. First synthesis of aza-calanolides—a new class of anti-HIV active compounds. *Tetrahedron*, **2003**, 59: 95–99.

60. Gurjar, M. K.; Sharma, G. V. M.; Ilangovan A; Narayanan, V. L. Indian Patent Application No. 1441/DEL/98; U.S. Patent 6,191,279, 2001.

# 9

# RECENT PROGRESS AND PROSPECTS ON PLANT-DERIVED ANTI-HIV AGENTS AND ANALOGS

DONGLEI YU AND KUO-HSIUNG LEE

*Natural Products Laboratory, School of Pharmacy, University of North Carolina, Chapel Hill, NC*

## 9.1 INTRODUCTION

Acquired immunodeficiency syndrome (AIDS), a degenerative disease of the immune system, is caused by the human immunodeficiency virus (HIV) and results in life-threatening opportunistic infections and malignancies.[1-3] As the world enters the third decade of the AIDS epidemic, this pandemic has spread rapidly through the human population and become the fourth leading cause of mortality worldwide.[4] The rapid spread of AIDS prompted an extensive search for chemotherapeutic anti-HIV agents. Large numbers of naturally and synthetically derived chemical entities were screened for potential anti-HIV activity. As a result, numerous compounds were found to have anti-HIV activity. Since then, 20 anti-HIV drugs have been approved by the U.S. Food and Drug Administration (FDA), including reverse transcriptase inhibitors (RTIs), protease inhibitors (PIs), and a fusion inhibitor.[5] Emergence of drug-resistant viruses and the side effects and toxicity of these agents have limited their benefits in AIDS treatment.[6-10] Therefore, the current development of new, effective, and less toxic anti-HIV agents is focusing on novel structures or new modes of action.

Figure 9-1 describes the modern drug research process. Our drug discovery program focuses on new lead discovery and lead optimization, using natural products as a reservoir of biologically active compounds. Many efforts have attempted to

*Medicinal Chemistry of Bioactive Natural Products* Edited by Xiao-Tian Liang and Wei-Shuo Fang
Copyright © 2006 John Wiley & Sons, Inc.

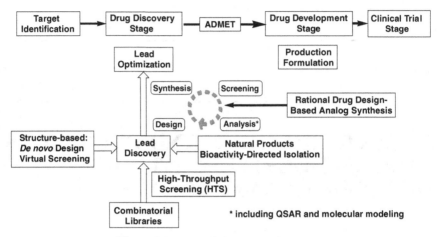

**Figure 9-1.** General drug research process.

find new anti-HIV compounds with unique structure characters or mechanisms of action.[11–14] Bioactivity-directed fractionation and isolation (BDFI) is a major approach for new lead generation. Rational drug design, synthesis, bioevaluation, and structure activity relationship (SAR) and quantitive SAR (QSAR) studies form a design circle, which optimizes the lead until it reaches a preclinical trial. Currently, driven by exciting technologic advances, enhanced efficiency in gathering absorption, distribution, metabolism, excretion, and toxicity (ADMET) data has been realized, which permits ADMET scientists to contribute more effectively to the drug discovery process as well.

In our long-term screening of plant extracts, particularly anti-infective or immunomodulative Chinese herbal medicines, we focus on lead identification followed by the structural modification of discovered leads. The author's laboratory has over 10 years of experience on plant-derived anti-HIV natural products, including polycyclic diones, saponins, alkaloids, terpenes, polyphenols, flavonoids, and coumarins.[11] This chapter will focus on the detailed investigation of structural modifications based on rational analog design, in vitro anti-HIV activity, SAR analysis, mechanisms of action, current status, and future prospects of the following natural product leads (categories): suksdorfin (khellactone coumarin analogs), dibenzocyclooocadiene lignans (biphenyl derivatives), betulinic acid (triterpene derivatives), and their derivatives and analogs.

## 9.2 KHELLACTONE COUMARIN ANALOGS AS ANTI-HIV AGENTS

### 9.2.1 Suksdorfin as a New Anti-HIV Agent

Anti-HIV coumarins have been found to inhibit viral adsorption, reverse transcription (RT), protease (PR) inhibition, and integrase (IN) in the HIV replication cycle.[12–14] A small branch of the coumarin family, khellactones, are notable for extensive bioactivities, including anti-HIV, antitumor promoting,[15] and antiplatelet aggregation[16] properties.

In 1994, Lee et al. reported on the anti-HIV activity of suksdorfin (**1**), a dihydroseselin type angular pyranocoumarin, isolated from the methanolic extract of *Lomatium suksdorfii* through BDFI. It suppressed viral replication in 11 separate acute HIV-1$_{\text{IIIB}}$ infections of H9 lymphocyte cells with an average EC$_{50}$ value of $2.6 \pm 2.1$ μM. It also suppressed acute HIV-1 infections in fresh peripheral blood mononuclear cells, monocyte/macrophages, and U937 cells, a promonocytic cell line.[17] Suksdorfin (**1**) represents a new class of potent anti-HIV agents that is structurally unique compared with other known anti-AIDS drugs; therefore, **1** has been chosen as lead compound for various structural modifications, following the principles of analog design.

Suksdorfin (**1**)

## 9.2.2 Pyrano-3′,4′ Stereoselectivity and Modification

### 9.2.2.1 3′,4′ Stereoselectivity

According to an almost universally accepted opinion, stereoisomers of an organic molecule should be anticipated to frequently exhibit widely different and unpredictable pharmacological effects.[18–20] The two chiral centers of suksdorfin (**1**), C-3′ and C-4′, both possess R configurations. Our first SAR effort was to elucidate the stereochemical requirement of these two positions, as a different configuration changes the orientation and shape of the whole molecule. Huang et al. synthesized a series of khellactone derivatives, including all four stereochemical isomers, 3′, 4′-*cis* or *trans* (**2–38**) (Figures 9-2 and 9-3).[21,22] Some *cis*-3′, 4′-substituted compounds showed activity against HIV-1 in H9 lymphocytes, which are listed in Table 9-1. Among them, 3′R,4′R-di-*O*-(−)-camphanoyl-(+)-*cis*-khellactone (DCK, **2**) had an EC$_{50}$ of $2.6 \times 10^{-4}$ μM and TI of 137,000, and it was much more potent than the remaining stereo-isomers, **3** (3′S,4′S, EC$_{50}$ 51 μM and TI 33), **4** (3′R,4′S, EC$_{50}$ 6.4 μM and TI 1), and **5** (3′S,4′R) (EC$_{50}$ 32 μM and TI 1). The three-dimensional (3-D) orientations of the four DCK stereochemical isomers (Figure 9-2), generated by Tripos Sybyl software,[23] showed different molecular orientations and shapes. Therefore, 3′R,4′R DCK (**2**) may fit the putative receptor, whereas the other three configurations do not. Thus, the low anti-HIV activity of most synthetic khellactone analogs may be attributed to racemization (Figure 9-3).[22] The rigid (−)-camphanoyl moiety in DCK (**2**) contains the isovaleryl group. An x-ray crystallographic analysis confirms that the isovaleryl group in **1** is more

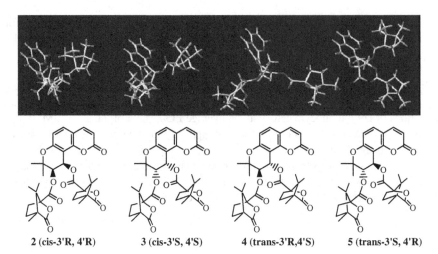

**2 (cis-3'R, 4'R)**     **3 (cis-3'S, 4'S)**     **4 (trans-3'R,4'S)**     **5 (trans-3'S, 4'R)**

**Figure 9-2.** Structures and 3-D orientation of DCK stereochemical isomers (**2–5**).

conformationally flexible, and its terminal carbon atoms are disordered over two orientations, whereas in **2**, the 4'-camphanoyl group neighbors another bulky camphanoyl substituent making the isovaleryl moiety more rigid.[24] The study data suggest that a rigid stereochemistry of 3' and 4'-configured khellactone derivatives is crucial for anti-HIV activity.

### 9.2.2.2   3'R, 4'R Modification
Several compounds with C-3'R,4'R small or bulky substituents were then designed and synthesized[25] (Figure 9-4). When the two (−)-camphanoyl groups in DCK were replaced with (+)-camphanoyl groups (**43**), the anti-HIV activity

**TABLE 9-1. Anti-HIV Activities of Khellactone Derivatives**[a,22]

| No. | $IC_{50}$ $(\mu M)^b$ | $EC_{50}$ $(\mu M)^c$ | $TI^d$ |
|---|---|---|---|
| Suksdorfin (**1**) | >52 | $2.6 \pm 2.1$ | $30.6 \pm 22.4$ |
| **2** | 35 | $2.56 \times 10^{-4}$ | $1.37 \times 10^5$ |
| **3** | 1700 | 51 | 33.3 |
| **4** | >6.4 but <32 | >6.4 but <32 | 1 |
| **5** | >32 | 32 | 1 |
| **8** | 14 | 7 | 2 |
| **10** | 101 | 7 | 14.4 |
| **11** | 16.5 | 4.7 | 3.5 |
| **13** | 3.1 | <1.4 | >2.2 |
| AZT | 1875 | 0.045 | 41,667 |

[a]Inhibitory of HIV-1 IIIB in H9 lymphocytes.
[b]Concentration that inhibits uninfected H9 cell growth by 50%.
[c]Concentration that inhibits viral replication by 50%.
[d]Therapeutic Index, $TI = IC_{50}/EC_{50}$.

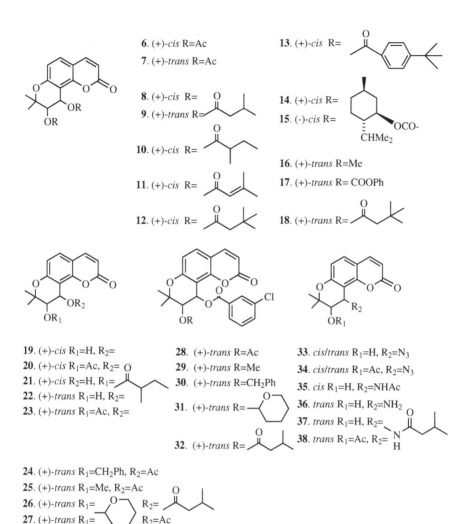

**Figure 9-3.** Modification of *cis*- and *trans*-khellactone derivatives (**6–38**).

(EC$_{50}$ 6.60 × 10$^{-3}$μM, TI > 2.38 × 10$^4$) was slightly lower than that of **2** [(−)-camphanoyl]. Compounds with aromatic acyl groups (**41** and **42**) or small acyl groups (**39** and **40**) on these positions did not show anti-HIV activity. Even with one bulky acyl group, **44, 46, 48,** and **49** did not suppress HIV-1 replication in H9 lymphocytes. 3′,4′-Disubstituted compounds with two different bulky acyl groups, 3′R-camphanoyl and 4′R-adamantanecarbonyl (**50**, EC$_{50}$ 0.60 μM, TI 35.5), exhibited anti-HIV activity but were much less active than DCK. These results indicate that the volume, size, and shape of the camphanoyl group are as or even more important than its absolute configuration.[25]

DCK (**2**) showed extremely increased potency in the anti-HIV screening; therefore, it became the new lead in the search for optimal synthetic method, SAR modifications,

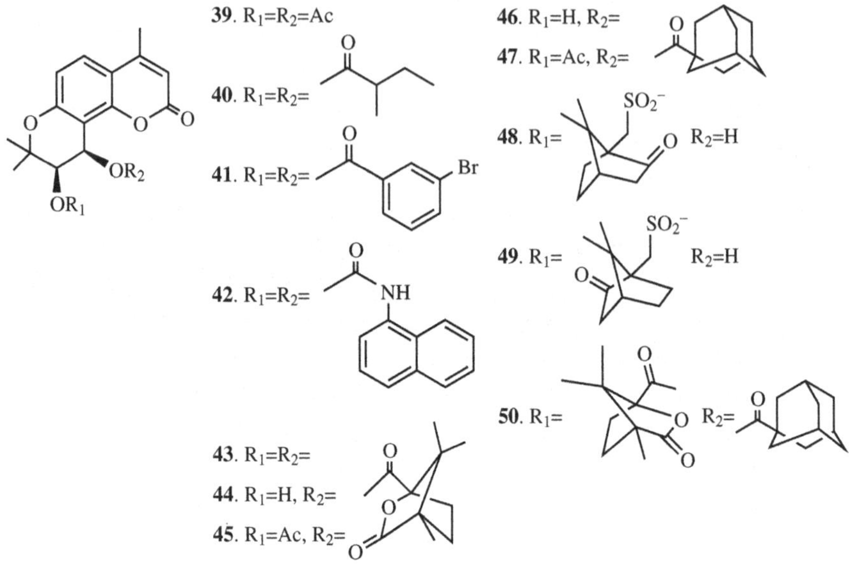

**Figure 9-4.** Khellactone 3'R,4'R-substitutions (**39–50**).

and better pharmaceutical properties. DCK was synthesized through a four-step route beginning with 7-hydroxycoumarin.[22] Osium oxide[26] and dioxane were used for the dihydroxylation of seselin, which produced racemic *cis*-khellactones. Isolation of pure 3'R,4'R-stereoisomer from the *cis*-racemic mixture is difficult and inefficient; therefore, we turned to asymmetric synthesis for the solution. Xie et al.[30] synthesized DCK (**2**) in up to 86% ee via a catalytic Sharpless asymmetric dihydroxylation (AD)[27,28] of seselin. Seven different chiral ligands were used for optimal enantioselectivity. Hydroquinidine (DHQD-R) type ligands[29] led to 3'S, 4'S-configuration, and hydroquinine (DHQ-R) type ligands gave 3'R, 4'R-configuration as the main product. (DHQ)$_2$PYR (hydroquinine 2,5-diphenyl-4,6-pyrimidine-diyl diether) gave the highest R,R stereoselectivity.[30] Methanesulfonamide (1 equiv.) was added to the reaction mixture to improve rate at 0 °C.[27] In a solid-phase synthetic route, 91% ee of the AD reaction was yielded by using OsO$_4$ as catalyst and (DHQ)$_2$-PHAL (hydroquinine 1,4-phthalazinediyl diether) as ligand at rt.[31] Scheme 9-1 describes the optimized synthetic route for DCK analogs.[25]

### 9.2.3 Coumarin Skeleton Modification

#### 9.2.3.1 Substituents on Coumarin Skeleton

9.2.3.1.1 *Monosubstituted Analogs*  Xie et al. synthesized a series of monosubstituted derivatives at the coumarin 3, 4, 5, and 6 positions.[25,32,33] The data in Table 9-2 show that mono-methyl substitution at the 3, 4, or 5 position (**51**, **53**, and **59**) greatly increased potency in comparison with **2**. 3- and 4-Methoxy DCK analogs (**52** and **54**) showed slightly less potency whereas 5-methoxy DCK (**60**) retained activity. However, although methylation or methoxylation at the 3, 4, and 5 positions did not affect the antiviral activity of DCK analogs, these same substitutions at

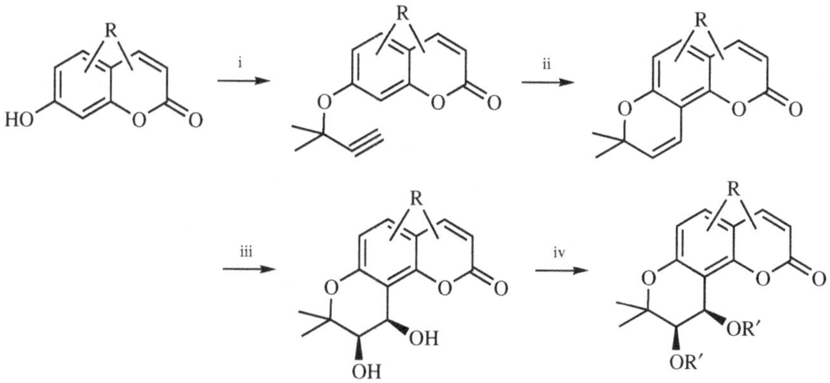

R=substitutents on coumarin

R′=camphonyl

i. 3-Chloro-3-methyl-1-butyne, $K_2CO_3$, KI, DMF/$N_2$;
ii. N, N-diethylaniline, reflux;
iii. $K_2OsO_2(OH)_4$, $K_3Fe(CN)_6$, DHQ-type ligan, $K_2CO_3$, ice-bath;
iv. Camphanoyl chloride, DMAP/$CH_2Cl_2$;

**Scheme 9-1.** General synthetic method for DCK analogs.

**TABLE 9-2. Structures and Anti-HIV Activities of Monosubstituted DCK Analogs (51–62)[a,25]**

| No. | 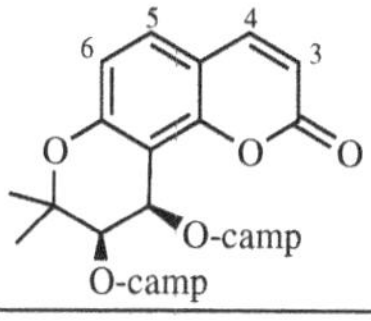 | | $IC_{50}$ $(\mu M)^b$ | $EC_{50}$ $(\mu M)^c$ | $TI^d$ |
|---|---|---|---|---|---|
| DCK(2) | | | 35.0 | $2.56 \times 10^{-4}$ | $1.37 \times 10^5$ |
| 51 | 3-substitution | $CH_3$ | >113 | $5.25 \times 10^{-5}$ | $> 2.15 \times 10^6$ |
| 52 | | $OCH_3$ | >153 | $2.38 \times 10^{-3}$ | $> 6.43 \times 10^4$ |
| 53 | 4-substitution | $CH_3$ | >126 | $1.83 \times 10^{-6}$ | $> 6.89 \times 10^7$ |
| 54 | | $OCH_3$ | >153 | $2.99 \times 10^{-3}$ | $> 5.12 \times 10^4$ |
| 55 | | $CH_2CH_2CH_3$ | >151 | $1.75 \times 10^{-2}$ | $> 8.63 \times 10^3$ |
| 56 | | $CH(CH_3)_2$ | >151 | $3.15 \times 10^{-2}$ | $> 4.79 \times 10^3$ |
| 57 | | $C_6H_5$ | >143 | 0.12 | $> 1.19 \times 10^3$ |
| 58 | | $CF_3$ | >145 | 1.81 | >80.1 |
| 59 | 5-substitution | $CH_3$ | >95 | $2.39 \times 10^{-7}$ | $> 3.97 \times 10^8$ |
| 60 | | $OCH_3$ | >153 | $1.92 \times 10^{-4}$ | $> 7.97 \times 10^5$ |
| 61 | 6-substitution | $CH_3$ | 33 | 0.151 | 218 |
| 62 | | $OCH_3$ | >153 | 15.8 | >9.68 |
| AZT | | | 1875 | 0.045 | 41,667 |

[a]Inhibitory of HIV-1 IIIB in H9 lymphocytes.
[b]Concentration that inhibits uninfected H9 cell growth by 50%.
[c]Concentration that inhibits viral replication by 50%.
[d]Therapeutic Index, TI = $IC_{50}/EC_{50}$.

the 6-position [6-methyl (**61**) and 6-methoxy (**62**)] dramatically decreased activity; thus, C-6 is not favorable for modification. More analogs with various substituents at C-4 were also synthesized. 4-Propyl (**55**), isopropyl (**56**), phenyl (**57**), and trifluoromethyl (**58**) DCK analogs were less potent or inactive (Table 9-2).[25] The larger volume of the propyl and isopropyl groups could interfere with target binding, whereas the phenyl substituent may affect compound planarity and not fit into the binding site. Although similar in size and shape to a methyl group, an electron-withdrawing trifluoromethyl group was unfavorable, based on the low activity of **58**. The extremely high anti-HIV potency of 4-methyl DCK (**53**) and 5-methyl DCK (**59**) indicates that a methyl group probably fits well into a hydrophobic cleft on the target's active surface and greatly increases both the agent's target affinity and the desired pharmacological response.[25]

*9.2.3.1.2 Disubstituted Analogs* Disubstituted DCK analogs were also developed and screened for inhibition of HIV replication in H9 lymphocytes (Table 9-3).[24] 4-Methyl-5-methoxy DCK (**69**) was the most promising compound and was similar to 4-methyl DCK (**53**) and much better than DCK (**2**) in the same assay. Dimethyl DCKs **63** (3,4-dimethyl), **67** (3,5-dimethyl), **68** (4,5-dimethyl), and **72** (4,6-dimethyl) showed about 20-fold lower activity compared with **2**. Among other 3,4-disubstituted compounds, 3-chloro-4-methyl (**64**) and 3,4-cyclohexano

**TABLE 9-3. Structures and Anti-HIV Activities of Disubstituted DCK Analogs (63–72)[a,24]**

| No. | | | $IC_{50}$ ($\mu M$)[b] | $EC_{50}$ ($\mu M$)[c] | $TI^d$ |
|---|---|---|---|---|---|
| | | 3,4-di-substitution | | | |
| 63 | 3-CH$_3$ | 4-CH$_3$ | >154 | $1.92 \times 10^{-3}$ | $>8.02 \times 10^4$ |
| 64 | Cl | CH$_3$ | 104 | $2.01 \times 10^{-3}$ | $5.17 \times 10^4$ |
| 65 | C$_6$H$_5$ | CH$_3$ | >140 | 43.7 | >3.20 |
| 66 | | -CH$_2$CH$_2$CH$_2$CH$_2$- | >148 | $2.12 \times 10^{-3}$ | $>6.98 \times 10^4$ |
| | | 3,5-di-substitution | | | |
| 67 | 3-CH$_3$ | 5-CH$_3$ | >154 | $9.10 \times 10^{-3}$ | $>1.69 \times 10^4$ |
| | | 4,5-di-substitution | | | |
| 68 | 4-CH$_3$ | 5-CH$_3$ | >154 | $4.19 \times 10^{-3}$ | $>3.68 \times 10^4$ |
| 69 | CH$_3$ | OCH$_3$ | >150 | $7.21 \times 10^{-6}$ | $>2.08 \times 10^7$ |
| 70 | CH(CH$_3$)$_2$ | CH$_3$ | >147 | No | suppression |
| 71 | CH$_3$ | OCH$_2$C$_6$H$_5$ | >135 | 1.54 | 87.7 |
| | | 4,6-di-substitution | | | |
| 72 | 4-CH$_3$ | 6-CH$_3$ | 3.75 | $4.69 \times 10^{-3}$ | $1.25 \times 10^3$ |

[a]Inhibitory of HIV-1 IIIB in H9 lymphocytes.
[b]Concentration that inhibits uninfected H9 cell growth by 50%.
[c]Concentration that inhibits viral replication by 50%.
[d]Therapeutic Index, TI = $IC_{50}/EC_{50}$.

(**66**) DCK analogs also showed potent anti-HIV activity, but 3-phenyl-4-methyl DCK (**65**) almost completely lost activity. For other 4,5-disubstituted DCKs, **70** (4-iso-propyl-5-methyl) was inactive and **71** (4-methyl-5-benzyloxy) showed weak HIV-1 suppression in the assay. These results indicate that anti-HIV activity can be maintained when two methyl or other aliphatic substituent(s) are placed on the khellac-tone coumarin nucleus and that an aromatic substituent on the nucleus is not favorable.[24] Comparison of the molecular 3-D orientations and torsional angles of these molecules provides an explanation that steric compression between the two bulky substituents at the 4- and 5- positions deforms the coumarin system. The [C(4)-C(4a)-C(5)-C(Me-5)] torsion angles are 42.9° and 29.5° in **70** and **68**, respectively, as computed by Sybyl software; thus, the 5-methyl groups are forced out of the resonance plane. However, the corresponding torsion angles in **69** and **53** are 3.3° and 2.0°, respectively. These data suggest that a planar system is an essential structural feature for potent anti-HIV activity in this compound class. Steric compression of the C(4) and C(5) coumarin substituents can reduce the overall planarity and, thus, resonance of the coumarin system, which results in decreased or completely lost anti-HIV activity.[24]

### 9.2.3.2 Bioisosteric Replacement

Bioisosteres are defined as groups or molecules that have chemical and physical properties producing broadly similar biological properties.[34] Bioisosteres likely affect the same receptor site or pharmacological mechanism.[18,35] This analog design principle was applied to our DCK modification, and thio-DCK and DCK lactam derivatives were developed.

*9.2.3.2.1 Thio-DCK Analogs* The thio bioisosteres of DCK contain *S* rather than *O* in the 2-ketone moiety and retain anti-HIV activity in comparison with the substituted DCKs.[36] Lawesson's reagent (LR)[37] converted the khellactones to the corresponding khellactone thiones (khelthiolactones) in 50% to 85% yields. Among the khelthiolactone analogs shown in Table 9-4, thioDCK (**73**), 4-methyl-thio (**75**), 3-methyl-thio (**74**), and 5-methyl-thio (**78**) analogs showed the highest potency in the H9 lymphocytes, but they were less active than correspondent DCK (**2**), 4-methyl (**53**), 3-methyl (**51**), and 5-methyl (**59**) DCKs. However, in the CEM-SS cells, **75** showed promising potency with $EC_{50} = 0.0635$ μM and TI >3149, and it was more potent than **2** and equipotent to **53**.[36]

Lawesson reagent (LR)

**TABLE 9-4. Structures and Anti-HIV Activities of Khelthiolactone Analogs (73–84)[36]**

| | | H9 Lymphocytes[a] | | CEM-SS Cell Line[b] | |
|---|---|---|---|---|---|
| Compound | Structure | $EC_{50}$ (μM) | TI | $EC_{50}$ (μM) | TI |
| 73 | $R_1 = R_2 = R_3 = R_4 = H$ | 0.029 | >5,390 | 0.307 | 13.7 |
| | **Mono-substitution** | | | | |
| 74 | $R_1 = CH_3, R_2 = R_3 = R_4 = H$ | 0.0119 | >12,900 | * | |
| 75 | $R_2 = CH_3, R_1 = R_3 = R_4 = H$ | 0.00718 | >21,300 | 0.0635 | >3,150 |
| 76 | $R_2 = C_3H_7, R_1 = R_3 = R_4 = H$ | 0.128 | >1,153 | 0.128 | >1,153 |
| 77 | $R_2 = $ phenyl, $R_1 = R_3 = R_4 = H$ | 2.48 | >56.6 | 2.48 | >56.6 |
| 78 | $R_3 = CH_3, R_1 = R_2 = R_4 = H$ | 0.0199 | >7,690 | * | |
| 79 | $R_3 = OCH_3, R_1 = R_2 = R_4 = H$ | 1.28 | >117 | * | |
| 80 | $R_4 = OCH_3, R_1 = R_2 = R_3 = H$ | N/S | | * | |
| | **Di-substitution** | | | | |
| 81 | $R_2 = R_3 = CH_3, R_1 = R_4 = H$ | 0.262 | >571 | * | |
| 82 | $R_2 = CH_3, R_3 = OCH_3, R_1 = R_4 = H$ | 1.46 | >1000 | * | |
| 83 | $R_1 = R_2, CH_3, R_3 = R_4 = H$ | 0.0334 | >4490 | * | |
| 84 | $R_1 = C_6H_5, R_2 = CH_3, R_3 = R_4 = H$ | 4.95 | 6.24 | * | |
| AZT | | 0.045 | 41,667 | 0.0038 | >263 |
| DCK (2) | | $2.56 \times 10^{-4}$ | $>1.37 \times 10^5$ | 0.14 – 0.26 | 100 |
| 53 | | $1.57 \times 10^{-7}$ | $>10^9$ | 0.0635 | >3,150 |

[a]Screened in H9 lymphocytes by Panacos, Inc.
[b]Screened in CEM-SS cells by the National Institutes of Health.

*9.2.3.2.2 DCK-Lactam Analogs* The DCK lactam analogs (**85–88**) were also synthesized asymmetrically and evaluated for antiviral activity against HIV-1 replication in H9 lymphocyte cells. As shown in Figure 9-5, the interchange of *O* and its similar-sized bioisostere *NH* retains the anti-HIV activity in DCK series derivatives because **87** and **88** had comparable activity with corresponding DCKs (**53** and **63**). Thus, the hydrogen bond acceptor ability of *NH* or *O* at position-1 is likely involved in receptor binding. Too bulky to fit and bind target, the *N-t*-BOC (butoxycarbonyl) intermediates (**85** and **86**) did not inhibit HIV-1 replication.[38]

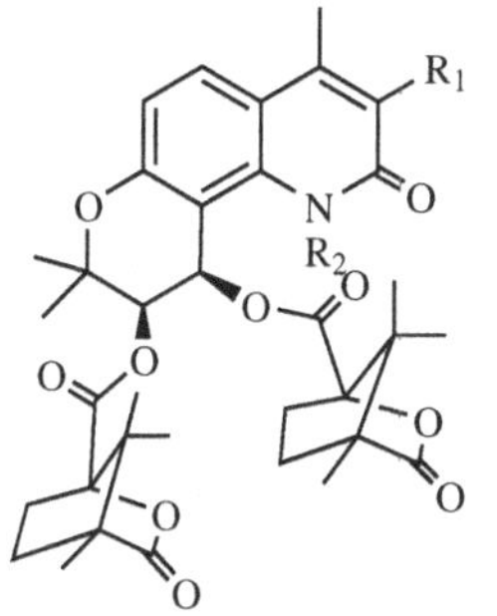

85, $R_1$ = H, $R_2$=BOC, no supression

86, $R_1$ = $CH_3$, $R_2$=BOC, no supression

87, $R_1$ =$R_2$ = H, $IC_{50}$=28 µM, $EC_{50}$ = 0.00024 µM, TI = 119,333

88, $R_1$ = $CH_3$, $R_2$ = H, $IC_{50}$=42 µM, $EC_{50}$ = 0.0046 µM, TI = 9,100

**Figure 9-5.** Structures and activities of the 4-methyl DCK lactam analogs (**85–88**).

### 9.2.3.3 Approach to Improve Water Solubility

An in vivo pharmacokinetics study showed that the poor water solubility of **53** is a major obstacle to its continued development as a drug candidate.[39] Introducing polar functional groups could improve water solubility and simultaneously provide the possibility of prodrugs. Scheme 9-2 shows our first approach: the parallel synthesis of 3'$R$, 4'$R$-di-$O$-cis-acyl-3-carboxyl khellactones (**89–94**) using Wang resin. However, no compounds, even the camphanoyl substituted **89**, showed promising activity, which indicates that a carboxylic acid is not favorable at 3-position.[31]

We therefore designed a synthetic route to introduce hydrophilic moieties into the coumarin skeleton. A hydroxy or amino group can form H-bonds with water to improve hydrophilicity or with an active site on a biological target to increase affinity. Scheme 9-3 shows the synthetic approaches to 3-substituted analogs (**95–99**).[40] 3,4-Disubstituted (**100–102**) and 4,6-disubstituted (**103–106**) DCK analogs were also synthesized using the same method beginning with corresponding dimethyl DCKs (**63** and **72**).[39] All target compounds and synthetic intermediates were screened against an HIV-1 strain in H9 lymphocytes. The data in Table 9-5 show that all analogs incorporating bromomethyl (**95** and **100**), hydroxymethyl (**97** and **102**), or acetoxymethyl (**96** and **101**) groups at C-3 showed similar or better activity than **2**, when tested in parallel in H9 lymphocytes; however, C-3 amino-methyl (**98**) and diethylaminomethyl (**99**) groups were not favored for anti-HIV activity. In contrast with the 3-substituted DCKs, 6-substituted analogs (**103–106**) showed only moderate antiviral activity and were much less potent than **2** and **53** in the $HIV_{IIIB}$/H9 assay.

Although similar in potency, 3-hydroxymethyl-4-methyl DCK (**102**, PA-344B) should be more water soluble than **53** and could be easily converted to a prodrug. The preclinical study of PA-344B is ongoing. Additional in vitro studies demonstrated that **102** exhibits significant activity against a panel of primary HIV-1 isolates propagated in primary peripheral blood mononuclear cells. The median $EC_{50}$ against these viruses was 0.024 µM, similar to those of the approved drugs AZT, nevirapine, and indinavir, which were tested in parallel. In addition, **102** exhibited moderate oral bioavailability (F = 15%) when administered orally to

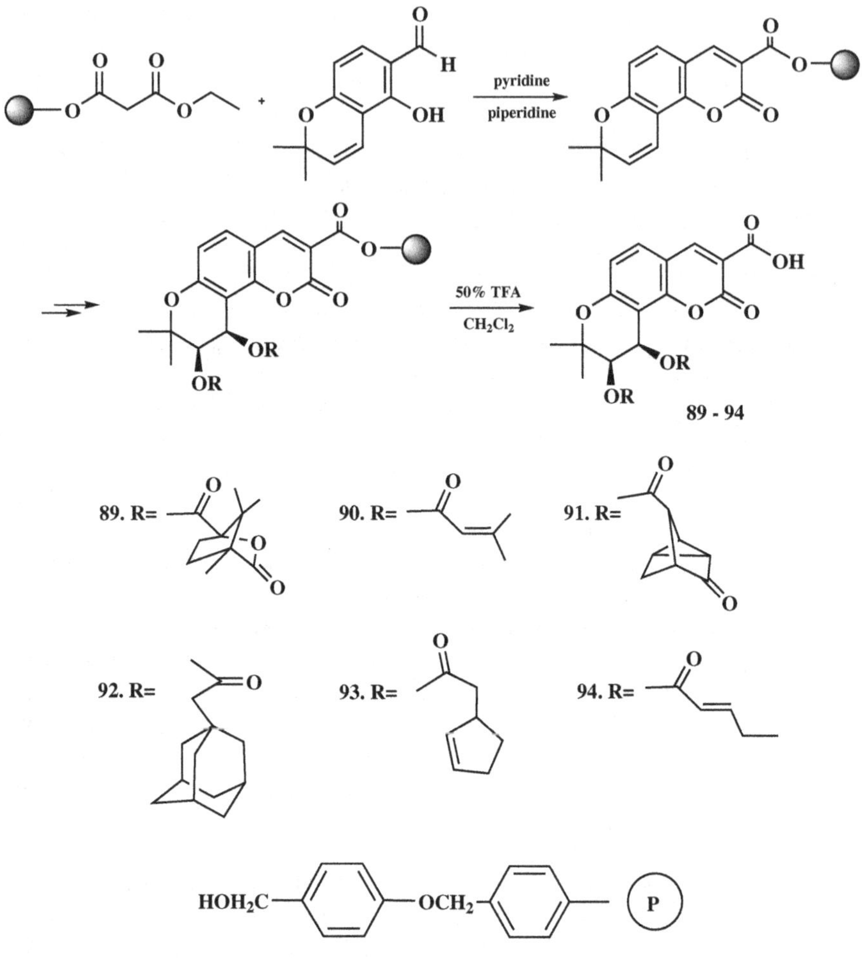

**Scheme 9-2.** Solid-phase parallel synthesis of 3′R, 4′R-di-O-cis-acyl-3-carboxyl khellactones (**89–94**).

rats as a suspension in carboxymethyl cellulose. The high systemic clearances also suggest that hepatic first-pass metabolism is a major factor in limiting the oral bioavailability of this series. Preliminary studies have indicated that DCK analogs are subject to rapid oxidative metabolism in human and rat liver microsomes, and P450 3A4 is the primary isoform that metabolizes **102**.[39]

### 9.2.3.4　Position Isomers
In 2003, a new screening system, HIV-1 RTMDR1 in MT-4 cell line, was introduced into our program. This viral strain is a multiple RT inhibitor-resistant strain,

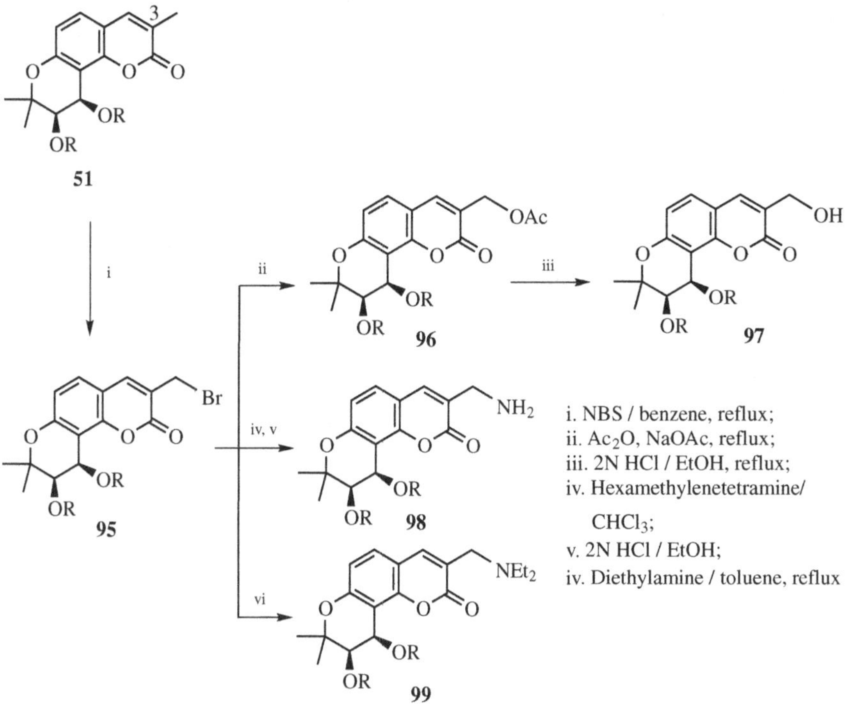

**Scheme 9-3.** Synthetic route to 3-substituted DCK analogs (**95–99**).

which contains several mutations in RT amino acid residues (M41L, L74V, V106A, and T215Y) and is resistant to AZT, ddI, nevirapine, and other non-nucleoside reverse transcriptase inhibitors (NNRTIs).[41] Most previously active DCK derivatives showed dramatically decreased activity or totally lost potency in this new screening system. It is not fully clear why DCK analogs show remarkably reduced activity against the drug-resistant viral strains. Mutations of the viral RT could cause conformational changes directly on the binding sites or other regions that affect binding. Accordingly, DCK might dissociate more rapidly from or not fit into the putative mutated target. Therefore, we turned to preparing ring position isomers, a common principle of analog design, as such a change may alter electron distribution in an aromatic ring system or affect the complimentarity toward in vivo receptors.[42,43]

A series of $3'R$, $4'R$-di-$O$-(−)-camphanoyl-$2',2'$-dimethyldihydropyrano[2,3-f]chromone (DCP, **107**) derivatives, which are positional isomers of DCK, were designed and synthesized.[42,43] All DCP (**107–117**) and $3'$-acyl-$4'$-camphanoyl (**118–124**) derivatives were tested against both the HIV-1$_{IIIB}$ non-drug-resistant strain in H9 lymphocytes and the HIV-1 RTMDR1 multi-RT inhibitor resistant viral strain in MT4 cells (Figure 9-6 and Table 9-6). Because the two screening assays used different protocols, virus strains and cell lines, the activity data cannot be compared between the two different assays. Therefore, our SAR analyses were summarized for each strain.

**TABLE 9-5. Structures and Anti-HIV Activities of DCK Analogs (95–106)[a,39]**

| | | 95–99 | 100–102 | 103–106 |

| Compound | | $IC_{50}$ $(\mu M)^{b}$ | $EC_{50}$ $(\mu M)^{c}$ | Therapeutic Index[d] |
|---|---|---|---|---|
| **95** | $R_3 = Br$ | >11.1 | 0.059 | >186 |
| **96** | $R_3 = OAc$ | 11.6 | 0.017 | 676 |
| **97** | $R_3 = OH$ | 23.0 | 0.029 | 806 |
| **98** | $R_3 = NH_2$ | >15.4 | 0.677 | >23 |
| **99** | $R_3 = NEt_2$ | >14.1 | 3.67 | >4 |
| **100** | $R_3 = Br$ | >20.9 | 0.00011 | >189,600 |
| **101** | $R_3 = OAc$ | 14.8 | 0.026 | 567 |
| **102** | $R_3 = OH$ | 24.9 | 0.0042 | 6,000 |
| **103** | $R_6 = Br$ | >13.7 | 0.156 | >88 |
| **104** | $R_6 = OAc$ | >14.1 | 0.544 | >26 |
| **105** | $R_6 = OH$ | >15.0 | 0.111 | >135 |
| **106** | $R_6 = NH_2$ | >15.0 | 0.148 | >102 |
| DCK $(2)^{e}$ | | >16.1 | 0.049 | >328 |
| **53**[f] | | >39.0 | 0.0059 | >6,600 |

[a] All data presented are averages of at least two separate experiments performed by Panacos Pharmaceutical Inc.
[b] Concentration that inhibits uninfected H9 cell growth by 50%.
[c] Concentration that inhibits viral replication by 50%.
[d] TI = $IC_{50}$ / $EC_{50}$ .
[e] $EC_{50} = 2.56 \times 10^{-4}$ and TI = $1.37 \times 10^{5}$ in previous screenings and publication.[22]
[f] $EC_{50} = 1.83 \times 10^{-6}$ and TI = $6.89 \times 10^{7}$ in previous screenings and publication.[25] Because of a systematic change of the Panacos screening method, we observed large differences in the results of previously active compounds in the screening. All compounds in Table 9-5 were screened in the new system, in parallel with DCK (2) and 4-methyl DCK (53).

In the assay against the non-drug-resistant strain, compounds **108** (3-methyl) and **111** (2-ethyl) exhibited extremely high anti-HIV activity with nanomolar $EC_{50}$ values (Table 9-6). Compounds **107** (DCP), **110** (2-methyl), and **116** (2,3-dimethyl) also showed better anti-HIV activity than DCK (2). Analogs **112–115** with a bigger 2- or 3-substituent (propyl, isopropyl, ethoxymethyl or phenyl) were less potent. Like the SAR of DCK analogs, **117** (6-t-butyl) lost almost all activity. A comparison of 3'-O-acyl compounds **118–120** indicates that an isovaleryl at the C-3' seems essential for high anti-HIV potency in H9 lymphocytes. The 2-phenyl analogs (**123** and **124**) showed weak activity. These data demonstrated that, in DCP analogs,

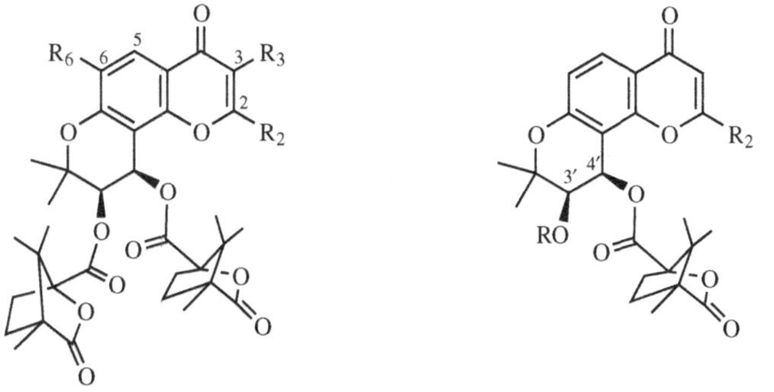

107. $R_2=R_3=R_6=H$, DCP
108. $R_2=R_6=H$, $R_3=CH_3$
109. $R_2=R_6=H$, $R_3=C_6H_5$
110. $R_2=CH_3$, $R_3=R_6=H$
111. $R_2=CH_2CH_3$, $R_3=R_6=H$
112. $R_2=CH_2CH_2CH_3$, $R_3=R_6=H$
113. $R_2=CH(CH_3)_2$, $R_3=R_6=H$
114. $R_2=CH_2OCH_2CH_3$, $R_3=R_6=H$
115. $R_2=C_6H_5$, $R_3=R_6=H$
116. $R_2=R_3=CH_3$, $R_6=H$
117. $R_2=CH_2CH_3$, $R_3=H$, $R_6=C(CH_3)_3$

118. $R_2=CH_3$, R=
119. $R_2=CH_3$, R=
120. $R_2=CH_3$, R=
121. $R_2=CH_2CH_3$, R=
122. $R_2=CH_2OCH_2CH_3$, R=
123. $R_2=$ph, R=
124. $R_2=$ph, R=

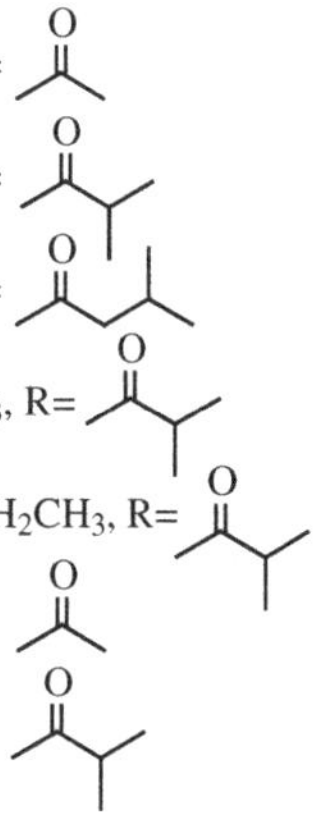

**Figure 9-6.** Substituted DCP derivatives **107–124**.

introducing a small alkyl substituent (one or two carbons) at the chromone 2- and/or 3-position retained anti-HIV activity, whereas a large substitution (over three carbons) at these positions significantly decreased activity against non-drug-resistant HIV in H9 lymphocytes, and substitution at 6-position is not favorable.

Interestingly, these compounds showed different SAR against the HIV-1 RTMDR1 strain. Analogs with a C-2 substituent showed activity. Among them, compound **111** (2-ethyl) exhibited the most promising antiviral activity, whereas **112** (2-propyl) and **113** (2-isopropyl) were slightly less potent. Compounds **110** (2-methyl), **115** (2-phenyl), and **116** (2,3-dimethyl) showed moderate activity. Compound **117**, with 6-*t*-butyl and 2-ethyl substitution, lost almost all activity; thus, similar to the SAR of DCK analogs, 6-substitution is not favorable. Interestingly, the DCP C-3 position is also not favorable against the resistant strain as the previous most potent compound in non-drug-resistant strain, **108**, showed only slight activity; and **109** (3-phenyl) was inactive at the testing concentration. The unsubstituted parent **107** lost potency against the resistant strain, as did **2** (DCK) and **53** (4-methyl DCK). Analogs with a 3′-isobutyryl or isovaleryl substituent exhibited comparable or slightly weaker potency than corresponding

**TABLE 9-6. Anti-HIV Activity of DCP Analogs (107–124)[43]**

| Compound | Non-drug-Resistant Strain[a] | | | Multiple RT Inhibitor Resistant Strain[b] | | |
|---|---|---|---|---|---|---|
| | $IC_{50}$ (μM) | $EC_{50}$ (μM) | TI | $IC_{50}$ (μM) | $EC_{50}$ (μM) | TI |
| **107**[c] | 14.20 | 0.0013 | 11,100 | | NS | |
| **108** | 14.70 | 0.00099 | 14,800 | 3.14 | 1.28 | 2.5 |
| **109** | >35.82 | 1.54 | 23.3 | | NS | |
| **110** | 27.30 | 0.0031 | 8,600 | 11.8 | 0.19 | 62.5 |
| **111** | 37.16 | 0.00032 | 116,200 | 43.08 | 0.06 | 718 |
| **112** | >37.65 | 0.020 | 1,860 | 37.65 | 0.139 | 272 |
| **113** | 33.4 | 0.07 | 483 | >15.04 | 0.14 | >111.1 |
| **114** | 15.1 | 0.1 | 151 | >12.5 | 0.37 | >34 |
| **115** | >35.82 | 0.129 | >277 | 12.2 | 0.17 | 71 |
| **116** | 11.3 | 0.007 | 1,500 | 4.72 | 0.31 | 15 |
| **117** | >35.4 | 1.62 | 22 | >14.15 | 7.36 | 1.9 |
| **118** | >50.15 | 2.29 | 21.9 | >40.16 | 2.2 | 18.2 |
| **119** | 40.68 | 0.88 | 46 | 17.1 | 0.30 | 56.3 |
| **120** | >32.15 | 0.0057 | 5,600 | 13.0 | 0.59 | 22 |
| **121** | >46.3 | 0.25 | 182 | >18.5 | 0.28 | >66.7 |
| **122** | >43.9 | 0.62 | 70 | >17.52 | 1.09 | >16.1 |
| **123** | >44.60 | 2.24 | 20 | 6.25 | 4.46 | 1.4 |
| **124** | >42.47 | 1.68 | 25.3 | | NS | |
| DCK (**2**)[d] | >16.1 | 0.049 | >328 | >16.1 | 12.06 | 1.3 |
| **53**[e] | >39.3 | 0.0059 | >6,660 | >15.7 | 9.43 | 1.7 |

All data presented in this table are averaged from at least two separate experiments.
[a]This assay was performed in H9 lymphocytes by Panacos, Inc., Gaithersburg, MD.
[b]This assay was performed in MT-4 cell line by Dr. Chin-Ho Chen, Duke University, NC.
[c]Previously obtained and published values: $EC_{50} = 0.00068$ μM and TI = 14, 500.[42]
[d]Previously obtained and published values: $EC_{50} = 0.000256$ μM and TI = $1.37 \times 10^5$.[22]
[e]Previously obtained and published values: $EC_{50} = 1.83 \times 10^{-6}$μM and TI = $> 6.89 \times 10^7$.[25]
NS, No Suppression at 10 μg/mL.

3′-camphanoyl-2-substituted compounds. Furthermore, from comparison of $EC_{50}$ and $IC_{50}$ values for compounds **108** (3-methyl) and **110** (2-methyl), substitution on the 2-position seems to increase the activity against a multi-RT inhibitor-resistant strain and decrease the cytotoxicity. Based on these data, a hydrophobic moiety, either aliphatic or phenyl, on C-2 is crucial for anti-HIV activity against the multi-drug-resistant HIV strain and may increase binding of these compounds to a putative hydrophobic cleft.[43]

### 9.2.3.5 Other Modifications
Homologation of an alkyl chain and alteration of ring size are also useful approaches in analog design. Therefore, 3′R,4′R-di-O-(−)-camphanoyl-2′,2′-dimethylnaphtho[5,6-a]pyran (**125**), 3′R,4′R-di-O-(−)-camphanoyl-2′,2′-dimethyl-dihydropyrano[2,3-e]-1-indanone (DCI, **126**), and three compounds with opened A ring (**127–129**) were designed and synthesized. These compounds vary

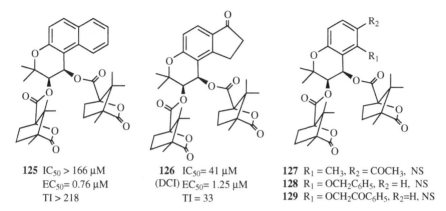

125 $IC_{50} > 166\ \mu M$
  $EC_{50} = 0.76\ \mu M$
  $TI > 218$

126 $IC_{50} = 41\ \mu M$
(DCI) $EC_{50} = 1.25\ \mu M$
  $TI = 33$

127 $R_1 = CH_3$, $R_2 = COCH_3$, NS
128 $R_1 = OCH_2C_6H_5$, $R_2 = H$, NS
129 $R_1 = OCH_2COC_6H_5$, $R_2 = H$, NS

**Figure 9-7.** Structures and activity of DCK related compounds **125–129**.

structurally in the A ring.[44,45] The naphthalene compound **125** and indanone analog **126** showed only slight activity against a non-drug-resistant HIV strain in H9 lymphocytes with $EC_{50}$ values of 0.76 and 1.25 μM, respectively, whereas opened ring compounds **127–129** were inactive (Figure 9-7). These results suggested that a bi-ring system is a requirement for anti-HIV activity.[45]

### 9.2.4 SAR Conclusions

The following conclusions were drawn from the SAR studies of DCK and DCP analogs and derivatives: (1) A bi-ring system is a requirement, and a planar system is probably an essential feature for potent anti-HIV activity. Steric compression between DCK C-4 and C-5 substituents or phenyl substituents on the skeleton can reduce the overall planarity and, thus, resonance of the system, which results in decreased or completely lost activity. (2) The stereochemistry at the 3' and 4' positions should be R-configured, and an isovaleryl structural feature is essential on these positions. Camphanoyl moieties are the most potent substituents. (3) Methyl or other aliphatic substitutions on the DCK C-3, C-4, and C-5 are favorable for anti-HIV activity against the non-drug-resistant strain, whereas aromatic substituents and 6-substitution are not. The DCK C-3 can tolerate polar but not charged or electron withdrawing substituents. (4) The bioisosteric isomers, thio-DCK, and DCK lactam retain activity. (5) The positional isomer, DCP, is even more promising because most DCP analogs are active against a multiple RT inhibitor resistant strain, whereas most DCK derivatives are ineffective against this RT inhibiter resistant strain. An appropriate alkyl substituent at position 2 is critical for the anti-HIV activity of DCP analogs against both viral strains. In addition, most 2-substituted DCP analogs are less toxic to the cells compared with DCP.[43,46]

### 9.2.5 Mechanism of Action

The preliminary mechanism of action studies indicated that DCK inhibits HIV after viral entry but before viral DNA integration.[22] To further elucidate the mechanism of action of this compound, a time-of-addition study was performed, where the point in viral replication inhibited by **102** was compared with the inhibitors acting at attachment (Chicago Sky Blue) or reverse transcription (AZT[47] and nevirapine[48]). This study indicated that **102** acts at a point in the virus life cycle immediately after the target for the RTIs AZT and nevirapine. Additional mechanism of action studies are ongoing, and preliminary data suggest that **102** inhibits the production of double-stranded viral DNA from the single-stranded DNA intermediate, in contrast to traditional RT inhibitors that block the generation of single-stranded DNA from the RNA template.[39]

Low water solubility, less potency against drug-resistant HIV strains, and fast metabolism are three main obstacles that limit the development of DCK analogs. Additional analog design is likely to use DCP as a new lead to increase the potency against drug-resistant viral strains and improve water solubility and pharmacological properties.

## 9.3 BIPHENYL DERIVATIVES AS ANTI-HIV AGENTS

### 9.3.1 SAR Analysis of Naturally Occurring Dibenzocyclooctadiene Lignans

In our previous search for natural products as anti-AIDS agents, an ethanolic extract of the stems of *Kadsura interior* showed significant activity. The BDFI led to the isolation of 12 lignans, including two new compounds. Seven of them showed HIV-1 suppression activity in H9 lymphocytes.[49,50] Among them, gomisin G (**135**, Figure 9-8) showed the most potent anti-HIV activity, with $EC_{50}$ and TI values of 0.011 μM and 300, respectively. Several other lignans also showed moderate activities; therefore, modifications were conducted on this compound type.

Natural products **130–140** are dibenzocyclooctadiene lignans (Figure 9-8). Structurally, they contain a linked biphenyl or phenyl/ketocyclohexadiene system, which can have either an *S*- (**130–136**) or *R*-configuration (**137–140**). Each ring is substituted at positions 1, 2, 3 and 12, 13, 14 with combinations of methoxy, methylenedioxy, or hydroxy groups. Four carbons C-6 ~ C-9 and the biphenyl linkage (C-5a, C-5, C-10, and C-10a) form a cyclooctadiene ring.[51,52] In all cases, positions 7 and 8 of this ring each bear one methyl group and they have α-methyl at C-8, but a different configuration at C-7. Heteroclitin F (**141**) has a tetrahydrofuran ring spanning the biphenyl linkage, and phenyl A is opened between C-2,3, which is called a C-2,3 seco-type.[50] The biphenyl structure feature is needed for the antiviral activity because **141** was inactive. However, the biphenyl configuration does not affect the activity. In Table 9-7, which lists the activities of lignans that showed $EC_{50}$ values lower than 10 μM, five compounds have an *S*-biphenyl and three possess *R*-configuration. Likewise, the configuration of the 7-methyl does not play a role in activity, because **130** and **132** (7α-methyl) showed similar activity to **136** (7β-methyl). Two C-6β substituents were found in these compounds, angeloyl and benzoyl (Bz). Compounds with an angeloyl group, **131** and **134**, did not

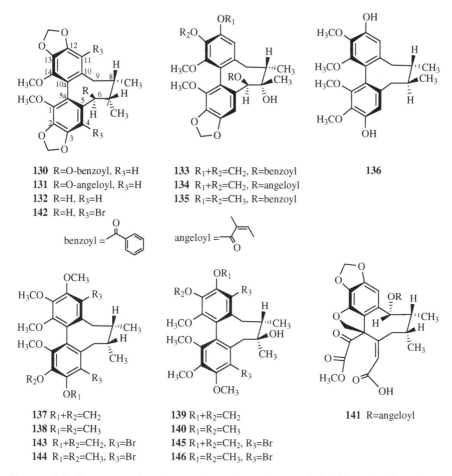

**Figure 9-8.** Structures of dibenzocyclooctadiene lignans **130–141** and dibrominated derivatives **142–146**.

show anti-HIV activity; however, compounds with Bz at C-6β, **130**, **133**, and **135**, did exhibit anti-HIV activity with $EC_{50}$ values of 5.9, 0.96, and 0.011 μM. The most potent compound, gomisin G (**135**), has $S$-biphenyl configuration and 6β-benzoyl-7β-methyl-7α-hydroxy substitutions. In contrast, a 7β-hydroxy substituent (**139** and **140**) diminished the activity. Compounds **132**, **136**, **137**, and **138** have different substituents on the aromatic rings, but no substituents on C-6. Although these four compounds showed similar $EC_{50}$ values, **136** and **138**, without a methylenedioxy group, showed much lower $IC_{50}$ values than **132** and **137**, which contain one or more methylenedioxy groups on the aromatic rings. These results showed that a C-6β benzoyl group, a C-7β hydroxy substituent, and methylenedioxy substitutions on the aromatic rings are important for enhanced anti-HIV activity.[50]

**TABLE 9-7. Anti-HIV Activities of Potent Dibenzocyclooctadiene Lignans and Biphenyl Derivatives**[49,50,53]

| Compound | Biphenyl Configuration | IC$_{50}$ ($\mu$M) | EC$_{50}$ ($\mu$M) | Therapeutic Index |
|---|---|---|---|---|
| **130** | S | 77.3 | 5.9 | 13 |
| **132** | S | 104 | 3.1 | 33.3 |
| **133** | S | 106 | 0.96 | 110 |
| **135** | S | 3.4 | 0.011 | 300 |
| **136** | S | 23.2 | 3.86 | 6 |
| **137** | R | 112 | 2.0 | 56 |
| **138** | R | 21.6 | 2.4 | 9 |
| **143** | R | 36 | 7.2 | 5 |
| **147** | | 13 | 6 | 2.2 |
| **151** | | >140 | 10 | >14 |
| **152** | | 4 | 2 | 2 |
| **153** | | 30 | 13 | 2.3 |
| **159** | | 10 | 5 | 2 |
| **160** | | >239 | 12.0 | >20 |
| **163** | | >201 | 1.05 | >190 |
| **164** | | >174 | 0.40 | >480 |
| **166** | | >166 | 3.3 | >50 |
| **176** | | >174 | 3.6 | >48 |

### 9.3.2 Structural Modifications

#### 9.3.2.1 Dibromination of Dibenzocyclooctadiene Lignans

Five dibrominated derivatives (**142–146**, Figure 9-8) were synthesized from compounds **132** and **137–140**, and evaluated for anti-HIV activity.[50] Only **143** had an EC$_{50}$ lower than 10 $\mu$M (EC$_{50}$ = 7.2 $\mu$M and TI = 5); however, it was still less potent than its parent compound **137** (EC$_{50}$ = 2.0 $\mu$M and TI = 56, Table 9-7).

#### 9.3.2.2 Biphenyl Derivatives

Because the biphenyl configuration does not affect the antiviral activity of dibenzocyclooctadiene lignans, to simplify the structure and achieve synthetic feasibility, we turned to modify biphenyl derivatives.[50,53] These derivatives were first designed and synthesized during the modification of anti-HIV tannins.[54–57] Each aromatic ring bears three variously substituted phenolic groups (position-4, 5, and 6) and, thus, are called hexahydroxydiphenyl derivatives (Figure 9-9). Among the benzoyl-substituted biphenyl derivatives (**147–157**), only **151** demonstrated inhibitory activity against HIV in acutely infected H9 lymphocytes (EC$_{50}$ = 10 $\mu$M) with low toxicity (IC$_{50}$ >140 $\mu$M). The two hydroxymethyl groups at C-2 and C-2′ seemed to be essential for specific HIV inhibition. Replacing them with CHO in **152** greatly increased the toxicity and slightly increased the activity. When bromomethyl at these positions (**153**), toxicity increased with no change of potency. No inhibition was observed with analogs that were mono- or dibrominated at C-3 and C-3′ and

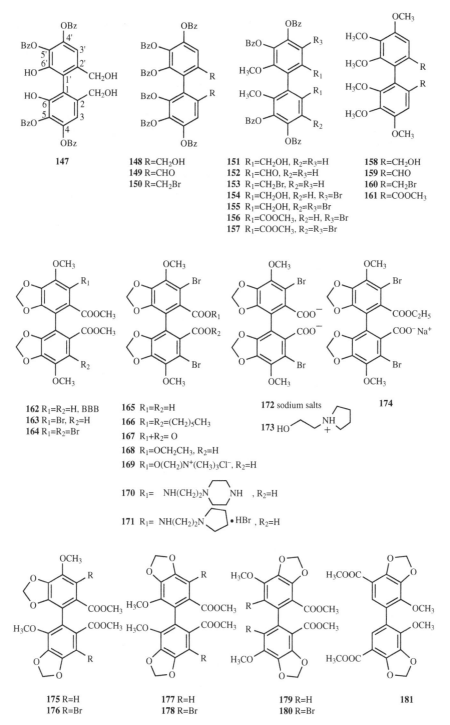

**Figure 9-9.** Structures of biphenyl derivatives (**147–181**).

had either hydroxymethyl (**154** and **155**) or methoxycarbonyl (**156** and **157**) groups on C-2 and C-2′. Simplification of the OBz to OMe reduced the activity (**158–161**).

The previously synthesized 4,4′-dimethoxy-5,6,5′,6′-bis(methylenedioxy)-2,2′-bis(methoxycarbonyl)biphenyl (BBB, **162**) is a known antiviral agent, which is used for the treatment of hepatitis in China, and a potential anti-HIV biphenyl compound.[58] Our preliminary evaluation of BBB demonstrated its anti-HIV activity, with an $EC_{50}$ value of 12 μM and a TI of >20. Therefore, we further modified biphenyl compounds as anti-HIV agents (Figure 9-9). The activities of potent biphenyl derivatives ($EC_{50} \leq 10$ μM) are listed in Table 9-7. Interestingly, **162**-derived compounds showed a different SAR from previous biphenyl compounds. Mono- or dibromination on C-3 and C-3′ greatly increased the activity of BBB. Compounds **163** and **164** are the most active biphenyl compounds, with $EC_{50}$ values of 1.05 and 0.40 μM and TIs of >190 and >480, respectively. The anti-HIV activity was retained if the methyl esters at C-2 and C-2′ were replaced with longer alkyl groups (**166**). However, hydrolysis of one or both esters at these positions dramatically reduced activity (**168** and **165**). Therefore, our approach to introducing water-soluble salts at C-2 and C-2′ was not successful, as **169–174** did not retain activity. The positions of methylenedioxy and methoxy groups also played a role in the anti-HIV activity. Compounds **175** and **176** have one methylenedioxy group moved to the 4,5-positions; they showed reduced activity relative to **162** and **164**, respectively. Compounds **177** and **178** have both methylenedioxy groups changed to the 4,5,4′,5′-positions; they lost almost all activity. Compounds **179** and **180**, which have 3,4,3′,4′-bismethylenedioxy groups, were not very potent, and as can be predicted, **181**, 2,2′-dimethoxy-3,4,3′,4′-bis(methylenedioxy)-5,5′-bis(methoxycarbonyl)biphenyl, did not show anti-HIV activity.

### 9.3.3  SAR Conclusions

The SAR of dibenzocyclooctadiene lignans and related biphenyl derivatives can be summarized as follows: (1) The biphenyl feature is needed for anti-HIV activity in dibenzocyclooctadiene lignans. A C-6β benzoyl group, C-7α hydroxy substituent, and methylenedioxy substitutions on the aromatic rings are important for enhanced anti-HIV activity. 7β-Hydroxylation or bromination on the biphenyl skeleton diminishes the activity. (2) In the case of Bz substituted hexahydroxy-biphenyl derivatives, the two hydroxymethyl groups at C-2 and C-2′ seem to be essential for specific HIV inhibition. (3) When the hexahydroxybiphenyl skeleton is substituted with methylenedioxy and methoxy groups, 3-bromination on one or both phenyls greatly increased the anti-HIV activity. C-2 methoxycarbonyl groups on both phenyls and the position of the methylenedioxy and methoxy groups are essential for the anti-HIV activity. Methoxy at C-4 and methylenedioxy at C-5,6 of each phenyl produced the best activity.

### 9.3.4  Mechanism of Action of Biphenyl Derivatives

The most active compounds **163** and **164** were investigated for RT-associated RT activity.[53] The RT-associated RT spectra of these compounds are template-primer dependent. The enzyme inhibition by **163** was approximately twice more sensitive with poly(rC)-oligo(dG) as template-primer than with poly(rA)-oligo(dT). In contrast, for **164**, the enzyme was about six times more sensitive with poly(rA)-oligo(dT). Nevirapine, an NNRTI, demonstrated more inhibitory activity using poly(rC)-oligo(dG) as template-primer than poly(rA)-oligo(dT).[59] Both compounds demonstrated potent inhibitory effects against HIV-1 RT-associated DNA polymerase activity using poly(dA)-oligo(dT) as a template-primer, with $IC_{50}$ values of 0.056 and 0.048 μM, respectively. They also displayed similar inhibition against HIV-RT-associated RNase H activity when poly(rG)-poly(dC) was used as the substrate and $Mn^{2+}$ (0.4 mM) was used as the divalent cation. However, when poly(rA)-poly(dT) was used as the substrate or $Mg^{2+}$ was used as the divalent cation, no inhibition was observed. Because RT plays an important role in HIV replication, the potent inhibitory activity of **163** and **164** against the enzyme-catalyzed DNA direct DNA synthesis might be related to their anti-HIV activity.[53]

## 9.4 TRITERPENE BETULINIC ACID DERIVATIVES AS ANTI-HIV AGENTS

Triterpenes represent a structurally varied class of natural products existing in various plant species. Thousands of triterpenes have been reported with hundreds of new derivatives described each year.[60,61] Some naturally occurring triterpenes exhibit moderate anti-HIV-1 activity and, therefore, provide good leads for further drug development because of their unique mode of action and chemical structures.[62] Anti-HIV triterpenes were found to block HIV entry, including absorption (glycyrrhizin[63,64]) and membrane fusion (RPR103611[65]), inhibit RT (mimusopic acid[66]) and PR (ganoderiol,[67] geumonoid,[68]), and act during viral maturation (DSB[69]). Here, we will mainly focus on the modification of betulinic acid and new mechanisms of action of RPR103611, IC9564, and DSB.

Lupane is a pentacyclic triterpene with a five-membered E ring. Two members of this group, 3β-hydroxy-lup-20(29)-en-28-oic acid (betulinic acid, BA, **182**) and platonic acid (**185**), were first reported to reduce $HIV_{IIIB}$ reproduction by 50% in H9 lymphocytes, with $EC_{50}$ values of 1.4 and 6.5 μM, respectively. With an additional hydroxy group at C-2, alphitolic acid (**183**) displayed much lower anti-HIV activity ($EC_{50} = 42.3$ μM) (Figure 9-10).[70] BA (**182**) has been widely investigated for significant chemical, spectral, and biological data. Several patents are related to the pharmaceutical application of BA for the treatment of cancer,[71,72] viral infections,[72,73] hair loss,[74,75] and other conditions.[76]

### 9.4.1 Betulinic Acid Derivatives as Entry Inhibitors

#### 9.4.1.1 Modification of Ring A

A group of scientists at the University Leuven, Belgium, observed weak inhibition of HIV-1 PR of betulinylglycine (**184**) from a high throughput screening; therefore, they modified the C-28 side chain.[77] Interestingly, no significant improvement of the anti-protease activity was achieved. However, some active analogs were

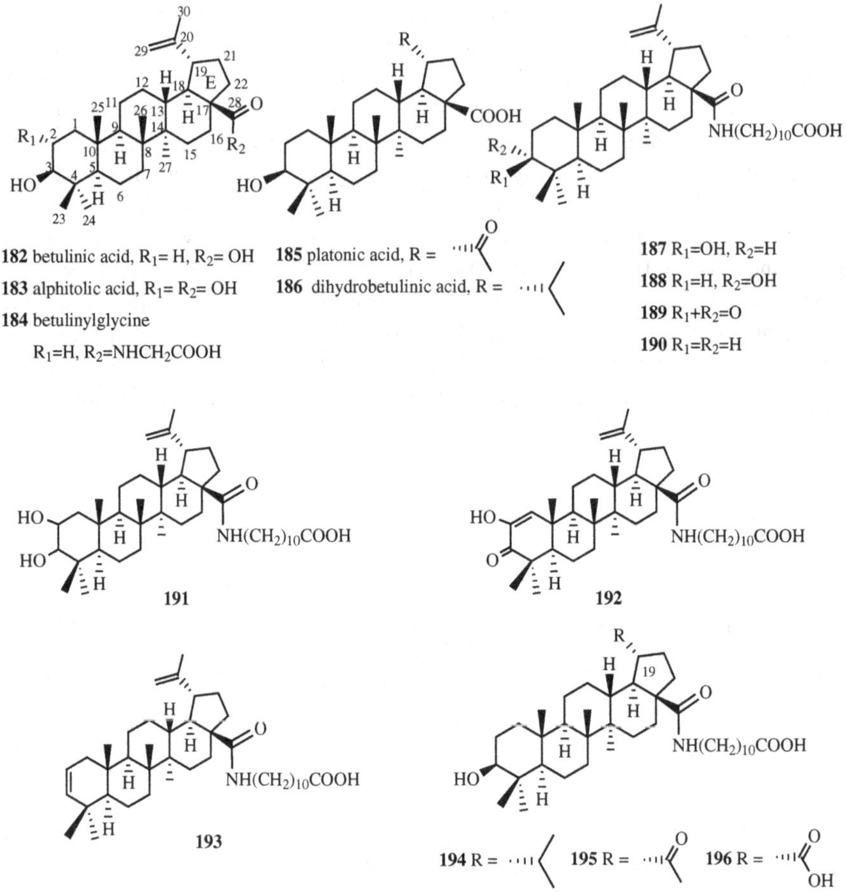

**182** betulinic acid, R$_1$= H, R$_2$= OH
**183** alphitolic acid, R$_1$= R$_2$= OH
**184** betulinylglycine
   R$_1$=H, R$_2$=NHCH$_2$COOH

**185** platonic acid, R =
**186** dihydrobetulinic acid, R =

**187** R$_1$=OH, R$_2$=H
**188** R$_1$=H, R$_2$=OH
**189** R$_1$+R$_2$=O
**190** R$_1$=R$_2$=H

**191**

**192**

**193**

**194** R =   **195** R =   **196** R =

**Figure 9-10.** Structures of lupane triterpenes and betulinic acid derivatives (**182–196**).

found to inhibit the cytopathogenicity of HIV-1 in CEM-4 cells, including betuli-nylaminoundecanoic acid (**187**) with IC$_{50}$ values of 0.23 and 0.44 μM in CEM and MT-4 cells, respectively. Changing the 3-hydroxy position from 3β (**187**) to 3α (**188**) led to a ten-fold decrease in activity (IC$_{50}$ = 3.0 and 2.52 μM). The 3-keto compound (**189**) exhibited intermediate activity. Interestingly, both the 3-deoxy analog (**190**) and the 3,4-dihydroxy derivative (**191**) lost activity. Furthermore, the antiviral activities of the 3,4-diketone (**192**) and 2,3-dehydro (**193**) derivatives were low. Therefore, chemical modification in ring A led to considerable loss of activity[77] (Figure 9-10).

### 9.4.1.2 Modification of the C-19 Side Chain
Hydrogenation of the isopropylidene double bond (**194**) reduced the activity of **187** by about five-fold. The replacement of the isopropylidene by an acetyl (**195**) or a carboxylic acid (**196**) led to virtually inactive molecules. However, various

substituents could be placed on the isopropylidene C-30 methyl group. The 30-hydroxy (**206**) and 30-(hydroxyethyl)thio (**197**) analogs retained the activity of **187**, whereas 30-(diethylaminoethyl)thio (**198**), aromatic (**200**), acetamide (**202**), and 30-pyrrolidine (**205**) derivatives showed decreased activity. Unlike **197**, the 30-(hydroxyethyl) amino analog (**203**) exhibited dramatically decreased activity. An unsubstituted amino moiety (**201**) and a free carboxyl (**199** and **204**) led to complete inactivation. In conclusion, the isopropylidene probably binds to a hydrophobic pocket on the HIV-1 molecular target. A certain lack of steric hindrance accommodates a variety of substituents without improvement of activity. However, amines and carboxylic acid moieties are not favorable (Table 9-8).[77]

### 9.4.1.3 Side-Chain Modification

Elongation of the glycine alkyl chain (one methylene) in **184** to an undecanoic acid in **187** (ten methylenes) resulted in greatly increased activity and a different mechanism of action. This result also suggested that position-28 might accommodate various substituents. Therefore, more compounds were designed to optimize the side chain at this position.

First, compounds (**207–214**, **187**, **215**, and **216**, Figure 9-11) with 1~12 methylenes were synthesized to determine the optimal chain length. These compounds

**TABLE 9-8. Structures and Activities of 30-Substituted BA Derivatives (197–206)[77]**

| Compound | R | CEM Cells | MT-4 Cells |
|----------|---|-----------|------------|
| 197 | SCH$_2$CH$_2$OH | 0.29 | 0.27 |
| 198 | SCH$_2$CH$_2$NHEt$_2$ | 1.0 | 0.63 |
| 199 | SCHCOOH | na | 1.16 |
| 200 | S⟨C$_6$H$_4$⟩F | 2.0 | 1.37 |
| 201 | NH$_2$ | na | Nt |
| 202 | NHCOCH$_3$ | 1.25 | 0.26 |
| 203 | NHCH$_2$CH$_2$OH | 5.75 | 2.0 |
| 204 | NHCH$_2$COOH | na | Nt |
| 205 | —N⟨pyrrolidine⟩ | 1.0 | 0.42 |
| 206 | OH | 0.20 | 0.38 |

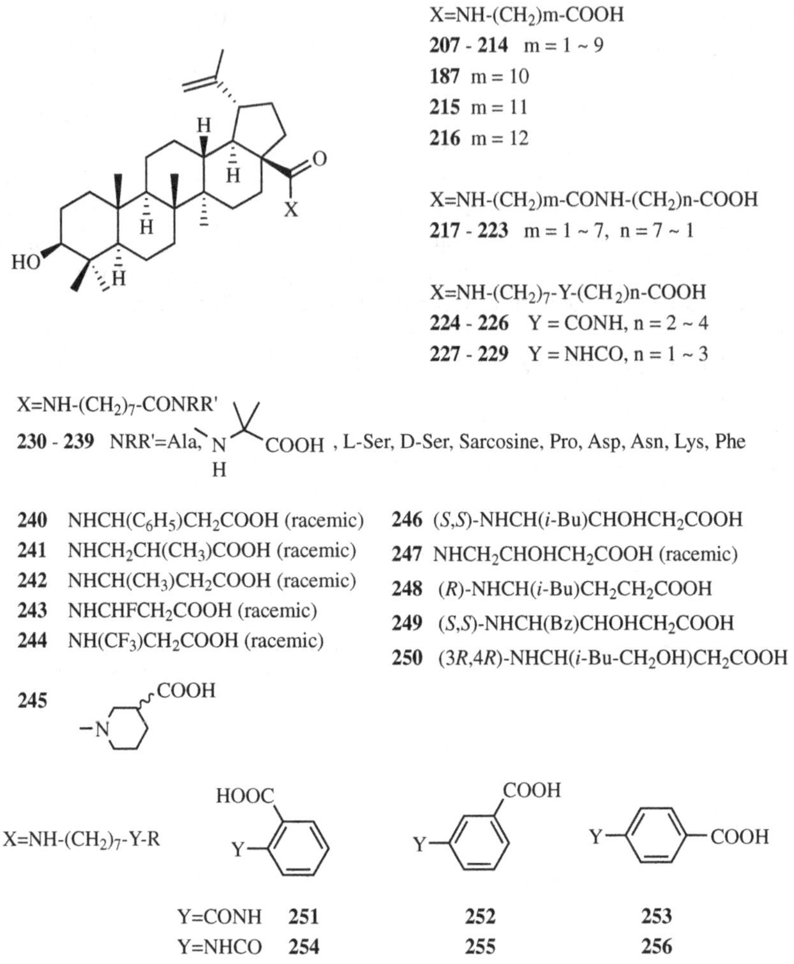

X=NH-(CH$_2$)m-COOH
**207 - 214**  m = 1 ~ 9
**187**  m = 10
**215**  m = 11
**216**  m = 12

X=NH-(CH$_2$)m-CONH-(CH$_2$)n-COOH
**217 - 223**  m = 1 ~ 7, n = 7 ~ 1

X=NH-(CH$_2$)$_7$-Y-(CH$_2$)n-COOH
**224 - 226**  Y = CONH, n = 2 ~ 4
**227 - 229**  Y = NHCO, n = 1 ~ 3

X=NH-(CH$_2$)$_7$-CONRR'
**230 - 239**  NRR'=Ala, $\overset{\displaystyle}{N}$ COOH , L-Ser, D-Ser, Sarcosine, Pro, Asp, Asn, Lys, Phe
                        H

**240**  NHCH(C$_6$H$_5$)CH$_2$COOH (racemic)
**241**  NHCH$_2$CH(CH$_3$)COOH (racemic)
**242**  NHCH(CH$_3$)CH$_2$COOH (racemic)
**243**  NHCHFCH$_2$COOH (racemic)
**244**  NH(CF$_3$)CH$_2$COOH (racemic)

**246**  (S,S)-NHCH(i-Bu)CHOHCH$_2$COOH
**247**  NHCH$_2$CHOHCH$_2$COOH (racemic)
**248**  (R)-NHCH(i-Bu)CH$_2$CH$_2$COOH
**249**  (S,S)-NHCH(Bz)CHOHCH$_2$COOH
**250**  (3R,4R)-NHCH(i-Bu-CH$_2$OH)CH$_2$COOH

**245**   COOH

X=NH-(CH$_2$)$_7$-Y-R

| | HOOC | COOH | |
|---|---|---|---|
| | Y=CONH  **251** | **252** | **253** |
| | Y=NHCO  **254** | **255** | **256** |

**Figure 9-11.** Structures of C-28 substituted betulinic acid derivatives (**207–256**).

were evaluated for inhibition of HIV-1-induced cytopathicity in CEM and MT-4 cells (Table 9-9). Based on the results, the activity is related to the length of the C-28 side chain. The optimal number of methylenes is between 7 and 11, and compound **187** ($m = 10$) showed the best activity. A second amide moiety (–CONH–) was introduced in the middle of the alkyl chain as shown in **217–223** (Figure 9-11), while keeping the total of the linking atoms [$m$, amide(=2), and $n$] at 10, the optimal number from the previous result. Only compound **223** ($m = 7$ and $n = 1$) was found to be more potent than its parent compound **212** ($m = 7$). Therefore, the chain length was extended by amidation of **212** with β-alanine (**224**, $n = 2$), 3-aminopropionic acid (**225**, $n = 3$), or 4-aminobutyric acid (**226**, $n = 4$), which results in increased activity (IC$_{50}$ values about 0.05 μM in CEM cells). Inverting the amide moiety within the chain (–NHCO–) led to compounds **227–229**, which showed two

**TABLE 9-9. Structures and Activity of Active C-28 Side-Chain Betulinic Acid Derivatives[65]**

NH(CH$_2$)m-Y-(CH$_2$)n-COOH

| Compound | m | n | Y | CEM Cells (IC$_{50}$ μM) | MT-4 Cells (IC$_{50}$ μM) |
|---|---|---|---|---|---|
| 210 | 5 | 0 | – | 2.3 | 2.4 |
| 211 | 6 | 0 | – | 12 | 4.3 |
| 212 | 7 | 0 | – | 0.75 | 0.47 |
| 213 | 8 | 0 | – | 0.42 | 3.4 |
| 214 | 9 | 0 | – | 0.6 | 0.42 |
| 187 | 10 | 0 | – | 0.23 | 0.44 |
| 215 | 11 | 0 | – | 0.55 | 0.25 |
| 219 | 3 | 5 | CONH | 5.5 | Nt |
| 220 | 4 | 4 | CONH | 4.5 | Nt |
| 221 | 5 | 3 | CONH | 3.0 | Nt |
| 222 | 6 | 2 | CONH | 4.5 | Nt |
| 223 | 7 | 1 | CONH | 0.15 | 0.16 |
| 224 | 7 | 2 | CONH | 0.1 | 0.94 |
| 225 | 7 | 3 | CONH | 0.044 | 0.13 |
| 226 | 7 | 4 | CONH | 0.061 | Nt |
| 227 | 7 | 1 | NHCO | 0.05 | 0.034 |
| 228 | 7 | 2 | NHCO | 0.095 | 0.053 |
| 229 | 7 | 3 | NHCO | 0.02 | Nt |

Compounds were tested for their ability to inhibit the cytopathic effect induced by HIV-1 infection. Nt, not tested.

to five times better activity than the corresponding **224–226** (Table 9-9). The (betulinylamino)-octanoic acid **212** was also reacted with small α-amino acids (**230–239**, Figure 9-11 and Table 9-10). The alanine derivatives **230** and **231** displayed improved activity compared with the parent glycine derivative **223**. The introduction of more bulky groups, such as Asp, Asn, Lys, and Phe, led to decreased activity. These results suggest that a small lipophilic space can accommodate one or two methyl groups.[65]

Compounds **240–250** are β-amino acid derivatives of **212**, and their activity profile indicated small substituents in the α- or β-position or even the more bulky phenyl did not significantly influence activity. Compounds **240** and **249** contain a substituted phenyl moiety and exhibited good activity. Therefore, *o*-, *m*-, and

**TABLE 9-10. Anti-HIV Activity of Additional C-28 Modified Betulinic Acid Derivatives[65]**

230–250                                        251–255

| Compound | CEM Cells IC$_{50}$ (μM) | MT-4 Cells IC$_{50}$ (μM) |
| --- | --- | --- |
| **230** NRR' = Ala | 0.092 | 0.062 |
| **231** NRR' = NHC(CH$_3$)$_2$COOH | 0.133 | 0.089 |
| **232** NRR' = L-Ser | 0.25 | 0.25 |
| **233** NRR' = D-Ser | 1.9 | Nt |
| **234** NRR' = Sarcosine | 0.25 | 0.33 |
| **235** NRR' = Pro | 0.64 | 0.058 |
| **240** | 0.085 | 0.047 |
| **241** | 0.13 | Nt |
| **242** | 0.13 | Nt |
| **243** | 0.095 | Nt |
| **244** | 0.155 | Nt |
| **245** | 0.185 | Nt |
| **246** | 0.05 | 0.04 |
| **247** | 0.2 | 0.033 |
| **248** | 0.052 | 0.085 |
| **249** | 0.13 | 0.033 |
| **250** | 0.05 | 0.044 |
| **252** | 0.08 | Nt |
| **253** | 0.105 | 0.04 |
| **255** | 0.07 | 0.026 |
| **256** | 0.11 | 0.022 |
| nevirapine | 0.084 | |

Nt, not detected.

p-aminobenzoic acid derivatives (**251–253**) and corresponding benzenedicarboxylic acid derivatives (**254–256**) were synthesized (Figure 9-11). The *ortho*-derivative (**251** and **254**) was inactive, probably because of steric hindrance. The *meta*- and *para*-derivatives (**252**, **253**, **255**, and **256**, Table 9-10) were active. The statine derivative **246** (RPR103611) was found to display the best overall activity, with an IC$_{50}$ value of 0.050 ± 0.026 μM in CEM cells and 0.040 ± 0.019 μM in MT-4 cells and low cytotoxicity (SI >100). Therefore, RPR103611 (**246**) was selected for continued investigations.[65]

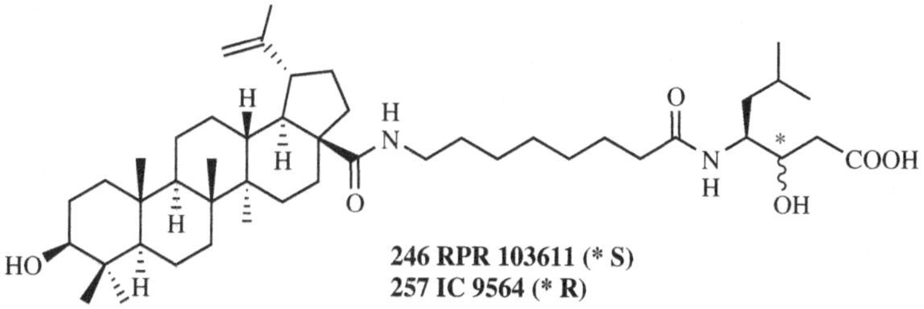

**246 RPR 103611 (* S)**
**257 IC 9564 (* R)**

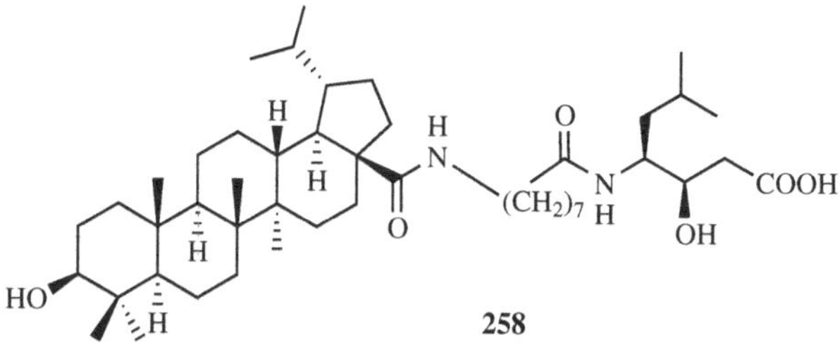

**258**

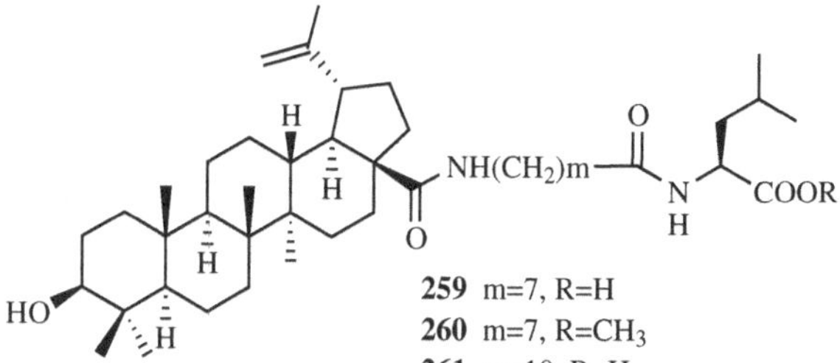

**259** m=7, R=H
**260** m=7, R=$CH_3$
**261** m=10, R=H

**Figure 9-12.** Structures of RPR103611 (**246**), IC9564 (**257**), and **258–261**.

IC9564 (**257**, Figure 9-12)[78] is an equipotent stereoisomer of **246**. The results confirmed that a free hydroxy group at C-3 and free carboxylic acid in the C-28 side chain are necessary for anti-HIV activity. The double bond of the isopropylidene group might not play a key role in the HIV inhibitory activity of this compound type, because **258** exhibited similar potency to **257**. Replacing the statine moiety of **257** with L-leucine yielded a four-fold less potent compound **259**. Interestingly, the methyl ester (**260**) of **259** showed increased potency ($EC_{50} = 0.52$ μM). Compound **261**, with ten methylenes, showed three-fold greater activity than **259**, which has seven methylenes. These results indicate that the statine group in the C-28 side chain of IC9564 (**257**) can be replaced by L-leucine and that the number of methylenes in the aminoalkanoic chain is crucial to optimal anti-HIV potency.

### *9.4.1.4 Mechanism of Action*

RPR103611 (**246**) was tested in a panel of isolates and cell lines, including the monocytic cell line U-937 and peripheral blood lymphocytes. The $IC_{50}$ values varied between 0.04 and 0.1 μM. However, no significant activity could be found against HIV-2 (ROD and EHO isolates). In addition, it was much less active against two isolates of Zairian origin, NDK and ELI. These results suggest that RPR103611 is a potent and selective inhibitor of HIV-1.[79]

The inhibitory effects of HIV RT, PR, or IN were not detected with RPR103611. Evidence showed that RPR103611 blocks virus infection at a postbinding step necessary for virus-membrane fusion. The postbinding molecular events leading to HIV-1 envelope-mediated membrane fusion are characterized by major conformational changes of the whole gp120/gp41 complex.[79] Sequence analysis of RPR103611-resistant mutants indicated that a single amino acid change, I84S, on the loop region of HIV-1 gp41 is sufficient to confer drug resistance. However, this I84S mutation did not occur in some naturally RPR103611-resistant HIV-1 strains such as NDK or ELI.[80] Continued investigation indicated that the antiviral efficacy of RPR103611-escape-mutant creates a conformational change that affects the stability and function of HIV-1 envelope glycoproteins.[81]

IC9564 (257) is the stereoisomer of RPR103611 (246), which also inhibited HIV-1 by blocking viral entry. These two compounds seem to be equally potent in their anti-HIV-1 and anti-fusion activity profiles. Analysis of a chimeric virus NLDH120 derived from exchanging envelope regions between IC9564-sensitive and IC9564-resistant viruses indicated that regions within gp120 and the N-terminal 25 amino acids (fusion domain) of gp41 are key determinants for the drug sensitivity. However, there is no change in the isoleucine at position 595, which corresponds to the cysteine at position 84 in Labrosse' et al.'s study.[81] The mutations G237R and R252K in gp120, which are located in the inner domain of the HIV-1 gp120 core that is believed to interact with gp41,[82] were found to contribute to the drug sensitivity in a drug-resistant mutant from the NL4-3 virus. These results confirm that HIV-1 gp120 plays a key role in the anti-HIV-1 activity of IC9564.[83] Additional evidence suggests that the bridging sheet is involved in IC9564 sensitivity, which is a critical structural motif involved in HIV-1 entry.[84]

### 9.4.2 Betulinic Acid Derivatives as Maturation Inhibitors

#### 9.4.2.1 Discovery of DSB (PA-457)

Our group modified the anti-HIV lead compounds BA (182) and dihydrobetulinic acid (186) to produce another series of compounds with substituents at C-3 and a free C-28 carboxylic acid. These 3-O-acyl derivatives (262–275) were obtained by treating the triterpenes with an acid anhydride and DMAP in pyridine or with an acid chloride in pyridine. The anti-HIV activities of active 3-O-acyl derivatives are listed in Table 9-11. Among them, 3-O-(3′,3′-dimethylsuccinyl)-betulinic acid (263, DSB, PA-457) and –dihydrobetulinic acid (268) both demonstrated extremely potent anti-HIV activity in acutely infected H9 lymphocytes, with EC$_{50}$ values of <0.35 nM, and remarkable TI values of >20,000 and >14,000, respectively. In contrast, 262 and 267, the 2′,2′-dimethyl isomers, showed significantly lower anti-HIV activity. Compounds 264–266, 269, and 270 were 10–100-fold less active than 263 and 268. Compounds without a carboxylic acid in the C-3 side chain

**TABLE 9-11. Structures and Activity of 3-O-Acyl-Betulinc Acid Derivatives (262–275)[69]**

|  | 262–266 | | | 267–275 | | |
| R | Compound | $EC_{50}(\mu M)$ | TI | Compound | $EC_{50}(\mu M)$ | TI |
| H | **182** | 1.4 | 9.3 | **186** | 0.9 | 14 |
| ![structure] COOH | **262** | 2.7 | 6.7 | **267** | 0.56 | 13.8 |
| ![structure] COOH | **263 (DSB)** | <0.00035 | >20000 | **268** | <0.00035 | >14000 |
| ![structure] COOH | **264** | 0.0023 | 1974 | **269** | 0.0057 | 1017 |
| ![structure] O COOH | **265** | 0.01 | 1172 | **270** | 0.0056 | 2344 |
| ![structure] COOH | **266** | 0.044 | 292 | **271** | 0.9 | 9 |
| ![structure] COOH | | | | **272** | 1.8 | 7.5 |
| ![structure] O O | | | | **273** | 0.5 | 2 |
| ![structure] | | | | **274** | 1.5 | 56 |
| ![structure] | | | | **275** | NS | |

The activity was screened in acutely infected H9 lymphocytes.
NS, no suppression at testing concentration.

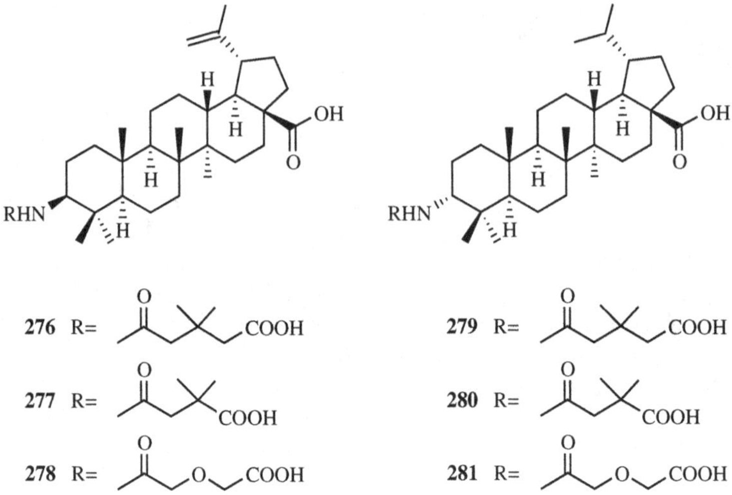

**Figure 9-13.** 3-*N*-acyl-betulinc acid derivatives (**276–281**).

(**273–275**) showed only slight activity. Several compounds with C-28 ester substitution and a free C-3 hydroxy or both C-3 and C-28 substituents did not exhibit high potency.[69,85] DSB (**263**) showed potent inhibition against both wild-type and drug-resistant HIV-1 isolates; therefore, it has been chosen as a drug candidate for the treatment of AIDS. Panacos Inc. filed an IND application on DSB (so-called PA-457) with the U.S. FDA in March 2004.

A series of 3-*N*-acyl-betulinc acid derivatives were synthesized to explore the SAR. The structures of 3β-*N*-acyl-betulinc acid derivatives (**276–278**) and 3α-*N*-acyl-betulinc acid analogs (**279–281**) are shown in Figure 9-13. None of them showed comparable activity with the corresponding 3β-*O*-acyl BA derivatives. Therefore, amino-acyl substitution at position 3 is not favorable for anti-HIV activity.[86]

### 9.4.2.2   Mechanism of Action of DSB

Although DSB (**263**) shares the BA skeleton of RPR103611 (**246**) and IC9564 (**257**), it does not target the HIV-1 entry process. Several targets for DSB were ruled out, including viral attachment, RT, IN, and PR. A time-of-addition study suggested that DSB blocks viral replication at a time point after the completion of viral DNA integration and *Tat* expression. Virions mature late in the virus replication cycle through cleavage of the *Gag* polyprotein by PR, which results in release of the virion proteins matrix (MA), capsid (CA), nucleocapsid (NC), p6, and two spacer peptides, SP1 and SP2.[87] Incomplete *Gag* processing, a ordered event,[88] results in virions that lack a functional core and are noninfectious.[89–91] Li et al. observed abnormal *Gag* protein processing in a virus generated in the presence of a compound. Radioimmunoprecipitation revealed analyses revealed that DSB specifically inhibited the conversion of p25 (CA-SP1) to p24 (CA), the last step of *Gag* processing, as increased levels of p25 were detected in both cell and virion lysates. Sequence analysis of DSB-resistant mutants indicated that a single mutation encoding an amino acid change (A to V) at the N terminus of SP1 was found. After introduction of this mutation A-to-V into the NL4-3 backbone to generate the NL4-3 SP1/A1V mutant, DSB had no effect on the processing of the mutant *Gag*, as no p25 was detected at 1 µg/mL. These results indicate that the A-to-V substitution at the p25 cleavage site confers DSB resistance.[92]

A similar study by Zhou et al.[93] indicates that a single mutation, L363F, at the P1 position of the CA-SP1 junction confers resistance to DSB. Additional evidence for the *Gag* protein as the target of DSB was obtained through substitution of amino acid residues flanking the CA-SP1 junction of HIV-1 or the simian immuodeficiency virus (SIV). This study takes advantage of the fact that DSB does not affect the replication of SIV. Alteration of three residues in the SIV CA-SP1 junction to the corresponding amino acids of HIV-1 resulted in an SIV mutant, with a similar degree of sensitivity as HIV-1 to DSB. On the other hand, two substitutions in HIV-1 with the corresponding residues of SIV resulted in resistance to DSB. These results suggest that the CA-SP1 junction is a key determinant of DSB sensitivity.[94] Therefore, DSB represents a unique class of anti-HIV compounds termed maturation inhibitors (MIs) that exploit a previously unidentified viral target, which provides additional opportunities for HIV drug discovery.

### 9.4.3 Bifunctional Betulinic Acid Derivatives with Dual Mechanisms of Action

The viral targets of BA derivatives vary, depending primarily on the side-chain structures of the compounds.[62] RPR103611 (**246**) and IC9564 (**257**) are entry inhibitors, and DSB (**263**) targets viral maturation. Therefore, the BA skeleton probably serves as a scaffold and binds to a common motif. Based on the molecular targets and chemical structures of IC9594 and DSB, Huang et al. reported on the synthesis, bioevaluation, and mechanism of action study of two new compounds, LH15 (**282**) and LH55 (**283**), which have a 3′,3′-dimethylsuccinyl ester at C-3, similar to DSB. LH-15 (**282**) is a leucine amide, and LH-55 (**283**) is an 11-aminoundecanoic amide derivative.[95] Their activities against several HIV strains are listed in Table 9-12. LH-55 (**283**) showed the best activities; its EC$_{50}$ values ranged from 0.03 to 0.006 µM against a wild-type strain (NL4-3), and two drug-resistant strains, PI-R and RTI-R. LH-15 (**282**) was two-to five-fold less potent than **283**, but more potent than the maturation inhibitor DSB (**263**) and the entry inhibitor IC9564 (**257**) screened side by side.[95]

The side chain at position-28 is critical for the antifusion activity of the BA derivatives, whereas a dimethylsuccinyl acid at position-3, as in DSB (**263**), results in inhibition of viral maturation rather than affecting the envelope-mediated membrane fusion. The key structural feature that enables LH55 (**283**) and its analogs to possess a dual mode of action is the presence of side chains at both C-3 and C-28. The unique biological activity of these compounds is that they not only inhibit viruses that are resistant to HIV-1 RT and PR inhibitors, but also they inhibit viruses that are resistant to compounds with side chains only at position-3 or

**TABLE 9-12. Structures and Activities of Bifunctional Betulinic Acid Derivatives[95]**

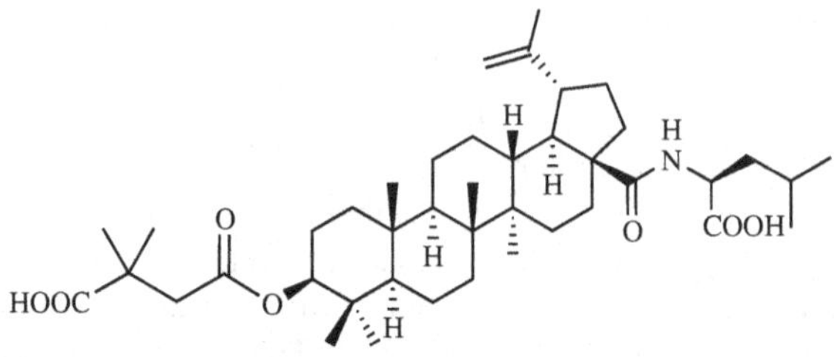

LH-15 (282)

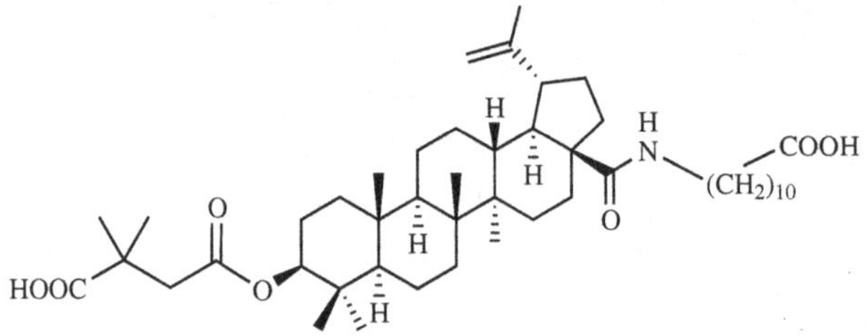

LH-55 (283)

| | EC$_{50}$ ($\mu$M) | | |
| --- | --- | --- | --- |
| | NL4-3 | PI-R | RTI-R |
| DSB (263) | 0.096 ± 0.012 | 1.71 ± 0.15 | 0.085 ± 0.008 |
| IC9564 (257) | 0.11 ± 0.02 | 0.12 ± 0.03 | 0.093 ± 0.016 |
| LH15 (282) | 0.0065 ± 0.0008 | 0.032 ± 0.004 | 0.01 ± 0.001 |
| LH55 (283) | 0.016 ± 0.003 | 0.049 ± 0.005 | 0.019 ± 0.003 |

NL4-3 is a wild-type strain; PI-R is a multiple PR inhibitor resistant strain, HIV-1 M46I, L63P, V82P, and I84V; and RTI-R is an HIV-1 strain, HIV-1 RTMDR1 (M41L, L74V, V106A, and T215Y, resistant to multiple HIV-1 RT inhibitors.

position-28. In a fusion assay, the potency of LH55 against NL4-3 envelope-induced membrane fusion was similar to that of IC9564 (257) and T20, the latter of which is the first FDA-approved fusion inhibitor for the treatment of AIDS. HIV-1 particles produced in the presence of LH55 were lysed and analyzed using Western blots. P25 accumulated in the virus was produced in the presence of LH55. The unique mode of action of DSB or LH55 is that these compounds affect the processing of only p25. These compounds do not affect the processing of other *Gag* proteins, such as p17. The antifusion activity allows LH55 to block HIV-1 before it enters the cell, whereas the antimaturation activity of this compound can combat any virus that does enter the cell. Because the molecular targets of these bifunctional anti-HIV betulinic acid derivatives are different from those of clinically available anti-HIV drugs, it is possible that LH55 (283) and derivatives might have the potential to become useful additions to current anti-HIV therapy.[94]

## 9.5 CONCLUSIONS

In a decade of extensive research, great progress has been achieved in the discovery of potential anti-HIV agents. Plants are a great reservoir of biologically active compounds and can provide good leads that are structurally unique or have new mechanisms of action. Our anti-HIV research program began from natural products and led to the discovery of potent khellactone analogs and related compounds, dibenzocyclooctadiene lignans and derived biphenyl components, and BA derivatives. In the whole process of drug discovery and development, we begin with bioactivity-directed isolation, followed by rational design-based structural modification and SAR analyses. Our goals include increasing the activity profile, overcoming absorption, distribution, metabolism, excretion and toxicity problems, generating new lead compounds, and finally entering clinical trials.

## ACKNOWLEDGMENTS

This investigation entitled "Anti-AIDS Agents 63" was supported by Grant AI-33066 from the National Institute of Allergies and Infectious Diseases awarded to K. H. Lee. We wish to express our sincerest thanks to Dr. Susan L. Morris-Natschke for her experienced editing. Also, we acknowledge the contributions from our collaborators, most of whom are cited in the references.[92-94]

## REFERENCES

1. Gottlieb, M. S.; Schroff, R.; Schanker, H. M.; Weisman, J. D.; Fan, P. T.; Wolf, R. A.; Saxon, A. Pneumocystis carinii pneumonia and mucosal candidiasis in previously healthy homosexual men. *N. Engl. J. Med.*, **1981**, 305: 1425–1431.

2. Barre-Sinoussi, F.; Chermann, J. C.; Rey, F.; Nugeyre, M. T.; Chamaret, S.; Gruest, J.; Dauguet, C.; Axler-Blin, C.; Vezinet-Brun, F.; Rouzioux, C.; Rozenbaum, W.; Montagnier, L. Isolation of a T-lymphotropic retrovirus from a patient at risk for acquired immune deficiency syndrome (AIDS). *Science*, **1983**, 220: 868–871.

3. Gallo, R. C.; Sarin, P. S.; Gelmann, E. P.; Robert-Guroff, M.; Richardson, E.; Kalyanaraman, V. S.; Mann, D.; Sidhu, G. D.; Stahl, R. E.; Zolla-Pazner, S.; Leibowitch, J.; Popovic, M. Isolation of human T-cell leukemia virus in acquired immune deficiency syndrome (AIDS). *Science*, **1983**, 220: 865–867.

4. UNAIDS; WHO. (2004) *AIDS epidemic update.* (online). Available: http://www.unaids.org.

5. FDA. (2005). *Antiretroviral drugs approved by FDA for HIV.* (online). Available: http://www.fda.gov/oashi/aids/virals.html.

6. Swanstrom, R.; Erona, J. Human immunodeficiency virus type-1 protease inhibitors: Therapeutic successes and failures, suppression and resistance. *Pharmacol. Ther.*, **2000**, 86: 145–170.

7. Loveday, C. Nucleoside reverse transcriptase inhibitor resistance. *JAIDS*, **2001**, 26: S10–S24.

8. Pellegrin, I.; Garrigue, I.; Caumont, A.; Pellegrin, J. L.; Merel, P.; Schrive, M. H.; Bonot, P.; Fleury, H. Persistence of zidovudine-resistance mutations in HIV-1 isolates from

patients removed from zidovudine therapy for at least 3 years and switched to a stavudine-containing regimen. *AIDS (London)*, **2001**, 15: 1071–1073.

9. Fellay, J.; Boubaker, K.; Ledergerber, B.; Bernasconi, E.; Furrer, H.; Battegay, M.; Hirschel, B.; Vernazza, P.; Francioli, P.; Greub, G.; Flepp, M.; Telenti, A. Prevalence of adverse events associated with potent antiretroviral treatment: Swiss HIV Cohort Study. *Lancet*, **2001**, 358: 1322–1327.

10. Lalezari, J.; Henry, K.; O'Hearn, M. et al. Enfuvirtide, an HIV-1 fusion inhibitor, for drug-resistant HIV infection in North and South America. *N. Engl. J. Med.*, **2003**, 348: 2175–2185.

11. Lee, K. H. Current developments in the discovery and design of new drug candidates from plant natural product leads. *J. Nat. Prod.*, **2004**, 67: 273–283.

12. Vlietinck, A. J.; De Bruyne, T.; Apers, S.; Pieters, L. A. Plant-derived leading compounds for chemotherapy of human immunodeficiency virus (HIV) infection. *Planta Med.*, **1998**, 64: 97–109.

13. Hudson, J.; Towers, G. H. N. Phytomedicines as antivirals. *Drugs Future*, **1999**, 24: 295–320.

14. Yang, S. S.; Cragg, G. M.; Newman, D. J.; Bader, J. P. Natural product-based anti-HIV drug discovery and development facilitated by the NCI developmental therapeutics program. *J. Nat. Prod.*, **2001**, 64: 265–277.

15. Nishino, H.; Okuyama, T.; Takata, M.; Shibata, S.; Tokuda, H.; Takayasu, J.; Hasegawa, T.; Nishino, A.; Ueyama, H.; Iwashima, A. Studies on the anti-tumor-promoting activity of naturally occurring substances IV. Pd-II [(+)anomalin, (+)praeruptorin B], a seselin-type coumarin, inhibits the promotion of skin tumor formation by 12-O-tetradecanoyl-phorbol-13-acetate in 7,12-dimethylbenz[a]anthracene-initiated mice. *Carcinogenesis (London)*, **1990**, 11: 1557–1561.

16. Tsai, I. L.; Wun, M. F.; Teng, C. M.; Ishikawa, T.; Chen, I. S. Anti-platelet aggregation constituents from Formosan Toddalia asiatica. *Phytochem.*, **1998**, 48: 1377–1382.

17. Lee, T. T.; Kashiwada, Y.; Huang, L.; Snider, J.; Cosentino, M.; Lee. K. H. Suksdorfin: an anti-HIV principle from *Lomatium suksdorfii*, its structure-activity correlation with related coumarins, and synergistic effects with anti-AIDS nucleosides. *Bioorg. Med. Chem.*, **1994**, 2: 1051–1056.

18. Cannon, J. G. In Wolff, M. E.; editors. *Burger's Medicinal Chemistry and Drug Discovery*, vol. 5. New York: Wiley, **1994**, 783.

19. Cannon, J. G.; Mohan, P.; Bojarski, J.; Long, J. P.; Bhatnagar, R. K.; Leonard, P. A.; Flynn, J. R.; Chatterjee, T. K. (*R*)-(−)-10-Methyl-11-hydroxyaporphine-a highly selective serotonergic agonist. *J. Med. Chem.*, **1988**, 31: 313–318.

20. Wikstroem, H.; Sanchez, D.; Lindberg, P.; Hacksell, U.; Erik Arvidsson, L.; Johnsson, A. M.; Thorberg, S. O.; Nilsson, J. L. G.; Svensson, K. Resolved 3-(3-hydroxyphenyl)-*N-n*-propylpiperidine and its analogs: Central dopamine receptor activity. *J. Med. Chem.* **1984**, 27: 1030-1036.

21. Huang, L.; Kashiwada, Y.; Cosentino, L. M.; Fan, S.; Lee, K. H. 3′,4′-Di-*O*-(−)-camphanoyl-(+)-*cis*-khellactone and related compounds: A new class of potent anti-HIV agents. *Bioorg. Med. Chem. Lett.*, **1994**, 4: 593–598.

22. Huang, L.; Kashiwada, Y.; Cosentino, L. M.; Fan, S.; Chen, C. H.; McPhail, A. T.; Fujioka, T.; Mihashi, K.; Lee, K. H. Anti-AIDS agents 15. Synthesis and anti-HIV activity of dihydroseselins and related analogs. *J. Med. Chem.*, **1994**, 37: 3947–3955 .

23. Tripos Associates. (**2004**). Sybyl (version 6.0). Louis, St. MO.

24. Xie, L.; Takeuchi, Y.; Cosentino, L. M.; McPhail, A. T.; Lee, K. H. Anti-AIDS agents 42. Synthesis and anti-HIV activity of disubstituted $(3'R,4'R)$-3',4'-di-$O$-($S$)-camphanoyl-(+)-$cis$-khellactone analogues. *J. Med. Chem.*, **2001**, 44: 664–671.

25. Xie, L.; Takeuchi, Y.; Cosentino, L. M.; Lee, K. H. Anti-AIDS agents 37. Synthesis and structure-activity relationships of $(3'R,4'R)$-(+)-$cis$-khellactone derivatives as novel potent anti-HIV agents. *J. Med. Chem.*, **1999**, 42: 2662–2672.

26. Beller, M.; Sharpless, K. B. Diols via catalytic dihydroxylation. In Cornils, B.; Herrmann, W. A. editors. *Applied Homogeneous Catalysis with Organometallic Compounds, vol. 3, 2nd Edition.* Weinheim, Germany: Wiley-VCH, **2002**, 1149–1164.

27. Mehltretter, G. M.; Dobler, C.; Sundermeier, U.; Beller, M. An improved version of the Sharpless asymmetric dihydroxylation. *Tetrahedron Lett.*, **2000**, 41: 8083–8087.

28. Kolb, H. C.; VanNieuwenhze, M. S.; Sharpless, K. B. Catalytic asymmetric dihydroxylation. *Chem. Rev.*, **1994**, 94: 2483–2547.

29. Becker, H.; Sharpless, K. B. A new ligand class for the asymmetric dihydroxylation of olefins. *Angew. Chem., Int. Ed. Engl*, **1996**, 35: 448–451.

30. Xie, L.; Crimmins, M. T.; Lee, K. H. Anti-AIDS agents 22. Asymmetric synthesis of 3',4'-di-$O$-(−)-camphanoyl-(+)-$cis$-khellactone (DCK), a potent anti-HIV agent. *Tetrahedron Lett.*, **1995**, 36: 4529–4532.

31. Xia, Y.; Yang, Z. Y.; Brossi, A.; Lee, K. H. Asymmetric solid-phase synthesis of $(3'R,4'R)$-di-$O$-$cis$-acyl 3-carboxyl khellactones. *Org. Lett.*, **1999**, 1: 2113–2115.

32. Takeuchi, Y.; Xie, L.; Cosentino, L. M.; Lee, K. H. Anti-AIDS agents 28. Synthesis and anti-HIV activity of methoxy substituted 3',4'-di-$O$-(−)-camphanoyl-(+)-$cis$-khellactone (DCK) analogs. *Bioorg. Med. Chem, Lett.*, **1997**, 7: 2573–2578.

33. Xie, L.; Takeuchi, Y.; Cosentino, L. M.; Lee, K. H. Anti-AIDS agents 33. Synthesis and anti-HIV activity of mono-methyl substituted 3',4'-di-$O$-(−)-camphanoyl-(+)-$cis$-khellactone (DCK) analogs. *Bioorg. Med. Chem, Lett.*, **1998**, 8: 2151–2156.

34. Thornber, C. W. Isosterism and molecular modification in drug design. *Chem. Soc. Rev.*, **1979**, 8: 563–580.

35. Krogsgaard-Larsen, P.; Hjeds, H.; Curtis, D. R.; Lodge, D.; Johnston, G. A. R. Dihydromuscimol, thiomuscimol and related heterocyclic compounds as GABA analogs. *J. Neurochem.*, **1979**, 32: 1717–1724.

36. Yang, Z. Y.; Xia, Y.; Xia, P.; Cosentino, L. M.; Lee, K. H. Anti-AIDS agents 31. Synthesis and anti-HIV activity of 4-substituted 3',4'-di-$O$-(−)-camphanoyl-(+)-$cis$-khellactone (DCK) thiolactone analogs. *Bioorg. Med. Chem, Lett.*, **1998**, 8: 1483–1486.

37. Cava, M. P.; Levinson, M. I. Thionation reactions of Lawesson's reagents. *Tetrahedron*, **1985**, 41: 5061–5087.

38. Yang, Z. Y.; Xia, Y.; Xia, P.; Brossi, A.; Cosentino, L. M.; Lee, K. H. Anti-AIDS agents 41. Synthesis and anti-HIV activity of 3',4'-di-$O$-(−)-camphanoyl-(+)-$cis$-khellactone (DCK) lactam analogues. *Bioorg. Med. Chem. Lett.*, **2000**, 10: 1003–1005.

39. Xie, L.; Yu, D.; Wild, C.; Allaway, G.; Turpin, J.; Smith, P. C.; Lee, K. H. Anti-AIDS agents 52. Synthesis and anti-HIV activity of hydroxymethyl $(3'R,4'R)$-3',4'-di-$O$-($S$)-camphanoyl-(+)-$cis$-khellactone derivatives. *J. Med. Chem.*, **2004**, 47: 756–760.

40. Xie, L.; Allaway, G.; Wild, C.; Kilgore, N.; Lee, K. H. Anti-AIDS agents 47. Synthesis and anti-HIV activity of 3-substituted 3',4'-di-$O$-($S$)-camphanoyl-$(3'R,4'R)$-(+)-$cis$-khellactone derivatives. *Bioorg. Med. Chem. Lett.*, **2001**, 11: 2291–2293.

41. Larder, B. A.; Kellam, P.; Kemp, S. D. Convergent combination therapy can select viable multidrug-resistant HIV-1 in vitro. *Nature*, **1993**, 365: 451–453.

42. Yu, D.; Brossi, A.; Kilgore, N.; Wild, C.; Allaway, G.; Lee, K. H. Anti-HIV agents 55. 3′R,4′R-di-O-(−)-camphanoyl-2′,2′-dimethyldihydropyrano[2,3-f]chromone (DCP), a novel anti-HIV agent. *Bioorg. Med. Chem. Lett.*, **2003**, 13: 1575–1576.

43. Yu, D.; Chen, C. H.; Brossi, A.; Lee, K. H. Anti-AIDS agents 60. Substituted 3′R,4′R-di-O-(−)-camphanoyl-2′,2′-dimethyldihydropyrano[2,3-f]chromone (DCP) analogs as potent anti-HIV agents. *J. Med. Chem.*, **2004**, 47: 4072–4082.

44. Xie. L. Design, synthesis, and biological evaluation of (+)-cis-khellactone derivatives as potent anti-HIV agents. Ph.D. Dissertation, **1999**, 124–147.

45. Yu, D.; Brossi, A.; Wild, C.; Allaway, G.; Lee, K. H. Design and synthesis of DCK-related compounds as potent anti-HIV agents. 44th Annual Meeting of the American Society of Pharmacognosy, July 12–16, **2003**, Chapel Hill, NC.

46. Yu, D.; Suzuki, M.; Xie, L.; Morris-Natschke, S. L.; Lee, K. H. Recent progress in the development of coumarin derivatives as potent anti-HIV agents. *Med. Res. Rev.*, **2003**, 23: 322–345.

47. Jones, B. C. N. M.; McGuigan, C.; O'Connor, T. J.; Jeffries, D. J.; Kinchington, D. Synthesis and anti-HIV activity of some novel phosphorodiamidate derivatives of 3′-azido-3′-deoxythymidine (AZT). *Antiviral Chem. Chemother.*, **1991**, 2: 35–39.

48. Grob, P. M.; Wu, J. C.; Cohen, K. A.; Ingraham, R. H.; Shih, C. K.; Hargrave, K. D.; Mctague, T. L.; Merluzzi, V. J. Nonnucleoside inhibitors of HIV-1 reverse transcriptase: nevirapine as a prototype drug. *AIDS Res. Hum. Retroviruses*, **1992**, 8: 145–152.

49. Chen, D. F.; Zhang, S. X.; Chen, K.; Zhou, B. N.; Wang, P.; Cosentino, L. M.; Lee, K. H. Two new lignans, interiotherins A and B, as anti-HIV principles from Kadsura interior. *J. Nat. Prod.*, **1996**, 59: 1066–1068.

50. Chen, D. F.; Zhang, S. X.; Xie, L.; Xie, J. X.; Chen, K.; Kashiwada, Y.; Zhou, B. N.; Wang, P.; Cosentino, L. M.; Lee, K. H. Anti-AIDS agents 26. Structure-activity correlations of gomisin-G-related anti-HIV lignans from Kadsura interior and of related synthetic analogs. *Bioorg. Med. Chem.*, **1997**, 5: 1715–1723.

51. Lee, K. H.; Morris-Natschke, S. L. Recent advances in the discovery and development of plant-derived natural products and their analogs as anti-HIV agents. *Pure Appl. Chem.*, **1999**, 71: 1045–1051.

52. Lee, K. H.; Zhang, S. X.; Morris-Natschke, S. L. The chemistry and pharmacology of "Wu Wei Zi" (*Schisandra chinensis* and *Kadsura Japonica*). *China Med. Coll. J.*, **1998**, 7: 41–78.

53. Xie, L.; Xie, J.; Kashiwada, Y.; Cosentino, L. M.; Liu, S. H.; Pai, R. B.; Cheng, Y. C.; Lee, K. H. Anti-AIDS agents 17. New brominated hexahydroxybiphenyl derivatives as potent anti-HIV agents. *J. Med. Chem.*, **1995**, 38: 3003–3008.

54. Nonaka, G.; Nishioka, I.; Nishizawa, M.; Yamagishi, T.; Kashiwada, Y.; Dutschman, G. E.; Bodner, A. J.; Kilkuskie, R. E.; Cheng, Y. C.; Lee, K. H. Anti-AIDS agents 2. Inhibitory effects of tannins on HIV reverse transcriptase and HIV replication in H9 lymphocyte cells. *J. Nat. Prod.*, **1990**, 53: 587–595.

55. Kilkuskie, R. E.; Kashiwada, Y.; Nonaka, G.; Nishioka, I.; Bodner, A. J.; Cheng, Y. C.; Lee, K. H. Anti-AIDS agents 8. HIV and reverse transcriptase inhibition by tannins. *Bioorg. Med. Chem. Lett.*, **1992**, 2: 1529–1534.

56. Lee, K. H.; Kashiwada, Y.; Nonaka, G.; Nishioka, I.; Nishizawa, M.; Yamagishi, T.; Bodner, A. J.; Kilkuskie, R. E.; Cheng, Y. C. Anti-AIDS agents 7. Tannins and related compounds as anti-HIV agents. In Chu, C. K.; Cutler, H. G. editors. *Natural Products as Antiviral Agents*. New York: Amer. Chem. Soc. Symposium Series, Plenum Press, **1992**, 69–90.

57. Kashiwada, Y.; Huang, L.; Kilkuskie, R. E.; Bodner, A. J.; Lee, K. H. Anti-AIDS agents 5. New hexahydroxydiphenyl derivatives as potent inhibitors of HIV replication in H9 lymphocytes. *Bioorg. Med. Chem. Lett.*, **1992**, 2: 235–238.

58. Xie, J.; Zhou, X. J.; Zhang, C. Z.; Yang, J. H.; Jin, H. Q.; Chen, J. Synthesis of schizandrin C analogs. II. Synthesis of dimethyl-4, 4'-dimethoxy-5,6,5',6'-dimethylenedioxybiphenyl-2,2'-dicarboxylate and its isomers. *Acta Pharm. Sinica*, **1982**, 17: 23–27.

59. Tramontano, E.; Cheng, Y. C. HIV-1 reverse transcriptase inhibition by a dipyridodiazepinone derivative: BI-RG-587. *Biochem. Pharmacol.*, **1992**, 43: 1371–1376.

60. Connolly, J. D.; Hill, R. A. Triterpenoids. *Nat. Prod. Rep.*, **2001**, 18: 131–147.

61. Connolly, J. D.; Hill, R. A. Triterpenoids. *Nat. Prod. Rep.*, **2001**, 18: 560–578.

62. Huang, L.; Chen, C. H. Molecular targets of anti-HIV-1 triterpenes. *Curr. Drug Targets: Infect. Disorders*, **2002**, 2: 33–36.

63. Sasaki, H.; Takei, M.; Kobayashi, M.; Pollard, R. B.; Suzuki, F. Effect of glycyrrhizin, an active component of licorice roots, on HIV replication in cultures of peripheral blood mononuclear cells from HIV-seropositive patients. *Pathobiol.*, **2003**, 70: 229–236.

64. Hirabayashi, K.; Iwata, S.; Matsumoto, H.; Mori, T.; Shibata, S.; Baba, M.; Ito, M.; Shigeta, S.; Nakashima, H.; Yamamoto, N. Antiviral activities of glycyrrhizin and its modified compounds against human immunodeficiency virus type 1 (HIV-1) and herpes simplex virus type 1 (HSV-1) in vitro. *Chem. Pharm. Bull.*, **1991**, 39: 112–115.

65. Soler, F.; Poujade, C.; Evers, M.; Carry, J. C.; Henin, Y.; Bousseau, A.; Huet, T.; Pauwels, R.; De Clercq, E. Betulinic acid derivatives: a new class of specific inhibitors of human immunodeficiency virus type 1 entry. *J. Med. Chem*, **1996**, 39: 1069–1083.

66. Sahu, N. P. Triterpenoid saponins of Mimusops elengi. *Phytochem.*, **1996**, 41: 883–886.

67. El Mekkawy, S.; Meselhy, M. R.; Nakamura, N.; Tezuka, Y.; Hattori, M.; Kakiuchi, N.; Shimotohno, K.; Kawahata, T.; Otake, T. Anti-HIV-1 and anti-HIV-1-protease substances from *Ganoderma lucidum*. *Phytochem.*, **1998**, 49: 1651–1657.

68. Xu, H. X.; Ming, D. S.; Dong, H.; But, P. P.-H. A new anti-HIV triterpene from *Geum japonicum*. *Chem. Pharm. Bull.*, **2000**, 48: 1367–1369.

69. Kashiwada, Y.; Hashimoto, F.; Cosentino, L. M.; Chen, C. H.; Garrett, P. E.; Lee, K. H. Betulinic acid and dihydrobetulinic acid derivatives as potent anti-HIV agents. *J. Med. Chem.*, **1996**, 39: 1016–1017.

70. Fujioka, T.; Kashiwada, Y.; Kilkuskie, R. E.; Cosentino, L. M.; Ballas, L. M.; Jiang, J. B.; Janzen, W. P.; Chen, I. S.; Lee, K. H. Anti-AIDS agents 11. Betulinic acid and platanic acid as anti-HIV principles from *Syzigium claviflorum*, and the anti-HIV activity of structurally related triterpenoids. *J. Nat. Prod.*, **1994**, 57: 243–247.

71. Ramadoss, S.; Jaggi, M.; Siddiqui, M. J. A.; Khanna, A. B. Betulinic acid derivatives for inhibiting cancer growth and process for the manufacture of betulinic acid. (Dabur Research Foundation). U.S. Patent 6,667,345, November 2, **1999**.

72. Pezzuto, J. M.; Kosmeder, II, J. W.; Xu, Z. Q.; Zhou, N. E. Synthesis of prodrugs of betulinic acid derivatives for treating cancers and HIV (The Board of Trustees of the

University of Illinois, and Advanced Life Sciences, Inc.). PCT Int. Appl. WO 0216395, February 28, **2002**.

73. Lee, K. H.; Kashiwada, Y.; Hashimoto, F.; Cosentino, L. M.; Manak, M. Betulinic acid derivatives and antiviral use (University of North Carolina, at Chapel Hill and Biotech, Research Laboratories). PCT Int. Appl. WO 9639033, December 12, **1996**.

74. Bradbury, J. B.; Schafer, S. J.; Kaczvinsky, J. R. Jr.; Bailey, D.; Gale, C. D. Method for regulating hair growth (The Procter & Gamble Company). PCT Int. Appl. WO0003749, January 27, **2000**.

75. Gardlik, J. M.; Severynse-Stevens, D.; Comstock, B. G. Compositions useful for regulating hair growth containing metal complexes of oxidized carbohydrates. (The Procter & Gamble Company) PCT Int. Appl. WO 2002007700, January 31, **2002**.

76. Cho, S. H.; Gottlieb, K.; Santhanam, U. Cosmetic compositions containing betulinic acid (Unilever Plc, U. K. and Unilever, N. V.). Eur. Pat. Appl. EP 071798395, June 6, **1996**.

77. Evers, M.; Poujade, C.; Soler, F.; Ribeill, Y.; James, C.; Lelievre, Y.; Gueguen, J. C.; Reisdorf, D.; Morize, I. Betulinic acid derivatives: a new class of human immunodeficiency virus type 1 specific inhibitors with a new mode of action. *J. Med. Chem.*, **1996**, 39: 1056–1068.

78. Sun, I. C.; Chen, C. H.; Kashiwada, Y.; Wu, J. H.; Wang, H. K.; Lee, K. H. Anti-AIDS agents 49. Synthesis, anti-HIV, and anti-fusion activities of IC9564 analogues based on betulinic acid. *J. Med. Chem.*, **2002**, 45: 4271–4275.

79. Mayaux, J. F.; Ousseau, A.; Pauwels, R.; Huet, T.; Henin, Y.; Dereu, N.; Evers, M.; Soler, F.; Poujade, C. Triterpene derivatives that block entry of human immunodeficiency virus type 1 into cells. *Proc. Natl. Acad. Sci. USA*, **1994**, 91: 3564–3568.

80. Labrosse, B.; Pleskoff, O.; Sol, N.; Jones, C.; Henin, Y.; Alizon, M. Resistance to a drug blocking human immunodeficiency virus type 1 entry (RPR103611) is conferred by mutations in gp41. *J. Virol.*, **1997**, 71: 8230–8236.

81. Labrosse, B.; Treboute, C.; Alizon, M. Sensitivity to a nonpeptidic compound (RPR103611) blocking human immunodeficiency virus type 1 Env-mediated fusion depends on sequence and accessibility of the gp41 loop region. *J. Virol.*, **2000**, 74: 2142–2150.

82. Kwong, P. D.; Wyatt, R.; Robinson, J.; Sweet, R. W.; Sodroski, J.; Hendrickson, W. A. Structure of an HIV gp120 envelope glycoprotein in complex with the CD4 receptor and a neutralizing human antibody. *Nature*, **1998**, 393: 648–659.

83. Holz-Smith, S. L.; Sun, I. C.; Jin, L.; Matthews, T. J.; Lee, K. H.; Chen, C. H. Role of human immunodeficiency virus (HIV) type 1 envelope in the anti-HIV activity of the betulinic acid derivative IC9564. *Antimicrob. Agents Chemother.*, **2001**, 45: 60–66.

84. Zhu, C. B.; Zhu, L.; Holz-Smith, S.; Matthews, T. J.; Chen, C. H. The role of the third β strand in gp120 conformation and neutralization sensitivity of the HIV-1 primary isolate DH012. *Proc. Natl. Acad. Sci. USA*, **2001**, 98: 15227–15232.

85. Hashimoto, F.; Kashiwada, Y.; Cosentino, L. M.; Chen, C. H.; Garrett, P. E.; Lee, K. H. Anti-AIDS agents 27. Synthesis and anti-HIV activity of betulinic acid and dihydrobetulinic acid derivatives. *Bioorg. Med. Chem.*, **1997**, 5: 2133–2143.

86. Kashiwada, Y.; Chiyo, J.; Ikeshiro, Y.; Nagao, T.; Okabe, H.; Cosentino, L. M.; Fowke, K.; Morris-Natschke, S. L.; Lee, K. H. Synthesis and anti-HIV activity of 3-alkylamido-3-deoxy-betulinic acid derivatives. *Chem. Pharm. Bull.*, **2000**, 48: 1387–1390.

87. Kraeusslich, H. G.; Paecke, M.; Heuser, A. M.; Konvalinka, J.; Zentgraf, H. The spacer peptide between human immunodeficiency virus capsid and nucleocapsid proteins is essential for ordered assembly and viral infectivity. *J. Virol.*, **1995**, 69: 3407–3419.

88. Vogt, V. M. Proteolytic processing and particle maturation. *Curr. Topics Microbiol. Immunol.*, **1996**, 214: 95–131.

89. Accola, M. A.; Hoglund, S.; Gottlinger, H. G. A putative a-helical structure which overlaps the capsid-p2 boundary in the human immunodeficiency virus type 1 Gag precursor is crucial for viral particle assembly. *J. Virol.*, **1998**, 72: 2072–2078.

90. Ashorn, P.; McQuade, T. J.; Thaisrivongs, S.; Tomasselli, A. G.; Tarpley, W. G.; Moss, B. An inhibitor of the protease blocks maturation of human and simian immunodeficiency viruses and spread of infection. *Proc. Natl. Acad. Sci. USA*, **1990**, 87: 7472–7476.

91. Wiegers, K.; Rutter, G.; Kottler, H.; Tessmer, U.; Hohenberg, H.; Krausslich, H. G. Sequential steps in human immunodeficiency virus particle maturation revealed by alterations of individual Gag polyprotein cleavage sites. *J. Virol.*, **1998**, 72: 2846–2854.

92. Li, F.; Goila-gaur, R.; Salzwedel, K.; Kilgore, N. R.; Reddick, M.; Matallana, C.; Castillo, A.; Zoumplis, D.; Martin, D. E.; Orenstein, J. M.; Allaway, G. P.; Freed, E. O.; Wild, C. T. PA-457: a potent HIV inhibitor that disrupts core condensation by targeting a late step in Gag processing. *Proc. Natl. Acad. Sci. USA*, **2003**, 100: 13555–13560.

93. Zhou, J.; Yuan, X.; Dismuke, D.; Forshey, B. M.; Lundquist, C.; Lee, K. H.; Aiken, C.; Chen, C. H. Small-molecule inhibition of human immunodeficiency virus type 1 replication by specific targeting of the final step of virion maturation. *J. Virol.*, **2004**, 78: 922–929.

94. Zhou, J.; Chen, CH.; Aiken, C. The sequence of the CA-SP1 junction accounts for the differential sensitivity of HIV-1 and SIV to the small molecule maturation inhibitor 3-O-{3′,3′-dimethylsuccinyl}-betulinic acid. *Retrovirology*, **2004**, 1: 15.

95. Huang, L.; Yuan, X.; Aiken, C.; Chen, C. H. Bifunctional anti-human immunodeficiency virus type 1 small molecules with two novel mechanisms of action. *Antimicrob. Agents Chemother.*, **2004**, 48: 663–665.

# 10

# RECENT PROGRESS ON THE CHEMICAL SYNTHESIS OF ANNONACEOUS ACETOGENINS AND THEIR STRUCTURALLY MODIFIED MIMICS

TAI-SHAN HU, YU-LIN WU, AND ZHU-JUN YAO
State Key Laboratory of Bioorganic and Natural Products Chemistry,
Shanghai Institute of Organic Chemistry, Chinese Academy of Sciences, Shanghai, China

## 10.1 INTRODUCTION

Annonaceous acetogenins are a large family of natural polyketides, whose number now reaches 400. In the past 15 years, it was recognized as an attractive and important area of phytochemistry and synthetic chemistry. For example, the *Journal of Natural Products* chose acetogenin as its cover for the first six issues in 1999. The first acetogenin, uvaricin, was isolated and characterized in 1982 by Jolad et al.,[1] from the plant *Uvaria acuminate*. Most annonaceous acetogenins have been reported to exhibit a variety of bioactivities, including anticancer activities and cytotoxicities against various cancer cells. A few review articles[2–12] were published in key peer-reviewed journals in recent years, covering studies on the phytochemistry, structural elucidation, bioactivity, and mechanism. Herein, we will conduct an overview on the progress of the total syntheses and related studies of the naturally occurring acetogenins since 1998.

So far, annonaceous acetogenins have only been found in various species of Annonaceae. Typical acetogenins present several unique structural characteristics

*Medicinal Chemistry of Bioactive Natural Products* Edited by Xiao-Tian Liang and Wei-Shuo Fang
Copyright © 2006 John Wiley & Sons, Inc.

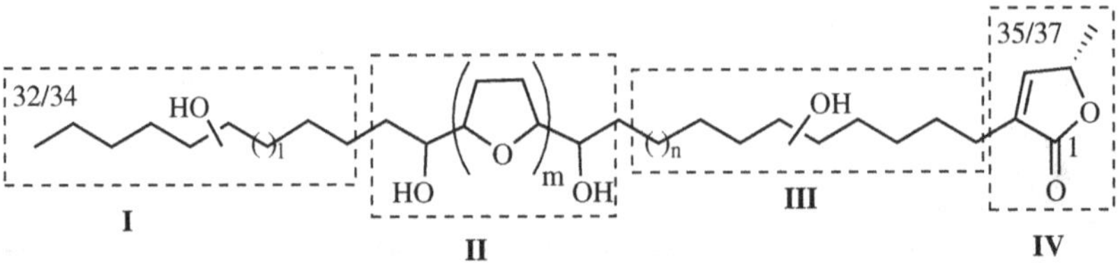

**Scheme 10-1.** General structure of annonaceous acetogenin.

(Scheme 10-1): (1) They contain one, two, or three 2,5-disubstituted THF ring(s) in the middle region of molecules (Scheme 10-1, II); (2) they have a 4-methyl γ-lactone (Scheme 10-1, IV) at one terminal of their linear structures; (3) they often contain several oxygenated functionalities along the hydrocarbon chains (Scheme 10-1, I and III), such as ketone, acetoxyl, and/or hydroxyl groups, as well as the double bonds. For varying positions of these functionalities and stereochemistries, the acetogenins present great molecular diversity and structural similarity. According to the statues of THF ring(s) in the molecules, six subclasses can be categorized: (1) *mono*-THF containing acetogenins, (2) adjacent *bis*-THF containing acetogenins, (3) nonadjacent *bis*-THF containing acetogenins, (4) THP-containing acetogenins, (5) tri-THF containing acetogenins, and (6) non-THF-containing acetogenins.

In addition, a broaden spectrum of activities caused by annonaceous acetogenins have been reported, such as anticancer, immunosuppressant, antimalarial, and pesticidal. Among these activities the most important is that many acetogenins present good-to-excellent toxicity against various tumor cells. For example, bullitacin, a bis-THF acetogenin, shows toxicity against the KB cell with $IC_{50}$ in the $10^{-12}$-μg/mL range.[13] Several reports have also appeared in the literatures to explain the mechanism of anticancer activities caused by acetogenins. The generally accepted mode of action is that the acetogenins inhibit the function of the complex I (NADH:ubiquinone redoxidase)[14–16] in mitochondria as well as the NADH oxidase[17] in the cancer cells. Those results indicate that the acetogenins decrease the biosynthesis level of ATP and lower cell proliferation, especially for those cancer cells with a high level of energy metabolism, and finally they lead to the death of cancer cells. Furthermore, some evidence shows that the acetogenins can inhibit the growth of some multidrug-resistance cancer cells.[18]

Although acetogenins were found in many existing species of plant annonaceae, their natural containments are rare and many stereoisomers sharing similar structures often exist together. The latter makes it difficult to obtain a pure sample of acetogenins [usually by high-performance liquid chromatography (HPLC)]. In addition, most acetogenins cannot be measured by x-ray crystallography because of their nature as a waxy solid. Therefore, the classic mission of total synthesis is becoming more important. It is not only a practical way to obtain an adequate amount of samples for additional biological investigations, but also it provides solid evidence to confirm the ambiguous absolute stereochemistries.

Considering the structural features of acetogenins, two major synthetic emphases are commonly concerned: One method is to establish the THF ring(s)

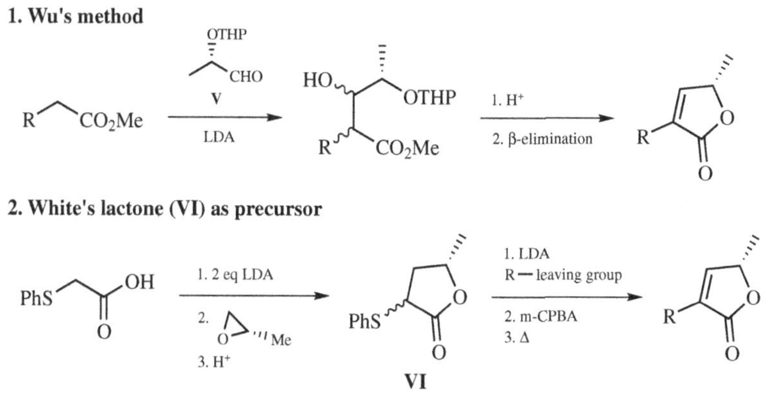

**1. Wu's method**

**2. White's lactone (VI) as precursor**

**Scheme 10-2.** Two popular methodologies for construction of the terminal lactone moiety.

with desired configurations, and the other method is to elaborate the terminal γ-lactone in a convenient way. For construction of 2,5-substituted THF ring(s), most existing protocols are either based on polyhydroxyl-containing starting materials or using proper oxidations [Sharpless asymmetric epoxidation (AE) or dihydroxylation (AD)] on the double bond(s)-containing precursor, followed by using Williamson etherification or intramolecular epoxide opening to form desired THF ring(s). For construction of the terminal lactone moiety, several methods have been adapted.[19–27] Among them, two protocols (Scheme 10-2) were frequently used: One protocol is the aldol strategy between an ester precursor and an O-protected lactal and subsequent lactonization, which was developed by us;[20,21] the other protocol uses β-phenylsulfurylactone as the unsaturated lactone precursor, which was originally developed by White et al.[27]

Because the acetogenins contain many stereogenic centers, and present great diversity and complexity, their total synthesis is still a challenging task to the organic chemist today. The following sections will introduce some representative total syntheses from recent years according to the order of different classes of acetogenins.

## 10.2 TOTAL SYNTHESIS OF MONO-THF ACETOGENINS

As one of the earliest groups involved into the total synthesis of acetogenins, we finished the total syntheses of 16,19,20,34-*epi*-corossolin, (10*RS*)-corossolin and corossolone in 1994 and 1995 (Scheme 10-3), respectively.[20,21] The aldol reaction-based lactonization strategy and the segment-coupling through epoxide-opening by terminal acetylene were developed in those syntheses, and both methods are now frequently used.

On basis of this work, we recently completed the total synthesis of five mono-THF acetogenins (Scheme 10-4), 4-deoxyannomonatacin,[28] tonkinecin,[29,30] (10*R*)- and (10*S*)-corossolin,[31] and annonacin.[32] These five acetogenins have a common *threo-trans-threo*-2,5-bishydroxymethyl substituted THF ring, whereas the length

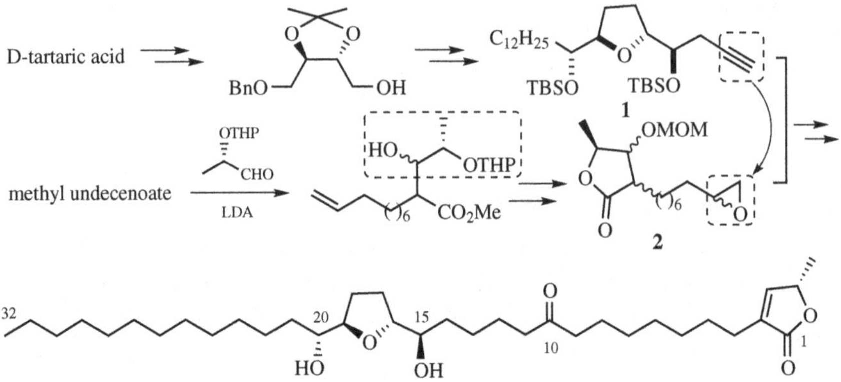

**Scheme 10-3.** Total synthesis of corossolone by Wu and Yao.

of hydrocarbon chains between the lactone unit and the THF ring, as well as the number, stereochemistries, and positions of hydroxyl groups along the hydrocarbon chains, are different from each other. It is noteworthy that these structural differences affect the bioactivities greatly in some cases. These synthetic studies not only provided us opportunities for biological investigations, but also they provided us solid evidence to confirm the absolute stereochemistry of those natural products, as well as some important structure activity relationship (SAR) information of naturally occurring acetogenins.

In those syntheses, some unambiguous stereogenic centers were introduced by chiron approaches, starting from inexpensive sugars and tartaric acid, and others were established by asymmetric reactions. For example, D-glucose was chosen as the starting material for 4-deoxyannomonatacin (Scheme 10-5).[28] The C-4 and C-5 of D-glucose were incorporated into the target as C-17 and C-18, and a Sharpless dihydroxylation furnished the stereochemistries of C-21 and C-22 in the target. Those steps finally established the key intermediate, vinyl iodide **6**. In the meantime, the other segment **9** was prepared through Jacobsen's hydrolytic kinetic resolution and subsequent epoxide opening reaction. Sonogashira coupling of both

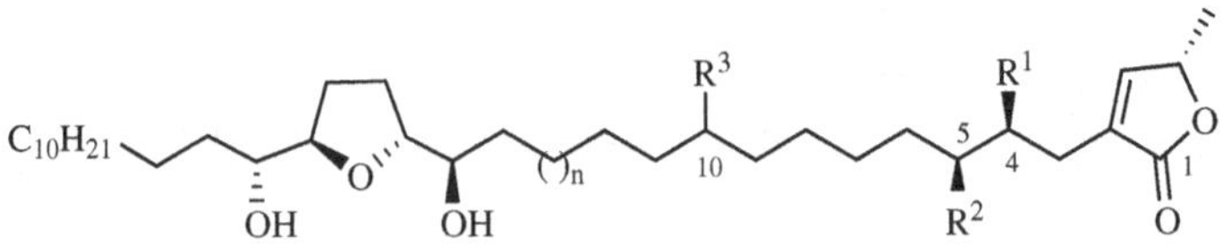

annonacin $R^1$ = OH, $R^2$ = H, $R^3$ = OH, n = 1
(10 *R*)-corossolin $R^1$ = H, $R^2$ = H, $R^3$ = OH, n = 1
(10 *S*)-corossolin $R^1$ = H, $R^2$ = H, $R^3$ = OH, n = 1
4-deoxyannomontacin $R^1$ = H, $R^2$ = H, $R^3$ = OH, n = 3
tonkinecin $R^1$ = H, $R^2$ = OH, $R^3$ = H, n = 3

**Scheme 10-4.** Five mono-THF acetogenins sharing common structural features.

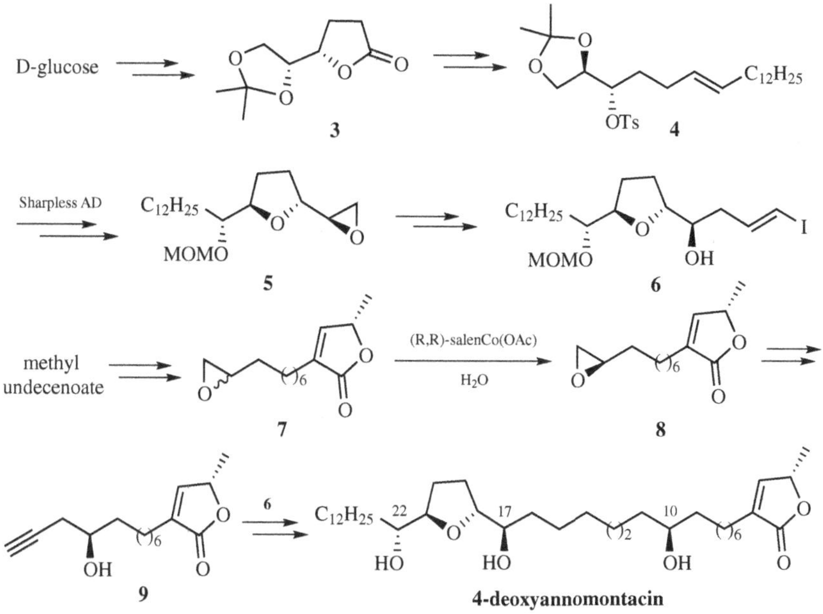

**Scheme 10-5.** Total synthesis of 4-deoxyannomonatacin by Wu et al.

segments was accomplished to afford the whole skeleton. The target was finally afforded after diimide-based selective reductions of double and triple bonds and deprotection of methoxymethyl (MOM) ethers.

Tonkinecin (Scheme 10-6) was an example of rare mono-THF acetogenins bearing a C-5 hydroxyl group. Starting from D-xylose, the ester **10** was prepared in

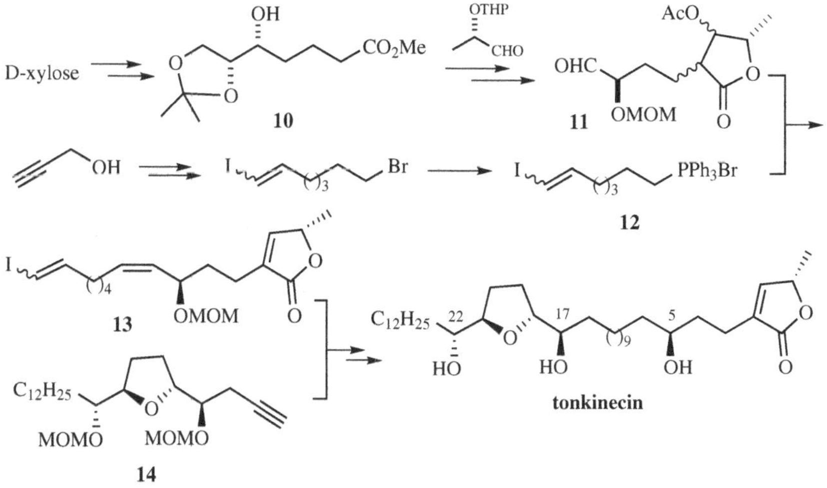

**Scheme 10-6.** Total synthesis of tonkinecin by Wu et al.

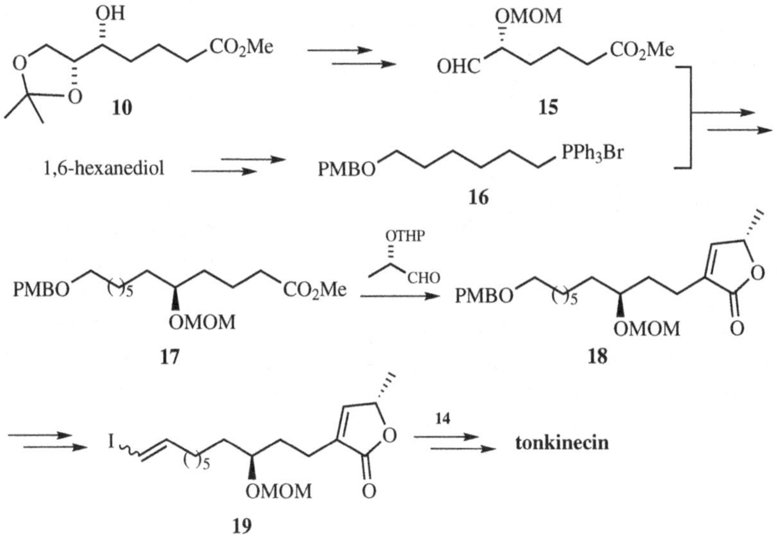

**Scheme 10-7.** Improved synthesis of tokinecin by Wu et al.

three steps. After hydroxyl group protection, an aldol condensation with lactal derivative and in situ lactonization and oxidative cleavage of terminal diol gave the aldehyde **11**. Wittig reaction of **11** with a ylide derived from **12** afforded vinyl iodide **13**. Sonogashira coupling of **13** with THF containing segment **14** established the whole skeleton, which was finally converted into tonkinecin after selective reductions and deprotections.[29] One advantage of using ester **10** is that this intermediate can be easily prepared on a large scale; it is of great value as a common building block for the synthesis of other acetogenins.

We later published an improved synthesis of tonkinecin[30] because of the low yield of the coupling reaction between **11** and **12** (only 30% previously) in a previous study. One major reason is that the vinyl iodide **12** was not stable enough to survive the basic reaction conditions. An alternative for the synthesis of right-side segment was then designed, in which the vinyl iodide functionality was introduced in the last step to avoid those basic conditions (Scheme 10-7). The Wittig reaction of aldehyde **15** with **16** and subsequent hydrogenation afforded ester **17**, which was converted to lactone **18** by aldol reaction with lactal followed by acid treatment and β-elimination. Selective deprotection, Dess–Martin oxidation, and Takai reaction (to introduce the vinyl iodide) afforded the precursor **19**. Similar treatments of **19** with **14**, as well as subsequent intermediates as described previously, finally provided tonkinecin.

Corossolin was isolated by a French group in 1991, and the absolute configuration of its C-10 hydroxyl group remained unknown until its total synthesis was achieved by us in 1999 (Scheme 10-8).[31] The first total synthesis of corossolin was achieved in 1996 by Makabe et al. in Japan.[34] However, the specific rotations and nuclear magnetic resonances (NMRs) of their two synthetic C-10 epimers are

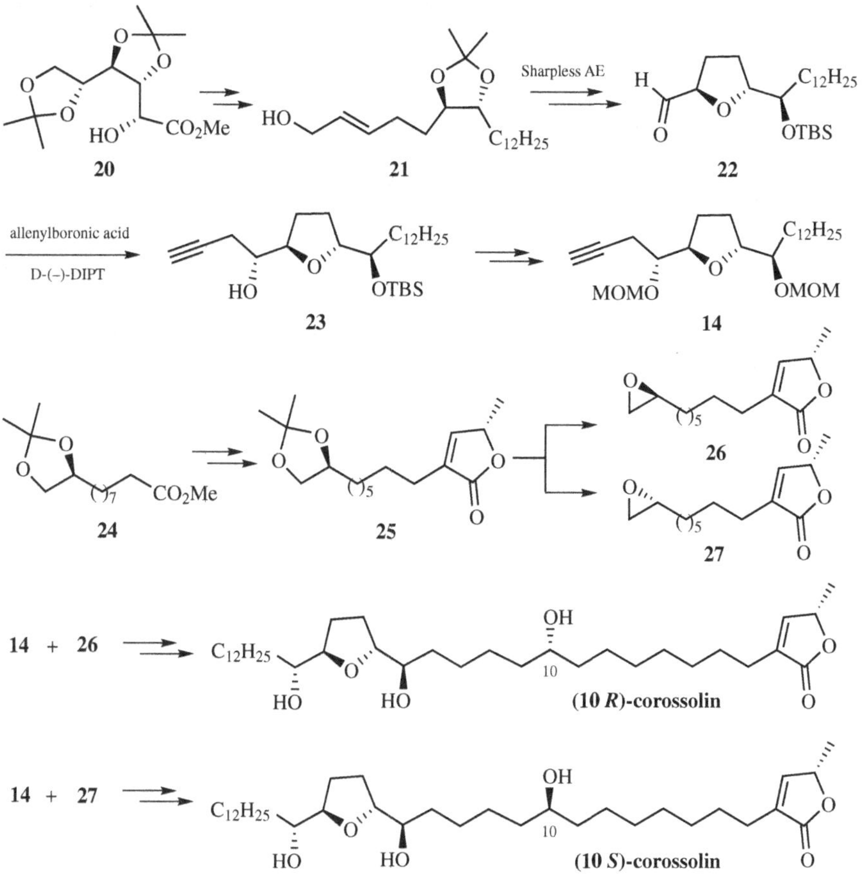

**Scheme 10-8.** Total synthesis of (10R)- and (10S)-corossolin by Wu et al.

almost the same. Therefore, they could not judge which one should be the natural product. In 1999, we also completed the synthesis of both C-10 isomers. The rotation comparison of our samples showed that (10R)-corossolin ($[\alpha]_D = 19.1$) was close to the natural product ($[\alpha]_D = 19$), whereas that of (10S)-corossolin ($[\alpha]_D = 24.6$) is more different. In addition, NMR data of our synthetic (10R)-corossolin was completely in accord with those of the natural one. So natural corossolin should have a 10R configuration. By comparison with physical data of known acetogenins, (10S)-corossolin was strongly suggested to be Howiicin C isolated by Zhang et al. in 1993.[35]

Our synthesis of corossolin started from the cheap sugar material, D-glucono-δ-lactone, whose C-3 and C-4 stereogenic centers were introduced into the target as C-19 and C-20, respectively. Subsequent introduction of C-15 and C-16 was achieved by Sharpless epoxidation of an allylic alcohol intermediate, and then the THF core **14** was established through a couple steps of transformations. On

406   CHEMICAL SYNTHESIS OF ANNONACEOUS ACETOGENINS

**TABLE 10-1. In Vitro Cytotoxicities Against B16BL6 by (10R)- and (10S)-Corossolin**

| Compound | $GI_{50}$ (µg/mL) | $LC_{50}$ (µg/mL) |
|---|---|---|
| (10R)-corossolin | 0.042 | ~7 |
| (10S)-corossolin | 0.77 | >10 |

the other hand, the epoxides **26** and **27** were prepared easily from a well-known chiron, (R)-isopropylidene glyceraldehyde. Coupling reactions of **14** with **26** and **27** in parallel, selective reduction of triple bonds and deprotections provided (10R) and (10S)-corossolin, respectively. It is noteworthy that the toxicity of (10R)-corossolin against cancer cells B16BL6 in vitro is more potent than that of (10S)-corossolin (Table 10-1).

A similar convergent synthetic strategy was applied in our synthesis of annonacin, using sequential assembly of the three key fragments, aldehyde **29**, phosphonium **30**, and alkyne **14** (Scheme 10-9). In this synthesis, the α,β-unsaturated-γ-lactone unit was constructed after completion of the linear skeleton. A modified method was applied into the synthesis of segment **3** (Scheme 10-5), in which the six carbons, as well as two chiral centers from D-glucono-δ-lactone, were all incorporated into the final target. Therefore, the efficiency of the whole

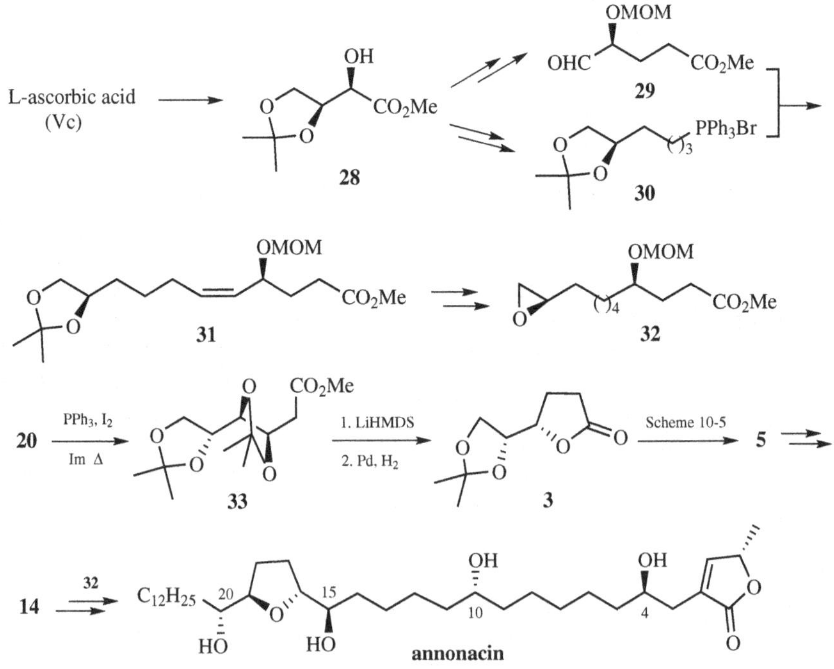

**Scheme 10-9.** Total synthesis of annoancin by Wu et al.

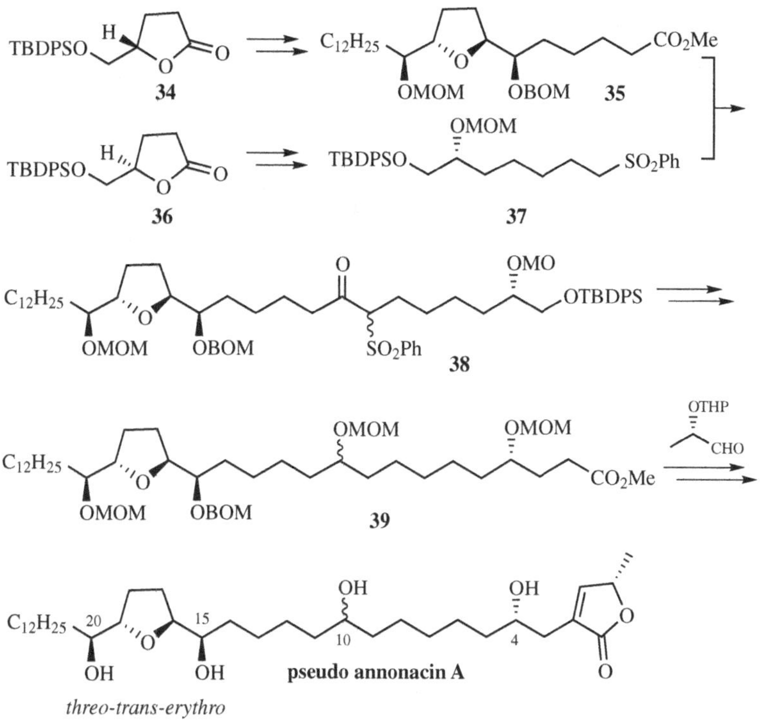

**Scheme 10-10.** Total synthesis of pseudo-annonacin A by Hanessian and Grillo.

route was significantly improved. The Wittig reaction of aldehyde **29** with ylide derived from compound **30**, and followed by two steps of transformations, provided epoxide **32**. The following epoxide-opening reaction of **32** by terminal alkyne **14** afforded the desired linear skeleton. The aldol-based protocol elaborated the lactone moiety on this intermediate, and annonacin was given after final deprotections.[31,33]

Hanessian and Grillo tried to identify the stereochemistries of the THF region of annonacin A by the total synthesis. Unfortunately, they only obtained pseudoannonacin A, a diastereomeric isomer of annonacin A (Scheme 10-10).[36] The key THF-containing segment **35** and sulfone **37** were prepared from L- and D-glutamic acid-derived chirons, respectively. Treatment of sulfone **37** with n-BuLi and then condensation with ester **35** afforded **38**. Reductive removal of sulfone followed by reduction of ketone functionality and two-carbon extension at the right side gave skeleton intermediate **39**. The lactone unit of target was introduced by the aldol strategy. Final deprotections gave a mixture of epimers at C-10, pseudoannonacin A. The relative configuration of the THF region of pseudoannonacin A is *threo-trans-erythro* (from left to right in Scheme 10-9).

A series of chiral stannic synthons was developed and applied successfully into the synthesis of acetogenins by Marshall Jiang. In the synthesis of longifolicin

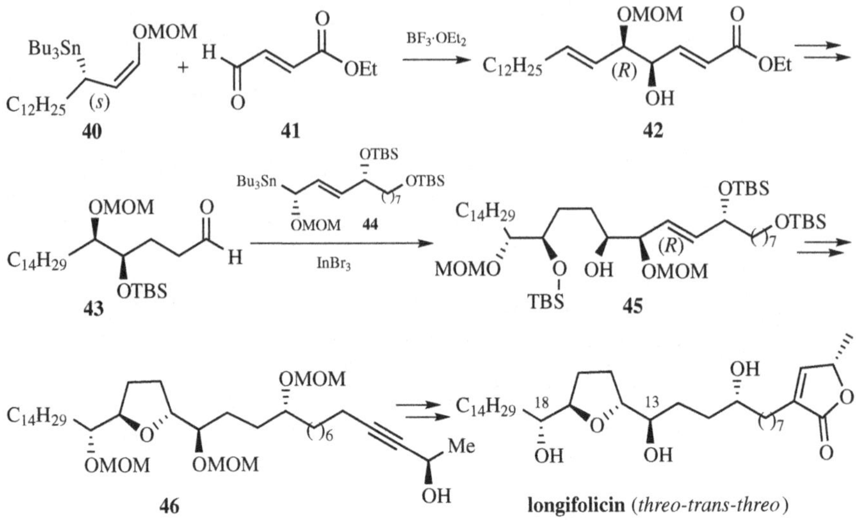

**Scheme 10-11.** Total synthesis of longifolicin by Marshall and Jiang.

(Scheme 10-11),[37] the allylic tin **40** was reacted with aldehyde **41** to afford the syn-adduct **42** in a 9 : 1 (syn/anti) ratio, under the promotion of BF$_3$•OEt$_2$. At a later stage, aldehyde **43** was treated with another allyic tin **44** (InBr$_3$ as the promoter), which gave the anti-adduct **45** in 1 : 9 (syn/anti) ratio. After Williamson etherification and other routine steps, the THF-containing intermediate **46** was synthesized. Introduction of terminal lactone was achieved by a Pd(0)-catalyzed carbonylation and lactonization. Final treatment with acid conditions removed the protecting groups and afforded longifolicin.

A pair of C-12 epimers of muricatetrocin were synthesized by Baurle et al.[38] This target was cut into three major segments, THF-containing phosphorium salt **51**, lactone-containing aldehyde **52**, and Grignard reagent **54** (Scheme 10-12). The intermediate **50** was synthesized by an enantioselective reaction of aldehyde **47** with zinc reagent **48** in the presence of a catalytic amount of cyclohexane-1,2-diamine derivative **49**. Williamson etherification of compound **50** furnished the THF segment and the corresponding phosphorium salt **51**. Wittig reaction between aldehyde **52** and **51**, selective hydrogenation, and oxidation of the primary alcohol afforded the aldehyde **53**. Grignard reaction of **53** and **54** gave only one product in a mechanism of chelation control. This product was further deprotected, affording (4R, 12R, 15S, 16S, 19R, 20R, 34S)-muricatetrocin. Using *ent*-**49** as the catalyst, the reaction between **47** and **48** afforded alcohol **55**, a C-12 diastereomer of **50**. After similar steps, (4R, 12S, 15S, 16S, 19R, 20R, 34S)-muricatetrocin was synthesized finally. It is noteworthy that the physical data of synthetic (4R, 12S, 15S, 16S, 19R, 20R, 34S)-muricatetrocin is exactly the same as those of natural muricatetrocin A (howiicin E). Although NMRs are the same as those of muricatetrocin B, the rotation value of synthetic (4R, 12R, 15S, 16S, 19R, 20R, 34S)-muricatetrocin is found smaller.

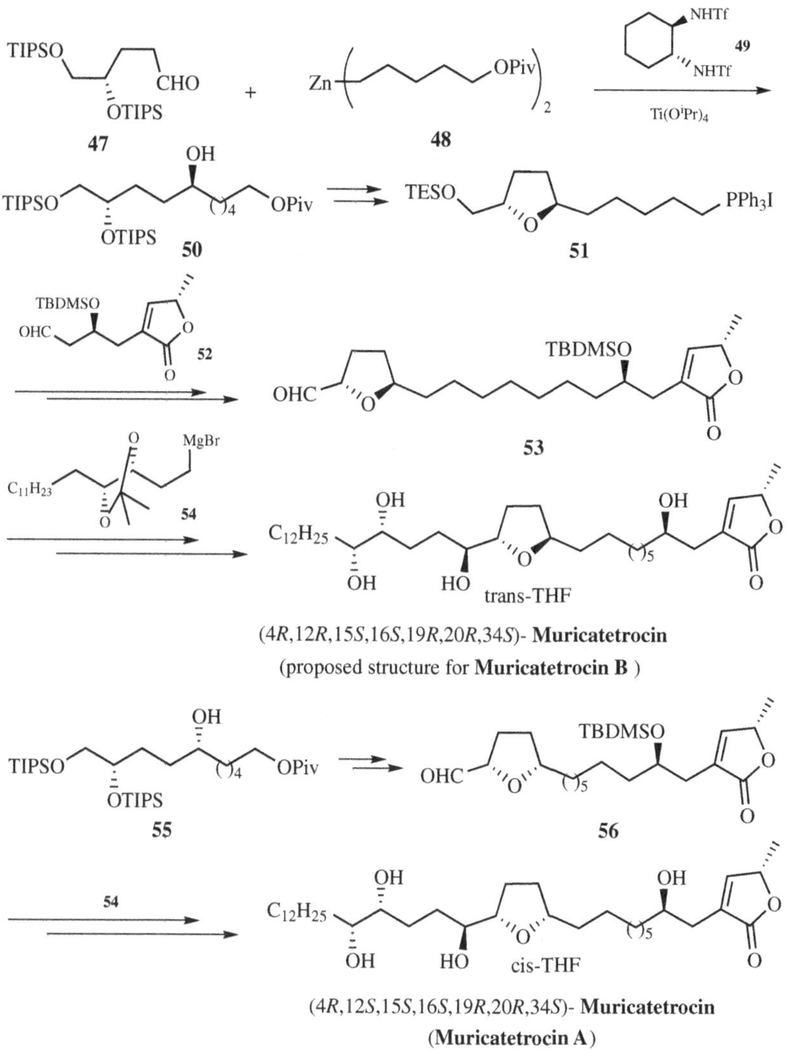

**Scheme 10-12.** Total synthesis of muricatetrocin A and B by Baurle et al.

A similar disconnection strategy was adopted by Dixon et al. in the United Kingdom in the synthesis of muricatetrocin C.[39,40] The retro-synthesis disconnected the target into three parts, aldehyde **59**, terminal alkyne precursor **60**, and vinyl iodide **64** (Scheme 10-13). The THF-containing aldehyde **59** was prepared from compound **58**, an O-C rearranged product from **57**. Compound **60** was synthesized from dimethyl L-tartrate. Compound **64** was prepared with a hetero Diels-Alder reaction. The terminal acetylene lithium derived from **60** was added to aldehyde **59**, which gave a major product with erythro configuration. This erythro product was converted into the threo product by alcohol oxidation and ketone reduction

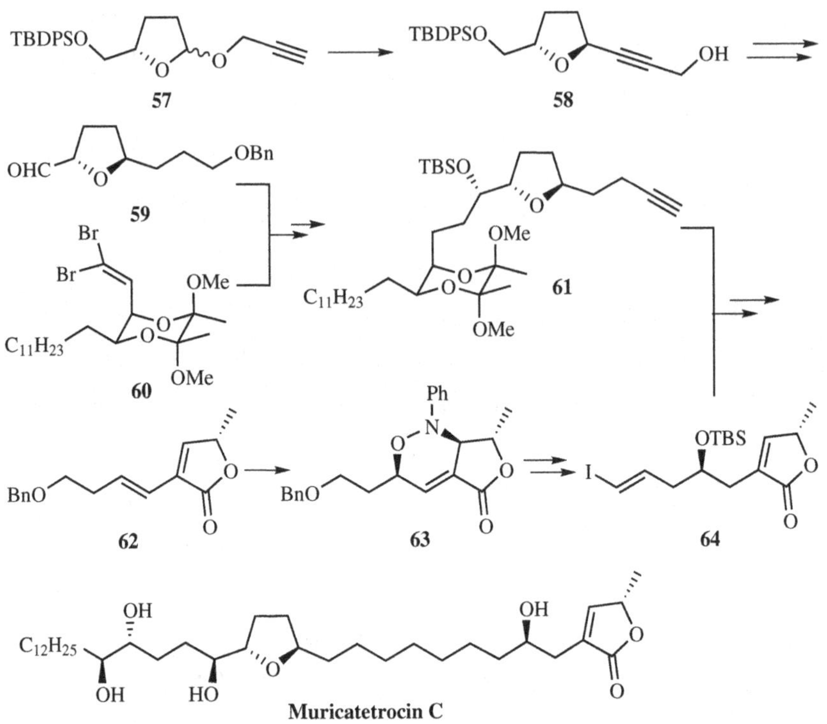

**Scheme 10-13.** Total synthesis of muricatetrocin C by Dixon et al.

with L-selectride. The intermediate **61** was given after hydroxyl group protection. Sonogashira coupling between **61** and **64**, followed by selective hydrogenation and deprotections, finally afforded muricatetrocin C.

Sharpless dihydroxylation, Co(II)-catalyzed oxidative *trans*-THF ring formation served as the key steps in Wang et al.'s route to gigantetrocin A,[41] which generated the THF-containing aldehyde **68** (Scheme 10-14). On the other hand, the ester **69** was prepared from L-glutamic acid. This ester was converted into the phosphorium salt **70** after the lactone unit was introduced. The molecular skeleton was established by Wittig reaction between **68** and **70**, although the chemical yield was less satisfactory. The total synthesis of gigantetrocin A was finally completed after several treatments, including β-elimination and deprotection of the hydroxyl groups.

Although the absolute configurations remained unknown, the relative configuration of the THF region of mosin B was known as *erythro-trans-threo*. In the synthesis published by Tanaka et al. (Scheme 10-15),[42,43] desymmetry of *meso*-cyclohexene-4,5-diol was achieved by mono-benzyl etherification, which gave **71** enantioselectively. The olefin **71** was then converted into allylic alcohol **73** by several steps including oxidative cleavage of the double bond and the Wittig reaction. Iodoetherification and base treatment furnished the *trans*-THF fragment **74** with

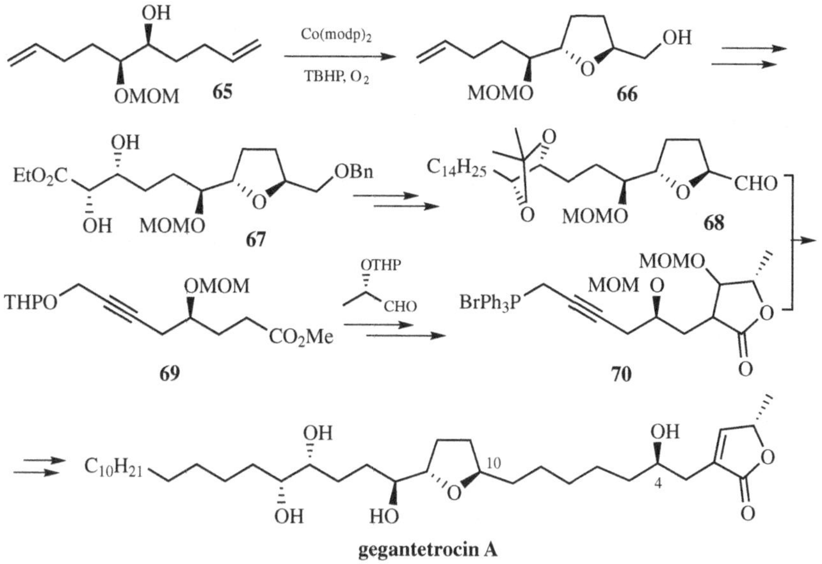

**Scheme 10-14.** Total synthesis of gegantetrocin A by Wang et al.

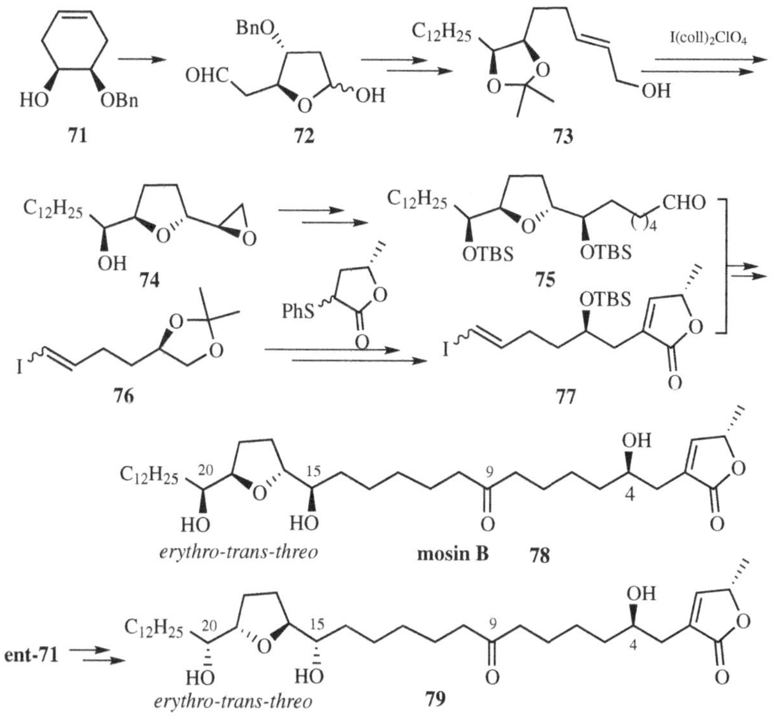

**Scheme 10-15.** Total synthesis of mosin B by Tanaka et al.

high selectivity. The Nozaki–Hiyama–Kishi reaction of aldehyde **75** (derived from **74**) with vinyl iodide **77**, followed by oxidation of the hydroxyl group, selective hydrogenation of the double bond, and deprotection of the *O*-TBS ethers, established the skeleton and finally afforded **78** (mosin B). In the meantime, the other isomer **79** was prepared from *ent*-**71**. Although both sets of NMRs of **78** and **79** are the same as those of the natural product, the rotation value comparisons suggests that natural mosin B is compound **78**.

As in the case of mosin B, the absolute configuration of *cis*-solamin (the relative configuration is *threo-cis-threo*) was also confirmed by total synthesis and comparison of rotations. Both Makabe et al.[44] and Cecil and Brown[45] finished its synthesis independently (Scheme 10-16). In the route developed by Makabe et al.,[44] the *cis*-THF-containing compound **81** was prepared from olefin **80**, using TBHP-VO (acac)$_2$-mediated epoxidation and an in situ intramolecular epoxide opening as key reactions. Coupling of alkyne **81** with vinyl iodide **82**, selective hydrogenation of double/triple bonds, and β-elimination to form unsaturated lactone provided *cis*-solamin. Cecil and Brown's protocol[45] used a chiral auxiliary to control the stereochemistry of KMnO$_4$-mediated 1,5-diene oxidation, which afforded

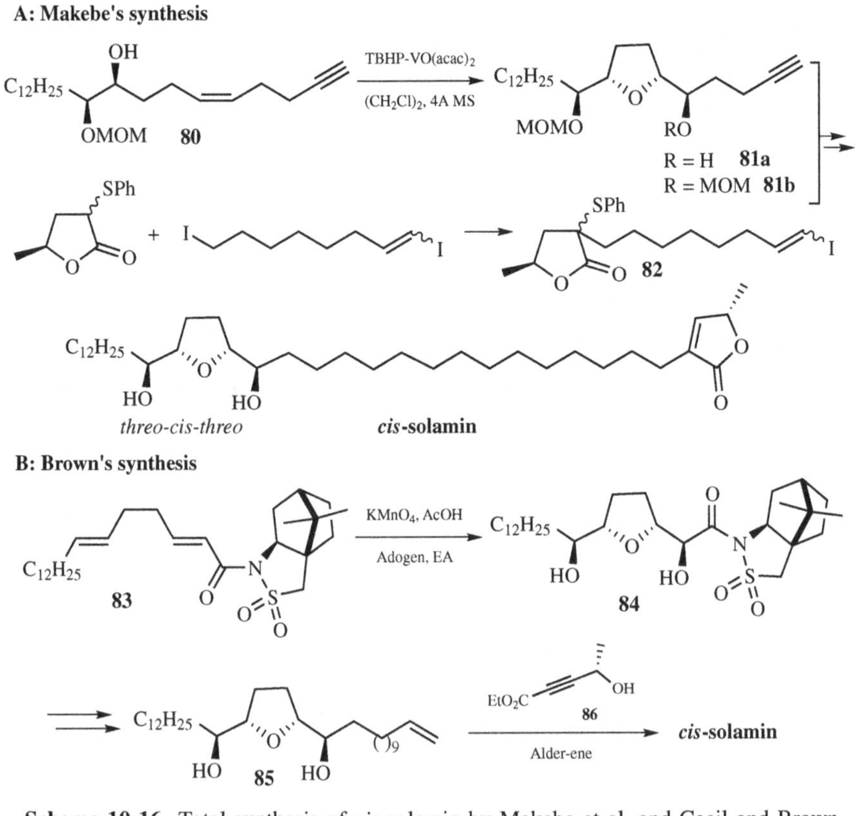

**A: Makebe's synthesis**

**B: Brown's synthesis**

**Scheme 10-16.** Total synthesis of *cis*-solamin by Makebe et al. and Cecil and Brown.

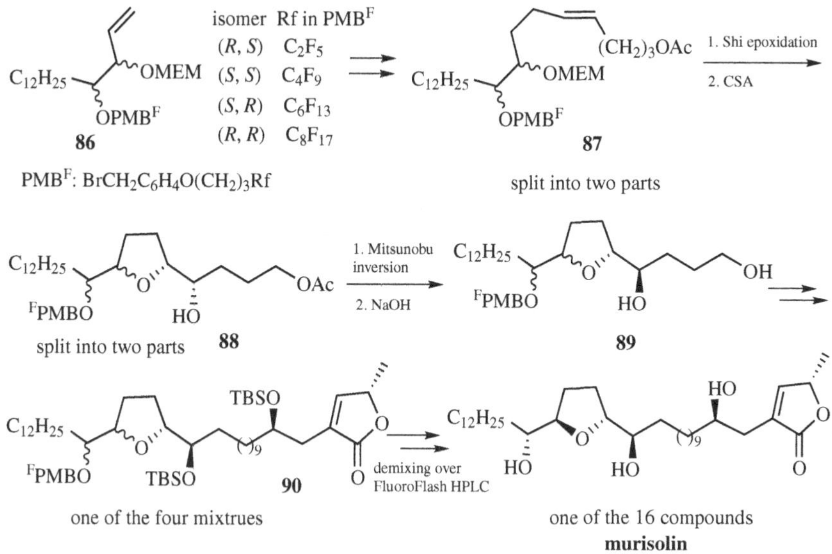

**Scheme 10-17.** Acetogenins library using perfluoroalkyl labeling protocols by Zhang et al.

*cis*-THF-containing intermediate **84**. Cleavage of the chiral auxiliary group and several subsequent transformations granted the epoxide, which was further opened by a Grignard reagent to afford terminal olefin **85**. Final introduction of lactone unit adopted an Alder-ene reaction previously developed by Trost.

A pioneer acetogenin library (16 members) was established by Zhang et al.[45] by mix-split protocols, using perfluoroalkyl groups as labels. The absolute configuration of murisolin was confirmed in this study (Scheme 10-17). Compound **86** is a mixture of four enantiopure isomers in equal amounts. The mixture **87** derived from **86** was split into two equal parts. One part of **87** was treated with Wang et al.'s epoxidation conditions and then acid to give **88**. Compound **88** was divided into two equal parts again, and one of the two parts was treated with Mitsunobu conditions to give **89**. After Koceinski–Julia coupling, **89** was converted into the procursor **90** as a mixture of four compounds, which could be separated on HPLC (fluorous silica gel as loading materials) to afford four pure single compounds. Removal of perfluoroalkyl labels and *O*-silyl ether protecting groups generated four acetogenins. Similarly, another 12 acetogenins were synthesized. From compounds **86** to **90**, the total steps for these four mixtures (each contains four isomers) is 39. If these compounds were synthesized individually, the total number of steps would be 156. Therefore, this strategy has a great advantage in synthetic efficiency.

## 10.3   TOTAL SYNTHESIS OF *BIS*-THF ACETOGENINS

Many naturally occurring *bis*-THF acetogenins have also been synthesized in recent years. Besides the *mono*-THF acetogenins, Marshall et al. also successfully applied

asiminocin $R^1 = OH, R^2 = R^3 = R^4 = H$     (5*S*)-uvarigrandin A   $R^3 = OH, R^1 = R^2 = R^4 = H$

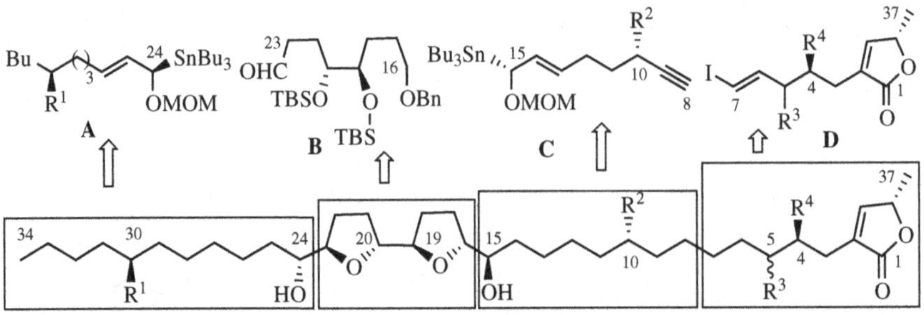

asimin      $R^2 = OH, R^1 = R^3 = R^4 = H$     (5*R*)-uvarigrandin A   $R^3 = OH, R^1 = R^2 = R^4 = H$

asimicin    $R^4 = OH, R^1 = R^2 = R^3 = H$

**Scheme 10-18.** General synthetic strategy for five *bis*-THF acetogenins developed by Marshall et al.

their useful chiral allylic tin or indium reagents to the syntheses of three *bis*-THF-containing acetogenin with a *threo-trans-threo-trans-threo* relative configuration (asimicin,[47,48] asiminocin, asiminecin,[49] and asimin[50]), one *bis*-THF-containing acetogenin with a *erythro-trans-threo-trans-threo* relative configuration [(30*S*)-bullanin[51]], and one *bis*-THF-containing acetogenin with a *threo-cis-erythro-trans-threo* relative configuration (trilobin[52]) in the *bis*-THF region. Very recently, they developed a more flexible and general strategy to synthesize these acetogenins with a *bis*-THF subunit (Scheme 10-18). Five acetogenins, asiminocin, asimin, asmicin, (5*S*)-uvarigrandin, and (5*R*)-uvarigrandin, could be cut into four fragments (A/B/C/D). Among these fragments, a masked di-aldehyde B is the common building block for these targets.

Scheme 10-19 shows the synthesis of asimicin.[58] The reaction of aldehyde **91** with allylic tin **92** in the presence of InBr$_3$ afforded the anti- (or erythro)-adduct **93**. Later, a similar reaction between aldehyde **94** and functionalized allylic tin **95** provided **96**. Both newly born hydroxyl groups in the above reactions were masked as tosylate immediately. The *threo-trans-threo-trans-threo bis*-THF rings (in compound **97**) were formed by treatment of the tosylate with TBAF. Coupling of the THF-containing segment **97** and lactone segment **98** was achieved with Sonogashira's conditions.

If fragment A in Scheme 17 is treated first with BF$_3$·OEt$_2$ and then subjected to aldehyde, the threo (syn) product will be produced predominantly. Thus, this methodology could be expanded for construction of those acetogenins with erythro (C-24/C-25) configuration. Scheme 10-20 illustrates the synthesis of bullanin[53] and (30*S*)-hydroxybullatacin.[54] In both cases, the reactions of allylic tins and aldehydes in the presence of InBr$_3$ or BF$_3$·OEt$_2$ gave anti- or syn-adducts, respectively. Again, the Sonogashira reaction was applied into the skeleton establishment.

Marshall et al. measured the activities of those *bis*-THF acetogenins synthesized against human cancer cells H-116 (Table 10-2). These results showed that the

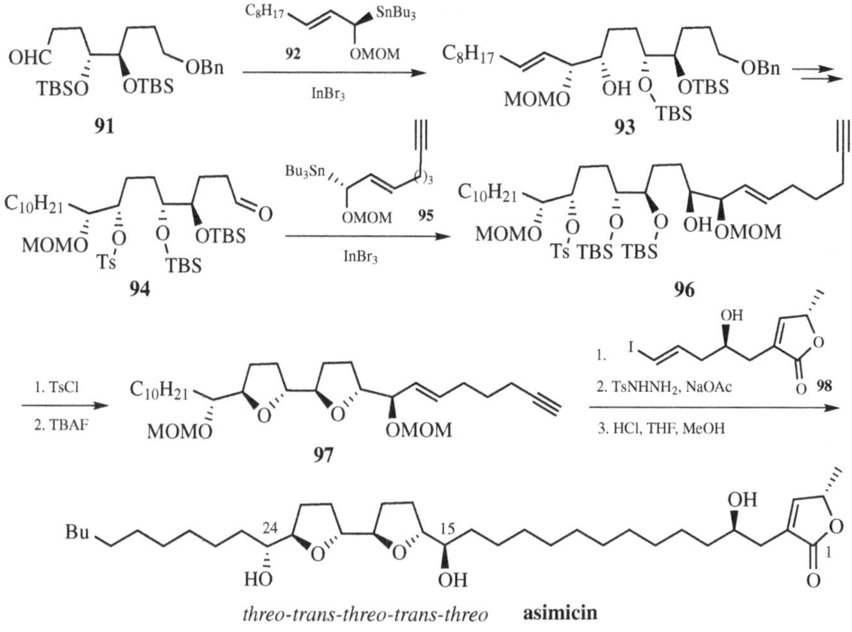

**Scheme 10-19.** Total synthesis of asimicin by Marshall et al.

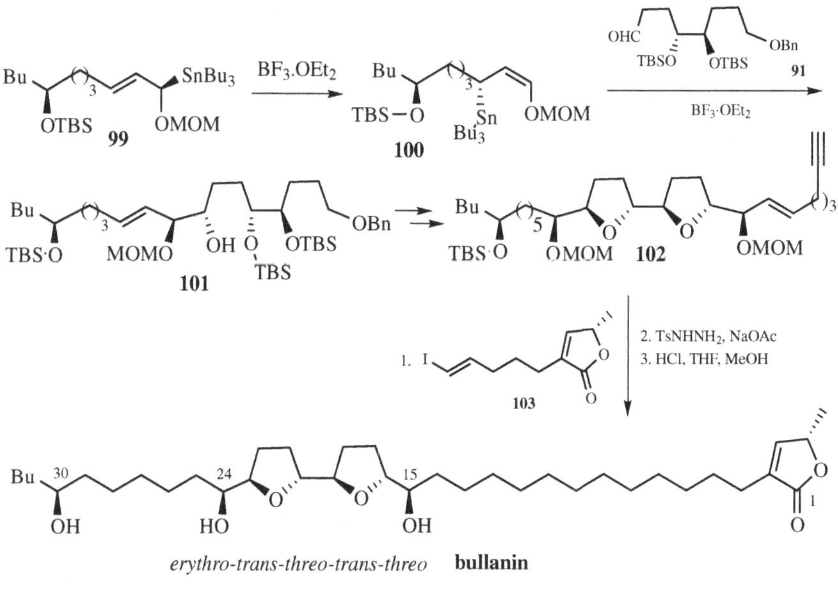

**Scheme 10-20.** Total synthesis of bullalin by Marshall et al.

**TABLE 10-2. In Vitro Cytotoxicities Against Cancer Cells by Asiminocin and Other *bis*-THF Acetogenins**

| Compound | H-116 (Marshall et al.'s results) | | | HT-29 (MacLaughlin et al.'s results) |
|---|---|---|---|---|
| | $IC_{50}$ (µg/mL) | $IC_{90}$ (µg/mL) | $IC_{90}/IC_{50}$ | $IC_{50}$ (µg/mL) |
| Asiminocin[a] | $5 \times 10^{-3}$ | 1.5 | 300 | $<10^{-12}$ |
| Asimicin[a] | $5 \times 10^{-4}$ | $2 \times 10^{-3}$ | 4 | $<10^{-12}$ |
| Dihydroasimicin[a] | $2 \times 10^{-4}$ | $1 \times 10^{-1}$ | 500 | |
| Asimin[a] | $3 \times 10^{-4}$ | $3 \times 10^{-3}$ | 10 | $<10^{-12}$ |
| Bullanin[a] | $3 \times 10^{-3}$ | $1 \times 10^{-1}$ | 33 | $5 \times 10^{-12}$ |
| 5-Fluorouracil[b] | $1.5 \times 10^{-1}$ | 1.5 | 10 | |

[a] 1 g/mL ≈ 1.5 M.
[b] 1 g/mL ≈ 7.0 M.

activities were more potent than that of 5-fluorouracil, a drug currently used for cancer chemotherapy. However, the Marshall et al. data are $10^8$–$10^9$ less potent than those reported previously by McLaughlin et al. by the same compounds. Additionally, the dihydroasimicin exhibited similar activities as asimicin, although its double bond of terminal lactone was saturated.

Elegant combinations of Sharpless dihydroxylations/epoixdation and Mitsunobu reaction on the unsaturated fatty acids were developed by Keinan et al., in constructing various THF rings in the syntheses of uvaricin,[55] squamotacin,[56] and trilobin.[57] Scheme 10-21 shows the syntheses of squamotacin and trilobin. During the course of converting γ-unsaturated lactone **104** into epoxide **108**, Sharpless dihydroxylations (AD-mix-α, AD-mix-β) were used twice and Sharpless epoxidation was used once. Epoxide opening of **108** with alkyne generated the phosphorium salt **109**, which was used for the following Wittig reaction to establish the whole skeleton. After two steps of routine transformations, squamotacin was finally synthezied. The synthesis of trilobin with *threo-trans-erythro-trans-threo bis*-THF configuration also used these two reactions as key steps. However, establishment of its skeleton used the epoxide opening reaction by a terminal alkyne. While synthesizing bullatacin and asimicin,[58] $Re_2O_7$-mediated oxidation of homoallylic alcohols was used as the key reaction.

$Re_2O_7$-mediated tandem oxidative polycyclization of polyenic *bis*-homoallylic alcohols was also developed by Keinan et al. in their synthesis of *bis*-THF acetogenins. Several important conclusions were drawn on $Re_2O_7$-mediated oxidations of ployalkene: (1) The first THF ring formed is always in *trans*-configuration. (2) If two oxygenated functional groups existing in the above THF containing intermediate are in *threo*-configuration, the next forming THF ring will be a *cis* one; otherwise, a *trans*-THF will form. These rules are well used in their synthesis of rollidecin C and rollidecin D[59] (Scheme 10-22). Epoxide **115** prepared by Sharpless AE was selectively reduced by Red-Al to give 1,3-diol, and then this diol was converted into the phophorium salt **116**. Wittig reaction of aldehyde **117** or **118** with

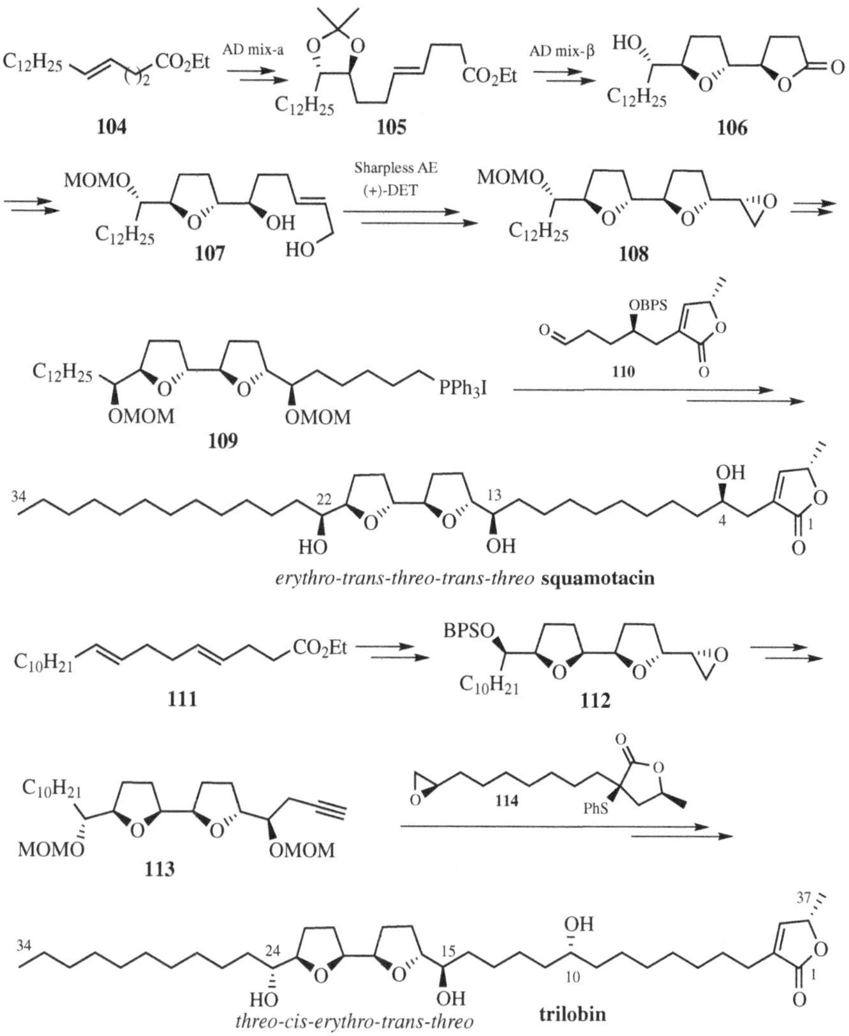

**Scheme 10-21.** Total synthesis of squamotacin and trilobin by Keinan et al.

**116** and selective deprotection of MOM ethers gave **119** and **120**, respectively. In the presence of $Re_2O_7$ and trifluroacetic acid anhydride, **119** and **120** were converted in only one step to *bis*-THF products, **121** and **122**, respectively. The final selective reduction of double bonds followed by removal of *O*-silyl ethers afforded rollidecin C and rollidecin D.

In Wang et al.'s synthesis of asimilobin[60,61] (Scheme 10-23), the key diol **123** was also prepared by Sharpless AD in high enantioselectivity. Oxidation of **123** furnished the *trans-threo-trans bis*-THF intermediate **124** in catalysis of $Co(modp)_2$. The $C_2$ symmetrical diol **124** was desymmetralized and converted

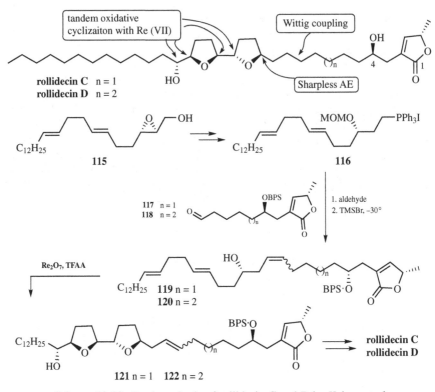

**Scheme 10-22.** Total synthesis of rollidecin C and D by Keinan et al.

into the aldehyde **125**, which was further subjected to the Wittig reaction with **70**. Final treatments of hydrogenation and deprotections, **126**, were obtained and characterized as natural asimilobin.

Grignard reactions were applied twice in Emde and Koert's synthesis of squamocin A and D,[62,63] to link the segments **127**, **128**, and **131** together and establish the

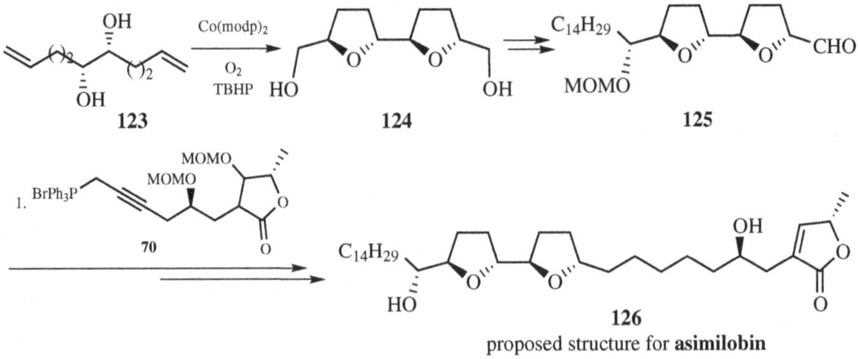

**Scheme 10-23.** Total synthesis of asimilobin by Wang et al.

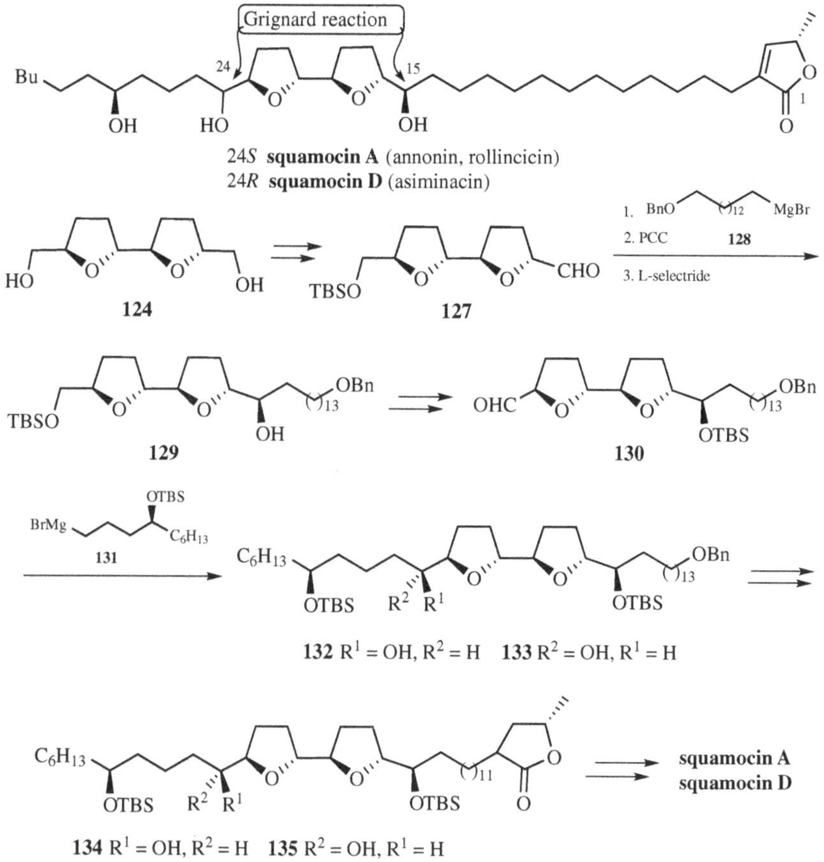

**Scheme 10-24.** Total synthesis of squamocin A and D by Emde and Koert.

skeletons (Scheme 10-24). In the second Grignard reaction, two diastereomers **132** and **133** were separated by chromatography. Both of them were used to synthesize squamocin A and D, respectively. The final procedures in both cases were to introduce the lactone subunit.

For those acetogenins having a $C_2$ symmetry in the THF region, two directional synthetic strategies are often adopted, although the efficiency of the desymmetrization step usually remains a problem. The iodoetherification was used in the synthesis of asimicin and trilobacin as the desymmertrization step to differentiate two chemically equal hydroxyl groups by Ruan and Mootoo[64] (Scheme 10-25). This step was carried out by using $I(coll)_2ClO_4$ as the reagent, affording the THF-containing intermediate **137**. Treatment of *mono*-THF containing iodide **138** with acid and base successively and reduction of double bond gave the known compound **139**. On the other hand, the iodide **138** was substituted by hydroxide and then converted into mesylate **140**. Compounds **139** and **140** were both advanced intermediates for trilobacin and asimicin in Keinan et al.'s routes, respectively.[65]

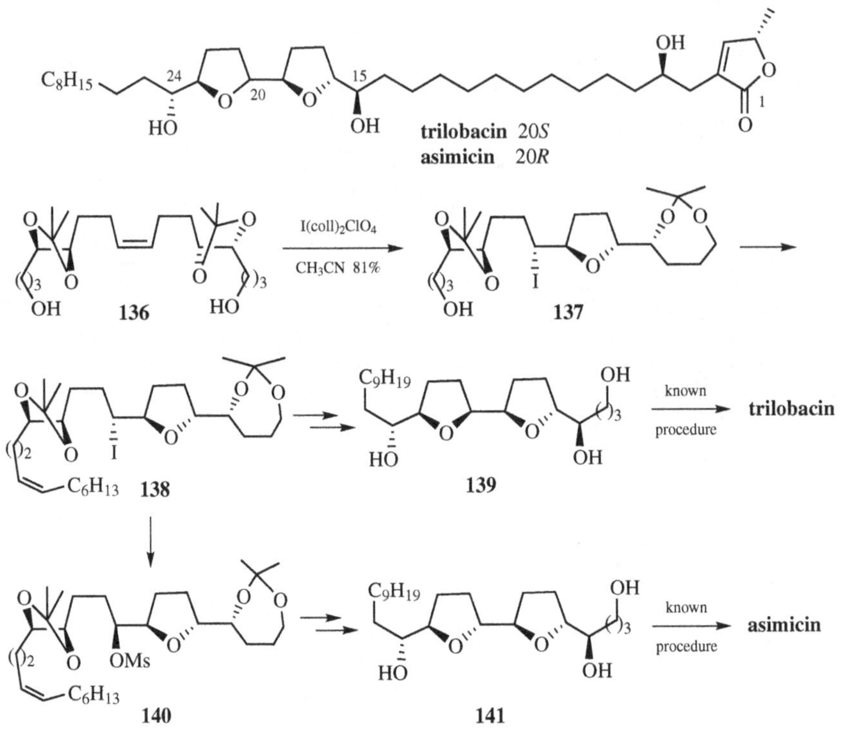

**Scheme 10-25.** Formal total synthesis of trilobacin and asimicin by Ruan and Mootoo.

Another route to uvaricin using a two-directional strategy was developed by Burke and Jiang[66] (Scheme 10-26). The linear diol **142** prepared from D-tartaric acid was converted into *bis*-THF-containing precursor **144** by Pd(0)-catalyzed reactions, using (*R,R*)-DPPBA as the ligand. One side of the di-olefin **144** was dihydroxylated by controlled Sharpless conditions, affording diol **145** in 70% yield after one recycle. The hydrocarbon chain was then introduced in the left side and yielded **146**, using the epoxide opening reaction by Grignard reagent. Again, the successful application of Sharpless AD upon the remaining olefin gave the diol **147**, which could be converted into uvaricin according to the Keinan et al. steps previously reported.[35]

Among those reported syntheses of acetogenins, few examples were working on the nonadjacent *bis*-THF acetogenins. These natural products often have two hydroxyl groups (1,4-dihydroxyl) between the separated THF rings, such as 4-deoxygigantecin (Scheme 10-27). In the route developed by Makabe et al.,[67–69] a *mono*-THF segment **148** was first synthesized, and then this compound was alkyalted by iodide **149** to yield **150**. Birch reduction of the triple bond afforded *trans*-olefin, which was then dihydroxylated to give a diol in a later satge. Triton B promoted intramolecular THF ring formation and furnished **153** with the desired stereochemistries. Finally, Sonogashira coupling between the terminal alkyne **153** and vinyl

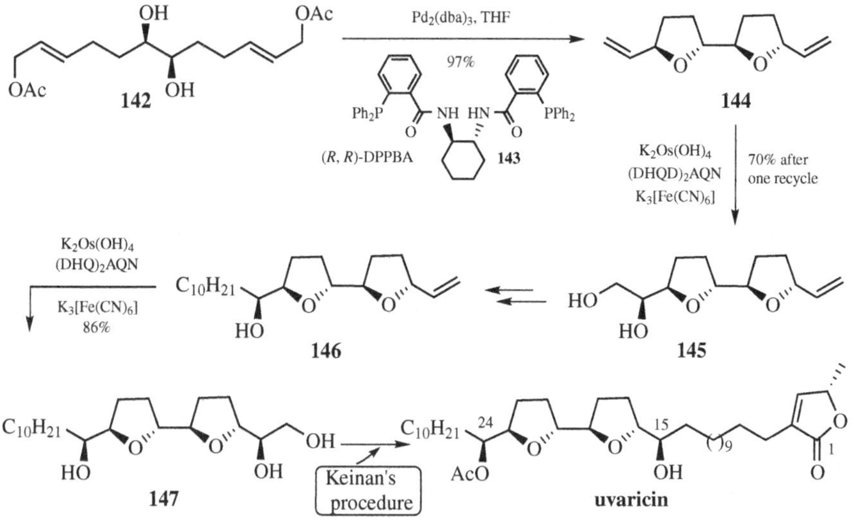

**Scheme 10-26.** Formal total synthesis of uvaricin by Burke and Jiang.

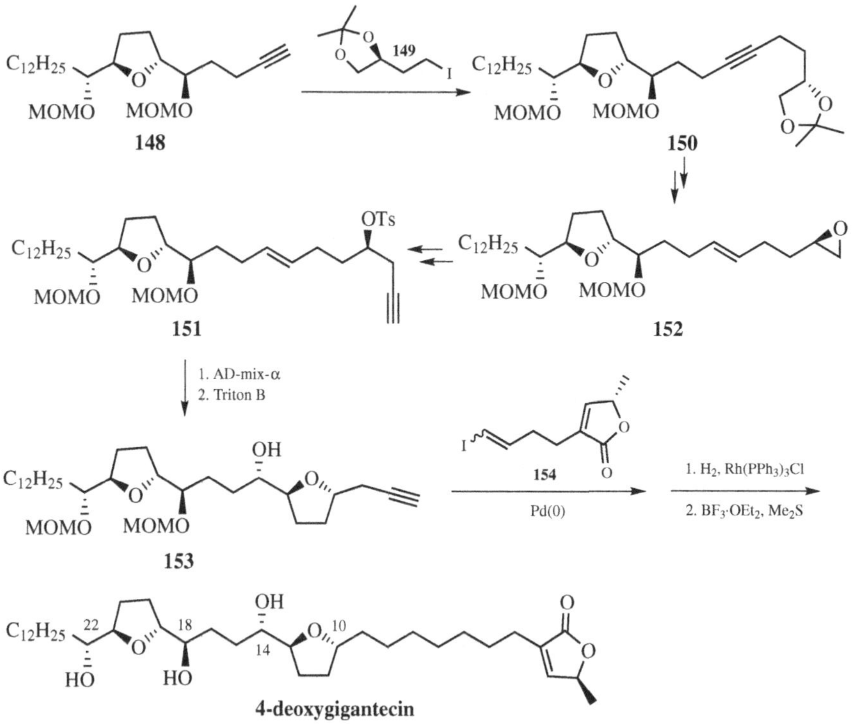

**Scheme 10-27.** Total synthesis of 4-deoxygigantecin by Makabe et al.

iodide **154** established the skeleton of 4-deoxygigantecin. The physical data confirmed the absolute stereochemistries of natural 4-deoxygigantecin.[67,68]

## 10.4   TOTAL SYNTHESIS OF THP-CONTAINING ACETOGENINS[70–76]

Among the more than 350 members of the acetogenin family characterized, only several acetogenins contain a tetrahydropyran ring. A typical member, mucocin, was isolated as the first example in 1995. Mucocin not only has a hydroxylayed THP moiety, but also it has a nonadjacent THF ring unit. The first total synthesis of mucocin was carried out by Keinan et al.[77] (Scheme 10-28). Similar to their previous synthesis of the *bis*-THF acetogenins, the first batch of stereochemistries was established by Sharpless AD and AE on the cheap unsaturated fatty acids. *Bis*-allylic alacohol **155** was prepared from (1*E*,5*E*,9*E*)-cyclododecane-triene through ozonolysis, Wittig–Hornor olefination, and DIBAL-H reduction of α,β-unsaturated esters. Sharpless epoxidation of **155** afforded epoxide alcohol **156**. Mono-protection of diol and olefination of the other side gave *cis*-alkene **157**. Selective dihydoxylation of the *trans*-olefins was achieved by Sharpless AD. The resulting tetraol **158** was then converted into key intermediate **159** with THP and THF rings installed. The final structure was again established by Sonogashira reaction.

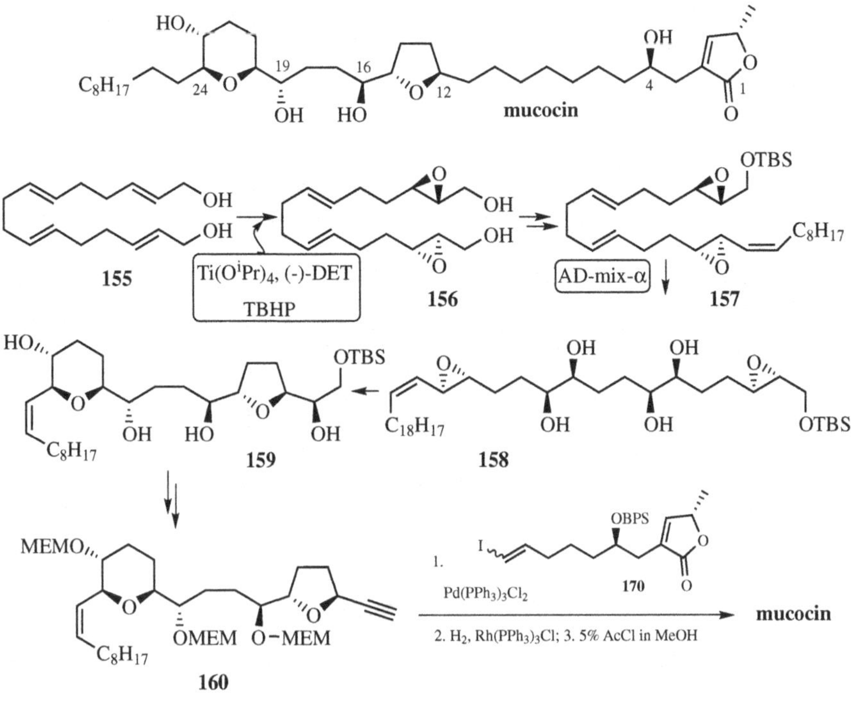

Scheme 10-28. Total synthesis of mucocin by Keinan et al.

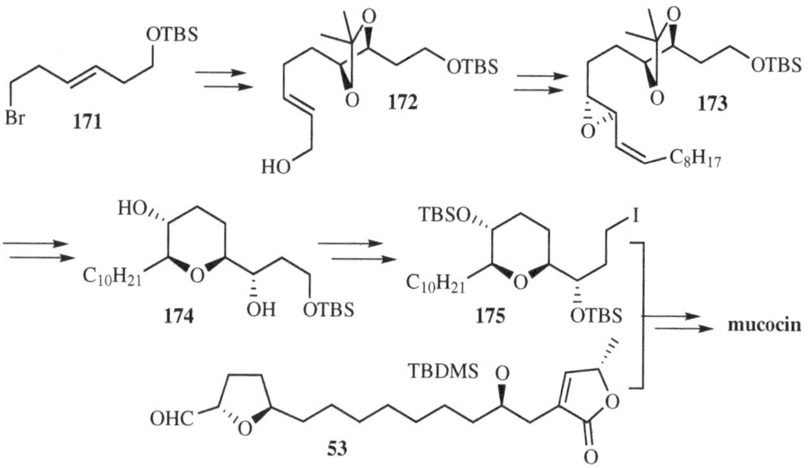

**Scheme 10-29.** Total synthesis of mucocin by Koert et al.

Koert et al.'s strategy[78,79] for mucocin disconnected the target into two major parts: One contains a THP moiety (diol **174** or its equivalent **175**), and the other contains a THF ring (aldehyde **53**). The coupling of those two segments was achieved by Grignard reaction (**175** and **53**). In this synthesis, the formation of the THP ring also used the *cis*-olefin site-direction effect on substrate **173**, which gives the 6-endo epoxide opening product **174** predominately (Scheme 10-29).

A chiron approach to mucocin was developed by Takahashi and colleagues[80,81] (Scheme 10-30). The simple addition of aldehyde **176** (prepared from D-lactose

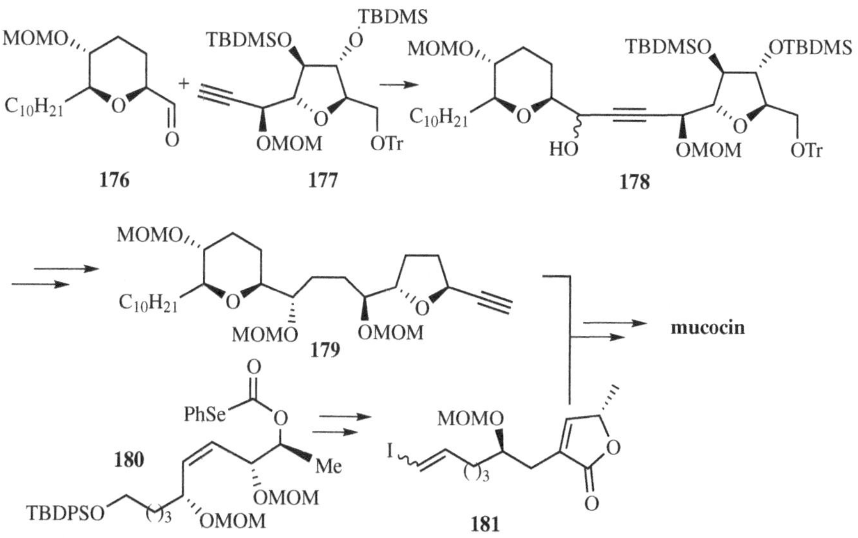

**Scheme 10-30.** Total synthesis of mucocin by Takahashi et al. (route one).

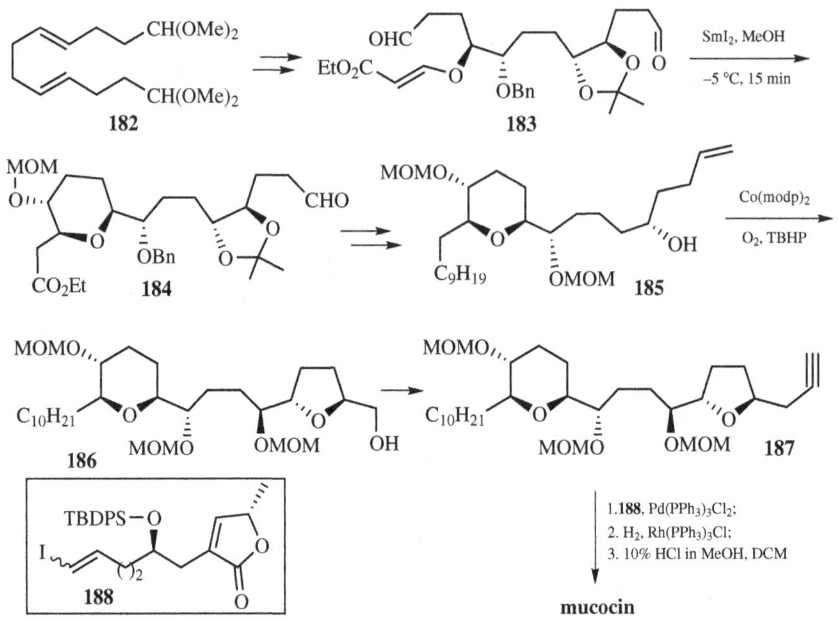

**Scheme 10-31.** Total synthesis of mucocin by Takahashi et al. (route two).

derivative) and terminal alkyne **177** (prepared from D-mannitol derived compound) furnished key compound **178**, which was a mixture of epimers at the newly formed chiral center. It was also found that $CeCl_3$ could greatly promote the reaction; however, the desired stereochemistry of new carbinol center was not favorable. After reduction of triple bond, this hydroxyl group was oxidized and then reduced by L-seletride, which gives the alcohol in correct configuration. Additional transformations gave one segment **179** with both THP and THF rings established. The other segment **180** was prepared from L-xylose and converted into the vinyl iodide **181** via a radical reaction. The coupling conditions and later elaboration are similar to that described in Scheme 10-28.

More recently, Takahashi et al. developed the second route to mucocin[82] (Scheme 10-31). A $SmI_2$-initiated free radical reaction was used to construct the *cis*-THP ring, and Co(II)-mediated oxidation furnished the *trans*-THF ring. The *trans/trans*-diene **182** (prepared from (1$E$,5$E$,9$E$)-cyclododecane-triene) was converted into di-aldehyde **183** by Sharpless AD and desymmetry step as well as other routine steps. Treatment of **183** with $SmI_2$ in a short time afforded *cis*-THP compound **184** in high yield, and the other aldehyde functionality showed good stability under such conditions. The second major reaction for THF ring formation was achieved by oxidation of **185** with the Co(II) complex. Sonogashira coupling of **187** with **188**, followed by couple routine transformations, completed the synthesis of mucocin smoothly.

The shortest route to mucocin to date was achieved by Evans et al. (Scheme 10-32)[83]; ring closing metathesis was elegantly used as a key step. In the presence of $InBr_3$ and tert-butyldimethylsilane, a reductive cyclization of **190** was achieved,

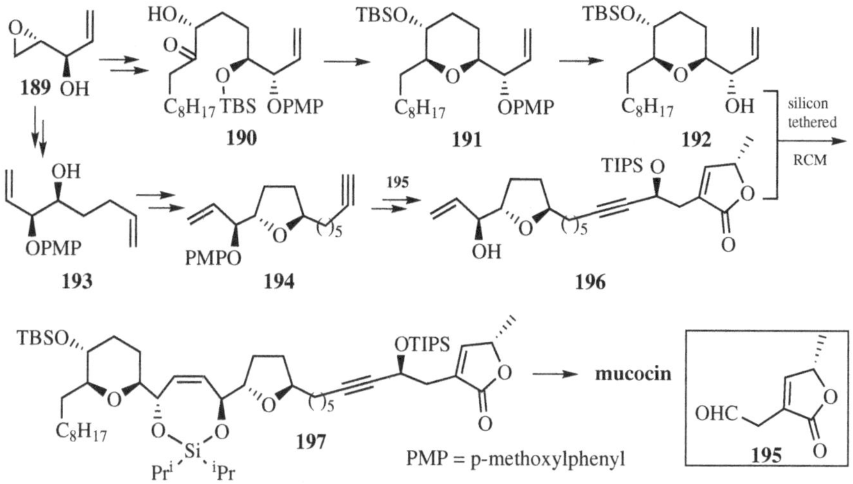

**Scheme 10-32.** Total synthesis of mucocin by Evans et al.

and THP intermediate **191** was obtained after in situ alcohol protection. Selective deprotection of PMP ether afforded the allylic alcohol **192**. In the meantime, oxidation of 5-hydroxyl olefin **193** with the Co(II) complex, followed by other conventional steps, elaborated THF-containing compound **194**. The enantioselective addition of aldehyde **195** with terminal acetylene **194** to give **196** was achieved in the presence of chiral amino alcohol and zinc triflate. Using a silyl ether as linking unit, both **192** and **196** were taken together. The key RCM reaction performed efficiently to generate **197** in 83% yield. The final modifications (deprotection of silyl ethers and reduction of the double bond in the middle region) on **197** afforded mucocin.

Another attractive molecule that contains both THF and THP rings is named muconin.[84] Different from mucocin, the THF and THP units in muconin are adjacent to each other. In the synthesis completed by Schans et al.,[85] four chiral building blocks (**200**, **201**, **202**, and **203**) were synthesized using salen(Co) **198**-catalyzed hydrolic kinetic resolution of terminal epoxide and salen(Cr) **199**-catalyzed hetero Diels–Alder reaction. The sequential assembly of these building blocks furnished the first total synthesis of muconin (Scheme 10-33). The ester derived from acid **200** and alcohol **201** performed an Ireland–Claisen rearrangement to provide diene **204**, which was further converted into key intermediate **205** by RCM, hydrogenation, and oxidation. Treatment of alykyne 206 with dicyclohexylborane and subsequent transmetallation afforded a vinylzinc reagent, which was coupled with the aldehyde **205** to give a precursor. The latter was then converted to muconin by a sequential of conventional procedures.

Yang and Kitahara's route[86,87] to muconin used D-glutamic acid as the only chiral starting material to introduce the stereogenic centers in the target (Scheme 10-34). That means the key segments in the synthesis, including bromide **208**, iodide **211**, and lactone **212**, were all prepared from D-glutamic acid. A palladium-catalyzed coupling reaction between **210** and **213** elaborated muconin skeleton efficiently.

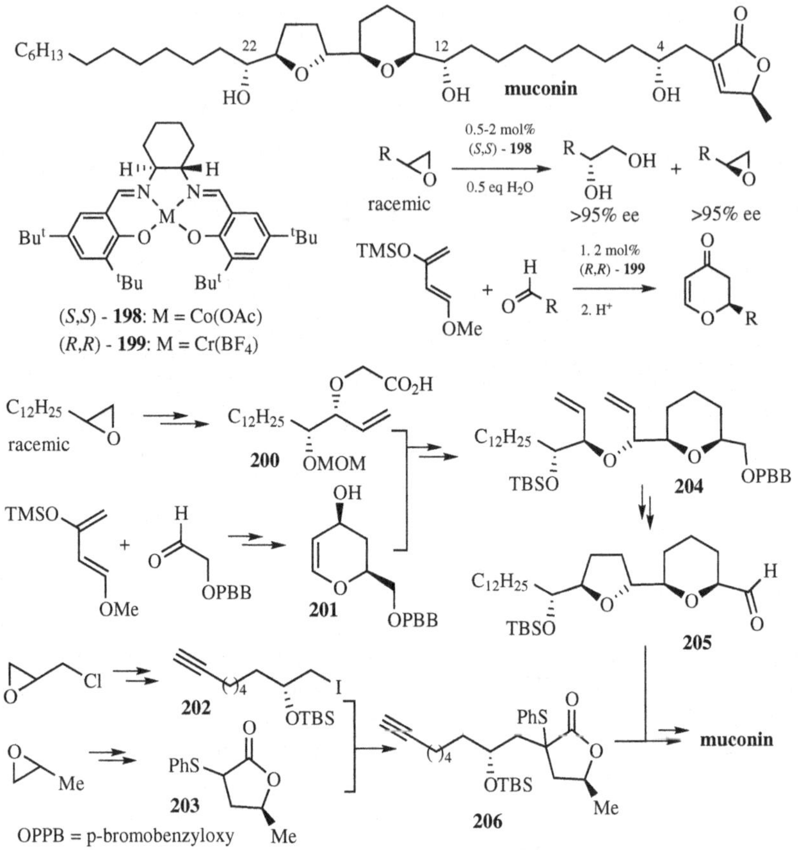

**Scheme 10-33.** Total synthesis of muconin by Schans et al.

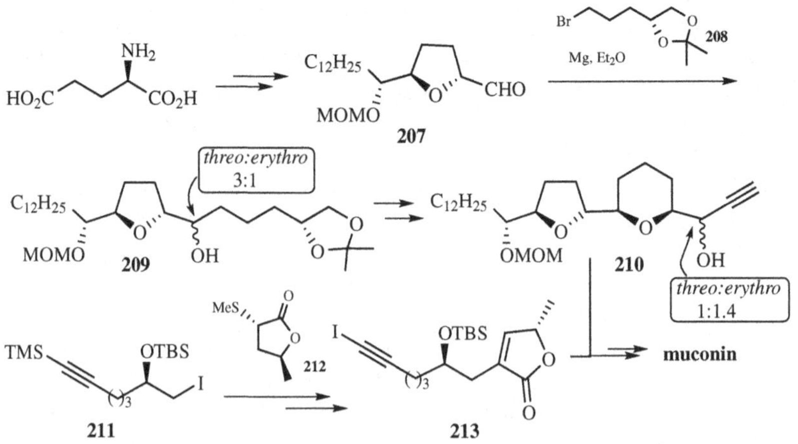

**Scheme 10-34.** Total synthesis of muconin by Yang and Kitahara.

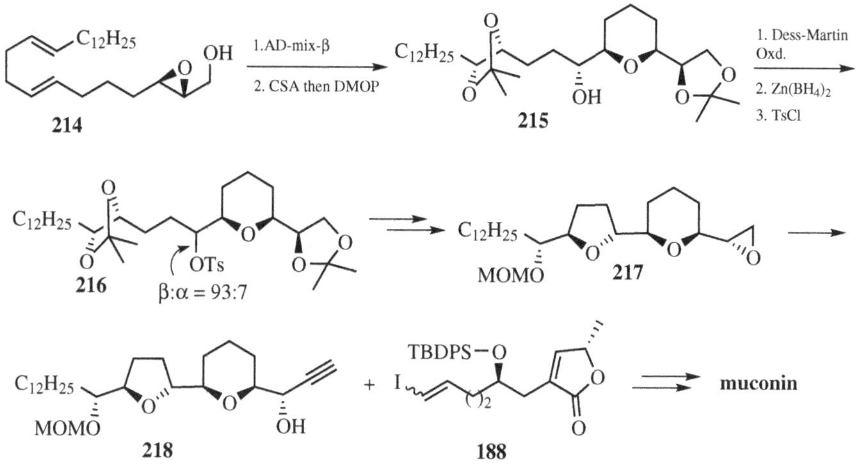

**Scheme 10-35.** Total synthesis of muconin by Takahashi et al.

Takahashi et al. also reported a route to muconin.[88,89] Their synthesis adopted Keinan et al.'s strategy to construct the stereochemistries by Sharpless AD and AE upon multiple olefin containing fatty acid (Scheme 10-35). The di-olefin 214 was subject to Sharpless AD conditions and then treated with acid, yielding a THP-containing diol. This diol was further protected as acetonide 215. The reversion of stereochemistry of alcohol 215 was achieved by Dess–Martin oxidation and Zn(BH₄)₂ reduction. Williamson etherification of tosylate 216 and epoxide formation afforded tri-ring intermediate 217. Opening with acetylene, 217 was converted into the terminal alkyne 218, which was coupled with vinyl iodide to finally give muconin.

Takahashi et al. also completed the total synthesis of other two acetogenins, jimenezin[90] (Scheme 10-36) and pyranicin[91] (Scheme 10-37), and they revised

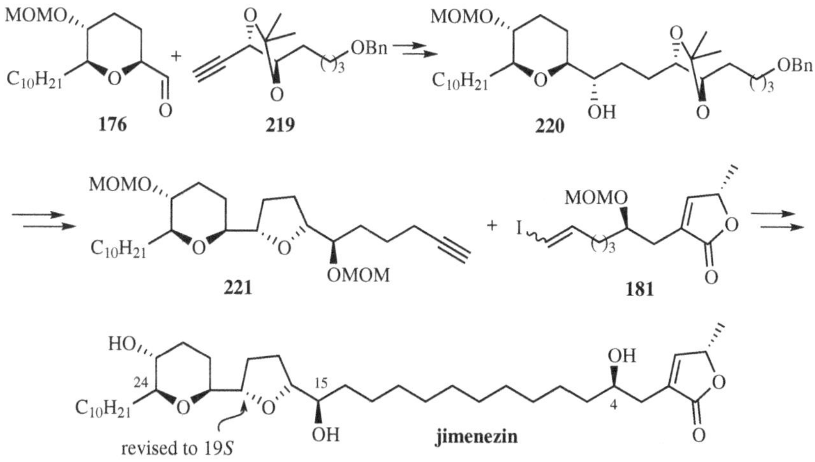

**Scheme 10-36.** Total synthesis of jimenezin by Takahashi et al.

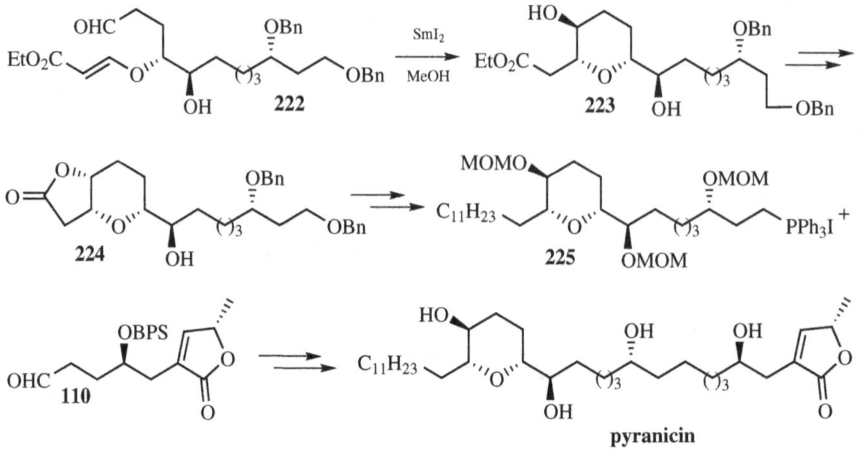

**Scheme 10-37.** Total synthesis of pyranicin by Takahashi et al.

the structure of jimenezin. In the synthesis of jimenezin, aldehyde **176** (Scheme 10-30) was used again as the key intermediate. The reaction of this aldehyde with terminal alkyne **219** gave the chelation-controlled product predominately (92 : 8 for the products ratio). After hydrogenation of triple bond, oxidation, and reduction, the configuration of hydroxyl group in **220** was reversed. The final coupling of two segments (**221** and **181**) again adopted Pd(0)-catalyzed protocol.

Total synthesis of pyranicin by Takahashi et al.[91] used a SmI$_2$-mediated reductive free radical reaction to construct the desired *cis*-THP ring **223** (Scheme 10-37). The stereochemistry of hydroxyl on the THP ring was reversed by an intramolecular S$_N$2 substitution, giving the lactone **224**. A Wittig reaction between **225** and known aldehyde **110** furnished the skeleton of the target.

## 10.5   DESIGN AND SYNTHESIS OF MIMICS OF ACETOGENINS

Although total synthesis to some extent alleviates the scarcity of acetogenins in nature, it is still a time-consuming work because of the structural complexities of these products. From a practical point of view, structurally simplified acetogenin mimics, preserving high cytotoxic potency, would be much more preferable. Thus, while synthesizing naturally occurring acetogenins, we initiated another project on design and synthesis of structurally simplified mimics of acetogenins, searching for medicinal leads with both potency and cytotoxic selectivity and a better understanding of SAR and the mode of action of acetogenin-like compounds.

It was reported[92-94] in 1994 that acetogenins could form a stable complex with Ca$^{2+}$, an important metal ion in the cell. However, whether this observed behavior of acetogenins was related to its biological activities was not clear. Nevertheless,

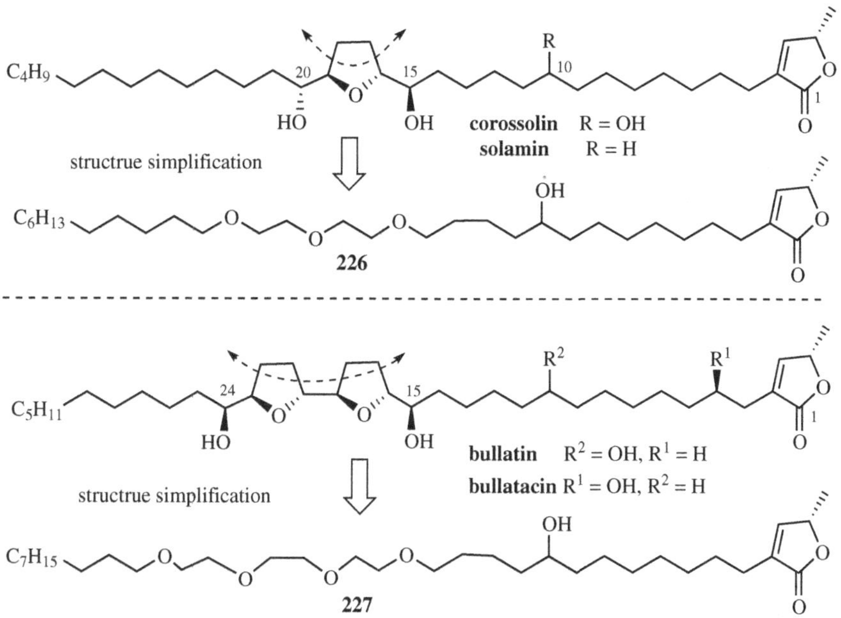

**Scheme 10-38.** Yao et al.'s first generation of mimics for *mono*- and *bis*-THF acetogenins.

we designed two acetogenin mimics, **226** and **227**, keeping in mind that oxygenated functionalities were important (Scheme 10-38). Compound **226** was a mimetic of *mono*-THF acetogenins, and two glycol ether units were used to simulate a hydroxy flanked *mono*-THF segment, whereas in compound 227, three glycol ether units were used as a substitute for a hydroxy flanked *bis*-THF segment commonly found in *bis*-THF acetogenins. This structural simplification would not only make synthesis much easier because of the reduction of chiral centers, but also it would maintain our ability to complex with $Ca^{2+}$. The synthesis of mimics **226** and **227** is shown in Scheme 10-39. Two-direction *O*-alkylation of diglycol with propargyl bromide afforded a dialkyne, which was treated with 1-bromooctane to give terminal alkyne **228**. Subsequent coupling of **228** with epoxide **2** led to **230**, which, after hydrogenation and β-elimination, furnished mimic **226**. Mimic **227** was prepared in a similar procedure. Much to our delight, preliminary bioactive assay indicated that these two mimics showed growth inhibitory activities against HL-60 and K562 comparable with that of natural acetogenins such as bullatacin (Table 10-3).[95]

Encouraged by these promising results, we set out to design and synthesize the second generation of mimics (Scheme 10-40). Here, *bis*-THF acetogenin was used as the template, and the four methylene carbons of the two THF rings were cut off as before. However, the two free hydroxy groups beside the *bis*-THF rings were kept intact because of the important role they may play in bioactivity; thus,

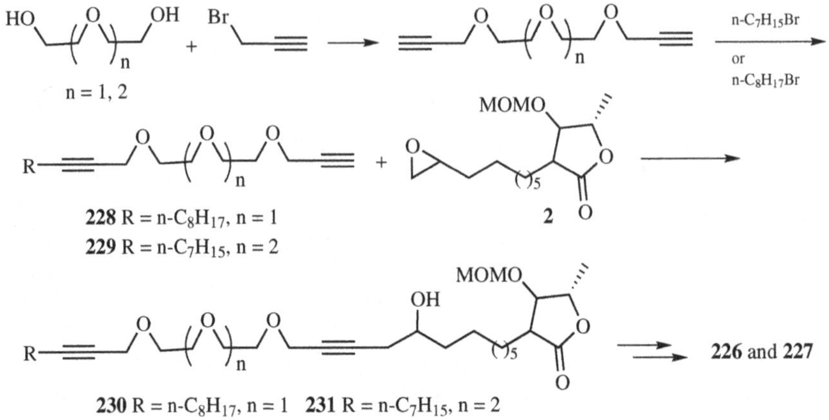
HO
OH
Br
n
n = 1, 2
n-C7H15Br
or
n-C8H17Br
MOMO
R
O
2
228 R = n-C8H17, n = 1
229 R = n-C7H15, n = 2
OH
226 and 227
230 R = n-C8H17, n = 1 231 R = n-C7H15, n = 2

**Scheme 10-39.** Synthesis of mimics **226** and **227** by Yao et al.

enantiopure diastereoisomers **232a–d** became the synthetic targets and *R*- and *S*-2,3-*O*-isopropylidene-glyceraldehyde were used as the chiral building blocks. From *S*-2,3-*O*-isopropylidene-glyceraldehyde diol, **233** and iodide **234b** were derived, whereas diol **235** and iodide **236b** were obtained in a similar way from *R*-2,3-*O*-isopropylidene-glyceraldehyde. The combination of two nucleophiles **233** and **235** and the two alkylating agents **234b** and **236b** would then give four linear esters, which after four-step conversions including a key aldol reaction, afforded **232a–d**, respectively. An example for the synthesis of **232c** is illustrated in Scheme 10-40. (Compounds **232a–d** are numbered according to their parent *bis*-THF acetogenin for easier comparison).[96,97]

The cytotoxic results of mimics **232a–d** toward some caner cells are shown in Table 10-4. As can be seen, maintenance of the two free hydroxy groups originally flanking *bis*-THF rings of acetogenin proved to be rewarding, and these mimics

**TABLE 10-3. Cytotoxicity of Mimics 226 and 227 toward HL-60 and K562 Cells**

| | IG % | | | | | |
|---|---|---|---|---|---|---|
| | HL-60 | | | K562 | | |
| cocn (μM) | 100 | 10 | 1 | 100 | 10 | 1 |
| cmp **226** | 100 | 50 | 0 | 31 | 18 | 0 |
| cmp **227** | 100 | 65 | 21 | 55 | 25 | 22 |
| Corossolone | 68 | 29 | 0 | 53 | 16 | 2 |
| (10*RS*)-Corossolin | 63 | 56 | 5 | 10 | 2 | 0 |
| Solamin | 24 | 8 | 0 | 59 | 39 | 29 |
| Bullatacin | 73 | 7 | 0 | 53 | 39 | 27 |

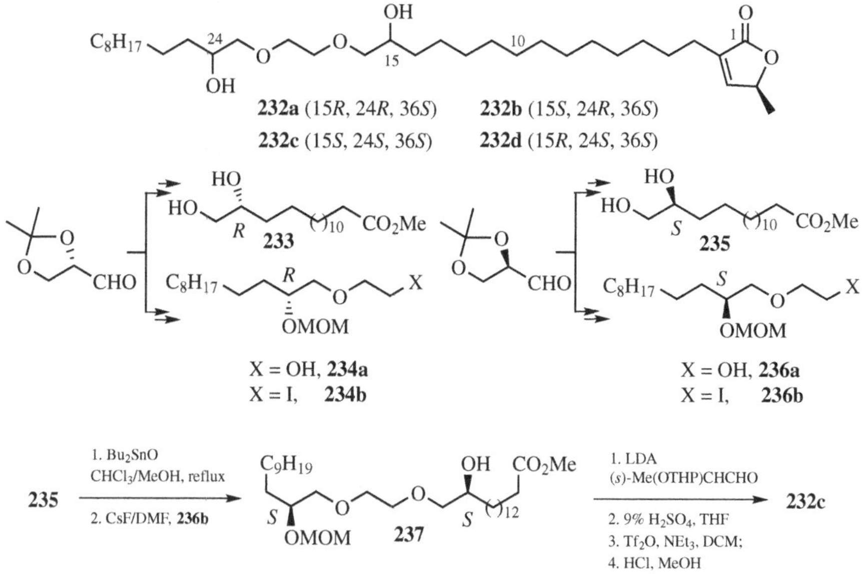

**Scheme 10-40.** Synthesis of second generation of mimics **232a–d** by Yao et al.

were active even under the μM scale. Most importantly, these mimics showed cytotoxic selectivity; they were active toward the HCT cell but inactive toward the KB and A2789 cells. In contrast, the control agent adriamycin displayed higher activities, but it did not differentiate these cells. It is noteworthy that these mimics had no cytotoxicity toward normal human cells. Among these four mimics, **232c** showed somewhat better activities, which might be attributed to their different stereochemistry.

To get more information about the SAR of acetogenin or its mimics, we synthesized **238a–d**, which have no butenolide unit, and butenolides **239**, which lack the THF ring or glycol ether unit. These compounds showed no or weak cytotoxicity, which indicated that both the butenolide segment and the glycol ether unit were indispensable.[96,97] The stereochemistry of the chiral center in the butenolide

**TABLE 10-4. Cytotoxicity of Mimics 232a–d Toward Four Kinds of Cells**

| Compound | $EC50$ ($\mu g$ $mL^{-1}$) | | | |
|---|---|---|---|---|
| | KB | A2780 | HCT-8 | HT-29 |
| **232a** | >1 | >1 | $6.6 \times 10^{-2}$ | $2.7 \times 10^{-1}$ |
| **232b** | >1 | >1 | $9.7 \times 10^{-2}$ | 1.1 |
| **232c** | >1 | >1 | $3.2 \times 10^{-2}$ | $1.1 \times 10^{-1}$ |
| **232d** | >1 | >1 | $6.5 \times 10^{-2}$ | 7.8 |
| Adriamycin | $2.89 \times 10^{-3}$ | $1.02 \times 10^{-3}$ | $4.65 \times 10^{-3}$ | $9.8 \times 10^{-4}$ |

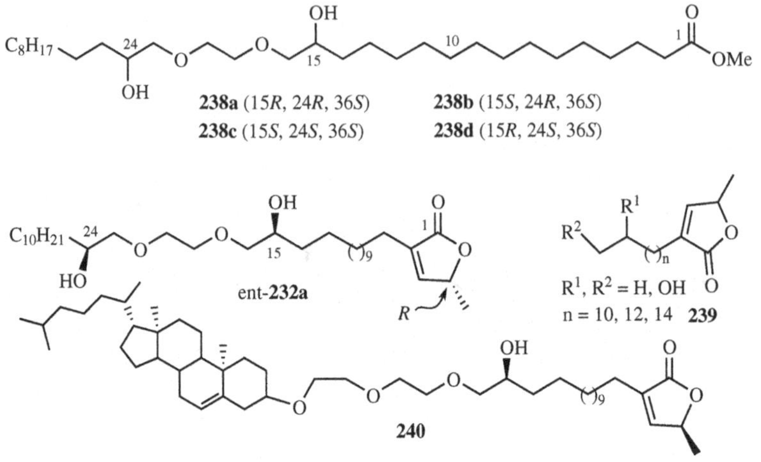

**Scheme 10-41.** Compounds **238**, **239**, **ent-232**, and steroid hybrid **240**.

segment seemed to have no effect on activities; **ent-232a** exhibited similar activities as **232a**. We also prepared a steroid hybrid **240** (Scheme 10-41) to examine the influence of lipophilic property on activities. Unfortunately, hybrid **240** showed little activities against the cancer cells screened, such as HCT-8.[98]

Inspection of the structures of acetogenins revealed that C4 and C10 were the two carbons where the hydroxy group was frequently located. We thus introduced, respectively, a hydroxy group to C4 and C10 of mimic **232c** to give two new mimics **243** and **246** (Scheme 10-42). Mimic **243** possessed one more free hydroxy group at C10 than did **232c** and had activities comparable with that of **232c**. In contrast, C4 hydroxy containing mimic **246** was 15 times as active as **232c** toward HT-29; furthermore, it had no cytotoxicity toward the normal human cell HELF (Table 10-5).[97,99]

From these results, we can see that it is possible to tune the activities and cytotoxic selectivities of acetogenin mimics by appropriate modification of their structures. To accelerate the space of searching for more potent acetogenin mimics with different cytotoxic selectivities, a systematic synthesis involving parallel fragment assembly strategy was developed. By using this strategy, a small library of ten mimics was established (Scheme 10-43).[100] Six of these ten mimics, **247a–f**, could be assembled from building blocks mesylate A, tartaric acid derivatives B, epoxide C, trimethylsilylacetylene D, and epoxide F, whereas the other four **247g–j** could be from A, B, C, 1,7-octadiyne E, and epoxide G. An example for the synthesis of **247d** is illustrated in Scheme 10-44. Two successive *O*-alkylations of **249** with mesylate **248** and R-epichlorohydrin afforded epoxide **251**, which was then converted to alkyne **252** via a three-step procedure: (1) epoxide opening reaction with lithium trimetylsilylacetylide, (2) hydroxy protection as a MOM ether, and (3) desilylation by TBAF. Finally, the coupling reaction of epoxide **253** and alkyne

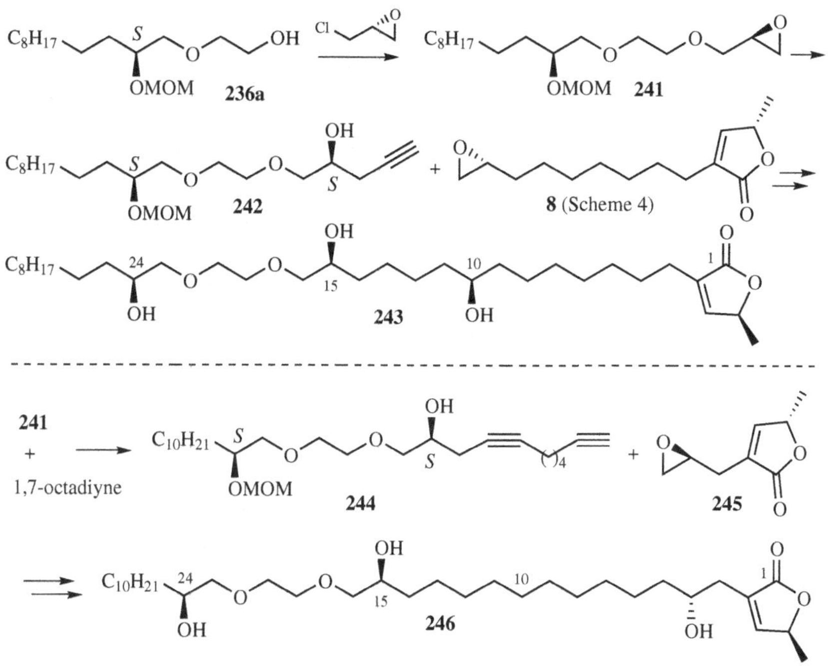

**Scheme 10-42.** Synthesis of mimics **243** and **246** by Yao et al.

**252**, followed by selective hydrogenation and global MOM ether cleavage, afforded mimic **247d**.

The bioassay results of mimics **247a–j** are shown in Table 10-6. As can be seen, these new mimics showed a different cytotoxic trend from **232c**. They were almost inactive toward the Bel-7402 cell. And those mimics with C4 hydroxy (**249g–j**) generally were more potent than those without the C4 hydroxy group (**249b–f**).

Besides our group, a few others are also interested in the synthesis of acetogenin analogs. Independently, Grée et al.[101–103] in France also envisioned using the glycol ether unit to replace the THF segment of acetogenin, and they synthesized a series

**TABLE 10-5. Cytotoxicity of Mimics 232c and 246 Toward Three Cancer Cells and One Normal Human Cell HELF**

| Compound | IC$_{50}$ ($\mu$g/mL) | | | |
|---|---|---|---|---|
| | HT-29 | HCT-8 | KB | HELF |
| **232c** | $2.4 \times 10^{-2}$ (15) | | | |
| **246** | $1.6 \times 10^{-3}$ (1) | $8.0 \times 10^{-2}$ | >10 | >10 |
| Adriamycin | $6.0 \times 10^{-2}$ (37.5) | $3.6 \times 10^{-2}$ | $7.6 \times 10^{-2}$ | 1.92 |

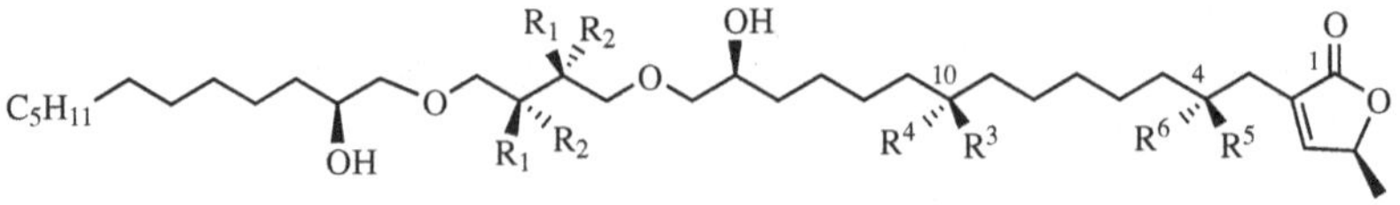

**249a** R¹=OH, R²=R³=R⁴=R⁵=R⁶=H  **249b** R²=OH, R¹=R³=R⁴=R⁵=R⁶=H
**249c** R¹=R³=OH, R²=R⁴=R⁵=R⁶=H  **249d** R¹=R⁴=OH, R²=R³=R⁵=R⁶=H
**249e** R²=R³=OH, R¹=R⁴=R⁵=R⁶=H  **249f** R²=R⁴=OH, R¹=R³=R⁵=R⁶=H
**249g** R¹=R⁶=OH, R²=R³=R⁴=R⁵=H  **249h** R¹=R⁵=OH, R²=R³=R⁴=R⁶=H
**249i** R²=R⁶=OH, R¹=R³=R⁴=R⁵=H  **249j** R²=R⁵=OH, R¹=R³=R⁴=R⁶=H

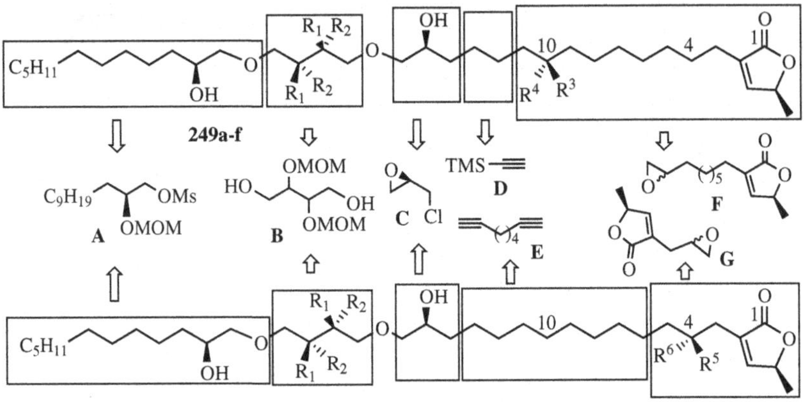

**Scheme 10-43.** Parallel fragment assembly strategy for a small library of acetogenin mimics (Yao et al.).

of mimics with a different tail (Scheme 10-45). Their cytotoxicity toward L1210 is summarized in Table 10-7. These analogs turned out to be much less cytotoxic than natural products; among these mimics, mimics with phenyl or piperidine as a side chain (**259b, c, h** and **260b, c**) showed lower activity than others.

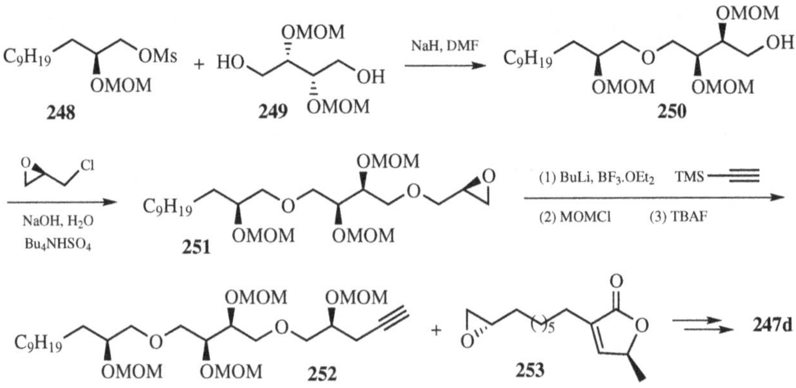

**Scheme 10-44.** A representative synthesis of mimics **247a–j**.

**TABLE 10-6. Bioassay Results of Mimics 247a–j**

| Compound | $IC_{50}$ ($\mu$g/mL) | | | |
|---|---|---|---|---|
| | KB | Bel-7402 | HT-29 | HCT-8 |
| 232c | 7.65 | 1.99 | 0.099 | 0.11 |
| 249a | 4.02 | >10 | 1.84 | 3.49 |
| 249b | 13.13 | >10 | 5.72 | 8.58 |
| 249c | 13.81 | >10 | 7.19 | 5.71 |
| 249d | 23.30 | >10 | 9.79 | 10.00 |
| 249e | 9.68 | >10 | 4.56 | 24.46 |
| 249f | 21.30 | >10 | 7.22 | 6.14 |
| 249g | 6.75 | >10 | 3.60 | 3.39 |
| 249h | 6.38 | >10 | 2.36 | 3.51 |
| 249i | 2.00 | >10 | 1.75 | 2.00 |
| 249j | 2.35 | 4.14 | 1.51 | 3.46 |

With quinone as the substitute for the butenolide unit of acetogenin, Koert et al.[104,105] synthesized two natural product hybrids, quinone-mucocin and quinone-squamocin D (Scheme 10-46). The complex I inhibitory activity of the two hybrids as well as some synthetic precursors and fragments is shown in Table 10-8. Interestingly, the quinone-mucocin hybrid was ten times more active than mucocin, whereas the quinone-squamocin hybrid lost some activity compared with the natural parent. Fragments **263–266** displayed weak activity; however, acid **277** showed comparable activity.

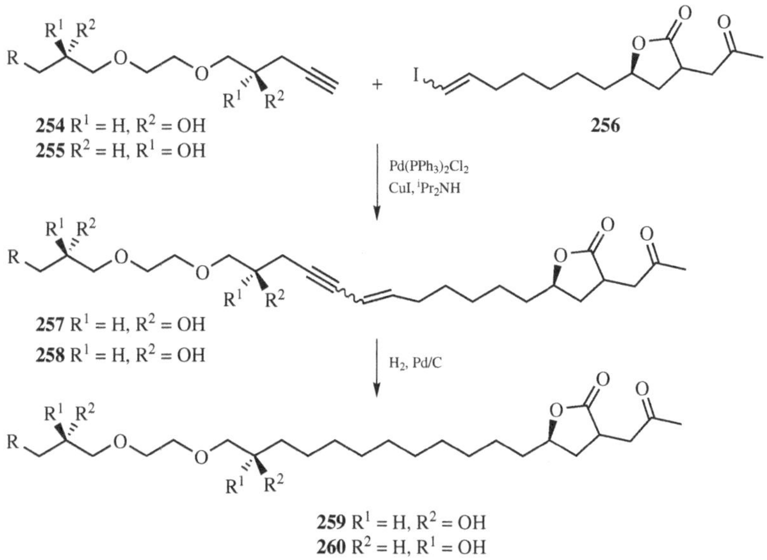

254 $R^1$ = H, $R^2$ = OH
255 $R^2$ = H, $R^1$ = OH

256

Pd(PPh₃)₂Cl₂
CuI, $^i$Pr₂NH

257 $R^1$ = H, $R^2$ = OH
258 $R^1$ = H, $R^2$ = OH

H₂, Pd/C

259 $R^1$ = H, $R^2$ = OH
260 $R^2$ = H, $R^1$ = OH

**Scheme 10-45.** Synthesis of acetognin mimics **259** and **260** by Grée et al.

**TABLE 10-7. IC$_{50}$ of Mimics 259 and 260 Toward L1210 Cell**

| R | Compound | IC$_{50}$ ($\mu$M) | Compound | IC$_{50}$ ($\mu$M) |
|---|---|---|---|---|
| $n$-C$_{10}$H$_{21}$ | 259a | 3.5 | 260a | 1 |
| Ph | 259b | 19.8 | 260b | 21.6 |
| $p$-MeOC$_6$H$_4$ | 259c | >10 | 260c | 32 |
| $p$-CF$_3$C$_6$H$_4$ | 259d | 3.3 | | |
| 2-Nph | – | | 260e | 3 |
| Bu$_2$N | 259f | 3.9 | 260f | 2.5 |
| Oct$_2$N | 259g | 3.7 | annonacin | 0.042 |
| $N$-piperidinyl | 259h | 28.7 | Bullatacin | 0.0004 |
| 3-$O$-cholesteryl | 259i | 2.8 | bullatacinone | 0.016 |

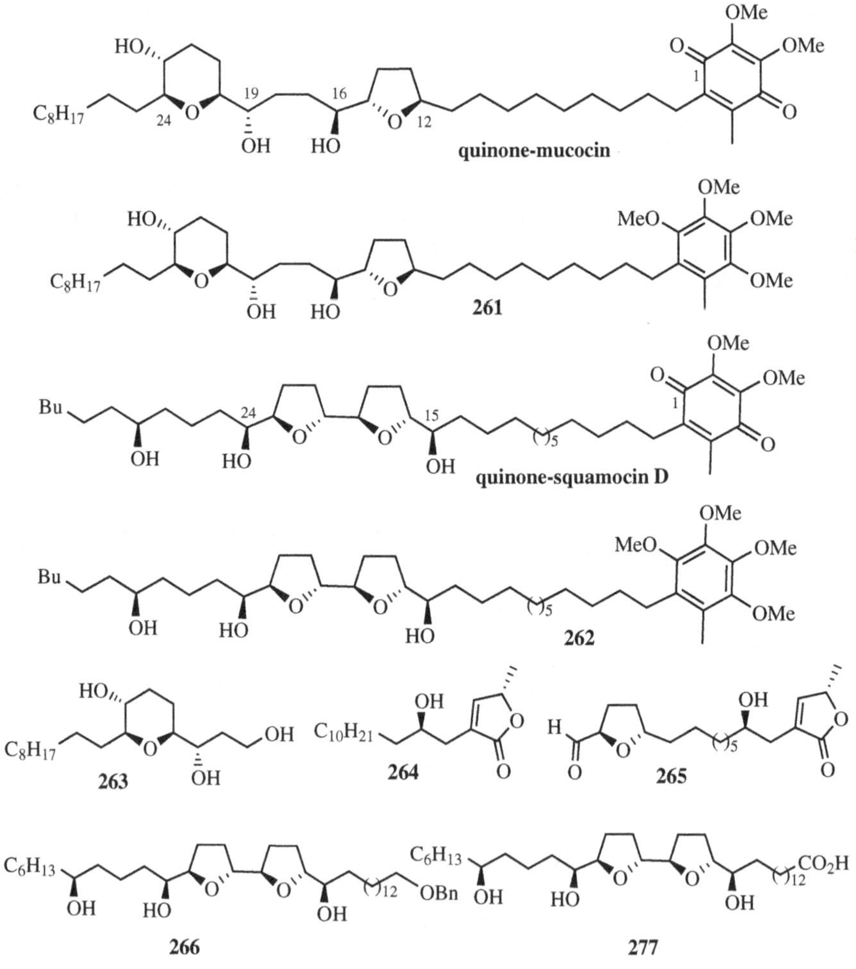

**Scheme 10-46.** Quinone-acetogenin hybrids (Koert et al.).

**TABLE 10-8. Inhibitory Activity of Quinone-Acetogenin Hybrids and Some Synthetic Fragments Toward Complex I**

| Compound | $IC_{50}$ (nM) | $IC_{50}$ ($\mu$mol mg$^{-1}$ protein) | $IC_{50}^a$ ($\mu$mol mg$^{-1}$ protein)[a] |
|---|---|---|---|
| Mucocin | 34 | 45 | 33.3 |
| Quione-mucocin | 3.6 | 4.9 | |
| **261** | 123 | 163 | |
| Squamocin D | | | 8.7 |
| Quinone-squamocin D | 1.7 | 2.3 | |
| **262** | 4.7 | 6.2 | |
| **263** | 32000 | | |
| **264** | 19500 | | |
| **265** | 2500 | | |
| **266** | 94000 | | |
| **267** | 173 | | |
| Retenone | 1.0 | 1.3 | |

[a] See Ref. 106.

## 10.6 SUMMARY

Annonaceous acetogenins are a large and famous class of natural products, which not only present unique and challenging structural features, but also exhibit an attractive wide spectrum of biological activities. The anticancer and cytotoxic activities of some members were reported more potent than the well-known drug, Taxol.[107–110] Although much attention has been paid to both the chemistry and biology investigations of acetogenins, little has been learned about the nature of those bioactive natural products, especially the origin and mechanism of activities. As mentioned, the endeavor made by the organic chemists through total synthesis developed many useful strategies and methodologies to construct these unique molecules. In addition, the fascinating work toward structure modification and simplification afforded a series of acetogenin mimics with potent activity and cytotoxic selectivity. These achievements not only exhibit again the power of modern organic chemistry, but also they provide more opportunities for a wide range of biological studies. Undoubtedly, these achievements and ongoing efforts by the chemical community will greatly promote the uncovering of the biological secrets of annonaceous acetogenins in the near future.

## REFERENCES

1. Jolad, S. D.; Hoffman, J. J.; Schram, K. H.; Cole, J. R.; Temesta, M. S.; Kriek, G. R.; Bates, R. B. *J. Org. Chem.*, **1982**, 47: 3151.
2. Rupprecht, J. K.; Hui, Y. H.; McLaughlin, J. L. *J. Nat. Prod.*, **1990**, 53: 237.
3. Fang, X. P.; Rieser, M. J.; Gu, Z. M.; McLaughlin, J. L. *Phytochem. Anal.*, **1993**, 4: 27.

4. Gu, Z. M.; Zhao, G. X.; Oberlies, N. H.; Zeng, L.; McLaughlin, J. In *Recent Advances in Phytochemistry, vol. 29.* New York: Plenum Press, **1995**, 249–310.

5. Zeng, L.; Ye, Q.; Oberlies, N. H.; Shi, G. E.; Gu, Z. M.; He, K.; McLaughlin, J. L. *Natural Product Reports*, **1996**, 275.

6. Alali, F. Q.; Liu, X.-X.; McLaughlin, J. L. *J. Nat. Prod.*, **1999**, 62: 504.

7. Cave, A.; Figadere, B.; Laurens, A.; Cortes, D. In *Progress in the Chemistry of Organic Natural Products.* Herz, W.; Kirby, G. W.; Moore, R. E.; Steglish, W.; Tamm, C., editors. New York: Springer-Verlag; **1997**, 81–287.

8. Zafra-Polo, M. C.; Figadere, B.; Gallardo, T.; Tormo, J. R.; Cortes, D. *Phytochemistry*, **1998**, 48: 1087.

9. Chen, Y.; Yu, D.-Q. *Yaoxue Xuebao*, **1998**, 37: 553.

10. Chen, R.-Y.; Yu, D.-Q. *Chinese J. Org. Chem.*, **2001**, 1: 1046.

11. Yao, Z.-J.; Wu, Y.-L. *Chinese J. Org. Chem.*, **1995**, 15: 120.

12. Casiraghi, G.; Zanardi, F.; Battistini, L.; Rassu, G.; Appendino, G. *Chemtracts*, **1998**, 11: 803.

13. Cortes, D.; Figadere, B.; Cave, A. *Phytochemistry*, **1993**, 32, 1467.

14. Londershausen, M.; Leicht, W.; Lieb, F.; Moeschler, H.; Weiss, H. *Pestic. Sci.*, **1991**, 33: 427.

15. Lewis, M. A.; Arnason, J. T.; Philogene, B. J.; Rupprecht, J. K.; McLaughli, J. L. *Pestic. Biochem. Physiol.*, **1993**, 45: 15.

16. Hollingworth, R. M.; Ahmmadsahib, K. I.; Gadelhak, G.; McLaughlin, J. L. *Biochem. Soc. Tran.*, **1994**, 22: 230 and references cited therein.

17. Morré, D. J.; de Cabo, R.; Farley, C.; Oberlies, N. H.; McLaughlin, J. L. *Life Sci.*, **1995**, 56: 343.

18. Oberlies, N. H.; Chang, C.-J.; McLaughlin, J. L. *J. Med. Chem.*, **1997**, 40: 2102 and references cited therein.

19. Hoye, T. R.; Hanson, P. R.; Kovelesky, A. C.; Ocain, T. D.; Zhang, Z. *J. Am. Chem. Soc.*, **1991**, 113: 9369.

20. Yao, Z.-J.; Wu, Y.-L. *Tetrahedron Lett.*, **1994**, 35: 157.

21. Yao, Z.-J.; Wu, Y.-L. *J. Org. Chem.*, **1995**, 60: 1170.

22. Hoye, T.; Humpal, P. E.; Jiménez, J.; Mayer, M.; Tan, L.; Ye, Z. *Tetrahedron Lett.*, **1994**, 35: 7517.

23. Koert, U. *Tetrahedron Lett.*, **1994**, 35: 2517.

24. Trost, B. M. Shi, Z. *J. Am. Chem. Soc.*, **1994**, 116: 7459.

25. Marshall, J. A.; Hinkle, K. W. *J. Org. Chem.*, **1997**, 62: 5989.

26. He, Y. T.; Yang, H. N.; Yao, Z. J. *Tetrahedron*, **2002**, 58: 8805.

27. White, D. J.; Somers, T. C.; Reddy, G. N. *J. Org. Chem.*, **1992**, 57: 4991.

28. Yu, Q.; Wu, Y.; Ding, H.; Wu, Y.-L. *J. Chem. Soc. Perkin Trans. I*, **1999**, 9: 1183.

29. Hu, T.-S.; Yu, Q.; Lin, Q.; Wu, Y.-L.; Wu, Y. *Org. Lett.*, **1999**, 1: 399.

30. Hu, T.-S.; Yu, Q.; Wu, Y.-L.; Wu, Y. *J. Org. Chem.*, **2001**, 66: 853.

31. Yu, Q.; Yao, Z.-J.; Chen, X.-G.; Wu, Y.-L. *J. Org. Chem.*, **1999**, 64: 2440.

32. Hu, T.-S.; Wu, Y.-L.; Wu, Y. *Org. Lett.*, **2000**, 2: 887.

33. Yu, Q.; Wu, Y.-K.; Xia, L.-J.; Tang, M.-H.; Wu, Y.-L. *Chem. Commun.*, **1999**, 2: 129.

34. Makabe, H.; Tanimoto, H.; Tanaka, A.; Oritani, T. *Heterocycles*, **1996**, 43: 2229.

35. Zhang, L. L.; Yang, R. Z.; Wu, S. J. *Acta Botanica Sinica*, **1993**, 35: 390.

36. Hanessian, S.; Grillo, T. A. *J. Org. Chem.*, **1998**, 63: 1049.

37. Marshall, J. A.; Jiang, H. *Tetrahedron Lett.*, **1998**, 39: 1493.

38. Baurle, S.; Peters, U.; Friedrich, T.; Koert, U. *Eur. J. Org. Chem.*, **2000**, 2207.

39. Dixon, D. J.; Ley, S. V.; Reynolds, D. J. *Angew. Chem. Int. Ed.*, **2000**, 39: 3622.

40. Dixon, D. J.; Ley, S. V.; Reynolds, D. J. *Chem. Eur. J.*, **2002**, 8: 1621.

41. Wang, Z.-M.; Tian, S.-K.; Shi, M. *Tetrahedron: Asymmertry*, **1999**, 10: 667.

42. Maezaki, N.; Kojima, N.; Sakamoto, A.; Iwata, C.; Tanaka, T. *Org. Lett.*, **2001**, 3: 429.

43. Maezaki, N.; Kojima, N.; Sakamoto, A.; Tominaga, H.; Iwata, C.; Tanaka, T.; Monden, M.; Damdinsuren, B.; Nakamori, S. *Chem. Eur. J.*, **2003**, 9: 390.

44. Makabe, H.; Hattori, Y.; Tanaka, A.; Oritani, T. *Org. Lett.*, **2002**, 4: 1083–1085.

45. Cecil, A. R. L.; Brown, R. C. D. *Org. Lett.*, **2002**, 4: 3715–3718.

46. Zhang, Q.; Lu, H.; Richard, C.; Curran, D. P. *J. Am. Chem. Soc.*, **2004**, 126: 36.

47. Marshall, J. A.; Hinkle, K. W. *J. Org. Chem.*, **1997**, 62: 5989.

48. Marshall, J. A.; Jiang, H. *J. Org. Chem.*, **1998**, 63: 7066.

49. Marshall, J. A.; Chen, M. *J. Org. Chem.*, **1997**, 62: 5996.

50. Marshall, J. A.; Jiang, H. *J. Nat. Prod.*, **1999**, 62: 1123.

51. Marshall, J. A.; Hinkle, K. W. *Tetrahedron Lett.*, **1998**, 39: 1303.

52. Marshall, J. A.; Jiang, H. *J. Org. Chem.*, **1999**, 64: 971.

53. Marshall, J. A.; Piettre, A.; Paige, M. A.; Valeriote, F. A. *J. Org. Chem.*, **2003**, 68: 1771.

54. Marshall, J. A.; Piettre, A.; Paige, M. A.; Valeriote, F. A. *J. Org. Chem.*, **2003**, 68: 1780.

55. Yazbak, A.; Sinha, S. C.; Keinan, E. *J. Org. Chem.*, **1998**, 63: 5863.

56. Sinha, S. C.; Sinha, S. C.; Keinan, E. *J. Org. Chem.*, **1999**, 64: 7067.

57. Sinha, A.; Sinha, S. C.; Sinha, S. C.; Keina, E. *J. Org. Chem.*, **1999**, 64: 2381.

58. Avedissian, H.; Sinha, C. S.; Yazbak, A.; Sinha, A.; Neogi, P.; Sinha, S. C.; Keinan, E. *J. Org. Chem.*, **2000**, 65: 6035.

59. D'souza, L. J.; Sinha, S. C.; Lu, S. F.; Keinan, E.; Sinha, S. C. *Tetrahedron*, **2001**, 57: 5255.

60. Wang, Z.-M.; Tian, S.-K.; Shi, M. *Tetrahedron Lett.*, **1999**, 40: 977.

61. Wang, Z.-M.; Tian, S.-K.; Shi, M. *Eur. J. Org. Chem.*, **2000**, 349.

62. Emde, U.; Koert, U. *Tetrahdron Lett.*, **1999**, 40: 5979.

63. Emde, U.; Koert, U. *Eur. J. Org. Chem.*, **2000**, 1889.

64. Ruan, Z.; Mootoo, D. R. *Tetrahedron Lett.*, **1999**, 40: 49.

65. Sinha, S. C.; Sinha, A.; Yazbak, A.; Keina, E. *J. Org. Chem.*, **1996**, 61: 7640.

66. Burke, S. D.; Jiang, L. *Org. Lett.*, **2001**, 3: 1953.

67. Makabe, H.; Tanaka, A.; Oritani, T. *Tetrahedron Lett.*, **1997**, 38: 4247.

68. Makabe, H.; Tanaka, A.; Oritani, T. *Tetrahedron*, **1998**, 54: 6329.

69. Mashall, J. A.; Jiang, H. *J. Org. Chem.*, **1998**, 63: 7066.

70. Sinha, S. C.; Sinha, A.; Sinha, S. C.; Keinan, E. *J. Am. Chem. Soc.*, **1997**, 119: 12074.

71. Sinha, S. C.; Sinha, A.; Sinha, S. C.; Keinan, E. *J. Am. Chem. Soc.*, **1998**, 120: 11279.

72. Baylon, C.; Prestat, G.; Heck, M. P.; Mioskowski, C. *Tetrahedron Lett.*, **2000**, 41: 3833.

73. Chandrasekhar, M.; Chandra, K. L.; Singh, V. K. *Tetrahedron Lett.*, **2002**, 43: 2773.

74. Konno, H.; Hiura, N.; Yanaru, M. *Heterocycles*, **2002**, 57: 1793.

75. Carda, M.; Rodriguez, S.; Gonzalez, F.; Castillo, E.; Villanueva, A.; Marco, J. A. *Eur. J. Org. Chem.*, **2002**, 2609.

76. Raghavan, S.; Joseph, S. C. *Tetrahedron: Asymmetry*, **2003**, 14: 101.

77. Neogi, P.; Doundoulakis, T.; Yazak, A.; Sinha, S. C.; Sinha, S. C.; Keinan, E. *J. Am. Chem. Soc.*, **1998**, 120: 11279.

78. Baurle, S.; Hoppen, S.; Koert, U. *Angew. Chem. Int. Ed. Engl.*, **1999**, 38: 1263.

79. Hoppen, S.; Baurle, S.; Koert, U. *Chem. Eur. J.*, **2000**, 6: 2382.

80. Takahahi, S.; Nakata, T. *Tetrahedron Lett.*, **1999**, 40: 723; *ibid.* **1999**, 40: 727.

81. Takahashi, S.; Nakata, T. *J. Org. Chem.*, **2002**, 67: 5739.

82. Takahashi, S.; Kubota, A.; Nakata, T. *Angew. Chem. Int. Ed. Engl.*, **2002**, 41: 4751.

83. Evans, P. A.; Cui, J.; Gharpure, S. J.; Polosukhin, A.; Zhang, H.-R. *J. Am. Chem. Soc.*, **2003**, 125: 14702.

84. Yoshimitsu, T.; Makino, T.; Nagaoka, H. *J. Org. Chem.*, **2004**, 69: 1993.

85. Schans, S. E.; Branact, J.; Jacobson, E. N. *J. Org. Chem.*, **1998**, 63: 4876.

86. Yang, W.-Q.; Kitahara, T. *Tetrahedron Lett.*, **1999**, 40: 7827.

87. Yang, W.-Q.; Kitahara, T. *Tetrahedron*, **2000**, 56: 1451.

88. Takahashi, S.; Kubota, A.; Nakata, T. *Tetrahedron Lett.*, **2002**, 43: 8661.

89. Takahashi, S.; Kubota, A.; Nakata, T. *Tetrahedron*, **2003**, 59: 1627.

90. Takahashi, S.; Maeda, K.; Hirota, S.; Nakata, T. *Org. Lett.*, **1999**, 1: 2025.

91. Takahashi, S.; Kubota, A.; Nakata, T. *Org. Lett.*, **2003**, 5: 1353.

92. Sasaki, S.; Naito, H.; Maruta, K.; Kawahara, E.; Maeda, M. *Tetrahedron Lett.*, **1994**, 35: 3337.

93. Sasaki, S.; Maruta, K.; Naito, H.; Sugihara, H.; Hiratani, K.; Maeda, M. *Tetrahedron Lett.*, **1995**, 36: 5571.

94. Peyrat, J.-P.; Mahuteau, J.; Figadere, B.; Cave, A. *J. Org. Chem.*, **1997**, 62: 4811.

95. Yao, Z.-J.; Wu, H.-P.; Wu, Y.-L. *J. Med. Chem.*, **2000**, 43: 2484.

96. Zeng, B.-B.; Wu, Y.-K.; Yu, Q.; Wu, Y.-L.; Li, Y.; Chen, X.-G. *Angew. Chem. Int. Ed.*, **2000**, 39: 1934.

97. Zeng, B.-B.; Wu, Y.; Jiang, S.; Yu, Q.; Yao, Z.-J.; Liu, Z.-H.; Li, H.-Y.; Li, Y.; Chen, X.-G.; Wu, Y.-L. *Chem. Eur. J.*, **2003**, 9: 282.

98. Jiang, S.; Wu, Y.-L.; Yao, Z.-J. *Chinese J. Chem.*, **2002**, 20: 692.

99. Jiang, S.; Liu, Z.-H.; Sheng, G.; Zeng, B.-B.; Chen, X.-G.; Wu, Y.-L.; Yao, Z.-J. *J. Org. Chem.*, **2002**, 67: 3404.

100. Jiang, S.; Li, Y.; Chen, X.-G.; Hu, T.-S.; Wu, Y.-L.; Yao, Z.-J. *Angew. Chem. Int. Ed.*, **2004**, 43: 329.

101. Huérou, Y. L.; Doyon, J.; Grée, R. L. *J. Org. Chem.*, **1999**, 64: 6782.

102. Rodier, S.; Huérou, Y.; Renoux, B.; Doyon, J.; Renard, P.; Pierré, A.; Gesson, J.-P.; Greé, R. *Bioorg. Med. Chem. Lett.*, **2000**, 10: 1373.

103. Rodier, S.; Huérou, Y.; Renoux, B.; Doyon, J.; Renard, P.; Pierré, A.; Gesson, J.-P., Greé, R. *Anti-Cancer Drug Design*, **2001**, 16: 109.

104. Hoppen, S.; Emde, V.; Friedrich, T.; Grubert, L.; Koert, U. *Angew. Chem. Int. Ed. Engl.*, **2000**, 39: 2099.

105. Arndt, S.; Emde, V.; Baurle, S.; Friedrich, T.; Grubert, L.; Koert, U. *Chem. Eur. J.*, **2001**, 7: 993.

106. Alfonso, D.; Johnson, H. A.; Colman-Saizarbitoria, T.; Presley, C. P.; McCabe, G. P.; McLaughlin, J. R. *Nat. Tox.*, **1996**, 4: 181.

107. Tormo, J. R.; Gallardo, T.; Aragon, R.; Cortes, D.; Estornell, E. *Chemico-Biological Interactions*, **1999**, 122: 171.

108. Duret, P.; Hocquemiller, R.; Gantier, J.-C.; Figadere, B. *Bioorg. Med. Chem.*, **1999**, 7: 1821.

109. Queiroz, E. F.; Roblot, F.; Duret, P.; Figadere, B.; Gouyette, A.; Laprevote, O.; Serani, L.; Hocquemiller, R. *J. Med. Chem.*, **2000**, 43: 1604.

110. Gallardo, T.; Zafra-Polo, M.; Tormo, J.; Gonzalez, M.; Franck, X.; Estornell, E.; Cortes, D. *J. Med. Chem.*, **2000**, 43: 4793 and references cited therein.

# INDEX

---

*Medicinal Chemistry of Bioactive Natural Products* Edited by Xiao-Tian Liang and Wei-Shuo Fang
Copyright © 2006 John Wiley & Sons, Inc.

**443**